Recent Progress in

HORMONE RESEARCH

The Proceedings of the Laurentian Hormone Conference

VOLUME 28

RECENT PROGRESS IN HORMONE RESEARCH

Proceedings of the
1971 Laurentian Hormone Conference

Edited by
E. B. ASTWOOD

VOLUME 28

COMMITTEE ON ARRANGEMENTS

E. Alpert	G. A. Grant
E. B. Astwood	R. O. Greep
G. D. Aurbach	E. C. Reifenstein, Jr.
R. W. Bates	H. J. Ringold
J. Beck	K. Savard
J. Fried	A. White

1972

ACADEMIC PRESS, New York and London
A Subsidiary of Harcourt Brace Jovanovich, Publishers

ACADEMIC PRESS, INC.
111 Fifth Avenue, New York, New York 10003

United Kingdom Edition published by
ACADEMIC PRESS, INC. (LONDON) LTD.
24/28 Oval Road, London NW1

LIBRARY OF CONGRESS CATALOG CARD NUMBER: Med. 47-38

Second Printing, 1973

PRINTED IN THE UNITED STATES OF AMERICA

CONTENTS

PREFACE

The interested readers of *Recent Progress in Hormone Research* will recognize this, the 28th volume, as another compendium of current work in some of the most active fields of endocrinological investigation. They will have long since noticed the growth in excellence of the contributions, a trend that is notably exemplified by the articles in this year's volume. The care with which the authors prepare their manuscripts, the choice of material to be included, and the excellence of the presentations leave little for the Editor to do beyond attending to clerical detail. The format of these volumes has not changed for some years, and its precise and pleasing aspect is a credit to the painstaking and expert attention of the Publisher.

The annual meeting of the Laurentian Hormone Conference, the proceedings of which make up this volume, took place at the Mont Tremblant Lodge, Mont Tremblant, Quebec, Canada, on August 28th to September 3rd, 1971. The members of the conference were again pleasantly surprised by the further improvements in the lodge and in the added amenities to the Chalet, where the meetings are held.

The papers were well received and, as usual, were well attended; periods of the discussion were as lively as ever and most of them profitably filled the allotted hours. It was to the credit of the experienced chairmen that seldom was the time monopolized by a single discussant. The members of the Committee on Arrangements are grateful to Drs. F. C. Bartter, L. L. Engel, F. C. Greenwood, R. Guillemin, C. W. Lloyd, P. L. Munson, D. N. Orth, W. H. Sawyer, and Jane E. Shaw for chairing the sessions.

It has often been said that the published discussion of the conference is the most interesting part of the printed proceedings. This would not be so were it not for the untiring efforts of our executive secretary, Miss Joanne Sanford, and her associates, Mrs. Mina Rano and Miss Lucy Passalapi, and the cooperation of the discussants in promptly editing their remarks.

Hamilton, Bermuda
May, 1972

E. B. ASTWOOD

Estrogen Action: An Inroad to Cell Biology[1]

Gerald C. Mueller, Barbara Vonderhaar,[2] Uh Hee Kim,
and Mary Le Mahieu

*McArdle Laboratory, University of Wisconsin,
Madison, Wisconsin*

I. Introduction

A description of the mechanism of estrogen action must account for
the rapid and diverse responses which this hormone can incite in a range
of tissues. To list just a few examples, estrogens have been shown to
effect: the suppression of gonadotropin production and release by an
action on the hypothalamic–pituitary axis (DaLage, 1966; Flerkó, 1957;
Schwartz, 1969), the growth and differentiation of mammary tissues
when combined with appropriate adrenal and pituitary hormones (Lyons
et al., 1958; Hilf *et al.*, 1967), accelerated growth and cornification
of vaginal epithelium (Bigger and Claringbold, 1954; Ladinsky *et al.*,
1968), the synthesis of phosphoproteins in the liver (Greengard *et al.*,
1965), the synthesis and accumulation of fat in adipose cells (Gassner
et al., 1958), and the growth and differentiation of uterine tissue
(O'Malley *et al.*, 1969; Mueller *et al.*, 1958; Oka and Schimke, 1969).
Looking for common denominators among the many studies dealing with
a broad spectrum of responses two observations stand out: (1) Cells,
which are responsive to estrogens, contain a receptor protein (or receptor
aggregate) that interacts specifically with certain structural features
of the estrogen molecule to form a high affinity, noncovalent complex
(Table I). (2) Many of the hormone responses (but not all) involve
and are highly dependent on a hormonal acceleration of the genetic
expression mechanisms in the sensitive cells (Table II). In the analysis
of estrogen action which follows, an attempt is made to relate these
two observations in molecular and biophysical terms. A concept of estrogen action is proposed in which: (a) the receptor protein is characterized
by a remarkable propensity to enter into distinctive associations or ag-
gregations with certain classes of macromolecules of the cell; (b) the
character, composition, or stability of protein components in these complexes is modified through some function of the receptor protein which
is activated by the binding of estrogen; (c) the character of the metabolic response which ensues is determined by the extent to which a

---

[1] The Gregory Pincus Memorial Lecture.

[2] *Present address:* National Institute of Arthritis and Metabolic Diseases, National
Institutes of Health, Bethesda, Maryland.

limited cellular process is affected by the availability, state, or activity of the specific altered component of such receptor complexes, either catalytically or genetically. In this concept the hormone is visualized as a modulator in the class expression of multiple genes through controlling the availability or state of specific proteins which are essential in the nuclear scene for the expression of these genes. It is also suggested that the same mechanism may as well regulate certain extragenomic processes.

TABLE I

Evidence for Estrogen Receptor Mechanism

Target tissue	References
Rat uterus	See Table III
Mammary gland	Sander (1968), Sander and Attramadal (1968), Puca and Bresciani (1969a)
Hormone-dependent mammary tumors	Jensen *et al.* (1967c), Mobbs (1968), Puca and Bresciani (1968b), Terenius (1968a), Stumpf (1969), Korenman and Dukes (1970), Lemon (1970)
Hypothalamus	Eisenfeld and Axelrod (1965), Kato and Villee (1967), Pfaff (1968), Stumpf (1968b), Anderson and Greenwald (1969), Kahwanago *et al.* (1969), Whalen and Maurer (1969), Woolley *et al.* (1969), Chader and Villee (1970), Eisenfeld (1970)
Pituitary	Eisenfeld and Axelrod (1966), Eisenfeld (1967, 1970), Kato and Villee (1967), Stumpf (1968c), Anderson and Greenwald (1969), Kahwanago *et al.* (1969), Kato *et al.* (1969, 1970), Leavitt *et al.* (1969)
Mouse uterus and vagina	Martin and Stone (1965), Stone and Baggett (1965a,b), Terenius (1966, 1968b), King *et al.* (1968), Folman and Pope (1969a,b), Harris (1971)
Chick oviduct	Jonsson and Terenius (1965), Terenius (1969)
Chicken liver	Arias and Warren (1971)

II. Some General Aspects of Nuclear—Cytoplasmic Interactions in Living Cells

Before taking up experiments dealing with estrogen action let us first consider some general properties of nuclei and nucleochromatin—particularly the manner in which the chromatin interacts with its environment in the control of gene expression. This subject is important in our analysis of estrogen action since the hormone appears to act chiefly by modifying the function of chromatin.

As revealed in the electron microscopic studies of Dr. Hans Ris, nucleochromatin is a highly complex molecular aggregate of DNA and proteins (Ris and Chandler, 1963; Ris, 1967). Using Kleinschmidt's tech-

nique to disrupt and gently spread nuclei on a water–air interface (Kleinschmidt *et al.*, 1961) he demonstrated that native chromatin consists of a complicated web of 200–250 Å nucleoprotein fibers (Fig. 1A–D). Treatment of the chromatin with chelating agents separates certain of these proteins and unravels the structure of the chromatin webs yielding 100 Å fibers with a "beaded" appearance. Progressive digestion with pronase removes the protein "beads" and reveals a core of 25 Å double-stranded DNA fibers in each branch of the web. It is apparent that

TABLE II

The Acceleration of Genetic Expression by Estrogens

Observation	References
Early general stimulation of uterine RNA synthesis	Mueller *et al.* (1958), Jervell *et al.* (1958), Gorski and Nicolette (1963), Wilson (1963), Hamilton *et al.* (1965), Means and Hamilton (1966a), Unhjem *et al.* (1968), Barry and Gorski (1971), Kapadia *et al.* (1971)
Estrogen induction of RNA polymerase	Gorski (1964), Nicolette and Mueller (1966b), Nicolette *et al.* (1968), Barker and Anderson (1968), Raynaud-Jammet and Baulieu (1969), Arnaud *et al.* (1971)
Inhibition of estrogen action by the inhibition of RNA synthesis	Ui and Mueller (1963), Szego and Lawson (1964), Nicolette and Mueller (1966a), DeAngelo and Gorski (1970)
Early general stimulation of uterine protein synthesis	Mueller (1953), Mueller *et al.* (1958), McCorquodale and Mueller (1958), Noteboom and Gorski (1963), Greenman and Kenney (1964), Szego and Lawson (1964)
Induced synthesis of uterine specific proteins by estrogens	Notides and Gorski (1966), Cecil and Bitman (1967), DeAngelo and Gorski (1970), Teng and Hamilton (1970), Cohen *et al.* (1970), Barker (1971)
Inhibition of estrogen action by inhibitors of protein synthesis	Mueller *et al.* (1961), Gorski and Axman (1964), Gorski and Morgan (1967), Cecil and Bitman (1967)

the webs of the native chromatin arise by the sticking together of protein-coated DNA structures and that in many cases a double-stranded chain of DNA may course its way into an arc of the web only to double back in the same arc, yielding what appears to be blind loops and blunt ends of nucleochromatin.

From these studies it is obvious that proteins play an important role in the structure of chromatin. In the analogy of the repressors of bacterial systems it appears quite likely that some proteins are highly specific in recognizing nucleotide sequences in DNA. An example of a transitory recognition of this type in bacterial systems is the initiation of RNA synthesis by RNA polymerase and the modifications of the tem-

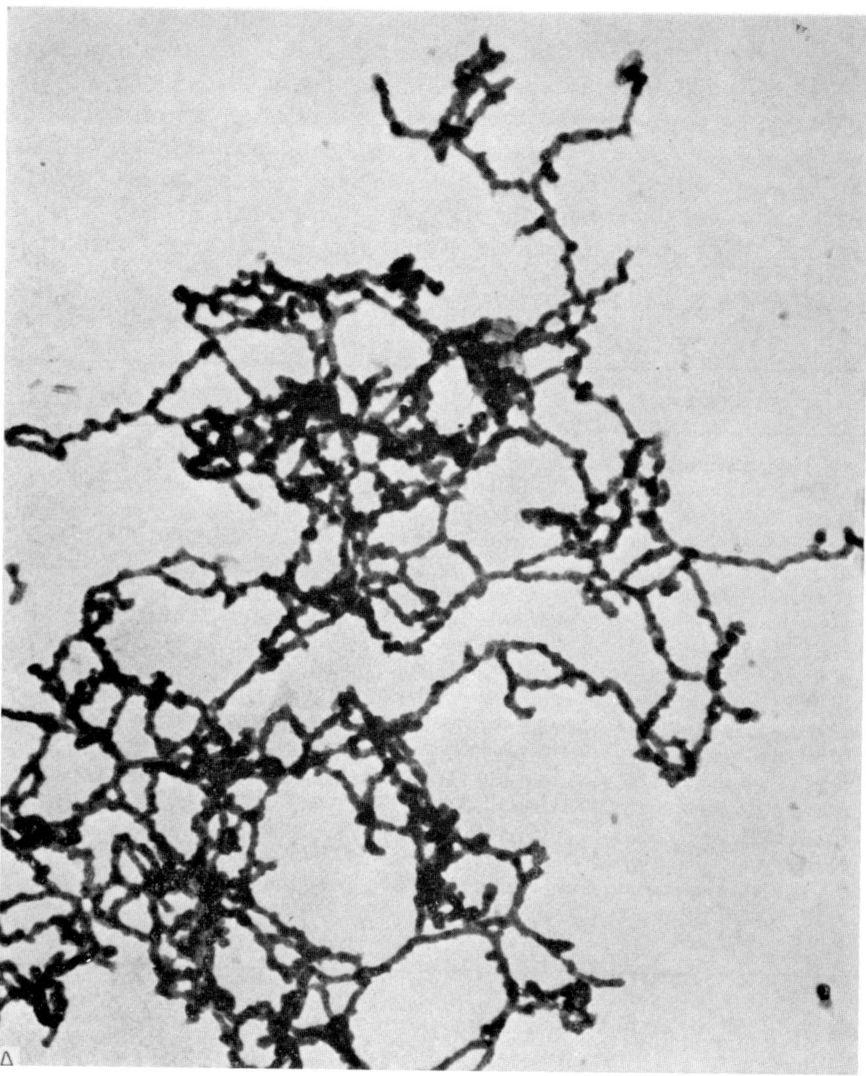

FIG. 1. The macromolecular character of nucleochromatin. Erythrocyte nuclei from the salamander *Triturus viridescens* have been spread with the indicated treatments at an air–water interface and examined electron microscopically.

(A) Nuclei have been spread directly at air–water interface, fixed in ethanol, dried from amyl acetate. Note that the webs are made up of 250 Å fibers. ×40,500.

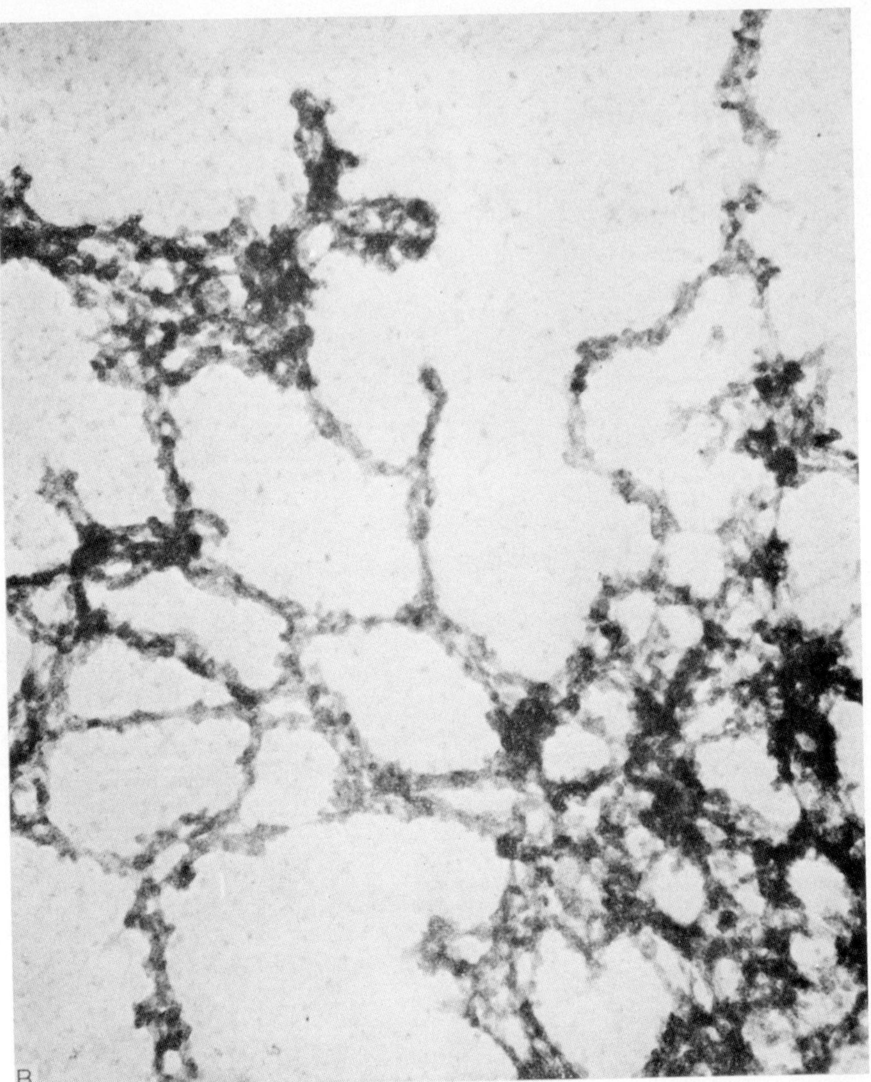

Fig. 1B. Nuclei have been treated for 10 seconds with 5 mM sodium citrate, fixed with 10% formalin, stained with 1% uranyl acetate, and dried from amyl acetate. Note that the 250 Å fibers consist of two 100 Å fibers which separate when treated briefly with chelating agents. ×96,000. See caption Fig. 1A.

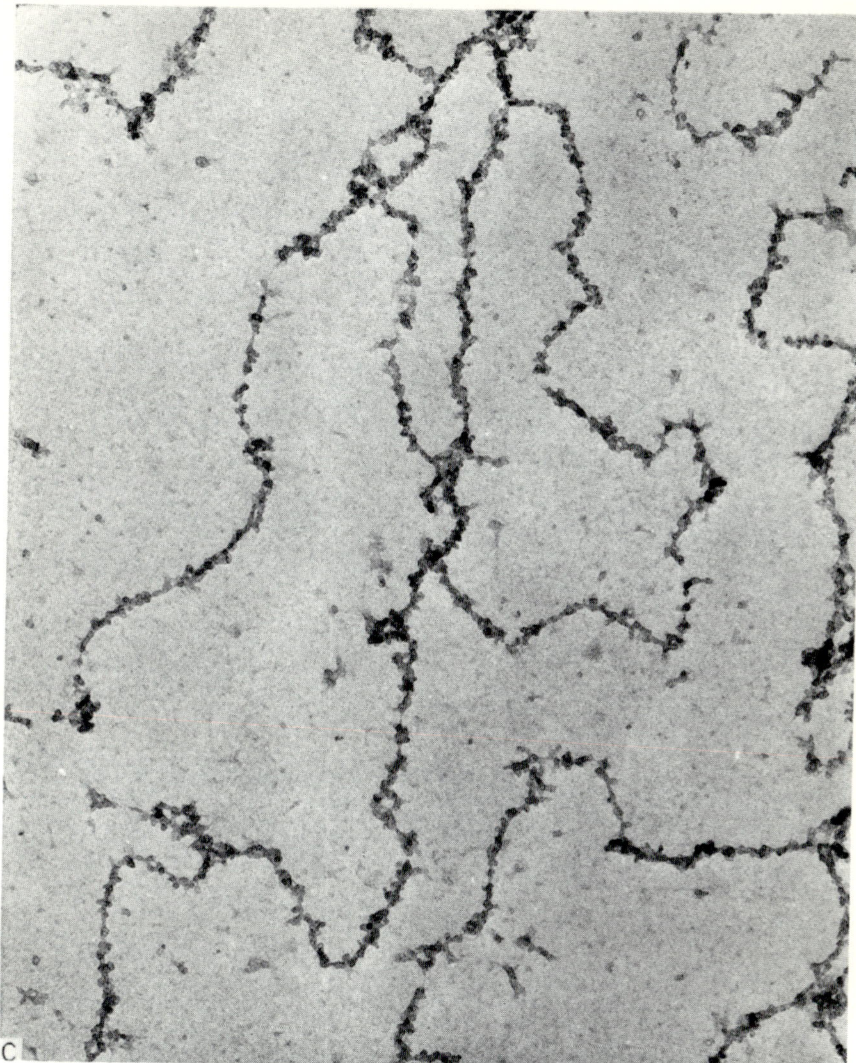

Fig. 1C.   Nuclei have been spread on 5 mM sodium citrate, stained with 1% uranyl acetate, fixed in ethanol, and dried from amyl acetate. Note the knobby projections and "moth-eaten" character of the 100 Å fibers after more extensive treatment with the chelating agents. ×96,000. See caption Fig. 1A.

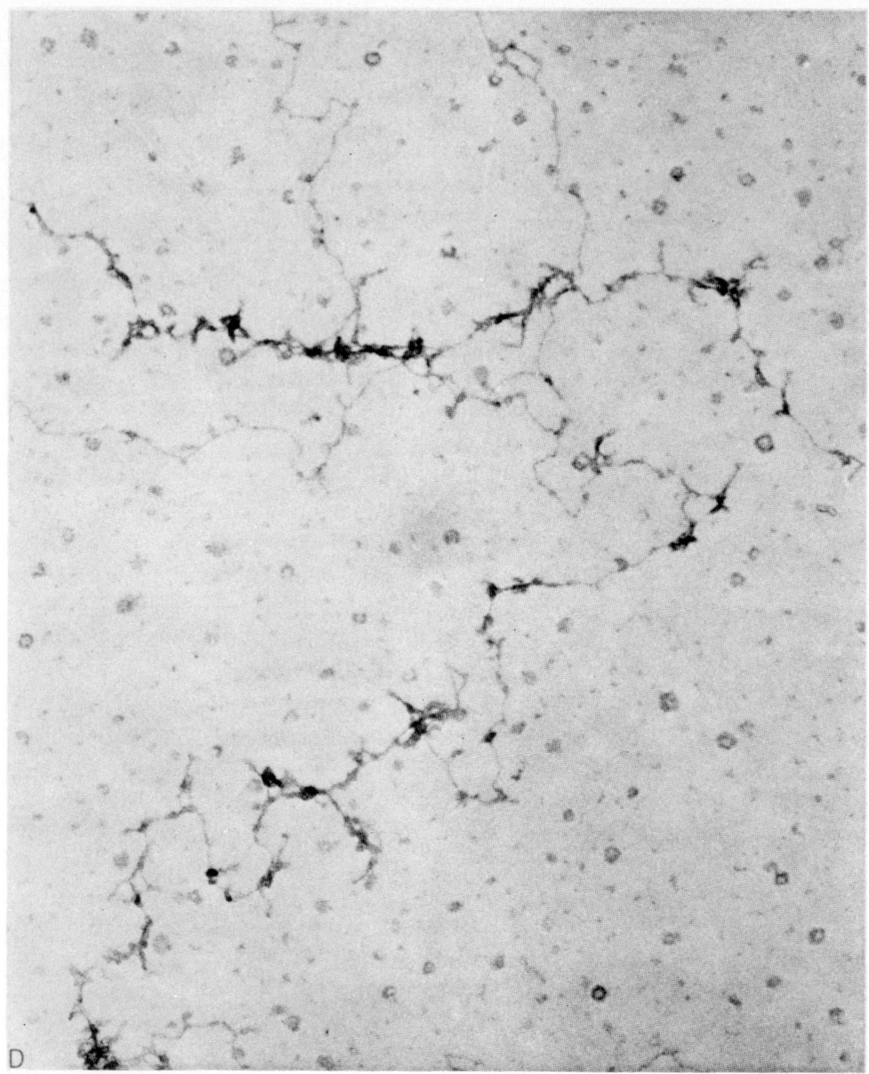

FIG. 1D. Nucleochromatin has been digested with pronase before fixation and staining as in (C). Note that digestion of the protein releases 20 Å double-stranded DNA. In many instances, the DNA folds back on itself to yield projections which appear to be residuals of the original web structure of the nucleochromatin. ×96,000. See caption Fig. 1A.

The author is very grateful to Professor Hans Ris, Department of Zoology, University of Wisconsin, for these electron microscopical studies and the permission to include the prints in this chapter. Further details of this study are present in his publication (Ris, 1967).

plate specificity of this enzyme by sigma factors (Burgess *et al.*, 1969; Travers and Burgess, 1969; Summers and Siegel, 1969; Crouch *et al.*, 1969). In eucaryotic nuclei a nonrandom distribution of certain proteins in chromatin is revealed in the distinctive nuclear distributions of antinuclear immunoglobulins from patients with systemic lupus erythematosis (Beck, 1961, 1963; Casals *et al.*, 1963). Similarly the reassociation experiments of Paul and Gilmour (1968) with acidic proteins and DNA attest to ordered relationships for some of these proteins with DNA.

Whereas the above observations point to the existence of unique nucleoprotein complexes within chromatin, it is quite likely that a large fraction of the proteins of chromatin recognize only general features of DNA such as the helical structure of the exposed chains of negatively charged phosphate groups. Histones appear to belong to this class. Still other proteins must recognize and interact with the proteins that already coat the DNA core to account for the bridging and webbing of the chromatin fibers seen in the electron-microscopical pictures. Whether or not RNA, ribonucleoproteins, complex lipids, and carbohydrates participate in the assembly or structure of chromatin remains to be studied. In any case chromatin is best viewed as the product of a self-assembly process in which the various protein subunits assemble along the chains of DNA according to their individual complementarities.

The importance of the complex and varied ultrastructure of chromatin in hormone action studies arises from its influence on gene expression. In order for given segments of DNA to be transcribed into RNA, the template must be accessible to RNA polymerase. Nucleochromatin, with its DNA existing as chains of DNA-protein aggregates, presents highly restricted templates for transcription. In a sense the chromatin system is a *molecular capacitor* in which information of the gene is stored in specific protein–DNA aggregates. The problem of releasing or discharging this information (i.e., the transcription of DNA into RNA by polymerase) is one of molecular mechanics involving the formation and dissolution of macromolecular aggregates amid the chromatin structure. It can be anticipated that the stability of each complex is a function of the degree of complementarity which exists between the molecular components of a particular aggregate. To change a molecular aggregate in preparation for transcription of the contained DNA it may be necessary to alter the existing complementarities in a specific manner; this requires the intervention of some outside force such as an approaching protein molecule. It is proposed that when an incoming molecule interacts with one of the components of the existing aggregate so as to induce a conformational change in this component, the bonding of the component to other units of the aggregate will be disturbed. A new comple-

mentarity will arise which in certain cases will expose the underlying template DNA or facilitate the transcription process.

In this field of molecular mechanics the incoming or perturbing molecules may be large or small, stable or unstable, and belong to any compositional class; the only important requirements are that they exhibit some complementarity to a component of the existing aggregate and induce some change in it through the interaction. The changes may be implemented simply through the mass of the incoming molecule in the case of large molecules or may result from localized enzyme catalysis. Selectivity of the response can be expected to be greater and more efficient as the degree of complementarity increases between the entering molecule and the interacting component of the chromosomal site.

In this picture of nucleochromatin it is obvious that structure and function are inseparable. It is also obvious that the control of gene expression is a function of the availability of the specific inducing or perturbing molecules in the extranuclear environment. To regulate this availability Nature has invented a second set of *molecular capacitors* which oppose the operation of the *nuclear molecular capacitor;* these are the membrane systems of the cell. The latter, like the nucleochromatin, are molecular aggregates, and also guided in their self-assembly by the complementarity of the constituent macromolecules. The stability and perturbability of the associations are subject to the same physical bonding principles which operate in the nucleochromatin or, for that matter, in any macromolecular aggregation. Existing in direct contact with both the cytoplasm and extracellular compartment, however, the membranes provide a bridge for moment to moment sensing and buffering of molecular changes in the environment. Through processes of sequestration, metabolism, and transport, the membranes act as a second molecular capacitor to control the availability of certain molecules that can interact with chromatin to regulate gene expression. Distinct equilibrium states are expected to arise between these two opposing systems; situations in which the molecules needed for the induction of specific gene expressions in the chromatin system are retained in specific molecular associations amid the membrane structures and are only released or transported in response to the entry of some molecule from outside the affected cell, such as a hormone.

Thus the living cell is portrayed as an interaction between two molecular capacitor systems: the nucleochromatin and the membrane of the cell. Since both capacitors record moment-to-moment changes in the cell, the interaction is of a cascading nature as depicted in Fig. 2. Accordingly, both systems are modulated by factors that may affect intermediately the translational steps in protein synthesis, nucleic acid matu-

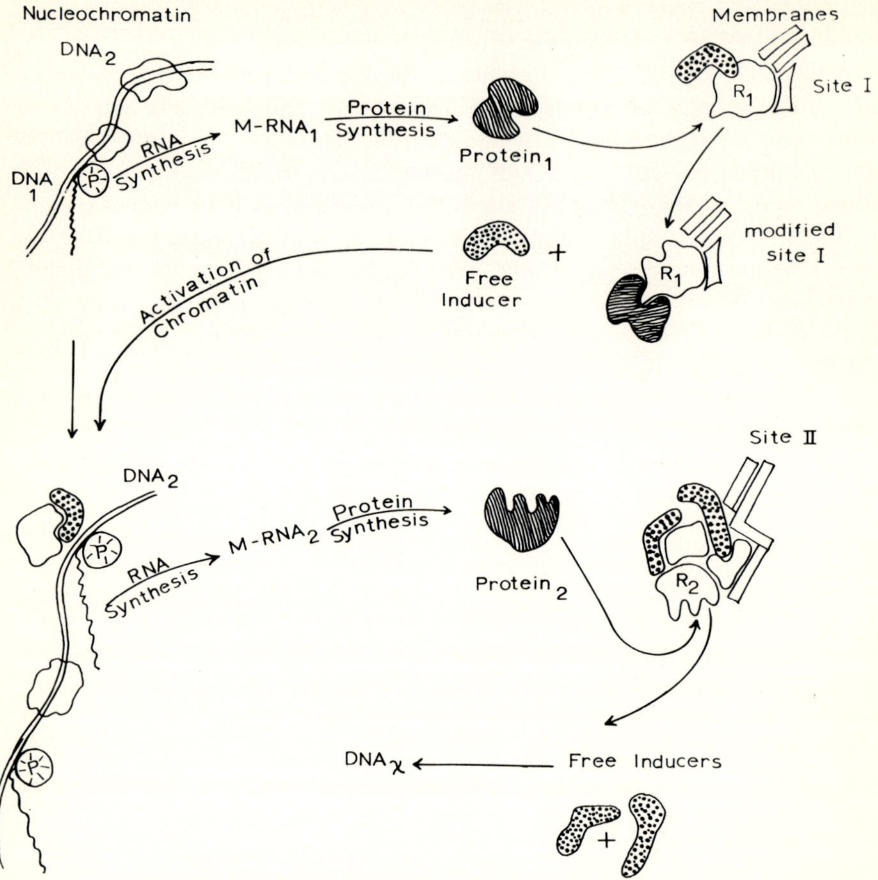

FIG. 2. A diagrammatic view of the interaction between nucleochromatin and the membrane systems of a cell during the sequential ordering of gene expressions. Nucleochromatin and membranes are visualized as two types of molecular capacitor systems which interact through the exchange of macromolecular components. In the diagram the synthesis of messenger RNA (M-RNA₁) gives rise to the synthesis of Protein₁ which in turn recognizes and associates with some specific site or component in the membrane (Site I). It is proposed that the perturbation which results leads to a localized activation of membrane metabolism and the release of a restrained inducer. By transport or diffusion the free inducer enters the nucleochromatin to react with a specific component of this structure to make the DNA available for transcription (messenger RNA, M-RNA₂). Carried through successive stages, a type of cascading control of gene expression ensures. The particular ordering rests in the character and stability of the macromolecular associations which can form in a given cell. Both the cell phenotype and the extracellular environment, i.e., factors from adjacent cells, influence the progression of gene expressions.

ration or metabolism, and any other processes that modify the character of existing macromolecules.

III. The Activation of Genetic Expression Mechanisms by Estrogens

In the framework of these concepts of cell biology, let us now examine and analyze the effects of estrogens in a highly responsive system, the rat uterus. Administration of a single physiological dose of a natural estrogen rapidly converts the atrophic uterus of the immature or ovariectomized female rat into an actively growing organ. Within minutes histamine levels decline (Szego, 1965) and cyclic AMP levels increase (Szego and Davis, 1967). As early as 1 hour after the administration of the hormone there is a generalized hyperemia of the tissue. (McLeod and Reynolds, 1938), followed by a general imbibition of fluid throughout the different cellular layers of this tissue. The water uptake reaches a maximum 4–6 hours after hormonal treatment (Astwood, 1938; Szego and Roberts, 1953). After 12 hours the increased dry weight of the organ becomes measurable and a second surge of water imbibition occurs which correlates with the striking acceleration of polymeric growth of the tissue (Mueller *et al.*, 1958; Telfer, 1953).

Compositional changes which attend this mobilization of growth are shown in Fig. 3. Where as the composition changes infer certain sequential effects of the hormone on the different synthetic pathways, incoporation studies with radioactive precursors demonstrate that each of these pathways (i.e., lipid, protein, and RNA synthesis) was accelerated with little or no lag period after the administration of estradiol (Mueller *et al.*, 1958). In confirmation of these findings, Hamilton (1968) and. Means and Hamilton (1966b) have reported an increased labeling of nuclear RNA as early as 2 minutes. Similarly Barnea and Gorski (1970) have documented the estrogen-induced synthesis of a single electrophoretically resolvable protein as early as 40 minutes after an injection of 17β-estradiol. In this case evidence for the synthesis of some RNA, the presumptive messenger for this protein, was obtained as early as 15 minutes (DeAngelo and Gorski, 1970).

Early experiments using puromycin to block protein synthesis (Mueller *et al.*, 1961) and using actinomycin D to block RNA synthesis (Ui and Mueller, 1963) demonstrated very clearly that the early estrogen response was highly dependent on the synthesis of both new RNA and protein. These results have been confirmed and extended by many laboratories. For example, Gorski and Axman (1964) obtained similar results using cycloheximide, which inhibits protein synthesis by a completely different mechanism. This group also showed that early estrogen effects on carbohydrate metabolism was prevented by cycloheximide (Nicolette and Gorski, 1964; Gorski and Morgan, 1967). They demonstrated further that the delayed administration of inhibitors of protein synthesis caused a rapid loss of the accelerated transport and phosphorylation of deoxyglucose which was achieved by prior estrogen treatment. This result, showing that the perpetuation of the hormonal response requires the continuous synthesis of some essential protein, suggests that the synthesis, metabolism, or transport of certain proteins are important targets

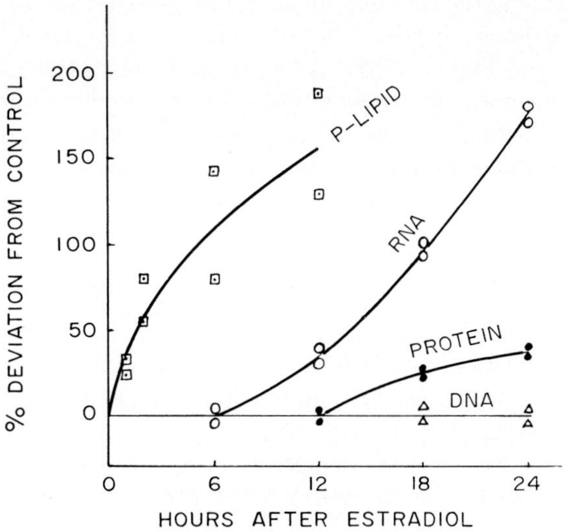

Fig. 3. Alterations in the composition of rat uteri following a single dose of estradiol (10 µg) injected at zero time. DNA was measured in micromoles of thymine per uterus. RNA was measured as micromoles of uridine and calculated as the ratio of uridine to thymine. Phospholipid (P-LIPID) was measured as micromoles of ethanolamine phosphate. All data are expressed as the percent deviation from the control during the first 24 hours after hormone treatment. Reprinted from Aizawa and Mueller (1961).

of the hormone action. Therefore, even though the hormones do exert extragenomic effects (Ui and Mueller, 1963; Lippe and Szego, 1965; Nicolette and Mueller, 1966a) a primary problem in elucidating estrogen action is to explain the role of protein synthesis in the control of genetic expression and the manner in which the hormone modulates this role.

IV. A Role for Protein Synthesis in the Estrogen Induction of and Maintenance of RNA Polymerase Activity

After the administration of estradiol, the DNA-dependent RNA polymerase activity of rat uterine nuclei rises rapidly (Gorski, 1964)

(Fig. 4). As in the case of the other *in vivo* estrogenic responses both the induction of this enzyme activity, as well as the maintenance of the induced activity, was abolished by levels of puromycin or cycloheximide which block protein synthesis in the intact rat (Gorski and Morgan, 1967; Gorski *et al.*, 1965). Thus protein synthesis appears to be required for both the induction and maintenance of this enzyme activity.

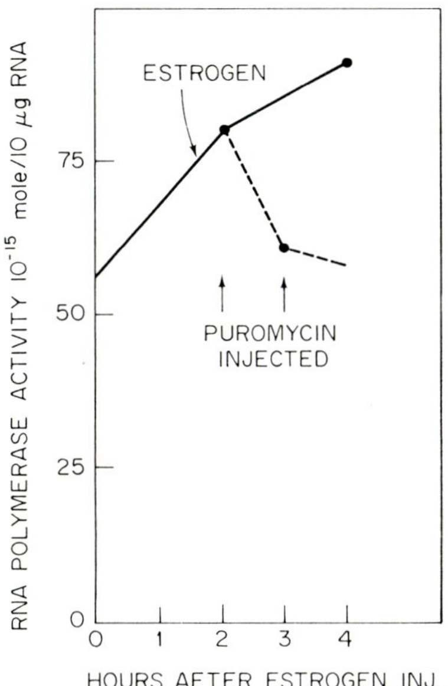

Fig. 4. The reversal by puromycin of the increase in RNA polymerase activity caused by estrogen. Estradiol injected at zero time into 21-day-old female rats. Puromycin (5 mg) was injected at 2 and in some groups again at 3 hours after estrogen. Rats were killed at times indicated in the figure and RNA polymerase assayed in uteri. Uteri from five rats were pooled for assays at each time point. Data are expressed as 10^{-15} mole cytidine-triphosphate-^{3}H incorporated per 10 μg of RNA released by ribonuclease treatment. Reprinted from Gorski (1964).

Analysis of the RNA products synthesized by isolated uterine nuclei has in turn revealed that the number of RNA chains in the process of synthesis per unit of DNA during the first hour of estrogen treatment is the same for both control and estrogen-treated rats; however, the rate of chain elongation is significantly enhanced (Barry and Gorski, 1971) (Fig. 5). These data, when added to the similarity of the product by nearest-neighbor frequency determinations (Mueller and LeMahieu, 1971) an hybridization studies (O'Malley *et al.*, 1969), suggests that the major initial effect of the hormone is to accelerate the rate of tran-

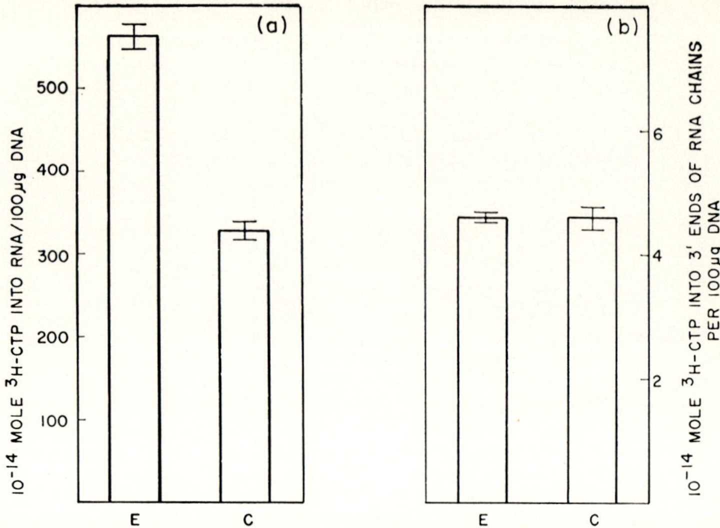

FIG. 5. Failure of early estrogen treatment to activate new polymerase sites. The incorporation of CTP-^{3}H into total RNA (a) and into 3' chain ends (b) by uterine nuclei from control and 1-hour estrogen-treated uteri. RNA synthesis was carried out for 10 minutes *in vitro,* and the product was subjected to alkaline hydrolysis. The amount of radioactivity in the cytidine-^{3}H liberated from the 3' ends of the RNA chains and the CMP-^{3}H of internucleotide linkages was separated by chromatography on PEI-cellulose and measured. Data are taken from a paper by Barry and Gorski (1971).

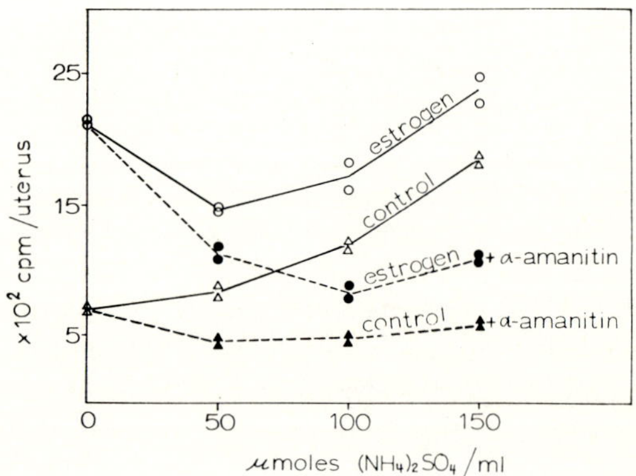

FIG. 6. The insensitivity of the estrogen-induced RNA polymerase activity to α-amanitin. Uterine nuclei from control (△) and estrogen-pretreated (○) (4 hour) rats were tested for their ability to synthesize RNA *in vitro* in the presence or absence of 5 μg/ml of α-amanitin at the specified concentration of (NH₄)₂SO₄ (Mueller and LeMahieu, (1971).

scription by RNA polymerase rather than to initiate new sites of synthesis. Furthermore, this acceleration is limited to the α-amanitin-insensitive fraction of RNA polymerase activity, which, in turn, is strikingly inhibited by low levels of $(NH_4)_2SO_4$ in the assay system (Figs. 6 and 7).

In an attempt to explore the relationship of protein synthesis to the induction and function of the estrogen-induced polymerase activity our laboratory has studied the maintenance of this enzyme in the nuclei of rat uterine segments incubated *in vitro*. For these experiments uteri from control or estrogen-treated immature female rats are transferred

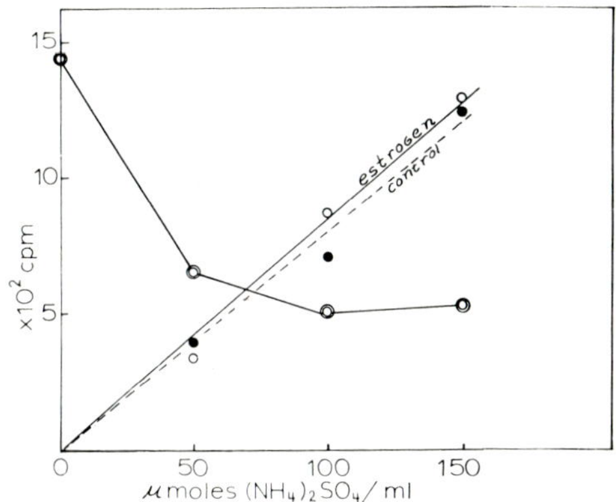

Fig. 7. The influence of salt concentration on the estrogen-induced RNA polymerase activity. All data are taken from Fig. 6. Amanitin-sensitive enzyme corresponds to the fraction of RNA polymerase activity which was sensitive to α-amanitin (5 μg/ml) at the specified concentrations of $(NH_4)_2SO_4$; this fraction was the same in both control ($\bullet$) and estrogen-activated ($\bigcirc$) nuclei. The fraction of RNA polymerase of control uterine nuclei that was insensitive to α-amanitin was then subtracted from the same value of estrogenized and plotted as the difference curve ($\triangle E\text{-}C$). The data reveal a selective depression of the estrogen-induced RNA polymerase activity by increasing concentrations of $(NH_4)_2SO_4$.

to small flasks containing a modified Eagle's medium with 10% bovine serum (Nicolette and Mueller, 1966b) and incubated with a 95% O_2–5% CO_2 gas phase at 37°C or at another temperature as indicated. At the specified times the uteri are removed and homogenized in Winnick and Winnick's buffer medium (Winnick and Winnick, 1960), and the nuclear fraction is isolated by differential contrifugation. The DNA-dependent RNA polymerase activity of these fractions is then assayed according to the method of Gorski (1964) as modified by Nicolette *et al.* (1968).

As shown in Fig. 8 the RNA polymerase activity of both control and estrogen-induced uteri is maintained reasonably well during a 2-hour incubation of the uteri in tissue culture medium; in fact nuclear polymerase survives well beyond 4 hours and even increases in activity in uteri incubated longer (Nicolette *et al.*, 1968; Nicolette, 1969). However, the addition of cycloheximide to the tissue culture medium, at a level that blocks protein synthesis in the surviving uteri, depressed the RNA polymerase activity of the estrogen-induced uteri to the level of the controls. This response is largely effected within 30 minutes. Treatment of control uteri with cycloheximide during the same interval had little or no effect on nuclear RNA polymerase. Similar results were obtained

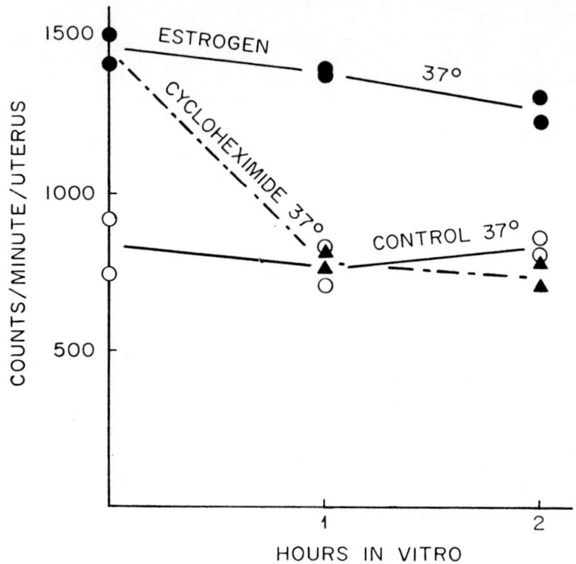

Fig. 8. Maintenance of estrogen induced RNA polymerase activity in uteri surviving *in vitro* and the effect of cycloheximade (25 μg/ml) ($\blacktriangle$---$\blacktriangle$) on this activity. Estradiol (5 μg) given *in vivo* 4 hours before *in vitro* incubation at 37°C; controls (○——○) received no estradiol. RNA polymerase was assayed in nuclei isolated at the indicated times by the method of Gorski (1964). Reprinted from Nicolette and Mueller (1966b).

when protein synthesis was limited by puromycin or by incubation of the surviving uterine segments in a medium deficient in certain essential amino acids.

Transfer of estrogenized uteri into fresh medium after a preliminary incubation in a medium containing cycloheximide was attended by a resumption of protein synthesis and restoration of the estrogen-induced RNA polymerase activity (Nicolette *et al.*, 1968). These results, which are in accord with the effects of cycloheximide in the living rat, also show that, once registered, the estrogenic state persists in the surviving uteri through temporary restrictions of protein synthesis which reversibly inactivate or depress the function of the estrogen-induced RNA polymerase activity. This observation suggests that establishment of the estrogenized state is separable from its function. Since cycloheximide treatment of the living rat can preclude the establishment of the estrogenized state (Mueller and LeMahieu, 1971; Gorski and Morgan, 1967; Smith and Gorski, 1968) there may be a requirement for the synthesis of more than one protein in the implementation of the early hormone response.

In the preliminary experiments it was noted that the maintenance of the estrogen-induced RNA polymerase activity in the nuclei of the

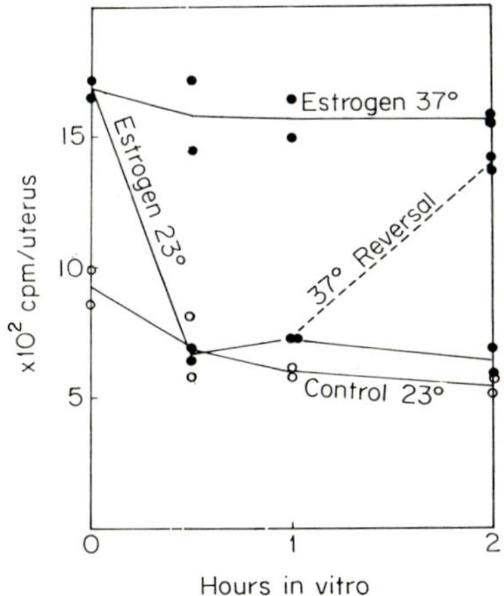

Fig. 9. The effect of *in vitro* incubation of estrogen-treated (●——●) and control (○——○) at 23°C on the RNA polymerase activity in isolated nuclei. Polymerase activity of estrogen-treated uteri incubated continuously at 37°C, or at 37°C after 1 hour at 23°C (●---●) were also determined. Reprinted from Nicolette and Mueller (1966b).

surviving uterine segments was unusually sensitive to the temperature of the incubation. Whereas incubation of the uteri *in vitro* at either 37°C or 0°C permitted the maintenance of the RNA polymerase activity of both control and estrogenized uteri, incubation at 23°C was attended by a rapid depression of the RNA polymerase activity in the nuclei of estrogenized uteri to the control level. Incubation of control uteri at 23°C had little or no effect on the RNA polymerase activity (Fig. 9) (Nicolette and Mueller, 1966b). Return of the surviving uteri to 37°C was associated with a very rapid rise of the RNA polymerase activity in the estrogenized uteri. In many cases the level of activity rose above the starting level of the freshly isolated uteri for a short time. Control uteri, which had been incubated *in vitro* at 23°C for a 3-hour period, also responded with an elevation of RNA polymerase activity when transferred to 37°C; however, this response was always quantitatively less than the estrogenized uteri and transitory in character. In both types of uteri the striking acquisition of polymerase activity was prevented when protein synthesis was blocked by the addition of cycloheximide (or puromycin) to the tissue culture medium (Fig. 10). In accord with the earlier results, the delayed addition of cycloheximide

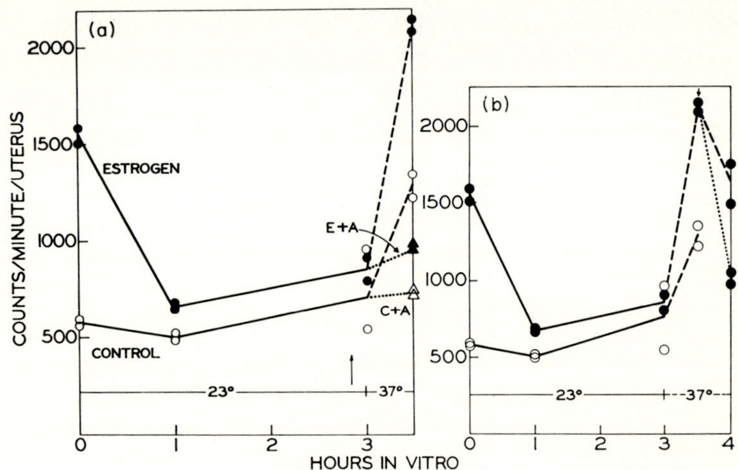

Fig. 10. The effect of cycloheximide on the temperature-induced reversal of uterine RNA polymerase activity. (a) Uteri from estrogen-treated (●——●) or control (○——○) animals incubated at 23°C for 3 hours, then transferred to 37°C (●--●, ○--○). Ten minutes before transfer to 37°C, 25 μg/ml cycloheximide added to appropriate flasks with estrogen-treated (○···▲) (E + A) and control (○···△) (C + A) uteri. (b) Same experiment and symbols as A; cycloheximide added 30 min after transfer to 37°C (●···●). Reprinted from Nicolette and Mueller (1966b).

caused a rapid and striking depression of the estrogen-induced polymerase which was regained by the shift of the uteri back to 37°C. In this situation cycloheximide blocked the reactivation response in both controls and estrogenized uteri.

The observations in these experiments showing that the estrogen-induced RNA polymerase was maintained in uteri incubated at 37°C or 0°C, but lost in uteri incubated at 23°C, suggests that at 37°C two opposing reactions are operating: one which inactivates the estrogen-induced polymerase, and a second which activates this enzyme. At 37°C the activation process appears to keep ahead of the inactivation mechanism; however, at 23°C the inactivation mechanism appears to dominate.

Since the reactivation of the RNA polymerase on transfer of the uteri from 23°C to 37°C is prevented by cycloheximide, it is suggested that the temperature-sensitive mechanism is concerned as well as the protein-requiring step in the control of RNA polymerase activity.

Considerable insight into the role of protein synthesis and the temperature-sensitive processes in the control of RNA polymerase in estrogenized uteri has been obtained through experiments with a reversible inhibitor of RNA synthesis, 2-mercapto-1-(β-4-pyridethyl)benzimidazole (MPB).

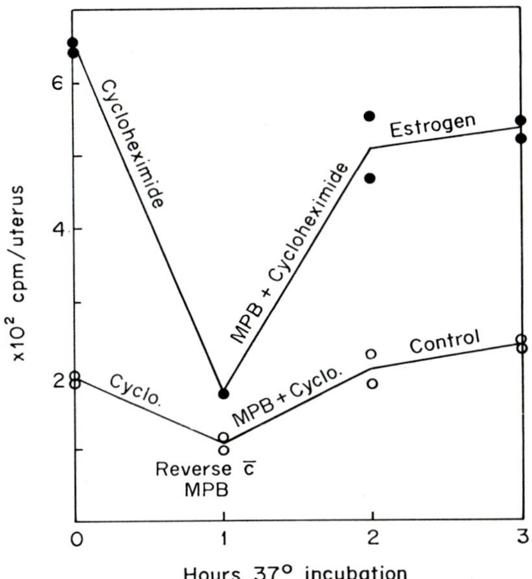

Fig. 11. Reversal of the cycloheximide depression of estrogen-induced RNA polymerase by 2-mercapto-1-(β-pyridethyl)benzimidazole (MPB). Estrogen-treated (●——●) and control uteri (○——○) were incubated in tissue culture medium containing cycloheximide (cyclo.) (1.0 μg/ml); after 1 hour, MPB (50 μg/ml) was added. Nuclei were isolated at the indicated times and assayed for RNA polymerase activity (Mueller and LeMahieu, 1971).

Contrary to the report of Nakata and Bader (1969), this agent is an excellent inhibitor of RNA synthesis in surviving rat uteri as well as in 4 strains of cells cultured in this laboratory (Mueller *et al.*, 1971). When this agent is added to the tissue culture medium, not only is RNA synthesis blocked *in situ*, but also the depressive action of cycloheximide on the level of the estrogen-induced polymerase activity which was described earlier. Even more interesting is the finding that the delayed addition of MPB to cycloheximide depressed uteri causes a restoration of the estrogen-induced RNA polymerase activity (Fig. 11). In contrast to these results, MPB is completely ineffective when the surviving uteri are incubated at 23°C.

When MPB is added along with cycloheximide at the onset of the incubation of the surviving uteri in tissue culture medium, a rapid increase in RNA polymerase activity was observed in the nuclei of both control and estrogenized uteri (Fig. 12); the surge of activity of the control uteri may even attain the level of the estrogenized uteri. This elevation of RNA polymerase activity in the control uteri is, however,

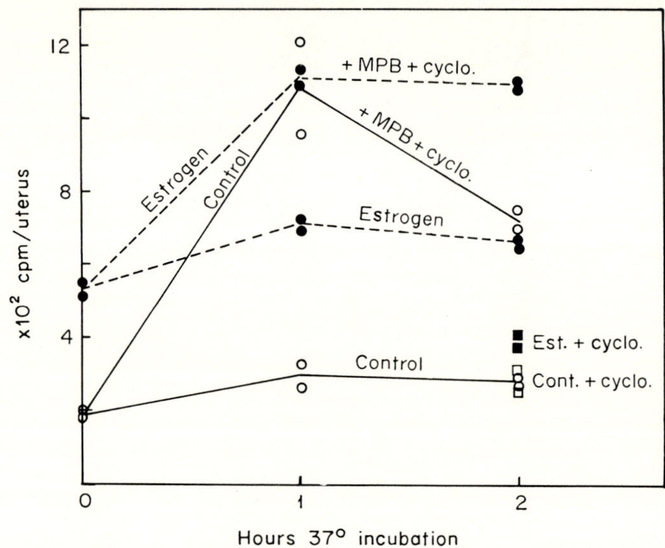

Fig. 12. Activation of nuclear RNA polymerase in surviving uteri incubated *in vitro* with 2-mercapto-1-(β-pyridethyl)benzimidazole (MPB). Estrogen-treated and control uteri were incubated in tissue culture medium ± the addition of 1.0 μg/ml cycloheximide (cyclo.). MPB (50 μg/ml) was added to cycloheximide-treated uteri as indicated. RNA polymerase activity was measured in nuclei isolated from the surviving uteri after the indicated treatments (Mueller and LeMahieu, 1971).

transitory and with time subsides toward the initial control level. The observation that the level of RNA polymerase may rise to the same high level in control and estrogenized uteri is in agreement with the conclusions of Barry and Gorski (1971) that the number of operating polymerase molecules in control and estrogenized uterine cells may be the same, but that they vary in their operating efficiency. Under the conditions of MPB treatment it would appear that at least for a short time this efficiency can be raised in both cases to the same level.

To interpret these findings, it is important to remember that RNA synthesis is blocked during the incubation with MPB and the isolation of the nuclei from such tissues reverses the inhibitory action of this compound. Thus, the assay of the RNA polymerase activity in nuclei isolated from such tissues reflects the amount of potential enzyme activity which has accrued under the specified conditions; as such it can be viewed as a summation of the positive and negative reactions that affect the functionality of the polymerase sites. The simple conclusion of the cycloheximide (or puromycin) experiments is that the estrogen-induced RNA polymerase activity depends on the availability of some protein which is used up in the process of RNA synthesis at certain polymerase sites. As shown in Fig. 6 these sites contain α-amanitin-insensitive polymerase units. When protein synthesis is blocked, this protein rapidly becomes limiting. This situation is reversible and can be corrected by allowing the tissue to resume protein synthesis.

Similarly, the effects of MPB can be explained in terms of this protein requirement. Assuming that a critical protein is actually used up in the cyclic operation of the polymerase, it would follow logically that slowing down RNA synthesis *in situ* with MPB would have a sparing action on this need; in uteri blocked with cycloheximide, this sparing action might even allow time for restoration of the polymerase sites to an active state by recovery of this protein from bound or inactive forms.

The temperature-sensitive step in the maintenance of the estrogen-induced polymerase activity appears most likely to involve a mechanism which supplies or activates the essential protein(s) for use by the polymerase sites. This role is compatible with the finding that the temperature-sensitive process completely overrides the stimulatory effect of MPB and depresses the polymerase activity by a process that is independent of protein synthesis. The observations that estrogenized uteri are distinguished from control uteri by the magnitude of this process (Figs. 9 and 10) suggests that this is also the hormone-sensitive step. Accordingly, the definition of estrogen action with respect to its effects on gene transcription becomes one of defining the process by which the hormone mediates the availability of some limiting proteins(s) at the RNA polymerase site. While the protein may function as a component of the polymerase, it appears more likely that it is involved in the processing or the transport of the newly synthesized RNA from the polymerase site.

The above metabolic studies and the emerging concept of estrogen action prompted a direct search for proteins which might facilitate RNA synthesis in isolated rat uterine nuclei.

Although these studies are still in a preliminary state, some evidence for the existence of such stimulatory protein has been obtained. As shown in Figs. 6 and 7 the estrogen-induced polymerase activity is sensitive to the presence of low levels of $(NH_4)_2SO_4$ in the assay system. In addition, it was found that prewashing the nuclei in 0.1 M $(NH_4)_2SO_4$ removed a large molecular weight factor which was important for polymerase function; nuclei of estrogenized uteri contained more of this entity than did the controls. When the estrogenized uteri were preincubated in tissue culture medium in the presence of MPB, both the polymerase activity and the fraction which was sensitive to prewashing of

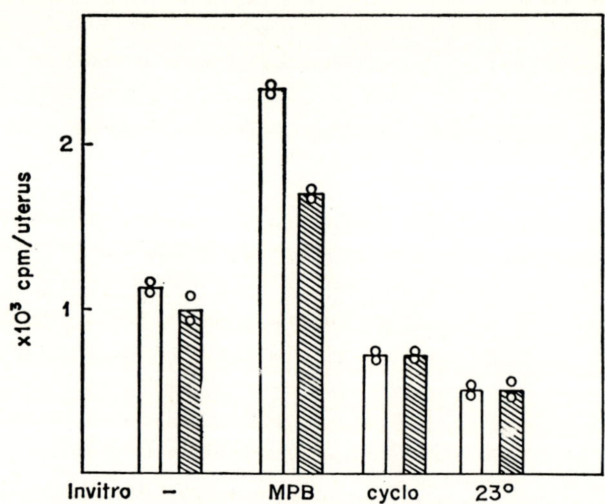

FIG. 13. The influence of salt extraction on the estrogen-induced RNA polymerase activity of nuclei from uteri incubated *in vitro*. Uterine segments from immature rats pretreated *in vivo* for 4 hours with estradiol were incubated for 1 hour in tissue culture at 37°C or 23°C as indicated. 2-Mercapto-1-(β-pyridethyl)benzimidazole (MPB) (50 μg/ml) or cycloheximide (cyclo.) (25 μg/ml) were added as shown. Nuclei were isolated and the RNA polymerase activity was measured directly (□) or after washing with 0.1M $(NH_4)_2SO_4$ (▨). Data are expressed as counts per minute of CTP-³H-incorporated into RNA (Mueller and LeMahieu, 1971).

the nuclei with 0.1 M $(NH_4)_2SO_4$ was increased (Fig. 13). Nuclei from uteri incubated in the presence of cycloheximide or at 23°C, treatments which depress the estrogen induced polymerase activity, were also insensitive to the $(NH_4)_2SO_4$ prewashing. These data suggest that when conditions are optimal for polymerase function, the nuclei contain an $(NH_4)_2SO_4$ releasable factor which is important for polymerase activity. In separate experiments this factor has been found to be high molecular weight and labile; it is presumed to be protein.

In other experiments the soluble protein fraction from uterine preliminary reports of Raynaud-Jammet and Baulieu (1969), Beziat homogenates has been prefractionated by gel filtration and then assayed for ability to stimulate the RNA polymerase activity in nuclei from rat uteri. These results are presented in Fig. 14. These data illustrate that the soluble protein fraction from rat uteri does contain a protein(s), or a related material fractionating like a protein, which increases the RNA polymerase activity of isolated uterine nuclei. The addition of estradiol to such fractions prior to gel filtration caused minor shifts

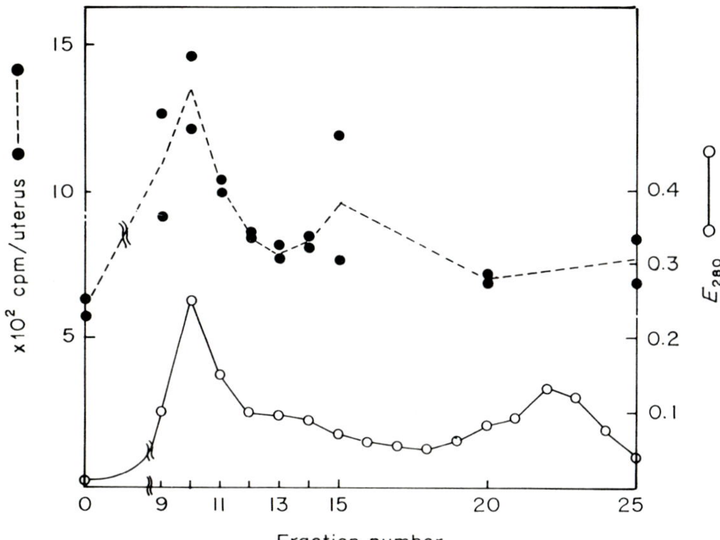

FIG. 14. Effect of soluble uterine proteins on RNA polymerase activity of uterine nuclei. A 20% homogenate of adult rat uteri (unstaged) was prepared in 0.1 M Tris, 0.0015 M EDTA (pH 7.4) and centrifuged for 1 hour at 105,000 g. The soluble protein fraction was passed over at a 6% agarose column, and 1 ml fractions were collected. Fractions (0.45 ml per 1.0 ml assay volume) were assayed for their influence on the RNA polymerase activity of control uteri (Mueller and LeMahieu, 1971).

in the elution character of the active fraction but had little effect on their activity.

Recently the soluble fractions of pig uteri have been treated for stimulatory effects on isolated nuclei from both rat and pig uteri. In this case preincubation of the uterine soluble protein fraction at 37°C for 15 minutes strikingly increases the stimulatory activity of the preparations. The addition of estradiol to the soluble protein fractions during the temperature activation step increases the activity significantly, but the results have been variable. The latter finding is in accord with the et al. (1970), Arnaud et al. (1971), and Mohla et al. (1971) on an estrogen-facilitated cytoplasmic stimulation of uterine RNA polymerase activity.

V. The Subunit Character of Estrogen Receptors and Their Possible Role in Estrogen Action

Through the pioneer studies of Jensen and Jacobson (1960) the presence of estrogen-binding proteins (i.e., estrogen receptors, or estrophiles) in estrogen-sensitive tissues has been established. Many laboratories have confirmed their initial discovery and expanded our knowledge of the properties and metabolic fate of estrogen receptors in target tissues (Table III). A summary of the overall picture that has emerged is presented in Fig. 15.

TABLE III

Estrogen Receptor Properties

Properties	References
Selectively binds active estrogens	Noteboom and Gorski (1965), Jensen *et al.* (1966), Toft and Gorski (1966), Brecher and Wotiz (1968), Talwar *et al.* (1968), Korenman (1969), Puca and Bresciani (1969b), Steggles and King (1970)
Hormone bound noncovalently	Jensen *et al.* (1966), Maurer and Chalkley (1967)
$K_{diss} = 0.4$–$2.0 \times 10^{-9}\ M$	Toft and Gorski (1967), Erdos *et al.* (1969, 1970), Puca and Bresciani (1969b), Shyamala and Gorski (1969), Vonderhaar *et al.* (1970a)
Destroyed by proteolytic enzymes but not by nucleases	Noteboom and Gorski (1965), Toft and Gorski (1966), Jensen *et al.* (1967a), Puca and Bresciani (1968a)
Binding sensitive to −SH reagents and iodination	Jensen *et al.* (1967b), Terenius (1967), Puca and Bresciani (1970), Steggles and King (1970)
Acidic character	Bresciani *et al.* (1969), DeSombre *et al.* (1969), King *et al.* (1969)
Soluble form sediments as 8–9 S at 0°	Toft and Gorski (1966, 1967), Jensen *et al.* (1967a)
Molecular weight of 200,000	Toft and Gorski (1966), DeSombre *et al.* (1969)
Dissociates to 4 S subunit with 0.3–0.4 M KCl	Erdos (1968), Korenman and Rao (1968), Rochefort and Baulieu (1968), Jensen *et al.* (1969), Vonderhaar *et al.* (1970b)
Transfers to nucleus at 37°	Brecher *et al.* (1967), Jensen *et al.* (1968, 1969), Stumpf (1968a), Brecher and Wotiz (1969), Rochefort and Baulieu (1969), Shyamala and Gorski (1969), Musliner *et al.* (1970), Giannopoulos and Gorski (1971)
Extracts from nucleus as 5 S form with 0.3–0.4 M KCl	Jensen *et al.* (1967a, 1969), Puca and Bresciani (1968a)

When estradiol enters the uterine cell it combines with the estrogen receptor located in the cytoplasm to form a noncovalent, high-affinity complex. In a low ionic strength sucrose gradient the receptor of the cytosol sediments as a 8–9 S complex; however, it is reversibly dissociated by 0.3–0.4 M KCl to yield a 4 S form of the receptor.

When the initial binding is done in surviving uterine segments at 0°C, the estrogen–receptor complex remains in the cytoplasm. However, when the incubation temperature is raised to 37°C, the estrogen, presumably in combination with the estrophile, transfers to the nucleus

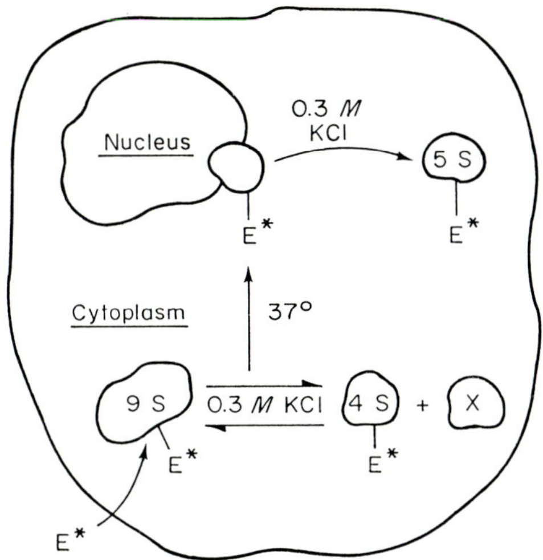

Fig. 15. A diagram of the interaction of estradiol with the receptor system in a hypothetical uterine cell. The entering estrogen molecule (E*) combines (by a temperature-insensitive process) with the 9 S salt-dissociable receptor of the cytoplasm. The hormone, presumably in combination with the receptor protein, translocates to the nucleus by a temperature-sensitive process. Salt treatment (0.3 M KCl) dissociates a 5 S bound hormone fraction from the nucleus.

(Jensen *et al.*, 1968, 1969; Shyamala and Gorski, 1969). This transfer occurs in the presence of actinomycin D, cycloheximide, or puromycin, agents which block much of the ensuing hormone response. Quantitative studies on the nature of this transfer reaction have shown that it is at the expense of the cytoplasmic estrogen–receptor complex and that the rate of depletion of the cytoplasmic receptor is dependent on the concentration of estradiol in the incubation medium (Giannopoulous and Gorski, 1971a).

The estrogen–receptor complexes can be released from nuclei by treatment with 0.3–0.4 M KCl (Jensen *et al.*, 1967a; Puca and Bresciani,

1968a) or by treatment at 0°C with pancreatic DNase (Harris, 1971; Fevold and Mueller, 1971). The salt released nuclear receptor sediments as a 5 S complex in a sucrose gradient containing 0.4 *M* KCl, but aggregates severely in low ionic media to yield heterogeneous rapidly sedimenting complexes (Fig. 16). In contrast the DNase released nuclear

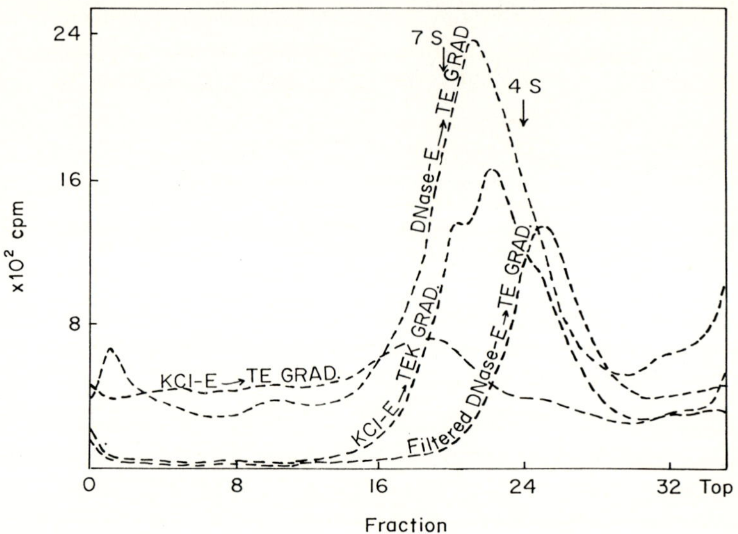

FIG. 16. Sedimentation properties of nuclear bound estrophiles after release by DNase or KCl. Immature rat uteri were incubated *in vitro* in Eagle's HeLa medium minus serum for 30 minutes at 37°C with ³H-estradiol to effect labeling of nuclear estrophiles. The nuclei were then isolated and extracted in the cold. Half of the nuclei were extracted for 30 minutes at 0°C with 10 μg DNase per nuclei from 12 uteri in 2 ml of 0.01 *M* Tris pH 7.4 and 0.001 *M* dithiothreitol solution. Another fraction of nuclei was extracted at 0°C with 0.4 *M* KCl. The released estrophiles were sedimented through 10–30% sucrose gradients prepared in 0.01 *M* Tris, 0.001 *M* EDTA (pH 7.4) (TEgrad.) or 0.01 *M* Tris, 0.001 M EDTA, 0.4 M KCl (TEK grad.). Sedimentation time was 16 hours in a SW 50.1 rotor at 40,000 rpm. Florescein-labeled immunoglobulins (7 S) and bovine serum (4 S) were included as markers in each sample. The distribution of receptor-bound ³H-labeled estradiol in the gradients was determined in fractions dripped from the bottom of each gradient. The filtered DNase-released estrophile preparation was diluted 10× with 0.01 M Tris, 0.001 *M* dithiothreitol and reconcentrated to 1.5× the original volume by filtration against an Amicon UM-10 filter prior to sedimentation.

bound receptor sediments in low ionic sucrose gradients as 6–7 S complexes. Treatment of the DNase-extracted receptors with 0.4 *M* KCl and sedimentation into a sucrose gradient containing 0.4 *M* KCl, however, yields a 5 S complex. While the latter sediments like the salt-released receptors in a high ionic sucrose gradient, it aggregates primarily to an 8–9 S size if sedimented instead into a low ionic sucrose gradient (Fig. 17). In the course of these studies it was also observed that dilution of the DNase-released receptor preparations, followed by reconcentration by filtration against an Amicon filter caused a shift in the sedimentation character from 6–7 S to a 4 S form (Fig. 16) suggesting that some aggregation promoting principle had been removed.

Summarizing these observations it appears that the estrogen receptor complex is bound to the nucleus primarily through electrostatic bonds.

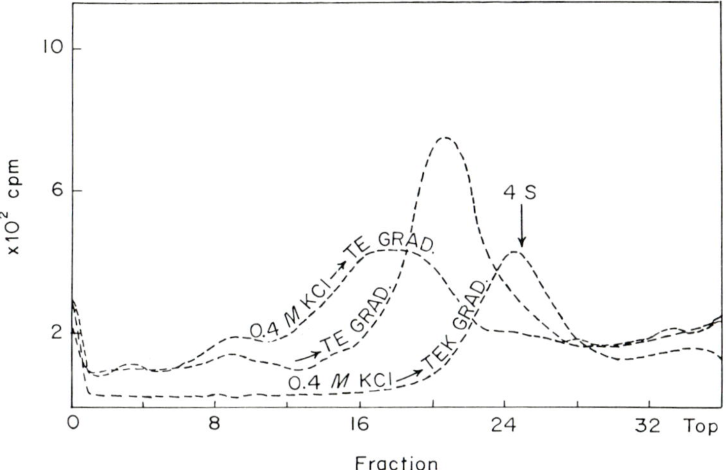

FIG. 17. The effect of salt treatment on the sedimentation properties of nuclear estrophiles released with DNase. DNase released nuclear estrophiles were prepared as described in Fig. 16, and sedimented directly into 10–30% sucrose gradients in 0.01 M Tris, 0.001 M EDTA pH 7.4 (→ TE grad.) or treated with 0.4 M KCl and then centrifuged into the TE or TEK gradients (0.01 M Tris–0.001 EDTA–0.4 M KCl pH 7.4). Sedimentation time was 16 hours in a SW 50.1 rotor at 40,000 rpm. The distribution of receptor-bound ³H-labeled estradiol was assayed in fractions dripped from the bottom of the gradient. Florescein-labeled bovine serum albumin (4 S) was added to each sample as a sedimentation marker.

Since a mild DNase treatment also releases the nuclear bound receptor, a close association with DNA is indicated; however, no evidence has been obtained for any specific associations of the receptor with DNA fragments. The demonstration that nuclei from a nontarget tissue (i.e., diaphragm) cannot substitute for uterine nuclei in the *in vitro* nuclear binding of the cytosol estrogen receptor suggests that specific nuclear receptor sites exist in the target tissue (Jensen *et al.*, 1969); little is known of these sites at present except that they may involve acidic proteins (O'Malley, personal communication). In any event, the released receptors exhibit a high tendency for aggregation in solution and the character of such aggregates, particularly in a low ionic medium, is very much dependent on the presence of other macromolecules in the extract.

The observation that the cytoplasmic binding and translocation of estradiol to the nucleus is not prevented or restricted by actinomycin D, cycloheximide, or puromycin, whereas these agents largely block the ensuing hormone response, suggests that the receptors play a primary role in estrogen action if related at all. It is evident, however, that the simple combination of the hormone with the receptor or its translocation to the nucleus are not equivalent to estrogen action since both of these steps take place in isolated uteri which are incubated *in vitro* without activation of the known estrogen-mediated responses. Accordingly, the estrogen receptor must mediate a second process if it is to be important in the hormone response. From the studies on the activation and maintenance of the estrogen-induced RNA polymerase activity, this process would appear to concern the supply of some essential protein to certain RNA polymerase sites. To explore this possibility our laboratory has undertaken a study of the dynamic properties of the cytoplasmic estrogen–receptor complexes. The following findings have emerged which appear significant for an understanding of estrogen action: (1) estrogen receptors constitute a family of macromolecular complexes; (2) they are composed of dissimilar subunits; and (3) their macromolecular association state is altered by the binding of the hormone with concomitant release of a small molecular component of the original receptor complex. In the final section of this paper these observations have been integrated into a working concept of estrogen action.

In early studies the dissociation of the 8–9 S estradiol receptors of the cytosol by 0.3–0.4 *M* KCl to yield 4 S forms was taken as evidence that the native cytosol receptor existed as a dimer in low ionic media (Jensen *et al.*, 1968). However, evidence that the picture was considerably more complex and that dissimilarity existed among the estrogen-receptor complexes was obtained in studies of estrogen-binding kinetics at different temperatures (Vonderhaar *et al.*, 1970a). As shown in Fig. 18, treatment of the cytosol fraction of rat uteri with estradiol-^{3}H at 0° resulted in only a fraction of the binding that can be obtained at 23°C. While the binding at 37°C is more rapid, there appears to be a concomitant destruction of certain receptor complexes following the combination with the hormone. At 0°C and 23°C the release of the bound estradiol is low and does not account for the differences in the forward binding reaction. The extra binding that was obtained at 23°C seems to involve a group of receptors that otherwise do not bind estradiol at 0°C. It also appears to involve some cooperative effect of the hormone and a temperature-sensitive process, since this binding does not occur if the hormone was added at 0° after a 23°C preincubation.

Further evidence for dissimilarity among the estrogen receptor complexes was obtained in experiments testing the effects of nucleotides and related compounds on the binding reaction (Vonderhaar *et al.*, 1970a). As shown in Fig. 19 the addition of ATP or GTP (10^{-3} M) also increased the binding capacity of the cytosol fraction. In these experiments the cytosol receptor solution was pretreated with unlabeled

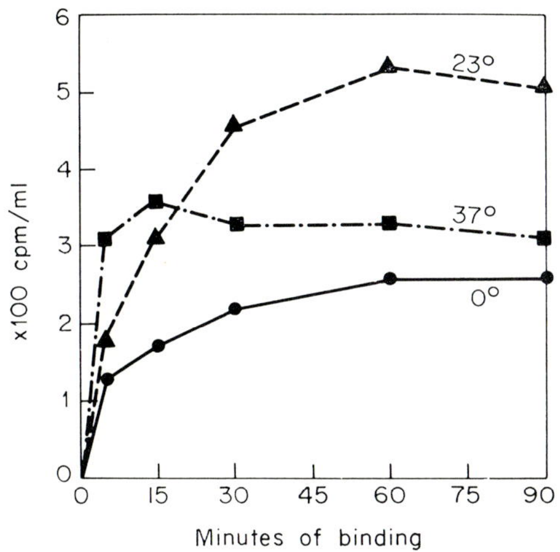

Fig. 18.   The effect of temperature on the kinetics of binding of estradiol- 6,7-^{3}H$_2$. To receptor solutions prepared by gel filtration on Biogel P-200, 10^{-4} μg of estradiol-^{3}H was added per milliliter and placed at 0°, 23°, or 37°C. At the indicated times 1-ml aliquots were assayed for bound hormone using the P-10 method described in the text. Reprinted from Vonderhaar *et al.* (1970a).

estradiol for 1.5 hours at 23°C to saturate available receptors. The extra cold hormone was then removed by gel filtration through a column of Biogel P-10. Estradiol-^{3}H was then added to the receptor solution and permitted to bind in the presence of added nucleotides as indicated. It appears that the gel filtration itself exposed some receptors in addition to removing the cold estradiol; however, the most striking effect was the extra binding obtained with purine nucleotides. Pyrimidine nucleotides proved to be relatively ineffective. In the case of the purine compounds, the nucleosides and free bases were similarly effective. In fact, a test of related compounds illustrated that the imidazole ring was the major determinant of activity (Vonderhaar *et al.*, 1970a). As in the case of the temperature-sensitive binding discussed above, the imidazole compounds had to be present during the exposure to estradiol-^{3}H in order to obtain the extra binding response.

The sensitivity of a fraction of potential estrogen receptor complexes to imidazole compounds presents an interesting, but unexplained phenomenon. Perhaps a fraction of estrogen receptors exists in a state which requires some conformational or hydrolytic change in order to close down in tight association with the incoming hormone molecule. The observations that certain fractions of receptors require the presence of imidazole compounds for binding whereas the other require only elevation of the

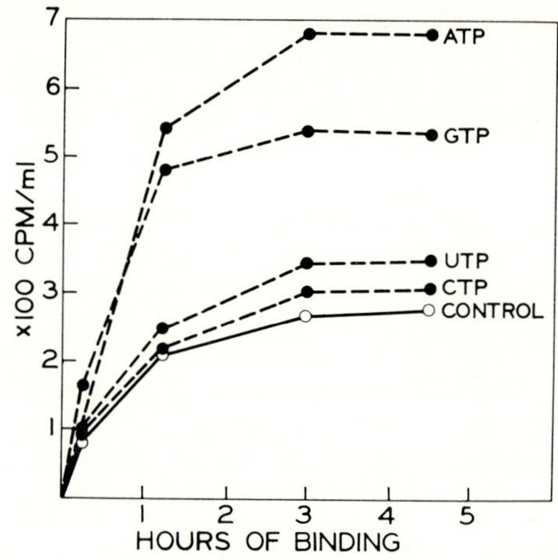

Fig. 19. The activation of a fraction of receptors in the P-200 eluate by certain nucleotides. Receptor solutions were incubated with 10^{-4} μg of unlabeled estradiol-17β for 1.5 hours at 23°C, and the free hormone was removed by gel filtration through a P-10 column. Aliquots of the eluate were then combined with 10^{-4} μg estradiol-^{3}H per milliliter (○——○) and the indicated nucleotide at 10^{-4} M (●--●) and incubated at 23°C. The amount of labeled hormone bound was determined at the indicated times by the P-10 method.

binding temperature to 23°C, attested to the heterogeneity of the receptors or receptor states and made improbable a simple dimer concept of the 8–9 S cytosol receptors.

Further evidence for the dissimilar nature of the subunits of the 8–9 S cytosol receptors was obtained in studies of the binding of the cytosol receptors to estradiol which had been immobilized to a modified polystyrene through a 17α-propyl side chain (Fig. 20) (Vonderhaar and Mueller, 1969). This resin is a highly efficient absorbent for estrogen receptors of the cytosol; however, great difficulty has been experienced in releasing the receptors again from the resin. To overcome this problem and to use the resins for isolation, attempts were made to bind the receptors in their assumed dimeric states, to titrate the sterically restricted subunits of the bound dimers with free estradiol-^{3}H and then dissociate the latter selectively with 0.3 M KCl. Surprisingly no estradiol-marked receptor could be dissociated; however, a receptor subunit was released, which as will be described later, modifies the ability of native 4 S receptors to reassociate. Additional attempts to associate native estradiol-^{3}H labeled 4 S estrophilic subunits with resin-bound estrophilic subunits also failed. These experiments, while not conclusive in themselves, support the view that the salt-dissociated subunits of the 8–9 S cytosol receptors are dissimilar and that only one type binds estradiol (Vonderhaar *et al.*, 1970b).

More compelling evidence of the heterogeneity of the subunits in the 8–9 S cytosol receptor complexes was obtained by first dissociating the

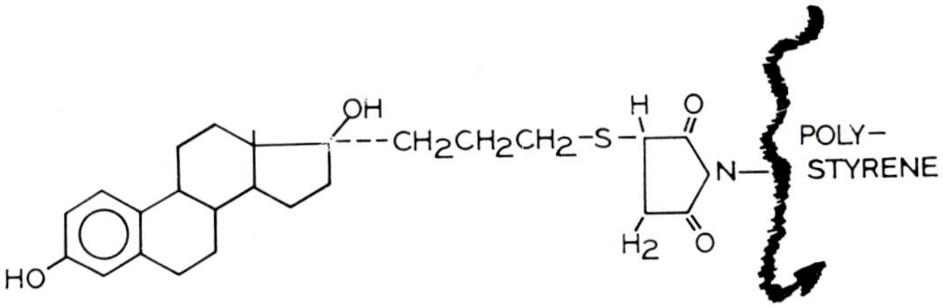

FIG. 20. Basic structure of PVM-estradiol. The 17α-propyl mercaptan derivative of estradiol was coupled to a polyvinyl (*N*-phenylenemaleimide resin) (Vonderhaar and Mueller, 1969).

receptors in 0.3 M KCl and then prefractionating the dissociated receptors by sedimentation in a sucrose gradient or by gel filtration over 6% agarose columns in the presence of 0.3 M KCl. In a characteristic experiment (Fig. 21) centrifugation of the salt-dissociated receptor solution yielded a symmetrical peak of 4 S estrogen receptors. Aliquots from the front of the peak (fraction 20) and the rear (fraction 23) were then recentrifuged in a low ionic strength sucrose gradient to test their reassociation characteristics. Whereas the front fraction yielded a 9 S peak on recentrifugation, the rear (or lighter) fraction reassociated only to a 7 S form and part may have failed to reassociate at all. It is apparent from these studies that the dissociated nonestrogen-binding subunits are dissimilar and can be separated in a sucrose gradient prepared in 0.3 M KCl. In fact the salt treatment appears to dissociate the nonestrophilic subunits into even smaller components, as witnessed by the formation of intermediate size (7 S) complexes in the reassociation study.

Further evidence was obtained for this view by filtration of the salt-dissociated receptors through a 6% agarose column and centrifugation of the prefractionated receptor peak in a low ionic sucrose gradient. In this case the reassociation was limited to the formation of 7 S com-

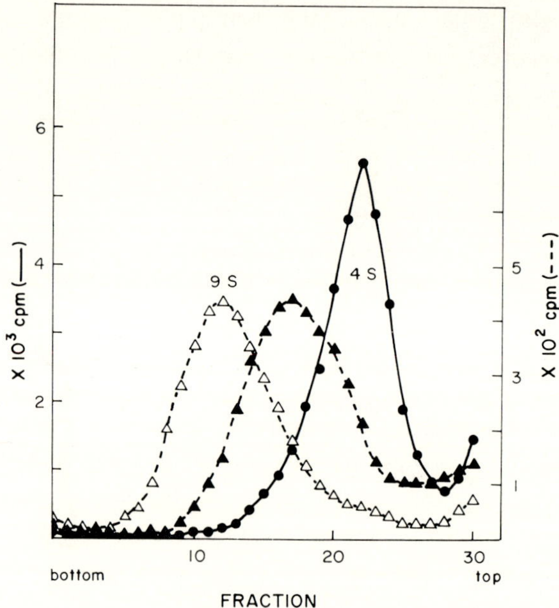

FIG. 21. The reassociation of receptor subunits in a low ionic environment. The 105,000 g supernatant prepared in 0.01 M Tris–0.0015 M EDTA buffer was combined with 0.3 M KCl and 10^{-3} μg estradiol-^{3}H per milliliter at 0°C for 15 minutes; 0.3-ml aliquots were layered on 4.3-ml gradients of 5–20% sucrose prepared in Tris–EDTA containing 0.3 M KCl. After centrifuging for 12 hours at 38,000 rpm (2°C) in a SW 39 rotor, the tubes were pierced and fractions collected by gravity flow. The equivalent fractions from these separate gradients were pooled, and 150-μl aliquots of fractions 20 ($\triangle$--$\triangle$) and 23 ($\blacktriangle$--$\blacktriangle$) were diluted with 300 μl and 250 μl of Tris–EDTA buffer, respectively. 0.3 ml aliquots of each were then layered on 4.3 ml gradients of 5–20% sucrose media in the Tris–EDTA buffer and recentrifuged for 12 hours at 38,000 rpm at 2°C in a SW 39 rotor. Fractions were collected by gravity flow and analyzed for estradiol-^{3}H. $\bullet$, High salt; $\triangle$, $\blacktriangle$, low salt. Reprinted from Vonderhaar *et al.* (1970b).

plexes, suggesting that the nonestrogen-binding subunit had been dissociated by the longer exposure to 0.3 M KCl in this procedure.

These experiments clearly show that the 8–9 S receptor complexes are made up of dissimilar subunits: a 4 S estrophilic unit (A subunit) and a nonestrogen binding unit (4–5 S) (B subunit). It also appears that the B subunits can dissociate further during prolonged exposures to 0.3 M KCl to yield smaller B′ subunits. Accordingly, the 8–9 S cytosol receptor is a macromolecular aggregate and the question arises as to what role such aggregates might play in the implementation of estrogen action. Some clues to the manner in which estrogen receptors may work were first obtained in studies of the effect of temperature on the physical character of the receptor complexes (Vonderhaar *et al.*, 1970a).

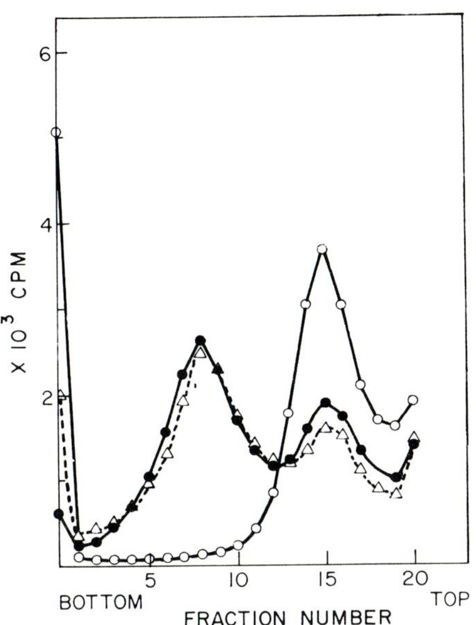

FIG. 22. The effect of temperature on the sedimentation character of the receptors. Aliquots of the 105,000 g extract of rat uteri were combined with 10^{-3} μg estradiol-³H per millitier at 0° (●——●), 23° (△---△), or 37°C (○——○) for 15 minutes. Samples (0.3 ml) were layered on 4.3 ml of a 5–20% sucrose gradient prepared in 0.01 M Tris, 0.01 M KCl, and 0.0015 M MgCl₂ (TKM buffer) at pH 7.4 and spun at 38,000 rpm for 12 hours at 2°C in a SW 39 rotor. Fractions were collected by gravity flow and assayed for radioactive estradiol. Data are expressed as counts per minute per fraction. Reprinted from Vonderhaar *et al.* (1970a).

In a series of experiments the cytosol receptors were exposed to estradiol-³H for 15 minutes at 0°, 23°, and 37°C at pH 7.4 and then centrifuged through a low-salt sucrose gradient at 0–4°C. In preparations subjected to binding at 0° or 23°C, the typical pattern with a major 8–9 S peak and a minor 4 S peak of receptor-bound hormone was obtained; however, at 37°C the 9 S component disappeared and could be accounted for by an increased amount of 4 S receptor (Fig. 22). This conversion of 9 S receptor to the 4 S form required both the 37°C treatment and the exposure to the hormone. A cytosol solution which had been preincubated for 15 minutes at 37°C in the absence of estradiol still contains receptors which sediment as 9 S complexes in a low ionic sucrose gradient. However, the preincubation labilizes the receptor complexes in some manner since the subsequent addition of estradiol-³H at 0°C causes a rapid conversion of the 9 S receptor to the 4 S form. This conversion of the 9 S receptor to the 4 S form is a function of the estrogen binding and proceeds without release of the bound hormone.

The intriguing aspects of this temperature-mediated conversion of the 8–9 S receptor to the 4 S form is the role of the hormone in the conversion and the observation that the new 4 S form is no longer able to hybridize or reassociate with native subunits of freshly dissociated cytosol receptors (Vonderhaar *et al.*, 1970a). In this respect the "temperature-modified" 4 S receptor complex is similar to the receptors dissociated from the nucleus. Additional evidence for a temperature modification of the cytosol receptor was reported subsequently by Brecher *et al.* (1970). They showed that temperature treatment of the receptor at pH 8.0, in contrast to treatment at pH 7.0, yields a 5 S form of the estrophilic component similar to the receptor released by salt extraction of labeled nuclei. Apparently, the pH during the temperature treatment determines whether a 4 S or a 5 S estrophilic component is obtained. In both cases the hormone is necessary for the conversion and the "modified" receptors are unable to hybridize with subunits obtained by salt dissociation of the native 8–9 S receptor complexes when centrifuged in a low ionic sucrose gradient.

A partial explanation for the temperature-mediated changes in the estrophilic subunit has been obtained in a study of the effects of prolonged salt treatment on the physical properties of the cytosol receptors. As cited earlier centrifugation or gel filtration of the cytosol receptors in the presence of 0.3 M KCl led to dissociation of the nonestrogen subunit (B) into smaller components (B′). To test whether or not this was a function of the aging in 0.3 M KCl a cytosol receptor preparation was deliberately aged for intervals up to 4 hours prior to testing for reassociation on centrifugation into a low ionic sucrose gradient. Comparison of the sedimentation patterns (Fig. 23) revealed that aging of the cytosol receptors in 0.3 M KCl led to a progressive conversion of 8–9 S forms to a nonassociating 4 S form. In further reassociation studies it was shown that the KCl aged 4 S forms, like the temperature-modified 4 S receptors, also failed to hybridize or reassociate with freshly dissociated components of 8–9 S cytosol receptors. These data suggest that the estrophilic subunit was converted in the salt treatment to a form similar to that obtained by the 37°C treatment.

This conversion now appears to be due to the release of an estrophile modifying factor (C-unit) during the salt treatment of the receptor preparations. Initial evidence for the existence of this principle was obtained in experiments studying adsorption and release of cytosol receptor subunits on the resin containing immobilized estradiol (PVM-estradiol)

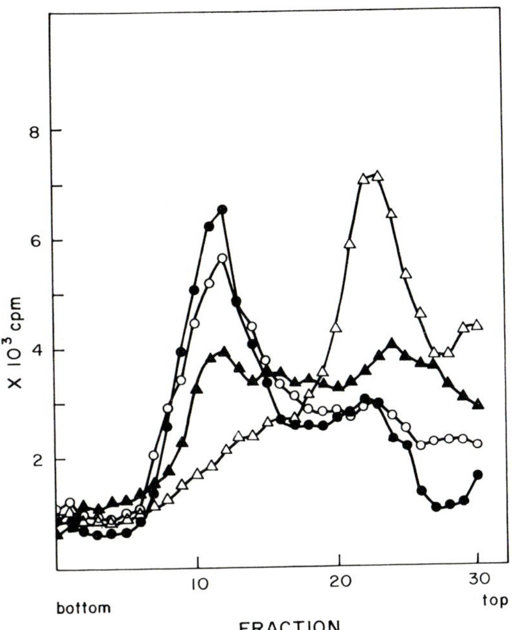

FIG. 23. The reassociation of receptor subunits aged in the presence of 0.3 M KCl. Aliquots of the 105,000 g uterine supernatant prepared in 0.01 M Tris–0.0015 M EDTA buffer at pH 7.4 were adjusted to 0.3 M KCl and aged at 0°C for either 0 minute (●——●), 15 minutes (○——○), 1.5 hours (▲——▲), or 4 hours (△——△). Fifteen minutes before centrifugation all samples received 10^{-3} μg of estradiol-³H per milliliter. Samples (0.3 ml) were then layered on 4.3 ml gradients of 5–20% sucrose prepared in 0.01 M Tris–0.0015 M EDTA at pH 7.4. Centrifugation was carried out at 2°C for 16 hours at 35,000 rpm in a SW 39 rotor. Fractions were collected by gravity flow and analyzed for estradiol-³H. Reprinted from Vonderhaar et al. (1970b).

(Vonderhaar et al., 1970b). This resin, after selectively adsorbing the estrogen receptors from a low ionic strength cytosol, was treated with 0.3 M KCl to dissociate the nonestrogen binding subunits from the bound estrophiles. Adding this extract to freshly dissociated cytosol receptor preparations precluded the reassociation of the latter in a low ionic strength sucrose gradient (Fig. 24) where as the corresponding extracts from the control resin PVM-BME had no effect on the reassociation

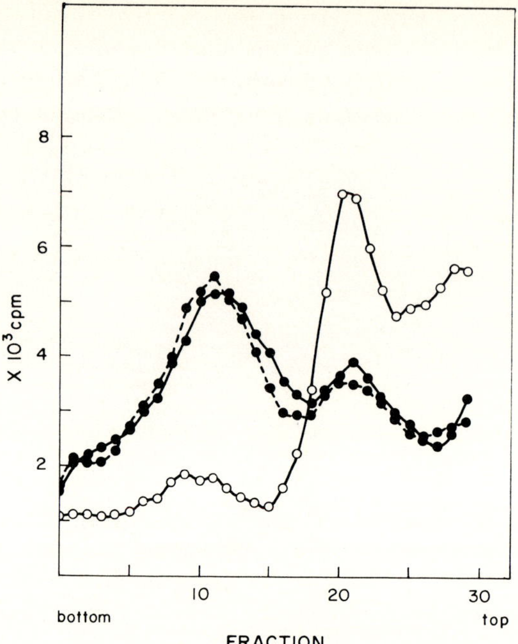

Fig. 24. The effect of a 0.3 *M* KCl extract of a PVM–estradiol–receptor complex on the reassociation of native receptor subunits. Of the 105,000 *g* uterine supernatant prepared in 0.01 *M* Tris–0.0015 *M* EDTA buffer at pH 7.4, 2.5 ml was combined with either 7.5 mg of PVM-estradiol or 7.5 mg PVM-β-ME for 15 minutes at 0°C. The supernatant was removed after centrifugation and the resins washed 2 times with 1-ml aliquots of the Tris–EDTA buffer. Each resin was then extracted with 0.5 ml of Tris-EDTA containing 0.3 *M* KCl. These extracts were then combined at 0° with 0.5-ml aliquots of the original receptor solution aged for 1 hour in the presence of 0.3 *M* KCl and labeled during the final 15 minutes with 10^{-3} μg of estradiol-^{3}H per milliliter. Samples (0.3 ml) were sedimented through 4.3-ml gradients of 5–20% sucrose prepared in 0.01 *M* Tris–0.0015 *M* EDTA at pH 7.4. Centrifugation was for 16 hours at 35,000 rpm (2°C) in a SW 39 rotor. Fractions were collected by gravity flow and analyzed for estadiol-^{3}H. ●——●, KCl 1 hour; ●---●, + PVM-β-ME extract; ○——○, + PVM-E₂ extract. Reprinted from Vonderhaar *et al.* (1970b).

process. Since the control resin does not adsorb cytosol receptors, it is suggested that the active principle is originally carried or generated by the adsorbed cytosol receptors. The C units which appear to be small, dialyzable, and ether soluble, act directly on KCl dissociated estradiol-^{3}H-labeled cytosol receptors to prevent their reassociation on centrifugation into a low ionic strength sucrose gradient (Fig. 25); they do not cause a release of estradiol once it has been bound.

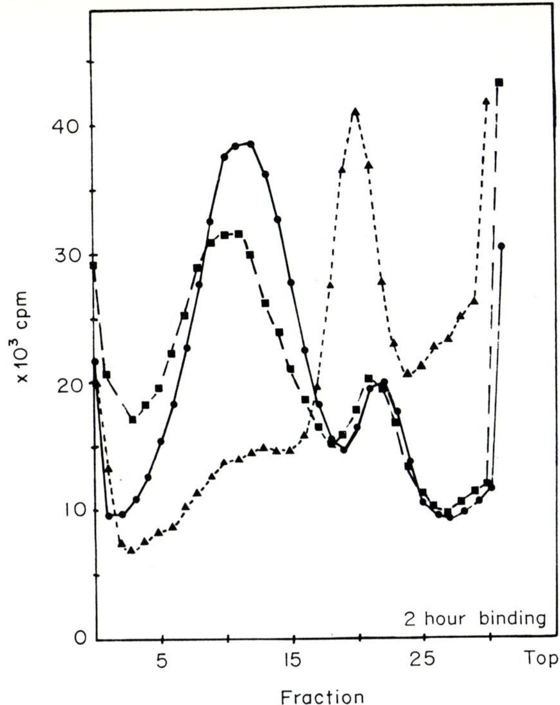

FIG. 25. The influence of C units on the reassociation of salt-dissociated cytosol receptors. A cytosol receptor solution was allowed to bind estradiol-^{3}H for 2 hours; the receptor solution was then treated with 0.3 M KCl (■) or 0.3 M KCl + C units (▲) to dissociate the 8–9 S receptor complexes. The control receptor solution (●, no salt treatment) and the KCl-dissociated preparations were then centrifuged into 5–20% sucrose gradients made in 0.01 M Tris–0.0015 M EDTA (pH 7.4). The distribution of receptor bound estradiol-^{3}H was determined in the fractions dripped from the bottom of the gradient (Vonderhaar *et al.*, 1971.)

VI. Summary: A Postulated Role for Receptors in Estrogen Action

In our studies with surviving uterine segments, blocking protein synthesis with cycloheximide, puromycin, or an amino acid deficiency, while allowing RNA synthesis to continue, led to a rapid, yet reversible, depression of the estrogen-induced RNA polymerase activity of nuclei; little or no effect was observed on the control level. In direct contrast, the reversible blocking of RNA synthesis in the surviving segments with MPB, increased the level of RNA polymerase activity. While this effect was observed in uteri from both estrogen-treated and control rats, the MPB effect in control uteri was transient and generally less. A remarkable feature of the MPB experiments was the ability of this agent to override or counteract the cycloheximide depression of the estrogen-induced RNA polymerase activity. Taken together these studies suggest the concept that the estrogen-induced polymerase activity requires a constant supply of some protein for efficient function of these polymerase units. They also suggest that this protein is used up in the course of RNA synthesis at these sites and is resupplied primarily from protein synthesis; however, in tissues in which RNA and protein synthesis are blocked simultaneously, it appears that the protein can also be regenerated or retrieved from some inactive form. The observations that a temperature-sensitive process controls the level of the estrogen-bound RNA polymerase activity in all cases, raises the additional possibility that the limiting protein is activated at 37°C prior to use and that at intermediate temperatures (i.e., 23°C) an inactivating process dominates.

These studies on the regulation of the estrogen-induced RNA polymerase activity of rat uterine nuclei clearly support the concept that estrogens facilitate the expression of certain genes by supplying the active proteins which are needed for the efficient function of RNA polymerase at these sites. In this concept the problem of estrogen action then becomes one of defining the molecular processes which mediate this supply and elucidating the manner in which the hormone affects their operation. It is proposed that estrogen receptors play an important role in this realm.

In the studies cited, estrogen receptors of rat uterine cytosol constitute a group of related molecular aggregates which extract and sediment as 8–9 S complexes in a low ionic environment (Fig. 26). In the presence of 0.3 M KCl these aggregates dissociate reversibly to yield a 4 S estrophilic subunit (A unit) which may be common to the different aggregates, and a 4–5 S subunit (B unit) which does not bind estradiol. On prolonged exposure to 0.3 M KCl or with treatment at 37°C the latter dissociates into smaller B′ subunits and a small C unit which appears to modify the 4 S estrophilic subunit so as to prevent its reassociation with native B units. Binding of the hormone to the estrophilic subunit of the 8–9 S cytosol receptors greatly facilitates this dissociation and release of the nonestrogen binding components.

In these associations the estrophilic subunit in the absence of estradiol appear to act as a nucleation center or a stabilizing corner stone in the formation of the 8–9 S macromolecular aggregates. Thus, in the absence of the hormone the receptor complexes may provide a mechanism for sequestering certain proteins or other macromolecules so as to make them unavailable for certain cellular processes. On the other hand, since the binding of the estrogen to the estrophilic subunit (A) quite obviously destabilizes these complexes the possibility is suggested that the estrophile, in the presence of an estrogen, may operate catalytically to modify B′ units for use at nuclear RNA polymerase sites.

In our current concept of estrogen action (Fig. 27) it is proposed that the estrophilic subunit of receptors recognizes a variety of nonestrogen binding subunits through a common surface entity (presumably C units), which originally is attached to a spectrum of B′ subunits. This variation in B′ subunits may also account for the fact that receptor complexes isolated in low ionic media vary in composition, size, and estrogen-binding character. It is proposed, that the binding of the estrogen molecule to the estrophilic subunit produces a conformational change

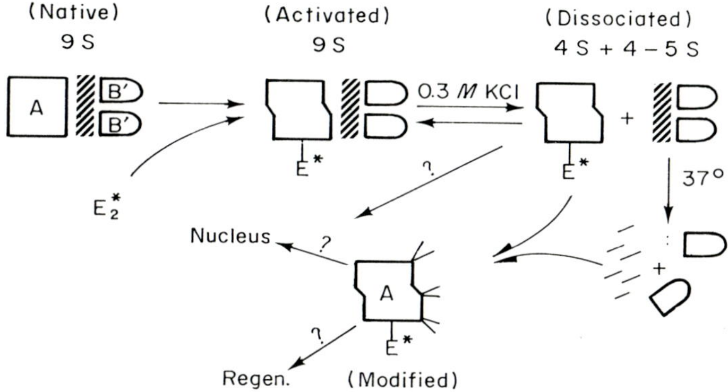

FIG. 26. A diagram of the subunit character of the cytosol receptors for estradiol. The native receptor (8–9 S) becomes labilized for dissociation (activated state) on binding estradiol to the estrophilic subunit (A). The treatment with 0.3 M KCl dissociates native or activated complexes to yield the 4 S estrophilic subunit (A) and a 4–5 S nonestrophilic unit (B). With aging in 0.3 M KCl at 0°C or incubation briefly at 37°C in the absence of KCl the B units dissociate further to yield B′ and C units; the latter react with A units to modify their ability to reassociate with native B units.

or converts the latter into an active catalytic unit, which cleaves the C component from the B′ subunit. In this way the receptor may play a working role in delivering active protein units to the nuclear scene to facilitate the transcription of certain genes. However, it is not necessary that the initial effects of receptor action be confined to genetic expression mechanisms as the active B′ units which are generated may as well affect catalytic processes or molecular associations at other cellular sites. In this way, the extragenomic effects of estrogens on water imbibition (Ui and Mueller, 1963) and nucleoside transport (Billing *et al.*, 1969) in rat uteri may be explained.

In this concept of estrogen action, the hormone in combination with its receptor modulates the interaction between the two molecular capacitor systems of the cell: the nucleochromatin and the cytoplasmic membranes. Operating within this conceptual framework, the estrogen–receptor mechanism, recognizing and processing different B′ units, provides not only for the progression and changing character of hormonal effects in a single organ, but accounts as well for the variation of estrogenic responses in different target organs. This working concept of estrogen receptors also provides a possible explanation for the different responses

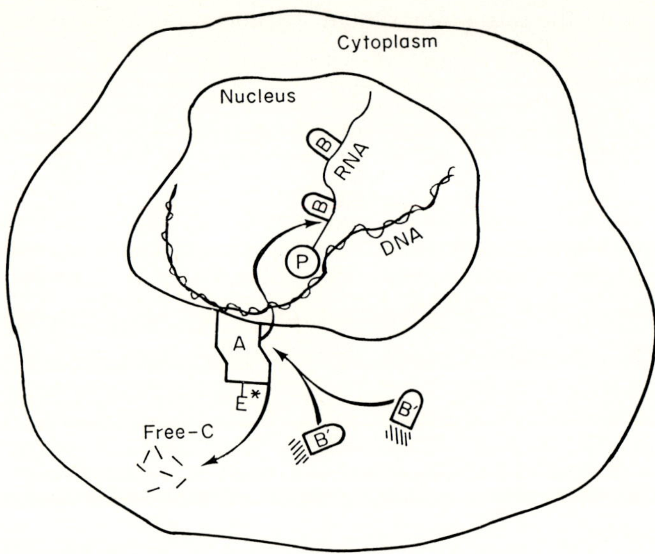

Fig. 27. A proposed role for estrogen receptors in genetic expression. In a hypothetical target cell the estrogen receptor (A) in combination with estrogen (E*) is visualized as combining with specific sites in or on the nucleus. In this position it enters into transitory combinations with B′ or B″ units to the polymerase sites (P) where they are used to facilitate RNA synthesis. In this process the receptor mechanism could as well provide activated B′, or B′ for extragenomic processes.

obtained with estrogen metabolites and antiestrogens since the character and stability of the receptor complexes would be expected to vary with the different estrogen analogues. Finally—since similar receptor systems appear to exist for many other steroid hormones, it is tempting to suggest that common principles of operation may apply—in this case the current studies on estrogen action may be predictive.

ACKNOWLEDGMENTS

Recent experimental work was supported by U.S. Public Health Service Grants TO-1-CA-5002-13 and CA-07175. The senior author (G.C.M.) is the recipient of a Career Research Award from the National Cancer Institute, U.S. Public Health Service.

REFERENCES

Aizawa, Y., and Mueller, G. C. (1961). *J. Biol. Chem.* **236**, 381.

Anderson, C. H., and Greenwald, G. S. (1969). *Endocrinology* **85**, 1160.

Arias, F., and Warren, J. C. (1971). *Biochim. Biophys. Acta* **230**, 550.

Arnaud, M., Beziat, Y., Guilleux, J. C., Hough, A., Hough, D., and Mousseron-Canet, M. (1971). *Biochim. Biophys. Acta* **232**, 117.

Astwood, E. B. (1938). *Endocrinology* **23**, 25.

Barker, K. L. (1971). *Biochemistry* **10**, 284.

Barker, K. L., and Anderson, J. M. (1968). *Endocrinology* **83**, 585.

Barnea, A., and Gorski, J. (1970). *Biochemistry* **9**, 1899.

Barry, J., and Gorski, J. (1971). *Biochemistry* **10**, 2384.

Beck, J. S. (1961). *Lancet* **1**, 1203.

Beck, J. S. (1963). *Brit. Med. Bull.* **19**, 192.

Beziat, Y., Guilleux, J., and Mousseron-Canet, M. (1970). *C. R. Acad. Sci.* **270**, 1620.

Bigger, J. D., and Claringbold, P. J. (1954). *J. Endocrinol.* **11**, 277.

Billing, R. J., Barbiroli, B., and Smellie, R. M. S. (1969). *Biochim. Biophys. Acta* **190**, 52.

Brecher, P. I., and Wotiz, H. H. (1968). *Proc. Soc. Exp. Biol. Med.* **128**, 470.

Brecher, P. I., and Wotiz, H. H. (1969). *Endocrinology* **84**, 718.

Brecher, P. I., Vigersky, R., Wotiz, H. S., and Wotiz, H. H. (1967). *Steroids* **10**, 635.

Brecher, P. I., Numata, M., DeSombre, E. R., and Jensen, E. V. (1970). *Fed. Proc. Fed. Amer. Soc. Exp. Biol.* **29**, 249.

Bresciani, F., Puca, G. A., Nola, E., Salvatore, M., and Ardovino, I. (1969). *Proc. 11th Congr. Ital. Soc. Pathol., Rome* p. 203.

Burgess, R. R., Travers, A. A., Dunn, J. J., and Bautz, E. K. F. (1969). *Nature (London)* **221**, 43.

Casals, S. P., Friou, G. J., and Teague, P. O. (1963). *J. Lab. Clin. Med.* **62**, 625.

Cecil, H. C., and Bitman, J. (1967). *Arch. Biochem. Biophys.* **119**, 105.

Chader, G. J., and Villee, C. A. (1970). *Biochem. J.* **118**, 93.

Cohen, S., O'Malley, B. W., and Stastny, M. (1970). *Science* **170**, 336.

Crouch, R. J., Hall, B. D., and Hager, G. (1969). *Nature (London)* **223**, 476.

DaLage, C. (1966). *In* "European Review of Endocrinology" (M. F. Jayle, ed.), Suppl. 2, Part 1, p. 79. Pergamon, Oxford.

DeAngelo, A. B., and Gorski, J. (1970). *Proc. Nat. Acad. Sci. U.S.* **66**, 693.

DeSombre, E. R., Puca, G. A., and Jensen, E. V. (1969). *Proc. Nat. Acad. Sci. U.S.* **64**, 148.

Eisenfeld, A. J. (1967). *Biochim. Biophys. Acta* **136**, 498.

Eisenfeld, A. J. (1970). *Endocrinology* **86**, 1313.

Eisenfeld, A. J., and Axelrod, J. (1965). *J. Pharmacol. Exp. Ther.* **150**, 469.

Eisenfeld, A. J., and Axelrod, J. (1966). *Endocrinology* **79**, 38.

Erdos, T. (1968). *Biochem. Biophys. Res. Commun.* **32**, 338.

Erdos, T., Bessada, R., and Fries, J. (1969). *FEBS (Fed. Eur. Biochem. Soc.) Lett.* **5**, 161.

Erdos, T., Best-Belpomme, M., and Bessada, R. (1970). *Anal. Biochem.* **37**, 244.

Fevold, H. R., and Mueller, G. C. (1971). Unpublished observations.

Flerko, B. (1957). *Arch. Anat. Microsc. Morphol. Exp.* **46**, 159.

Folman, Y., and Pope, G. S. (1969a). *J. Endocrinol.* **44**, 203.

Folman, Y., and Pope, G. S. (1969b). *J. Endocrinol.* **44**, 213.

Gassner, F. X., Reifenstein, E. C., Algeo, J. W., and Mattox, W. E. (1958). *Recent Progr. Horm. Res.* **14**, 183.

Giannopoulous, G., and Gorski, J. (1971). *J. Biol. Chem.* **246**, 2524.

Gorski, J. (1964). *J. Biol. Chem.* **239**, 889.

Gorski, J., and Axman, M. C. (1964). *Arch. Biochem. Biophys.* **105**, 517.

Gorski, J., and Morgan, M. S. (1967). *Biochim. Biophys. Acta* **149**, 282.

Gorski, J., and Nicolette, J. A. (1963). *Arch. Biochem. Biophys.* **103**, 418.

Gorski, J., Noteboom, W. D., and Nicolette, J. A. (1965). *J. Cell Comp. Physiol.* **66**, Suppl. 1, 91.

Gorski, J., Toft, D., Shyamala, G., Smith, D., and Notides, A. (1968). *Recent Progr. Horm. Res.* **24**, 45.

Greengard, O., Sentenac, A., and Acs, G. (1965). *J. Biol. Chem.* **240**, 1687.

Greenman, D. L., and Kenney, F. T. (1964). *Arch. Biochem. Biophys.* **107**, 1.

Hamilton, T. H. (1968). *Science* **161**, 649.

Hamilton, T. H., Widnell, C. C., and Tata, J. R. (1965). *Biochim. Biophys. Acta* **108**, 168.

Harris, S. G. (1971). *Nature (London)* **231**, 246.

Hilf, R., Michel, I., and Bell, C. (1967). *Recent Progr. Horm. Res.* **23**, 229.

Jensen, E. V., and Jacobson, H. I. (1960). *In* "Biological Activities of Steroids in Relation to Cancer" (G. Pincus and E. P. Vollmer, eds.), p. 161. Academic Press, New York.

Jensen, E. V., Jacobson, H. I., Flesher, J. W., Saha, N. N., Gupta, G. N., Smith, S., Colucci, V., Shiplacoff, D., Neumann, H. G., DeSombre, E. R., and Jungblut, P. W. (1966). *In* "Steroid Dynamics" (G. Pincus, T. Nako, and J. F. Tait, eds.), p. 133. Academic Press, New York.

Jensen, E. V., DeSombre, E. R., Hurst, D. J., Kawashima, T., and Jungblut, P. W. (1967a). *Arch. Anat. Microsc. Morphol. Exp.* **56**, 547.

Jensen, E. V., Hurst, D. J., DeSombre, E. R., and Jungblut, P. W. (1967b). *Science* **158**, 385.

Jensen, E. V., DeSombre, E. R., and Jungblut, P. W. (1967c). *In* "Endogenous Factors Influencing Host-Tumor Balance" (R. W. Wissler, T. L. Dao, and S. Wood, Jr., eds.), p. 15. Univ. of Chicago Press, Chicago, Illinois.

Jensen, E. V., Suzuki, T., Kawashima, T., Stumpf, W. E., Jungblut, P. W., and DeSombre, E. R. (1968). *Proc. Nat. Acad. Sci. U.S.* **59**, 632.

Jensen, E. V., Suzuki, T., Numata, M., Smith, S., and DeSombre, E. R. (1969). *Steroids* **13**, 417.

Jervell, K. F., Diniz, C. R., and Mueller, G. C. (1958). *J. Biol. Chem.* **231**, 945.

Jonsson, C. E., and Terenius, L. (1965). *Acta Endocrinol. (Copenhagen)* **50**, 289.

Kahwanago, I., Heinrichs, W. L., and Hermann, W. L. (1969). *Nature (London)* **223**, 313.

Kapadia, G., Means, A. R., and O'Malley, B. W. (1971). *Cytobios* **3**(9), 33.

Kato, J., and Villee, C. A. (1967). *Endocrinology* **80**, 1133.

Kato, J., Inaba, M., and Kobayashi, T. (1969). *Acta Endocrinol. (Copenhagen)* **61**, 585.

Kato, J., Atsumi, Y., and Muramatsu, M. (1970). *J. Biochem. (Tokyo)* **67**, 871.

King, R. J. B., Gordon, J., and Martin, L. (1968). *J. Endocrinol.* **41,** 223.

King, R. J. B., Gordon, J., and Stegglcs, A. W. (1969). *Biochem. J.* **114,** 649.

Kleinschmidt, A., Lang, D., Plescher. C., Hellman, W., Haass, J. Zahn, R. K., and Hagedorn, A. (1961). *Z. Naturforsch. B* **16,** 730.

Korenman, S. G. (1969). *Steroids* **13,** 163.

Korenman, S. G., and Dukes, B. A. (1970). *J. Clin. Endocrinol. Metab.* **30,** 639.

Korenman, S. G., and Rao, B. R. (1968). *Proc. Nat. Acad. Sci. U.S.* **61,** 1028.

Ladinsky, J. L., Grunchow, H. W., and Peckman, B. M. (1968). *J. Endocrinol.* **41,** 161.

Leavitt, W. W., Friend, J. P., and Robinson, J. A. (1969). *Science* **165,** 496.

Lemon, H. M. (1970). *Cancer* **25,** 423.

Lippe, B. M., and Szego, C. M. (1965). *Nature (London)* **207,** 272.

Lyons, W. R., Li, C. H., and Johnson, R. E. (1958). *Recent Progr. Horm. Res.* **14,** 219.

McCorquodale, D. J., and Mueller, G. C. (1958). *J. Biol. Chem.* **232,** 31.

McLeod, J., and Reynolds, S. R. M. (1938). *Proc. Soc. Exp. Biol. Med.* **37,** 366.

Martin, L., and Stone, G. M. (1965). *Steroids* **6,** 473.

Maurer, H. R., and Chalkley, G. R. (1967). *J. Mol. Biol.* **27,** 431.

Means, A. R., and Hamilton, T. H. (1966a). *Proc. Nat. Acad. Sci. U.S.* **56,** 686.

Means, A. R., and Hamilton, T. H. (1966b). *Proc. Nat. Acad. Sci. U.S.* **56,** 1594.

Mobbs, B. G. (1968). *J. Endocrinol.* **41,** 339.

Mohla, S., DeSombre, E. R., and Jensen, E. V. (1971). *Fed. Proc. Fed. Amer. Soc. Exp. Biol.* **30,** 1214.

Mueller, G. C. (1953). *J. Biol. Chem.* **204,** 77.

Mueller, G. C., and Le Mahieu, M. (1971). Unpublished observations.

Mueller, G. C., Herranen, A. M., and Jervell, K. F. (1958). *Recent Progr. Horm. Res.* **14,** 95.

Mueller, G. C., Gorski, J., and Aizawa, Y. (1961). *Proc. Nat. Acad. Sci. U.S.* **47,** 164.

Mueller, G. C., Le Mahieu, M., Nishigori, H., and Kajiwara, K. (1971) Unpublished obseivations.

Musliner, T. A., Chader, G. J., and Villee, C. A. (1970). *Biochemistry* **9,** 4448.

Nakata, Y., and Bader, J. P. (1969). *Biochim. Biophys. Acta* **190,** 250.

Nicolette, J. A. (1969). *Arch. Biochem. Biophys.* **135,** 253.

Nicolette, J. A., and Gorski, J. (1964). *Arch. Biochem. Biophys.* **107,** 279.

Nicolette, J. A., and Mueller, G. C. (1966a). *Endocrinology* **79,** 1162.

Nicolette, J. A., and Mueller, G. C. (1966b). *Biochem. Biophys. Res. Commun.* **24,** 851.

Nicolette, J. A., Le Mahieu, M. A., and Mueller, G. C. (1968). *Biochim. Biophys. Acta* **166,** 403.

Noteboom, W. D., and Gorski, J. (1963). *Proc. Nat. Acad. Sci. U.S.* **50,** 250.

Noteboom, W. D., and Gorski, J. (1965). *Arch. Biochem. Biophys.* **111,** 559.

Notides, A., and Gorski, J. (1966). *Proc. Nat. Acad. Sci. U.S.* **56,** 230.

Oka, T., and Schimke, R. T. (1969). *J. Cell Biol.* **43,** 123.

O'Malley, B. W., McGuire, W. L., Kohler, P. O., and Korenman, S. G. (1969). *Recent Progr. Horm. Res.* **25,** 105.

Paul, J., and Gilmour, R. S. (1968). *J. Mol. Biol.* **34,** 305.

Pfaff, D. W. (1968). *Endocrinology* **82,** 1149.

Puca, G. A., and Bresciana, F. (1968a). *Nature (London)* **218,** 967.

Puca, G. A., and Bresciani, F. (1968b). *Cancer* 3, 475.
Puca, G. A., and Bresciani, F. (1969a). *Endocrinology* 85, 1.
Puca, G. A., and Bresciani, F. (1969b). *Nature (London)* 223, 745.
Puca, G. A., and Bresciani, F. (1970). *Nature (London)* 225, 1251.
Raynaud-Jammet, C., and Baulieu, E. E. (1969). *C. R. Acad. Sci.* 268, 3211.
Ris, H. (1967). *In* "Regulation of Nucleic Acid and Protein Biosynthesis" (V. V. Koningsberger and L. Bosch, eds.), p. 11. Elsevier, Amsterdam.
Ris, H., and Chandler, B. (1963). *Cold Spring Harbor Symp. Quant. Biol.* 28, 1.
Rochefort, H., and Baulieu, E. E. (1968). *C. R. Acad. Sci.* 267, 662.
Rochefort, H., and Baulieu, E. E. (1969). *Endocrinology* 84, 108.
Sander, S. (1968). *Acta Endocrinol. (Copenhagen)* 58, 49.
Sander, S., and Attramadal, A. (1968). *Acta Endocrinol. (Copenhagen)* 58, 235.
Schwartz, N. B. (1969). *Recent Progr. Horm. Res.* 25, 1.
Shyamala, G., and Gorski, J. (1969). *J. Biol. Chem.* 244, 1097.
Smith, D. E., and Gorski, J. (1968). *J. Biol. Chem.* 243, 4169.
Steggles, A. W., and King, R. J. B. (1970). *Biochem. J.* 118, 695.
Stone, G. M., and Baggett, B. (1965a). *Steroids* 5, 495.
Stone, G. M., and Baggett, B. (1965b). *Steroids* 6, 277.
Stumpf, W. E. (1968a). *In* "Radioisotopes in Medicine: *In Vitro* Studies" (R. L. Hayes, F. A. Goswitz, and B. E. P. Murphy, eds.), p. 633. USAEC, Oak Ridge, Tennessee.
Stumpf, W. E. (1968b). *Science* 162, 1001.
Stumpf, W. E. (1968c). *Z. Zellforsch. Mikrosk. Anat.* 92, 23.
Stumpf, W. (1969). *Endocrinology* 85, 31.
Summers, W. C., and Siegel, R. B. (1969). *Nature (London)* 223, 1111.
Szego, C. M. (1965). *Fed. Proc. Fed. Amer. Soc. Exp. Biol.* 24, 1343.
Szego, C. M., and Davis, J. S. (1967). *Proc. Nat. Acad. Sci. U.S.* 58, 1711.
Szego, C. M., and Lawson, D. A. (1964). *Endocrinology* 74, 372.
Szego, C. M., and Roberts, S. (1953). *Recent Progr. Horm. Res.* 8, 419.
Talwar, G. P., Sopori, M. L., Biswas, D. K., and Segal, S. J. (1968). *Biochem. J.* 107, 765.
Telfer, M. A. (1953). *Arch. Biochem. Biophys.* 44, 111.
Teng, C. S., and Hamilton, T. H. (1970). *Biochem. Biophys. Res. Commun.* 40, 1231.
Terenius, L. (1966). *Acta Endocrinol. (Copenhagen)* 53, 611.
Terenius, L. (1967). *Mol. Pharmacol.* 3, 423.
Terenius, L. (1968a). *Cancer Res.* 28, 328.
Terenius, L. (1968b). *Mol. Pharmacol.* 4, 301.
Terenius, L. (1969). *Acta Endocrinol. (Copenhagen)* 60, 79.
Toft, D., and Gorski, J. (1966). *Proc. Nat. Acad. Sci. U.S.* 55, 1574.
Toft, D., and Gorski, J. (1967). *Proc. Nat. Acad. Sci. U.S.* 57, 1740.
Travers, A. A., and Burgess, R. R. (1969). *Nature (London)* 222, 537.
Ui, H., and Mueller, G. C. (1963). *Proc. Nat. Acad. Sci. U.S.* 50, 256.
Unhjem, O., Attramadal, A., and Sölna, J. (1968). *Acta Endocrinol. (Copenhagen)* 58, 227.
Vonderhaar, B., and Mueller, G. C. (1969). *Biochim. Biophys. Acta* 176, 629.
Vonderhaar, B. K., Kim, U. H., and Mueller, G. C. (1970a). *Biochim. Biophys. Acta* 208, 517.

Vonderhaar, B. K., Kim, U. H., and Mueller, G. C. (1970b). *Biochim. Biophys. Acta* **215**, 125.
Vonderhaar, B. K., Kim, U. H., and Mueller, G. C. (1971).
Whalen, R. E., and Maurer, R. A. (1969). *Proc. Nat. Acad. Sci. U.S.* **63**, 681.
Wilson, J. D. (1963). *Proc. Nat. Acad. Sci. U.S.* **50**, 93.
Winnick, R. E., and Winnick, T. (1960). *J. Biol. Chem.* **235**, 2657.
Woolley, D. E., Holinka, C. F., and Timiras, P. S. (1969). *Endocrinology* **84**, 157.

DISCUSSION

C. W. Bardin: The stimulation of RNA polymerase during treatment with MPB is somewhat similar to the "superinduction" phenomenon with actinomycin D. One way that the superinduction data have been explained is that actinomycin D inhibits protein breakdown. Have you determined whether the increase in polymerase during MPB is due to increased protein inactivation?

G. C. Mueller: With respect to the action of the inhibitor (MPB), I want to make it clear that we are carrying out these experiments with surviving uterine segments and that during the treatment we are blocking RNA synthesis in the intact cells of this tissue. We then isolate the nuclei and measure the polymerase activity which has been stored up or accumulated in a potentially active form during that interval of time. I do not think that any kind of superinduction phenomenon is occurring since the activation of RNA polymerase is able to take place in the presence of cyclohexamide or puromycin. Under these conditions we blocked protein synthesis as well as RNA synthesis. What seems more likely is that there is a redistribution of some protein within the cell to supply RNA polymerase sites in the nucleus with a protein required for RNA synthesis. Our extraction studies with ammonium sulfate support this concept in that they demonstrated that activated nuclei contain some extractable protein which facilitates RNA polymerase function. For these reasons I do not think superinduction phenomenon is occurring.

C. W. Bardin: Using sucrose gradients, we have identified multiple forms of progesterone "receptor" in guinea pig uterine cytosol. These include a 7 S receptor which can be dissociated with KCl to a 3.5 S form. This latter binding protein does not reassociate to the 7 S form when recentrifuged in a salt-free gradient. When uterine cytosol labeled with progesterone-^{3}H is filtered through sephadex G-25, the protein–progesterone complex then sediments at 2 S. These observations suggest that other steroid receptors may be heterogeneous.

G. C. Mueller: This interesting observation is certainly in accord with our studies on the estrogen receptors as cited.

A. White: One is impressed with the heterogeneity which you and Dr. Bardin have stressed. You showed on one of your slides evidence for heterogeneity on the basis of the electron micrograph of a section of uterine tissue in which ^{3}H-labeled estradiol had accumulated. You expressed surprise that all of the diverse cells in the uterine tissue seem to have taken up the estrogen into their nuclei. Is it possible that the experimental studies you have described are made somewhat more difficult to interpret by working with whole uterine tissue, which is a mixture of cell types? Would it be practical and/or useful to dissociate the cells of uterine tissue by techniques which have been utilized for other tissues and then subject these cells to gradient centrifugation in order to obtain homogeneous populations of cells and then compare the cytoplasmic receptors of estrogen present in each class of cells? This might afford an approach to the question of the wide variety and the heterogeneity of estrogen receptors that you and others have described. Perhaps under these circumstances one might also see differential cell type responses to estrogen.

G. C. Mueller: I think that would be a very good thing to do, but it is technically very difficult; instead we are attempting to cultivate uterine cells in tissue culture. In preliminary experiments we have found that uterine cells from the pig grow quite nicely and you can obtain both epithelial and fibroblastic cell types. I think that this might be a step in the proper direction. With respect to the heterogeneity of receptors, I think that as you go from cell type to cell type one will find that the B' subunits which interact with the estrophile will indeed vary according to the cell type. I think that the estrophiles, on the other hand, may be rather constant and interact with B' units through some common group which is present, and are dealing with some kind of class reaction on a variety of proteins; this recognition may be quite analogous to recognition of carbohydrate chains on proteins by concanavalin or phytohemagglutinin.

E. B. Astwood: Could you briefly tell us to what extent one can see actions of estrogens *in vitro?*

G. C. Mueller: In surviving rat uterine segments one can obtain estrogenic activations of RNA polymerase as high as 100% on occasion, but unfortunately it is not a predictable event; our usual experience is a 15–30 activation of RNA polymerase. In these experiments the estrogen is added to the tissue under incubation in tissue culture medium. After a specified period, the nuclei are isolated and the RNA polymerase activity is measured. In another type of experiment Dr. Mousseron-Canet (France) claims upwards of 80% stimulation of the nuclei when a solution of the estrogen receptor plus estradiol is added directly to the nuclei. While we have obtained 15–30% stimulation doing similar experiments, we have not been able to duplicate the large effects in our laboratory. I know that Dr. Jensen has also had difficulty in this type of experiment; however, he recently reported that stimulations of RNA polymerase activity approximating 40%. In our hands, it is easy to obtain cytoplasmic protein fractions which stimulate nuclear RNA polymerase activity; the difficulty is to demonstrate a dependence on estrogens. Things are moving in the right direction, but we have not reached the predictable state as yet.

R. Chatterton: G. A. Puca and F. Bresciani [*Endocrinology* 85, 1 (1969)] reported that the retention of estradiol by different organs was qualitatively similar but that there was only a quantitative difference. Possibly working with the rat uterus which has diverse tissues may not be as much of a problem as it would be if there were not this qualitatively similar response.

G. C. Mueller: One is dealing with an additional problem when one considers hormone retention in tissues; this is the nature of the release mechanism. Actually, very little is known as to the fate of receptors or how long it retains the hormone. If you look at the data in Barry and Gorski's paper you see that the higher levels of the estrogen appear to deplete the amount of estrogen receptor both in the nucleus and in the cytoplasm. Thus, it is quite probable that the estrogen receptor itself is subject to some metabolic activation and inactivation steps. We are currently looking for cyclic alterations in the estrophile which might provide for a hormone release mechanism. In any case, in evaluating hormone retention in tissues, I think we have to allow for both a heterogeneity of receptors and varying functionality in different cells.

R. Chatterton: Is it possible that the C units are also steroids?

G. C. Mueller: Yes, it is possible; I cannot at this point exclude that they may be some kind of an estrogen derivative, but, since they can affect the association state of receptor subunits which have already been titrated with estradiol, they appear to differ from the usual estrogen.

I. L. Schwartz: I gather that you are suggesting that the second level receptor in your system; i.e., the entity with which the cytoplasmic receptor is reacting within the nucleus, is RNA polymerase or its product, RNA. However, it has been suggested recently that this second level (nuclear) receptor may be an acidic nucleoprotein. Has this possibility been examined by your group or others with the estrogen system?

G. C. Mueller: I think that the estrophile is not working directly with the polymerase like a sigma factor in the bacterial polymerase systems. It may, however, exist in close association with DNA because if you treat nuclei with RNase in low magnesium medium a major fraction of the nuclear bound receptor is dissociated from the nucleus very quickly. Since we have not been able to show that there is DNA associated with the released receptor, one cannot be sure whether the estrophile is attached directly to DNA or to some entity such as an acidic protein existing in the chromatin. On the other hand, we do know that receptors from the cytoplasm are very easily precipitated by basic proteins or protamine; however, this latter phenomenon appears too unspecific to account for the estrogen localization in target cell nuclei.

I. L. Schwartz: Do you feel that it is excluded or merely unlikely that the steroid itself might be active in the nucleus rather than its cytoplasmic receptor?

G. C. Mueller: Attempts to bind estradiol to a specific nuclear receptor by direct addition of the hormone to the nuclear fraction have not been successful. Instead, the evidence favors the concept that the hormone remains in combination with the cytosol receptor as it enters or attaches to the nucleus. The possibility that estradiol, once in the nucleus, can dissociate from the receptor to affect another site or reaction as a free steroid cannot be rules out, but the lack of direct effects of the free hormone on nuclear function makes this unlikely.

I. L. Schwartz: How do you assess the evidence from Dr. O'Malley's laboratory suggesting that acidic chromatin protein might function as a nuclear receptor for steroids or steroid-cytoplasmic receptor complexes?

G. C. Mueller: With respect to acidic nuclear proteins, I think that the data of O'Malley are interesting in several respects. First, they suggest that acidic proteins in the nucleus may be concerned with the nuclear binding of cytosol receptors for both progesterone and estradiol. Second, he presents evidence that the nuclear binding of the cytosol receptors appears to change with progressive hormone treatment. This suggests that the chromatin system is not constant and can be changed by the accumulative action of the hormone. These changes are understandable if one postulates that the hormone activates a mechanism that interacts with a specific prosthetic group on a variety of different proteins to deliver these proteins in turn to their complementary sites amid the chromatin structure as discussed in this paper.

A. Segaloff: I gather that in most of the plots which were not labeled you are still plotting just the counts from estradiol. In your Fig. 27 it looks as though you were only binding on the primary site and the pieces you were breaking off did not have estrogen. Yet from you plot, it seemed to me that they had to. How do you join them to tie the two together?

G. C. Mueller: In all estrogen receptor studies, the only marker we have for the estrophile is estradiol-^{3}H. For example, the influence of C units on the physical state of the estrogen receptor (i.e., 4 S versus the 8 S forms) is still measured by the distribution of the receptor bound estradiol-^{3}H in a sucrose gradient after centrifugation. As yet we have no good way to measure the protein itself, but it is clear that such methodology must be developed. Recently we have been approaching this problem by using the resins' to selectively adsorb the receptor complexes, which are subsequently eluted, labeled with iodine-125, and resolved electrophoretically. This is helpful and has shown that at least 3 proteins are present in the receptors of pig uteri; however, progress awaits the development of specific immune assays for each of the subunits of the estrogen receptors.

H. G. Friesen: Is it really essential that estrogen enter the cytoplasm to exert an effect? Have you or other laboratories any data on the biological effect of estrogen covalently linked to polystyrene? If this estrogen conjugate were active and mimicked the effect of estrogen in solution, would this not necessitate considerable revision of the mechanism of action of estrogens which you proposed?

G. C. Mueller: While I cannot exclude that estrogens can activate cells by acting at their surfaces as shown for insulin and prolactin, I think that the localization of estrogens in or on the nucleus argues against this possibility. It is possible, however, that some of the effects of the hormone on permeability which appear to be independent of RNA synthesis, might be implimented at this site without entry of the hormone into the cell. This would be still in accord with the proposed mechanism (test Figs. 26 and 27).

J. Weisz: You have postulated, I think, that homogeneity in your receptor proteins is based on the fact that under the different conditions that you used, e.g., different temperature, the number of counts taken up differed. Could these differences not be explained on the basis of conformational changes induced in the protein receptor by the different conditions?

G. C. Mueller: As I mentioned earlier, we have begun to study the proteins directly and have found at least 3 proteins associated with the estrogen receptors of pig uteri. I believe these data, in addition to the centrifugation studies, argue for the heterogeneity of the receptor complexes on a primary compositional basis. Our thesis is that conformational changes in the estrophile do indeed underlie the observed physical changes, but that the major effects concern the size and composition of these macromolecular aggregates.

C. H. Monder: Over the years there has been considerable debate about whether the estrogen-binding proteins in the cytoplasm are required for the transport of estradiol into the nucleus or whether estradiol is required for the transport of the protein into the nucleus. Considerable evidence has accumulated which supports the latter type of dependence. In this regard, estradiol has been described as an "outboard motor," propelling the receptor to or into the nucleus. With the introduction of the C component, do you visualize the estradiol as having an "outboard motor" type of function or does it appear now to have some other one in the entire system. The imagery that comes to my mind as I look at your model is one in which the estradiol-A component acts as a "launching pad" for the B component in the nucleus.

G. C. Mueller: At the present time, I subscribe to the concept that the estrogen, as a molecule, is required primarily to activate the estrophilic subunit of receptor complexes. Once activated, the estrophilic units no longer form stable associations with B subunits, but localize instead in or on nuclei. This appears to be due to specific receptors for the estrophile in the target cell nuclei rather than the presence of different estrogen binding proteins in the nuclei. In this new location the cytosol receptors appear to regulate the availability of some protein which is used by α-amanitin-insensitive RNA polymerase sites in the synthesis or processing of RNA. Our studies suggest that this is a catalytic role and may be concerned with the activation or transport of existing protein molecules as well as newly synthesized ones. Our concept proposes that the proteins (i.e., B′ subunits) may differ, but that a recognition unit or prosthetic group (i.e., C subunit) may be common among them. In a sense we do suggest that the A subunit of the estrogen–receptor complex acts as a "launching" system to deliver active proteins to dependent sites. In immature or estrogen-deficient uteri, one of the most responsive sites for such proteins is the nucleus; however, our concept of estrogen action allows that the same fundamental action of the receptor may regulate the availability of active proteins for other cellular processes.

R. B. Billiar: Does the C factor actually cause the 9 S complex to go to the 4 S component? Have you any information about the affinity constants of the different complexes such as the 4 S, 7 S, and 9 S for estradiol?

G. C. Mueller: Our data indicate that C units favor the shift from 9 S to a 4 S form of the receptor, but most studies have dealt with the prevention of the reassociation of salt-dissociated 9 S complexes. The affinities of the different derived units have not been studied exhaustively but appear to be of the same order.

S. L. Cohen: You spoke of estrogen frequently. Does stilbestrol give a similar picture to estradiol? Also, do the antiestrogens act before the estrogen gets into the cell or do they cause the splitting off of the estrogen from the receptor protein?

G. C. Mueller: We have used estradiol almost exclusively in these experiments. In a few cases we have used estrone and estriol, and in a very few cases diethylstilbestrol. With respect to the binding of those estrogens to the receptor, diethylstilbestrol binds very effectively; estrone and estriol bind rather poorly. Many people have studied different estrogen derivatives including the so-called estrogen antagonists. In the case of the estrogen antagonists, it appears that they are able to get onto the receptor and preferentially prevent the natural estrogen from associating with it. While it has also been reported that antiestrogens can facilitate the dissociation of estrogens already bound to estrogen–receptor complexes, our experiments along this line have not been successful. At present it appears that both estrogen receptors of the cytosol combines with all types of estrogens as well as the so-called antiestrogens. The subsequent function of the receptor, however, may depend on the nature of the associated estrogen. This is an interesting possibility as it might provide an explanation for the different biological responses which are seen with certain estrogen metabolites and synthetic analogs.

Prostaglandins in Luteal Function

B. B. Pharriss, S. A. Tillson, and R. R. Erickson

ALZA Corporation, Palo Alto, California

I. Introduction

Since their discovery in the early 1930's (Goldblatt, 1933; von Euler, 1934), prostaglandins have been associated with reproductive processes in many mammalian species. At present, this association has progressed to the point that these unsaturated, hydroxylated fatty acids are being explored as pharmaceutical agents with implications ranging from preventing or interrupting pregnancy (Karim and Filshie, 1970; Embrey, 1970; Wiqvist *et al.*, 1970) to enhancing fertility in the male (Bygdeman and Samuelsson, 1966; Hawkins, 1968). However, two early studies are of most interest to us here. von Euler and Hammarström (1937) described the presence of prostaglandin-like material in the human ovary but did not suggest its role in this gland. Somewhat later, Pickles and associates, in a series of works (Pickles, 1957, 1959; Pickles and Clitheroe, 1960; Pickles and Hall, 1963; Pickles *et al.*, 1965), reported on the presence of prostaglandins in human menstrual fluid and hypothesized on its role in menstruation. These reports have been repeatedly confirmed from various aspects, and we will draw upon these early discoveries to discuss a physiological role for prostaglandins for each of these tissues, ovary and uterus.

II. Luteolysis

Loeb (1923) described a phenomenon which was the subject of little concern until 1960, when it became one of the most intensively pursued problems in the area of reproductive physiology. Loeb's discovery that the uterus played an important role in control of ovarian function presented no theoretical problems to the scientific community but was of interest in that the uterus was considered to be the end recipient of ovarian hormonal control but certainly not involved in the cyclic functioning of the ovary. However, du Mesnil du Buisson (1961) reported on his experiments, which created a controversy that still persists. He showed not only that the uterus has luteolytic properties in pigs but that this is a local effect in that each uterine horn in this bicornuate system controls the corpus luteum of only its adjacent ovary (Fig. 1). This, of course, implies some sort of local portal-type vascular connection or a unilateral nervous reflex by which a uterine horn could control or influence its ipsilateral ovary.

                    B. B. PHARRISS ET AL.

Since that time, we have been treated to a conglomeration of experiments in which the uteri and ovaries and their accompanying vasculature, nerves, and connective tissue have been severed, ligated, reflexed, distended, transplanted, denervated, perfused, and generally insulted as no other system in the body. No common mammalian experimental species have escaped this malignment, and even some rare laboratory animals have been included. Out of these various experiments have come the following general conclusions:

1. There is no "vascular portal system" similar in nature to the portal system that exists between the small intestine and liver.

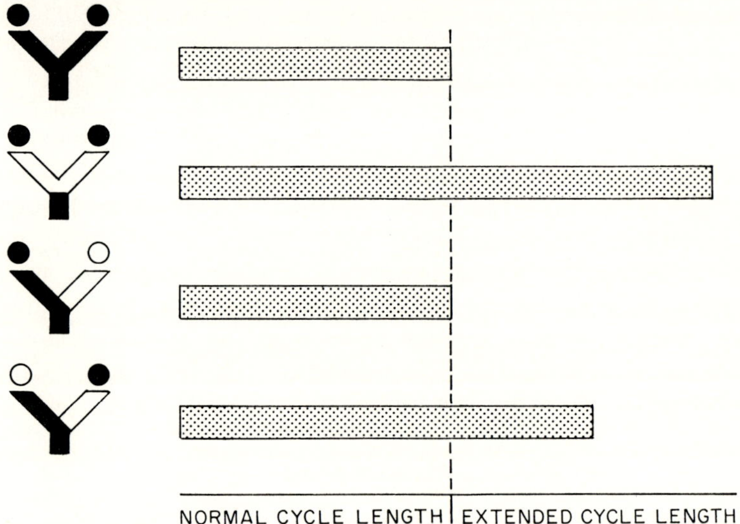

Fig. 1. Effect of hysterectomy on cycle length.

2. No evidence can be obtained for a lymphatic connection which could transport material from the uterus to the ovary.

3. Nervous connection to the uterus, ovary, or between these two tissues is not important in the uterine luteolytic effect.

4. Material is not transported from the uterus to the ovary by way of the oviduct, as this latter tube is unnecessary for luteolysis to occur.

5. The unknown factor is elusive but can be harvested from the endometrium and gives positive assays when injected into various species.

6. No matter what the pathway in the intact animal when uterine tissue is transplanted distant to its normal position, it still will terminate luteal activity although its efficiency is somewhat reduced.

These last two points, coupled with the uterine and ovarian vasculature arrangement, formed the basis of a hypothesis for uterine control of luteal function.

Caldwell *et al.* (1967) showed that homologous transplantation of uterine or endometrial tissue to the cheek pouch of the hamster could partially reverse the effect of hysterectomy (Table I). Others working in the pseudopregnant rat observed not only the localized effect of unilateral hysterectomy in this species, but also that an intact uterine horn on one side of an animal could appreciably shorten the life expectance of luteal activity on the contralateral ovary (Barley *et al.*, 1966). These

TABLE I

Systemic Mediation of Uterine Luteolysis

Treatment	Mean length of pseudopregnancy (days)
A. *Effect of hysterectomy and ovariectomy on the length of pseudopregnancy in the rat[a]*	
Sham operated	13.6 ± 0.3
Unilateral hysterectomy and ovariectomy on same side	13.3 ± 0.2
Unilateral hysterectomy and ovariectomy on opposite sides	17.4 ± 0.7
Bilateral hysterectomy intact ovary	21.5 ± 0.3
B. *Effect of uterine transplantation on ovarian activity in hamsters[b]*	
Sham operated	9.2 ± 0.03
Total hysterectomy	18.1 ± 0.27
Hysterectomized with uterine segment transplanted to cheek pouch	13.2 ± 0.83
Hysterectomized with endometrium alone transplanted to cheek pouch	13.0 ± 0.23

[a] From Barley *et al* (1966).
[b] From Caldwell *et al.* (1967).

studies suggested that whatever the luteolytic factor was, it had ready access to the systemic circulation and that the substance was effective even when the two tissues were spatially separated. Since the uterine luteolytic material has such ready access to the circulation, it could be argued that this is probably an important part of its mechanism. For this to be true, it would be necessary either for the uterine venous blood to reach the ipsilateral ovary or that there is some intermediate target tissue that could in turn affect the ovary. This line of reasoning led to a further examination of the venous vascular bed in this region. In most, if not all species under consideration here, the ovary and uterus share a final vein in common. It was suggested that the uterine luteolytic material is a venoconstrictor agent that is continually released from the endometrium during the crucial period. The constant exposure of the utero-ovarian vein to the venoconstrictor material would result in a reduction in total flow. Since this is the only effluent for ovarian blood, the perfusion of this organ is depressed. Also, since this is a period of peak ovarian activity, a reduction in blood flow would be more acutely felt at this time. Prostaglandin $F_{2\alpha}$ ($PGF_{2\alpha}$) was selected as the test material for examining this hypothesis as it had been described as being a potent venoconstrictor (DuCharme *et al.*, 1968) and, as was mentioned previously, it is present in endometrial tissue (Pickles, 1967). Subsequently, $PGF_{2\alpha}$ was tested and found to be luteolytic in several species.

This then, left us with two questions: Is prostaglandin the endogenous luteolytic factor? What is the mechanism whereby luteolysis is induced by uterine prostaglandin? During the investigation into the latter question, a potential role of prostaglandins in the ovary was uncovered. On the following pages, we will deal with these subjects in an attempt to establish the physiological role of prostaglandins in the ovary.

III. Consequences and Interpretations of Prostaglandin Experiments

A. Is $PGF_{2\alpha}$ THE UTERINE LUTEOLYTIC FACTOR?

$PGF_{2\alpha}$ was tested initially by infusion into pseudopregnant rats (Pharriss and Wyngarden, 1969). Infusions were done continuously on days 7 and 8 of pseudopregnancy; the ovaries were then removed and their progesterone contents were quantitated. It was shown that when infused into either the uterine lumen or right heart, $PGF_{2\alpha}$ could cause a shift in the steroid pattern from that seen during active luteal activity to one of a cycling ovary (Table II). This information was substantiated by evidence that $PGF_{2\alpha}$ injected subcutaneously would shorten the pattern of vaginal cornification seen with pseudopregnancy from 17 days to 8 days (Pharriss and Wyngarden, 1969). Furthermore, when given to rabbits during days 4 through 8 of pseudopregnancy or pregnancy, there was a complete morphological regression of luteal tissue while large follicles developed normally (Pharriss, 1970). It was obvious from these experiments that $PGF_{2\alpha}$ was luteolytic, but there was little evidence to suggest it as the endogenous luteolytic factor. For this to be the case, five criteria would have to be satisfied.

1. Prostaglandin $F_{2\alpha}$ must be luteolytic in all species where the uterus is implicated in luteal function.

$PGF_{2\alpha}$ has been reported to terminate luteal activity in rats (Pharriss and Wyngarden, 1969), hamsters (Gutknecht *et al.*, 1971b), rabbits (Pharriss, 1970), guinea pigs (Blatchley and Donovan, 1969), and sheep

(McCracken *et al.*, 1970). Furthermore, no negative results have been reported with prostaglandins in laboratory or farm animals. Positive results have even been reported in the rhesus monkey, where the uterus is not thought to be functional in luteolysis (Kirton *et al.*, 1970). Attempts to exhibit the luteolytic effect of $PGF_{2\alpha}$ have been fruitless in humans; however, conditions similar to that in the rhesus monkey have not been selected (Wiqvist *et al.*, 1970).

TABLE II

Effect of Prostaglandin $F_{2\alpha}$ ($PGF_{2\alpha}$) Infusion on the Concentration of Progesterone and 20α-Dihydroprogesterone in Ovaries of Pseudopregnant Rats

Treatments	No. of rats	Ovarian weight (mg)	Steroid concentration[a] (μg/gm of tissue)		P:OHP ratio
			Progesterone	20α-Dihydro-progesterone	
Experiment 1					
Saline infusion (2.06 ml/day) in uterus	2	137	11.7	1.2	9.75
$PGF_{2\alpha}$ infusion	3	281	3.6	16.2	0.22
(1 mg/kg/day) in uterus	2	184	2.1	7.8	0.27
			$\bar{x} = 2.8$	12.0	
Experiment 2					
Saline infusion (2.06 ml/day) in uterus	3	474	4.4	5.2	0.85
	3	484	6.7	4.6	1.46
			$\bar{x} = 5.6$	4.9	
$PGF_{2\alpha}$ infusion	3	296	0.70	11.4	0.06
(1 mg/kg/day) in uterus	3	426	0.64	8.7	0.07
			$\bar{x} = 0.67$	10.1	
$PGF_{2\alpha}$ infusion	3	631	0.32	5.5	0.06
(1 mg/kg/day) in right heart	2	274	0.89	8.3	0.11
			$\bar{x} = 0.61$	6.9	

[a] Each value is the average of duplicate determinations on pooled ovaries from 2 or 3 rats.

2. Prostaglandins must cause luteolysis when administered systemically and on a local basis.

The uterus can exhibit a localized effect on an ovary, yet endometrial preparations cause a bilateral response when injected systemically. $PGF_{2\alpha}$ must have the same potential if it is to be the luteolytic factor. Many studies have been reported in which $PGF_{2\alpha}$ was given by intravenous injection (Pharriss *et al.*, 1970), intrauterine and intracardiac infusion (Pharriss and Wyngarden, 1969), subcutaneous injection

(Gutknecht *et al.*, 1971b), and even intravaginal application (Pharriss, 1969), all of which resulted in luteolysis. Few reports exist demonstrating a localized effect. Goding and associates (1972) have shown that infusion of $PGF_{2\alpha}$ into the ovarian artery and uterine vein of sheep will cause luteolysis at doses that are below threshold when given systemically. This particular argument needs to be strengthened by testing the local effectiveness of $PGF_{2\alpha}$ in other species.

3. $PGF_{2\alpha}$ must be present in or released from the endometrium at a time consistent with uterine induced luteolysis.

Evidence for this point is substantial and supports the role of $PGF_{2\alpha}$ in cases where luteolysis is induced via the uterus and during normal cycling. Poyser *et al.* (1971) and Blatchley *et al.* (1971) have examined this point in the guinea pig under two different situations. In the study reported by Poyser *et al.* (1970), uterine horns were removed from guinea pigs and distended *in vitro* by filling with fluid. Uterine distension has been shown to induce luteolysis in guinea pigs when carried out *in situ* (Donovan and Traczyk, 1962). The medium bathing the distended horns was then examined for prostaglandin. Solvent extraction and thin-layer separation of the polar lipid extract was subjected to mass spectroscopy, and $PGF_{2\alpha}$ was identified as a major substance released from the uterus by distension. Subsequently, Blatchley *et al.* (1971) applied a different procedure to the guinea pig. Injections of estradiol causes luteolysis in this species, presumably by causing release of uterine luteolytic factor (ULF) since estradiol is ineffective in hysterectomized animals. These authors injected 10 μg of estradiol benzoate per day for 3 days and collected blood from the uterine vein for 60–90 minutes. This blood was processed much as was the bathing fluid in the previous experiment. The results showed an increase in $PGF_{2\alpha}$ ($>300\%$) released by the estrogen-injected animals.

There is a crucial period after which, when hysterectomy is carried out, the corpus luteum will regress at its usual time. Presumably, this timing represents the "luteolytic factor surge" from the uterus. If this is the case, then the endometrial and uterine venous content of luteolytic substance should mirror this surge. Bland *et al.* (1971) have taken advantage of this thinking and examined ovine uterine venous blood throughout the cycle and bioassayed for $PGF_{2\alpha}$ activity with conformation by mass spectrometry. Levels of 3 ng/ml were reported through day 13 of the cycle, at which point there was a sharp increase to greater than 8 ng/ml on day 14. The crucial time for uterine effect is days 13–15, depending upon method of assessment (see discussion: Wilson *et al.*, 1972). The uterine effect was followed by a return to the post-ovulatory levels. Wilson and co-workers (1972) in the sheep, examined endometrial tissue on days 3, 5, 11, and 14 for $PGF_{2\alpha}$ and found maximal levels at day 14. These workers also showed that insertion of an intra-uterine device in sheep causes a significant increase in concentration of endometrial $PGF_{2\alpha}$ 5 days post insertion (day 7 of the cycle). Such a treatment has previously been demonstrated to cause premature luteal regression in sheep (Ginther *et al.*, 1966).

4. Pregnancy should counteract the presence or action of $PGF_{2\alpha}$.

In species that show a true ovarian cycle, i.e., guinea pig, bovine, ovine, and porcine, pregnancy results in a prolongation of the function-ality of the corpus luteum. Whether this is due to elimination of the luteolytic factor or overcoming its lytic effect via a chorionic gonado-tropin is not known for certain in these species. However, it has been shown that processed fetal material will maintain the corpus luteum in sheep when placed in the uterus (Rowson and Moor, 1967).

To date, there have been no reports concerning the effect of pregnancy on endometrial $PGF_{2\alpha}$ levels, nor has anyone reported on the luteolytic potential of pregnant endometrium. Several experiments have been described, however, in which it has been shown that various gonado-tropins can overcome $PGF_{2\alpha}$-induced luteolysis. Johnston and Hunter (1970) have reported that the luteotrophic complex (FSH + prolactin) will block the effect of $PGF_{2\alpha}$ in hamsters. LH and prolactin tend to overcome the effect of $PGF_{2\alpha}$ in the rat, and in the rabbit estradiol, probably the ultimate luteotropin in this species, will completely over-come $PGF_{2\alpha}$ induced luteolysis (Gutknecht *et al.*, 1971a; this paper).

Therefore, although this argument for $PGF_{2\alpha}$ as the luteolytic factor is not strong, there is supportive evidence.

5. The luteolysis induced by $PGF_{2\alpha}$ should mimic spontaneous luteoly-sis, biochemically, morphologically, and functionally.

In the rat, one of the earliest indications that the luteal tissue is re-gressing is a shift from progesterone to 20α-dihydroprogesterone dominance in the ovary (Lidner and Shelesnyak, 1967). This shift is also seen when $PGF_{2\alpha}$ is administered to the rat (Pharriss and Wyn-garden, 1969). Also, within 24 hours after treatment of pregnant ham-sters with $PGF_{2\alpha}$, initiation of the estrous cycle occurs, as shown by cyclic changes of progesterone concentration both in the ovary and the venous blood (Table III). Furthermore, photomicrographs of the treated ovaries showed irregularly shaped luteal cells, pycnotic nuclei, and a decrease in luteal cell number and size—in short, a disorganization typical of luteal degeneration.

Corpora lutea of rabbit ovaries undergo rapid degeneration to corpora albicans after exposure to $PGF_{2\alpha}$. Within 7 days after initiation of treat-ment, all signs of luteal tissue are absent, and the ovaries contain large follicles (Pharriss, 1970). Termination of luteal activity with $PGF_{2\alpha}$ in hamsters (Labhsetwar, 1971) and monkeys (Kirton *et al.*, 1970) results immediately in a new ovulatory cycle which is fertile. The $PGF_{2\alpha}$ effect seems to be selective in the ovary, as only luteal function is impaired and the ovary renews its cycling activity in a manner not atypical to that following.spontaneous luteolysis.

Thus, there is much evidence to suggest that $PGF_{2\alpha}$ is the endogenous luteolytic substance. In fact, there is only one type of evidence which

TABLE III

Effect of $PGF_{2\alpha}$ on Progesterone Levels in the Pregnant Hamster

Day of pregnancy	Control		$PGF_{2\alpha}$ treated	
	$\bar{x}$ $\pm$ SEM[a]		$\bar{x}$ $\pm$ SEM	
A. *Plasma progesterone*[b]				
5	11.4	2.27	1.7	0.57
7	16.0	5.71	4.2	3.54
8	17.2	4.20	7.2	2.85
9	17.3	3.72	9.2	3.83
10	13.8	4.29	4.5	3.97
13	13.6	7.34	7.7	4.73
B. *Ovarian progesterone*[c]				
5	702	90.49	360	31.48
7	1039	132.60	753	272.00
8	1359	99.48	621	136.31
9	1492	184.39	375	102.70
10	944	231.79	91	43.46
13	514	32.24	243	127.51

[a] Standard error of the mean.
[b] In ng/ml.
[c] In ng/100mg.

argues against this point, and that is the description of the chemical nature of the active material extracted from bovine endometrium. Lukaszewaska and Hansel (1970) and Caldwell *et al.* (1968) have reported that the molecular weight of natural ULF is quite high, suggesting a protein or other polymer which would eliminate $PGF_{2\alpha}$. However, it has been shown by Shaw (1971) that the prostaglandins have a high binding affinity for many proteins. Therefore, it is possible that the active fraction of these endometrial preparations is actually a protein-bound $PGF_{2\alpha}$.

It is also possible that the assay used for the endometrial preparation is not specific for the luteal effects and may be a less specific assay. In general, the hysterectomized hamster has served as the assay animal (Caldwell *et al.*, 1968) for detecting luteolysis. Conceivably, this test animal would also be sensitive to hypothalamic-pituitary alterations which could block secretion of the luteotrophins. The hamster is also the species most sensitive to prostaglandin-induced luteolysis of any examined, and Ramwell and Shaw (1971) as well as others have shown that many nonspecific stresses cause an outpouring of prostaglandins from tissues in general.

B. Mechanism of $PGF_{2\alpha}$-Induced Luteolysis

The evidence presented establishes without doubt that $PGF_{2\alpha}$ is luteolytic in several species, and there is strong evidence to argue for this substance as the endogenous uterine luteolytic factor. Although this evidence in itself is quite satisfying, it does little to explain the mystery associated with the unilateral or local relationship between uterine cornu and ovary. If we assumed that $PGF_{2\alpha}$ is or closely resembles the endogenous lytic substance, then an examination as to its mechanism of action could go far in explaining the mystery. To understand the mechanism(s) of $PGF_{2\alpha}$ in this action, we will break the system down into 5 areas for discussion. These are the hypothalamic-hypophysial axis, the uterus (release of endogenous factor), the ovarian parenchyma, effectiveness of circulating luteotropins, and ovarian blood flow.

In many of the following studies, termination of pregnancy has been used as an index of $PGF_{2\alpha}$ termination of luteal progesterone secretion. Justification of this approach is found from the following experiments.

At the time when pregnancy is terminated in rats, hamsters, and rabbits, the corpus luteum is necessary for continuation of that pregnancy. In the rat and hamster, $PGF_{2\alpha}$ causes a reduction in circulating and ovarian progesterone levels (Gutknecht *et al.*, 1971b; Pharriss and Wyngarden, 1969). Replacement of the missing progesterone will maintain pregnancy in hamsters and rats (Table IV). Synthetic progestogens (6-medroxyprogesterone acetate) will also reverse the effect of $PGF_{2\alpha}$ in rats. The pregnancy support is superior to that seen with progestogen support in ovariectomized animals, thus suggesting that other ovarian secretions (estrogens) are not affected by $PGF_{2\alpha}$ treatments (Gutknecht *et al.*, 1970a).

Data of this type suggested that inadequate progesterone was the cause of pregnancy termination but left open the possibility that $PGF_{2\alpha}$ was antiprogestational at the uterus. Therefore, $PGF_{2\alpha}$ was examined for its effect on progesterone-induced endometrial proliferation.

In this experiment, twenty-seven female Dutch Belted rabbits were ovariectomized on the same day (day of ovariectomy = day 0). The fallopian tubes of these animals were not removed.

From day 8 through day 13, all rabbits received daily subcutaneous injections of 7 μg of 17β-estradiol in 10% ethanol saline. During days 14 through 18 animals received either vehicle, progesterone 75 μg/kg, progesterone 200 μg/kg, progesterone 400 μg/kg, progesterone 75 μg/kg plus PGF$_{2\alpha}$ 1 mg/kg, or progesterone 75 μg/kg plus PGF$_{2\alpha}$ 5 mg/kg. Intramuscular injections of progesterone in propylene glycol were given

TABLE IV

Effect of PGF$_{2\alpha}$ on Pregnancy in Ovariectomized Rats Maintained on Progesterone[a]

Treatment[b]	No. animals pregnant/ No. animals treated	Average No. live implants/ average No. dead and resorbing implants[c]
Progesterone (2.5 mg)	0/7	0/2
Progesterone (5 mg)	9/9	6/2
Progesterone (10 mg)	8/8	8/1
PGF$_{2\alpha}$ (0.8 mg)	7/12	10/[d]
PGF$_{2\alpha}$ (1.6 mg)	2/16	2/[d]
Progesterone (5 mg) + PGF$_{2\alpha}$ (0.8 mg)	4/8	5/7
Progesterone (5 mg) + PGF$_{2\alpha}$ (1.6 mg)	2/7	2/6
Progesterone (10 mg) + PGF$_{2\alpha}$ (0.8 mg)	7/7	5/5
Progesterone (10 mg) + PGF$_{2\alpha}$ (1.6 mg)	4/8	3/5

[a] Rats ovariectomized on day 8 of pregnancy.

[b] PGF$_{2\alpha}$ reported as total daily doses which were given b.i.d. on days 9 and 10 of pregnancy. Progesterone given daily from day 8 to day 19.

[c] Rats not pupping were autopsied on day 22.

[d] Indistinguishable resorption sites.

once a day and subcutaneous injections of PGF$_{2\alpha}$ in 10% ethanol saline were administered b.i.d. On day 19, animals were sacrificed, then uteri were removed, sectioned, and stained with hematoxylin and eosin. Table V reports the average of three independent evaluations of the degree of proliferation which shows that pregnancy termination is not due to the antiprogestational effects of PGF$_{2\alpha}$. In fact, if PGF$_{2\alpha}$ has any action, it tends to enhance the effect of exogenous progesterone in estrogen-primed ovariectomized rabbits rather than inhibit it. Therefore, it seems that pregnancy termination in rats can be used as an acceptable assay for PGF$_{2\alpha}$-induced luteolysis in this species.

The 5 possible areas where $PGF_{2\alpha}$ could be exerting its primary effect are depicted in Fig. 2.

Since the pituitary gland is important in the maintenance of luteal activity in laboratory rodents, this gland was naturally suspected as the target for $PGF_{2\alpha}$. Several lines of evidence, however, suggest that this is not the case, suggestions to the contrary notwithstanding (Labhsetwar, 1970).

Although the pituitary is necessary in the various species, the actual luteotropic complex varies. For example, LH is necessary in the rabbit, prolactin in the rat, and prolactin and FSH in the hamster. Since prolactin is under a different type of control than LH and FSH, it would

TABLE V

Antagonism of Progesterone-Induced Endometrial Proliferation in Ovariectomized Rabbits by $PGF_{2\alpha}$

Treatment[a]	No. of animals	McPhail Index
Vehicle	6	0
Progesterone (75 μg/kg)	6	1[+]
Progesterone (200 μg/kg)	3	3[+]
Progesterone (400 μg/kg)	3	4[+]
Progesterone (75 μg/kg) + $PGF_{2\alpha}$ (1 mg/kg)	5	1[+]
Progesterone (75 μg/kg) + $PGF_{2\alpha}$ (5 mg/kg)	4	2[+]

[a] Estrogen given 7 μg/day in 10% ethanol saline as a daily subcutaneous injection for 6 days prior to further treatment. Progesterone given once a day in propylene glycol and $PGF_{2\alpha}$ given b.i.d. in 10% ethanol saline as subcutaneous injections for 5 days following estrogen priming.

have to be argued that $PGF_{2\alpha}$ either totally blocks the pituitary or has the ability to selectively inhibit the luteotropin characteristic of each species. Both arguments seem untenable. A second argument is that $PGF_{2\alpha}$ is not effective until corpora lutea have reached a certain stage of maturity. Gutknecht *et al.* (1969) showed that if $PGF_{2\alpha}$ was given prior to day 4 of pregnancy in the rat then pregnancy was maintained and 100% pregnancy termination did not occur until the animal was treated on day 6 or later. Shelesnyak (1957) has shown that ergocornine, which is luteolytic in the rat by virtue of blocking prolactin release from the pituitary, is more effective prior to luteal maturity.

Gutknecht *et al.* (1969) observed the lactation was not affected by $PGF_{2\alpha}$ whereas ergocornine caused cessation of milk production in rats. Pharriss, Wyngarden, and Gutknecht (1968) treated ovariectomized rats with PGF$_{2\alpha}$ and assayed the pituitaries for LH. Using the ovarian ascorbic acid depletion assay, they showed that the LH content was identical in the PGF$_{2\alpha}$ and vehicle-treated animals. It has also been reported (Gutknecht *et al.*, (1969) that PGF$_{2\alpha}$ will decrease luteal progesterone content when hypophysectomized rats were maintained on prolactin.

Ovarian arterial infusions of PGF$_{2\alpha}$ have been shown to induce luteolysis at doses which are ineffective when given systemically (Goding *et al.*, 1972). Thus, based on the above arguments, it is suggested that

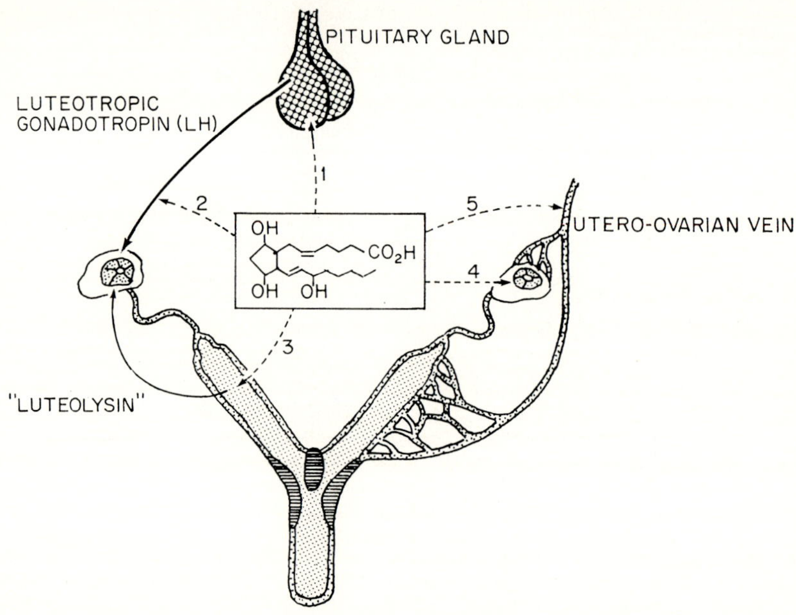

FIG. 2. Possible mechanisms of prostaglandin F$_{2\alpha}$ in luteolysis. *1,* Direct feedback on pituitary gland; *2,* antigonadotropic effect; *3,* stimulation of uterus to produce luteolysin; *4,* direct toxicity on corpus luteum; *5,* constriction of utero-ovarian vein. From Behrman *et al.*, *Ann. N.Y. Acad. Sci.* **180,** 437 (1970).

the hypothalamus and pituitary are not directly involved in PGF$_{2\alpha}$-induced luteolysis.

Another mechanism by which PGF$_{2\alpha}$ could induce luteolysis is by way of the uterus. Prostaglandin is a potent smooth-muscle stimulator and could excite the uterus to contract and release endogenous uterine luteolysin, much as does oxytocin in the bovine (Ginther *et al.*, 1967). However, Blatchley and Donovan (1969) treated hysterectomized guinea pigs with PGF$_{2\alpha}$ for 7 days and examined the ovaries histologically.

All six of the treated animals had corpora that were smaller than the controls, and five exhibited an advanced state of regression. Similar responses have been observed with PGF$_{2\alpha}$ in hysterectomized hamsters (Mazer and Hansel, 1970).

A third possible area for PGF$_{2\alpha}$ action would be a direct toxic effect of the molecule on the corpus luteum itself. Reasons for rejection of this mechanism are based on the stimulation to progesterone synthesis seen in the ovary when exposed to PGF$_{2\alpha}$ *in vitro*. Details of this argument will be presented in the third section of this report.

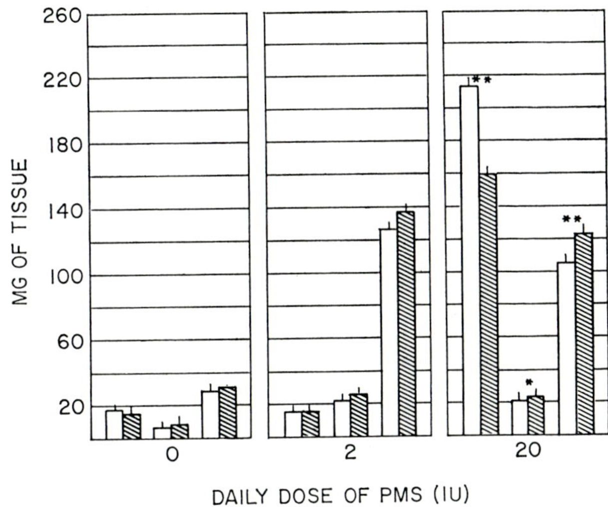

FIG. 3. The effect of (PGF$_{2\alpha}$) on rat uterine and ovarian weight at dose levels of 0, 2, and 20 IU of pregnant mare's serum. □, No PGF$_{2\alpha}$; ▨, 500 μg PGF$_{2\alpha}$; *, $P < 0.05$; **, $P < 0.01$. In each panel, the left-hand pair of bars indicate ovarian weight; the center pair, uterine dry weight; right-hand pair, uterine wet weight.

Even though the pituitary per se had been eliminated as a site of action, there still existed the possibility that PGF$_{2\alpha}$ exerted an antigonadotropic effect. The interaction of PGF$_{2\alpha}$ and gonadotropins could take place in the circulation or at the receptor site on the corpus luteum. This antigonadotropic effect does appear to be important, as studies by Pharriss and Hunter (1971) and data reported here (Fig. 3) have shown that the ovarian effects of HCG and PMS can be overcome by simultaneous injections of PGF$_{2\alpha}$.

Pharriss and Hunter (1971) tested PGF$_{2\alpha}$ for its ability to substitute for, or to antagonize, the gonadotropin induction of ovarian weight gain and ovulation in the rat. PGF$_{2\alpha}$ would not supply FSH or LH components in these two assays. However, this prostaglandin did inhibit FSH-like and LH-like activities of pregnant mare's serum and human chorionic gonadotropin. The PMS stimulation of ovarian weight gain was inhibited 30–60% by $PGF_{2\alpha}$. HCG-induced ovulations were almost totally blocked by this prostaglandin.

Figure 3 shows the antagonism between $PGF_{2\alpha}$ and PMS in the rat. Of special interest is the stimulation of uterine weight gain seen even though there is a suppression of the ovarian response. This is interpreted to mean that the $PGF_{2\alpha}$ has a suppressive effect only on the luteal tissue and may, in fact, stimulate estrogen secretion from other cells.

TABLE VI

17β-Estradiol Antagonism of $PGF_{2\alpha}$-Induced Luteolysis in Rabbits

Treatment[a]	No. of animals	No. with corpora lutea
Vehicle	5	5
$PGF_{2\alpha}$ (1 mg/kg)	7	3
$PGF_{2\alpha}$ (5 mg/kg)	5	0
Estradiol (2 μg/kg)	6	6
$PGF_{2\alpha}$ (1 mg/kg) + estradiol (2 μg/kg)	5	4
$PGF_{2\alpha}$ (5 mg/kg) + estradiol (2 μg/kg)	5	0
$PGF_{2\alpha}$ (5 mg/kg) + estradiol (4 μg/kg)	5	5
$PGF_{2\alpha}$ (8 mg/kg) + estradiol (4 μg/kg)	5	1

[a] $PGF_{2\alpha}$ and estradiol given subcutaneously in 10% ethanol saline are reported as total daily doses. Injections given on afternoon of day 6 and b.i.d. on day 7 of pregnancy.

Table VI shows a similar relationship between $PGF_{2\alpha}$ and estradiol on luteal activity in the rabbit. It can be seen that there is a competitive antagonism between these substances. Similar results have been reported by Gutknecht *et al.* (1971a). This effect, however, does not hold true for the rat (Table VII), as LH and prolactin will cause only a partial reversal of $PGF_{2\alpha}$ in this species.

Johnston and Hunter (1970) measured the ability of the luteotropic complex (FSH + prolactin) in the hamster to overcome the pregnancy-terminating effect of $PGF_{2\alpha}$. $PGF_{2\alpha}$-induced luteolysis had previously been shown to be the mechanism whereby this prostaglandin terminated pregnancy in the hamster (Gutknecht *et al.*, 1969). Johnston and Hunter (1970) reported that the FSH–prolactin complex antagonized 200 μg, but not 500 μg, of $PGF_{2\alpha}$ in terminating pregnancy, thereby suggesting a competitive inhibition. Behrman *et al.* (1970) also suggested competitive inhibition of gonadotropins and $PGF_{2\alpha}$ on secretion from rat ovaries during pseudopregnancy. Therefore, it is suggested that antagonism to luteotropins is a possible mechanism whereby $PGF_{2\alpha}$ exerts its luteolytic effect.

The final area for discussion of mechanisms is ovarian blood flow. Although this potential mechanism was the subject which initiated the luteolytic studies, adequate tests for the role of altered blood flow in ovarian physiology still have not been carried out. Three studies have been reported, however, which suggest that ovarian perfusion is reduced after $PGF_{2\alpha}$ treatment. Pharriss *et al.* (1970) cannulated the utero-

TABLE VII

Antagonism of $PGF_{2\alpha}$-Induced Termination of Pregnancy in Rats by 17β-Estradiol and Luteotropins

Treatment[a]	No. of animals	Percent terminated
Vehicle	8	0
$PGF_{2\alpha}$ (0.8 mg)	23	65
$PGF_{2\alpha}$ (1.6 mg)	18	90
$PGF_{2\alpha}$ (0.8 mg) + estradiol (2 µg)	9	100
$PGF_{2\alpha}$ (0.8 mg) + estradiol (5 µg)	5	100
$PGF_{2\alpha}$ (0.8 mg) + estradiol (10 µg)	6	100
$PGF_{2\alpha}$ (0.8 mg) + estradiol (20 µg)	5	100
$PGF_{2\alpha}$ (1.6 mg) + LH (10 µg)	8	100
$PGF_{2\alpha}$ (1.6 mg) + prolactin (2 IU)	8	100
$PGF_{2\alpha}$ (1.6 mg) + LH (10 µg) + prolactin (2 IU)	8	75

[a] Total daily doses per rat are reported. $PGF_{2\alpha}$ and 17β-estradiol given b.i.d. and LH and prolactin given once a day. Treatment was on days 9 and 10 of pregnancy.

ovarian vein at its juncture with the inferior vena cava in rats and rabbits and measured blood flow by timed collection. In rats, after a single dose of $PGF_{2\alpha}$ (300–600 µg/kg iv) there was an immediate drop in blood flow to 50–60% of the control levels, lasting about 25 minutes. The depression was more severe (10–20% of control blood flow) in rabbits which responded to doses as small as 50 µg/kg and lasted longer than 45 minutes. Gutknecht *et al.* (1970) reported similar observations in rabbits using a more sophisticated method. These investigators employed a hydrogen desaturation technique for measuring blood flow and compared the ovarian perfusion to that of the kidney. Blood flow was significantly depressed in ovarian tissue at doses of 200–400 µg/kg of $PGF_{2\alpha}$, while renal blood flow was not altered. These doses of $PGF_{2\alpha}$ also were effective in depressing ovarian and plasma progesterone concentrations in these rabbits. McCracken *et al.* (1970) measured ovarian blood flow in sheep in which the uterus, ovary, and intact vascular supply had been transplanted into the neck. They found that $PGF_{2\alpha}$ caused a reduction in ovarian blood flow, but there was a depression in progesterone output prior to this decrease. In this investigation, the ovary was subject to large doses (up to 100 μg $PGF_{2\alpha}$ per ovary per hour) as the prostaglandin was infused directly into the ovarian artery. The antiluteotropic effect of this high concentration of $PGF_{2\alpha}$ probably caused the depression in progesterone secretion. Aldridge and associates (1970), using a similar sheep preparation, found that PGE_1 caused a significant depression in progesterone secretion with no consistent alteration in blood flow. Nevertheless, it does appear that $PGF_{2\alpha}$ does have the ability to decrease ovarian perfusion.

The mechanism whereby prostaglandins are luteolytic is still unanswered but two hypothesis are supportable: gonadotropin antagonism and alteration of ovarian blood flow. Both of these can be used in arguments for the localized effect of luteolysis. The blood flow hypothesis fits with what is known about the vascular anatomy in species which exhibit the unilateral effect; the antagonism to luteotropins fits the hypothesis forwarded by Goding and his collaborators. Goding *et al.* (1971) and McCracken *et al.* (1971) have described a possible countercurrent exchange between the uterine vein and ovarian artery in the sheep. This would allow uterine prostaglandins direct access to the ipsilateral ovary and their antigonadotropic effect could be exerted.

We believe the vascular congestion hypothesis still offers the best explanation to this problem since there are some observations that fit this mechanism but do not fit with a countercurrent-luteotropin inhibition pattern.

Two observations, one in the guinea pig (Culiner, 1944) and the other in sheep (Morris and Sass, 1966), have been made on an increase in lymph flow from the ovary as peak luteal activity is approached. Both reports suggest the importance in venous blood flow in such a phenomenon. No mention has been made here as to the final mechanism of luteal degradation, but if one considers the possibility of an inadequate blood flow, then the result in the tissue would be one of stagnation or congestion. There would not be an optimal supply of nutrients and precursors to the cells, and likewise their ability to eliminate end products would be hampered. Consistent with this line of reasoning is the fact that progesterone is luteolytic in ewes (Ginther, 1968), heifers (Woody *et al.*, 1967), and guinea pigs (Ginther, 1969) and that LH is luteolytic in rats (Rothchild, 1965). Both these treatments would aggravate the situation that exists in our model. Keyes and Weiner (1971) have shown a local negative feedback of progesterone metabolites on steroidogenesis in the corpus luteum which might also be active in blocking luteal progesterone production.

Finally, there are the reports concerning the induction of two generations of corpora on the same ovary and their pattern of regression (Neill and Day, 1964; Caldwell *et al.*, 1969; Deanesly and Perry, 1969). If the second generation is induced shortly after the first, then both regress at the same time, the first generation showing the expected life span. If the second luteal induction is separated by a week or more, these corpora will continue to secrete progesterone after the first generation has undergone regression. This information suggests that there are some inherent qualities associated with stage of development that renders corpora lutea susceptible to lysis. It has been reported that $PGF_{2\alpha}$ is relatively ineffective in terminating luteal function in the rat during the first 5 days after ovulation but that after this time the corpora are quite susceptible. This is also the time when luteal tissues reach maturity in these species and are synthesizing progesterone at high rates. It may be that a reduced blood flow will not affect the growing corpora because they are not burdened with excreting high levels of steroids, but once they reach maturity, elimination of these products is necessary for a healthy existence.

C. ROLE OF PROSTAGLANDINS IN THE OVARY

Most, if not all, of the actions of the prostaglandins which have been described are associated with cyclic 3′,5′-adenosine monophosphate cyclic AMP (cAMP). The postaglandins (depending upon the tissues being studied) either mimic the action of cAMP, probably by stimulating the adenyl cyclase system (Butcher and Baird, 1968), or antagonize cAMP-mediated responses, conjecturally by inhibiting the formation of this nucleotide (Steinberg *et al.*, 1964). At present, there is no evidence that the prostaglandins are antagonistic to cAMP on a competitive basis (Orloff *et al.*, 1965; Steinberg, 1966).

As mentioned earlier, initial attempts to show a direct inhibiting effect of $PGF_{2\alpha}$ to luteal steroidogenesis *in vitro* resulted in a stimulation to progesterone synthesis (Pharriss *et al.*, 1968; Speroff and Ramwell, 1970b). It was, in fact, reported that $PGF_{2\alpha}$ mimicked several of the effects of LH in the rat (Pharriss *et al.*, 1968). In the ovarian incubation studies, 10 μg/ml of the $PGF_{2\alpha}$ resulted in a 33–150% increase in progesterone synthesized in 4 hours. Cholesterol was probably the progesterone precursor during this stimulation, since acetate-1-^{14}C was not incorporated into the progesterone. These results were quite similar to those reported for LH stimulation of rat ovaries *in vitro* by Armstrong *et al.* (1964). In an attempt at further comparisons, $PGF_{2\alpha}$ and LH were tested together at concentrations of 10 μg/ml each. This resulted in no further increase in progesterone synthesis than was seen when these substances were tested separately at the same concentrations, indicating that they were acting on the same metabolic pathway or were at least both restricted by the same rate-limiting step. Another test for LH activity, ovarian ascorbic acid depletion, was carried out with $PGF_{2\alpha}$. $PGF_{2\alpha}$ at 10–250 μg resulted in an 18–32.5% depletion in ovarian ascorbic acid. Higher doses failed to increase the response. While this assay is not specific for LH activity, the correlation of this response with the simulation of progesterone synthesis argued strongly for the possibility of similarities in action between $PGF_{2\alpha}$ and LH.

Bedwani and Horton (1968) attempted to measure an inhibitory response of prostaglandins to gonadotropins in minced rabbit ovaries. Their results, however, indicated that in the presence of human chorionic gonadotropin and pregnant mare serum, PGE_2 (1 μg/ml) caused an increase in the formation of 20α-hydroxypregn-4-en-3-one with no change in progesterone synthesis. The significance of an increase in the production of this inactive metabolite of progesterone is unknown, but it does show that prostaglandins will stimulate steroid formation in this preparation.

More recently, Speroff and Ramwell (1970b) have reported on the prostaglandin stimulation of steroidogenesis in bovine corpora lutea. In this study, PGE_2 was found to be more active than PGE_1, $PGF_{2\alpha}$, or PGA_1 in increasing progesterone synthesis. PGE_2 was approximately half as active as LH on a molar basis in their preparations. The stimulation was quite similar in character to the effect of LH, in that the incorporation of acetate-1-^{14}C was increased in the same manner. The incorporation of acetate is an important point here because there seems to be a species difference between rats and cows in response to LH, and this difference is mirrored by prostaglandins. Speroff and Ramwell (1970b) also reported that the time-response curves for PGE_2 and LH were similar and that there was no additive effect when prostaglandins were added to luteal slices incubated with saturating doses of LH or HCG.

Snellner and Wickersham (1970) found that the effect of PGE_1, PGE_2, and $PGF_{2\alpha}$ on progesterone formation in bovine luteal slices was dose related. They also stated that the mean response to a combination of prostaglandin and LH was greater than to prostaglandin alone, presumably at a submaximal dose of both substances.

It has been shown in both bovine corpora lutea and luteinized rabbit ovaries that the response to LH is probably mediated through cAMP (March and Savard, 1966; Marsh *et al.*, 1966; Dorrington and Baggett, 1969). In an attempt to further associate the actions of prostaglandins and LH, Erickson and Pharriss (1970) have demonstrated that one of the methylxanthines (theophylline) will potentiate the increase in steroidogenesis in luteinized rat ovaries caused by $PGF_{2\alpha}$ and LH (Table VIII). By measuring cAMP directly, Marsh (1970) has specifically shown this to be the case. In homogenates of luteal tissues obtained from pregnant cows, PGE_2 caused an activation of the adenyl cyclase system and an increase in the concentration of cAMP. The increase in progesterone formation correlated well with the increase in cAMP. Behrman and co-workers (1970) have been able to stimulate ovarian

TABLE VIII

Effects of $PGF_{2\alpha}$, LH, Theophylline and Combinations of Each on Progesterone and 20α-Dihydroprogesterone Synthesis in Vitro in 6-Day Pseudopregnant Rat Ovaries during a 4-Hour Incubation

Treatment	De novo synthesis (μg/gm tissue)
A. *LH*	
Control	$7.4 \pm 0.69^{a*}$
Theophylline	$9.6 \pm 0.82^{b*}$
LH	$22.2 \pm 1.23^{c*}$
LH + theophylline	$34.1 \pm 3.91^{d*}$
B. *$PGF_{2\alpha}$*	
Control	$26.1 \pm 2.94^{a**}$
Theophylline	$25.6 \pm 2.15^{a**}$
$PGF_{2\alpha}$	$31.8 \pm 2.19^{b**}$
$PGF_{2\alpha}$ + theophylline	$36.7 \pm 1.97^{c**}$

* b different from a, $p < 0.05$; c different from a and b, $p < 0.001$; d different from c, $p < 0.025$.

** b different from a, $p < 0.05$; c different from b, $p < 0.025$.

progesterone synthesis with prostaglandins *in vivo*. When $PGF_{2\alpha}$ (10 μg) was injected into pseudopregnant rats, there was a 33% decrease in progesterone secretion into the ovarian vein. Concomitant injection of LH and $PGF_{2\alpha}$ returned the progesterone concentration to normal. When this procedure was carried out in hypophysectomized rats, however, $PGF_{2\alpha}$ injection resulted in an increase in progesterone secretion. The control secretion rate for progesterone was decreased 50% by hypophysectomy and $PGF_{2\alpha}$ caused a 23% increase. LH (20 μg) brought the secretion rate back to the intact level but $PGF_{2\alpha}$ and LH together were no more effective than $PGF_{2\alpha}$ alone.

Fried and co-workers (1969) have shown that 7-oxa-13-prostynoic acid antagonizes the action of PGE_1 and PGE_2 on cAMP formation in a competitive manner. Kuehl *et al.* (1970) reported that this prostynoic acid analog would not only inhibit prostaglandin stimulation of cAMP formation in the mouse ovary but also would inhibit this response to LH. Kinetic studies revealed the possibility of a single LH-related prostaglandin receptor in ovarian tissue and that its activation was an essential requirement for LH stimulation of steroidogenesis.

In some recent work in our laboratory (Chasalow and Pharriss, 1972), we have been able to show an effect of LH on prostaglandin synthetase activity. Ovaries from pregnant rats were homogenized and incubated for 30 minutes with ^3H-labeled arachidonic acid. Ether extracts were chromatographed in 2% acetic acid in ethyl acetate and eluted; the percentage of ^3H incorporation into PGE_2 fractions was measured. If the rats were injected with anti-LH antiserum 30 minutes prior to sacrifice, there was a 36% decrease in incorporation. Concomitant injections of LH or addition of LH to the homogenation media caused a stimulation above that seen in the control animals. Since the percentage incorporation of added arachidonic acid was increased with LH, it appears that the gonadotropin works on the synthetase system directly.

Although the picture is still not as complete with prostaglandins as it is with cAMP, it is beginning to appear that these lipids are important intermediates in the response of luteal tissue to LH. If this is the case, then the data suggest that PGE_2 fits in between the LH contact with the luteal cell and the stimulation to adenyl cyclase activity. PGE_2 and $PGF_{2\alpha}$ are implicated as the important prostaglandin in the ovary because of the high sensitivity and reproducibility seen with PGE_2 and the preference for PGE_2 synthesis following LH stimulation of the ovary.

IV. Summary

The prostaglandins create various responses in most tissues that have been studied. This, coupled with their ubiquitous presence, suggests they must play an important role in biochemical and physiological systems at least in mammals. The reproductive processes seem especially sensitive to these agents, and many functions have been suggested for the prostaglandins in this area (Speroff and Ramwell, 1970a). At present, two functions seem better documented than others, these being the subject of this paper. $PGF_{2\alpha}$ has the qualities and effectiveness to be the long sought uterine luteolytic factor and fits the current hypothesis for the mechanism of uterine directed luteolysis. Prostaglandins in general, and PGE_2 specifically, seem important in mediating LH stimulation of luteal steroidogenesis. This latter role is complicated and requires further experimentation, but we predict that it will ultimately be shown to be necessary in LH control of ovarian steroidogenesis.

ACKNOWLEDGMENT

The authors wish to acknowledge the technical assistance of P. Chiau, J. Wong, E. Pollard, and M. Lee.

REFERENCES

Aldridge, R. R., Barrett, S., Brown, J. B., Funder, J. W., Goding, J. R., Kaltenbach, C. C., and Mole, B. J. (1970). *J. Reprod. Fert.* **21,** 360.

Armstrong, D. T., O'Brien, J., and Greep, R. O. (1964). *Endocrinology* **75,** 488.

Barley, D. A., Butcher, R. L., and Inskeep, E. K. (1966). *Endocrinology* **79,** 119.

Bedwani, J. R., and Horton, E. W. (1968). *Life Sci.* **7,** 389.

Behrman, H. R., Yoshinaga, K., and Greep, R. O. (1970). *Ann. N.Y. Acad. Sci.* **180,** 426.

Bland, K. P., Horton, E. W., and Poyser, N. L. (1971). *Life Sci.* **10,** 509.

Blatchley, F. R., and Donovan, B. T. (1969). *Nature (London)* **221,** 1065.

Blatchley, F. R., Donovan, B. T., Poyser, N. L., Horton, E. W., Thompson, C. J., and Los, M. (1971). *Nature (London)* **230,** 243.

Butcher, R. W., and Baird, C. E. (1968). *J. Biol. Chem.* **243,** 1713.

Bygdeman, M., and Samuelsson, B. (1966). *Clin. Chim. Acta* **13,** 465.

Caldwell, B. V., Mazer, R. S., and Wright, P. A. (1967). *Endocrinology* **80,** 477.

Caldwell, B. V., Moor, R. M., and Lawson, R. A. S. (1968). *J. Reprod. Fert.* **17,** 567.

Caldwell, B. V., Moor, R. M., Wilmut, I., Polge, C., and Rowson, L. E. A. (1969). *J. Reprod. Fert.* **18,** 107.

Chasalow, F., and Pharriss, B. B. (1972). Prostaglandins (in press).

Culiner, A. (1944). *Anat. Rec.* **90,** 217.

Deanesly, R., and Perry, J. S. (1969). *J. Reprod. Fert.* **20,** 503.

Donovan, B. T., and Traczyk, W. J. (1962). *J. Physiol. (London)* **161,** 227.

Dorrington, J. H., and Baggett, B. (1969). *Endocrinology* **84,** 989.

DuCharme, D. W., Weeks, J. R., and Montgomery, R. G. (1968). *J. Pharmacol. Exp. Ther.* **160,** 1.

du Mesnil du Buisson, F. (1961). *C.R. Acad. Sci.* **253,** 727.

Embrey, M. P. (1970). *Brit. Med. J.* **2,** 258.

Erickson, R. R., and Pharriss, B. B. (1970). Unpublished observations; also, Erickson, R. R. (1970). Ph.D. Thesis. Western Michigan Univ.

Fried, J., Santhanakrishnan, T. S., Himizu, J., Lin, C. H., Ford, S. H., Rubin, B., and Grigas, E. O. (1969). *Nature (London)* **223,** 208.

Ginther, O. J. (1968). *Endocrinology* **83,** 613.

Ginther, O. J. (1969). *Amer. J. Vet. Res.* **30,** 261.

Ginther, O. J., Pope, A. L., and Casida, L. E. (1966). *J. Anim. Sci.* **25,** 472.

Ginther, O. J., Woody, C. O., Mahajan, S., Janakiraman, K., and Casida, L. E. (1967). *J. Reprod. Fert.* **14,** 225.

Goding, J. R., Baird, D. T., Cumming, I. A., and McCracken, J. A. (1971). *Karolinska Symp. Res. Meth. Reprod. Endocrinol. 4th; Perfusion Techniques.*

Goding, J. R., Cain, M. D., Cerini, J., Cerini, M., Chamley, W., and Cumming, I. A. (1972). *Biol. Reprod.* In press.

Goldblatt, M. W. (1933). *Chem. Ind.* **52**, 1056.

Gutknecht, G. D., Cornette, J. C., and Pharriss, B. B. (1969). *Biol. Reprod.* **1**, 367.

Gutknecht, G. D., Duncan, G. W., and Wyngarden, L. J. (1970). *Mtg. Amer. Physiol. Soc., Bloomington, Indiana, 1970.*

Gutknecht, G. D., Duncan, G. W., and Wyngarden, L. J. (1971a). *Biol. Reprod.* **5**, 87.

Gutknecht, G. D., Wyngarden, L. J., and Pharriss, B. B. (1971b). *Proc. Soc. Exp. Biol. Med.* **136**, 1151.

Hawkins, D. F. (1968). *In* "Prostaglandins" (P. W. Ramwell and J. E. Shaw, eds.), Symposium Worcester Foundation for Experimental Biology, New York.

Johnston, J. O., and Hunter, K. K. (1970). *Mtg., Amer. Physiol. Soc., Columbus, Ohio, 1970.*

Karim, S. M. M., and Filshie, G. M. (1970). *Lancet* **1**, 157.

Keyes, P. L., and Weiner, M. (1971). *Proc. Endocrine Soc. Mtg.; Abstr. No.* **340**.

Kirton, K. T., Pharriss, B. B., and Forbes, A. D. (1970). *Proc. Soc. Exp. Biol. Med.* **133**, 314.

Kuehl, F. A., Humes, J. L., Tarnoff, J., Cirillo, V. J., and Ham, E. A. (1970). *Science* **169**, 883.

Labhsetwar, A. P. (1970). *J. Reprod. Fert.* **23**, 155.

Labhsetwar, A. P. (1971). *Nature (London)* **230**, 528.

Lidner, H. R., and Shelesnyak, M. E. (1967). *Acta Endocrinol.* **56**, 27.

Loeb, L., (1923). *Proc. Soc. Exp. Biol. Med.* **20**, 441.

Lukaszewaska, J. H., and Hansel, W. (1970). *Endocrinology* **86**, 261.

McCracken, J. A., Baird, D. T., and Goding, J. R. (1971). *Recent Progr. Horm. Res.* **27**, 537.

McCracken, J. A., Glew, M. E., and Scaramuzzi, R. J. (1970). *J. Clin. Endocrinol. Metab.* **30**, 544.

Marsh, J. M. (1970). *FEBS Lett.* **7**, 283.

Marsh, J. M., and Savard, K. (1966). *Steroids* **8**, 133.

Marsh, J. M., Butcher, R. W., Savard, K., and Sutherland, E. W. (1966). *J. Biol. Chem.* **241**, 5436.

Mazer, R., and Hansel, W. L. (1970). Personal communication.

Morris, B., and Sass, M. B. (1966). *Proc. Roy. Soc. Ser. B.* **164**, 577.

Neill, J. D., and Day, B. N. (1964). *Endocrinology* **74**, 355.

Orloff, J., Handler, J. S., and Bergström, S. (1965). *Nature (London)* **205**, 397.

Pharriss, B. B. (1969). Unpublished observation.

Pharriss, B. B. (1970). *Perspect. Biol. Med.* **13**, 434.

Pharriss, B. B., and Hunter, K. K. (1971). *Proc. Soc. Exp. Biol. Med.* **136**, 503.

Pharriss, B. B., and Wyngarden, L. J. (1969). *Proc. Soc. Exp. Biol. Med.* **130**, 92.

Pharriss, B. B., Wyngarden, L. J., and Gutknecht, G. D. (1968). *In* "Gonadotropins" (E. Rosemberg, ed.), p. 121. Geron-X, Los Altos, Calif.

Pharriss, B. B., Cornette, J. C., and Gutknecht, G. D. (1970). *J. Reprod. Fert. Suppl.* **10**, 97.

Pickles, V. R. (1957). *Nature (London)* **180**, 1198.

Pickles, V. R. (1959). *J. Endocrinol.* **19**, 150.

Pickles, V. R. (1967). *Int. J. Fert.* **12**, 335.

Pickles, V. R., and Clitheroe, H. J. (1960). *Lancet* **2**, 959.

Pickles, V. R., and Hall, W. J. (1963). *J. Reprod. Fert.* **6**, 315.

Pickles, V. R., Hall, W. J., Best, F. A., and Smith, G. N. (1965). *J. Obstet. Gynaecol. Brit. Commonw.* **72**, 185.

Poyser, N. L., Horton, E. W., Thompson, C. J., and Los, M. (1971). *Nature (London)* **230**, 526.

Ramwell, P. W., and Shaw, J. E. (1971). *Ann. N. Y. Acad. Sci.* **180**, 10.

Rothchild, I. (1965). *Vitam. Horm. (New York)* **23**, 209.

Rowson, L. E. A., and Moor, R. M. (1967). *J. Reprod. Fert.* **13**, 511.

Shaw, J. E. (1971). Personal communication.

Shelesnyak, M. C. (1957). *Rec. Progr. Horm. Res.* **13**, 269.

Snellner, R. G., and Wickersham, E. W. (1970). *J. Anim. Sci.* **31**, 230.

Speroff, L., and Ramwell, P. W. (1970a). *Amer. J. Obstet. Gynecol.* **107**, 1111.

Speroff, L., and Ramwell, P. W. (1970b). *J. Clin. Endocrinol. Metab.* **30**, 345.

Steinberg, D. (1966). *Pharm. Rev.* **18**, 217.

Steinberg, D., Vaughn, M., Nestel, P. S., Strand, O., and Bergström, D. (1964). *J. Clin. Invest.* **43**, 1533.

von Euler, U. S. (1934). *Arch. Exp. Pathol. Pharmakol.* **175**, 78.

von Euler, U. S., and Hammarström, S. (1937). *Skand. Arch. Physiol.* **77**, 96.

Wilson, L., Jr., Cenedella, R. J., Butcher, R. L., and Inskeep, E. K. (1972). *J. Anim. Sci.* **54**, (1), 93.

Wiqvist, N., Bygdeman, M., and Kirton, K. T. (1970). *Nobel Symp., Reprod. Mech.* In press.

Woody, C. O., First, N. L., and Pope, A. L. (1967). *J. Anim. Sci.* **26**, 139.

DISCUSSION

J. E. Shaw: The presentation by Dr. Pharriss emphasizes the potential multiplicity of effects exerted by prostaglandins in physiological test systems, for here we are just looking at the effects of these compounds on the ovary. I think we must be careful to dissociate the obvious pharmacological actions from the possible physiological effects of the endogenous compounds. I think what we have heard outlined is perhaps the first indication of the involvement of the endogenous prostaglandins in a physiological process.

H. Behrman: There now appears to be little doubt that prostaglandin is luteolytic in many species, and Dr. Pharriss has presented several possible mechanisms whereby this effect may be produced. We have data that indicate that a possible vascular effect of $PGF_{2\alpha}$ is not the sole cause of luteolysis, and additional data on the nature of the intracellular lesion sites induced by $PGF_{2\alpha}$.

Dr. Larry Demers and I have successfully cultured functional rat corpora lutea (CL) explants for several days [L. Demers, H. Behrman, and R. Greep, *Endocrine Soc., Abstr.* (1971)]. This method consists of culturing 6–12 CL in Trowell's T-8 complete nutritive medium supplemented with fetal calf serum and glucose (400 mg/100 ml%). The medium, changed each day, contained a mixture of ^{14}C-labeled amino acids (25 μCi/ml), acetate-1-^{14}C (2 μCi/ml), and either $PGF_{2\alpha}$ (10 μg/ml) or prolactin (NIAMD rat prolactin; 0.5 IU/ml). At the indicated times the CL were removed, then extracted with chloroform:methanol to isolate the steroid fraction; the residue was treated with perchloric acid (1.6 N) for 24 hours. The protein was then treated with trichloroacetic acid (5%), extracted twice with ether:ethanol (1:1) and once with ether, and filtered; the radioactivity was determined (Fig. A). After 24 hours in culture, amino acid incorporation into luteal protein was severely reduced by PGF$_{2\alpha}$, an effect expressed at each time interval compared to either control of prolactin-treated CL.

Progesterone was isolated by two-dimensional thin-layer chromatography [H. Behrman *et al., Can. J. Biochem.* **48,** 881 (1971)], and the radioactivity associated with this fraction was determined (Fig. B). Prostaglandin produced a significant decrease in acetate incorporation into material that ran with progesterone.

The combined effects of PGF$_{2\alpha}$ on amino acid incorporation into protein and acetate incorporation into progesterone indicate that active luteolysis was induced by PGF$_{2\alpha}$ where no change in blood flow was possible. This is the first time that prostaglandin has been shown to be luteolytic *in vitro,* and this method may help solve the enigma of the luteolytic effect of PGF$_{2\alpha}$ *in vivo* and the apparent luteotropic effect of PGF$_{2\alpha}$ in very short-term incubations. The mechanism by which

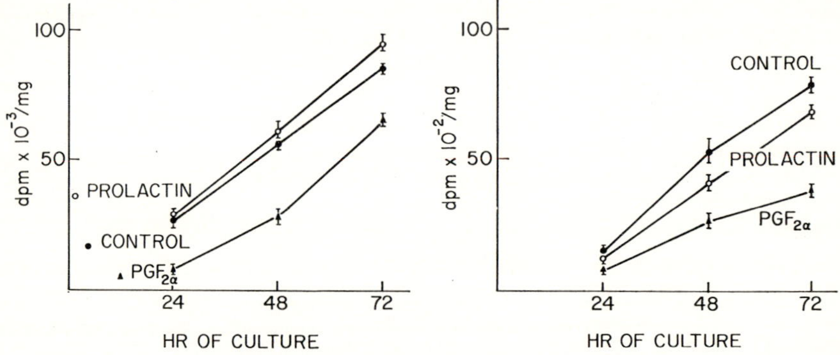

FIG. A. (left). Incorporation of ^{14}C-labeled amino acid into protein of rat corpora lutea. ○, prolactin; ●, control; ▲, prostaglandin F$_{2\alpha}$.

FIG. B. (right). Incorporation of active-^{14}C into progesterone-associated material in organ culture of rat corpora lutea. ●, Control; ○, prolactin; ▲, prostaglandin F$_{2\alpha}$.

PGF$_{2\alpha}$ induced luteolysis under these conditions is not known, but these results point to a direct action on or in the luteal cell.

We have additional data which we think point to the site of action of PGF$_{2\alpha}$ in producing luteolysis There is an enzyme system in the rat luteal cell which we believe is fundamental to the regulation of steroidogenesis and the function of this system is simply to provide adequate storage and release of cholesterol for conversion to steroids. Two enzymes are involved here: cholesterol ester synthetase, which catalyzes the synthesis of cholesterol esters, and cholesterol esterase, which catalyzes the hydrolysis of cholesterol esters thereby releasing free cholesterol. We have shown that LH activates cholesterol esterase [H. Behrman, and D. Armstrong, *Endocrinology* **85,** 474 (1969)] and this action of LH is closely linked to the rapid increase in progesterone formation produced by LH. Over longer periods, however, both these enzymes are maintained by prolactin [H. Behrman *et al., Endocrinology* **87,** 1251 (1970)]. What I wish to demonstrate here is that, when administered *in vivo,* PGF$_{2\alpha}$ markedly reduced the specific activity of these enzymes, particularly the synthetase enzyme; PGF$_{2\alpha}$ prevented the maintenance of synthetase activity by prolactin when administered to hypophysectomized rats;

LH and prolactin together were able to partially reverse the detrimental action of $PGF_{2\alpha}$ on synthetase esterase activity; $PGF_{2\alpha}$ *in vivo* exerted a luteolytic action in the absence of the pituitary, ruling out an effect of $PGF_{2\alpha}$ on the pituitary as the cause of luteolysis (Table A). From these data $PGF_{2\alpha}$ appears to neutralize the tropic expression of prolactin. This effect could be expressed directly within the luteal cell by inhibiting enzyme synthesis ($PGF_{2\alpha}$ had no direct effect on synthetase activity in a partially purified enzyme extract of luteal tissue) or may in some manner alter gonadotropin receptor interaction and thereby prevent hormone expression.

J. A. McCracken: I would like to support Dr. Pharriss' contention that prostaglandin $F_{2\alpha}$ is indeed the uterine luteolytic factor. At this meeting last year [J. A. McCracken, D. T. Baird, and J. R. Goding, *Rec. Progr. Horm. Res.* **27,** 537 (1971)] data were presented describing the isolation of a luteolytic factor from the uterine blood of sheep. This was achieved by cross-circulating uterine blood from ewes bearing utero-ovarian transplants as donors into ewes with the ovary transplanted alone as recipients. It was found that uterine venous blood of the donor infused into the ovarian arterial supply of the recipient contained a luteolytic factor at the time of normal luteal regression (day 15). This uterine blood factor was mimicked by infusions of small doses of $PGF_{2\alpha}$ into the ovarian arterial supply of the recipient ewes. Evidence was also presented in the form of ligation and separation experiments, and by one experiment with tracer amounts of $PGF_{2\alpha}$-^{3}H, that $PGF_{2\alpha}$ was transferred from the uterine vein to the adherent ovarian artery by means of a countercurrent exchange mechanism. Futher experiments with ^{3}H-labeled $PGF_{2\alpha}$ have now been performed which confirm our original report.

Figure C illustrates one of the more recent experiments which shows the time course of the transfer of $PGF_{2\alpha}$ from the uterine vein to the adjacent ovarian artery; 0.1 μCi per minute of $PGF_{2\alpha}$-^{3}H was infused into the uterine vein for 1 hour on day 14 of the cycle. Ovarian arterial blood was collected continuously from a branch of the ovarian artery near the hilus of the ovary and simultaneously from the adjacent iliac artery (as a peripheral control). The radioactivity associated with the $PGF_{2\alpha}$ after extraction, solvent partition, and two TLC separations, was compared in the ovarian and iliac arterial samples. The radioactivity associated with the $PGF_{2\alpha}$ zone in the ovarian arterial blood showed a significant increase at about 30 minutes after the infusion was begun. The radioactivity in ovarian arterial blood continued to rise throughout the infusion and rather surprisingly reached a peak value about 35 minutes after the infusion was stopped. On the other hand, the radioactivity associated with the $PGF_{2\alpha}$ zone isolated from the iliac arterial blood remained low throughout the experiment indicating that the clearance rate for $PGF_{2\alpha}$ in the peripheral circulation must be extremely high [P. Piper, J. R. Vane, and J. H. Wyllie, *Nature (London)* **225,** 600 (1970)]. From these experiments one can say the following: (a) The transfer of $PGF_{2\alpha}$ does occur by a countercurrent mechanism. (b) The transfer proceeds relatively slowly. (c) The high levels found in ovarian arterial blood are not due to recirculation.

It is possible to make a minimum estimate of the amount of $PGF_{2\alpha}$-^{3}H transferred from the uterine vein to the ovarian artery if one takes the peak value of radioactivity found in the ovarian arterial blood as representative of the equilibrated state. In the experiment described, only about 10% of the ovarian arterial flow was collected. The recovery of $PGF_{2\alpha}$-^{3}H from plasma was about 50%, and the counting efficiency for tritium was also about 50%. From these figures it was calculated that an amount in excess of 2% of the infused $PGF_{2\alpha}$-^{3}H was trans-

TABLE A

Effect of $PGF_{2\alpha}$ in Vivo on Corpus Luteum Function in the Rat (Mean $\pm$ SEM)[a]

Treatment	Cholesterol ester synthetase (CE cpm/min/mg protein)	Cholesterol esterase (% hydrolysis/min/mg protein)	Cholesterol ester (mg/gm tissue)	*In vitro* Progesterone synthesis[e] (μg/gm tissue)
Intact pituitary	855 $\pm$ 85	0.18 $\pm$ 0.03	11.0 $\pm$ 2.4	57.4 $\pm$ 7.6
Intact + $PGF_{2\alpha}$[b]	234 $\pm$ 10	0.10 $\pm$ 0.02	3.1 $\pm$ 0.6	23.3 $\pm$ 2.8
APX	119 $\pm$ 11	0.06 $\pm$ 0.02	2.3 $\pm$ 0.1	28.5 $\pm$ 6.7
APX + PL[c]	544 $\pm$ 87	0.15 $\pm$ 0.01	4.3 $\pm$ 1.3	38.4 $\pm$ 4.4
APX + PL + $PGF_{2\alpha}$	160 $\pm$ 10	0.11 $\pm$ 0.02	1.7 $\pm$ 0.4	11.6 $\pm$ 1.8
APX + PL + LH + $PGF_{2\alpha}$[d]	318 $\pm$ 21	0.14 $\pm$ 0.01	2.0 $\pm$ 0.2	14.5 $\pm$ 4.3

[a] H. R. Behrman *et al.*, *Lipids* (1971) in press.

[b] Superovulated rats treated 3 days following HCG treatment (25 iu; s.c.) with $PGF_{2\alpha}$ (0.5 mg/kg; b.i.d.), for 2 days.

[c] NIH-P-S9 (2 IU/rat; b.i.d.).

[d] NIH-LH-S16 (10 μg/rat; b.i.d.).

[e] Ovaries were removed, sliced, and incubated for 2 hours at 37°C. All treatments were administered *in vivo*. Progesterone content was measured after extraction and thin-layer chromatography by gas-liquid chromatography. [H. R. Behrman *et al.*, *Can. J. Biochem.* **48,** 881 (1970)].

ferred from the uterine vein to the adherent ovarian artery. However, as shown in Fig. C, the peak value occurred some 30–35 minutes after stopping the infusion. It is therefore likely that this peak value is not representative of the equilibrated state so that the figure of 2% represents only a very minimal estimate. Since we have found that 1–2 μg/hour is about the lowest rate of infusion of cold $PGF_{2\alpha}$ that will induce complete luteal regression in the transplanted ovary, and since it is known that the uterus releases up to 25 μg of $PGF_{2\alpha}$/hour into the uterine vein (see below), then the true value for the amount transferred by the countercurrent

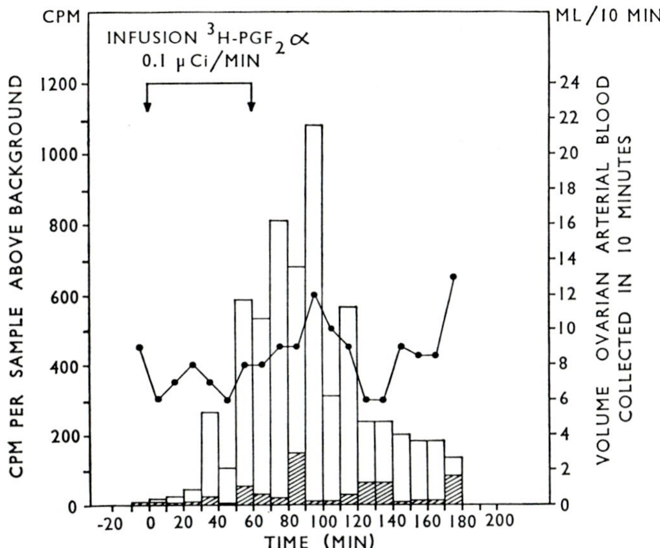

Fig. C. Time course of countercurrent transfer of ³H-labeled prostaglandin $F_{2\alpha}$ from utero-ovarian vein to adherent ovarian artery following 1-hour infusion into the adjacent uterine vein. □, Cpm, ovarian artery; ▨, cpm, iliac artery; ●——●, volume of blood from branch of ovarian artery. From J. A. McCracken, J. C. Carlson, M. E. Glew, J. R. Goding, and D. T. Baird, K. Gréen, and B. Samuelsson, *Nature (London)* (1972) in press.

exchange is likely to be in the range of 5–10%, i.e., two to three times higher than our preliminary estimate.

Longer-term infusion experiments of $PGF_{2\alpha}$-³H are now underway to determine the average value of radioactivity in ovarian arterial blood obtained during a state of equilibrium. In this way it should be possible to determine the quantitative aspects of the countercurrent exchange phenomenon with greater accuracy and precision.

The probability that $PGF_{2\alpha}$ was the luteolytic factor was further examined as a result of a collaborative study with Professor B. Samuelsson and Dr. K. Gréen in Stockholm [K. Gréen, B. Samuelsson, J. C. Carlson, and J. A. McCracken *Rec. Progr. Horm. Res.* **27**, 519 (1971)]. Single samples of ovine uterine blood at different stages of the estrous cycle were collected from a series of individual sheep. In addition peripheral arterial bloods were collected simultaneously. $PGF_{2\alpha}$ was identified and measured in uterine venous and iliac arterial blood by GLC/mass spectrometry [V. Axen, K. Gréen, D. Horlin, and B. Samuelsson, *Biochem. Biophys. Res. Commun.* **45,** 519 (1971); K. Gréen and B. Samuelsson, *Analytical Biochem.* (1972) in press]. Figure D shows the levels of $PGF_{2\alpha}$ in uterine venous blood during the course of luteal regression in a series of individual sheep. At the same time the

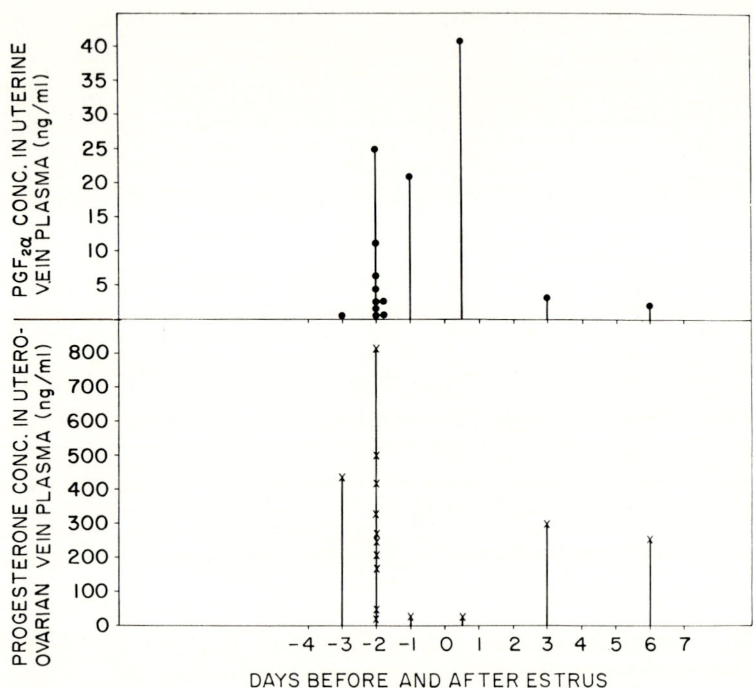

FIG. D. Gas–liquid chromatography/mass spectrometry measurements of prostaglandin $F_{2\alpha}$ concentration in uterine vein blood during luteal regression in a series of individual sheep. ●, Prostaglandin $F_{2\alpha}$ concentration; ✕, progesterone concentration. From K. Gréen, B. Samuelsson, J. C. Carlson, and J. A. McCracken, K. Gréen, and B. Samuelsson, *Nature (London)* (1972) in press.

concentration of progesterone in venous blood draining the ovary containing the corpus luteum was measured since the secretion of progesterone is known to be the most reliable index of luteal function [H. W. Deane, M. F. Hay, R. M. Moor, L. E. A. Rowson, and R. V. Short, *Acta Endocrinol.* **51,** 245 (1966); R. M. Moor, M. F. Hay, R. V. Short, and L. E. A. Rowson, *J. Reprod. Fert.* **21,** 319 (1970)]. Figure D shows a 10- to 20-fold rise in the concentration of $PGF_{2\alpha}$ in uterine venous plasma concomitant with luteal regression (indicated by the fall in progesterone concentration in uteroovarian venous plasma). The levels of $PGF_{2\alpha}$ in the plasma from iliac arterial blood collected simultaneously were generally below 2.0

ng/ml. These preliminary data were collected from a series of individual sheep, which is not perhaps as satisfactory as a series of samples from one individual animal.

In order to confirm that these levels of $PGF_{2\alpha}$ were of physiological significance, experiments were carried out in which the levels of $PGF_{2\alpha}$ in uterine vein blood associated with luteal regression were experimentally created in the uterine vein

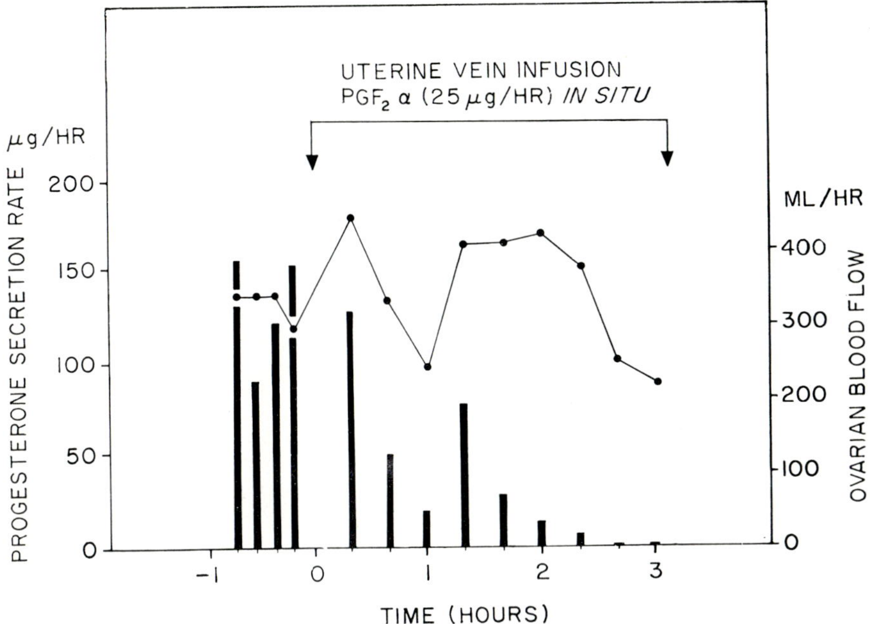

Fig. E. Luteal regression induced on day 10 by the infusion of 25 μg of prostaglandin $F_{2\alpha}$ per hour for 3 hours into the uterine vein of sheep ewe No. 156. ■, Progesterone secretion rate; ●——●, ovarian blood flow. From K. Gréen, B. Samuelsson, J. C. Carlson, and J. A. McCracken, *Nature (London)* (1972) in press.

at a time prior to the onset of normal luteal regression. It was calculated that, based on a maximum level of 42 ng $PGF_{2\alpha}$/ml and a blood flow of 15 ml/min through the uterine vein, the uterus of the sheep was releasing 25 μg $PGF_{2\alpha}$ per hour into the uterine vein. Figure E shows the results obtained when $PGF_{2\alpha}$ was infused *in situ* into the right uterine vein of a ewe on day 10 of the estrous cycle (i.e., 5 days before expected luteal regression). The secretion rate of progesterone and ovarian blood flow from the right ovary were measured by means of a cannula in the major vein draining the right ovary. Within 1 hour of beginning the infusion of $PGF_{2\alpha}$ into the uterine vein there was a marked fall in progesterone secretion rate, which continued to drop until progesterone secretion was undetectable at the end of the experiment. This type of experiment shows that $PGF_{2\alpha}$ is capable of inducing luteal regression by mimicking the physiological levels of $PGF_{2\alpha}$ at the time of luteal regression in the sheep. Since 25 μg PGF$_{2\alpha}$ per hour for 6 hours is without effect on the CL when given systematically to the sheep [J. A. McCracken, *Ann. N.Y. Acad. Sci.* **180**, 456 (1971)], then this experiment also shows that the PGF$_{2\alpha}$ must be operating via a countercurrent mechanism.

B. B. Pharriss: This suggestion of the countercurrent method that Dr. Goding developed some time ago I think is very exciting. The fact that it has been described in the sheep and in Dr. McCracken's work shows a concentrating effect in the ovarian artery that probably negates the need for restricting blood for hypothesy. I would like to suggest, however, that if you had an antagonism to blood flow out of the utero-ovarian vein it might help to accentuate the concentrating mechanism of the luteolytic agent across this by effectively slowing down blood flow from the uterus.

B. V. Caldwell: We have been interested in trying to determine what sequence of hormones may be responsible for inducing the release of the F prostaglandins. In ovariectomized animals, progesterone was administered every other day for a period of 10 days, followed 2 days later by a single injection of estradiol to control animals and animals immunized against estradiol. During the period of progesterone injections, F prostaglandin levels in the nonimmunized animals rose gradually over the 1-day period, rising significantly by the latter phase of this schedule. However, by 3 hours after the injection of estradiol there was a dramatic rise which often increased 10-fold in the peripheral levels of F prostaglandin by 6–12 hours post estrogen. This supports the view that if prostaglanding is "luteolysin" it is released under the sequence of hormonal events which is reportedly normal for the sheep estrous cycle. In the animals immunized against estradiol, there were no detectable levels of prostaglandin throughout the entire period.

In another experiment we infused extracts of day 14 uterine vein blood into sheep on day 8 of the estrous cycle. The prostaglandin levels in the uterine vein blood were 7–12 times higher than in day 8 blood and luteal regression was induced, demonstrating that the uterine vein blood is able to effect luteolysis [B. V. Caldwell and R. M. Moor, *J. Reprod. Fert.* **26**, 133 (1971)].

I would like to ask Dr. McCracken to explain how 60–90 minutes after stopping an infusion of radioactive prostaglandins he found extremely high levels of radioactivity maintained in the ovarian artery.

J. A. McCracken: We, too, were surprised that there was such a long delay in the transfer of PGF$_{2\alpha}$-^{3}H from the uterine vein to the adjacent ovary. It takes a good 30 minutes for the transfer to begin (see Fig. C). It would appear that PGF$_{2\alpha}$ is passing across a concentration gradient; i.e., from a region of high concentration in the uterine vein to a level of lower concentration in the ovarian artery against the hydrostatic difference in pressure between the ovarian artery and the uterine vein which is likely to be about 100 mm Hg. The mass of PGF$_{2\alpha}$ infused was $<$ 1 μg/hour (i.e., about a 2-fold increase over endogenous levels). It is possible that the transfer rate could be dose dependent. Presumably the reason why the level persists or even increases after the infusion is stopped is that the amount of PGF$_{2\alpha}$ contained between the endothelium of the uterine vein and the endothelium of the ovarian artery is not exposed to catabolic processes and simply continues to pass on into the artery in an unchanged form. Since the ovary of the sheep is likely to contain arteriovenous shunts [P. E. Mattner and G. D. Thorburn, *J. Reprod. Fert.* **19**, 547 (1969)], the possibility exists that some of the PGF$_{2\alpha}$ may recycle back into the uterine vein in an unchanged form, thus causing the level in the ovarian artery to persist longer than one might expect.

B. V. Caldwell: We have found that the leved of PGF$_{2\alpha}$ is virtually undetectable in the blood within 5 minutes after stopping an infusion of PGF$_{2\alpha}$ at 50 μg per minute for the induction of labor at term.

Another result pertinent to the concept of PGF$_{2\alpha}$ as the luteolysin was observed in two rabbits which were hyperimmunized against PGF$_{2\alpha}$. The normal length of pseudopregnancy is 15–16 days in normal rats; however, in these animals the pseudopregnancy was extended to 25–28 days, which is about the same duration as hysterectomized pseudopregnancy. This demonstrates neutralization of circulating PGF$_{2\alpha}$ and shows the same prolongation of luteal function observed following the removal of the uterus.

J. R. Goding: Dr. Caldwell seemed surprised that PGF$_{2\alpha}$-^{3}H failed to decay rapidly in the ovarian artery after the infusion into the uterine vein was terminated.

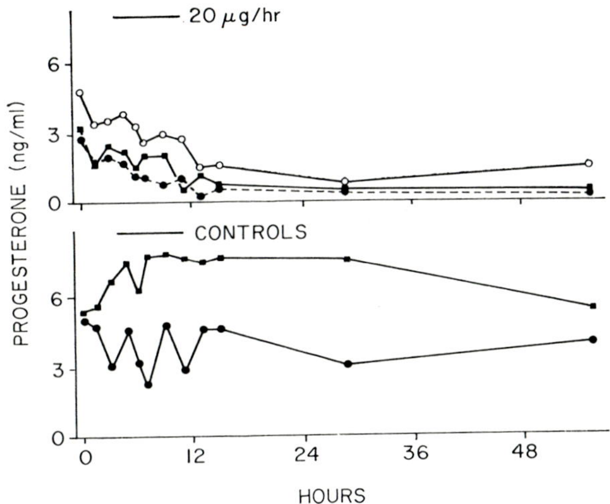

FIG. F. Infusion of prostaglandin F$_{2\alpha}$ (20 μg/hr) into the uterine vein on the same side as the corpus luteum in ewes on day 7 of the cycle. Control and saline infusions are of similar duration.

The delay in reaching a plateau after the injection starts presumably represents the transit time across the walls of both vessels, and once in this situation the lungs are effectively out of the circuit and have little effect on the rate of entry into the ovarian artery. Dr. Caldwell's statement on the rapid decay of PGF$_{2\alpha}$ from the peripheral circulation, however, has important physiological implications. I would like to know whether anyone has exact figures on the metabolic clearance rate of PGF$_{2\alpha}$.

As an extension of the experiments which showed transfer of PGF$_{2\alpha}$-^{3}H from the uterine vein to the ovarian artery, we have looked to see whether cold PGF$_{2\alpha}$ could get from the uterine vein into the ovarian circulation at a rate adequate to cause luteolysis. We had been getting luteolysis reliably after infusing PGF$_{2\alpha}$ into the ovarian arterial circulation at 40 μg/hr, so we infused PGF$_{2\alpha}$ into the uterine vein at the rate of 1000 μg/hr. Rapid and complete luteolysis resulted. We then reduced the rate of infusion to 200 μg/hr (Fig. F). Again, luteal regression was complete, and the animal went on to a new cycle. We then aimed for a dose below the one that would cause complete regression and infused 20 μg/hr for 9 hours. To our surprise, complete luteal regression ensued and a fresh ovulation resulted. Only when we reduced the infusion time to 7 hours did we begin to get an occasional, incomplete luteolysis. In one sheep, luteolysis was incomplete. Dr. Jeremy O'Shea reported that the CL showed areas of complete regression alongside other large areas typical of active secretory tissue with foamy luteal cells and numerous mitotic figures. All these infusions were carried out on day 7 of the cycle to obviate the complication of an early, spontaneous regression. None of the control animals showed luteal regression. Thus, incomplete regression caused by $PGF_{2\alpha}$ was accompanied by the histological picture of patchy infarction of the CL, which suggests that prostaglandin may indeed cause luteolysis by interfering with the blood supply to the CL.

After completing these experiments I was pleased to learn from Dr. McCracken that Gréen and Samuelsson had found 25 ng/ml $PGF_{2\alpha}$ in uterine vein blood at the time of luteolysis. This finding virtually completes the evidence necessary to prove that $PGF_{2\alpha}$ is "the" luteolysin in sheep. I am, however, not satisfied with some of the data published on blood levels of $PGF_{2\alpha}$. I believe that GLC mass spectrometry can be used as a sort of OK phase. In view of the obvious importance of exact values and knowledge of the compounds concerned, I would like to see a full validation of the methods used, including specificity, sensitivity, precision, accuracy, recoveries, and blanks.

J. E. Shaw: The Swedish group at the Karolinska Institutet, headed by Dr. B. Samuelsson, measured $PGF_{2\alpha}$ by combined GLC and mass spectrometry. Though to date we have little information as to the precision and accuracy of this method, I think identification can be regarded as definitive. The specificity of radioimmunoassay and protein-binding methods is still, however, open to question. It would be interesting to see the results of assays of identical samples in different laboratories using varying techniques.

J. R. Goding: I accept unreservedly that the Swedes are capable of measuring what they say they are measuring. My problem was with the recent paper by K. P. Bland, E. W. Horton, and N. L. Poyser [*Life Sci.* **10**, 509 (1971)], where the evidence was not nearly as clear-cut as I would have liked.

J. E. Shaw: In the sheep, do you actually get decreased blood flow in association with luteolysis induced by $PGF_{2\alpha}$?

J. R. Goding: If you give massive doses of $PGF_{2\alpha}$ you will observe a substantial decrease in ovarian blood flow. J. A. McCracken, M. E. Glew, and R. J. Scaramuzzi [*J. Clin. Endocrinol.* **30**, 544 (1970)] showed that an intra-arterial infusion of 100 μg/hour caused luteolysis accompanied by a large fall in blood flow. If we stick to what we may now define as physiological doses, we do not see any appreciable change in ovarian blood flow. I would not say that there was none, as we cannot be certain of measuring blood flow to much better than 10% accuracy. However, I believe that there is now evidence that $PGF_{2\alpha}$ causes redistribution of the circulation within the ovary itself. Dr. Mattner and colleagues at Prospect have been looking at the microcirculation of the ovine ovary and have evidence that arteriovenous shunts account for a considerable proportion of the blood flow through the ovary. Hence prostaglandins may act by diverting blood which should go to the CL into these shunts.

L. Speroff: J. P. O'Grady working on his thesis in our laboratory has developed a modified organ culture system for studying luteal tissue. Incubation of rabbit corpora lutea for 6 hours results in a significant increase in progestin content. Total progestin was assayed by radioimmunoassay utilizing an antibody to progesterone conjugated with BSA at the 11 position. The cross-reaction with 20-dihydroprogesterone was only 1%, therefore, the assay measured almost exclusively progesterone, since 17-hydroxyprogesterone is not a significant product of rabbit luteal tissue. Incubation of this tissue with prostaglandin $F_{2\alpha}$ in the medium at a concentration of 10 μg/ml results in approximately 50% inhibition of progesterone production. Incubating prostaglandin with HCG completely overcomes the inhibitory effect in this *in vitro* system. Along with Dr. Behrman's recent *in vitro* work with rat tissue, this is the first demonstration of *in vitro* inhibition of luteal steroidogenesis by prostaglandins. We have not figured out why this *in vitro* system differs from others, but the results argue against the blood flow hypothesis as a sole explanation for the mechanism of action.

We are also studying the effect of prostaglandins on ovarian steroidogenesis in the monkey. Prostaglandin $F_{2\alpha}$ has been given, with similar results, both through a peripheral vein infusion and via a catheter threaded up the femoral artery to the point of origin for the ovarian arteries, as verified by Hypaque injection and X-ray. The animals are stimulated with Pergonal and HCG, and progesterone is measured by radioimmunoassay in blood collected by catheterization of an ovarian vein.

A dose-response study of prostaglandin $F_{2\alpha}$ has shown stimulation of progesterone output at doses of up to 1 μg/min systemically and 10 ng/min arterially. When the infusion rate is increased to 50 μg/min systemically and 100 ng/min arterially, a rapid and marked fall in progesterone output is seen. If the high-dose infusion is maintained and HCG is given intravenously, the inhibition can be overcome. At no time have we detected a change in the blood flow, but admittedly measurement of blood flow by timed collection via a catheter is not very precise.

We were motivated to measure PGF levels in the ovarian vein to try to get an idea of the collective levels of prostaglandin $F_{2\alpha}$. The arteriovenous difference across the ovary was different by a factor of 5–10 times; i.e., the PGF content in th ovarian vein was 5–10 times greater than the content in the aorta at the origin of the ovarian arteries. Preliminary data show that administration of HCG markedly increases the content of PGF in the ovarian vein, suggesting that there is active synthesis of prostaglandin in the ovary itself, which is responsive to gonadotropin.

In one animal studied prior to ovulation, we have demonstrated stimulation of estradiol production. At the same infusion rate which effectively shuts off luteal steroidogenesis, no inhibition of estrogen steroidogenesis was seen.

Excision of an ovarian wedge at termination of the prostaglandin infusion and examination of electron and light microscopy have not revealed any morphological degeneration of the tissue.

K. T. Kirton: We have been interested in the effects of $PGF_{2\alpha}$ in the cycling, nonpregnant rhesus monkey. Daily injections of 15 mg/day on days 22 and 23 in 8 animals shortened the cycle from 27.9 ± 0.7 to 25.0 ± 0.8 days ($P < 0.05$). The cycles that were shortened were associated with a premature decrease in progesterone levels in peripheral blood. Glen Gutknecht in our laboratory has measured blood flow through the ovary of monkeys after an SQ injection of 15 mg of $PGF_{2\alpha}$ by using the hydrogen desaturation technique. Blood flow in treated animals was depressed to less than 30% of pretreatment values, for up to 4 hours. I would like to point out that this is a pharmacological effect of exogenously administered

PGF$_{2\alpha}$, and that we are not suggesting it is a physiological effect. However, this change in blood flow through the ovary cannot be accounted for by change in peripheral blood pressure alone.

B. B. Pharriss: Even though these were pharmacological doses, they are doses associated with luteal regression in the animal when given by this route. So we do get these blood flow changes at minimal doses which induce luteolysis.

J. E. Shaw: At the recent meetings of the reproductive society in Boston, results reported from Dr. Inskeep's laboratory indicated that the main prostaglandin in the sheep endometrium is PGF$_{2\alpha}$. We have heard here today that PGF$_{2\alpha}$ is the main prostaglandin released. The knowledge that both PGE$_2$ and PGF$_{2\alpha}$ come from the same precursor (arachidonic acid) implies that in this tissue there is some control mechanism which on stimulation of prostaglandin synthesis keeps the arachidonic acid converting into PGF$_{2\alpha}$ rather than PGE$_2$. Does PGE$_2$ cause regression of the corpus luteum?

B. B. Pharriss: Yes. In one experiment on the synthesis of prostaglandins by ovarian tissue we get measurable amounts of ^{3}H-labeled arachidonic acid incorporated into prostaglandins only if we define the medium to preferentially select for E synthesis over that of the F's. If you look back on the data on stimulation of steroidogenesis or stimulation of adenyl cyclase activity, the E's are much more consistently effective and much more potent in this system than are the F's. Therefore, while the F's look like the major prostaglandins coming from the endometrium, I believe it will probably be the E's that turn out to be important as mediators within the ovary.

H. A. Robertson: As a result of sequentially measuring the concentration of progesterone in the plasma of large numbers of cows I have come across a number of animals which show a considerable decrease in plasma progesterone levels over the period 20–25 days post conception. This is the period at which most of the "early embryonic mortality" is considered to occur. The progesterone levels of these animals subsequently recover and a normal pregnancy ensues. This led me to speculate upon the possibility of secondary ovulations and secondary corpora lutea occurring in these animals at this time. Could these effects on the plasma progesterone levels occurring early in gestation in the pregnant animal be due to a prostaglandin? I should like to ask Dr. Pharriss if he ever observed a decrease in plasma progesterone followed by a recovery to a normal level in animals given a subluteolytic dose of prostaglandin.

B. B. Pharriss: We have data which suggest that we might be doing this in several species. If we give prostaglandins as a single injection to the rat once a day, we have been unable to terminate luteal activity, and these are dosages greater than 10 times the effective dose as determined by twice daily injections. A continuous infusion of PGF$_{2\alpha}$ will cause another significant lowering of the effective dose. It appears that the suppression brought about by prostaglandins must be maintained for a certain period of time before it is effective and that the progesterone synthesis or luteal activity in the rat cannot be terminated acutely by prostaglandins.

H. A. Robertson: I should mention that, in one of the cows I referred to, the plasma progesterone fell to 0.4 ng/ml on day 25. The progesterone level recovered by day 30, and the animal subsequently calved normally. The level quoted is comparable to the level found during estrus when no corpus luteum is present.

M. R. Henzl: We are presently investigating whether prostaglandin E$_2$ is luteolytic in humans during the normal menstrual cycle. To a small group (4) of women we have administered prostaglandin E_2 at various periods during the corpus luteum phase in doses known to induce abortion during early pregnancy.

Figure G shows the results in a woman on days 20 and 21 of the menstrual cycle. Prostaglandin E_2 was administered in doses of 5–10 μg/minute for 8 hours. Blood for progesterone determinations was sampled before the infusion and several times during infusion on both days of treatment. As might be observed, progesterone blood levels fluctuated only slightly during PGE_2 administration. No substantial decrease was noted, and premature menstrual bleeding did not occur. The graph also shows the presence of side effects; the increase of temperature and pulse rate shortly after the infusion of PGE_2 started; and variations of blood pressure.

In another case, (Fig. H) PGE_2 was given on days 16, 17, and 18 of the cycle, soon after ovulation. During the infusion there were some fluctuations in the progesterone blood levels; however, the trend to continouously increase progesterone blood levels was maintained. The influence on temperature, blood pressure, and pulse was similar to that seen in the previous case.

We have summarized (Table B) the progesterone blood levels from day 16 to day 21. The values before each daily infusion resemble the pattern of a normal

TABLE B

Progesterone Blood Levels during Prostaglandin Administration

Cycle day	Pretreatment	Infusion of PGE_2			Posttreatment
		2 Hours	4 Hours	6 Hours	
16	0.9	3.2	2.6	3.4	1.9
17	2.7	3.2	0.6	5.4	3.6
18	7.8	2.7	6.9	5.4	2.8
20	17.4	14.1	13.0	18.0	17.7
21	17.2	16.0	11.9	13.9	17.7

luteal phase. During the PGE_2 infusion there are some fluctuations; however, we are unable to say whether they are due to the PGE_2 administration or whether they are spontaneous. The cycle lengths were unaffected, and menstrual bleeding occurred at the expected time, i.e., on day 28 or 29 of the cycle.

It is unlikely that the fluctuations in progesterone levels are due to experimental error. They are certainly real but not profound or lasting enough to represent luteolysis. This appears to be different from what has been reported here on experimental animals and indicates species difference. The cause of the fluctuations is uncertain. Do they represent spontaneous variations in progesterone secretion? Is progesterone secreted in a series of brief episodes, as is cortisol? Are there circadian variations that may partially explain our findings? Another possibility is that the fluctuations of progesterone represent transient vascular effects of PGE_2. At the height of the luteal phase, one of our women experienced slight spotting during the infusion, which may be a sign of an endometrial vascular effect.

B. B. Pharriss: These same general data have been reported for the human by Wiqvist and Bygdeman. They could induce an early menses in cycling women which appeared as normal menses about 5–6 days earlier than during normal men-

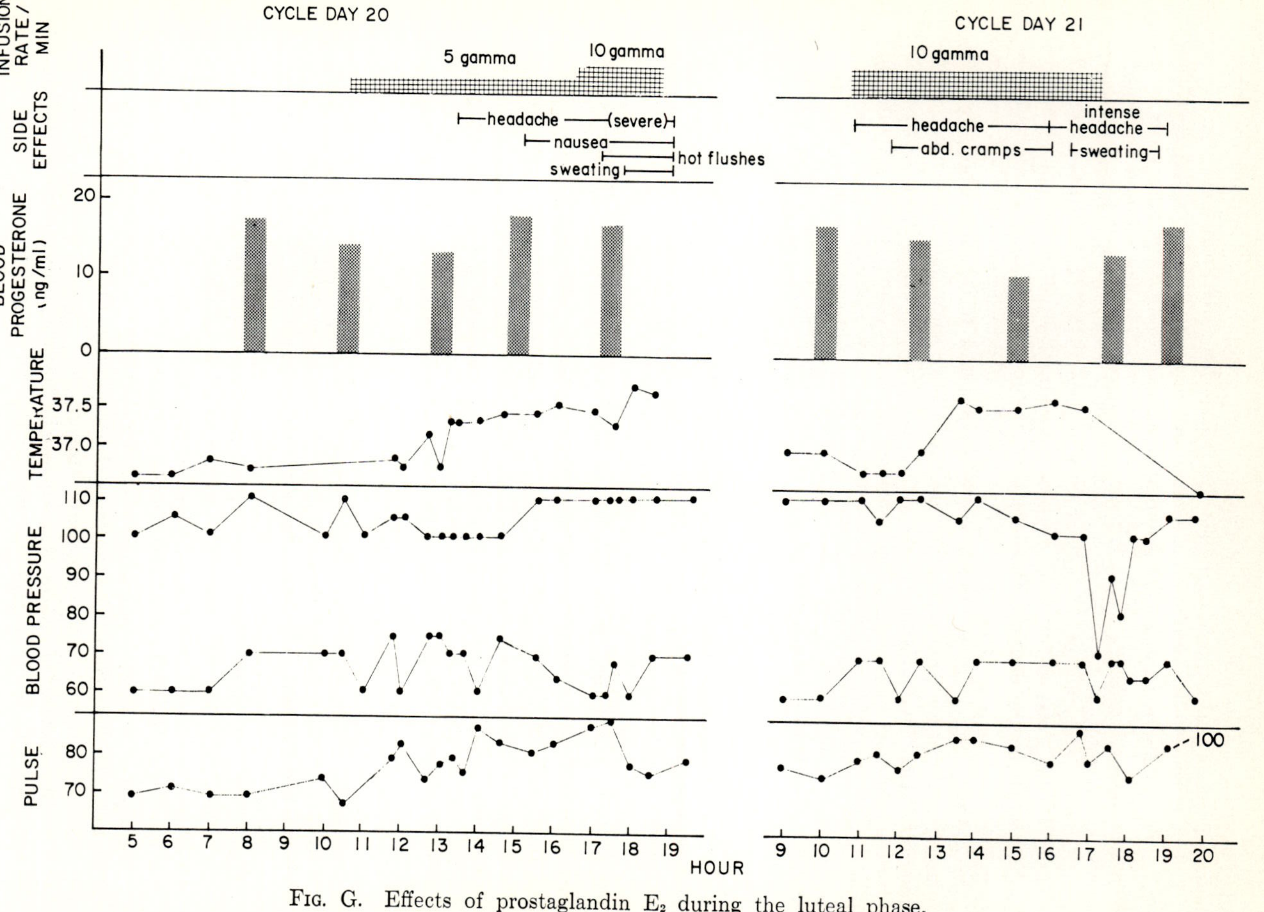

Fig. G. Effects of prostaglandin E₂ during the luteal phase.

struation. At the time of expected menses they again menstruated. The peripheral progesterone levels were essentially unaffected so this appeared to be a direct effect on the endometrium. Attempts to show luteolysis in the cycling rhesus monkey have also been fruitless. However, there is a time when the rhesus ovary seems sensitive to luteolysis induced by prostaglandin, and that is in the early pregnant state. Similar studies have been reported in humans.

K. Savard: May we discuss a few of the data that Dr. Pharriss presented? I refer to the effects in the *in vitro* assessment of steroidogenesis, and particularly

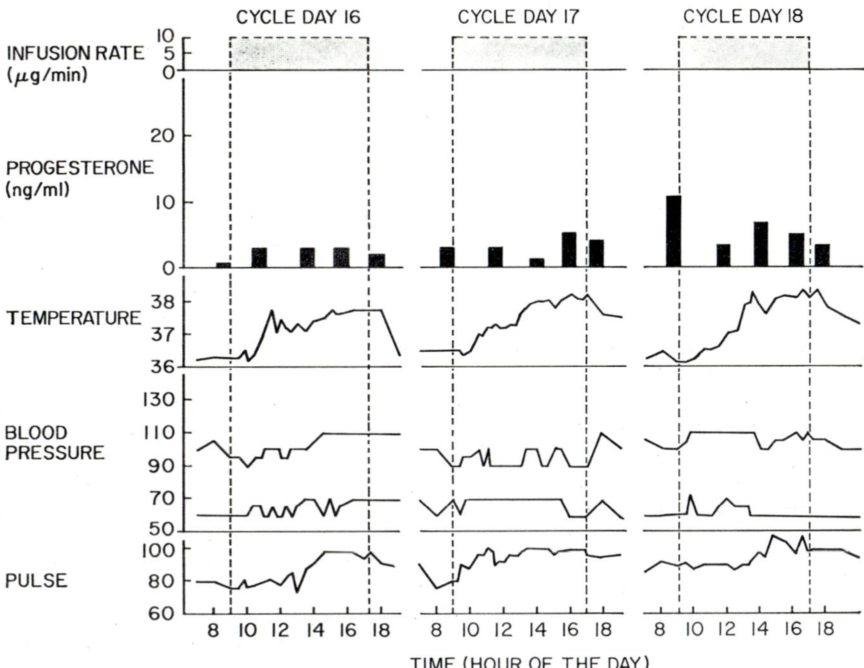

FIG. H.   Effects of prostaglandin E_2 during the early luteal phase.

to the effect of LH on the endogenous concentrations of prostaglandin. If it is to play a role in activating adenyl cyclase, it is most critical that concentration changes of prostaglandins be observed in a time sequence that precedes the activation of adenyl cyclase.

B. B. Pharriss: We have not become sophisticated enough to look at the timing or the adenyl cyclase stimulation in these preparations. We are solely measuring the incorporation of tritiated arachidonic acid into prostaglandins over a 30-minute period. Besides the effect of the anti-LH antiserum, we see a stimulation of prostaglandin synthetase activity when the rats are pretreated with amino glutethimide. We are not sure what this means at the present time.

K. Savard: It may be necessary to expend more effort in order to observe changes in prostaglandin concentrations in an *in vitro* system. As you know, LH activates adenyl cyclase in a matter of seconds—less than 1 minute. Therefore, the concentration of prostaglandin may have risen and fallen in the time interval before your observations were made and certainly would have done so in a prolonged incubation system.

B. B. Pharriss: I am not sure what the interpretation should be on those particular data. The adenyl cyclase activity has been reported in the total ovary. It may be that LH is effective in stimulating adenyl cyclase in tissues in the ovary other than the luteal cells. We know that it stimulates estrogen secretion from cells other than the luteal cells, and it may be that that is the portion of the ovary where the adenyl cyclase is responding. We are directing the work with prostaglandins strictly to the luteal cell in our interpretations.

K. Yoshinaga: You mentioned that traumatization of the endometrium causes a release of prostaglandins. Have you any information on whether the amount of prostaglandins released from the uterus after traumatization of the endometrium is enough to shorten the duration of pseudopregnancy in rats?

B. B. Pharriss: I think that Dr. Goding would be better able to answer your question.

J. R. Goding: Certainly in our control experiments, on intravenous infusion of $PGF_{2\alpha}$ when we handled the uterus and its vessels in the same way, but infused saline only, no consistent fall in progesterone resulted. Hence trauma to the uterus was not a major factor in these experiments. However, I have seen trauma to the uterus, incidental to moderately heavy-handed laparotomy, cause some increase in the incidence of early embryonic loss.

So far as the quantitative aspects of prostaglandins are concerned, I would like to put a few things in perspective. We can get reliable luteolysis in sheep with ovarian autotransplants after an arterial infusion of as little as 2 μg/hour of $PGF_{2\alpha}$ [W. A. Chamley, J. M. Brown, M. D. Cain, J. C. Cerini, M. E. D. Cerini, I. A. Cumming, and J. R. Goding, *J. Reprod. Fert.* **28**, 153, Abst. (1971)], and the ovarian blood flow was about 600 ml/hour. This represents an arterial concentration of $PGF_{2\alpha}$ of about 3 ng/ml. This is in complete contrast to the *in vitro* experiments in which 10 μg/ml was used. Even more startling, though, was the finding of Harrison *et al.* [F. A. Harrison, R. B. Heap, E. W. Horton, and N. L. Poyser, *J. Endocrinol.* (1972) in press] that in some cases of hydrops uteri following ovarian autotransplantation, there were enormous amounts of $PGF_{2\alpha}$ in the uterus. There were no prostaglandins of the E series at all, and in one case the uterus contained 7.6 mg of $PGF_{2\alpha}$.

A. R. Fuchs: I have some data on the effects of prostaglandin $F_{2\alpha}$ on pregnant rats that seem to support the hypothesis that the luteolytic action of $PGF_{2\alpha}$ is mediated by an interference with the action of gonadotropins on the corpus luteum rather than with the ovarian blood flow.

$PGF_{2\alpha}$ was administered as a slow intravenous infusion to pregnant rats. Autopsy was performed 1 or 3 days later, the uterine contents were examined for implantation sites, and the ovaries were removed and studied for their 20α-hydroxy steroid dehydrogenase activity. We observed a differential effect of $PGF_{2\alpha}$ depending on the stage of gestation. When infused on day 10, a total dose of 250 μg per rat was 100% effective in terminating gestation, and 125 μg was effective in 50% of the rats.

On the other hand, when infused on day 14 the dose of 250 μg of $PGF_{2\alpha}$ did not interfere with gestation. Increasing the dose to up to 1000 μg of $PGF_{2\alpha}$ had no effect in 50% of the rats and caused the resorption of only part of the litter in the remaining rats.

These effects are similar to those seen in pregnant rats after treatment with anti-LH serum; or after active immunization against LH. Before day 12 anti-LH treatment terminates pregnancy in rats, whereas after day 12 it has very little effect on rat pregnancy.

Furthermore, in our hands local infusion of $PGF_{2\alpha}$ into the uterine horn on day 7 or day 10 of gestation has been completely inefficient in terminating pregnancy. Nor have we seen a unilateral effect on the ovary ipsilateral to the infused horn when looking at the 20α-hydroxysteroid dehydrogenase activity in the ovaries of these rats. The ovaries of normal pregnant rats do not show this activity during most of the gestation. Ovaries of rats infused with prostaglandin $F_{2\alpha}$ on day 10 show intense activity on the following day if the infusion is given systemically, but when the same amount of prostaglandin $F_{2\alpha}$ was given into the uterine horn no activity was seen in either ovary. When we tried to overcome the effect of prostaglandin infusions, we observed, like Dr. Pharriss, that progesterone is effective in overcoming the effect on termination of gestation, whereas prolactin even in large doses (2 mg b.i.d) was ineffective. However, either 25 μg or 50 μg of LH given 2 times daily in gelatin vehicle prevented the resorption of implantations following the infusion of 250 μg of prostaglandin $F_{2\alpha}$ on day 10, which is 100% effective in terminating pregnancy in otherwise untreated rats.

B. B. Pharriss: We have also attempted to show a unilateral effect of infusing prostaglandins into the uterine horn with no clear-cut results. Generally we had a bilateral effect or no effect at all.

The Effect of Melanocyte-Stimulating Hormone on Coat Color in the Mouse

Irving I. Geschwind, Robert A. Huseby, and Richard Nishioka

*Department of Animal Science, University of California, Davis, California;
American Medical Center at Denver, Spivak, Colorado; and Department
of Zoology and Its Cancer Research Genetics Laboratory,
University of California, Berkeley, California*

I. Introduction

The research to be reported had its inception in 1963 when a single hybrid mouse of a C3H $\times$ A cross, fed a diet containing 0.2 μg of stilbestrol per gram of diet for 11 months in order to produce mammary cancer, was found to have a darker coat than other mice similarly treated. At necropsy a pituitary tumor was found, which upon transplantation into mice of the same genotype fed the stilbestrol-containing diet also produced darkening of the hosts. Microscopic investigation of dorsal hairs from control and darkened animals revealed that the typical agouti pattern, of black hairs with subterminal yellow bands, had been obliterated in the tumor-bearers. We have previously reported (Geschwind and Huseby, 1966) the correlation between darkening of coat color, as measured reflectometrically, and plasma melanocyte-stimulating activity levels determined by bioassay, the concentrations of such activity in the tumors and urine of the mice, and the provisional identification of α-MSH as the melanocyte-stimulating principle in the tumors. Subsequently, we were able to demonstrate that injection of α-MSH into nontumor bearing mice of the same genotype duplicated the darkening effect of the tumor on newly growing hair (Geschwind, 1966).

After the publication of the second report we ran into a problem. We had stated in our original report that the tumor-bearing animals became obese as the tumor enlarged, and we soon found that with a subline of the tumor, as well as with later transplant generations of the tumor, the mice were becoming obese earlier and were not darkening. At necropsy grossly enlarged adrenal glands were found in these mice. Moreover, assay of circulating melanocyte-stimulating activity levels revealed them to be as high as in mice which had darkened. Other mice bearing a tumor of the same subline were then adrenalectomized, and shortly thereafter they began to darken. We then turned to mice carrying the original tumor line. We had originally reported that the darkening of such animals plateaus from 3 to 5 months after the tumor is transplanted, the time at which the animals started to become obese (cf. Figs. 1 and 2 of Geschwind and Huseby, 1966). When plateaued animals were adrenalectomized, further darkening always ensued. We also found that, if the tumors of plateaued animals were excised, then as the remnants regrew the animals darkened further.

When additional studies revealed the failure of hair regrowth in intact obese tumor-bearers, we found that we had at least partially solved our problem. As adrenal stimulation occurred either late in the growth of our original tumor line or early in the growth of the subline, hyper-adrenocorticism[1] had an inhibitory effect on the normal hair cycle of the mouse. Change in hair color does not take place in preexisting hair, but occurs only in hair growing during the period of elevated melano-cyte-stimulating activity. Thus, with no hair regrowth, darkening could not be observed. This could be overcome either by adrenalectomy or, in the original tumor line, by excision of the tumor, since that operation led to the removal of the adrenal-stimulating agent. In the latter case, as the tumor remnants began to regrow, they produced melanocyte-stim-ulating, but not adrenal-stimulating, activity. We also considered the possibility that a glucocorticoid might be inhibiting tyrosinase, and since that time Pomerantz and Chuang (1970) have reported that cortisol produces a partial inhibition of the enzyme activity, which may be over-ridden by MSH. Such an inhibition may have also contributed to the effect.

With that particular problem out of the way, we turned our attention to other matters.

II. Genetic Studies

Wolfe and Coleman (1966) reported that there are approximately 70 genes at 40 different loci which affect coat color or spotting in mice. We had used black agouti animals in all our studies, and we considered it worthwhile to determine the genetic specificity of the MSH effect. We had no desire to see whether we could modify the expression of 70 genes, so we restricted ourselves to alleles at seven different loci. At the agouti locus, we investigated the action of MSH on the expression of agouti (A), viable yellow (A^{vy}), lethal yellow (A^y), black and tan (a^t), and tanoid (a^{td}) alleles; and at the color locus, we looked at mice homozygous for the albino (c), chinchilla (c^{ch}), or himalayan (c^h) trait. At five other loci we concentrated on single recessive traits: brown (b),

---

[1] Whereas intact animals fattened tremendously, had no new hair growth cycles, demonstrated enlarged islets of Langerhans and fatty metamorphosis of the liver, these did not occur in tumor-bearing animals that had been adrenalectomized. Also the hyperphagia, polydipsia, and polyuria (so noticeable in intact, tumor-bearing animals) was not seen in adrenalectomized recipients. This is in accord with pre-vious reports concerning hyperadrenocorticism in rats (Bahn *et al.*, 1957). For biochemical evidence of the hyperadrenocorticism, see Section XIII.

dilute (*d*), yellow (*e*), pink-eyed dilute (*p*), and pallid (*pa*). Alleles at every other locus other than that being investigated were usually wild type.

Circulating MSH levels were raised in mice in these experiments either by injection of MSH or by transplantation of the tumor. In the former group, hair was plucked from a dorsal area of mice 60–90 days of age, of the desired genotype. Plucking serves to stimulate new hair growth, and starting 5 days later MSH was injected once daily for 10 days. Natural or synthetic α-MSH (3 mg/ml) was suspended in beeswax–peanut oil and injected subcutaneously in the interscapular region in a volume of 50 μl. For the tumor experiments to be successful, rejection of the tumor by hosts of different genotypes had to be minimized, and this problem was circumvented by passing fragments of the tumor through organ culture (Jacobs and Huseby, 1967). When growth of the fragment indicated that the tumor was established, the dorsal hair of the host was either plucked or shaved. Changes in the intensity or quality of pigmentation were followed by gross observation and by reflectometry until the end of the experiment, at which time freshly-plucked hair samples were treated according to Russell's procedure (1946), and skin biopsies were fixed in ethanol–formaldehyde–acetic acid; the samples and histological sections were then examined microscopically. To be certain that the growing tumor was functional in the host, levels of melanocyte-stimulating activity in blood obtained from an orbital plexus were assayed by our modification (Geschwind and Huseby, 1966) of the frog skin reflectometric method of Shizume *et al.* (1954). This modification allows the detection of activity in 50 μl of normal mouse plasma and has a sensitivity of 2 U/ml (100 pg α-MSH equivalent per milliliter).

By these end points of altered pigmentation intensity or type, both approaches yielded concordant results: the only effect of MSH was on the expression of the agouti locus, where the most obvious effect of the hormone was to convert the pigmentation to the nonagouti type (e.g., black and brown agoutis grew hair which was black or brown, respectively). In those animals whose genotypes were a/a (nonagouti), with substitution at another locus (e.g., b/b, c/c, d/d, or p/p) MSH had no visible effect.[2] The most marked effect was on the A^y mouse, whose hair

---

[2] Steelman *et al.* (1956) had shown that transplantable ACTH-secreting tumors, which had been produced in LAF₁ mice by Furth and his co-workers by ionizing radiation, contained very large amounts of MSH. In their experiments, the LAF₁ mice were implanted with the tumor, and 3 weeks later the mice were bilaterally adrenalectomized. After having reached a size of 1–6 gm, the tumors were excised and assayed. The report is notable in that no mention was made of any coat color change in the brown nonagouti hosts (a/a,b/b).

is normally yellow. MSH was injected into animals of the genotype A^y/a,B/B, whereas the tumors were implanted in animals of the genotype A^y/a,b/b, producing new black hair in the former and new brown hair in the latter.

The effect in the A^y is so remarkable that one should note the effects of the hormone in two other types of yellow mice, the pink-eyed dilute agouti, A/-,B/-,p/p, and the yellow recessive a/a,e/e. Hair of the pink-eyed dilute agouti is basically gray with a subterminal yellow band, which gives an overall yellow appearance to the coat. New hairs growing in during the period of MSH administration are totally gray, characteristic of the pink-eyed dilute nonagouti, a/a,B/-,p/p. On the other hand, the coloration of the yellow recessive is completely unaltered by MSH administration or by the presence of the tumor (Figs. 1 and 2). The failure of animals of this genotype to respond to MSH stimulated us sufficiently to try to produce mice of the genotype A^y/a,B/B,e/e, i.e., with two different sets of genes for yellowness, and to test their response to MSH.

A^y (A^y/a,E/E) mice were bred to yellow recessives (a/a,e/e) and all yellow progeny (A^y/a,E/e) were saved. These animals were tested with MSH and were found to be responsive. They were then back-crossed to a/a,e/e mice, and the yellow progeny (3/4) were again saved. These

Fig. 1. A^y and e/e mice, each bearing the transplantable pituitary tumor, with comparable blood melanocyte stimulating activity levels. A small area on the back of each animal was shaved; only hair regrowing in the A^y mouse is black. Note also the masklike pattern of black pigment in the A^y and the darkening around the ears. No hair was shaved in these regions; the hormone is simply marking the position of the growth wave of the hair.

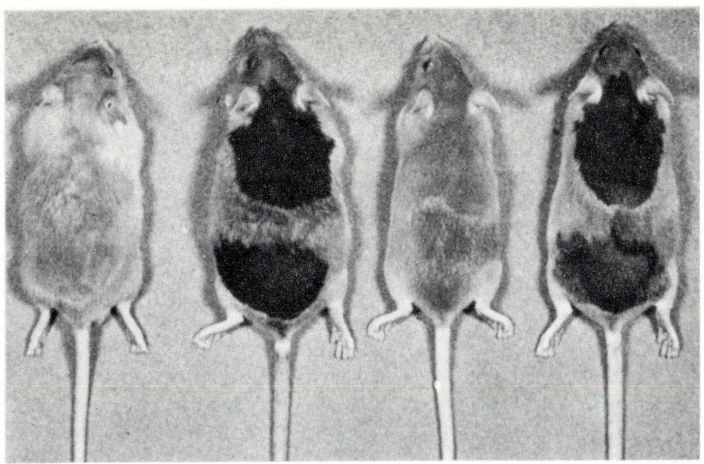

FIG. 2. From left to right: e/e mouse which had been partially plucked, and starting 5 days later had received a series of 10 daily injections of MSH; A^y mouse, similarly treated; A^y mouse, similarly treated but receiving medium rather than MSH; A^y mouse, similarly treated but receiving MSH only on days 8 and 9. All mice were 90 days old, and were photographed 20 days after plucking.

animals should be of the A^y/a,E/e: A^y/a,e/e: and a/a,e/e genotypes, theoretically in a 1:1:1 ratio. When tested with MSH, only 13 of the 35 progeny were found to respond. Since the A^y/a,E/e should have responded, and the a/a,e/e should not have, a 2:1 or 1:2 ratio of responders to nonresponders was expected. The results with MSH indicated that A^y/a,e/e animals were nonresponders, *if* the actual ratio of genotypes was similar to theoretical. To test this, 34 were bred to C57Bl mice (a/a,E/E). Mice of the genotypes A^y/a,E/e and A^y/a,e/e would be expected to have progeny which was 50% yellow, whereas a/a,e/e mice would have no yellow progeny by this cross. Of the 34, 16 mice which had not responded to MSH gave all black litters, and were presumably a/a,e/e; 13 who responded to MSH were presumably A^y/a,E/e; and the 5 who failed to respond but had partially yellow litters were presumably the desired genotype, A^y/a,e/e. Thus the ratio of 13:5:16 was far from the theoretical 1:1:1, and the close approximation of the ratio of responders to nonresponders to the theoretical, was fortuitous.

To confirm the above assignment, the same mice were bred to a/a,e/e mice. Here the division is different, for A^y/a,E/e would be expected to have 25% black progeny by this cross, whereas A^y/a,e/e and a/a,e/e would not be expected to have any black progeny. From these matings 13 partially black litters were obtained, and every animal had been in the group which originally had responded to MSH, thus confirming the assignments. As a final check the progeny from the cross of two of the suspected $A^y/a,e/e$ with $a/a,e/e$ were tested with MSH, but none of the 15 responded. Since 50% of the progeny would be expected to be $A^y/a,e/e$ and the other 50% $a/a,e/e$, this is additional proof that $A^y/a,e/e$ does not respond to MSH. It must be emphasized that by only one of the tests that we have employed can $A^y/a,e/e$ be distinguished from $a/a,e/e$ but in that test, the mating with mice of the genotype $a/a,E/E$, the production of yellow progeny clearly distinguishes the two genotypes.

The finding of a lack of response in $A^y/a,e/e$ animals was unanticipated. Hauschka *et al.* (1968), who found the yellow recessive mutant, stated that the "size, round shape and concentration of the medullary pigment granules were much the same for the two kinds of yellow. However, the cortex of e/e hairs contained far less pigment than the A^y/a cortex." They also found that immersion in 10% KOH for 15 minutes extracted all the pigment from both types of hairs, leaving them colorless, a finding which is consistent with what Prota has observed (see below). That A^y and e/e can influence pheomelanin formation independently of the other (i.e., both $A^y/-,E/E$ and $a/a,e/e$ genotypes are yellow) is obvious from their nonallelism, but why the presence of e/e prevents the action of MSH is not known. How MSH may affect the pigment change in $A^y/-,E/-$ animals is the further subject of this paper.

Before leaving these genetic studies, it should be pointed out that an allele of the e locus, sombre (E^{so}), expresses itself in intensely black animals even in the presence of the agouti locus. Thus, the presence of either e/e or E^{so} apparently will override some a locus-dependent pigmentary phenomena. It seemed possible that E^{so} may exert its effect by increasing MSH levels, but assays of plasma from $A/-,E^{so}/-$ mice revealed no elevation of melanocyte stimulating activity levels when they were compared to levels in $A/-,E/E$ mice.

III. A Digression: The *a* Locus

The peculiar banding pattern of agouti hair has fascinated mouse geneticists for a long time (for reviews see Wolfe and Coleman, 1966; Silvers, 1961). Now classical experiments by Silvers and Russell (1955) and Silvers (1958) showed that the pigmentary response of melanocytes in the hair follicle was dependent upon the agouti locus genotype of the follicle rather than upon their own. Thus, if the skin of an albino mouse ($a/a,c/c,E/-$) is grafted onto the back of a histocompatible black mouse of *either* the agouti ($A/-,B/-,C/-,E/-$) or nonagouti ($a/a,B/-,C/-,$ $E/-$) genotype, melanoblasts from the surrounding skin of the host invade the graft, take up residence in the follicles, and contribute pigment to the production of completely black hairs at the periphery of the graft. However, if the graft is obtained from an albino donor of the genotype A/-,c/c,E/-, then melanocytes of the two aforementioned genotypes will produce pigment for an agouti pattern. When these findings are taken together with the claims of Markert and Silvers (1956) that extrafollicular melanocytes of agouti mice synthesize eumelanin (black) exclusively (however, see below), it appears that the eumelanin-synthesizing machinery of the melanocyte is normally redirected to pheomelanin (yellow) production only by the follicular environment of the mouse with the agouti genotype. To explain the production of all yellow hair, the "A^y genotype" can be substituted for "agouti genotype" in the above discussion.

In the growth of agouti hair a switching mechanism operates at the beginning and termination of yellow band formation. The switches are "thrown," according to Cleffmann (1963a), on days 4 and 6.5 in the newborn mouse, and as reported by Galbraith (1964), 6.5 and 9 days after plucking older mice; thus the yellow band is formed during a 2.5-day period. Both Cleffmann (1954) and Galbraith (1964) also demonstrated that the production of yellow pigment in agouti mouse skin was by the very same cells which had been producing black pigment.

A number of theories have been offered regarding the nature of the switching signals. Cleffmann (1954, 1963a,b, 1964) has for many years been an advocate of changing sulfhydryl group levels. He has found that melanocytes producing yellow pigment give a very strong cytochemical reaction for free sulfhydryls (Cleffmann, 1954), and that there is a greater specific uptake of labeled cysteine and reduced glutathione into yellow pigment cells (A^y or yellow phase of agouti) than into black pigment cells (Cleffmann, 1964). But by far his most interesting discoveries have been in experiments with mouse skin cultures. He reported in 1954 that when skin from a black mouse was cultured, a phenocopy of the agouti pattern could be obtained by addition of glutathione for a limited period of time to the culture. Subsequently (Cleffmann, 1963a,b) he showed that in the absence of glutathione from the culture medium, melanocytes of A^y follicles will produce only black pigment, and in the presence of sufficient glutathione melanocytes of black nonagouti animals will produce only yellow pigments. No theory as to how elevated sulfhydryl levels might stimulate yellow pigment formation was suggested, but Cleffmann (1954) was intrigued by the finding that the yellow banding occurred at the time of greatest growth of the hair and at the peak of mitotic activity in the hair follicle (Cleffmann, 1963a,b). Subsequently, he suggested that the agouti gene controlled the amount of sulfhydryl compounds in the pigment cell and that these compounds were probably concerned with both pheomelanin formation and inhibition of tyrosinase activity, the latter by virtue of the chelation of tyrosinase copper by the sulfhydryl compounds (Cleffmann, 1964).

Galbraith (1964), on the basis of results obtained after intradermal injection of tyrosine and colchicine, originally favored the idea that pheomelanin formation at the time of most rapid hair growth resulted from a diversion of tyrosine, the substrate for eumelanin formation, into hair protein synthesis. He also suggested that the renewal of eumelanin formation on day 9 could result from an increased blood supply to the follicle or to a decrease in the mitotic activity of the bulb. For us, theories invoking increased mitotic activities, rapid hair growth, diversion of tyrosine, and/or changes in blood flow have seemed inadequate, for they scarcely contribute to the problem of the all-yellow hair of the A^y animal, and recently Galbraith (1971) measured the mitotic activities in the hair follicles of A^y, A, and nonagouti mice and found them similar.

In contrast to these latter theories, those stressing the importance of sulfhydryl groups have received considerable support from the chemical findings of Prota, Nicolaus, and their colleagues. These have revealed that 5-S-cysteinyl DOPA and 2-S-cysteinyl DOPA, formed from a condensation of cysteine and DOPA-quinone, are intermediates in pheomelanin synthesis in chicken feathers, with the 5-S intermediate quantitatively the important one (Misuraca *et al.*, 1969); that these intermediates cyclize to form dihydrobenzothiazine derivatives during enzymatic oxidation *in vitro* (Prota *et al.*, 1970); and that dihydrobenzothiazines by oxidative coupling give rise to trichosiderin pigments (Prota *et al.*, 1969), characteristic of red and yellow hairs and feathers. These findings, incidentally, with their emphasis on the nonenzymatic diversion of DOPA-quinone, an important intermediate of eumelanin synthesis, should lay to rest at least some of the speculation (e.g., different enzymes, and tryptophan as a substrate) which has grown up concerning the mechanism of formation of pheomelanin (for earlier reviews of this subject, see Fitzpatrick *et al.*, 1958; Wolfe and Coleman, 1966). However, Takeuchi (cited by Holstein *et al.*, 1971) has recently suggested a new variation of an older idea, that tyrosinase activity is altered by the A allele, and has postulated the production of an "A" factor which inhibits the tyrosinase-catalyzed conversion of DOPA to DOPA-quinone. According to this suggestion, tyrosine and DOPA are the critical intermediates which are diverted to pheomelanin formation.

IV. Dynamics of Darkening of the A^y Mouse Treated with MSH

Most of our early studies on the A^y mouse were with animals between 75 and 90 days of age which had been plucked 8 days earlier. Histological examination of skin biopsies revealed a gradual shift from production of pheomelanin to that of eumelanin in follicular melanocytes 12–24 hours after a subcutaneous injection of MSH in beeswax–peanut oil, or after a series of three consecutive hourly injections of MSH in saline. In animals receiving their hormone in beeswax–peanut oil, eumelanin granule content appears to be maximum approximately 36 hours after the injection (Figs. 3 and 4), and the black granules predominate until some time between the 72nd and 96th hour post-injection, at which time the MSH effect has worn off and the melanocytes have resumed making pheomelanin. It seems probable that the presence of eumelanin granules for so long a time is due to a continued release of the hormone from its site of injection, to the relatively long life of the machinery which MSH has set into motion, and to the length of time required by a melanocyte to transfer completely all the eumelanin granules formed earlier in the presence of high levels of MSH. Microscopic inspec-

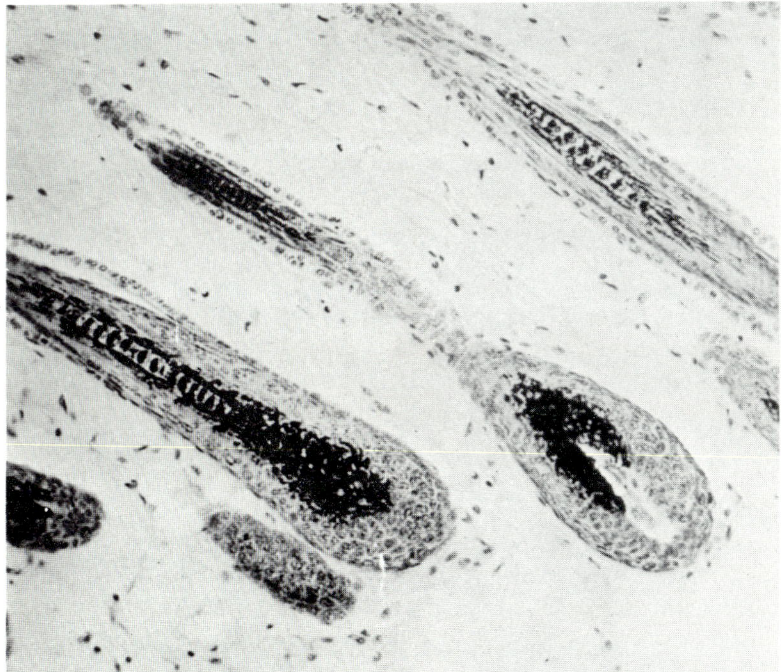

FIG. 3. Histologic section of the dorsal skin from an A^y mouse which had been partially plucked and had received a single injection of MSH on day 8, with sacrifice 48 hours later. Note the presence of black pigment in melanocytes in the bulb and in proximal hair shafts. Fixed in FAA and stained with Nuclear Fast Red ×200.

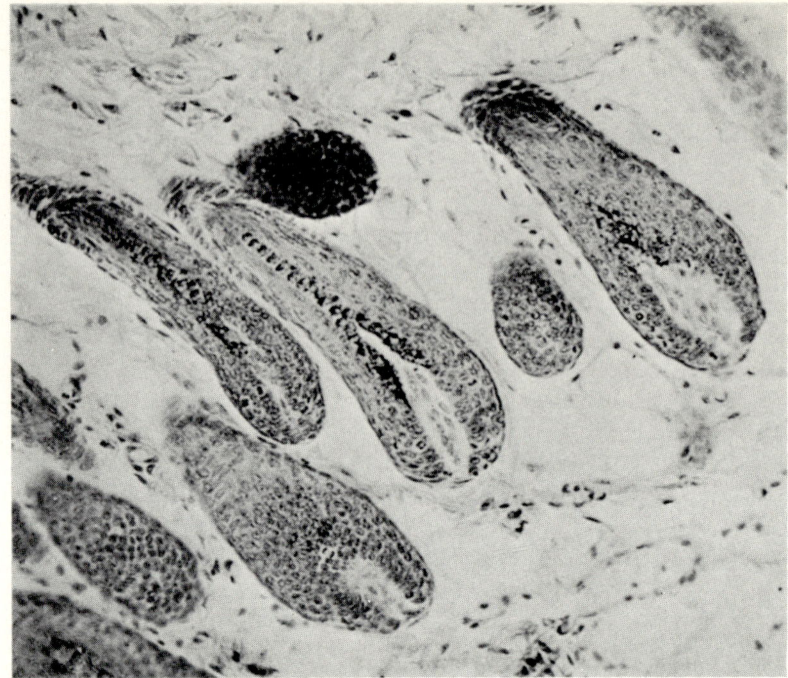

Fig. 4.   Similar to Fig. 3, but the mouse received medium alone. Note bulbs and proximal hair shafts. The pigment was yellow both grossly and when viewed through the microscope.

tion, under oil, clearly demonstrates at the times when reversal of synthesis takes place (i.e., between the 12th and 18th, and between the 72nd and 96th hours) the presence of both eumelanin and pheomelanin granules in the same melanocyte, proving that it is one and the same melanocyte which produces both pigments, and confirming Cleffman and Galbraith's conclusions concerning the source of eu- and pheomelanin in the agouti mouse.

More recently we have examined the effect of the hair growth phase on the sensitivity to MSH. Mice 73 to 88 days of age were paired for a close match of coat shade, and on the appropriate day after plucking one member was given a single subcutaneous injection of β-seryl MSH (600 μg) in beeswax–peanut oil, while the other received an injection of medium alone. Three pairs were used for each time period, and animals were sacrificed 30 hours post-injection, except for one extra group of animals which was sacrificed 96 hours after a day 9 injection. The days referred to below are those on which the injections were given. Plucking and injection schedules were staggered to permit comparison of effects at a single sitting.

An initial series demonstrated no response at 3 days, but a positive response at 6 days. The response at 9 days was very strong, and certainly more intense than at 6 days. Ninety-six hours after the day 9 injection the skins were very dark and the proximal portions of the hair shaft were also black. Good responses were also seen at days 12 and 15. A second series revealed no response on day 4, and a weak positive response on day 5, establishing this day as the first sensitive one. Fitzpatrick *et al.* (1958) had shown in the C57Bl mouse that melanocytes could first be identified in the hair bulb on day 4, and that tyrosinase activity was weakly positive on that day. They rated tyrosinase activity "positive" on day 6 and "strongly positive" on days 8 and 14. Burnett *et al.* (1969) rating tyrosinase activity in mice of the same strain, found it to be negative on days 1 to 3, plus 1 on day 5, plus 2 on days 7 and 8, and plus 3 on days 9 to 12. By day 15 the activity had returned to plus 1. They also noted that in their experience melanocytes can be detected as early as day 3.

On day 16 the darkening response was patchy, and by day 17 we graded the response weak to very weak. No response was seen on day 18 or 24, as the growth phase terminates. Fitzpatrick *et al.* (1958) could no longer identify melanocytes in the bulb or detect tyrosinase activity on day 24 of their studies, and Burnett *et al.* (1969) made similar findings on days 19 and 24. Thus the responsiveness to MSH and the naturally occurring tyrosinase activity follow the same pattern during the hair cycle. This would indicate that the hormone is incapable of hastening the development of melanocytes or of their pigment synthesizing machinery at the beginning of a new hair cycle, or of prolonging the life of melanocytes beyond their natural span.

A comparable study of more limited scope in very young mice, has indicated that the 4-day-old responds much like the adult 6 days after plucking. In these baby A^y mice we have also determined the minimum effective dose of MSH by inspecting the inner surface of the skin under a dissecting microscope and looking for blackening of the follicles, and have found it to be 50 ng when administered as a single injection in beeswax–peanut oil 24–30 hours earlier. At low doses, both in adult and baby mice, the effect is spotty and is probably the result of local, rather than systemic effects of the hormone.

Microscopic investigation of dorsal skin, nipple, tail, and scrotum employing histological sections has revealed two effects of the hormone other than that on coat color. In dorsal skin, even after a few injections, it is easy to observe a marked increase in the *number* of pigment granules in the follicular melanocytes of the treated mice. As was mentioned above, it has been reported than in A^y mice pheomelanin is produced only in hair follicles while the melanin in other areas is eumelanin (Markert and Silvers, 1956). In our studies, histologic examination of untreated A^y mice revealed a few melanocytes in the dermis of the tail and scrotum as well as in the connective tissue of the nipple where the glandular epithelium transforms to squamous epithelium. In none of these locations were melanin granules present in the epithelium, and the granules in the melanocytes definitely were *yellow*, i.e., pheomelanin. As these areas darkened with the growth of the tumor, or after a lengthy series of injections, the number of melanocytes increased very significantly in number, the granules darkened, and in all three areas dark granules could readily be found in the epithelium. It is also in these latter groups that one finds marked darkening of the ventrum as well (Figs. 5 and 6). Melanocytes were found in the same sites in e/e mice. The melanin granules were also yellow and confined to the melanocytes. Their number, color, and distribution were unchanged by the growth of the tumor.

Occasionally, in A^y mice which have been partially plucked and given a short series of MSH injections or in animals bearing the tumor, localized darkening of a nonplucked area some distance from the site of injection is observed. MSH, in such circumstances, is marking the position of the normal growth wave of the hair, and it may serve as a useful tool for that purpose (Fig. 1).

Finally, it must be emphasized that after a series of injections of MSH has terminated, the darkened coat color persists only until the next growth wave leads to a falling out of the existing hair and its replacement with hair of the color normal for the animal. This effect is shown vividly in Figs. 7 and 8, which portray two A^y mice, one of whose coats was plucked in a "UC" (University of California) pattern. Both animals received multiple injections of MSH after plucking, to produce the patterns shown in Fig. 7. About 10 weeks later the patterns showed definite erosions as the new growth waves irregularly overtook them (Fig. 8).

V. Chemistry of the Pigments of A^y Hair and the Effect of MSH

The general solubility characteristics of the pigments in the hair of A^y mice—for example, their solubility in alkali, which distinguishes them

FIG. 5. A^y mice at both sides of the photograph received MSH on alternate days, for 10 injections, starting on day 2. The mouse in the middle is a control littermate which received medium alone. Photographed 48 hours after the last injection.

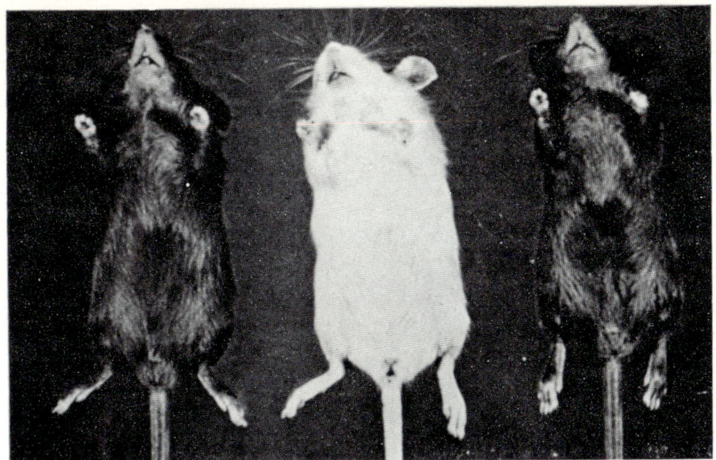

Fig. 6. Ventral view of the mice in Fig. 5.

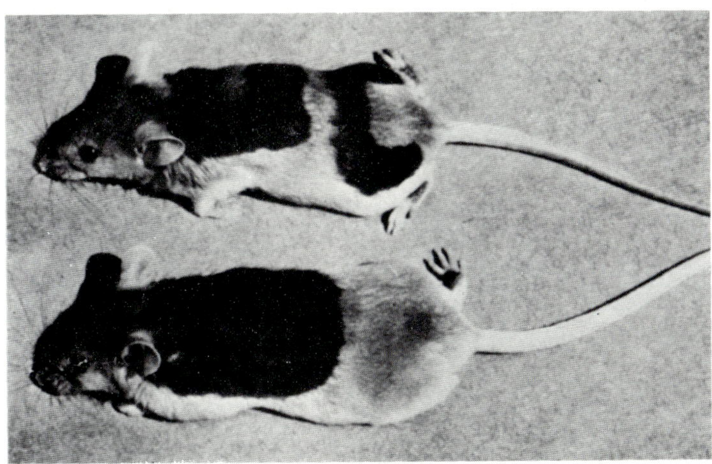

Fig. 7. Two A^y mice were partially plucked (the upper in a "UC" pattern), and starting on day 5 after plucking they were given 10 daily injections of MSH. Photographed about 1 week after the last injection.

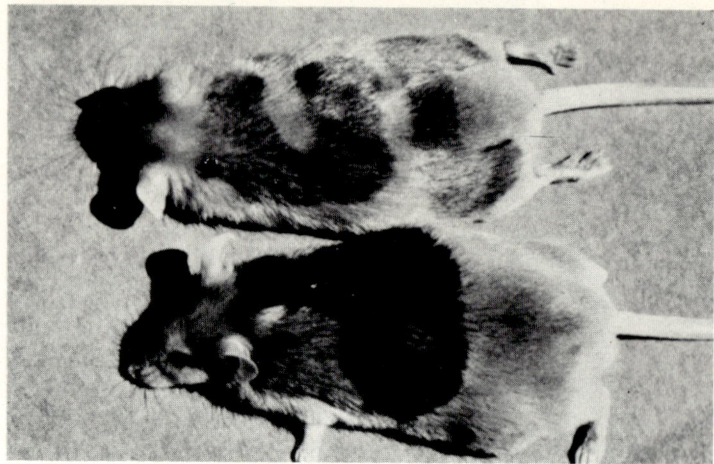

FIG. 8. Same animals as in Fig. 7, but photographed about 10 weeks later.

from eumelanin—place the pigments in the pheomelanin class. However, before it could be accepted that MSH was promoting a conversion from pheomelanin synthesis to that of eumelanin, chemical evidence for the existence of pheomelanin in A^y hairs was required, especially since according to Professor R. A. Nicolaus of the Institute of Organic Chemistry of the University of Naples, a number of melanins appearing to be red or yellow have turned out to be eumelanins which had been deposited in an unusual form. Through the good offices of Professor Joseph T. Bagnara of the University of Arizona, then on leave at the Zoological Station in Naples, hair harvested from A^y mice receiving 10 injections on alternate days starting on day 2, of either MSH or medium alone (Figs. 5 and 6) and from e/e mice injected only with medium, were sent to Dr. Giuseppe Prota of the University of Naples for chemical analyses.

The quantity of hair was small (shavings of 6 mice per group) so that Dr. Prota chose to limit his analysis to the trichosiderins since they are easily extracted in alkali, have characteristic absorption spectra, and can be easily identified by paper chromatography. The 0.1 N NaOH extracts from the control A^y and e/e mice were yellow, and after removal of acid-insoluble material and fractionation on a cation-exchange resin, the acidified eluates were subjected to paper chromatography. From both the control A^y and e/e hairs, two red-purple pigment spots with chromatographic properties identical with those of trichosiderins E1 and F1 were obtained. Both had been previously identified and their structures determined by Prota and his colleagues (1969) in extracts of the feathers of New Hampshire chickens. In addition, a yellow spot, which does not correspond to any known trichosiderin, was present on the chromatograms from both sources.

In contrast to the above, extracts of the hair of A^y mice treated with MSH were almost colorless; after they were fractionated by the procedures outlined above, *no* trichosiderins could be detected chromatographically.

These results indicate that trichosiderins, and presumably pheomelanins which always seem to be associated with them, are indeed present in hair of the A^y mouse; that trichosiderins E1 and F1 and the same unknown yellow pigment are present in hairs from both A^y and e/e mice, suggesting that the end products of pigment formation are the same for both genotypes; and that MSH does alter a switching mechanism by which an intermediate of tyrosine metabolism is shunted to pheomelanin production.

VI. Failure of Melatonin to Affect the Pigmentary Response to MSH

The tryptophan derivative melatonin (N-acetyl-5-methoxytryptamine) has been shown to be a potent paling agent in amphibian larvae. Its action there, on the dermal melanophore, is to cause aggregation of melanin granules, thus counteracting the effect of the MSH's, which are powerful melanin granule-dispersing peptides. The only known effect of melatonin in a mammalian pigmentary system has been reported by Rust and Meyer (1969), who found that implantation of melatonin in beeswax in the short-tailed weasel, *Mustela erminea*, prevented the acquisition of the brown coat typical of the summer phase. Both brown and white weasels molted and grew a new white coat whereas controls retained or acquired a brown coat. However, melatonin had no effect on the growth of new brown pelage in weasels with pituitary autografts under the kidney capsule. Such autografts presumably produce large quantities of MSH which stimulate the production of a brown coat.

The darkening of the A^y mouse by MSH seemed to offer an attractive system for testing the effect of melatonin in another mammal. Mice were plucked and 8 days later were divided into 5 groups, three of which were given a single injection of MSH in beeswax–peanut oil (0 time). Mice in one of these three groups received three injections of 25 μg of melatonin in saline at zero time and 6 and 12 hours later.

A second group received a single subcutaneous injection of 25 μg of melatonin in beeswax–peanut oil at zero time. The third MSH-treated group served as a control. The two remaining groups were controls which received melatonin in saline and melatonin in beeswax–peanut oil. Inspection of the skins of all MSH-treated animals 24 hours later showed a typical MSH response, while hair bulbs in both melatonin control groups were yellow.

Rust and Meyer employed a *weekly* dose of 1 mg of melatonin for 100–200 gm weasels, and we believe that the 25 μg *daily* dose (and certainly the 75 μg total daily dose used in one experiment) injected into 25–30 gm mice should have been entirely adequate to produce a response if one was forthcoming. The failure to obtain a melatonin effect in the mice simply extends the list of mammals (man and short-tailed weasels) in which melatonin has not shown any interaction with MSH.

VII. The Effect of MSH on Skin Tyrosinase Activity

The enzyme responsible for melanin synthesis in melanocytes is tyrosinase, and it seemed logical to follow our histological studies with measurements of skin tyrosinase activity. To this end, skin samples have been obtained from mice in our various experiments and assayed for the enzyme activity, through the kindness of Dr. Walter Chavin, by the procedure of Chen and Chavin (1965).

Our first study was designed to correlate the enzyme activity present in the skin of young adult A^y mice with the type of pigment present in follicular melanocytes, as observed in histological sections. Ninety-day-old mice were partially plucked and 8 days later received subcutaneously three consecutive hourly injections of MSH in saline. In two separate experiments, at approximately 3-hour intervals from 6 to 24 hours after the initial MSH injection, individual mice were sacrificed. The dorsal skin which had been plucked was excised, a small piece was fixed for histology, and the remainder was rapidly cleaned, frozen om dry ice, and individually wrapped in aluminum foil for shipment to Dr. Chavin's laboratory. In two other experiments a single injection of MSH in beeswax–peanut oil was administered to 4-day-old A^y mice, and skins were taken solely for tyrosinase estimation at a single time interval, 24 hours later. (Still another kinetic study for the period 24 to 48 hours post-injection is reported below as part of an experiment with cycloheximide—Table III.)

The results shown in Table I for tyrosinase activities are generally consistent with the kinetics of appearance of eumelanin in the follicles and suggest that the lag period of about **9** hours before the first eumelanin granules are found may be an indication of the length of time required to form a new melanosome (it may be shorter, but it is certainly not longer). The marked increase in activity seen at **9** hours in experiment **1** is probably the earliest detectable response that we have so far observed, and we had switched over to injections in saline in the hope of obtaining an earlier response. However, in the second experiment we could not duplicate the early tyrosinase effect, and even at **12** hours, when we observed the appearance of some eumelanin granules in the melanocytes, the tyrosinase levels were still at the control

TABLE I

A^y Mouse Skin Tyrosinase Activity Following the Injection of MSH

Group	Hours after initial MSH injection	Type of melanin present		Tyrosinase specific activity[a]	
		Expt. 1	Expt. 2	Expt. 1	Expt. 2
Control	9 to 24	Pheo	Pheo	5.3 ± 0.5	
Experimental	6	Pheo	Pheo	5.8	6.2
	9	Mostly pheo	Pheo	16.0	6.5
	12	Pheo + eu	Pheo + eu	12.2	6.8
	15	Mostly eu	Pheo + eu	19.7	12.4
	18	—	Mostly eu	—	11.2
	24	—	Eu	—	15.3

[a] Picomole of L-tyrosine converted to melanin per microgram of skin protein nitrogen.

level. Both the control-like level found here at **12** hours and some of the anomalous values reported in Table II (e.g., in experiment **2**, the value of 13.3 for the skin of an MSH-treated animal which had grossly darkened) probably reflect problems which investigators frequently find with tyrosinase methods (A. B. Lerner, personal communication). Nevertheless, the results (Tables I and II) reveal that a 2.5- to 5-fold increase in tyrosinase activity can be produced in adult mouse skin within **24** hours by injection of MSH, and increases of **2** to **3** times are to be found in the normally elevated levels of the unplucked baby *A^y* mouse.

In attempting to analyze the mechanism of action of MSH in producing darkening of the *A^y* mouse, we have used the injected e/e mouse as a type of control. Unfortunately, our results with the e/e have been far from consistent, and as this is written we await assay results from Dr. Chavin for what we would like to think is a definitive experiment.

In an experiment similar to those listed in Table I, control values of 6.3 and 11.2 were found, while at 24 hours post-injection the levels were 10.0 and 10.1. In a more recent experiment, similar to that shown in Table III (see below), skins of beeswax–peanut oil controls had tyro-

TABLE II

Tyrosinase Activity in the Skins of 5-Day-Old A^y
Mice 24 Hours after an MSH Injection

	Tyrosinase specific activity	
	Expt. 1	Expt. 2
Control	11.5	14.2
	13.9	14.9
	—	17.5
	—	23.1
Experimental	20.6	13.3
	20.8	31.6
	23.8	40.1
	—	44.1

sinase specific activities of 10.6, 11.7, and 30.7(!), while 24 hours after MSH the activities were 11.9, 12.2, and 14.4. The control specific activities ought to have been approximately the same in both experiments (cf. controls in Table I), and if so, the values of 6.3 and 30.7, surely the latter, are aberrant. Certainly, from the second experiment the effect of MSH on tyrosinase levels in the e/e mouse seems to be minimal.

Most of these results were reported at the Seventh International Pigment Cell Conference in 1969 (Geschwind and Huseby, 1970). They have since been supplemented by Pomerantz and Chuang (1970), who found that injections of 50 μg of β-MSH in beeswax–sesame oil, starting on day 1 or 2, into C57 black and brown mice produced a 45–50% increase in tyrosinase levels. In contrast to some of our results with older animals, reported above, these workers found that "the MSH-treated black mice were darker than untreated controls. However, brown mice treated with MSH were not visibly darker than controls even though they did have more tyrosinase."[3]

The limited results we have obtained with e/e mice would indicate that they may be exceptional in their lack of a tyrosinase response to MSH administration. *If* what is required for darkening of a yellow

---

[3] See, also, footnote 2.

coat is simply an increase in tyrosinase activity, then the failure of e/e to darken is explained. We shall consider this matter further, below.

There is evidence from a number of sources that the tyrosinase activity of the skin of an A^y mouse is less than that of the skin of a black or brown pigmented animal (for reviews of earlier reports, see Fitzpatrick *et al.*, 1958; Wolfe and Coleman, 1966). The most striking recent evidence is found in the reports of Holstein *et al.* (1967, 1971), who have detected the presence of 3 bands of tyrosinase activity in acrylamide gel electropherograms of extracts of skin of non-yellow-pigmented species. Extracts of A^y skin, however, give rise to but a single band, and its intensity is attenuated when compared to that of other genotypes, indicating "marked quantitative and perhaps qualitative differences between the tyrosinases associated with eumelanogenic and pheomelanogenic melanocytes" (Holstein *et al.*, 1971). We have confirmed these findings, but have detected with certainty only 2 bands in non-yellow-pigmented species and have also noted both bands in extracts of skin of MSH-treated A^y mice. Although we have not, as yet, electrophoresed extracts of e/e skin, there is some evidence that total tyrosinase activity is greater than in A^y skin. This evidence comes from a comparison of the tyrosinase specific activities in control 90-day-old A^y (Tables I and III) and e/e mouse skins. The values reported above for the latter, averaging approximately 11 to 12, are just about double those found for A^y's. It must be emphasized, however, that the tyrosinase assay of Chen and Chavin measures the activity in *homogenates* of skin, and Chavin (1969) has demonstrated that most of the activity is associated with a particulate fraction, whereas in the procedure of Holstein *et al.* the tyrosinases are soluble, and are either free, or associated with some protein (perhaps from the particle from which they have been released). Therefore, the two techniques are probably not measuring the same thing.

VIII. Studies with Inhibitors of Protein and RNA Synthesis

The MSH-induced increase in tyrosinase activity and the change in the quality of pigment may be dependent upon *de novo* RNA and/or protein synthesis. Studies with inhibitors have been useful in solving similar problems in the past, and we carried out a number of experiments with them. We appreciated the fact that such inhibitors would be expected to affect hair growth, but histological observation of the melanocyte's pigment content circumvents that particular problem. More important, the formation of the protein matrix of the developing melanin granule (premelanosome) would be prevented by inhibitors of protein synthesis, unless replacement comes from a large pool of soluble protein monomeric units. The same would hold true for replacement of dendrites inserted and left in keratinocytes when pigment granules are transferred. On the other hand, failure of an inhibitor to affect the response might favor an activation or a permeability phenomenon as an explanation for the MSH effects. However, since the effects (darkening and increase in tyrosinase activity) we are measuring have a relatively long lag period before they become evident, one must be concerned about the adequacy of the dose of inhibitor and the persistence of its effect. These remain today incompletely solved problems.

The most striking and consistent effects have been obtained with an inhibitor of protein synthesis, cycloheximide. Both the darkening and tyrosinase responses were observed 24 and 36 hours after the injection of MSH in beeswax–peanut oil into 90-day-old animals, 8 days after plucking. Simultaneous with the MSH injection the animals received an intraperitoneal injection of 2 mg of cycloheximide, and the latter was repeated 7 and 14 hours later. The 24-hour darkening responses were investigated either in skins obtained from animals sacrificed at that time or in skin biopsies.

A number of darkening experiments were initially run, and in these the usual MSH effect was displayed by control animals, with the follicles alone blackened at 24 hours, and with extension of the blackening to the proximal hair shafts at 36 hours. Cycloheximide control animals were similar to medium controls, and when MSH was administered at 36 hours, the animals proceeded to darken normally over the next 24 hours. The experimental animals, receiving both MSH and cycloheximide, showed absolutely no darkening at 24 hours, and histologically, the follicular melanocytes were devoid of eumelanin granules. However, at 36 hours, 22 hours after the last cycloheximide injection, large numbers of eumelanin granules were evident in the melanocytes. Thus, as the inhibition wore off, the hormone, which had been injected 36 hours earlier, began to exert its effect. Identical results were obtained in a number of experiments, including one in which tyrosinase activity was also followed. The results of that experiment are shown in Table III; the typical MSH effect, the failure of cycloheximide alone to alter normal levels of enzyme activity, and the inhibition of the MSH effect at 24 hours by the cycloheximide are all evident. However, in this experiment only one of two mice of the experimental group may have begun to show an enzyme effect at 36 hours although histology clearly revealed the overwhelming presence of black bulbs in these same skins.

The persistence of MSH at its target, or the long biological half-life of its effect, noted in earlier darkening experiments and reported above, is also evident in the cycloheximide studies in the form of the delayed response.

TABLE III

Effect of Cycloheximide Administration on the MSH-Induced Increase in Skin Tyrosinase Activity

	Treatment		Hours after MSH injection	Tyrosinase specific activity
Group	MSH	Cycloheximide		
1	−	−	36	5.2, 5.7
2	+	−	24	10.0, 11.0
3	+	−	36	10.7, 11.9
4	+	−	48	15.0, 16.4
5	−	+	24	5.9, 6.7
6	+	+	24	5.7, 5.9
7	+	+	36	4.9, 8.2

Experiments with another inhibitor of protein synthesis, puromycin, have not shown any inhibition of the darkening response, even when given at the same dose and the same time intervals as was cycloheximide. Changing the injection schedule to 0, 5, and 10 hours was no more successful. Since the possibility exists that the duration of any inhibition produced by puromycin is shorter than that of cycloheximide, injections were given at 0, 5, 10, and 15 hours, but no animal has survived that insult.

Actinomycin D is an inhibitor of DNA-dependent RNA synthesis, and we have attempted to determine its effect in our system. However, the material has been lethal at levels of about 1 μg per gram body weight, when more than a single injection is given over a 10-hour period. A single injection, given either 0.5 hour before, at the same time as, or 9 hours after the administration of MSH has not visibly altered the darkening process, as observed macroscopically or microscopically 24 hours after the hormone injection. Hair regrowth, observed 4 days later, had been definitely inhibited in some of the same animals, so that a long-lasting effect on at least one system was evident.

The failure of actinomycin to affect the darkening process may be interpreted as meaning that no new RNA need be synthesized for it, while the cycloheximide results suggest a requirement for *de novo* protein synthesis in order to obtain the darkening response. A number of possibilities can be considered to explain its effect. Most obvious among these is that MSH regulates a translational process involved in tyrosinase synthesis which is inhibitable by cycloheximide. Another possiblity is that MSH activates existing enzyme when the latter is associated with a premelanosome, and that inhibition of premelanosome matrix formation by cycloheximide prevents that effect. We cannot choose among these and other possibilities at present.

IX. Cyclic AMP, Adenyl Cyclase, and the Darkening Process

The discovery by Bitensky and Burstein (1965) of the ability of cyclic AMP to mimic MSH in promoting dispersion of melanin granules in frog skin melanophores, and the subsequent demonstration by Abe *et al.* (1969) that MSH induced an increase in cyclic AMP levels in dorsal, but not in ventral frog skin which is essentially devoid of melanophores, served to place MSH in the ever-enlarging camp of those substances which depend upon this nucleotide to act as a second messenger in their mechanism of action. Recently, Bitensky and Demopoulos (1970) reported that MSH also activated *in vitro* the adenyl cyclase of both amelanotic and melanotic mouse melanomas, and suggested the involvement of the nucleotide in melanin synthesis.

For the past three years we have been attempting to mimic the effect of MSH in the A^y mouse by injecting either cyclic AMP, N,O'-dibutyryl cyclic AMP (DBC), theophylline, or a combination of one of the two nucleotides with theophylline. When the compounds were administered to 80- to 90-day-old mice, 8 days after plucking, as single injections in beeswax–peanut oil or as multiple injections in saline, in doses as high as 5 mg of DBC and 4.8 mg of theophylline, in very few instances was there any indication of darkening, and even then microscopic inspection of histological sections was necessary to confirm the effect which has always been very weak and largely questionable (Table IV). Experiments with baby mice have produced slightly more consistent results (Table IV) although the doses given were usually lethal to some of the animals, and all combinations of the nucleotide and theophylline which we employed killed all the babies into which they were injected. When eumelanin was found in any of the experiments listed at the bottom of Table IV it was always in the form of extremely fine granules. However, a word of caution is in order here. In our breeding program for A^y mice we note quite a bit of variation in coat color, and attempts for about 5 years to select for a homogeneous bright yellow coat have been unsuccessful. In histological sections of 4- to 6-day-old control mouse skin we frequently observe follicles that appear to contain eumelanin granules, and although we select pairs of mice of closely matched shade to be experimental and control subjects, this does not seem to have entirely solved the problem of questionable responses. The problem presented itself during our attempts to determine a minimum effective dose of the hormone (see above) and is especially important in the cyclic AMP studies, for at no time has more than a weak response

TABLE IV

Cyclic AMP and Darkening of A^y Mouse Skin

Age group	Expt. No.	Injections[a] (mg)				Schedule	Result
		cAMP	DBC	Theoph.	Medium[b]		
80–90 days	1	—	2.5	—	BWPO	Single inj.	Both animals yellow
	2	—	—	3.6	Saline	2×/day for 4 inj.	Both animals yellow
	3	1.9	—	2.4	Saline	2×/day	4 animals yellow
		—	1.9	2.4	Saline	2×/day	1 of 4, some darkening
	4	—	3.8	4.8	BWPO	Single inj.	1 of 2, some darkening
		—	—	4.8	BWPO	Single inj.	Both animals yellow
	5	—	3.8	4.8	BWPO	Single inj.	1 of 2, some darkening
		3.8	—	4.8	BWPO	Single inj.	Both animals yellow
	6	—	5.0	—	Saline	1×/2 hr for 4 inj.	3 animals yellow
		—	—	1.0	Saline	1×/2 hr for 4 inj.	3 animals yellow
		—	5.0	1.0	Saline	1×/2 hr for 4 inj.	1 of 2, slight darkening
4–5 days	7	—	0.60	—	Saline	1×/3 hr for 3 inj.	Some darkening in 3 of 3, but 2 of 3 controls also show some degree of darkening
	8	0.50	—	—	Saline	1×/8 hr for 3 inj.	Some darkening in 1 of 2
		—	0.66	—	Saline	1×/8 hr for 3 inj.	Some darkening in 2 of 2
	9	—	0.75	—	Saline	1×/3 hr for 3 inj.	Very slight darkening in 1 of 5

[a] cAMP, cyclic AMP; DBC, dibutyryl cyclic AMP; Theoph., theophylline.

[b] BWPO, beeswax–peanut oil; all injections were given subcutaneously except in Expt. 6, where the last injection was given intraperitoneally.

ever been obtained. Attempts to use the same baby mouse as its own control, by obtaining a skin biopsy prior to the first injection, have not been successful because the animal operated upon is rejected by its mother and is not suckled, and because the skin changes tremendously from day to day during this period of rapid growth.

The inconsistent, weak, and questionable responses that we have observed have left the question of cyclic AMP mediation of the MSH effect in doubt. We therefore sought to determine whether cyclic AMP levels and adenyl cyclase activities were increased in the skins of mice treated with MSH. Homogenates and extracts of skins from a number of experiments have been assayed by Dr. Lewis Chase, to whom we are indebted for the analyses by the methods of Steiner *et al.* (1969)

TABLE V
Cyclic AMP Levels[a] in Control and MSH-Treated Mouse Skins[b]

Genotype	16 Hr		24 Hr	
	MSH	Control	MSH	Control
A^y/-	0.063 ± 0.003	0.056 ± 0.002	0.051 ± 0.006	0.040 ± 0.006
e/e	0.028 ± 0.002	0.039 ± 0.010	0.046 ± 0.008	0.028 ± 0.004

[a] Nanomole per gram tissue.

[b] Mice (4 per group), 80–90 days of age, were partially plucked and 8 days later were given a single injection of MSH in beeswax–peanut oil or medium alone. At sacrifice 16 and 24 hours later, only skin from the previously plucked area was removed and frozen for assay. Blackening was quite evident in the 24-hour samples. *No differences were significant at the 5% level.*

for cyclic AMP and of Chase and Aurbach (1968) for adenyl cyclase. Adenyl cyclase levels were barely above background and did not differ in MSH and control groups. However, all could be stimulated by addition of fluoride. Although cyclic AMP levels were measurable, no significant differences between control and experimental groups have been found (Table V). The concentrations of the nucleotide in mouse skin are very low in comparison to other tissues and to frog skin (Abe *et al.*, 1969). This comparison is of some importance, for it is possible that since follicular melanocytes make up probably far less than 1% of the total cell population of the skin, even a 100-fold increase in their adenyl cyclase activity might not be observed against a very large background of unchanging activity contributed by the other cells. However, the very low activities found would indicate either that the cell population as a whole has very low enzyme activity, or that a very large proportion of the weight of the skin represents cellular products (e.g., hair and collagen) and dying cells (stratum corneum), effectively diluting the specific enzyme activity.

In all probability the only real test of the participation of cyclic AMP in the MSH-induced darkening response and/or increase in tyrosinase activity will come from experiments with mouse skin in culture where higher concentrations of the nucleotide may be employed and theophylline added without as much concern for lethality.

X. MSH and the Ultrastructure of the Melanosome

Early in the development of a mature eumelanin granule (melanosome) an ellipsoidal particle, the premelanosome, is formed containing within a membrane an array of helical longitudinal fibers which are cross-linked (Figs. 9 and 10). Tyrosinase is located at regular intervals on these parallel fibers and serves as a focus for melanin deposition

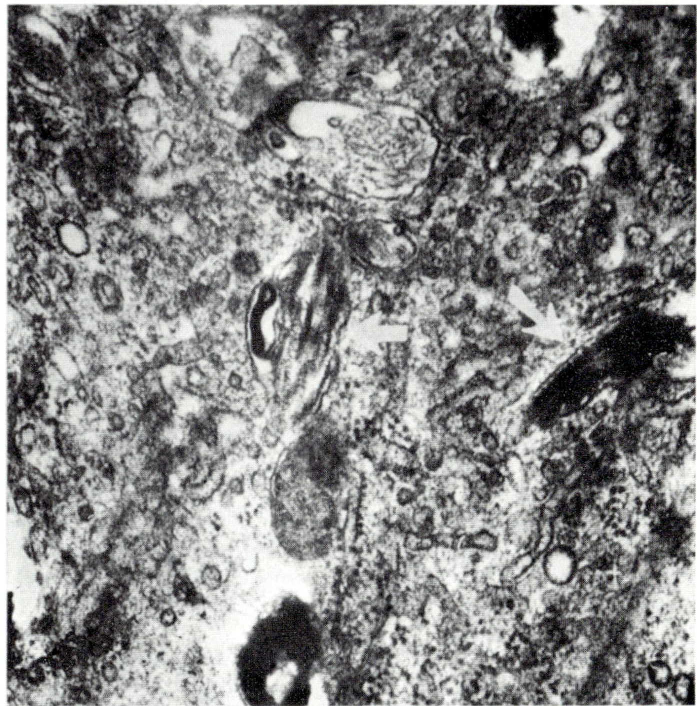

FIG. 9.

FIGS. 9 and 10. Electron micrographs illustrating premelanosomes (arrows) in a black nonagouti mouse (a/a,B/B,C/C). Similar magnification (the scale line is 0.5 μ).

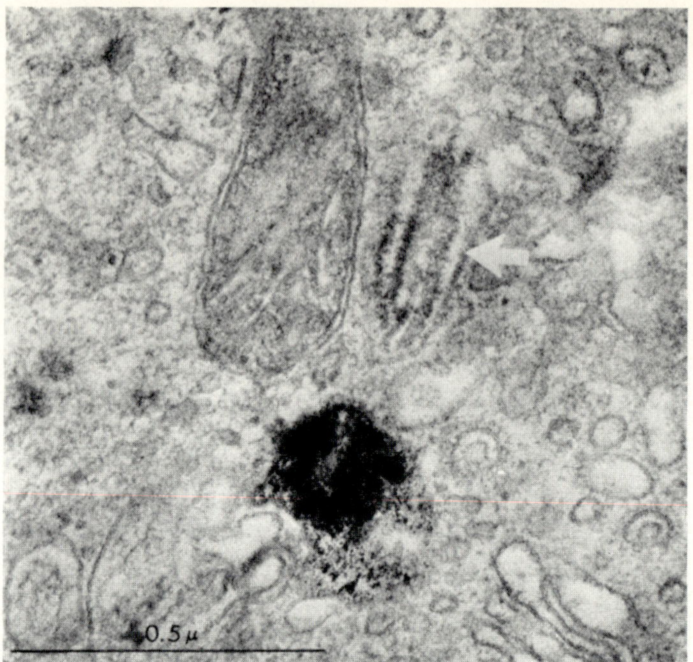

FIG. 10.

on a matrix which "appears to be a lattice of parallel fibers spaced approximately 100 Å apart and cross-linked by fibers approximately 35 Å in diameter" (Moyer, 1966). Deposition continues until the particle takes on the appearance of the homogeneous mature melanin granule.

The ultrastructure of the elementary particle which is to contain pheomelanin differs from that for eumelanin. Moyer (1966) was the first to report the electron microscopic appearance of the prepheomelanosome of the A^y mouse: "The ontogeny of these granules is quite different from that of granules producing eumelanin. No organized matrix is formed and the pheomelanin is deposited in discrete but randomly distributed areas on a tangled mat of extremely fine fibers that are almost translucent to the electron beam. There is no ordered aggregation of fibers nor is there any organized cross-linking."

It was of considerable interest to us to determine whether MSH, in darkening the A^y mouse, also affected the ultrastructure of the premelanosome. The problem has considerable interest both for the solution of the question of the mechanism of MSH action and for the more basic question of the relationship between particle ultrastructure, tyrosinase activity and the nature of the pigment synthesized.

Skin was obtained from 3- to 6-day-old mice either injected 24 hours earlier with MSH in beeswax–peanut oil, or which had received three injections of MSH in saline during the previous 24 hours. The skin was immediately fixed in paraformaldehyde-glutaraldehyde, buffered with pH 7.5 cacodylate. The fixative was changed once during a period of 2–4 days. The material was then postfixed in OsO_4 buffered with cacodylate, block-stained with uranyl acetate, embedded in Epon 812, and sectioned with a glass or diamond knife on a Porter-Blum MT2 microtome. The sections were stained with lead citrate and examined

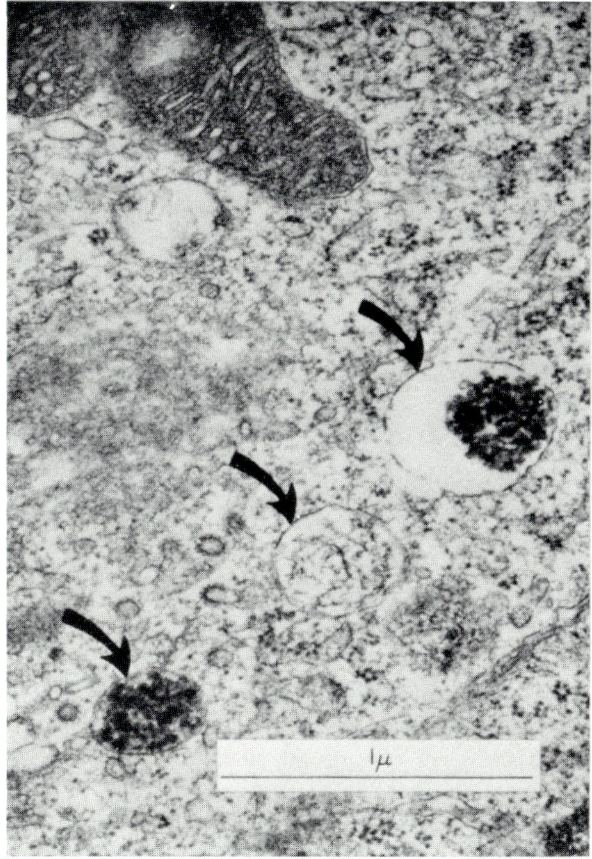

FIG. 11.

FIGS. 11 and 12. Electron micrographs of premelanosomes in skins of control A^y mice 4 and 6 days old. The uppermost particle in Fig. 12 is not typical, but does illustrate the greatest degree of strand helicality we have found in control A^y mice. See Fig. 17 for detail. The scale line is 1 μ.

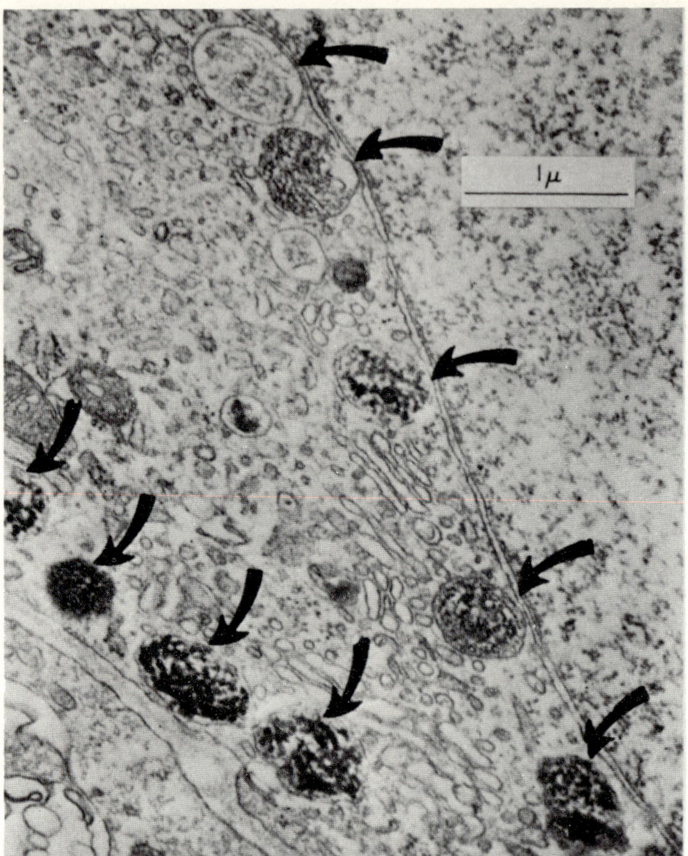

Fig. 12.

in an RCA EMU-3F or Siemens Elmiskop I electron microscope at 8000–40,000 magnification. Figures 11 and 12 are electron micrographs of melanocytes from control mice injected with beeswax–peanut oil alone on day 3 (Fig. 11) and day 5 (Fig. 12). Inspection of a large number of such fields reveals very few premelanosome structures which resemble those found in the skin of black nonagouti (a/a,B/B,C/C) mice (Figs. 9 and 10). One of the prepheomelanosomes shown in Fig. 12, although it can hardly be considered typical, reveals an ultrastructure which to some extent approximates the appearance of preeumelanosomes (Figs. 9 and 10), with some evidence of a few disorganized helical fibers. These premelanosomes are to be compared to those found in Figs. 13–17, which are found in great number in A^y mice treated with MSH. Oriented longitudinal fibers are present in a "zigzag" (helical) pattern, hardly a "tangled mat of extremely fine fibers . . . almost translucent to the electron beam."

It will also be noted that many of these premelanosomes have an ellipsoidal shape rather than the spherical shape seen in the control micrographs. Moyer (1966) has pointed out that eumelanosomes are rod-shaped, whereas pheomelanosomes are spherical. We have looked at the sphericality of mature granules, examining skin obtained from baby A^y mice which had been injected with MSH in beeswax–peanut oil on days 3, 5, 7, and 9 and sacrificed 24 hours after the last injection. Most pigment granules seen in the skin must have been formed in the presence of elevated levels of MSH, which is not true of the animal receiving a single injection of the hormone. Not only are the granules more ellipsoidal in skins from the experimental animal, but their size

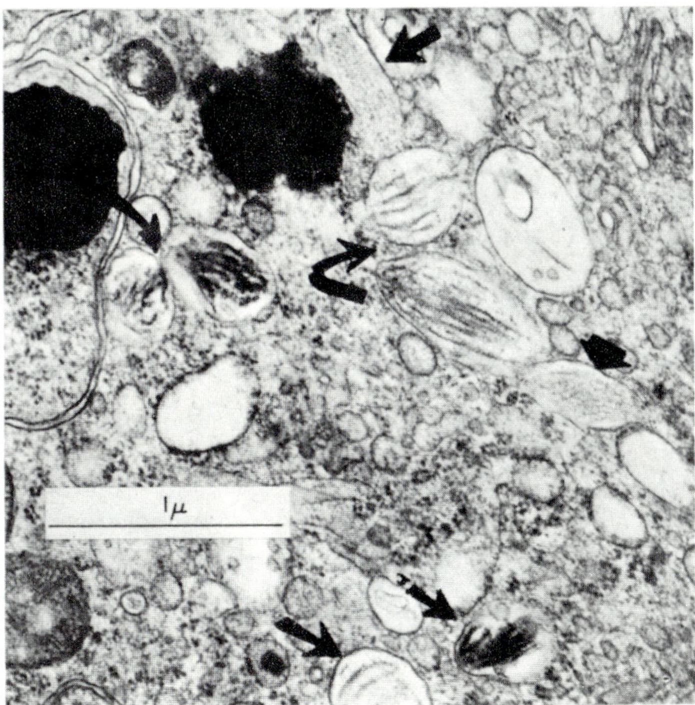

FIG. 13.

FIGS. 13–16.   Electron micrographs of premelanosomes from four A^y mice, 4 and 5 days old, given a single injection of MSH in beeswax–peanut oil 24 hours before sacrifice. Note the organization of the longitudinal strands, the "zigzag" appearance of the strands, and the ellipsoidal shape of the particles. See Fig. 17 for details. Compare to Figs. 11 and 12. The scale line is 1 μ.

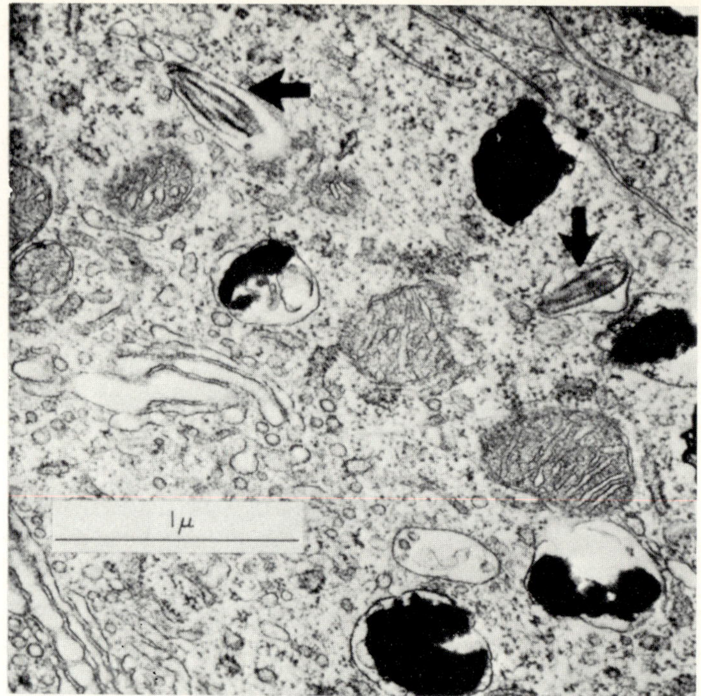

FIG. 14.

distributions appear to be much more uniform than in controls. Controls also frequently possess granules 2 to 3 times the maximum size seen in MSH-treated mice.

Electron micrographs of the skin of baby e/e mice show many more examples of ellipsoidal particles with organized fibers, although the predominant pattern is similar to that found in the A^y. However, we are uncertain whether the melanocytes are producing pheomelanin, since in all baby mice of this genotype we find large areas of darker hair on the dorsum, and it is our experience that it is only after 1 or 2 waves of hair growth that a more homogeneous yellow coat is found. MSH injection seems to have little, if any, effect on the premelanosomes of these mice.

XI. MSH and Coat Color

An effect of MSH on mammalian pigmentation was first shown in man by Lerner and McGuire (1961), but the hyperpigmentation was of the epidermis rather than of hair, and resulted from an increase in the number of free melanin granules in keratinocytes (Lerner *et al.,*

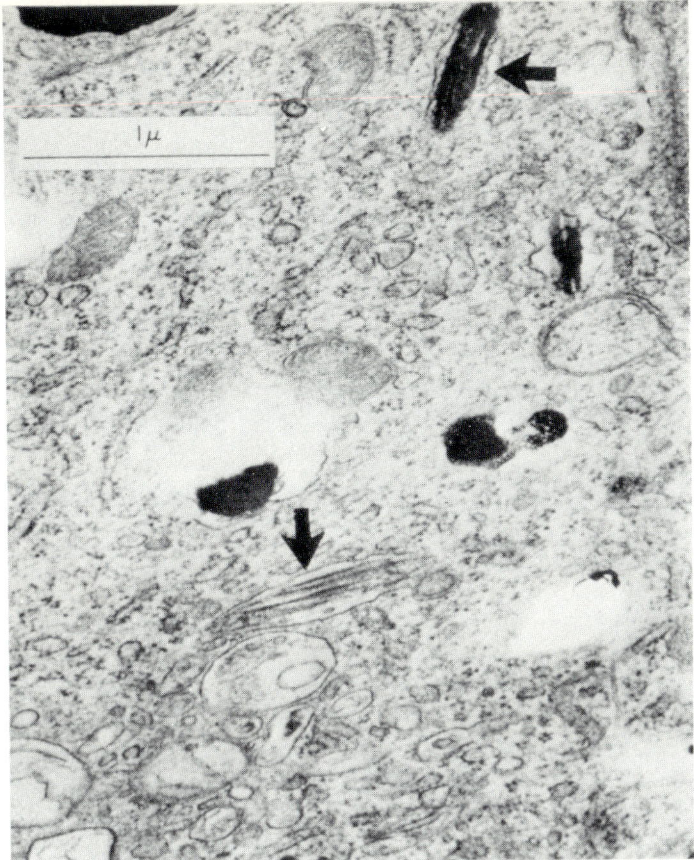

Fig. 15.

1966). A similar effect has been reported in the guinea pig (Snell, 1964) and has been seen in the A^y mouse subjected to persistently elevated levels of MSH, as has been reported above.

Effects on coat color have also been reported. Rust (1965) reported that when MSH was injected into partially plucked hypophysectomized short-tailed weasels with a white coat, the regrown hair was brown. In these experiments the animals were receiving 12 hours of light and so would normally have been brown. Thus, MSH was functioning as replacement therapy for the absent pituitary. Later, Rust and Meyer (1968) observed darkening of the coat in weasels with pituitary auto-grafts under the kidney capsule, presumably secreting large amounts of MSH, even when the animals were subjected to short day lengths which induce the growth of a white coat in normal animals. Adrenalec-

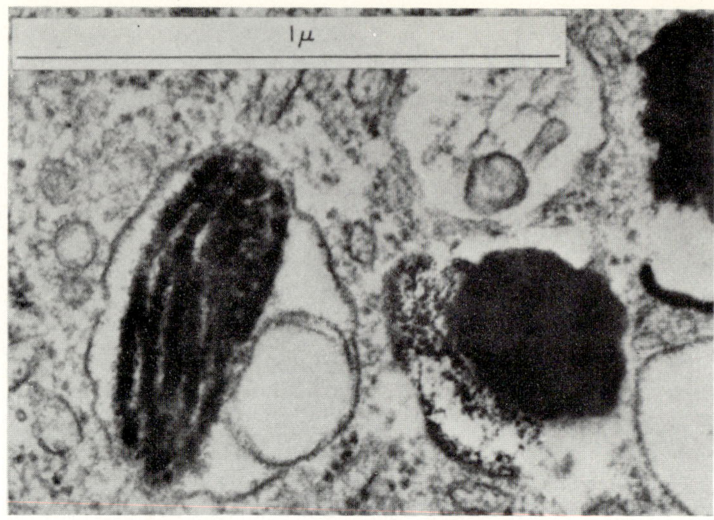

Fig. 16.

tomy of the prairie deer mouse, *Peromyscus maniculatus bairdii,* leads
to a marked darkening of the coat over a period of 1–3 months (Bronson
and Clarke, 1966), which is accompanied by a striking increase in blood
melanocyte-stimulating activity (Bronson *et al.,* 1969). In the guinea
pig treated with MSH the percentage of dark colored hairs increases,
and the effect is greater in black animals than in red. In black animals
light-colored hair is gray, and about two-thirds of anterior abdominal
wall hair in untreated animals is gray (the proportion is apparently
somewhat greater on the scalp), whereas in the red animals light-colored
hair is light red, and represents about 75–80% of the total hair.

To the effects observed in these three species must be added those
observed in the laboratory mouse of the proper genotype. Unlike the
prairie deer mouse, however, the laboratory mouse (A^y, for example)
does not darken after adrenalectomy (Geschwind and Huseby, 1968;
Bronson *et al.,* 1969), nor is there any increase in plasma melanocyte-
stimulating activity (Geschwind and Huseby, 1968). Furthermore, the
time lag before a response to elevated MSH levels is observed is 14–17
days in the weasel, 1–3 months in the deer mouse, and approximately
2 months in the guinea pig, compared to 24 hours or less in the A^y
mouse. In these other species, however, the follicular melanocytes were
not examined microscopically, and a response would certainly be detected
earlier if such an examination were carried out.

On the basis of the failure of the hypophysectomized or adrenalectomized laboratory mouse to alter its coat color, and of our inability

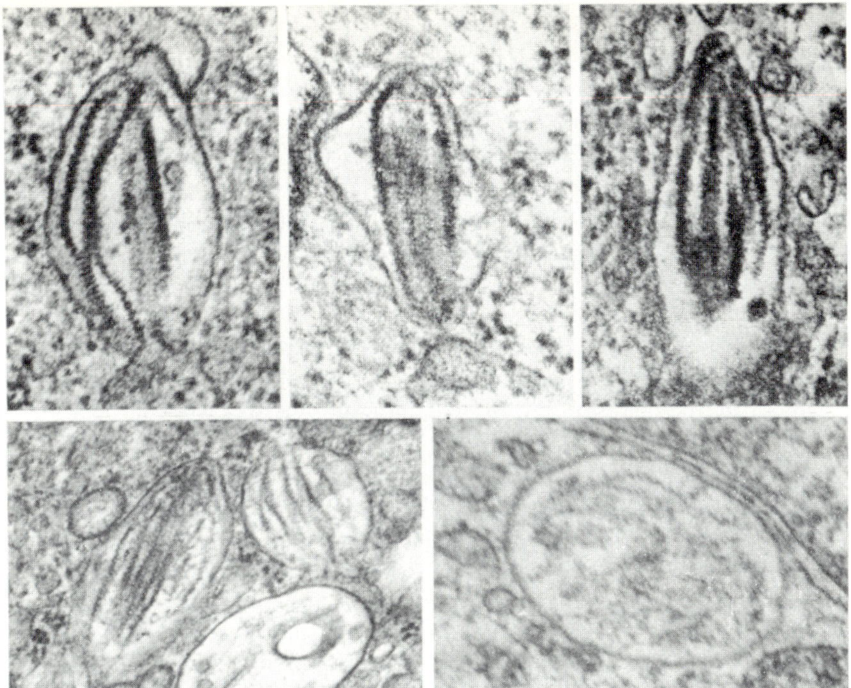

Fig. 17. Electron micrographs of premelanosomes. All but the lower right figure, which is the particle referred to in the legend of Figs. 11 and 12, are from the MSH-treated A^y baby mice (Figs. 13–16). The details of the strands are more obvious at this magnification.

to detect differences in the circulating melanocyte-stimulating activity levels among all the genotypes we employed in our genetic studies, we have come to the conclusion that MSH has no normal physiological function in determining coat color of the laboratory mouse (Geschwind and Huseby, 1971). The hormone may certainly have an effect when blood levels are elevated, as for example, in the presence of a pituitary tumor secreting MSH, or in the animal with a pituitary transplanted away from the hypothalamus, and we have previously shown that only a small increase in blood hormone levels was sufficient to darken the agouti mouse (Geschwind, 1966). The situation is reminiscent of the effect of ACTH-MSH on pigmentation in normal subjects and in Addisonian patients.

XII. Mechanism of Action of MSH

The melanocyte producing pheomelanin differs in at least three ways from that producing eumelanin (or if it is in the agouti animal, from

itself when it is producing eumelanin): its premelanosomes are different,
its tyrosinase activity is decreased, and the chemistry of the pigment
it produces is altered. In the agouti animal a small change in some
parameter brings on yellow pigment synthesis, and presumably by reversal, shuts it off, allowing black pigment production to resume. As
mentioned earlier sulfhydryl groups have been implicated in the production of yellow pigment, but unfortunately their effect is not unique.
Takeuchi (1970), for example, has found that skin from a 2-day-old
agouti mouse in culture will not form pheomelanin, confirming Cleffmann's findings, but if DOPA is added to the medium yellow pigment
is formed. On the other hand, he finds that skin from a 3-day-old does
produce pheomelanin in culture, but this is reversed if excess tyrosine
is added to the medium. Moreover, Flesch (1970), working with New
Zealand red rabbits induced black pigmentation simply by vigorously
rubbing them. Thus, a number of factors are capable of producing a
switch from red or yellow to black or in the reverse direction.

As mentioned earlier, Cleffmann has favored the idea that the agouti
locus controls the amount of sulfhydryl compounds in pigment cells
and has shown that such compounds are incorporated to a greater extent
during yellow pigment formation than during black. When coupled with
the discovery of 5-S-cysteinyl DOPA and the elucidation of the structures of two trichosiderins, the increased incorporation is readily explained. Since in the yellow phase incorporation of DOPA and tyrosine
is less than in the black phase, an inhibition of enzyme activity was
also invoked, but the importance of that inhibition in yellow pigment
formation was not mentioned by Cleffmann. Nicolaus and his colleagues
can hardly consider the effect unimportant, for although the formation
of cysteinyl DOPA occurs nonenzymatically, the formation of the intermediate DOPA-quinone depends upon the enzyme. Holstein et al. (1971)
feel that their own data on the electrophoretic variants of tyrosinase
suggest that in the A^y mouse the absence of a normal complement of
tyrosinases may explain pheomelanin production, perhaps as a loss of
some control function exerted by the missing variants. However, in none
of their papers do Holstein and his colleagues consider the possibility
that activity of the enzyme may have been altered by interaction *in
vivo* with sulfhydryl groups. They do propose, though, as one of a number of possibilities, that the variants may represent "a single enzyme
moiety bound to structural components of the melanosomal matrix"
(Holstein *et al.*, 1971).

Support for Cleffmann's ideas has also come from Brumbaugh (1968),
who has studied the ultrastructural differences between eumelanin and
pheomelanin formation in fowl. He has found disorganized matrices in

prepheomelanosomes, similar to Moyer's observations in the mouse, and decreased enzymatic activity. On the basis of these findings he has offered an alternate hypothesis, suggesting that "sulfhydryls react with the matrix subunits so that crosslink and strand formation is reduced . . . (and) the resulting pheomelanin premelanosome is less able to oxidize and polymerize melanin precursors because of its disorganized structure." This suggestion is a variation of one offered by Moyer (1966), who, as a result of Cleffmann's work, hypothesized that the agouti locus somehow alters the melanocyte redox potential, and the change "interferes with orderly aggregation of the protein subunits and causes an altered enzymatic activity leading to the production of pheomelanin . . . ." Either suggestion could offer the basis for the absence of multiple forms of the enzyme reported by Holstein *et al.* (1967, 1971), if their proposal about the origin of the variants, referred to above, is correct.

At this point we reemphasize the findings reported in this paper: MSH alters the three parameters which differ in yellow pigment-forming and black pigment-forming melanocytes. Thus, in the A^y, and probably in the e/e mouse, in the agouti mouse in its yellow phase, in the fowl, and in A^y and e/e mice treated with MSH, the three phenomena seem to behave as parts of a single coordinated unit. The only exceptions that we have seen are in the correlation between darkening (i.e., eumelanin formation) and enzyme activity in some experiments (Tables I–III), but here the deficiencies may lie in the methodology for tyrosinase determination.

If the three phenomena are behaving as a single unit, they they represent most probably either a linear array of cause and effect relationships (e.g., maximum tyrosinase activity depends upon a highly organized premelanosome structure and eumelanin is formed only when tyrosinase activity is high; or the suggestions of Moyer and Brumbaugh), or discrete phenomena independently affected by a single regulator (e.g., a sulfhydryl compound, cellular redox potential, or cyclic AMP). The "cause and effect" suggestions of Moyer and Brumbaugh were offered before the chemistry of yellow pigments was developed, and so could not anticipate the importance of cysteine residues in pigment formation. On the other hand, if all that is required for pheomelanin formation is an increase in intracellular free cysteine molecules, either a complex machinery would have had to be present to release the bound residues from glutathione in Cleffmann's experiments, or yellow pigments were formed from the coupling of glutathione itself to DOPA-quinone, or glutathione (alone or together with a glutathione reductase) only served to maintain a reducing potential so that intracellular free cysteine was not oxidized. The first two alternatives, but not the last, are consistent with Cleffmann's (1964) findings of an increased incorporation of the label of glutathione-^{35}S into the yellow band of agouti hair. From these experiments, incidentally, it does not seem probable that an active uptake mechanism is affected by the A locus, since one might expect some discrimination between the transport of cysteine and glutathione, but none is here evident. The role played by cysteine must be critical in pheomelanin formation, yet it is obvious that the melanocytes must be competing with the keratinocytes for this amino acid which is so important in keratin, and therefore in hair formation.

A critical experiment for the interpretation of these data lies in an electron microscope investigation of the premelanosomes in melanocytes of A^y and black nonagouti skins cultured in the presence and absence of added glutathione. As for MSH, since it acts upon the unit as a whole, the simplest explanation would be to suggest that it directly opposes the A locus effect. Whether MSH can stimulate the production of eumelanin in skin cultured in the presence of glutathione remains to be determined, and the answer obtained will unquestionably contribute to our understanding of its mechanism of action.

XIII. Other Possible Effects of MSH

It has been reported that MSH influences thyroid function (Bowers *et al.*, 1964). The 20-hour uptake of ^{131}I by the thyroid of ten control and nine tumor-bearing adrenalectomized C3H $\times$ A F_1 female mice was determined. No significant difference was noted. Female animals of this cross were hypophysectomized when 6 weeks of age. Two weeks later synthetic α-MSH, 50 μg (1×10^6 U), was injected twice daily in beeswax–peanut oil. Injections were continued for 2 weeks with the animals being sacrificed 2 hours after the last injection. Although ^{131}I uptake studies were not carried out, histologic examination of the thyroids of five injected animals showed no evidence of stimulation when compared to those of six control animals.

Since the adrenals of later transplant generations of tumor-bearing animals consistently were enlarged and demonstrated histologic evidence of cortical hyperactivity, *in vitro* studies of cortical function were carried out employing the adrenals of the MSH-treated hypophysectomized animals referred to in the ^{131}I experiments, plus 10 animals that had received 2 IU of ACTHAR Gel subcutaneously each day for the same period as those receiving α-MSH. In all instances, homogenates equivalent to five adrenals per incubation flask were employed using a final incubation volume of 10 ml of phosphate buffer containing 25 nmoles of progesterone-^{14}C and concentrations of cofactors previously described

(Dominguez and Huseby, 1968). Duplicate evaluations were carried out, and the extraction of steroids as well as their identification were also similar to that previously described.

The injection *in vivo* of α-MSH had no effect on spleen or thymus weights, or on the *in vitro* conversion by the adrenals of progesterone to either corticosterone or deoxycorticosterone while ACTH was most effective, increasing the former from 0.1% to 16.0%, and the latter from 0.6% to 31.3%, and markedly reducing the weights of both spleen and thymus. Six incubations were carried out employing adrenal homogenates from tumor-bearing hypophysectomized animals under the same *in vitro* conditions outlined above. Splenic weights decreased and the thymus were so small as to make it impossible to dissect and weigh with any accuracy. The *in vitro* conversion of progesterone to corticosterone and deoxycorticosterone varied considerably, probably relative to the size of the tumors, but the conversion to corticosterone was in the range of that of the adrenals of hypophysectomized animals receiving large doses of ACTH, but lower in respect to conversion to deoxycorticosterone.

ACKNOWLEDGMENTS

We are pleased to acknowledge the contributions made by many people to this project: Drs. Eric Bradford, Elizabeth S. Russell, Gordon Eaton, Theodore S. Hauschka, and George L. Wolff for supplying the mice of the various genotypes used in this investigation; Drs. Virginia Upton, Saul Lande, Aaron B. Lerner, Andrew V. Schally, and W. Rittel, for supplying natural or synthetic α-MSH preparations; Dr. Joe Fisher of the Armour Co. for a gift of porcine posterior lobe powder which enabled us to prepare our own MSH preparations; Drs. Walter Chavin and Y. M. Chen for the tyrosinase assays, and Dr. Lewis Chase for the cyclic AMP and adenyl cyclase assays; Dr. Giuseppe Prota for allowing us to cite his unpublished results; and Mr. Robert Dewey, Mr. Oskar Lang, Mrs. Sally Thurlow, Mrs. Eva Vlajk, and Miss Susan Hamamoto for expert technical assistance.

Finally, we acknowledge our indebtedness to the National Institutes of Health for Grants HD 00394 and CA 05191, which partially supported the research of two of us (I. I. G. and R. A. H., respectively), and to the National Science Foundation for Grant GB 6424 (awarded to Dr. Howard Bern) which supported the research of our third member (R. N.).

REFERENCES

Abe, K., Butcher, R. W., Nicholson, W. E., Baird, C. E., Liddle, R. A., and Liddle, G. W. (1969). *Endocrinology* **84**, 362.

Bahn, R., Furth, J., Anderson, E., and Gadsden, E. (1957). *Amer. J. Pathol.* **33**, 1075.

Bitensky, M. W., and Burstein, S. R. (1965). *Nature (London)* **208**, 1282.

Bitensky, M. W., and Demopoulos, H. B. (1970). *J. Invest. Dermatol.* **54**, 83.

Bowers, C. Y., Redding, T. W., and Schally, A. V. (1964). *Endocrinology* **74**, 559.

Bronson, F. H., and Clarke, S. H. (1966). *Science* **154**, 1349.

Bronson, F. H., Eleftheriou, B. E., and Dezell, H. E. (1969). *Proc. Soc. Exp. Biol. Med.* **130,** 527.

Brumbaugh, J .A. (1968). *Develop. Biol.* **18,** 375.

Burnett, J. B., Holstein, T. J., and Quevedo, W. C., Jr. (1969). *J. Exp. Zool.* **171,** 369.

Chase, L. R., and Aurbach, G. D. (1968). *Science* **159,** 545.

Chavin, W. (1969). *Amer. Zool.* **9,** 505.

Chen, Y. M., and Chavin, W. (1965). *Anal. Biochem.* **13,** 234.

Cleffmann, G. (1954). *Z. Naturforsch. B* **9,** 701.

Cleffmann, G. (1963a). *Ann. N.Y. Acad. Sci.* **100,** 749.

Cleffmann, G. (1963b). *Wilhelm Roux' Arch. Entwicklungsmech. Organismen* **154,** 239.

Cleffmann, G. (1964). *Exp. Cell Res.* **35,** 590.

Clive, D., and Snell, R. (1967). *J. Invest. Dermatol.* **49,** 314.

Dominguez, O. V., and Huseby, R. A. (1968). *Cancer Res.* **28,** 348.

Fitzpatrick, T. B., Brunet, P., and Kukita, A. (1958). *In* "Biology of Hair Growth" (W. Montagna and R. A. Ellis, eds.), pp. 255–334. Academic Press, New York.

Flesch, P. (1970). *J. Invest. Dermatol.* **54,** 87.

Galbraith, D. B. (1964). *J. Exp. Zool.* **155,** 71.

Galbraith, D. B. (1971). *Genetics* **67,** 559.

Geschwind, I. I. (1966). *Endocrinology* **79,** 1165.

Geschwind, I. I., and Huseby, R. A. (1966). *Endocrinology* **79,** 97.

Geschwind, I. I., and Huseby, R. A. (1968). *Excerpta Med. Found. Int. Congr. Ser.* **157,** 70.

Geschwind, I. I., and Huseby, R. A. (1970). *J. Invest. Dermatol.* **54,** 87.

Geschwind, I. I., and Huseby, R. A. (1971). *In* "Pigmentation: Its Genesis and Biologic Control" (V. Riley, ed.), Appleton, New York.

Hauschka, T. S., Jacobs, B. B., and Holdridge, B. A. (1968). *J. Hered.* **59,** 339.

Holstein, T. J., Burnett, J. B., and Quevedo, W. C., Jr. (1967). *Proc. Soc. Exp. Biol. Med.* **126,** 415.

Holstein, T. J., Quevedo, W. C., Jr., and Burnett, J. B. (1971). *J. Exp. Zool.* **177,** 173.

Jacobs, B. B., and Huseby, R. A. (1967). *Transplantation* **5,** 410.

Lerner, A. B., and McGuire, J. S. (1961). *Nature (London)* **189,** 176.

Lerner, A. B., Snell, R. S., Chanco-Turner, M. L., and McGuire, J. S. (1966). *Arch. Dermatol.* **94,** 269.

Markert, C. L., and Silvers, W. K. (1956). *Genetics* **41,** 429.

Misuraca, G., Nicolaus, R. A., Prota, G., and Ghiara, G. (1969). *Experientia* **25,** 920.

Moyer, F. H. (1966). *Amer. Zool.* **6,** 43.

Pomerantz, S. H., and Chuang, L. (1970). *Endocrinology* **87,** 302.

Prota, G., Scherillo, G,. Petrillo, D., and Nicolaus, R. A. (1969). *Gazz. Chim. Ital.* **99,** 1193.

Prota, G., Crescenzi, S., Misuraca, G., and Nicolaus, R. A. (1970). *Experientia* **26,** 1058.

Russell, E. S (1948). *Genetics* **31,** 327.

Rust, C. C. (1965). *Gen. Comp. Endocrinol.* **5,** 222.

Rust, C. C., and Meyer, R. K. (1968). *Gen. Comp. Endocrinol.* **11,** 548.

Rust, C. C., and Meyer, R. K. (1969). *Science* **165,** 921.

Shizume, K., Lerner, A. B., and Fitzpatrick, T. B. (1954). *Endocrinology* **54,** 553.
Silvers, W. K. (1958). *J. Exp. Zool.* **137,** 189.
Silvers, W. K. (1961). *Science* **134,** 368.
Silvers, W. K., and Russell, E. S. (1955). *J. Exp. Zool.* **130,** 199.
Snell, R. S. (1964). *J. Invest. Dermatol.* **42,** 337.
Steelman, S. L., Kelly, T. L., Norgello, H., and Weber, G. F. (1956). *Proc. Soc. Exp. Biol. Med.* **92,** 392.
Steiner, A. L., Kipnis, D. M., Utiger, R., and Parker, C. (1969). *Proc. Nat. Acad. Sci. U.S.* **64,** 364.
Takeuchi, T. (1970). *J. Invest. Dermatol.* **54,** 98.
Wolfe, H. G., and Coleman, D. L. (1966). *In* "Biology of the Laboratory Mouse" (E. L. Green, ed.), 2nd Ed., pp. 405–425. McGraw-Hill, New York.

DISCUSSION

G. A. Bray: The A^y gene is associated not only with yellow coat color, but with an increased body fat content and obesity. Do you know whether your yellow mice treated from birth with MSH showed any differences in the body fat content as compared with the untreated dominant yellow? Also, is there any change in the percentage of carcass fat associated with the other forms of yellow coat color in the e/e or the pink-eyed agouti?

I. Geschwind: As I recall, the e/e has very little effect on body fat. I really cannot answer your first question because those babies we showed which had been injected with MSH from day 2 received only 10 injections on alternate days. We do not have enough MSH to inject in animals up to about day 90, when the A^y animal ordinarily fattens.

A. Bartke: Since you mentioned that MSH blood levels were the same in yellow and black animals, would it be fair to assume that the effect of dominant yellow genes is to yield melanocytes relatively insensitive to endogenous MSH, and you can overcome this with MSH from tumor or with injected hormone? In the case of the e/e animal, you would have an absolute failure to respond rather than a relative lack of sensitivity.

I. Geschwind: I think it is not a question of lack of sensitivity. As I mentioned, an increase in tyrosinase activity was found by Pomerantz in black and brown animals after MSH administration. We found in injection experiments that all one had to do was to increase the circulating level of MSH from about the normal level of 3–6 U/ml to about 25 U/ml in order to get darkening. This is really a very small increase, and the animal does not encounter this sort of increase during its lifetime. Under pathological conditions, or in conditions where you might implant the pituitary under the kidney capsule, you may be able to see such high levels.

G. C. Mueller: You administered cycloheximide and at the same time administered MSH and got no effect until after a period of about 36 hours, when you began to get the effects in the absence of any further administration of MSH. Is that correct?

I. Geschwind: That is correct. We administered cycloheximide and MSH together at zero time. Then at 7 and 14 hours we gave a second and third injection of the cycloheximide. Ten hours later, at 24 hours, we looked for the typical 24-hour effect of MSH and found none. However, if we waited another 12 hours, we began to see the effect. It is good at 36 hours, better at 48 hours, and it's beautiful at 60 hours after the single injection of MSH in beeswax–peanut oil. As I mentioned, in our kinetic experiments we demonstrated that maximum black pigment formation takes place at 36 hours after an injection, and you can still see the effect after 72–96 hours.

G. C. Mueller: In the case of puromycin, is there absolutely no effect upon the response picture?

I. Geschwind: That is right. It is nice and black at 24 hours.

G. C. Mueller: Does this not point to quite a distinction between the action of these two chemicals in terms of protein synthesis? The cycloheximide is very effective in preserving the structure of a polysome, whereas puromycin actually has a tendency to disperse polysomes. Are you not dealing with some kind of preserved particulate structure here which is made in the presence of MSH but which can finally evolve for later action?

I. Geschwind: With puromycin we get a beautiful MSH effect. Puromycin does not inhibit the MSH effect, so if it had some effect on the ribosomes I presume we would not have seen the MSH effect. I believe it has to do with the persistence of the puromycin effect as compared to that of cycloheximide in terms of the levels we have given. We gave it at 0, 5, and 10 hours and then tried some animals with a follow-up injection at 15 hours, but they were all dead in a few hours.

V. L. Gay: I understand that the structure of the MSH inhibitory factor has been determined and the factor should be available in large quantities. Have there been any experiments in which you have attempted to control the MSH-secreting tumor or to demonstrate an influence of inhibitory factors on coat color in your animals?

I. Geschwind: We have not done any experiments with it yet.

If you hypophysectomize a black mouse, the new hair which is going to come will be black. It does not need MSH. If you hypophysectomize a yellow mouse, the new hair which comes in will be yellow. It is not going to be changed. There are no effects that we can see as far as normal coat color determination is concerned on the part of MSH. These facts would also serve to refute Dr. Bartke's suggestion above about differences in sensitivity to MSH.

I. L. Schwartz: Your finding that cyclic AMP does not mimic MSH effects in the system you are studying means that either some other membrane-bound secondary messenger system is involved or that this peptide hormone is acting directly at an intracellular locus. May I ask therefore whether anyone has attempted to decide between these possibilities by localizing labeled MSH at the subcellular level or by studying the activity of MSH covalently coupled to Sepharose beads?

I. Geschwind: Not to my knowledge, but I believe this has been done with ACTH, and it is a membrane localization with ACTH. The difference here would be a difference between 39 amino acids and 18 with portions of the same sequence.

R. O. Greep: In some cases you shaved the animals, and in others you plucked the hair. Does this make a difference?

I. Geschwind: Plucking is a very good stimulus to new hair growth; therefore, you can start your experiment on the day of plucking as day 0 and follow it over the next 19 days. With shaving you are hoping some of the hairs have not completed their growth and will continue to grow after shaving. I prefer to pluck, and Dr. Huseby prefers to shave because it is easier on the fingers.

Oxytocin Analogs in the Analysis of Some Phases of Hormone Action

J. RUDINGER, V. PLIŠKA, AND I. KREJČÍ

*Institute of Molecular Biology and Biophysics, Swiss Federal Institute of
Technology, Zürich, Switzerland; Institute of Pharmacy and
Biochemistry, Prague, Czechoslovakia*

I. Introduction

For the past fifteen years or so peptide chemists have been busy synthesizing vast numbers of analogs of the peptide hormones. At the same time physiologists and endocrinologists have been studying the action of peptide hormones in numerous and varied biological systems. Unfortunately, most of the synthetic analogs have been examined in only one or a few standard biological assay preparations, and conversely only the natural hormones or at best a few analogs have been studied in most biological systems. Partly this situation no doubt results from simple limitations of time and resources; however, we believe that there has also been a lack of appreciation of the useful information that can be obtained by the judicious use of synthetic hormone analogs in diverse biological systems.

This criticism is perhaps less valid for the neurohypophysial hormones and their analogs than for most other areas of peptide endocrinology, and it is the purpose of this paper to examine the way in which synthetic analogs have been, or may be, used to analyze hormone action in all its stages. At the very outset we would emphasize that we commend such work not as an alternative, but as an adjunct to other, more purely biological or biochemical ways of studying hormone action.

Figure 1 shows a schematic representation of the life, deeds, and death of a generalized peptide hormone, divided into what we may term the phases of hormone action. The definition of these phases is, of course, to some extent arbitrary but it will serve for the purposes of this discussion. As is indicated in the figure, an exogenous hormone or hormone analog will bypass some of these phases; and it is also obvious that many biological preparations will by their very nature eliminate some or most of them. Indeed, one is tempted to design biological preparations that would isolate these phases one at a time and permit their separate study. However, two dangers attend all such attempts. One is inherent in the assumption that an anatomically simple system—say, an isolated target tissue of the hormone—is in fact "simple" also in the sense of the scheme in Fig. 1; and the second in the assumption that phenomena observed in the simplified, and therefore necessarily disintegrated, biological preparation are relevant to hormone action under physiological conditions. We shall encounter both these problems in the subsequent discussion.

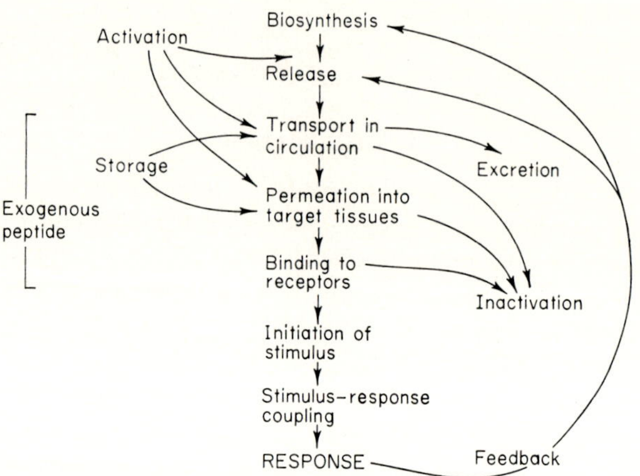

Fig. 1. Generalized scheme of action of a peptide hormone.

II. Interaction with Tissue Receptors

A. A Receptor Model

The interaction of hormones with their specific tissue receptors is the linchpin of the scheme and has also exerted the greatest fascination over investigators. Early interpretations of the results obtained with peptide hormone analogs tended to assume, explicitly or implicitly, that the potency of an analog acting on an assay preparation was a measure of its affinity for the hormone receptors of the target tissue. Such interpretations tended to ignore not only the contribution of other factors, but also the distinction which pharmacologists were by that time making between the "affinity" of a drug for the receptor and its "efficacy" (Stephenson, 1956) or "intrinsic activity" (Ariëns *et al.*, 1956) as factors contributing to potency. Rushing in where angels, apparently, feared to tread, we applied the pharmacological technique of recording full dose–response curves to the action of oxytocin and its analogs on the isolated rat uterus and found that peptide hormones did, in fact, behave like "lesser" drugs. Under the experimental conditions used, [2-*O*-methyltyrosine]-oxytocin (methyloxytocin: Fig. 4e) acted as a competitive inhibitor and other analogs modified in sequence position 2, such as [2-leucine]-oxytocin and [2-phenylalanine]-oxytocin (Fig. 4b) showed the properties of "partial agonists," eliciting a maximal response of the uterus which was lower than the maximal response to oxytocin itself (Rudinger and Krejčí, 1962). It therefore seemed that the action of peptide hormones at the receptor level can usefully be discussed in terms of the model of pharmacological receptor theory which distinguishes between the binding of a drug to its receptor (characterized by the "affinity") and the ability of the drug–receptor complex to generate a primary stimulus[1] (expressed in its "efficacy" or "intrinsic activity").

However, a closer study of the properties of methyloxytocin and related analogs revealed features that were difficult to reconcile with this relatively simple, "classical" receptor model. Thus it turned out that methyloxytocin could show oxytocin-like uterotonic action, or inhibitor properties, or intermediate behavior according to the experimental conditions (ionic composition of the medium, temperature, etc.; see Section III) (Krejčí *et al.*, 1964, 1966, 1967b). These and related findings, which will be discussed in more detail below, could be interpreted either by assuming that such changes in experimental conditions alter relations at the receptor level or, in some cases more plausibly, that they intervene in the stimulus–response coupling at a stage subsequent to the primary event at the receptor (Krejčí *et al.*, 1966, 1967b; Rudinger, 1966; Rudinger and Krejčí, 1968). Further consideration showed that the latter interpretation would require reference to a model in which the response was a nonlinear function of the stimulus, and which in consequence would show the phenomena of receptor reserve capacity and threshold (Krejčí *et al.*, 1970). Eggena *et al.* (1970) have independently resorted to a model with receptor reserve and threshold to rationalize their findings with the same series of analogs acting on the toad bladder.

To facilitate subsequent discussion, some properties of this model are described here. In later sections, it will then be used for the interpretation of the experimental results.

When a response is not a linear function of the stimulus, a stimulus sufficient to elicit the maximal response of a given biological preparation may already be generated when only a certain fraction of the available receptors is occupied by the drug (curves *1* in Fig. 2). The occupation

---

[1] There is a good deal of semantic confusion in the literature of pharmacological receptor theory because some terms are defined (alternatively or promiscuously) as features of a conceptual model, or phenomenologically, or mathematically. In what follows we shall use "efficacy" with reference to stimulus generation (Stephenson, 1956; van Rossum and Ariëns, 1962), i.e., that primary event (whatever it might be) which follows or attends the formation of the drug-receptor complex and is linked by a chain of further events ("stimulus-effect coupling") to the observed response.

of further receptors (the "receptor reserve") then generates "surplus" stimulus which is no longer translated into an overt effect. Such models have been considered by Stephenson (1956), van Rossum and Ariëns (1962), and Furchgott (1955, 1966). They have shown that in a system with a receptor reserve, a drug with decreased efficacy (curves *2* in Fig. 2a) may yet elicit the same maximal response as the reference drug (curves *1* in Fig. 2a). Even though such a drug may have the same

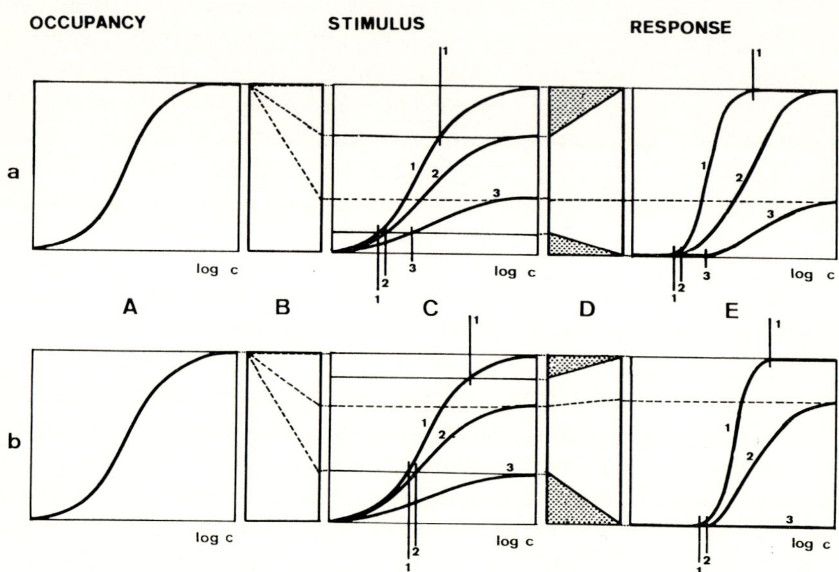

FIG. 2. Receptor model with variable stimulus–response coupling. Panels: A, receptor occupancy *vs* log concentration; B, "black box" of stimulus generation; C, stimulus *vs* log concentration; D, "black box" of stimulus–response coupling; E, response *vs* log concentration. Curves: *1*, for the reference drug ("full" efficacy); *2* and *3* for analogs with progressively lower efficacy; (a) represents a situation with more efficient, (b) with less efficient, stimulus–response coupling. The stippled areas represent the threshold and receptor reserve.

affinity for the receptors as the reference drug (as assumed in Fig. 2; see panel A) its log-dose–response curve will yet appear shifted to higher concentrations (panel E in Fig. 2a). In appearance, such log-dose–response curves simulate those which in the simpler system without receptor reserve result from a decrease in affinity only. Not until, in the system with a receptor reserve, the efficacy of a drug is so low that it can no longer generate the stimulus required for maximal response even at full receptor occupancy does the log-dose–response curve take on the appearance characteristic of "partial agonism" (curve *3* in Fig. 2a, panel E), with a maximal overt response lower than that of which the system is capable.

It has already been shown theoretically, and demonstrated experimentally, that the progressive decrease in the number of receptors in such a system (e.g., by blocking with irreversible competitive inhibitors) will also result in an apparently parallel shift of the log-dose–response curves for a given substance and may reveal the "partial agonism" of analogs with decreased efficacy (Furchgott, 1955, 1966; Nickerson, 1956).

In all these discussions, the stimulus–response coupling was assumed to be described by a function which was nonlinear but which was invariably taken to be constant. Our results with methyloxytocin and related analogs prompted us to consider the effect of varying the efficiency of the stimulus–response coupling in a model such as that described above. Mere inspection (compare Figs. 2a and 2b) shows that a decrease in the effectiveness of stimulus–response coupling is likely to have a similar effect, phenomenologically, as a decrease in the number of receptors or of the efficacy. In Fig. 2b, a greater stimulus must enter the "black box" of stimulus–response coupling (D) than in Fig. 2a to give the same response. Consequently the maximal stimulus now produced by the analog *2* at full receptor occupancy is no longer sufficient to elicit the maximal response, and the analog is now clearly seen to be a partial agonist (Fig. 2b, panel E).

Similarly, the threshold value of the stimulus which is capable of giving a measurable response is raised by a decreased efficiency of stimulus–response coupling. Under the conditions of Fig. 2a, the stimulus generated by analog *3* at full receptor occupancy exceeds the threshold value and therefore causes a measurable response (panel E). The same stimulus is no longer sufficient to exceed the threshold in Fig. 2b, and the analog therefore causes no measurable response; however, since the analog is still occupying the receptors it acts as a competitive inhibitor of compounds such as *1* and *2*.

More quantitative treatment, using various nonlinear functions to represent the stimulus–response coupling, confirms the essential features of the model described qualitatively in Fig. 2 (Pliška and Rudinger, 1972).

Lest it should appear that such "model building" is not related to the title of this review, we would again emphasize that the development of this receptor model was a direct consequence of the need to explain the peculiar properties of a series of oxytocin analogs—the transition from agonism to "partial agonism" and to inhibition with changes in the experimental conditions (Krejčí *et al.*, 1970; Eggena *et al.*, 1970). We believe, however, that the model will prove more generally applicable.

B. Binding to the Receptors

A series of analogs modified in sequence position 3 (Fig. 3) in numerous experiments on the isolated rat uterus under a variety of conditions (Rudinger and Krejčí, 1962; Krejčí *et al.*, 1964; Krejčí and Poláček, 1968) invariably gave log-dose–response curves parallel to that for oxytocin and merely displaced from it along the concentration axis. It was therefore concluded (Rudinger and Krejčí, 1962; Nesvadba *et al.*, 1963; Krejčí *et al.*, 1964) that these differences in potency reflect differences in binding (affinity) alone.

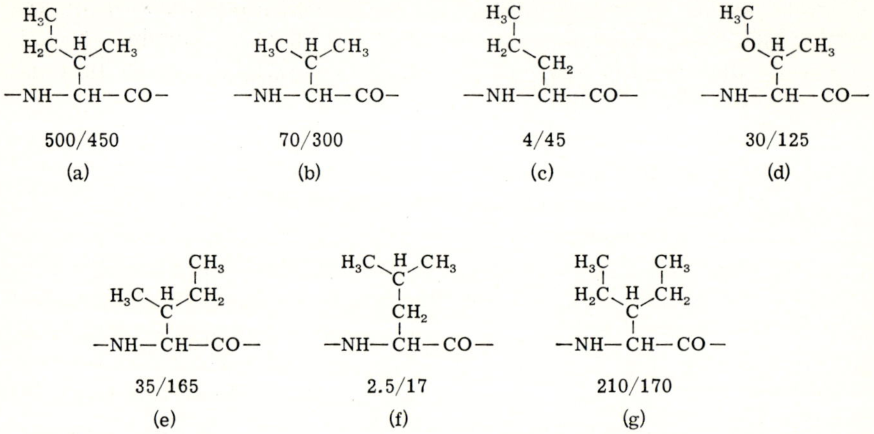

500/450	70/300	4/45	30/125
(a)	(b)	(c)	(d)

35/165	2.5/17	210/170
(e)	(f)	(g)

Fig. 3. Some replacements in sequence position 3 of oxytocin: (a) isoleucine (oxytocin); (b) valine; (c) norvaline; (d) O-methylthreonine; (e) alloisoleucine; (f) leucine; (g) β,β-diethylalanine. Figures give uterotonic potency *in vitro* (IU/μmole) in van Dyke-Hastings medium without/with 0.5 mM magnesium.

Recent developments require a reexamination of this conclusion. The implicit assumption that the analogs do not differ appreciably in their distribution or disposition in the isolated uterine tissue still seems plausible: The structural changes made have a rather small effect on the gross properties of the molecule, and there is no evidence that position 3 is a site of enzymatic attack. However, the interpretation of the parallel log-dose–response curves becomes inconclusive since the system has been shown to have a receptor reserve capacity (see above) so that the "parallel shift" of the log-dose–response curves might, in fact, be a reflection of differences in efficacy rather than, or as well as, differences in affinity. Against this we may argue that no decrease in the maximal response to these analogs relative to oxytocin has been observed even under experimental conditions when analogs of similar potency modified in position 2 clearly behaved as "partial agonists" or inhibitors. Finally, it has been found that replacement of the isoleucine in position 3 of oxytocin by phenylalanine does give analogs with decreased efficacy (Walter *et al.*, 1968, 1969) ; however, this may well be a specific property of the aromatic side chain of phenylalanine, as against the aliphatic side chains in Fig. 3, and has in fact been so interpreted in a recent discussion (Walter *et al.*, 1971).

If, as we believe, the analysis of the potencies of these analogs in terms of affinity is still justified, some conclusions may be drawn about the properties of the receptor. From the loss of potency which occurs when a methylene group is omitted (in [3-valine]- and [3-norvaline]-oxytocin; Fig. 3b, c) it was concluded that lipophilie interaction of this side chain with the receptor contributes to binding (Nesvadba *et al.*, 1963). This conclusion was borne out by the properties of [3-*O*-methylthreonine]-oxytocin (Fig. 3d), an analog in which one (lipophilic) methylene group of the isoleucine side chain is replaced by a (hydrophilic) oxygen and which shows the same decreased potency as the [3-valine] analog, although it must be nearly isosteric with oxytocin (Chimiak and Rudinger, 1965).

On the other hand, it is not lipophilicity generally which is important at this site, but rather lipophilic groups distributed in a particular steric pattern: Placement of additional methyl groups in the side chain as in the [3-alloisoleucine]-, [3-leucine]-, or [3-β,β-diethylalanine]-oxytocins (Fig. 3e, f, g) *decreases* potency, and presumably binding, evidently by steric hindrance (Nesvadba *et al.*, 1963; Eisler *et al.*, 1966; Rudinger, 1968). We may therefore conclude that the oxytocin receptor of the uterus—and, by analogy, of the rat mammary gland myoepithelium (Poláček *et al.*, 1967; Poláček and Krejčí, 1969)—has a lipophilic, sterically rather strictly defined, region whose interaction with the side chain in sequence position 3 makes an important contribution to hormone binding.

An even more extensive and well graded series of substitutions has been made in position 4 of oxytocin (for summaries, see Flouret and du Vigneaud, 1969; Sawyer and Manning, 1971; Rudinger, 1971). In the one case where full dose–response relations have been reported for an analog of this type (Chan and Kelley, 1967), there was no evidence for a decreased maximal response. In the absence of indications to the contrary, the structure potency relations within this group may be discussed in terms of requirements for receptor binding (Sawyer and Manning, 1971; Rudinger, 1971). However, more evidence on this point would certainly be welcome; a note of warning is sounded by the finding that the [4-leucine] analogs of oxytocin and mesotocin may act as antagonists of the natural hormones in the kidney and amphibian bladder (Chan *et al.*, 1968; Eggena *et al.*, 1970; Chiu and Sawyer, 1970; Chan and du Vigneaud, 1970).

In general, discussion of structure–activity relations in terms of receptor affinity will require more detailed pharmacological analysis than has hitherto been customary. It might be considered that so far such studies have given a poor yield of information laboriously achieved; however, at present such "probing" of the receptor site with properly designed analogs is the only available approach to this problem.

C. Generation of the Stimulus

The mechanism by which the hormone–receptor complex, once formed, generates the stimulus which initiates the response is a central problem of molecular endocrinology. According to one conceptual model of drug action, the quantum of action corresponds to the event of receptor binding, the receptors once occupied by the drug being inactive ("rate" receptor theory; see, e.g., Paton and Rang, 1966). On the other hand, the classical "receptor occupation" theory and its various developments regard the drug–receptor complex as active throughout the duration of its existence. Intermediate models are also possible (Gosselin, 1970). The following discussion will be based on the occupation model, although admittedly there are no compelling reasons for excluding a "rate" mechanism.

Antagonists which act as competitive inhibitors are in the nature of things the most informative about the phase of stimulus generation. Their properties imply that they have retained the structural elements required for binding to the receptor but lost a structural feature necessary for initiation of the stimulus.

As a general conclusion from structure–activity relations among neurohypophysial hormone analogs (see, e.g., Berde and Boissonnas, 1968; Rudinger, 1969; Pickering, 1970), it follows that no single group of oxytocin so far examined is *essential* for biological activity. In particular, the systematic work of du Vigneaud and his school (see du Vigneaud, 1965) has demonstrated that none of the chemically reactive substituents are critically required. Even when the disulfide bridge is omitted, some slight hormonelike activity is retained (Huguenin and Guttmann, 1965; Poláček *et al.*, 1970); and when the sulfur atoms are replaced by methylene groups rather than omitted, the resulting analogs are highly active (Rudinger and Jošt, 1964; Schwartz *et al.*, 1964; Yamanaka *et al.*, 1970; Jošt and Šorm, 1971a).

On the other hand, quite a few structural changes are known by now which lead to the appearance of antagonistic or "partial agonist" properties. In the assay on the isolated rat uterus, these include omission or substitution of the tyrosine hydroxyl group (Law and du Vigneaud, 1960; Rudinger and Krejčí, 1962, 1968; Chan and Kelley, 1967; Smyth, 1967a,b); omission of the carboxamide group in position 5 (Chan and Kelley, 1967); and omission or (formal) hydrolysis of the C-terminal carboxamide group (Dutta *et al.*, 1966; Chan and Kelley, 1967). Potent inhibitors have been found in [1-penicillamine]-oxytocin and its deamino derivative (Schulz and du Vigneaud, 1966; Chan *et al.*, 1967). Hormone analogs with phenylalanine in position 3 have "partial agonist" prop-

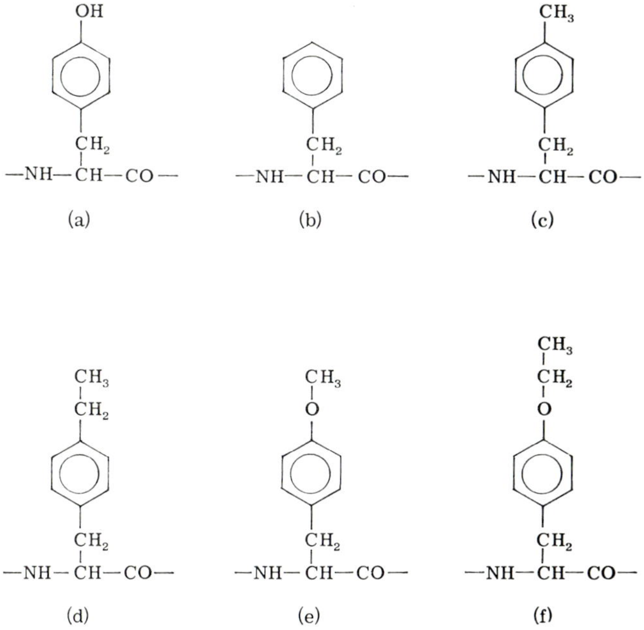

FIG. 4. Some replacements in sequence position 2 of oxytocin: (a) tyrosine (oxytocin); (b) phenylalanine; (c) *p*-methylphenylalanine; (d) *p*-ethylphenylalanine; (e) *O*-methyltyrosine; (f) *O*-ethyltyrosine.

erties on the uterus (Walter *et al.*, 1968, 1969) as have the acyclic analogs, [1,6-dialanine]- and [1,6-diserine]-oxytocin (Poláček *et al.*, 1970). A more general review of inhibitors has been given elsewhere (Rudinger and Krejčí, 1968).

The most thoroughly studied group of inhibitors and "partial agonists" are compounds in which the hydroxyl group of the tyrosine has been replaced as shown in Fig. 4. Several other analogs in which the L-tyrosine in position 2 has been substituted by leucine (Jošt *et al.*, 1963a; Rudinger and Krejčí, 1962), D-tyrosine (Drabarek and du Vigneaud, 1965), *p*-amino-phenylalanine (Rudinger, 1965; Rudinger and Krejčí, 1968; Rudinger *et al.*, 1968b), or a tyrosyl-tyrosine sequence (Guttmann *et al.*, 1957) have also shown antagonistic or transitional properties, as have the *O*-methyl and *O*-ethyl derivatives of lysine vasopressin (Vogel and Hergott, 1963; Zaoral *et al.*, 1965; Krejčí *et al.*, 1967a, 1971). However, these compounds have been investigated in less detail.

The properties of the analogs in various biological systems are summarized in Figs. 5 and 6; more quantitative results from work with

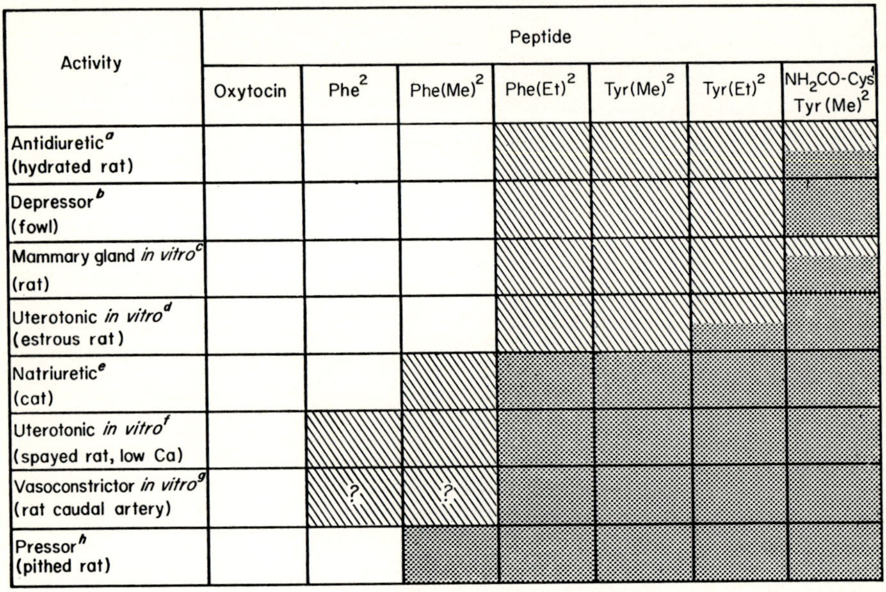

Activity	Peptide						
	Oxytocin	Phe^2	Phe(Me)^2	Phe(Et)^2	Tyr(Me)^2	Tyr(Et)^2	NH$_2$CO-Cys^1 Tyr(Me)^2
Antidiuretic[a] (hydrated rat)							
Depressor[b] (fowl)							
Mammary gland *in vitro*[c] (rat)							
Uterotonic *in vitro*[d] (estrous rat)							
Natriuretic[e] (cat)							
Uterotonic *in vitro*[f] (spayed rat, low Ca)							
Vasoconstrictor *in vitro*[g] (rat caudal artery)		?	?				
Pressor[h] (pithed rat)							

FIG. 5. Responses of various biological preparations to oxytocin analogs modified in sequence position 2. White fields mark oxytocin-like action; black inhibition; gray, intermediate properties, such as decreased maximal response ("partial agonism"), marked tachyphylaxis, dual response, or variable response (agonism or antagonism in different experiments). Superscript letters in column 1 indicate references: [a] Jošt *et al.* (1963a), Zhuze *et al.* (1964), Pliška (1971); [b] Krejčí *et al.* (1972), Bisset *et al.* (1970); [c] Poláček *et al.* (1967), Bisset *et al.* (1970); [d] Krejčí *et al.* (1967b, 1972), Bisset *et al.* (1970); [e] Cort *et al.* (1966); [f] Krejčí *et al.* 1972); [g] Krejčí *et al.* (1971), Bisset *et al.* (1970); [h] Krejčí *et al.* (1967a, 1971, 1972), Bisset *et al.* (1970).

the isolated rat uterus are given in Fig. 7. It will be seen that the tendency to antagonism or related anomalous behavior (decreased maximal response, pronounced tachyphylaxis) varies in two dimensions. In the "biological" dimension, there is a graded tendency of various target tissues, or of the same tissue under different experimental conditions, to reveal the anomalous or inhibitor properties of a given analog. In the other, "structural" dimension it is found that mere omission of the hydroxyl group in [2-phenylalanine]-oxytocin already leads to the appearance of anomalous properties, including full antagonism under suitable conditions[2] (see Fig. 5). However, the inhibitor properties become more pronounced as the size of the *para*-substituent in the aromatic ring is increased (Zhuze *et al.*, 1964; Rudinger, 1965; Cort *et al.*, 1966; Rudinger and Krejčí, 1968). On the other hand, the pD_2 values measured under conditions under which there is no receptor reserve for the analogs are rather similar, as are the pA_2 values in experiments in which the same analogs act as inhibitors (Fig. 7).

Activity	Peptide			
	Tyr^2 (Oxytocin)	$Tyr(Me)^2$	$Tyr(Et)^2$	$NH_2CO\text{-}Cys^1$ $Tyr(Me)^2$
Galactobolic[a] (rat)				
Uterotonic *in situ*[b] (rat)		?	?	
Natriferic[c] (frog skin)				
Hydroosmotic[d] (amphibian bladder)				
Lipogenesis[e] (rat fat cells)				

Fig. 6. Response of further biological preparations to some oxytocin analogs modified in position 2. Significance of shading as in Fig. 5. Superscript letters in column 1 indicate references: [a] Bisset (1964), Bisset *et al.* (1970); [b] Krejčí *et al.* (1967b), Bisset *et al.* (1970); [c] Jard *et al.* (1970); [d] Eggena *et al.* (1970), Jard *et al.* (1970); [e] Braun *et al.* (1969).

The tendency toward inhibition can be further accentuated by substituting the α-amino group *in addition to* modifying the side chain in position 2. Thus introduction of a carbamoyl group into methyloxytocin (Bisset and Smyth, 1965; Chimiak *et al.*, 1968) affords an analog showing some inhibitor properties in all the biological preparations so far examined (see Figs. 5 and 6). The pA_2 value for this inhibitor in the isolated rat uterus under comparable conditions (7.1) is, however, probably somewhat lower than that for the analogs modified in position

---

[2] Smyth (1970), arguing somewhat obliquely from results obtained with the isolated rat uterus by a different experimental approach, concluded that [2-phenylalanine]- and [2-*p*-methylphenylalanine]-oxytocin have the same capacity to initiate a stimulus (i.e., efficacy) as oxytocin itself, and assigns to the hydroxyl group a role mainly in receptor binding. Our results clearly show this conclusion to be untenable.

2 alone. From studies on related compounds (Smyth 1967a,b, 1970; Jošt and Šorm, 1971b), it appears that the structural modification in position 2 is the primary cause of reduced efficacy, with N-substitution accentuating this effect (see Bisset *et al.*, 1970).

What do these findings tell us about stimulus generation at the receptor? First of all, they suggest that the molecular mechanism of this

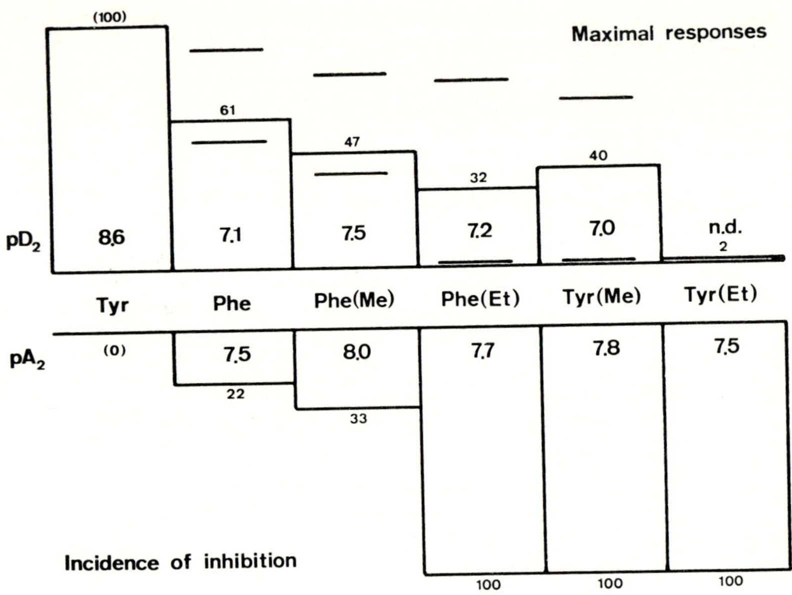

FIG. 7. Effects of oxytocin analogs modified in sequence position 2 on the isolated rat uterus under two sets of experimental conditions. Top: Holton's medium with 0.3 m*M* Ca and no Mg, cumulative dose procedure, isotonic recording. Blocks and small figures show mean maximal responses as percentages of the maximal response to oxytocin; horizontal lines, highest and lowest maximal response recorded in 10–12 experiments; mean values of pD_2 are given. Bottom: Uteri from spayed rats washed free of Ca and restored to van Dyke-Hastings medium with 0.2 m*M* Ca and no Mg. Blocks and small figures show incidence of inhibition (percent of experiments) in 9–20 experiments with each analog; mean values of pA_2 are given. From Krejčí *et al.* (1972).

process is similar, at least in that feature in which the tyrosine is concerned, in all the diverse tissues on which the hormone acts. The differences in the *degree* to which the decreased efficacy is apparent are presumably due to differences in the extent of the receptor reserve (see Sections II, A and III).

Second, they indicate that loss of efficacy is due to interference with some function of the tyrosine hydroxyl group itself rather than to an indirect effect of conformational perturbation which results in the "misalignment" of some other, functionally important group [for a recent discussion of this general problem with reference to a specific conformational model of oxytocin, see Walter *et al.* (1971) and Walter (this volume, p. 167)]. It is unlikely that substitution of the hydroxyl group both by smaller and by larger groups would have the same conformational effect, and that such a conformational perturbation would leave the binding properties largely unchanged, as we know them to be from a comparison of pD_2 and pA_2 values within this series of analogs. To account for the effect of substituent size on the efficacy, we have suggested (Rudinger, 1965; Cort *et al.*, 1966; Rudinger and Krejčí, 1968) that while some loss of efficacy is in the first instance associated with a functional role of the hydroxyl group, substituents replacing this hydroxyl group may in proportion to their size interfere with an event necessary for stimulus generation (but not for binding) which takes place in the vicinity of this site in the normal hormone–receptor complex.

Finally, the results may be relevant to a choice between what may be called a "participation" model of stimulus generation and an allosteric model. In gross terms, the first type of model assumes that the drug molecule itself, or some portion of it, participates in the molecular event of stimulus generation (e.g., by completing an enzymatically active site or a metal chelating site whose other groups preexist in the receptor, etc.). In the allosteric model, the binding of the drug induces a conformational change in the receptor which results in its "activation"; the molecule of the drug does not itself participate directly in the molecular event which generates the stimulus (see, e.g., Changeux *et al.*, 1970). Now in a "participation" model, inhibitor properties would imply that the groups contributing to the biological functionality of the drug–receptor complex are absent or misaligned. In terms of the allosteric model, the inhibitor would bind yet fail to induce the conformational change required for receptor activation, or "lock" the receptor in an inactive form; the efficacy of an analog would be given by the ratio of its affinities for the "active" and "inactive" forms of the receptor. We suggest that in the case under discussion the loss of efficacy when the hydroxyl group is replaced both by a smaller and by larger substituents, and in particular the relative constancy of the pD_2 and pA_2 values is in better accord with a "participation" model than with an allosteric model, although of course the evidence is not compelling.

III. Stimulus–Response Coupling

At first sight it would seem (Fig. 1) that the participation of the hormone in the scheme of regulation ends at the stage of receptor interaction. Certainly beyond the receptor the nature of the signal is the same no matter which analog generated the stimulus; but the influence of stimulus–response coupling on the character of the response to some analogs, as discussed in Section II, A, does in fact link the phenomenology of dose–response relations to processes in the "black box" of stimulus–response coupling.

In the case of methyloxytocin and related analogs, some of the factors that make for the appearance of "anomalous" or inhibitor properties in experiments with the rat uterus *in vitro* are a low calcium concentration in the medium; the absence of magnesium; low temperature; and the use of uteri from spayed rats (Krejčí *et al.*, 1964, 1966, 1967b, 1972; Rudinger *et al.*, 1968b). These conditions, then, decrease or eliminate the receptor reserve and increase the threshold value of the stimulus required for response in one of the ways discussed in Section II, A. The effect of the hormonal status of the uterus is probably most plausibly interpreted in terms of the actual number of receptors: Estrogen treatment might well increase, withdrawal from steroid hormone control (ovariectomy) decrease, the actual number of receptors and thereby alter the receptor reserve [see Braun and Hechter (1970) for a somewhat analogous case involving corticotropin, fat cells, and glucocorticoids]. On the other hand, calcium ions are certainly involved in the actual stimulus–response coupling in smooth muscle, and temperature has been shown to affect the compartmentation and mobility of calcium in the uterus (Poláček and Pliška, 1969); it is reasonable to assume that temperature may also affect other biochemical processes involved in coupling.

The role of magnesium appears to be complex. On the one hand, the structure-related potentiation by magnesium (see Fig. 3) of the analogs modified in sequence position 3 (which are thought to have the same efficacy as oxytocin; see Section II, B) has been interpreted as an effect an affinity, and as indicating participation of the ion in peptide–receptor binding (Krejčí and Poláček, 1968; Rudinger, 1968; see also Bentley, 1965; Chan and Kelley, 1967; Schild, 1969). On the other hand, it has been shown (Krejčí *et al.*, 1964, 1967b, 1972; Walter *et al.*, 1968, 1969; Somlyo and Somlyo, 1970; Yamanaka *et al.*, 1970) that the addition of magnesium may increase the maximal response to analogs with "partial agonist" properties, and even dramatically change the antagonist properties shown by deamino-[2-*O*-methyltyrosine]-oxytocin under standard assay conditions to oxytocin-like uterotonic activity (Krejčí *et al.*, 1969, 1972). Somlyo and Somlyo (1970) have offered an explanation of magnesium potentiation based on the assumption that magnesium makes available additional hormone receptors in the tissue. Alternatively, the metal might increase the true efficacy of the peptide–receptor complexes

(Schild, 1969; Somlyo and Somlyo, 1970). Finally, it might also play a role in processes following stimulus generation. At present it seems likely to us that magnesium acts both at the binding stage and in stimulus–response coupling.

As an example of the use of a hormone analog as a "probe" in examining stimulus–response coupling, we may consider an experiment (Krejčí *et al.*, 1966) in which a rat uterus was washed with calcium-free solution until it no longer responded to oxytocin and then was exposed to stepwise increasing calcium concentrations. The response to methyloxytocin was examined for each calcium concentration. It was found that methyloxytocin still acted as an antagonist rather than an agonist when the calcium concentration in the medium had reached 0.8 mM, a concentration at which under standard experimental conditions the analog invariably had a uterotonic action. Only when the calcium-deprived uterus was placed in 1.2 mM calcium did a contractile response to methyloxytocin appear. It was concluded (Krejčí *et al.*, 1966) that the calcium involved in stimulus–response coupling was in a special compartment in relatively slow equilibrium with the interstitial calcium (see, e.g., Daniel, 1965). The transition of methyloxytocin from antagonism to agonism here signals the achievement of a critical calcium concentration in this particular compartment.

For the action of the neurohypophysial hormones on water transport in the toad bladder and kidney, the biochemical situation in stimulus–response coupling is somewhat more transparent since in these organs the adenyl cyclase–3′,5′-AMP system is known to mediate the response. The work of Walter, Schwartz, and Hechter and their colleagues (Eggena *et al.*, 1970; Bär *et al.*, 1970) suggests that in these systems 3′,5′-AMP may be produced in excess of the amount required to give the maximal permeability response; that part of the stimulus which produces the "excess" 3′,5′-AMP might then represent the "receptor reserve capacity" (Bär *et al.*, 1970; Schwartz, this volume, p. 166).

Most of the results discussed here were obtained in the course of the development of the receptor model described in Section II, A, rather than from experiments already based on this model. Further work will show to what extent experiments deliberately designed to this end and using suitable analogs can contribute to our knowledge of this phase of hormone action.

IV. Transport in the Circulation

Hardly anything has been done to study the behavior of oxytocin analogs in the circulation. Yet the finding (Smith and Ginsburg, 1961) that [3-valine]-oxytocin has a half-life in the circulation some three times longer than oxytocin, whereas the half-life of the [3-phenyl-alanine] analog is similar to that of the hormone itself, suggests intriguing possibilities. The binding of the hormones to the serum proteins is apparently small in extent, or weak, or both (see Lauson, 1970); however, most of the information on this point has been obtained under nonequilibrium conditions so that no quantitative estimates of binding constants or binding capacities are available. Pharmacokinetic considerations suggest that differences in binding constants could make an appreciable difference to the rate of disappearance of the peptide from the circulation even if the binding is weak, provided the binding capacity is sufficiently high. Investigation of representative hormone analogs both by binding studies *in vitro* and by pharmacokinetic measurements would certainly be of interest.

There is one pathological condition in which hormone binding in the blood is presumably of critical importance: the antibody-induced lack of response to endogenous or exogenous vasopressin. It has been shown (Miller and Moses, 1969) that rabbits immunized with a vasopressin conjugate develop mild to moderate diabetes insipidus concurrently with the appearance of antibodies in the circulation. The serum of a patient with diabetes insipidus who had developed resistance to vasopressin was shown to bind radioiodinated vasopressin strongly (Roth *et al.*, 1966). In a similar case, injection of the patient's serum or of its γ-globulin fraction into assay rats was found to reduce or abolish their antidiuretic response to vasopressin (Bisset *et al.*, 1971). In all these cases it may be assumed that the circulating antibodies bind the endogenous or injected vasopressin and prevent it from reaching the target tissue.

It is evident that the specificity of such antibodies may be examined either by conventional immunochemical procedures (Glick *et al.*, 1968; Miller and Moses, 1969; Chard *et al.*, 1970) or with "passively immunized" assay animals as used by Bisset *et al.* (1971). Analogs which do not cross-react with the antivasopressin antibodies but have sufficiently high antidiuretic potency should be able to overcome the immunological block to the antidiuretic response. Clinical experiments have in fact shown that deamino-[8-D-arginine]-vasopressin (Zaoral *et al.*, 1967) will evoke an excellent antidiuretic response in patients with hypothalamic diabetes insipidus even when these patients have become resistant to lysine and arginine vasopressins (Vávra *et al.*, 1968), presumably by developing antibodies.

V. Inactivation

The typical responses to the neurohypophysial hormones set in rapidly but, unless hormone levels are maintained by continuous release of endogenous or infusion of exogenous peptide, they are of short duration. This time course seems to be a feature of their physiological role and, taken in conjunction with the ubiquity of proteolytic enzymes, suggests that the phases of excretion and inactivation will play an important role in the overall scheme of hormone action. For this reason the enzymatic inactivation of oxytocin and the vasopressins attracted early attention both in biochemical and in synthetic work. The extensive work on the metabolism of the hormones has been summarized in excellent reviews (Lauson, 1967, 1970; Tuppy, 1968; Ginsburg, 1968), and we shall be concerned with it only insofar as it involves synthetic analogs.

Two approaches can be distinguished in the work with synthetic oxytocin analogs on hormone inactivation. In the first, conclusions about possible mechanisms of inactivation are drawn from the biological properties of analogs (which may or may not have been designed with this aim in mind). In the second approach, the specificity of inactivation by tissues, slices, homogenates, or extracts is examined using a range of synthetic analogs essentially as substrates; bioassay here merely serves as a convenient way of measuring rates of destruction, and it may be supplemented by chemical analysis of the degradation products. Obviously the two approaches are complementary and should be most fruitful when used in conjunction.

Even a careful evaluation of conventional pharmacological assay results can sometimes give pointers to the intervention of metabolic factors. In particular, protracted action and gross differences in potencies measured on the same target tissue *in vitro* and *in vivo* may be such pointers.

Protracted action may sometimes be seen merely by an inspection of time–response curves. However, even when there are no obvious differences in the time course of the response, differences in the rate of disposition may be revealed by the determination of formal elimination constants [the exponential rate constants for the decrease in concentration in the receptor compartment, as measured by the decay of the response; see Pliška (1966)]. Such elimination constants may be determined either by following the decay of the response from a steady state reached by infusion (Pliška, 1969) or from a comparison of log-dose–response curves constructed with the total (time-integrated) effects as a measure of the response (Pliška, 1966; Beránková-Ksandrová *et al.*, 1966). A steeper slope of such a log-dose–response curve signifies a slower elimination. Conventionally, antidiuretic responses are assayed in terms of the total effect, and it is for this reason that differences in the slope of log-dose–response curves presumably associated with differences in metabolism are most frequently noticed in this assay [classically, the difference between arginine and lysine vasopressins; see, e.g., Berde and Cerletti (1961)]. By a slight modification of the rat pressor assay—definition of the pressor response by the total area under the blood pressure curve rather than the highest blood pressure reached, as in the conventional assay—similar information may be obtained for the pressor activity. Pliška and Krejčí (1966) have shown that [4-asparagine, 8-lysine]-vasopressin (Zaoral, 1965), which has a steeper log-dose-response curve than lysine vasopressin in the antidiuretic assay, exhibits the same "anomaly" also in its pressor activity provided the response is expressed in the same way.

Deamino-[8-D-arginine]-vasopressin, which from its structure might well be resistant to inactivating enzymes of aminopeptidase and trypsin-like specificity, similarly shows a markedly steeper dose dependence of the antidiuretic effect than arginine vasopressin (Zaoral *et al.*, 1967) and has been reported to have protracted antidiuretic action at higher doses in rats, dogs, and human patients with diabetes insipidus (Vávra *et al.*, 1968).

In neither of these cases have the metabolic factors which presumably underlie the prolonged action been studied by more direct pharmacological or biochemical methods.

The second approach suggested above—comparison of activities *in vitro* and *in vivo*—is based on the following reasoning: If an organ other than the target organ is responsible for a large share of the total inactivation, and if an analog should be resistant to inactivation in that organ, a larger proportion of the analog than of the hormone would reach the target tissue, and a lower dose of the analog would therefore be required to match the response to a given dose of hormone *in vivo* than *in vitro;* the potency *in vivo* would thus be appreciably higher than that determined *in vitro*. Unfortunately, the conditions conventionally used for the assay of oxytocin and its analogs *in vitro*, particularly the ionic composition of the medium, differ considerably from conditions *in vivo* and since magnesium and calcium are known to potentiate different analogs to a different extent, a straightforward interpretation of the potency ratios obtained by standard assays *in vitro* and *in vivo* is hardly possible. However, a striking example of potency differences *in vitro* and *in vivo* due to metabolic factors is provided by some analogues of bradykinin (see Rudinger, this volume p. 171), and in the special case of the "hormonogen" analogs of oxytocin and vasopressin (see Section VI) similar reasoning (applied to activation rather than inactivation) is probably valid.

Quite early in the work with synthetic oxytocin analogs, deliberate attempts were made to design enzyme-resistant molecules (Jošt *et al.*, 1961, 1963b). By that time the pregnancy serum "oxytocinase" had been defined as an aminopeptidase (Tuppy and Nesvadba, 1957), and it also appeared that the most conspicuous inactivating system found in liver cell sap involved aminopeptidase action (Rychlík, 1964). Analogs were therefore prepared which were expected to be resistant to aminopeptidases, by replacing the amino-terminal L-hemicystine of oxytocin with the D-isomer or with N-methylhemicystine (Jošt *et al.*, 1961, 1963b; Hope *et al.*, 1963), or the tyrosine with N-methyltyrosine (Jošt *et al.*, 1961; Huguenin and Boissonnas, 1961). All these analogs had very low activity and, moreover, showed no signs of protracted effects when tested *in vivo* (Rychlík, 1964; Beránková-Ksandrová *et al.*, 1966). In deaminooxytocin (Fig. 8a) Hope *et al.* (1962) prepared a highly potent analog

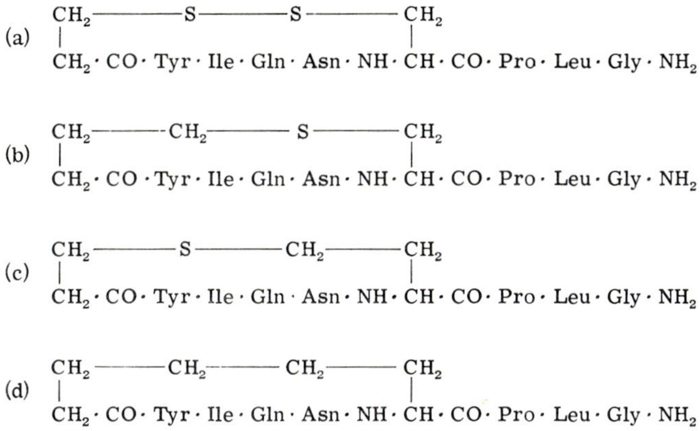

(a)
$$CH_2\text{———}S\text{———}S\text{———}CH_2$$
$$CH_2 \cdot CO \cdot Tyr \cdot Ile \cdot Gln \cdot Asn \cdot NH \cdot CH \cdot CO \cdot Pro \cdot Leu \cdot Gly \cdot NH_2$$

(b)
$$CH_2\text{———}CH_2\text{———}S\text{———}CH_2$$
$$CH_2 \cdot CO \cdot Tyr \cdot Ile \cdot Gln \cdot Asn \cdot NH \cdot CH \cdot CO \cdot Pro \cdot Leu \cdot Gly \cdot NH_2$$

(c)
$$CH_2\text{———}S\text{———}CH_2\text{———}CH_2$$
$$CH_2 \cdot CO \cdot Tyr \cdot Ile \cdot Gln \cdot Asn \cdot NH \cdot CH \cdot CO \cdot Pro \cdot Leu \cdot Gly \cdot NH_2$$

(d)
$$CH_2\text{———}CH_2\text{———}CH_2\text{———}CH_2$$
$$CH_2 \cdot CO \cdot Tyr \cdot Ile \cdot Gln \cdot Asn \cdot NH \cdot CH \cdot CO \cdot Pro \cdot Leu \cdot Gly \cdot NH_2$$

FIG. 8. Analogs of oxytocin with the amino group omitted and the disulfide bridge replaced: (a) deamino-oxytocin; (b) deamino-1-carba-oxytocin; (c) deamino-6-carba-oxytocin; (d) deaminodicarba-oxytocin.

which must also be resistant to aminopeptidases; yet again, the avian depressor, rat pressor, or uterotonic activity *in vivo* showed no evidence for protracted action (Chan and du Vigneaud, 1962). Even deamino-1-carba-oxytocin (Fig. 8b) and deaminodicarba-oxytocin (Fig. 8d) (Jošt and Rudinger, 1967; Yamanaka *et al.*, 1970; Jošt and Šorm, 1971a), analogs which are resistant not only to aminopeptidase attack, but also to reductive cleavage of the ring, showed the same time course for the antidiuretic response as oxytocin itself (Pliška *et al.*, 1968).

Two possible explanations for this situation come to mind: Either aminopeptidase degradation is not, in fact, a physiologically important mode of inactivation, or for pharmacokinetic reasons decreased inactivation does not necessarily result in a prolonged effect (Rudinger *et al.*, 1968a; Pliška, 1968).

Pharmacokinetic computations were based on the three-compartment model shown in Fig. 9 (Pliška, 1968). Some parameters of the system were obtained from experimental results (rate of disappearance of the hormone from the circulation, excretion in urine); others were computed. The output was plotted as the time course of the response (obtained from the calculated time course of the concentration in the receptor compartment); the computed curves could be made to mimic the experimental time course, e.g., of the antidiuretic response, quite closely.

In this model, variation in the rate constants for inactivation in the circulation or in the "residual" compartment turned out to have very

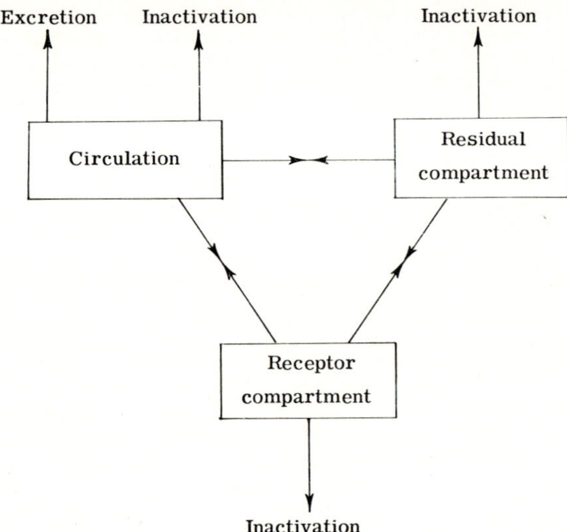

FIG. 9.   Scheme of three-compartment model used in pharmacokinetic calculations (Pliška, 1968).

little effect on the time course or *duration* of the response; rather, a decrease in the rate of inactivation particularly in the "residual" compartment increased the maximal *intensity* of the response to a given dose. On the other hand, even slight changes in the rate coefficient for inactivation in the receptor compartment markedly changed the time course of the response.

These results establish at least the theoretical possibility that analogs resistant to even physiologically significant inactivating systems located outside the receptor compartment may not show protracted effects but may show relatively high potency. It therefore becomes important to identify not only the inactivating systems responsible for physiological disposition in the whole organism (kidney, liver, lactating mammary gland; see, e.g., Ginsburg, 1968), but also the inactivating enzymes in the target tissues which, although they may dispose of only a small fraction of the circulating peptide, may determine the duration of the effect.

The neurohypophysial hormones are stable in normal plasma but inactivated by the "oxytocinase" of primate pregnancy serum. Deamino-oxytocin (Fig. 8a) has, as expected, proved resistant to this enzyme (Golubow *et al.*, 1963); some inactivation of the analog by human pregnancy plasma has been observed and ascribed to reversible reduction of the disulfide bond (Branda *et al.*, 1968). Saameli (1964) has shown that the potency of deamino-oxytocin relative to oxytocin is the same for the cat uterus *in situ* and for the human uterus near term. This finding argues against an important role of pregnancy serum oxytocinase in the physiological disposition, or regulation, of oxytocin since this enzyme, being absent in the cat but present in the human, would be expected to inactivate oxytocin but not deamino-oxytocin only in the human, and thereby increase the relative potency of deamino-oxytocin in the human. Pharmacokinetic experiments with a purified enzyme preparation in rats have led to a similar conclusion (Pliška *et al.*, 1967).

The rate of inactivation of oxytocin by hog kidney homogenates was found to be about 2.5 times greater than the rate of inactivation of deamino-oxytocin (Fig. 8a) or deamino-1-carba-oxytocin (Fig. 8b). The particulate (microsomal) fraction of the homogenate inactivated oxytocin 20–40 times more rapidly than the analogs (Barth *et al.*, 1969). Similar results were obtained with homogenates of rat kidney (Suska-Brzezińska *et al.*, 1972) and rat liver (see Fig. 10; Pliška *et al.*, 1972b). The particulate fractions of the homogenates did not inactivate the deaminocarba or deaminodicarba analogs appreciably whereas the cytoplasmic fraction destroyed all the peptides (Table I). Koida *et al.* (1971) have partially purified an enzyme from the soluble fraction of rat kidney homogenates, which was also found to inactivate the deaminocarba analogs as well as oxytocin. From examination of a range of synthetic analogs and analysis of the cleavage products, they concluded that the enzyme has a specificity resembling that of chymotrypsin and attacks oxytocin and the analogs at the leucyl-glycine bond.

Both oxytocin and deamino-oxytocin are inactivated by homogenates of human placenta (Branda and Ferrier, 1971) and by homogenates of pregnant and nonpregnant rat uterus (Chan and Wahrenburg, 1968). In the latter case, it has been concluded that different enzyme systems are responsible. In homogenates of nonpregnant rat uterus, the particulate fraction destroys oxytocin but not a deaminocarba analog whereas the soluble fraction inactivates both (Suska-Brzezińska *et al.*, 1972)

(Table I). Again, a substantially purified enzyme which cleaves both oxytocin and deamino-oxytocin has been prepared from the same source, and its specificity has been characterized by means of hormone analogs and sequence fragments (Glass *et al.*, 1970).

The relevance of all these results obtained with tissue homogenates and their fractions to the physiological disposition mechanism unfortunately remains unclear since we do not know which of the enzymes

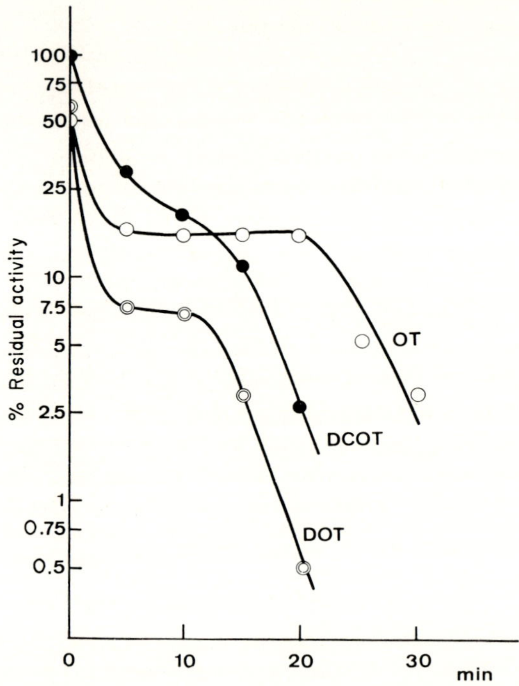

FIG. 10. Rate of disappearance of oxytocin (OT), deamino-oxytocin (DOT), and deamino-1-carba-oxytocin (DCOT) on incubation with a rat liver homogenate containing 1.3 mg/ml protein. Peptide concentrations 0.5 μM, pH 7.5, 37°C; ordinate: % residual activity, log scale. From Pliška *et al.* (1972b).

in fact have access to the hormone in the intact tissue. Comparisons of suitable analogs in clearance experiments with perfused intact organs would be most welcome at this stage—and are in fact long overdue.

As a possible alternative approach to this problem, we are examining the simple and effective "oil-bath" technique of Kalsner and Nickerson (1968). In experiments with this technique, a strip of smooth muscle is contracted by exposure of the agent to be examined and the medium in the organ bath is then replaced with mineral oil. Under suitable conditions, the rate of relaxation of the muscle strip then indicates the rate

TABLE I

Relative Rates of Inactivation of Deaminocarba Analogs of Oxytocin by Particulate and Soluble Fractions of Tissue Homogenates[a,b]

	Uterus		Kidney		Liver	
Analog	Particulate	Soluble	Particulate	Soluble	Particulate	Soluble
Deamino-1-carba-	—	—	0	0.20	0	0.35
Deamino-6-carba-	0	0.67	0	0.11	0	0.22
Dicarba-	—	—	0	0.38	0	0.30

[a] Data from Suska-Brzezinska *et al.* (1972) and Pliška *et al.* (1972b).
[b] Rate constants of inactivation relative to oxytocin = 1.

of (physiologically significant) disposition of the active agent. We have shown (Rudinger *et al.*, 1972) that the depolarized rat uterus can be used in such experiments and have examined the rates of relaxation after contraction by oxytocin, deamino-oxytocin, and deamino-1-carba-oxytocin. An example of such an experiment is recorded in Fig. 11, and

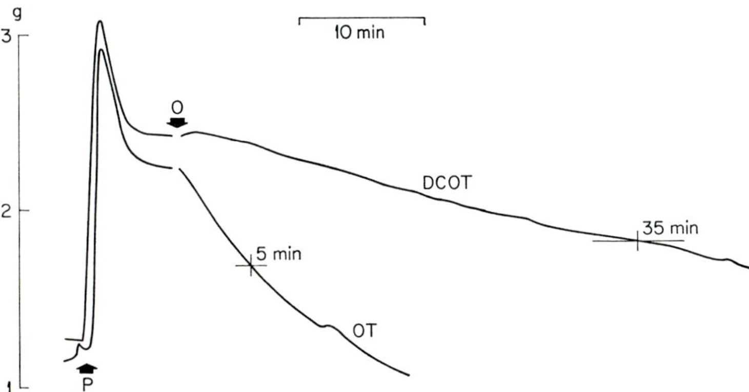

FIG. 11. Relaxation of a depolarized rat uterine strip exposed at *P* to oxytocin (OT) or deamino-1-carba-oxytocin (DCOT), after replacement of the aqueous medium with mineral oil at *O*. Aqueous medium: Ca-containing K_2SO_4–Ringer (Schild, 1969) with 0.5 mM $MgCl_2$. Tension recorded isometrically. Points of half-relaxation from plateau tension are marked with crosses and the time required.

a logarithmic plot of the decrease in effective concentration of the peptides in the receptor compartment with time (calculated from curves such as those in Fig. 11) is shown in Fig. 12. The preliminary results show that the disposition of both deamino-oxytocin and the deaminocarba analog is about 5–7 times slower than that of oxytocin.

This finding again raises the question whether, or to what extent, the remarkably high potency of deamino-oxytocin even *in vitro* is due to metabolic factors rather than to increased affinity for the receptors. The concentration of a peptide in the receptor compartment (or the "biophase") will be a steady-state value determined, among other things, by the rate of accession from the bulk solution, which would be expected to be similar for oxytocin and the deamino analog, and the rate of inactivation, which is now shown to be different. For a given concentration in the organ bath the analog therefore may reach a higher concentration at the tissue receptors than oxytocin, and part or all of the

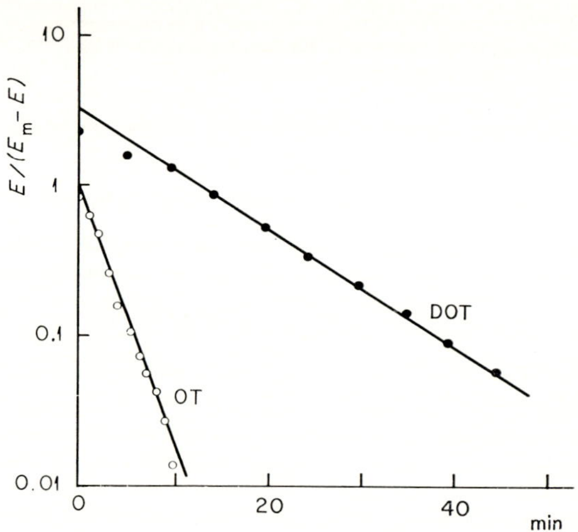

Fig. 12. Rate of decrease in peptide concentration in the receptor compartment after exposure of a depolarized rat uterus strip to oxytocin (OT) and deamino-oxytocin (DOT) and replacement of the aqueous medium by oil. Experimental conditions as for Fig. 11. The concentration in the receptor compartment is taken as proportional to $E/(E_m - E)$ (ordinate, logarithmic scale) where E is the response at a given time, E_m the maximal response (as plateau tension) of the preparation to oxytocin.

higher potency may be due to this effect. Experiments and calculations designed to test this point are under way.

The pharmacokinetic considerations set out on page 150 suggest that a changed rate of inactivation in the receptor compartment should lead to protracted action. It is therefore of interest to note that with both deamino-oxytocin (Chan and Kelley, 1967) and with deaminodicarba-oxytocin (Yamanaka *et al.*, 1970), there is a tendency for the contractions of the isolated uterus to persist during wash-out after exposure to relatively high doses of the peptides. There are also indications of protracted uterotonic effects *in vivo* (Krejčí, 1971).

Our results indicate that an aminopeptidase does, indeed, participate in the disposition of oxytocin in the uterus. However, there must obviously be at least one other mechanism, responsible for the disposition of the deamino analogs. One candidate for this role is certainly the enzyme characterized in uterus homogenates by Glass *et al.* (1970). However, another possibility would be an attack at the tyrosine side chain: Deamino-[2-*O*-methyltyrosine]-oxytocin (Kasafírek *et al.*, 1969) has been found to have a prolonged effect on the uterus *in situ* (Bisset and Clark, 1971) and also *in vitro* (Krejčí *et al.*, 1969, 1972), observations suggesting that substitution of the hydroxyl group may block some additional mode of inactivation. The persistence of the activity even after repeated washing indicates, however, that storage or compartmentation processes may also be involved.

VI. Activation: Synthetic Hormonogens

The phase of "activation" allowed for in the scheme of Fig. 1 is confined in the case of the neurohypophysial hormones to biosynthesis within the neurons (see, e.g., Sachs, 1967); there is no evidence for an activation step subsequent to release. In order to make good this deficiency of

TABLE II

Some Properties of Hormonogen Derivatives of Oxytocin[a]

X-Cys-Tyr-Ile-Gln-Asn-Cys-Pro-Leu-Gly-NH_2

X	Uterotonic activity			Antidiuretic activity	
	In vivo[b]	*In vitro*[b]	Ratio	Potency[b]	IP[c]
Leu	18	5	3.5	9.0	2.9
Phe	10	0.6	17	6.8	2.6
Gly	0.8	0.07	11	4.4	3.2
Gly-Gly	1.5	0.16	9	0.83	4.0
Pro	1.5	0.13	11	0.17	4.3
Leu-Gly-Gly	4.0	0.05	80	0.13	5.0
Sar	0.34	0.46	0.8	0.83	1.0

[a] From Beránková-Ksandrová *et al.* (1966).
[b] Relative to oxytocin, percent.
[c] Index of persistence; for definition see text.

Nature we have introduced a purely man-made activation phase by constructing analogs of oxytocin and vasopressin capable of behaving as "hormonogens" and releasing the free hormones by enzyme action *in vivo*. The essential feature of the hormonogens (see Tables II and

III) is attachment of an additional amino acid or short peptide chain to the terminal amino group of the hormone (Jošt *et al.*, 1961, 1963b; Kasafírek *et al.*, 1965, 1966; Zaoral and Šorm, 1965). The first analog of this structural type, N^α-glycyloxytocin, was prepared by du Vigneaud *et al.* (1960) and found by them to have protracted action and inhibitor properties.

An examination of the biological actions of the aminoacyl and peptidyl derivatives (Tables II and III) showed that they had all the expected

TABLE III

Some Properties of Hormonogen Derivatives of Lysine Vasopressin[a]

X-Cys-Tyr-Phe-Gln-Asn-Cys-Pro-Lys-Gly-NH$_2$

X	Antidiuretic activity		Pressor activity	
	Potency[b]	IP[c]	Potency[b]	IP[c]
Leu	19	1.1	15	1.0
Tyr	15	1.1	12	2.2
Gly-Pro	9.1	1.2	14	1.2
Lys	2.8	2.1	4.0	4.9
Phe	1.4	3.2	2.5	2.6
Ala	—	—	2.3	2.2
Gly	1.2	4.3	0.80	3.2
Gly-Gly-Gly	1.1	4.9	0.84	5.1
Gly-Gly	0.68	3.9	0.36	3.9
Pro	0.16	2.9	0.20	4.7
Sar-Gly	—	—	0.10	3.2
Trp	0.04	2.4	0.05	3.0
Sar	0.29	1.4	0[d]	—

[a] Kynčl *et al.* (1972).

[b] Relative to lysine vasopressin, percent.

[c] Index of persistence; for definition see text.

[d] No pressor activity; inhibits and prolongs the response to vasopressin.

features of hormonogens and some unexpected ones (Rychlík, 1964; Bisset, 1964; Beránková-Ksandrová *et al.*, 1964, 1966; Bisset *et al.*, 1966; Kasafírek *et al.*, 1966; Kynčl *et al.*, 1969, 1972; Kynčl and Rudinger, 1970). The uterotonic, antidiuretic, avian depressor, rat pressor, and milk ejection responses *in vivo* generally developed more slowly than the responses to oxytocin or the vasopressins, and also faded more slowly. The uterotonic potency *in vivo* of the oxytocin derivatives was invariably greater than the potency *in vitro*, indicating a share of tissues other than the target tissue in producing active peptide. However, some of the analogs also elicited gradually increasing responses when acting on the uterus or mammary-gland strip *in vitro,* suggesting that the target tissues were also capable of liberating the active hormone (Beránková-Ksandrová *et al.,* 1966). When the extended-chain peptides were incubated with tissue extracts, pregnancy serum, or aminopeptidase preparations, the hormonelike activity of the incubates initially increased before finally decreasing (Beránková-Ksandrová *et al.,* 1964; Pliška *et al.,* 1972a). Finally, it could be shown that after the administration of N^α-glycylglycylglycyl-[8-lysine]-vasopressin (triglycylvasopressin) to rats or cats an antidiuretic substance was excreted in the urine which corresponded in properties with vasopressin (Kynčl and Rudinger, 1970).

An unexpected property of some hormonogens was their ability to inhibit the response to challenge doses oxytocin in the avian depressor or milk-ejection assay during, and for some time after, the response to the analog, or when the analog was given in subthreshold doses. However, it turned out that this was a case of oxytocin tachyphylaxis rather than inhibition by the analog since very similar effects could be evoked by infusing oxytocin at suitable rates (Beránková-Ksandrová *et al.,* 1966) and so this property of the analogs provides evidence for, rather than against, the proposed mechanism of action.

The two-phase effect seen in some experiments was interpreted as a superposition of the response to the hormonogen itself in its quality as an oxytocin analog, upon a time course similar to that for oxytocin, and the slowly developing and fading response to the released oxytocin (Beránková-Ksandrová *et al.,* 1966; Pliška, 1968; Rudinger *et al.,* 1968a). On this assumption, such responses could also be mimicked (Fig. 13) by computation based on the three-compartment pharmacokinetic model (Fig. 9) (Pliška, 1968).

The structure dependence of the activities within this group provides some evidence about the metabolic processes involved. Derivatives which would not be expected to undergo enzymatic conversion to the hormone (sarcosyloxytocin, D-leucyloxytocin) had very low potency and, in the case of the sarcosyl derivative, no protracted action[3] (Beránková-Ksandrová *et al.,* 1966).

The "protractedness" of the response can be expressed quantitatively by the "index of persistence," defined as the ratio of the formal elimination constant (see page 147) for the reference compound (in this case oxytocin) and the analog (Pliška, 1966). For a series of oxytocin hor-

---

[3] The very low, protracted activity of the D-leucyloxytocin may be due to contamination with traces of the L-leucyl derivative (see Beránková-Ksandrová *et al.,* 1966).

monogens, the index of persistence proved to be inversely proportional to the logarithm of the potency, with a good degree of approximation. This was interpreted on the "hormonogen" hypothesis by assuming that a rapid conversion to hormone gave rise to a relatively high hormone concentration and therefore high potency, but led to rather rapid exhaustion of the hormonogen; slow conversion to the hormone conversely gave a more protracted effect but lower potency. The dependence of the rates of activation on the nature of the added amino acid or peptide chain was compatible with activation largely by an aminopeptidase with a specificity similar to that of leucine aminopeptidase (Beránková-Ksandrová *et al.*, 1966). The properties of the hormonogen derivatives of lysine vasopressin (Kasafírek *et al.*, 1966; Kynčl *et al.*, 1972) sum-

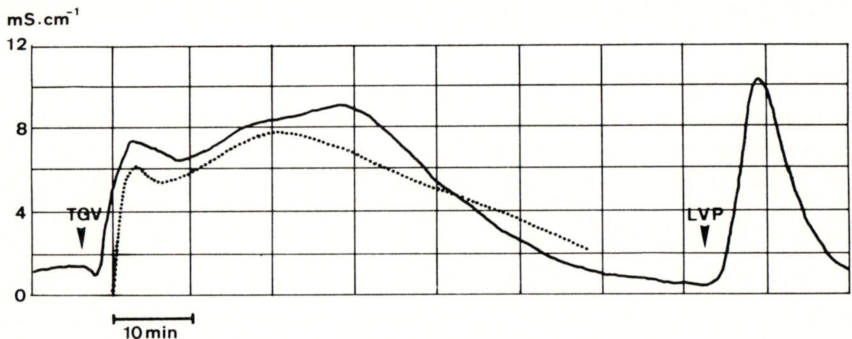

Fig. 13. Time course of the antidiuretic response (measured as urine conductivity in millisiemens/cm) to N^{α}-glycylglycylglycyl-[8-lysine]-vasopressin (TGV) and to lysine vasopressin (LVP). The dotted line shows a curve (response and time in machine units) computed for the three-compartment model of Fig. 9 assuming the inherent potency of the analog to be 2% of the potency of lysine vasopressin. After Pliška (1968).

marized in Fig. 14 show that these conclusions require some modification. The hormonogen properties of the sarcosylglycyl and glycylprolyl derivatives suggest that endopeptidases might also be operative: A sarcosyl peptide is unlikely to be attacked by the common aminopeptidases (as witness the properties of sarcosyloxytocin) and prolylvasopressin cannot be an intermediate in the activation of the glycylprolyl derivative since the latter has a less protracted action than the former. Furthermore, at the low-potency end of the series the index of persistence decreases again. This may be rationalized by the more complete metabolic scheme given in Fig. 15. Like the hormones themselves (see Section V), the hormonogens must be inactivated by enzymes other than aminopeptidases and also excreted in the urine, and these "unproductive" modes of elimination compete with the "productive" conversion to the hormone. If conversion is very slow, the rate of unproductive elimination becomes decisive for the time course of the response.

The deviation of triglycylvasopressin and lysylvasopressin from the trend shown by the other analogs cannot at present be explained. Differences in distribution might conceivably be responsible.

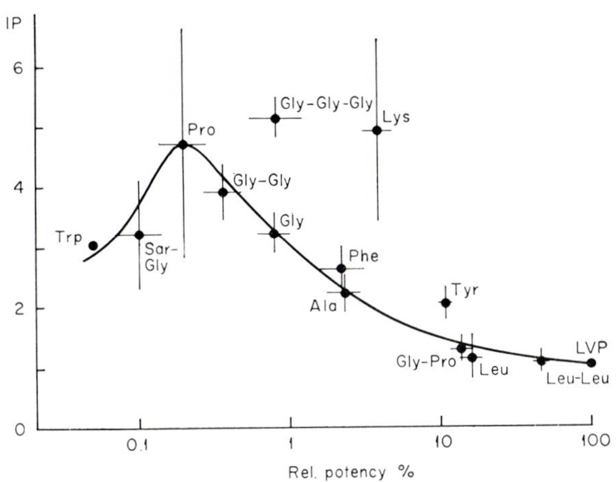

FIG. 14. Index of persistence (IP) plotted against the logarithm of the potency (% of lysine vasopressin) for the pressor response to some hormonogen derivatives of lysine vasopressin. The amino acid or peptide sequence attached to the terminal amino group of lysine vasopressin is indicated for each derivative. Bars show standard deviations. Results from Kynčl et al. (1972).

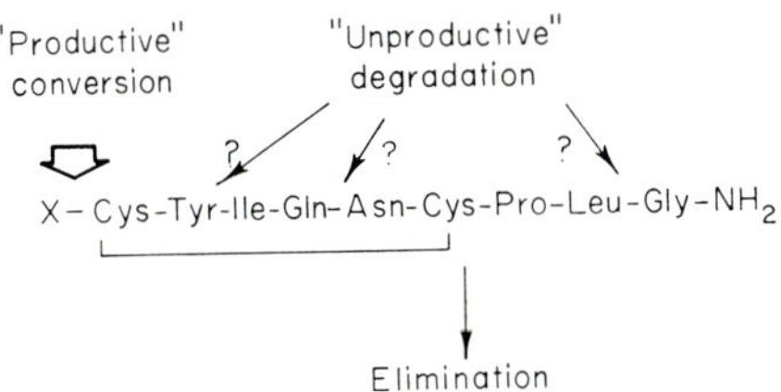

FIG. 15. Scheme of metabolic conversion for a hormonogen derivative of oxytocin.

A lower limit for the "productive" conversion of triglycylvasopressin can be calculated from the amount of lysine vasopressin excreted in the urine after an injection of the hormonogen, as determined by antidiuretic assay on the urine. In this way it has been found that at least 3.2–6.3% of the triglycyl derivative is converted to lysine vasopressin in the cat (Kynčl and Rudinger, 1970).

Although, as has been emphasized, such activation processes do not constitute a normal part of the scheme of hormone action for the neurohypophysial peptides, they may serve as a model for the generalized scheme—this in addition to the intrinsic interest attaching to the hormonogens as pharmacological tools and potential drugs.

VII. Storage

During storage in the neurosecretory granules of the neurohypophysis the hormones are present in a complex with specific proteins, the neurophysins. Deamino-oxytocin and deamino-[8-arginine]-vasopressin are not bound by the neurophysins, showing that the α-amino group is of critical importance for the peptide–protein interaction (Stouffer et al., 1963; Breslow and Abrash, 1966; Hollenberg and Hope, 1967; Ginsburg and Jayasena, 1968b). A more detailed examination with a series of analogs (Breslow and Abrash, 1966) indicated that the side chains in position 2 and 3 also contribute to binding. However, most of these results were obtained with a mixture of neurophysins whose individual components might well differ in their detailed binding specificity.

The physiological function of the neurophysins is still in doubt. If they play a role in storage of the hormones before release, their varying ability to bind synthetic analogs is unlikely to be of significance in vivo.[4]

However, Ginsburg and Jayasena (1968a,b) have found that extracts of kidney, uterus, and mammary gland cross-react with antibodies against neurophysin (see also Legros and Franchimont, 1971) and have purified a protein from kidney which binds vasopressin but does not bind deaminovasopressin. The characteristics of binding are similar as for the neurophysins.

Independent evidence for the existence of tissue proteins capable of binding the hormone and their analogs was obtained in inactivation experiments with homogenates (Section V). When oxytocin was incubated with uterus homogenates, there was a rapid initial decrease in activity followed by a second, slower phase of inactivation. The first, rapid process has been interpreted as binding to tissue constituents in the homogenate and is much less marked with deamino-oxytocin (Chan and Wahrenburg, 1968). In similar studies with oxytocin, deamino-oxy-

---

[4] Although labeled oxytocin and lysine vasopressin are taken up by the neurohypophysis from the circulation (Aroskar et al., 1964; Willumsen and Bie, 1969) or from the medium during incubation in vitro (Pliška et al., 1971; Edwards, 1971), it seems that at most very small amounts of the hormones enter the nerve endings intact with the possibility of being "re-stored" (Pliška et al, 1971; Edwards, 1971).

tocin, and deamino-1-carba-oxytocin and homogenates of rat kidney, uterus, and liver the same feature has been observed (Suska-Brzez ńska *et al.*, 1972; Pliška *et al.*, 1972b). The example in Fig. 10 shows that there is also a lag phase before the second, slower phase of disappearance which might possibly be due to a "buffering" effect of the protein-bound peptide (Pliška *et al.*, 1972b) ; again, this is less marked for the deamino analogs.

The presence and structural specificity of binding sites (other than the receptors) in target tissues of the hormones will require further study. In particular, it must be clarified whether such sites are accessible to the hormones or analogs *in vivo* and whether the binding characteristics are such that the binding might substantially affect the concentration of the peptide at the receptor or its time course, or the availability of the peptide for translocation or inactivation. Only if these requirements are met can a physiological function of storage be considered for such sites.

VIII. Other Phases

No conspicuous feedback mechanisms, other than those operated through the normal physiological responses, have been detected in work with the neurohypophysial hormones. The potentially important phase of distribution has not been examined with any of the hormone analogs; the properties of some analogs [e.g., the unexpectedly low potency of [4-glutamic acid]-oxytocin in contrast to other derivatives modified in sequence position 4 (Photaki and du Vigneaud, 1965)] lead one to suspect that this phase might be of some importance and would merit more attention. Also, while the excretion of the natural hormones has been studied quite intensively, no information about the elimination of analogs in the urine is available.

IX. Conclusions

Disappointment and skepticism have often been expressed about the results achieved in work with synthetic peptide analogs. This review has attempted to show that these judgments are perhaps shortsighted; and to the extent that they are justified, one of the reasons is an under-exploitation of the possibilities offered by such analogs as physiological or pharmacological probes. As these possibilities become better appreciated we are likely to see a revival of interest in structure–activity relationships at this intellectually higher level.

ACKNOWLEDGMENT

The recent work reported in this paper was supported in part by Swiss National Science Foundation Grants Nos. 3.372.70 and 3.424.70.

REFERENCES

Ariëns, E. J., van Rossum, J. M., and Simonis, A. M. (1956). *Arzneim.-Forsch.* **6**, 282.

Aroskar, J. P., Chan, W. Y., Stouffer, J. E., Schneider, C. H., Murti, V. V. S., and du Vigneaud, V. (1964). *Endocrinology* **74**, 226.

Bär, H.-P., Hechter, O., Schwartz, I. L., and Walter, R. (1970). *Proc. Nat. Acad. Sci. U.S.* **67**, 7.

Barth, T., Hütter, H. J. Pliška, V., and Šorm, F. (1969). *Experientia* **25**, 646.

Bentley, P. J. (1965). *J. Endocrinol.* **32**, 215.

Beránková-Ksandrová, Z., Rychlík, I., and Šorm, F. (1964). *Proc. 2nd Int. Pharmacol. Meet., Prague, 1963* **10**, 181.

Beránková-Ksandrová, Z., Bisset, G. W., Jošt, K., Krejčí, I., Pliška, V., Rudinger, J., Rychlík, I., and Šorm, F. (1966). *Brit. J. Pharmacol. Chemother.* **26**, 615.

Berde, B., and Boissonnas, R. A. (1968). *In* "Handbook of Experimental Pharmacology. Vol. 23: Neurohypophysial Hormones and Similar Polypeptides" (B. Berde, ed.), pp. 802–870. Springer-Verlag, Berlin and New York.

Berde, B., and Cerletti, A. (1961). *Helv. Physiol. Pharmacol. Acta* **19**, 135.

Bisset, G. W., (1964). *Proc. 2nd Int. Pharmacol. Meet., Prague, 1963* **10**, 21.

Bisset, G. W., and Clark, B. J. (1971). Unpublished observations.

Bisset, G. W., and Smyth, D. G. (1965). Unpublished observations. See Smyth (1967b).

Bisset, G. W., Haldar, J., and Lewin, J. E. (1966). *Mem. Soc. Endocrinol.* **14**, 185.

Bisset, G. W., Clark, B. J., Krejčí, I., Poláček, I., and Rudinger, J. (1970). *Brit. J. Pharmacol.* **40**, 342.

Bisset, G. W., Jones, N. F., and Hilton, P. J. (1971). Unpublished observations.

Branda, L. A., and Ferrier, B. M. (1971). *Amer. J. Obstet. Gynecol.* **109**, 943.

Branda, L. A., Ferrier, B. M., Archimaut, G., Marchelli, E. A., and Rucanski, B. (1968). *Science* **160**, 81.

Braun, T., and Hechter, O. (1970). *Horm. Metab. Res. Suppl.* **2**, 11.

Braun, T., Hechter, O., and Rudinger, J. (1969). *Endocrinology* **85**, 1092.

Breslow, E., and Abrash, L. (1966). *Proc. Nat. Acad. Sci. U.S.* **56**, 640.

Chan, W. Y., and du Vigneaud, V. (1962). *Endocrinology* **71**, 977.

Chan, W. Y., and du Vigneaud, V. (1970). *J. Pharmacol. Exp. Ther.* **174**, 541.

Chan, W. Y., and Kelley, N. (1967). *J. Pharmacol. Exp. Ther.* **156**, 150.

Chan, W. Y., and Wahrenburg, M. (1968). *Endocrinology* **82**, 475.

Chan, W. Y., Fear, R., and du Vigneaud, V. (1967). *Endocrinology* **81**, 1267.

Chan, W. Y., Hruby, V. J., Flouret, G., and du Vigneaud, V. (1968). *Science* **161**, 280.

Changeux, J.-P., Blumenthal, R., Kasai, M., and Podleski, T. (1970). *In* "Molecular Properties of Drug Receptors" (R. Porter and M. O'Connor, eds.), Ciba Found. Symp., pp. 197–214. Churchill, London.

Chard, T., Forsling, M. L., James, M. A. R., Kitau, M. J., and Landon, J. (1970). *J. Endocrinol.* **46**, 533.

Chimiak, A., and Rudinger, J. (1965). *Collect. Czech. Chem. Commun.* **30**, 2592.

Chimiak, A., Eisler, K., Jošt, K., and Rudinger, J. (1968). *Collect. Czech. Chem. Commun.* **33**, 2918.

Chiu, P. J. S., and Sawyer, W. H. (1970). *Amer. J. Physiol.* **218**, 838.

Cort, J. H., Rudinger, J., Lichardus, B., and Hagemann, I. (1966). *Amer. J. Physiol.* **210**, 162.

Daniel, E. E. (1965). *In* "Muscle" (W. M. Paul, E. E. Daniel, C. M. Kay, and G. Monckton, eds.), pp. 295–313. Pergamon, Oxford.

Drabarek, S., and du Vigneaud, V. (1965). *J. Amer. Chem. Soc.* **87**, 3974.

Dutta, A. S., Anand, N., and Kar, K. (1966). *Indian J. Chem.* **4**, 488.

du Vigneaud, V. (1965). *Proc. Robert A. Welch Found. Conf. Chem. Res., 1964* **8**, 133.

du Vigneaud, V., Fitt, P. S., Bodanszky, M., and O'Connell, M. (1960). *Proc. Soc. Exp. Biol. Med.* **104**, 653.

Edwards, B. A. (1971). *J. Endocrinol.* **50**, 669.

Eggena, P., Schwartz, I. L., and Walter, R. (1970). *J. Gen. Physiol.* **56**, 250.

Eisler, K., Rudinger, J., and Šorm, F. (1966). *Collect. Czech. Chem. Commun.* **31**, 4563.

Flouret, G., and du Vigneaud, V. (1969). *J. Med. Chem.* **12**, 1035.

Furchgott, R. F. (1955). *Pharmacol. Rev.* **7**, 183.

Furchgott, R. F. (1966). *Advan. Drug Res.* **3**, 21.

Ginsburg, M. (1968). *In* "Handbook of Experimental Pharmacology. Vol. 23: Neurohypophysial Hormones and Similar Polypeptides" (B. Berde, ed.), pp. 286–371. Springer-Verlag, Berlin and New York.

Ginsburg, M., and Jayasena, K. (1968a). *J. Physiol.* (*London*) **197**, 53.

Ginsburg, M., and Jayasena, K. (1968b). *J. Physiol.* (*London*) **197**, 65.

Glass, J. D., Dubois, B. M., Schwartz, I. L., and Walter, R. (1970). *Endocrinology* **87**, 730.

Glick, S. M., Kumaresan, P., Kagan, A., and Wheeler, M. (1968). *In* "Protein and Polypeptide Hormones" (M. Margoulies, ed.), Proc. Int. Symp., pp. 81–83. Excerpta Med. Found., Amsterdam.

Golubow, J., Chan, W. Y., and du Vigneaud, V. (1963). *Proc. Soc. Exp. Biol. Med.* **113**, 113.

Gosselin, R. E. (1970). *Brit. J. Pharmacol.* **39**, 214P.

Guttmann, S., Jaquenoud, P.-A., Boissonnas, R. A., Konzett, H., and Berde, B. (1957). *Naturwissenschaften* **44**, 632.

Hollenberg, M. D., and Hope, D. B. (1967). *Biochem. J.* **105**, 921.

Hope, D. B., Murti, V. V. S., and du Vigneaud, V. (1962). *J. Biol. Chem.* **237**, 1563.

Hope, D. B., Murti, V. V. S., and du Vigneaud, V. (1963). *J. Amer. Chem. Soc.* **85**, 3686.

Huguenin, R. L., and Boissonnas, R. A. (1961). *Helv. Chim. Acta* **44**, 213.

Huguenin, R. L., and Guttmann, S. (1965). *Helv. Chim. Acta* **48**, 1885.

Jard, S., Rajerison, R. M., and Montegut, M. (1970). *Biochim. Biophys. Acta* **196**, 85.

Jošt, K., and Rudinger, J. (1967). *Collect. Czech. Chem. Commun.* **32**, 1229.

Jošt, K., and Šorm, F. (1971a). *Collect. Czech. Chem. Commun.* **36**, 234.

Jošt, K., and Šorm, F. (1971b). *Collect. Czech. Chem. Commun.* **36**, 297.

Jošt, K., Rudinger, J., and Šorm, F. (1961). *Collect. Czech. Chem. Commun.* **26**, 2496.

Jošt, K., Rudinger, J., and Šorm, F. (1963a). *Collect. Czech. Chem. Commun.* **28**, 1706.

Jošt, K., Rudinger, J., and Šorm, F. (1963b). *Collect. Czech. Chem. Commun.* **28**, 2021.

Kalsner, S., and Nickerson, M. (1968). *Can. J. Physiol. Pharmacol.* **46**, 719.

Kasafírek, E., Jošt, K., Rudinger, J., and Šorm, F. (1965). *Collect. Czech. Chem. Commun.* **30**, 2600.

Kasafírek, E., Rábek, V., Rudinger, J., and Šorm, F. (1966). *Collect. Czech. Chem. Commun.* **31**, 4581.

Kasafírek, E., Eisler, K., and Rudinger, J. (1969). *Collect. Czech. Chem. Commun.* **34**, 2848.

Koida, M., Glass, J. D., Schwartz, I. L., and Walter, R. (1971). *Endocrinology* **88**, 633.

Krejčí, I. (1971). Unpublished observations.

Krejčí, I., and Poláček, I. (1968). *Eur. J. Pharmacol.* **2**, 393.

Krejčí, I., Poláček, I., Kupková, B., and Rudinger, J. (1964). *Proc. 2nd Int. Pharmacol. Meet., Prague, 1963* **10**, 117.

Krejčí, I., Poláček, I., Rudinger, J. (1966). *Mem. Soc. Endocrinol.* **14**, 171.

Krejčí, I., Kupková, B., and Vávra, I. (1967a). *Brit. J. Pharmacol. Chemother.* **30**, 497.

Krejčí, I., Poláček, I., and Rudinger, J. (1967b). *Brit. J. Pharmacol. Chemother.* **30**, 506.

Krejčí, I., Poláček, I., and Rudinger, J. (1969). *Abstr. 4th Int. Congr. Pharmacol. Basel* p. 208.

Krejčí, I., Pliška, V., and Rudinger, J. (1970). *Brit. J. Pharmacol.* **39**, 217P.

Krejčí, I., Kupková, B., Vávra, I., and Rudinger, J. (1971). *Eur. J. Pharmacol.* **13**, 65.

Krejčí, I., Poláček, I., and Rudinger, J. (1972). In preparation.

Kynčl, J., and Rudinger, J. (1970). *J. Endocrinol.* **48**, 157.

Kynčl, J., Jelínek, V., and Rudinger, J. (1969). *Acta Endocrinol. (Copenhagen)* **60**, 369.

Kyncl, J., Řežábek, K., Pliška, V., and Rudinger, J. (1972). In preparation.

Lauson, H. D. (1967). *Amer. J. Med.* **42**, 713.

Lauson, H. D. (1970). In "International Encyclopedia of Pharmacology and Therapeutics. Sect. 41, Vol. 1: Pharmacology of the Endocrine System and Related Drugs; The Neurohypophysis" (H. Heller and B. T. Pickering, eds.), pp. 377–397. Pergamon, Oxford.

Law, H. D., and du Vigneaud, V. (1960). *J. Amer. Chem. Soc.* **82**, 4579.

Legros, J. J., and Franchimont, P. (1971). In "Radioimmunoassay Methods" (K. E. Kirkham and W. M. Hunter, eds.), pp. 156–158. Churchill Livingstone, Edinburgh and London.

Miller, M., and Moses, A. M. (1969). *Endocrinology* **84**, 798.

Nesvadba, H., Honzl, J., and Rudinger, J. (1963). *Collect. Czech. Chem. Commun.* **28**, 1691.

Nickerson, M. (1956). *Nature (London)* **178**, 697.

Paton, W. D. M., and Rang, H. P. (1966). *Advan. Drug Res.* **3**, 57.

Photaki, I., and du Vigneaud, V. (1965). *J. Amer. Chem. Soc.* **87**, 908.

Pickering, B. T. (1970). In "International Encyclopedia of Pharmacology and Therapeutics. Sect. 41, Vol. 1: Pharmacology of the Endocrine System and Related Drugs; The Neurohypophysis" (H. Heller and B. T. Pickering, eds.), pp. 81–110. Pergamon, Oxford.

Pliška, V. (1966). *Arzneim.-Forsch.* **16**, 886.

Pliška, V. (1968). *Farmaco, Ed. Sci.* **23**, 623.

Pliška, V. (1969). *Eur. J. Pharmacol.* **5**, 253.

Pliška, V. (1971). Unpublished observations.

Pliška, V., and Krejčí, I. (1966). *Arch. Int. Pharmacodyn. Ther.* **161**, 289.

Pliška, V., and Rudinger, J. (1972). In preparation.

Pliška, V., Barth, T., and Rychlík, I. (1967). *Experientia* **23**, 196.

Pliška, V., Rudinger, J., Douša, T., and Cort, J. H. (1968). *Amer. J. Physiol.* **215**, 916.

Pliška, V., Thorn, N. A., and Vilhardt, H. (1971). *Acta Endocrinol. (Copenhagen)* **67**, 12.

Pliška, V., Cort, J. H., Barth, T., Hütter, H. J., Jošt, K., Vítek, A., and Rudinger, J. (1972a). In preparation.

Pliška, V., Gasparović, I., and Rudinger, J. (1972b). In preparation.

Poláček, I., and Krejčí, I. (1969). *Eur. J. Pharmacol.* **7**, 85.

Poláček, I., and Pliška, V. (1969). *Eur. J. Pharmacol.* **6**, 143.

Poláček, I., Krejčí, I., and Rudinger, J. (1967). *J. Endocrinol.* **38**, 13.

Poláček, I., Krejčí, I., Nesvadba, H., and Rudinger, J. (1970). *Eur. J. Pharmacol.* **9**, 239.

Roth, J., Glick, S. M., Klein, L. A., and Petersen, M. J. (1966). *J. Clin. Endocrinol.* **26**, 671.

Rudinger, J. (1965). *Proc. 2nd Int. Congr. Endocrinol., London, 1964* p. 1202–1206. Excerpta Med. Found., Amsterdam.

Rudinger, J. (1966). *Mem. Soc. Endocrinol.* **14**, 183.

Rudinger, J. (1968). *Proc. Roy. Soc., Ser. B* **170**, 17.

Rudinger, J. (1969). *In* "Progress in Endocrinology" (C. Gual, ed.), Proc. 3rd Int. Congr. Endocrinol., pp. 419–424. Excerpta Med. Found., Amsterdam.

Rudinger, J. (1971). *In* "Drug Design" (E. J. Ariëns, ed.), Vol. 2, pp. 319–419. Academic Press, New York.

Rudinger, J., and Jošt, K. (1964). *Experientia* **20**, 570.

Rudinger, J., and Krejčí, I. (1962). *Experientia* **18**, 585.

Rudinger, J., and Krejčí, I. (1968). *In* "Handbook of Experimental Pharmacology. Vol. 23: Neurohypophysial Hormones and Similar Polypeptides" (B. Berde, ed.), pp. 748–801. Springer-Verlag, Berlin and New York.

Rudinger, J., Pliška, V., Rychlík, I., and Šorm, F. (1968a). *In* "Pharmacology of Hormonal Polypeptides and Proteins" (N. Back, L. Martini, and R. Paoletti, eds.), pp. 66–72. Plenum, New York.

Rudinger, J., Krejčí, I., Poláček, I., and Kupková, B. (1968b). *In* "Protein and Polypeptide Hormones" (M. Margoulies, ed.), Proc. Int. Symp., p. 217. Excerpta Med. Found., Amsterdam.

Rudinger, J., Pliška, V., and Furrer, J. (1972). *In* "Structure-Activity Relationships of Protein and Polypeptide Hormones. Part 3" (M. Margoulies and F. C. Greenwood, eds.), Proc. Int. Symp. Excerpta Med. Found., Amsterdam (in press).

Rychlík, I. (1964). *Proc. 2nd Int. Pharmacol. Meet., Prague, 1963* **10**, 153.

Saameli, K. (1964). *Brit. J. Pharmacol. Chemother.* **23**, 176.

Sachs, H. (1967). *Amer. J. Med.* **42**, 687.

Sawyer, W. H., and Manning, M. (1971). *J. Endoc.inol.* **49**, 151.

Schild, H. O. (1969). *Brit. J. Pharmacol.* **36**, 329.

Schulz, H., and du Vigneaud, V. (1966). *J. Med. Chem.* **9**, 647.

Schwartz, I. L., Rasmussen, H., and Rudinger, J. (1964). *Proc. Nat. Acad. Sci. U.S.* **52**, 1044.

Smith, M. W., and Ginsburg, M. (1961). *Brit. J. Pharmacol. Chemother.* **16**, 244.

Smyth, D. G. (1967a). *J. Biol. Chem.* **242,** 1579.

Smyth, D. G. (1967b). *J. Biol. Chem.* **242,** 1592.

Smyth, D. G. (1970). *Biochim. Biophys. Acta* **200,** 395.

Somlyo, A. P., and Somlyo, A. V. (1970). *Pharmacol. Rev.* **22,** 249.

Stephenson, R. P. (1956). *Brit. J. Pharmacol. Chemother.* **11,** 379.

Stouffer, J. E., Hope, D. B., and du Vigneaud, V. (1963). *In* "Perspectives in Biology" (C. F. Cori *et al.,* eds.), pp. 75–80. Elsevier, Amsterdam.

Suska-Brzezińska, E., Fruhaufová, L., Barth, T., Rychlík, I., Jošt, K., and Šorm, F. (1972). In preparation.

Tuppy, H. (1968). *In* "Handbook of Experimental Pharmacology. Vol. 23: Neurohypophysial Hormones and Similar Polypeptides" (B. Berde, ed.), pp. 67–129. Springer-Verlag, Berlin and New York.

Tuppy, H., and Nesvadba, H. (1957). *Monatsh. Chem.* **88,** 977.

van Rossum, J. M., and Ariëns, E. J. (1962). *Arch. Int. Pharmacodyn. Ther.* **136,** 385.

Vávra, I., Machová, A., Holeček, V., Cort, J. H., Zaoral, M., and Šorm, F. (1968). *Lancet* **i,** 948.

Vogel, G., and Hergott, J. (1963). *Arzneim.-Forsch.* **13,** 415.

Walter, R., Dubois, B. M., and Schwartz, I. L. (1968). *Endocrinology* **83,** 979.

Walter, R., Dubois, B. M., Eggena, P., and Schwartz, I. L. (1969). *Experientia* **25,** 33.

Walter, R., Schwartz, I. L., Darnell, J. H., and Urry, D. W. (1971). *Proc. Nat. Acad. Sci. U.S.* **68,** 1355.

Willumsen, N. B. S., and Bie, P. (1969). *Acta Endocrinol. (Copenhagen)* **60,** 389.

Yamanaka, T., Hase, S., Sakakibara, S., Schwartz, I. L., Dubois, B. M., and Walter, R. (1970). *Mol. Pharmacol.* **6,** 474.

Zaoral, M. (1965). *Collect. Czech. Chem. Commun.* **30,** 1853.

Zaoral, M., and Šorm, F. (1965). *Collect. Czech. Chem. Commun.* **30,** 2812.

Zaoral, M., Kasafírek, E., Rudinger, J., and Šorm, F. (1965). *Collect. Czech. Chem. Commun.* **30,** 1869.

Zaoral, M., Kolc, J., and Šorm, F. (1967). *Collect. Czech. Chem. Commun.* **32,** 1250.

Zhuze, A. L., Jošt, K, Kasafírek, E., and Rudinger, J. (1964). *Collect. Czech. Chem. Commun.* **29,** 2648.

DISCUSSION

I. L. Schwartz: I would like to expand Dr. Rudinger's consideration of stimulus–response coupling by calling attention to recent observations in a target organ which is ideally suited for analysis of this phenomenon, namely, the urinary bladder of the toad. It is possible in the toad bladder system to measure the affinity and intrinsic activity of neurohypophysial hormonal peptides in terms of an early step (adenylate cyclase activation) and a late or terminal step (hydroosmosis) in the biological reaction sequence which constitutes the action of these hormones on membrane permeability.

Drs. Kirchberger, Walter, and I, in collaboration with Drs. Douša, Bär, and Hechter, have tested the relative stimulatory effects of a set of agonistic neurohypophyseal hormones (and analogs) and the relative inhibitory effects of a set of antagonistic analogs, and we found a striking parallelism in the activities of all compounds at the level of the adenylate cyclase system of the toad bladder epithelial cells and at the level of the final biological effect (increased water transport along an osmotic gradient superimposed across the intact bladder). In other words all compounds exhibited the same relative order of potency (affinity) whether tested on a broken-cell preparation of adenylate cyclase or on the intact bladder. From this finding we were able to conclude that the potential for *selective* quantitative modulation of the ultimate hormonal response is realized *only* at the initial hormone–receptor interaction.

In sharp contrast to the parallelism observed when affinities are measured in both cyclase and intact systems, the intrinsic activities of agonistic peptides, i.e., the ability of saturating concentrations of these agents to evoke a maximal response, did not correlate when the results of studies of adenylate cyclase activation and of the final response of the intact toad bladder were compared. For example, it was found that each of the peptides studied at saturating concentrations evokes a different maximal response in the cyclase system whereas most of these compounds evoke the same (maximal) response in the intact system. Thus if the peptide which evokes the greatest maximal response in the cyclase system is defined as exhibiting an intrinsic activity of 100%, the other compounds will exhibit variably lower intrinsic activities in the cyclase system. On the other hand, most of the peptides will exhibit maximal (100%) intrinsic activity in terms of the final hydroosmotic effect in the intact system. It is only when intrinsic activity, as measured in the cyclase system, decreases below 30% that we begin to see any fall whatever in intrinsic activity as measured in the intact system. In other words the maximal response capacity at the early cyclase step must fall by 70% before we can detect any decrease in the maximal response capacity at the final step in hormone action. Thus most of the neurohypophysial hormone analogs that we have studied can stimulate adenylate cyclase to a much greater level of activity than is required to elicit a maximal response in the final effector system. These observations are consistent with our earlier proposal that the toad bladder has a "receptor reserve" with respect to neurohypophysial hormone-induced changes in membrane permeability and with the concept of stimulus response coupling which Dr. Rudinger has described.

R. Walter: I feel that with the elucidation of the X-ray structure of insulin [M. J. Adams, T. L. Blundell, E. J. Dobson, G. G. Dobson, M. Vijayan, E. N. Baker, M. M. Harding, D. C. Hodgkin, B. Rimmer, and S. Sheat, *Nature* (*London*) **224,** 491 (1969)] and the proposal of a solution conformation for the neurohypophysial hormones oxytocin and vasopressin [D. W. Urry and R. Walter, *Proc. Nat. Acad. Sci. U.S.* **68,** 956 (1971); R. Walter, *in* "Structure-Activity Relationships of Protein and Polypeptide Hormones" (M. Margoulies and F. C. Greenwood, eds.), p. 181. Excerpta Med. Found., Amsterdam, 1971]—and certainly other hormones will follow suit shortly—we have entered a new era of structure–activity analysis: a type of analysis which transcends the classical approach of considering single or multiple amino acid replacements in terms of vicinal effects within the hormone sequence. We now consider a structural modification in terms of its effects on the intramolecular stabilization of the peptide backbone, or amino acid side chains, and on any changes of intermolecular interactions connected with the binding to receptor molecules or the expression of the "hormonal information" of the peptide. From a conformational point of view, chemical modifications of neurohypophysial hormones can be grouped into three categories, each of which can be correlated with specific changes in biological activity: "(a) those affecting the stabilization of the backbone of the peptide, which would extensively perturb the spatial relationships among all the constituent amino acids and, hence, affect both affinity and intrinsic activities uniformly; (b) those which, while retaining the stability of the backbone conformation, alter the steric environment and charge distribution of limited surface areas, and thereby can affect affinity and intrinsic activity differentially; and (c) those changing the steric and electronic requirements of moieties comprising the active surface of the neurohypophyseal peptide, without perturbing the peptide backbone of the hormone molecule and, hence, affecting intrinsic activity without altering affinity" [R. Walter, I. L. Schwartz, J. H. Darnell, and D. W. Urry, *Proc. Nat. Acad. Sci. U.S.* **68**, 1355 (1971)].

Certainly, as progress is being made to efficiently determine amino acid side chain conformations and the relative populations of their different rotomers, and how various "biophases" perturb the hormonal topography, our biofunctional interpretation will be refined. Nevertheless it is to be appreciated that conformational considerations are not limited to questions of relative hormonal "affinity" and "intrinsic activity" but also can be the foundation for meaningful interpretations in such diverse areas as hormonal evolution, biosynthesis, enzymatic degradation, immunologic, and various other properties.

M. Manning: Dr. William Sawyer and I have been trying to track down a postulated intermediate between the neurohypophsial hormones containing 4-serine and those containing 4-glutamine. It had been postulated by Geschwind, among others, that this intermediate might be an analog of oxytocin which contains proline in place of glutamine in the 4 position. Using the Merrifield solid phase method we have synthesized three such proline-containing analogs. They are all totally inactive. Continuing the search for this suggested intermediate we next substituted threonine for glutamine in the 4 position of oxytocin. This led to the discovery of a series of synthetic analogs possessing most interesting and surprising properties.

Figure A shows the rat uterus, fowl vasodepressor, and milk ejection potencies of ocytocin and [4-threonine]-oxytocin together with their deamino analogs. As you can see, the substitution of threonine for glutamine in the oxytocin molecule brings about a striking enhancement of these characteristic oxytocinlike properties. Yet amazingly, the identical substitution in deamino-oxytocin leads to a drastic reduction in these same potencies.

Figure B shows what happens to the vasopressin-like effects of [4-threonine]-oxytocin and deamino-[4-threonine]-oxytocin. There is a remarkable diminishment in both analogs of pressor and antidiuretic potencies. Here, we also see an interesting difference between [4-threonine]-oxytocin and deamino-oxytocin, both of which are more potent oxytocic agents than oxytocin itself. In the former case there is a diminishment of antidiuretic activity, yet in the latter there is a marked enhancement of this activity. I might add, that we have obtained findings similar to these on substituting threonine for glutamine in the 4 position of related peptides both with and without the N-terminal amino group. In searching for clues as to why we obtained such a curious spectrum of activities for these peptides, we looked at a series of other 4-substituted analogs of oxytocin.

Table A shows the side chains of each amino acid in the 4 position together with the rat uterus potency of each analog. Most of these peptides were synthesized and pharmocologically evaluated by other investigators. The purpose of this table is to try to bring together some of the factors that we think influence the degree of activity which each of these analogs possesses. By taking a close look at the side chains in position 4, we were able to come up with some kind of a generalization in terms of what structural features might be involved. The critical features seem to be: (1) The length of the side chain, (2) the degree of branching, (3) the possession of both hydrophilic and lipophilic characteristics, and (4) the nature of the hydrophilic substituent. By taking these criteria into account, we have attempted to explain why we are getting an enhancement of activity by substituting threonine for glutamine in oxytocin. Simply stated, the explanation goes something like this: Glutamine and threonine both possess side chain characteristics which

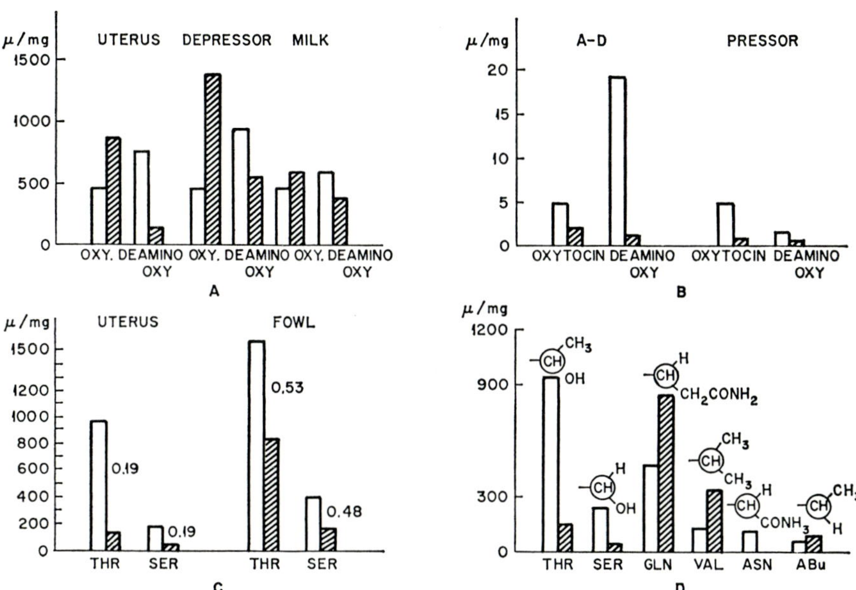

FIG. A.   Effect of threonine (▨) and glutamine (☐) interchange on rat uterus, fowl vasodepressor, and milk ejection potencies of oxytocin and deamino-oxytocin.

FIG. B.   Effect of threonine (▨) and glutamine (☐) interchange on pressor and antidiuretic (A-D) potencies of oxytocin and deamino-oxytocin.

FIG. C.   Effects of deamination of [4-threonine]- and [4-serine]-oxytocin analogs with NH₂ (☐) and without NH₂ (▨).

FIG. D.   Effect on rat uterus activity of deamination of 4-substituted oxytocins

$$\text{NH}_2\text{-Cys-Tyr-Ile-}\boxtimes\text{-Asn-Cys-Pro-Leu-Gly-NH}_2$$

☐, With NH₂; ▨, without NH₂.

optimize criteria 1–4, with one major difference: glutamine possesses a carboxamide group whereas threonine possesses a hydroxyl group. Interestingly enough, the sole difference between aspargine and serine is a carboxamide/hydroxyl interchange in their side chains. 4-Substituted analogs containing these amino acids, asparagine and serine, could thus serve as models to determine what the relative effects of the carboxamide group versus the hydroxyl group in position 4 of oxytocin might be. Both of these analogs had already been synthesized, and thus their rat uterus activities, as shown here, were known. The oxytocic activity of [4-serine]-

oxytocin is *twice* that possessed by [4-asparagine]-oxytocin, thus indicating that the hydroxyl group exerts a 2-fold greater effect relative to the effect of the carboxamide group in mediating the oxytocic response.

From these and other considerations we proposed the following *theory* to account for the 2-fold enhancement of oxytocic activity observed in [4-threonine]-oxytocin relative to oxytocin: The side chains of threonine and glutamine possess approximately equal amounts of lipophilic character. Therefore, by analogy with the serine/asparagine interchange as noted above, the enhancement of oxytocic activity in [4-threonine]-oxytocin is due solely to the greater effect of the hydroxyl group in threo-

TABLE A

Correlation of Side-Chain Structures with Rat Uterus Activities
of 4-Substituted Oxytocin Analogs

Amino acid in 4-position	Structure of side chain	Rat uterus activity
Threonine	$-\overset{\displaystyle CH_3}{\underset{\displaystyle |}{CH}}-OH$	900
Glutamine	$-CH_2-CH_2-\overset{\displaystyle O}{\overset{\displaystyle \|}{C}}-NH_2$	450
Serine	$-CH_2-OH$	195
Valine	$-\overset{\displaystyle CH_3}{\underset{\displaystyle |}{CH}}-CH_3$	140
Asparagine	$-CH_2-\overset{\displaystyle O}{\overset{\displaystyle \|}{C}}-NH_2$	108
α-Aminobutyric acid	$-CH_2-CH_3$	72
Norvaline	$-CH_2-CH_2-CH_3$	61
Isoleucine	$-\overset{\displaystyle CH_3}{\underset{\displaystyle |}{CH}}-CH_2-CH_3$	37
Alanine	$-CH_3$	36
Leucine	$-CH_2-\overset{\displaystyle CH_3}{\underset{\displaystyle |}{CH}}-CH_3$	13

nine relative to the carboxamide group of glutamine. If this is true, it should be easy to design an analog that would help to substantiate this-idea by simply replacing the carboxamide group of glutamine with an OH group to give [4-homoserine]-oxytocin. We have encountered a number of interesting problems in trying to make this analog, but we are still hopeful of success. In the meantime, we turned again to the peculiar effect of deamination on [4-threonine]-oxytocin. If this were due to the presence of the OH group in position 4, then by analogy we wondered what effect deamination would have on the characteristic potencies of [4-serine]-oxytocin.

We obtained here a remarkable correlation between what happens on deamination of [4-threonine]-oxytocin and deamination of [4-serine]-oxytocin. We got identical potency ratios between the unaminated and the aminated analogs in both instances in the rat uterus assay and in the fowl vasodepressor assay (Fig. C).

This gives us the opportunity to construct something like what appears on Fig. D, which shows the effect of removing the amino group on the rat uterus activity of a number of 4-substituted oxytocin analogs. The presence of an OH group in the side chain at position 4 leads to a drastic reduction in potency upon removal of the amino group. Removal of the amino group from the other analogs shown here leads to an enhancement of rat uterus potency. The deamino derivative of [4-asparagine]-oxytocin has not yet been synthesized, but one could almost predict from this series that it will possess increased activity.

The point that I would like to emphasize by way of conclusion is that in many instances it is very difficult to make predictions on the basis of structure–activity data. One can make a change in one molecule and get a specific effect, yet an identical change in another closely related molecule may result in a totally opposite effect, as we have seen with oxytocin and deamino-oxytocin. We take oxytocin, substitute threonine in the 4 position and obtain an enhancement of activity. We take deamino-oxytocin, substitute threonine in the 4 position, and get diminishment of activity. We know that the OH group in the 4-position is in some way responsible for these striking differences but have absolutely no clear ideas as to how it might be involved. I think that this is where the type of conformational studies and interpretations that Dr. Walter and Dr. Schwartz are doing would help very much to clarify what is going on and I await with interest their analysis of this series of analogs.

J. Rudinger: I would certainly agree, and I anticipate that the experimentally derived conformational model which is now available should be very valuable in rationalizing the structural aspects of structure–activity relations, more so than the essentially intuitive models that we have used from time to time on the basis of structure–activity relations alone.

I might perhaps make one additional point about the studies of hormone inactivation. What has been conspicuously lacking is work with isolated organs perfused *in situ* or *in vitro*, or with specific circulatory regions. Such experiments have been carried out with angiotensin and with bradykinin [e.g., J. R. Vane, *Brit. J. Pharmacol.* **35**, 209 (1969); J. R. Ryan, J. Roblero, and J. M. Stewart, *Biochem. J.* **110**, 795 (1968); J. M. Stewart, *in* "Structure-Activity Relationships of Protein and Polypeptide Hormones" (M. Margoulies and F. C. Greenwood, eds.), pp. 23–30. Excerpta Med. Found., Amsterdam, 1971]. It has turned out that the properties of some bradykinin analogs *in vivo* are largely determined by their metabolic behavior, particularly in the lung, which is the main site of inactivation of bradykinin [Stewart (1971) see above].

It is interesting to note that several years earlier some of these analogs had been found to be very much more active (relative to bradykinin) when assayed on the rabbit blood pressure than when assayed on the isolated rat uterus or guinea pig ileum [E. Schröder, *Experientia* **21**, 271 (1965)]. Apparently nobody thought this result sufficiently interesting to do experiments with isolated rabbit vascular muscle to find whether the effect was due to an organ or species specificity, or represented a difference between the situations *in vitro* and *in vivo*. If this finding had been followed up, it would surely have led to an earlier appreciation of the metabolic component in the overall action of bradykinin.

R. Walter: Dr. Rudinger pointed to the great desirability of carrying studies concerned with the enzymatic degradation of peptide hormones from the "homogenate work" to the *in vivo* situation. At least in one instance we seem to have obtained such data with neurohypophyseal hormones which make an interpolation between these two levels of complexity possible. Earlier we reported that homogenates of rat kidney inactivate oxytocin more rapidly than arginine vasopressin and that different kidney enzymes are responsible for the inactivation of each hormone. Moreover, the most active oxytocin-inactivating principle was purified and it was found to cleave the glycinamide residue from oxytocin. Since this particular type of inactivating activity was found to be highest in kidney tissue, it was suggested that this organ also in the *in vivo* situation is capable of inactivating oxytocin by removing the C-terminal glycinamide residue [M. Koida, J. D. Glass, I. L. Schwartz, and R. Walter, *Endocrinology* **88**, 633 (1971)]. In line with this notion Walter and Shlank (Endocrinology, October issue, 1971) found that the intravenous administration of a single dose of oxytocin, specifically radioactively labeled in the glycine residue [R. Walter and R. T. Havran, *Experientia* **27**, 645 (1971)], yields only two detectable radioactive components in the urine, e.g., oxytocin and glycinamide. With the perfused rat kidney system developed by R. H. Bowman [*J. Biol. Chem.* **245**, 1604 (1970)], we were able to confirm that the kidney is highly efficient in cleaving glycinamide from oxytocin. Moreover, we found that the perfused kidney also removes the glycinamide moiety from arginine vasopressin, although somewhat less effectively than from oxytocin (R. Walter and R. H. Bowman, unpublished). Based in part on the above *in vitro* and *in vivo* inactivation profile of oxytocin by kidney, we were able to design neurohypophysial hormone analogs which exhibit in the rat significantly protracted uterotonic responses, suggesting that renal inactivation processes of oxytocin—or processes which in enzymatic specificity are identical to that observed in kidney— play a role in determining in part the time course of the response of the uterus to oxytocin.

P. L. Munson: I should like to reemphasize Dr. Rudinger's point that when the biological properties of a series of related compounds are compared it is not enough merely to run them through a standardized assay system and put down a series of numbers indicating relative potencies, no matter how much they may be dignified by statistical paraphernalia. We should actually study them in more detail and compare dose-response curves, then go further and try to analyze the reasons behind the differences in the dose-response curves.

Clinical Experience with Hypothalamic Releasing Hormones
Part 1. Thyrotropin-Releasing Hormone

Carlos Gual, Abba J. Kastin, and Andrew V. Schally

Department of Endocrinology, Instituto Nacional de la Nutrición, Mexico City, Mexico; and Endocrinology Section of the Medical Service and Endocrine and Polypeptide Laboratories, Veterans Administration Hospital, and Department of Medicine, Tulane University School of Medicine, New Orleans, Louisiana

I. Introduction

Although the hypothalamus has long been known to control the release of thyrotropin from the anterior pituitary gland by means of neurohumoral agents (Harris, 1955; Greer, 1957; D'Angelo, 1963; Reichlin, 1963), the first unequivocal demonstration of the presence of a substance in hypothalamic tissue with thyrotropin-releasing activity was reported a decade ago by Guillemin *et al.* (1962) and Schreiber *et al.* (1962). This substance was initially designated as thyrotropin-releasing factor (TRF), but more recently Schally *et al.* (1968) suggested the term thyrotropin-releasing hormone (TRH), which will be used in this report.

TRH has been purified from hypothalamic tissue containing the pituitary stalk and the median eminence region from a variety of sources including pig (Schally *et al.*, 1966a,c), sheep (Guillemin *et al.*, 1966), cow (Schally *et al.*, 1966b), and man (Bowers *et al.*, 1965; Schally *et al.*, 1967). Its biological activity has been demonstrated in a number of *in vitro* (Guillemin, 1965, 1967; Guillemin *et al.*, 1965; Schally *et al.*, 1966c; Schally and Redding, 1967; Mittler *et al.*, 1969), and *in vivo* experiments (Guillemin, 1965, 1967; Guillemin *et al.*, 1965, 1966; Schally *et al.*, 1966b,c, 1968; Bowers *et al.*, 1967). The structure of TRH of porcine origin (Folkers *et al.*, 1969; Bøler *et al.*, 1969; Schally *et al.*, 1969, 1970b; Enzmann *et al.*, 1971) and of ovine origin (Burgus *et al.*, 1969a,c, 1970a) has been shown to be L-pyroglutamyl-L-histidyl-L-proline amide. TRH has been synthesized in several laboratories (Bøler *et al.*, 1969; Burgus *et al.*, 1969b; Nair *et al.*, 1970; Gillessen *et al.*, 1970; Baugh *et al.*, 1970; Flouret, 1970; Inouye *et al.*, 1971); this synthetic tripeptide has shown biological activity similar to that of natural TRH obtained from porcine (Bowers *et al.*, 1970a) and ovine (Burgus *et al.*, 1970b) hypothalami. It has been postulated (Schally *et al.*, 1970a; Bowers *et al.*, 1970a) that TRH of several other mammalian species may have the same chemical structure and that species specificity may be minimal or absent. If this is correct, it is probable that bovine and human TRH will share the above-mentioned structure. *In vitro* and *in vivo* studies have demonstrated that TRH of human origin can release

173

thyroid stimulating hormone (TSH) when tested in assay systems employing rats and mice (Bowers *et al.*, 1965; Schally *et al.*, 1970a). Also, it has been shown (Bowers *et al.*, 1968) that when a highly purified preparation of porcine TRH was injected intravenously into humans, a significant increase in plasma TSH was observed, thus indicating a certain lack of species specificity, since this porcine preparation was active also in the mouse, rat, and nutria.

The availability of pure synthetic TRH and the potential clinical usefulness of this material, prompted a number of investigators to test its biological activity in man (Fleischer *et al.*, 1970; Hall *et al.*, 1970; Bowers *et al.*, 1970b,c; Sakoda *et al.*, 1970; Hershman and Pittman, 1970). These studies demonstrated very clearly that the intravenous or oral administration of synthetic TRH elevates plasma TSH levels without significantly altering the levels of corticotropin, gonadotropins, or growth hormone.

In this report, the effects of synthetic TRH on TSH release in normal subjects, as well as in documented cases of hypothalamic–pituitary–thyroid disorders and in other endocrinopathies, are presented. Most of these data have been previously reported (Bowers *et al.*, 1968; Gual *et al.*, 1971) or have been submitted for publication (Gual *et al.*, 1972b,c).

II. Materials and Methods

A. Subjects

Eighteen normal young female and male volunteers underwent a detailed clinical examination before the administration of TRH. In none of them was evidence of pituitary or thyroid disease observed. The subjects were fasted overnight and remained seated throughout the test, which was performed between 8:00 AM and 11:00 AM. In all cases informed signed consent was obtained before the TRH administration, and the safety and scientific relevance of all procedures were reviewed in accordance with the standard practice at the "Instituto Nacional de la Nutricion" in Mexico City.

In addition to normal subjects, TRH was administered to hospitalized patients or outpatient volunteers attending the Endocrine Clinic of the Instituto Nacional de la Nutricion. As previously outlined, the test was performed in three patients with primary hypothyroidism, seven patients with secondary or tertiary hypothyroidism, four patients with untreated hyperthyroidism, nine patients with functional and nonfunctional pituitary tumors, eight patients with galactorrhea–amenorrhea syndromes, four patients with galactorrhea and normal menses, four patients with documented primary or secondary hypothalamic amenorrhea, two patients with gonadal dysfunction, four hypogonadotropic men, and two pituitary dwarfs (idiopathic isolated growth hormone deficiency).

B. TRH Stimulation Test

The synthetic TRH used in this study was supplied by M. S. Anderson and W. F. White (Abbott Laboratories, Co., North Chicago, Illinois) in sterile ampoules containing 500 μg of pyroglutamyl-histidyl-proline amide (TRH) in 1.0 ml of isotonic sodium chloride (lots Nos. 843-8928 and 844-8903). The intravenous TRH test was performed after an overnight fast, through an indwelling catheter inserted into a forearm vein at approximately 8:00 AM, and blood samples were withdrawn at minus 5 minutes, zero time, and 15, 30, 60, and 120 minutes later. After the blood sample was obtained at zero time, 500 μg was injected intravenously over a period of 15–30 seconds. In some cases 500 μg TRH was infused in 100 ml of saline over a period of 30 minutes. In three normal men TRH was injected intravenously at weekly intervals in doses of 50 μg, 500 μg, and 1500 μg. One fasting subject was given a single oral dose of 10 mg of TRH in 100 ml of water, and blood samples were drawn at minus 5 minutes, zero time, and 30, 60, 120, 180, 240, and 300 minutes later. In all cases blood samples were collected into heparinized tubes and the separated plasma kept frozen at —20°C until assayed.

C. Assays

Plasma TSH was measured in duplicate by double antibody radioimmunoassay according to the method of Odell *et al.* (1967) and expressed as microunits per milliliter of plasma, in terms of the human thyrotropin Standard International Reference Preparation A (Division of Biological Standards, National Institute for Medical Research, Mill Hill, London). Plasma FSH and LH were determined in duplicate samples by radioimmunoassay methods according to Midgley (1966, 1967), and results are expressed in terms of units of biological activity (mIU), of the Second International Reference Preparation for Human Menopausal Gonadotropin (HMG 2nd IRP). Radioimmunoassayable plasma prolactinlike substance was measured in duplicate by methods of Midgley (1971) and expressed as microliters per milliliter of standardized pooled lactating women serum preparation. The assay is a homologous ovine–ovine radioimmunoassay which appears to be capable of quantitating prolactin in human serum. It does not cross react appreciably with HCG, HPL, TSH, growth hormone, LH, or FSH.

III. Clinical Studies with Natural TRH

The only recorded study with natural TRH in man was that by Bowers *et al.* (1968), who used a highly purified preparation of porcine thyrotropin-releasing hormone in three untreated cretins. Since release of TSH was high in these patients, a single small 25 μg dose of T_3 was given orally 24 hours before the studies were performed, to lower existing high plasma TSH levels. The time course of the TSH rise in two of the three cretins is shown in Fig. 1. A detectable rise is seen within 3 minutes, with a further rise between 6 and 30 minutes, and a peak at 15–30 minutes, followed by a gradual fall over the next 120 minutes. It was also found that 300 μg of porcine TRH, equivalent to 20 μg of pure TRH, gave a rise of more than 400% over the control TSH values at 30 minutes in the most responsive case. From these early results it was concluded that TRH prepared from pig hypothalami probably lacks species specificity since it is active in the mouse and rat as well as in man, and it was suggested that its intravenous administration in humans should prove a useful test of pituitary TSH reserve.

IV. Clinical Studies with Synthetic TRH

A. TSH Response to Intravenous TRH in Normal Subjects

1. Normal Women

The normal range of plasma TSH control values obtained in 24 subjects (13 men and 11 women) was 2.1–12.0 μU/ml, mean 6.0 μU/ml, standard error ±0.38, and no significant difference between the basal TSH levels in males and females was found. The intravenous administration of 500 μg of synthetic TRH consistently increased plasma TSH in all normal subjects. Figure 2 depicts the responses in seven normal women injected with 500 μg of TRH intravenously. Peak values for plasma TSH occurred at 15–30 minutes after TRH injection with a gradual fall over the next 90 minutes. The mean TSH level was 5.5 μU/ml at time zero, 19.0 μU/ml at 15 min, 21.5 μU/ml at 30 minutes, 16.6 μU/ml at 60 minutes, and 10.3 μU/ml at 120 minutes. At 30 minutes the peak values ranged from 15.5 to 25.8 μU/ml. Similar responses, after intravenous TRH in doses ranging from 100 to 800 μg, injected into 13 normal women, were observed by Bowers *et al.* (1970b, 1971) and Beckers *et al.* (1971b). However, lower peak plasma TSH values (Ormston *et al.*, 1971b,c) or higher values (Haigler *et al.*, 1971a; Karlberg *et al.*, 1971b; Raptis *et al.*, 1971; Rothenbuchner *et al.*, 1971a,b)

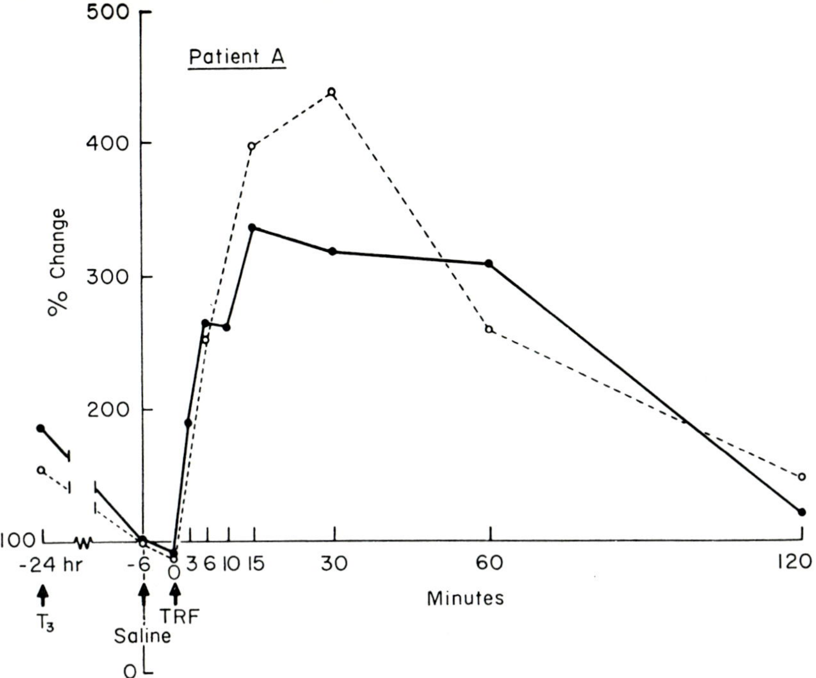

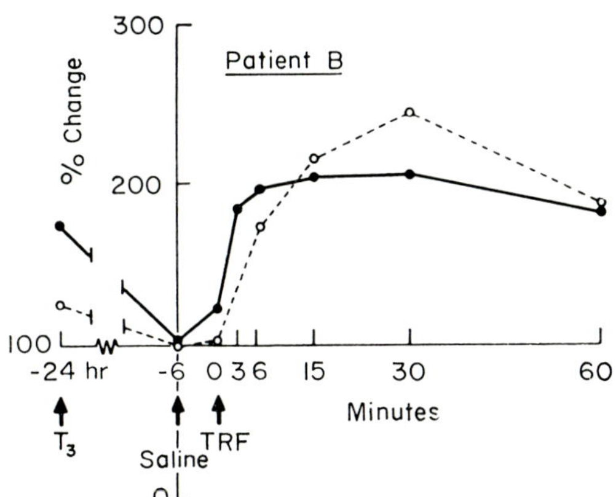

Fig. 1. Changes in plasma thyroid-stimulating hormone (TSH) levels before and after the injection of thyrotropin-releasing factor (TRF) to cretins: at —24 hours 25 μg of T_3 was given orally; at time —6 minutes, 2 ml of saline was given intravenously (iv); and at 0 time, 300 μg of porcine TRF was given iv. Plasma samples were measured by both bioassay (●——●) and radioimmunossay (○---○). Results recorded were obtained by comparing each plasma TSH value with those obtained at time —6 minutes and calculating the percentage change. From Bowers *et al.* (1968). Courtesy of the *Journal of Clinical Endocrinology.*

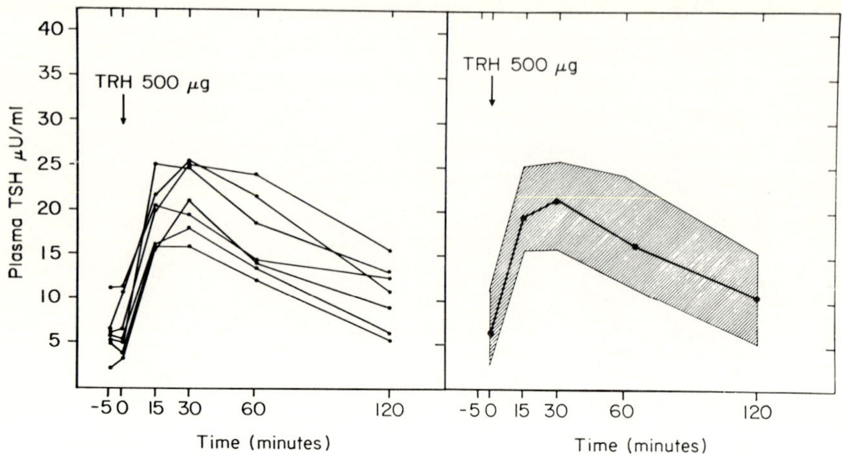

Fig. 2. Effect of 500 μg thyrotropin-releasing hormone (TRH) given intravenously on plasma thyroid-stimulating hormone (TSH) levels in 7 normal women. The shaded area on the right indicates the range of responses in normal women, and the filled circles connected by lines depict the mean of individual responses. From Gual *et al.* (1972b).

have also been reported. In any case, most of the data available to date point out that the rapid intravenous administration of TRH into normal women produces a transient rise in plasma TSH levels, with peak responses within 15 to 30 minutes after injection in most of tested subjects.

Differences in the response to TRH administration by a prolonged (30 minutes) infusion of 500 μg of TRH in 100 ml of saline were investigated in three normal women (Fig. 3). A peak plasma TSH value was observed in all cases immediately after the end of the TRH infusion, which was followed by a gradual fall to nearly basal values over the next 150 minutes. These responses were not significantly different from those seen in normal women after rapid intravenous injection 500 μg TRH and confirm previously reported data by Hershman and Pittman (1970).

When 500 μg TRH was infused over an 8-hour period (1.25 μg/ml), plasma TSH steadily rose from a control value of 8.7 μU/ml to a peak of 20.2 μU/ml at 6 hours, followed by a decline in the next 2 hours in spite of the continuous TRH infusion. Although these results may indicate a diminished sensitivity of the pituitary due to a rise of circulating thyroxine and triiodothyronine, induced by the elevated TSH levels

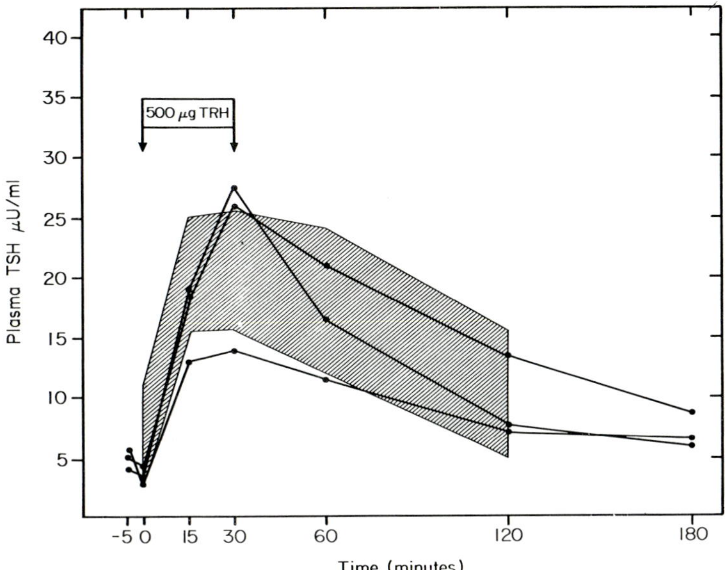

FIG. 3. Effect of 500 µg of thyrotropin-releasing hormone (TRH) infused in 100 ml of saline over a period of 30 minutes in 3 normal women. The shaded area indicates the range of responses seen in normal women rapidly given 500 µg of TRH intravenously. From Gual et al. (1972b).

after the TRH infusion, further studies will be necessary to confirm this lack of response.

2. Normal Men

Responses to rapid intravenous administration of 500 µg of TRH in seven normal men are shown on the left of Fig. 4. The greatest elevation in plasma TSH levels occurred 15–30 minutes after injection, and in most cases the peak values at 30 minutes were less than 15 µU/ml. The average responses at 15 and 30 minutes were significantly lower than those observed in normal women. Analogous results have been reported for men by a number of authors (Anderson et al., 1971a,b; Beckers et al., 1971a; Bowers et al., 1970b,c, 1971; Fleischer et al., 1970, 1971a,b; Hall et al., 1970; Hershman and Pittman, 1970, 1971; Karlberg et al., 1971a,b; Köbberling et al., 1971; Ormston et al., 1971a; Sakoda et al., 1970). It has been stated that the response to TRH in normal females appears to be greater than in males (Bowers et al., 1971; Haigler et al., 1971b; Ormston et al., 1971b,c). In this regard, Adams and Maloof (1970) have shown that ethynylestradiol causes a rise in serum TSH in normal males and hypogonadal females, which is suppressible with triiodothyronine. Concurring findings, have been published more recently by Glatstein *et al.* (1971) and Hall *et al.* (1971). Despite these observations, we failed to induce a rise in plasma TSH responses to 500 µg of TRH, given intravenously, by pretreating six of the same normal men (right side of Fig. 4) with daily divided doses of 150 µg of ethynylestradiol for 6 consecutive days.

A logical explanation for this lack of response cannot be deduced from the present data; nevertheless the concurrent increase in the blood

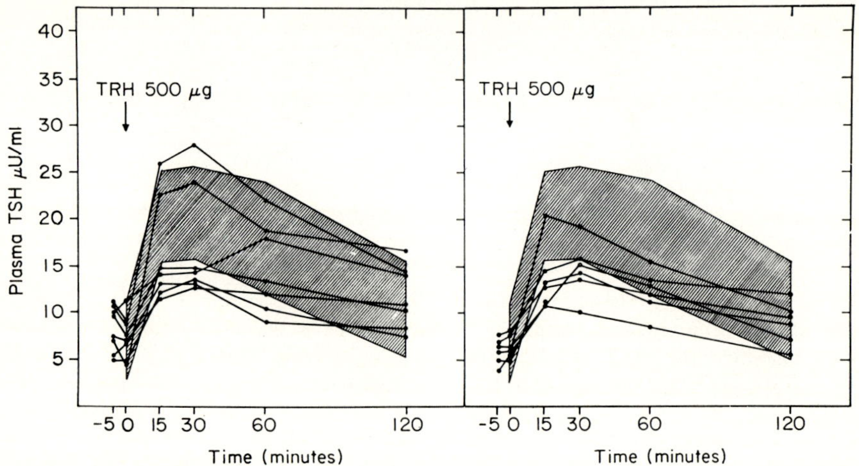

FIG. 4.   On the left is shown the effect of 500 µg of thyrotropin-releasing hormone (TRH) administered intravenously on plasma thyroid-stimulating hormone (TSH) levels in 7 normal men. On the right is depicted the TSH response obtained in the same men pretreated with daily divided doses of 150 µg of ethynylestradiol for 6 consecutive days. The shaded area indicates the range of responses seen in normal women receiving the same intravenous TRH dose. From Gual *et al.* (1972b).

levels of thyroid hormone induced by the repeated TRH administration to the same individual may in part be responsible for the decrease in the pituitary response. In support of this hypothesis, Odell *et al.* (1967), have shown that withdrawal of thyroxine therapy in patients with primary hypothyroidism gives a delayed recovery of plasma TSH of 10–14 days. To examine this question, we injected increasing doses of TRH, at weekly intervals, to three of the same normal men. As shown in Fig. 5, a dose of 1500 µg of TRH was not capable of provoking higher TSH responses than those seen with 50 µg. Comparable results were seen in normal subjects in whom repeated injections on the same day of increasing doses of TRH, ranging from 400 to 800 μg, caused less or equal changes than those obtained with 25 μg (Bowers *et al.*, 1971). Moreover, Haigler *et al.* (1971a), reported that 500 μg of TRH daily blunted the TSH rise in 2 of 3 subjects under study, and Ormston *et al.* (1971c) also found that one man, who 5 days before the TRH test had finished a 1-week course of T_3 (120 μg daily), showed no rise in

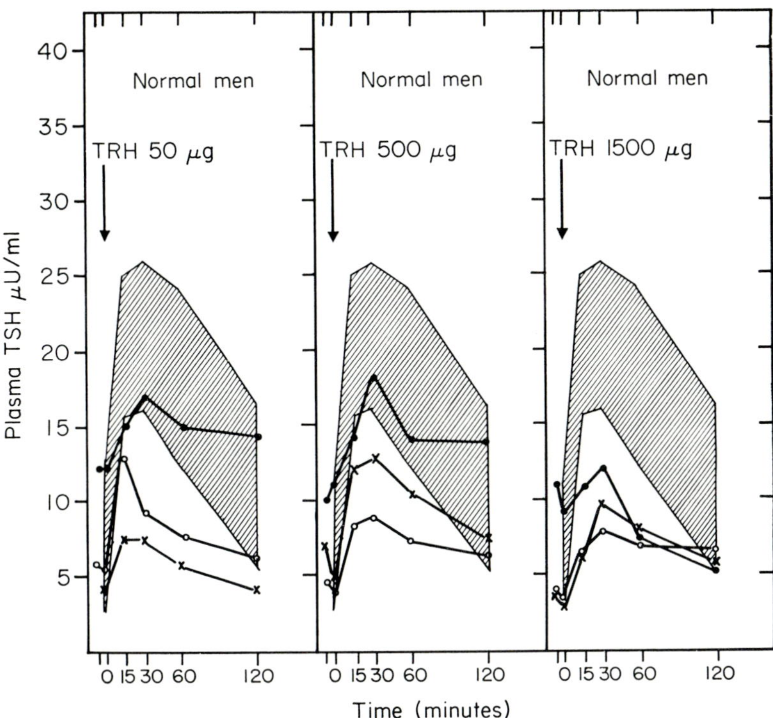

FIG. 5. Effect of administering increasing doses of thyrotropin-releasing hormone (TRH) at weekly intervals to normal men. The shaded area indicates the range of responses seen in normal women. From Gual *et al.* (1972b).

TSH in spite of the fact that he had previously responded well to TRH. However, a number of authors have reported that increasing intravenous doses of TRH were able to provoke significantly more elevated TSH responses when injected into different normal subjects (Bowers *et al.*, 1970b, 1971; Fleischer *et al.*, 1970; Haigler *et al.*, 1971a,b; Köbberling *et al.*, 1971; Ormston *et al.*, 1971c; Wagner *et al.*, 1971). These results further support the idea that repeated TRH stimulus on the pituitary gland may cause in normal subjects a transient refractory condition that is thyroid hormone dependent, and that a resting period of at least 10–14 days is required before a normal response can be restored.

B. TSH RESPONSE TO ORAL ADMINISTRATION OF TRH

Immediately after TRH synthesis was achieved, it was reported that its oral administration in microgram or submicrogram quantities to mice resulted in acute stimulation of the release of TSH (Vale *et al.*, 1970). These preliminary experiments were promptly confirmed in the human by Hershman and Pittman (1970, 1971) and subsequently by others

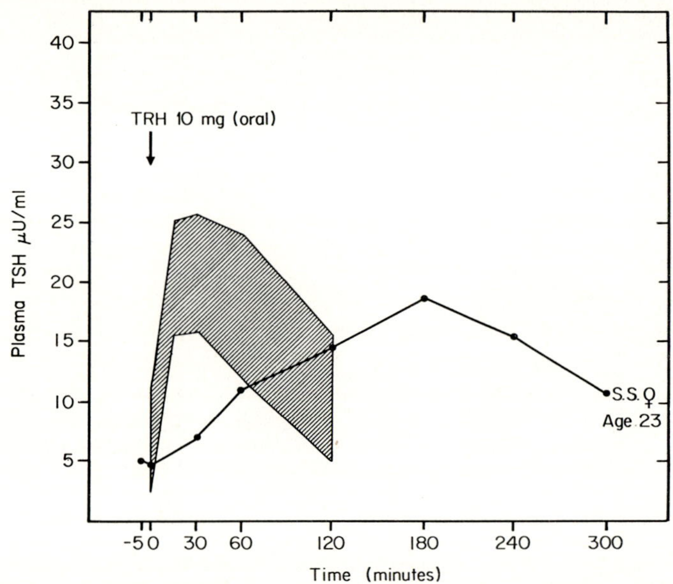

FIG. 6. Effect of oral dose of 10 mg of thyrotropin-releasing hormone (TRH) on plasma thyroid stimulating hormone (TSH) levels in a normal woman. The shaded area indicates the range of responses seen in normal women receiving 500 μg of TRH intravenously. From Gual *et al.* (1972b).

(Bowers *et al.*, 1971; Beckers *et al.*, 1971a); Köbberling *et al.*, 1971; Ormston *et al.*, 1971a,c). In general it was shown that a single oral dose of TRH caused more prolonged elevation of TSH levels than those seen after intravenous administration, with a dose-dependent TSH peak 2–5 hours after TRH ingestion. Interestingly enough, sublingual TRH administration also causes a significant TSH increase (Wagner *et al.*, 1971). Figure 6 depicts our experience with a single oral administration of 10 mg TRH to a 23-year-old woman. A TSH peak response was seen at 180 minutes with gradual decrease over the following 2 hours, thus confirming previously reported data. The prolonged rise in the TSH level stimulated by oral administration of TRH is highly suggestive that this may be the method of choice to achieve sustained clinical effects.

C. Toxic and Other TRH Hormonal Effects

As most of the authors have already reported, no significant toxic effects were noted in any individual. Anderson *et al.* (1971b) have reported "No objective toxicity was encountered in any subjects as judged by careful assessment of hematological, renal, and hepatic status." Pulse and blood pressure showed no change in any of our normal subjects and only mild and transient symptoms, consisting most commonly of nausea, a "queazy feeling" in the abdomen, a feeling of facial flushing, and an urge to urinate, were observed in two-thirds of the subjects tested. These symptoms began within 1 minute after TRH injection and ιasted in most cases less than 2 minutes. The occurrence of side effects was definitely not dose related, nor was it related to the magnitude of plasma TSH response. Nevertheless, Hall *et al.* (1970) reported that no side effects were noted with doses lower than 50 μg.

Other hormonal effects of TRH on plasma or serum levels of human growth hormone (HGH), LH, FSH, cortisol, insulin, glucose, protein-bound iodine (PBI), thyroxine, and triiodothyronine resin sponge up-take, have been investigated. A delayed HGH increase to TRH administration has been observed by several authors (Anderson *et al.*, 1971b; Bowers *et al.*, 1970b, 1971; Karlberg *et al.*, 1971b; Raptis *et al.*, 1971; Rothenbuchner *et al.*, 1971b; Saito *et al.*, 1971); however, a lack of significant responses of these hormones to TRH has also been reported (Fleischer *et. al.*, 1970; Kastin *et al.*, 1971; Ormston *et al.*, 1971a,c). In any case, most authors agree that the effect of TRH on HGH release may be considered to be an indirect one caused by unknown mechanisms. From available data it appears that TRH has no effect on the release of serum LH and FSH (Anderson *et al.*, 1971b; Bowers *et al.*, 1970b, 1971; Kastin *et al.*, 1971; Ormston *et al.*, 1971a,c). In regard to the effect of TRH on plasma or serum cortisol levels, there are contradictory reports, since some authors claim a decrease in circulating cortisol (Anderson *et al.*, 1971b; Rothenbuchner *et al.*, 1971b), while others have found no significant blood changes of cortisol (Fleischer *et. al.*, 1970; Kastin *et al.*, 1971; Ormston *et al.*, 1971a,c; Saito *et al.*, 1971), or even slight or irregular increases (Bowers *et al.*, 1971; Karlberg *et al.*, 1971b). Further studies should be done before any definite conclusions are reached. Reports on the insulin response to TRH indicate no significant changes (Karlberg *et al.*, 1971b; Saito *et al.*, 1971), although a decrease in blood glucose levels has been reported by Köbberling *et al.* (1971). Concerning blood thyroxine and T_3 resin uptake responses to TRH administration, it has been reported that slight or no responses occur after intravenous administration (Anderson *et al.*, 1971b; Sakoda *et al.*, 1970). On the other hand, PBI determinations seem to be a better index for clinical evaluation since a sustained rise after oral or intravenous TRH administration has been consistently observed (Karlberg *et al.*, 1971b; Köbberling *et al.*, 1971; Ormston *et al.*, 1971a,c; Sakoda *et al.*, 1970).

D. Pituitary TSH Reserve in Primary and Secondary Hypothyroidism

1. Primary Hypothyroidism

Patients with clinical findings and laboratory data of primary hypothyroidism were stimulated with 500 μg TRH given intravenously. Figure 7 depicts results obtained in three untreated patients, in whom peak TSH responses were induced at 30–60 minutes after injection. In one patient a very slow TSH decrease was noticed over the following 60 minutes, yet in the other 2 cases a plateau was sustained. These findings are in accordance with the prolonged half-life of the plasma TSH found by Odell *et al.* (1967) in patients with primary myxedema. The percentage rise in plasma TSH concentration was no greater than in control subjects, but the absolute increase in plasma TSH concentration was much higher, as in patient M.E.Z. with juvenile myxedema in whom the control plasma TSH levels rose from 1725 μU/ml to a record 4875 μU/ml, 60 minutes after TRH injection. Comparable results have been reported by others (Bowers *et al.*, 1968; Fleischer *et al.*, 1971b; Hershman and Pittman, 1970; Haigler *et al.*, 1971a,b; Job *et al.*, 1971; Karlberg *et al.*, 1971b; Ormston *et al.*, 1971b; Saito *et al.*, 1971). In one patient (M.D.T.) a second injection of 500 μg TRH was given after 1 month of daily therapy of 1 grain of desiccated thyroid; even though the control TSH plasma values were diminished by thyroid therapy, they were still elevated, and the second response was greater than the first one. Patient C.M. was maintained on 2 grains of desiccated thyroid therapy for a 2-month period before repeated intravenous injections of 500 μg TRH at 6-hour intervals were given on the same day. As shown in Fig. 8, the thyroid replacement therapy not only caused a marked diminution of the previous control values of plasma TSH, but also blunted the TSH response to the repeated administration of TRH. These results further support previous observations on the inhibitory effect of thyroid hormones on the pituitary TSH response to TRH stimulation.

Seven patients with secondary hypothyroidism were also studied, five of them being diagnosed as panhypopituitarism secondary to either Sheehan's syndrome or a pituitary tumor and two with isolated TSH deficiency in the absence of an organic destructive lesion of the pituitary. As expected most of the patients with Sheehan's syndrome did not re-

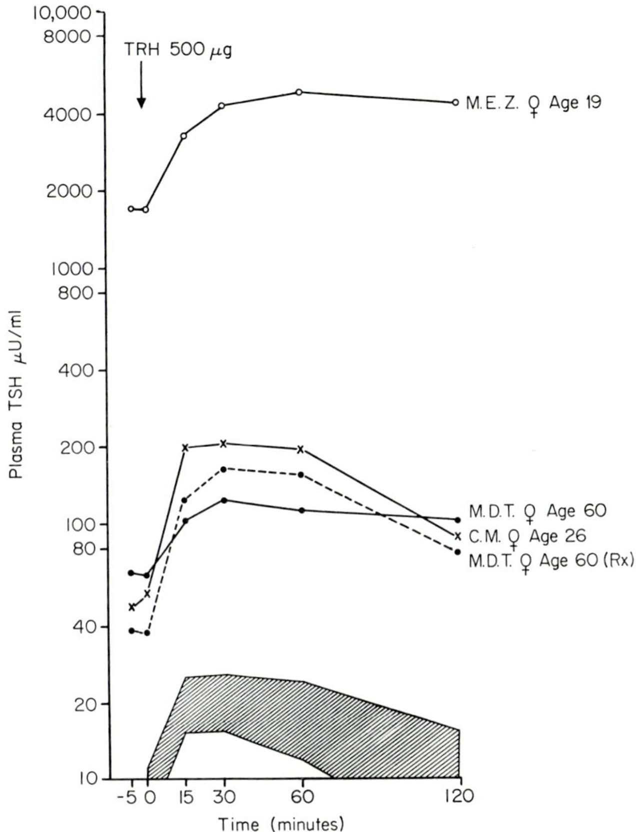

Fig. 7. Plasma thyroid stimulating hormone (TSH) after intravenous injection of 500 μg of thyrotropin-releasing hormone (TRH) in three patients with untreated primary hypothyroidism; the interrupted line shows the TSH response in one patient (M.D.T.) after 40 days of therapy with 1 grain of desiccated thyroid daily. From Gual *et al.* (1972b).

spond to the TRH test, confirming previous reports in the literature (Fleischer *et al.*, 1970, 1971b; Hershman and Pittman, 1970; Anderson *et al.*, 1971b; Meyer *et al.*, 1971; Karlberg *et al.*, 1971b; Saito *et al.*, 1971). To our surprise one of the patients with well documented Sheehan's syndrome responded normally to the TRH stimulus, as did two patients with pituitary tumors, who presented a diminished TSH response. It is not easy to explain these findings, but it is likely that residual functional pituitary tissue may exist, as has been reported in cases of a patient with Sheehan's syndrome who became pregnant after several years of disease (Jackson *et al.*, 1969). The two patients with a diagnosis of isolated TSH deficiency responded normally to TRH administration. At present the exact nature of the failure in TSH secretion

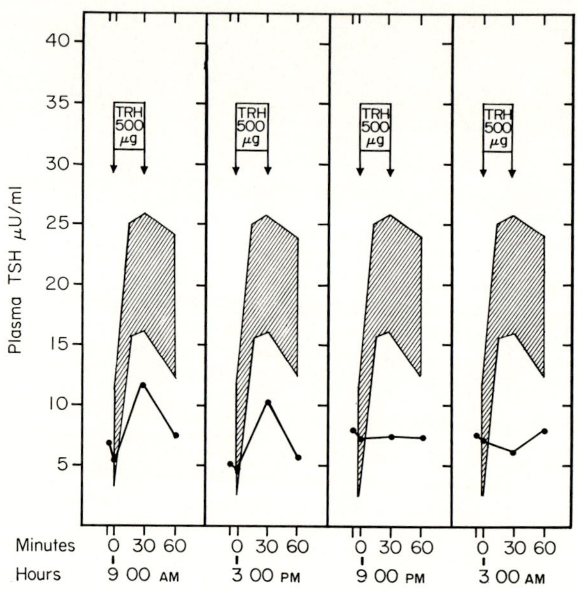

FIG. 8. Plasma thyroid-stimulating hormone (TSH) responses to four consecutive intravenous doses (6-hour interval) of 500 μg of thyrotropin-releasing hormone (TRH) each for a period of 30 minutes, in a female primary myxedema patient (C.M., age 26) under thyroid therapy. From Gual *et al.* (1972b).

is not clear, but it is conceivable that a primary hypothalamic defect in the secretion of TRH may be the cause for the isolated TSH pituitary deficiency.

E. Inhibitory Effect of Hyperthyroidism, Thyroxine and Triiodothyronine on the TRH-Induced TSH Release

Reports by Bowers *et al.* (1967), Vale *et al.* (1967), and Averill (1969) have shown in mice and rats that exogenous thyroxine (T_4) and triiodothyronine (T_3) inhibited the release of TSH produced by TRH and that T_4 or T_3 blockade can be overcome by increasing the TRH stimulus. These reports on T_4 and T_3 inhibition of the TSH response after TRH administration in animals have been partially confirmed in humans (Bowers *et al.*, 1970b, 1971; Fleischer *et al.*, 1970; Ormston *et al.*, 1971c). In addition to previous observations on the blocking effect of desiccated thyroid on the TSH response to TRH in a patient with primary myxedema, we have tested four hyperthyroid patients (Graves' disease) with 500 μg of TRH given intravenously. Figure 9 depicts a lack in TSH response in all tested patients. When the same patients were tested

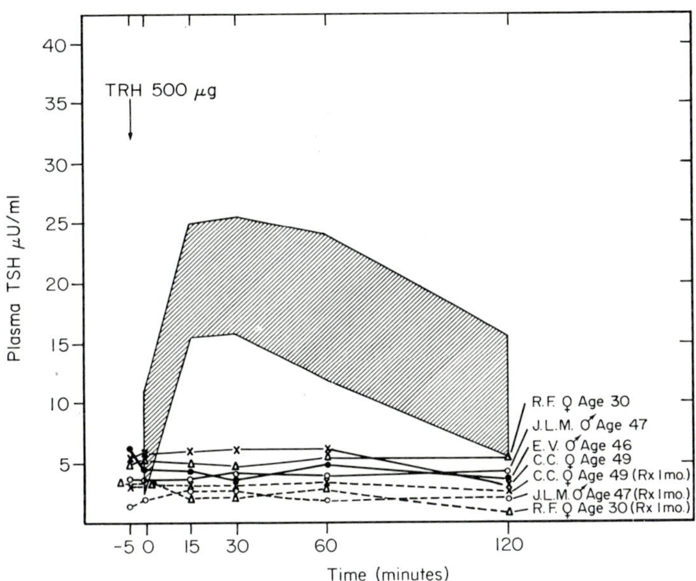

Fig. 9. Plasma thyroid-stimulating hormone (TSH) after injection of 500 μg of thyrotropin-releasing hormone (TRH) given intravenously in patients with untreated hyperthyroidism; interrupted lines show the TSH response in some of these patients after 1 month of antithyroid drug (PTU) therapy. From Gual *et al.* (1972b).

after 1 month of propylthiouracil therapy, again no response to TRH stimulus was obtained. These results have been comprehensively documented by reports from Hershman and Pittman (1970, 1971), Sakoda *et al.* (1970), Beckers *et al.* (1971b), Haigler *et al.* (1971a), Job *et al.* (1971), von zur Mühlen *et al.* (1970, 1971), Ormston *et al.* (1971b), Saito *et al.* 1971), and Wagner *et al.* (1971). Using this TRH test, von zur Mühlen *et al.* (1971) showed that 62% of euthyroid patients presented a partially exaggerated TSH response after antithyroid drug therapy, but in 38% of the cases no TSH could be released by TRH even several months after attainment of a euthyroid state. Similar findings have recently been reported by Lawton *et al.* (1971), suggesting that a decision on treatment should not be taken on the basis of the TRH test alone.

F. Pituitary TSH Reserve in Euthyroid Patients with Pituitary Tumors

Since the TSH response to intravenous TRH in euthyroid patients with different types of pituitary tumors did not follow a single pattern,

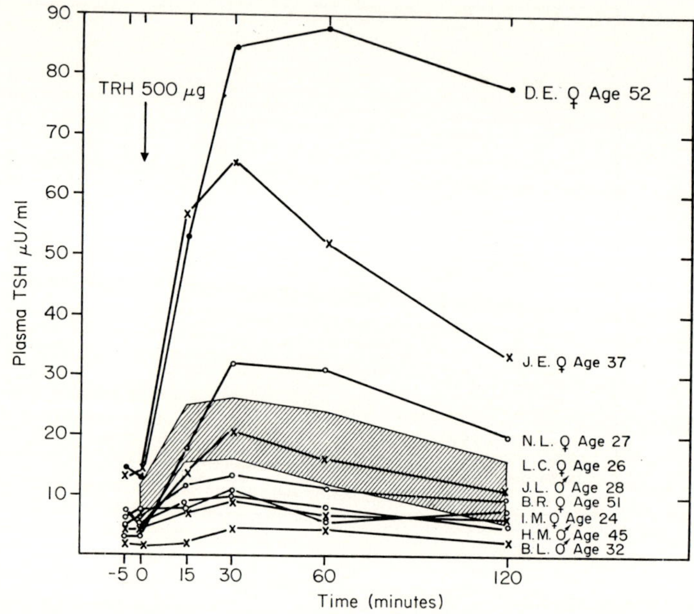

Fig. 10. Effect of 500 µg of thyrotropin-releasing hormone (TRH) given intravenously on plasma thyroid-stimulating hormone (TSH) levels in euthyroid patients with functional and nonfunctional pituitary tumors. The shaded area indicates the range of responses in normal women. From Gual *et al.* (1972b).

they will be individually analyzed. Figure 10 depicts the pituitary TSH reserve test in 4 acromegalic subjects, 4 patients with post adrenalectomy Nelson's syndrome, as well as in 1 patient with a suspected craniopharyngioma. After routine laboratory tests in all patients, no thyroid dysfunction was found.

1. Acromegaly

As depicted in Fig. 10 the TSH response to intravenous TRH in subject N.L. with active acromegaly was above normal limits, whereas subject H.M. in an inactive phase of the disease presented a response which was diminished but within the normal range for men. Subject J.L., who had undergone a selective transsphenoidal surgical removal of the pituitary tumor, presented postoperatively a normal male TSH response, and finally subject B.R. who was under thyroid replacement therapy, gave a diminished pituitary TSH reserve test. From these data it is concluded that TSH response to intravenous administration of TRH in patients with eosinophilic pituitary tumors depends on several factors, such as the clinical course of the disease and the type of therapy employed. These results confirm and extend previous reports of other investigators (Fleischer *et al.*, 1970; Hershman and Pittman, 1970; Haigler *et al.*, 1971a,b; Kastin *et al.*, 1971; Karlberg *et al.*, 1971b; Saito *et al.*, 1971).

2. Nelson's Syndrome

Postadrenalectomized patients who had later developed Nelson's syndrome, presented different types of pituitary response to TRH administration. In two patients under corticoid therapy, opposite responses were observed. In subject J.E. an exaggerated response to TRH stimulus was obtained, whereas in subject I.M. only a moderate response below normal limits was observed. A TRH test within normal limits was found in subject L.C. after selective surgical removal of the pituitary tumor, indicating an adequate TSH pituitary reserve. On the contrary, in subject B.L. a lack of TSH response to TRH stimulus was observed after conventional radiation therapy. Since, to our knowledge these results represent the first experience with the use of TRH in patients with Nelson's syndrome, more tests need to be done to allow a final conclusion; however, it should be pointed out that, as in patients with acromegaly, the response may be conditioned by the therapeutic procedures employed.

3. Craniopharyngioma

Although craniopharyngiomas originate from cells unrelated to the pituitary, it was felt to be of interest to test one patient with suspected craniopharyngioma. As shown in Fig. 10, subject D.E. responded to the TRH stimulus with a marked increase in plasma levels of TSH, thus demonstrating high TSH pituitary reserve due to increased functional pituitary tissue. Comparable but lower responses have been also reported by Fleischer *et al.* (1970), Hershman and Pittman (1970), and Sakoda *et al.* (1970).

G. Galactorrhea–Amenorrhea Syndromes

The clinical features of patients with galactorrhea–amenorrhea syndromes associated with either pituitary tumors or primary myxedema (Kinch *et al.*, 1969) have been well established. In order to investigate the TSH pituitary reserve in patients with galactorrhea–amenorrhea associated with hypothyroidism, 8 subjects received intravenously 500 μg of synthetic TRH and their TSH responses were evaluated. The results obtained in this study are shown in Fig. 11 in which solid lines represent patients with primary hypothyroidism and dashed lines represent patients with normal thyroid function. All patients with primary hypothyroidism presented high control levels of plasma TSH as well as an

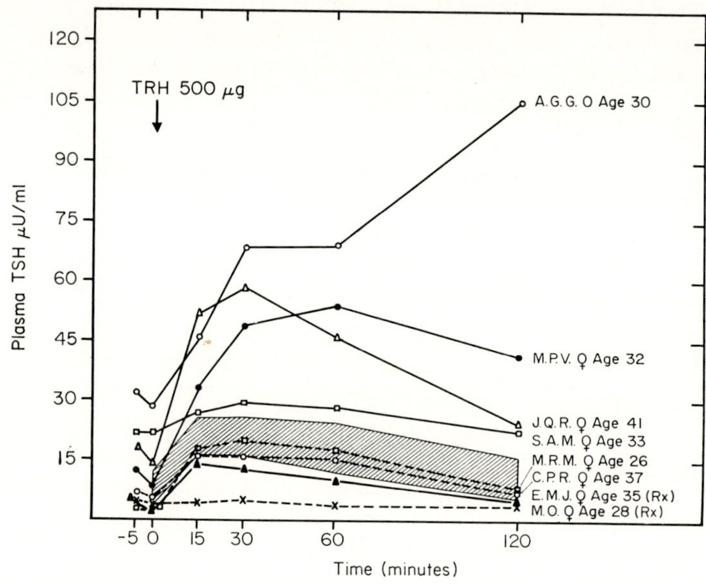

Fig. 11. Effect of 500 μg of thyrotropin-releasing hormone (TRH) given intravenously on plasma thyroid-stimulating hormone (TSH) levels in patients with galactorrhea-amenorrhea syndromes and high plasma prolactin. Solid lines depict responses in patients with clinical hypothyroidism; dashed lines depict responses in clinically euthyroid patients. Shaded area indicates the range of responses seen in normal women. From Gual et al. (1972c).

increased response to TRH stimulus, with the exception of subject E.M.J., who had a diminished TRH response probably caused by prolonged thyroid replacement therapy. On the other hand, the group of euthyroid patients had a normal pituitary response to the administered TRH; however, subject M.O., under desiccated thyroid therapy, failed to respond to TRH stimulation. In all these patients the serum levels of prolactin were well above the normal limits, but patients with hypothyroidism showed lower prolactinlike levels than those found in patients with normal thyroid function or under thyroid replacement therapy

(Table I). Further investigations were undertaken in order to explore the pituitary TSH response to TRH administration in patients with galactorrhea, normal menses, and normal plasma prolactin and in subjects with documented hypothalamic amenorrhea, no galactorrhea, and normal plasma prolactin levels. Figure 12 depicts the TSH response to TRH stimulus in 4 patients with galactorrhea, normal menses, and normal plasma prolactin. The TSH response in 3 subjects was within the normal range; only subject C.P. demonstrated a lack of pituitary response while on thyroid therapy, but after its withdrawal a normal

TABLE I

Prolactin and TSH Plasma Levels in Galactorrhea Syndromes

Diagnosis	Subject	Age	Prolactin (μl/ml)	TSH (μU/ml)
Galactorrhea, normal menses	C.P.	32	50.8	4.2
	C.P.	32	61.4	3.6
	M.G.	25	107.8	7.1
	M.T.P.	30	113.0	5.0
	M.A.P.	29	69.9	3.0
Galactorrhea–amenorrhea	C.P.R.	37	717.5	6.0
	M.O.	28	963.8	4.9
	M.R.M.	26	587.6	3.6
	E.M.J.	35	718.5	3.5
	M.P.V.	32	546.0	10.6
	A.G.G.	30	320.2	30.3
	J.Q.R.	41	172.5	16.4
	S.A.M.	33	267.2	21.2

response was seen, thus confirming our previous findings. Figure 13 shows the results of pituitary stimulation with TRH in patients with primary or secondary hypothalamic amenorrhea, no galactorrhea, and normal prolactin values. In this group of patients, the hypothalamic origin of the amenorrhea was established by the demonstration of normal ovarian and gonadotropic pituitary function after induction of ovulation by exogenous gonadotropin administration (HMG and HCG) as well as by the increase of circulating levels of both LH and FSH after the intravenous administration of LH-RH/FSH-RH (Gual *et al.*, 1971, 1972a). As can be seen in Fig. 13, the four tested subjects had a normal pituitary TSH response, thus indicating normal pituitary reserve not only for gonadotrophins, but also for TSH.

At present, it is difficult to establish the relationship between prolactin secretion and thyroid function; however, it is clear that galactorrhea in patients with primary hypothyroidism is associated with an increased prolactin secretion. The exact nature of this phenomenon remains to be elucidated. However, Tashjian *et al.* (1971) have demonstrated that TRH *in vitro* stimulates prolactin production in cultured pituitary cells, and Jacobs *et al.* (1971) have found 10-fold plasma prolactin rises, after TRH administration in normal volunteers, to be greater in magnitude than the concomitant TSH elevations. We are in the process of

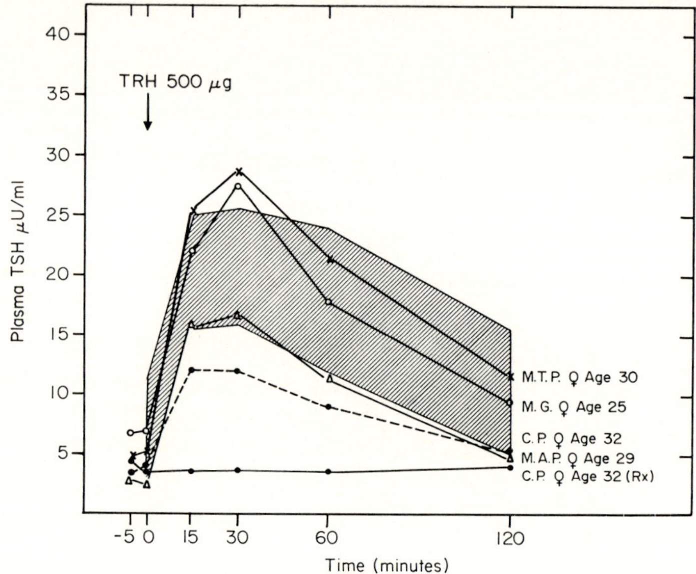

FIG. 12. Effect of 500 μg of thyrotropin-releasing hormone (TRH) given intravenously on plasma thyroid-stimulating hormone (TSH) levels in patients with galactorrhea, normal menses, and normal plasma prolactin. In one patient (C.P.), TRH was given during desiccated thyroid therapy (solid line) and after 1 month of withdrawal (dashed line). The shaded area indicates the range of responses in normal women. From Gual *et al.* (1972c).

evaluating the effects of TRH upon prolactin secretion in the subjects reported herein, and preliminary results support this concept of a significant response in normal subjects, as well as in patients with galactorrhea–amenorrhea syndromes. If these data are confirmed, they may explain the high prolactin values observed in patients with primary hypothyroidism in whom endogenous TRH secretion may be increased. The higher plasma prolactin levels in patients with normal thyroid function cannot be explained on this basis, and further clinical experience is required to elucidate these mechanisms.

H. Pituitary TSH Reserve in Miscellaneous Endocrine Disorders

In this group, 2 patients with genetic gonadal disorders, 4 males with hypogonadotrophic hypogonadism, and 2 male pituitary dwarfs with isolated growth hormone deficiency and normal thyroid function, were tested.

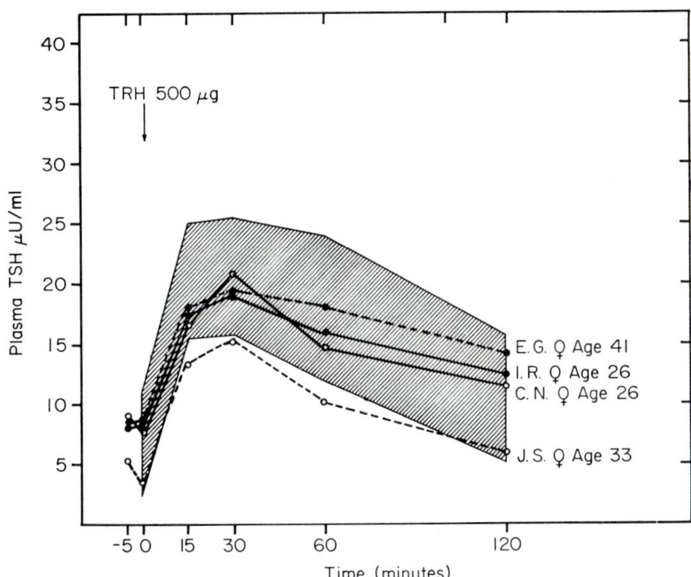

Fig. 13. Plasma thyroid-stimulating hormone (TSH) responses to 500 μg of thyrotropin-releasing hormone (TRH) given intravenously, in patients with suspected hypothalamic amenorrhea and normal plasma prolactin. Ovulation was induced in all cases by HMG and HCG, but no ovarian response was seen after clomiphene citrate therapy. The shaded area indicates the range of responses in normal women. From Gual *et al.* (1972c).

1. Gonadal Dysgenesis

A 20-year-old patient with pure gonadal dysgenesis (46, XX chromosomal pattern) and one 29-year-old patient with Klinefelter's syndrome (47, XXY) received 50 μg iv of TRH for evaluation of TSH response. Both subjects had basal TSH levels within the normal range. The gonadal dysgenesis patient exhibited a sharp increase of plasma TSH up to 36 μU/ml at 30 and 60 minutes after TRH stimulation with a gradual fall to 20 μU/ml at 120 minutes, whereas in the Klinefelter's syndrome patient, a much lower TSH response within the lower limits of normal male responses was obtained. No definitive conclusions can be established from these limited experiences, but it looks as though a normal TSH pituitary reserve does exist in hypergonadotrophic syndromes.

2. Hypogonadotropic Hypogonadism

As depicted in Fig. 14, TSH responses were variable in the four male subjects tested. In two patients a brisk TSH response within the upper range for normal females was observed, but in the other two patients subnormal responses were found. Although, a limited number of cases was studied, it may be possible to conclude that a normal TSH pituitary reserve is present in hypogonadotropic patients.

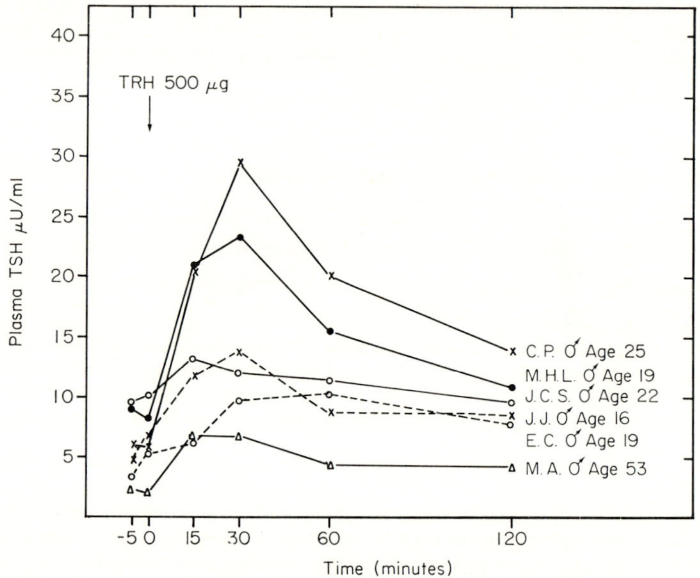

Fig. 14. Plasma thyroid-stimulating hormone (TSH) responses to 500 μg of thyrotropin-releasing hormone (TRH) given intravenously in patients with hypo-gonadotropic hypogonadism (——) and in pituitary dwarfs (idiopathic isolated growth hormone deficiency) (---). From Gual *et al.* (1972b).

3. Pituitary Dwarfism

In the two euthyroid pituitary dwarfs a diagnosis of idiopathic isolated growth hormone deficiency was established by clinical features, by a negative response of plasma growth hormone to provocative stimulation with arginine and insulin, and by normal metopyrapone and ACTH tests. Plasma TSH response to TRH stimulation was found to be within the values found in normal males (Fig. 14). These results confirm that in these dwarfs with normal thyroid function, a normal pituitary TSH reserve is found. Concurring results have been reported very recently by Costom *et al.* (1971), who also have studied hypothyroid dwarfs and concluded that in these patients functional thyrotrophs in the anterior hypophysis are present and suggest that TRH deficiency may be the cause of the TSH deficiency and secondary hypothyroidism.

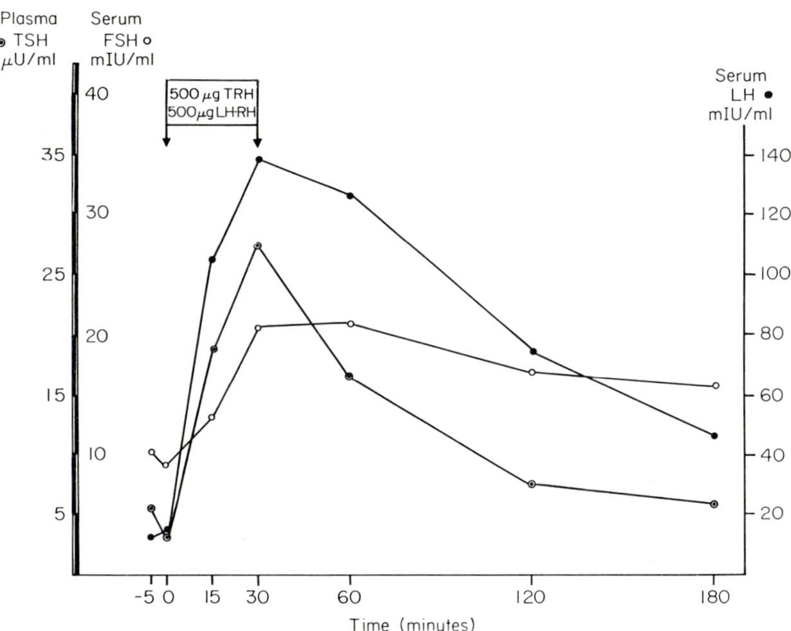

Fig. 15. Plasma thyroid-stimulating hormone (TSH) (⊙), luteinizing hormone (LH) (●), and follicle-stimulating hormone (FSH) (○) responses to a mixture of 500 μg of TRH and 200 μg of synthetic LH-RH, infused in 100 ml of saline over a period of 30 minutes as a clinical test of pituitary reserve in a normal woman (age 18).

I. Simultaneous Administration of Synthetic TRH and LH-RH as a Pituitary Reserve Test

The availability of synthetic TRH and LH-RH made it possible to design a pituitary TSH, LH, and FSH reserve test. In Fig. 15 are depicted the results obtained in a normal female who received a mixture of 500 μg of TRH and 200 μg of LH-RH by a 30-minute intravenous infusion; as can be noted, the TRH/LH-RH stimulus gave a significant rise in plasma TSH, LH, and FSH levels. These results clearly indicate that this test may have an important role in both the diagnosis and in the understanding of the physiology of the hypothalamus-pituitary axis.

V. Summary

The studies described here show that natural and synthetic TRH induce in human subjects a significant plasma TSH increase after its oral or intravenous administration. Responses to intravenous TRH in normal females was significantly greater than in normal males in most studies, but pretreatment with ethynylestradiol in normal males has not consistently increased the TSH response in all investigations. Thyroxine therapy as well as repeated and increasing doses of TRH caused a blockade in the pituitary TSH response. Patients with primary myxedema showed an exaggerated pituitary TSH as compared with hypothalamic hypothyroid patients. No response was obtained when the TRH test was performed in pituitary hypothyroid subjects. Euthyroid patients with pituitary tumors showed variable responses, often related to previous therapeutic procedures. When hypothyroid women with galactorrhea and amenorrhea syndromes were tested with intravenous TRH, the pituitary TSH reserve was found to be like that seen in primary myxedema patients, whereas normal responses were detected in euthyroid galactorrhea–amenorrhea subjects. A normal TSH response to TRH injection was observed in women with normal menses and galactorrhea as well as in women with primary or secondary amenorrhea without galactorrhea. Elevated plasma prolactinlike activity was found only in patients with galactorrhea–amenorrhea syndromes. A normal pituitary TSH response was shown in patients with gonadal dysgenesis, hypergonadotropic hypogonadism, and isolated growth hormone-deficient dwarfs. On the basis of these studies, it is suggested that the TRH test alone or combined with LH-RH may be an important diagnostic tool as well as a means for better understanding of the hypothalamic–pituitary physiology.

ACKNOWLEDGMENTS

This work was supported by a Grant (650-025B) from the Ford Foundation. The authors would like to express their appreciation to Dr. Michael S. Anderson and Dr. Wilfred F. White of the Abbott Laboratories, North Chicago, Illinois, for the generous supply of synthetic TRH. The authors also are grateful to Dr. John F. Wilber of Northwestern University, Medical School, Chicago, Illinois, for plasma TSH determinations and to Dr. A. Rees Midgley of the University of Michigan School of Medicine, Ann Arbor, Michigan, for plasma LH, FSH, and prolactin determinations. We thank Dr. Michael S. Anderson, Dr. Bruce H. Costom, Dr. Norman Fleischer, and Dr. David Haigler for permitting us to quote data still in press.

REFERENCES

Adams, L., and Maloof, F.   (1970).   *J. Clin. Invest.* **49,** 1.
Anderson, M. S., Bowers, C. Y., Kastin, A. J., Schalch, D. S., Schally, A. V.,

Utiger, R. D., Snyder, P. J., Wilber, J. F., and Wise, A. J. (1971a). *Clin. Res.* **19**, 366.

Anderson, M. S., Bowers, C. Y., Kastin, A. J., Schalch, D. S., Schally, A. V., Snyder, P. J., Utiger, R. D., Wilber, J. F., and Wise, A. J. (1971b). *New Engl. J. Med.* **285**, 1279.

Averill, R. L. W. (1969). *Endocrinology* **85**, 67.

Baugh, C. M., Krumdieck, C. L., Hershman, J. M., and Pittman, J. A., Jr. (1970). *Endocrinology* **87**, 1015.

Beckers, C., Cornette, C., and Maskens, A. (1971a). *Program 53rd Meet., Amer. Endocrine Soc., San Francisco* p. 168 (abstr.).

Beckers, C., Maskens, A., and Cornette, C. (1971b). *Ann. Endocrinol.* **32**, 214.

Bøler, J., Enzmann, F., Folkers, K., Bowers, C. Y., and Schally, A. V. (1969). *Biochem. Biophys. Res. Commun.* **37**, 705.

Bowers, C. Y., Redding, T. W., and Schally, A. V. (1965). *Endocrinology* **77**, 609.

Bowers, C. Y., Schally, A. V., Reynolds, G. A., and Hawley, W. D. (1967). *Endocrinology* **81**, 741.

Bowers, C. Y., Schally, A. V., Hawley, W. D., Gual, C., and Parlow, A. (1968). *J. Clin. Endocrinol. Metab.* **28**, 978.

Bowers, C. Y., Schally, A. V., Enzmann, F., Bøler, J., and Folkers, K. (1970a) *Endocrinology* **86**, 1143.

Bowers, C. Y., Schally, A. V., Schalch, D. S., Gual, C., Kastin, A., Castañeda, E., and Folkers, K. (1970b). *Program 52nd Meet., Amer. Endocrine Soc., St. Louis* p. 41 (abstr.).

Bowers, C. Y., Schally, A. V., Schalch, D. S., Gual, C., Kastin, A. J., and Folkers, K. (1970c). *Biochem. Biophys. Res. Commun.* **39**, 352.

Bowers, C. Y., Schally, A. V., Kastin, A., Arimura, A., Schalch, D. S., Gual, C., Castañeda, E., and Folkers, K. (1971). *J. Med. Chem.* **14**, 477.

Burgus, R., Dunn, T. F., Desiderio, D., and Guillemin, R. (1969a). *C. R. Acad. Sci.* **269**, 1870.

Burgus, R., Dunn, T. F., Desiderio, D., Vale, W., and Guillemin, R. (1969b). *C. R. Acad. Sci.* **269**, 226.

Burgus, R., Dunn, T. F., Ward, D. N., Vale, W., Amoss, A., and Guillemin, R. (1969c). *C. R .Acad. Sci.* **268**, 2116.

Burgus, R., Dunn, T. F., Desiderio, D., Ward, D. N., Vale, W., and Guillemin, R. (1970a). *Nature (London)* **226**, 321.

Burgus, R., Dunn, T. F., Desiderio, D. M., Ward, D. N., Vale, W., Guillemin, R., Felix, A. M., Gillessen, D., and Studer, R. O. (1970b). *Endocrinology* **86**, 573.

Costom, B. H., Grumbach, M. M., and Kaplan, S. L. (1971). *J. Clin. Invest.* **50**, 2219.

D'Angelo, S. A. (1963). *Advan. Neuroendocrinol., Proc. Symp., Miami, 1961* p. 158.

Enzmann, F., Bøler, J., Folkers, K., Bowers, C. Y., and Schally, A. V. (1971). *J. Med. Chem.* **14**, 469.

Fleischer, N., Burgus, R., Vale, W., Dunn, T., and Guillemin, R. (1970). *J. Clin. Endocrinol. Metab.* **31**, 109.

Fleischer, N., Lorente, M., Hauger-Klevene, J., and Calderon, M. (1971a). *Program 53rd Meet., Amer. Endocrine Soc., San Francisco* p. 87 (abstr.).

Fleischer, N., Lorente, M., Kirkland, J., Kirkland, R., Clayton, G., Calderon, M., and Hauger-Klevene, J. (1971b). Submitted for publication.

Flouret, G. (1970). *J. Med. Chem.* **13**, 843.

Folkers, K., Enzmann, F., Bøler, J., Bowers, C. Y., and Schally, A. V. (1969). *Biochem. Biophys. Res. Commun.* **37**, 123.

Gillessen, D., Felix, A. M., Lergier, W., and Studer, R. O. (1970). *Helv. Chim. Acta* **53**, 63.

Glatstein, E., McHardy-Young, S., Brast, N., Eltringham, J. R., and Kriss, J. P. (1971). *J. Clin. Endocrinol. Metab.* **32**, 833.

Greer, M. (1957). *Recent Progr Horm Res.* **18**, 67.

Gual, C., Kastin, A. J., Midgley, A. R., Jr., and Flores, F. (1971). *Program 53rd Meet., Amer. Endocrine Soc., San Francisco,* p. 128 (abstr.).

Gual, C., Schally, A. V., Kastin, A. J., Midgley, A. R., Jr., and Flores, F. (1972a). Submitted for publication.

Gual, C., Wilber, J. F., Tello, C., and Rios, E. (1972b). *Rev. Invest. Clin.* **24** (in press).

Gual, C., Villanueva, A., Wilber, J. A., and Midgley, A. R., Jr. (1972c). Submitted for publication.

Guillemin, R. (1965). *Proc 23rd Int. Congr. Physiol. Sci.* p. 284.

Guillemin, R. (1967). *Annu. Rev. Physiol.* **29**, 313.

Guillemin, R., Yamazaki, E., Jutisz, M., and Sakiz, E. (1962). *C. R. Acad. Sci.* **255**, 1018.

Guillemin, R., Sakiz, E., and Ward, D. N. (1965). *Proc. Soc. Exp. Biol. Med.* **118**, 1132.

Guillemin, R., Burgus, R., Sakiz, E., and Ward, D. N. (1966). *C. R. Acad. Sci.* **262**, 2278.

Haigler, E. D., Jr., Hershman, J. M., and Pittman, J. A., Jr. (1971a). *Clin. Res.* **19**, 373.

Haigler, E. D., Jr., Pittman, J. A., Jr., Hershman, J. M., and Baugh, C. M. (1971b). *J. Clin. Endocrinol. Metab.* **33**, 573.

Hall, R., Amos, J., Garry, R., and Buxton, R. L. (1970). *Brit. Med. J.* **ii**, 274.

Hall, R., Amos, J., and Ormston, B. J. (1971). *Brit. Med. J.* **i**, 582.

Harris, G. W. (1955). "Neural Control of the Pituitary Gland." Arnold, London.

Hershman, J. M., and Pittman, J. A., Jr. (1970). *J. Clin. Endocrinol. Metab.* **31**, 457.

Hershman, J. M., and Pittman, J. A., Jr. (1971). *Ann. Intern. Med.* **74**, 481.

Inouye, K., Namba, K., and Otsuka, H. (1971). *Bull. Chem. Soc. Jap.* **44**, 1689.

Jackson, I. M. D., Whyte, W. G., and Garrey, M. M. (1969). *J. Clin. Endocrinol. Metab.* **29**, 315.

Jacobs, L., Schneider, P., Wilber. J. F., Daughaday, W. H., and Utiger, R. D. (1971). *J. Clin. Endocrinol. Metab.* **33**, 996.

Job, J. C., Milhaud, G., Binet, E., Rivaille, P., and Moukhtar, M. S. (1971). *Rev. Eur. Etud. Clin. Biol.* **16**, 537.

Karlberg, B., Almqvist, S., and Werner, S. (1971a). *Acta Endocrinol. (Copenhagen) Suppl* **155**, Abstr. 7.

Karlberg, B., Almqvist, S., and Werner, S. (1971b). *Acta Endocrinol. (Copenhagen)* **67**, 288.

Kastin, A. J., Schally, A. V., Gonzalez-Barcena, D., Schalch, D. S., Lee, L., and Villalpando, S. (1971). *Clin. Res.* **19**, 374.

Kinch, R. A. H., Plunkett, E. R., and Devlin, M. C. (1969). *Amer. J. Obstet. Gynecol.* **105**, 766.

Köbberling, J., von zur Mühlen, A., and Emrich, D. (1971). *Acta Endocrinol. (Copenhagen) Suppl.* **155**, Abstr. 1.

Lawton, N. F., Ekins, R. P., and Nabarro, J. D. N. (1971). *Lancet* **ii**, 14.

Meyer, W. J., Smith, H. C., and Bartter, F. C. (1971). *Program 53rd Meet., Amer. Endocrine Soc., San Francisco* p. 88 (abstr.).

Midgley, A. R., Jr. (1966). *Endocrinology* **79**, 10.

Midgley, A. R., Jr. (1967). *J. Clin. Endocrinol. Metab.* **27**, 295.

Midgley, A. R., Jr. (1971). To be published.

Mittler, J. C., Redding, T. W., and Schally, A. V. (1969). *Proc. Soc. Exp. Biol. Med.* **130**, 406.

Nair, R. M. G., Barrett, J. F., Bowers, C. Y., and Schally, A. V. (1970). *Biochemistry* **9**, 1103.

Odell, W. D., Wilber, J. F., and Utiger, R. D. (1967). *Recent Progr. Horm. Res.* **23**, 47.

Ormston, B. J., Amos, J., Garry, R., Kilborn, J. R., and Hall, R. (1971a). *J. Endocrinol.* **49**, 10.

Ormston, B. J., Cryer, R. J., Garry, R., Besser, G. M., and Hall, R. (1971b). *Lancet* **ii**, 10.

Ormston, B. J., Kilborn, J. R., Garry, R., Amos, J., and Hall, R. (1971c). *Brit. Med. J.* **ii**, 199.

Raptis, S., Rothenbuchner, G., Birk, J., Loos, V., Schleyer, M., and Pfeiffer, E. F. (1971). *Acta Endocrinol. (Copenhagen) Suppl.* **155**, Abstr. 182.

Reichlin, S. (1963). *New Engl. J. Med.* **269**, 1182.

Rothenbuchner, G., Birk, J., Raptis, S., Loos, U., Schleyer, M., and Pfeiffer, E. F. (1971a). *Acta Endocrinol. (Copenhagen) Suppl.* **155**, Abstr. 2.

Rothenbuchner, G., Vanhaelst, L., Birk, J., Golstein, J., Loos, U., Raptis, S., Winkler, G., Schleyer, M., and Pfeiffer, E. F. (1971b). *Eur. Soc. Clin. Invest.* **1**, 389.

Saito, S., Abe, K., Yoshida, H., Kaneko, T., Nakamura, E., Shimizu, N., and Yanaihara, N. (1971). *Endocrinol. Jap.* **18**, 101.

Sakoda, M, Otsuki, M., Hiroshige, N., Kanzo, K., Yagi, A., and Honda M. (1970). *Endocrinol. Jap.* **17**, 541.

Schally, A. V., and Redding, T. W. (1967). *Proc. Soc. Exp. Med.* **126**, 320.

Schally, A. V., Bowers, C. Y., and Redding, T. W. (1966a). *Proc. Soc. Exp. Biol. Med.* **121**, 718.

Schally, A. V., Bowers, C. Y., and Redding, T. W. (1966b). *Endocrinology* **78**, 726.

Schally, A. V., Bowers, C. Y., Redding, T. W., and Barrett, J. F. (1966c). *Biochem. Biophys. Res. Commun.* **25**, 165.

Schally, A. V., Muller, E. E., Arimura, A., Bowers, C. Y., Saito, T., Redding, T. W., Sawano, S., and Pizzolato, P. (1967). *J. Clin. Endocrinol. Metab.* **27**, 755.

Schally, A. V., Arimura, A., Bowers, C. Y., Kastin, A. J., Sawano, S., and Redding, T. W. (1968). *Recent Progr. Horm. Res.* **24**, 497.

Schally, A. V., Redding, T. W., Bowers, C. Y., and Barrett, J. F. (1969). *J. Biol. Chem.* **244**, 4077.

Schally, A. V., Arimura, A., Bowers, C. Y., Wakabayashi, I., Kastin, A. J., Redding, T. W., Mittler, J. C., Nair, R. M. G., Pizzolato, P., and Segal, A. J. (1970a). *J. Clin. Endocrinol Metab.* **31**, 291.

Schally, A. V., Nair, R. M. G., Barrett, J. F., Bowers, C. Y., and Folkers, K. (1970b). *Fed. Proc., Fed. Amer. Soc. Exp. Biol.* **29**, 470.

Schreiber, V., Rybak, M., Eckertova, A., Jirgl, V., Koci, J. Franc, Z., and Kmentova, V. (1962). *Experientia* **18**, 388.

Tashjian, A. H., Jr., Barowsky, N. J., and Jensen, D. K.   (1971).   *Biochem. Biophys. Res. Commun.* **43,** 516.
Vale, W., Burgus, R., and Guillemin, R.   (1967).   *Proc. Soc. Exp. Biol. Med.* **125,** 210.
Vale, W., Burgus, R., Dunn, T. F., and Guillemin, R.   (1970).   *J. Clin. Endocrinol. Metab.* **30,** 148.
von zur Mühlen, A., Hesch, R. D., Embrich, D., and Creutzfeldt, W.   (1970).   *Deut. Med. Wocherschr.* **95,** 2623.
von zur Mühlen, A., Hesch, R. D., Köbberling, J., and Emrich, D.   (1971).   *Acta Endocrinol. (Copenhagen) Suppl.* **155,** Abstr. 6.
Wagner, H., Hrubesch, M., Bökel, K., Vosberg, H., Junge-Hülsing, G., and Hauss, W. H.   (1971).   *Acta Endocrinol. (Copenhagen) Suppl.* **155,** Abstr. 3.

[Discussion follows Part 2 of this chapter, page 217.]

Clinical Experience with Hypothalamic Releasing Hormones

Part 2. Luteinizing Hormone-Releasing Hormone and Other Hypophysiotropic Releasing Hormones

ABBA J. KASTIN, CARLOS GUAL, AND ANDREW V. SCHALLY

Endocrinology Section of the Medical Service and Endocrine and Polypeptide Laboratories, Veterans Administration Hospital, and Department of Medicine, Tulane University School of Medicine, New Orleans, Louisiana; and Instituto Nacional de la Nutricion, Mexico City, Mexico.

I. Luteinizing Hormone-Releasing Hormone

A. INTRODUCTION

After it was clearly established in laboratory animals that the hypothalamus controls the release of luteinizing hormone (LH) by means of a neurohormone called LH-releasing hormone (LH-RH), it became possible to start a series of clinical studies with LH-RH (Schally *et al.*, 1968, 1970b; Schally and Kastin, 1970). Since most of the natural LH-RH obtained after extraction of a huge number of hypothalamic (stalk median eminence) fragments had to be used for studying the structure of LH-RH, the amount available for clinical investigation was restricted. An attempt was made, therefore, to obtain the most information from the small amount available.

The results of the injection of porcine LH-RH into 53 subjects are summarized herein. No other subjects have been tested with highly purified LH-RH, but there were two studies in which crude hypothalamic extracts were used. In the first of these, Igarashi and his co-workers (1968) used beef hypothalami to increase serum follicle-stimulating hormone (FSH) values in 3 of 4 women, and in the second study Root and his co-workers (1969) injected over 100 mg of ovine hypothalamic extract to release LH from 3 abnormal children.

At the time of our first study, it had not been established whether LH-RH obtained from the pig would be effective in releasing LH in the human being. Then, because of the limited amount of material available, it was desirable to determine if there existed a particular type of subject who was especially sensitive to the LH-releasing effect of LH-RH. After this problem was examined, it was possible to start investigating certain theoretical and practical questions concerning control of the release of LH in man.

Although the studies were organized on this type of problem-solving basis, the 53 subjects tested with porcine LH-RH arbitrarily have been regrouped into three sections for this review. The first group started with LH values which were probably within the normal range (Table I). There were 9 untreated men, 2 also injected with putrescine; 12 untreated women, 6 of whom had secondary amenorrhea; 2 untreated boys; 2 untreated girls; and 6 patients with pituitary tumors.

In the second group of subjects, LH values were suppressed either by ethynylestradiol, as in 4 men, 2 of whom received the LH-RH subcutaneously, or by an oral contraceptive, as in 4 women, 2 being postmenopausal. Hypogonadotropic hypogonadism was present in a family of 3 other subjects, two of whom had decreased ability to smell. It is pos-

TABLE I
Subjects Receiving Porcine Luteinizing Hormone-Releasing Hormone (LH-RH)

Group I: Normal resting levels of LH and FSH
    1. 9 untreated men (2 injected with putrescine)
    2. 12 untreated women (6 with secondary amenorrhea)
    3. 2 untreated boys
    4. 2 untreated girls
    5. 6 pituitary tumors (4 acromegalics)
Group II: Suppressed resting levels of LH and FSH.
    1. by ethynylestradiol: 4 men (2 injected subcutaneously)
    2. by oral contraceptive: 4 women (2 postmenopausal)
    3. hypogonadotropic hypogonadism with anosmia: 1 woman, 2 men
    4. ? some of group I-5 (pituitary tumors)
Group III: Elevated resting levels of LH and FSH
    1. by menopause: 2 women
    2. by clomiphene: 6 men
    3. Turner's syndrome: 2
    4. Klinefelter's syndrome: 1

sible that the LH values of some patients with pituitary tumors also belong in this group, but this is difficult to distinguish with the presently available assays.

Another group of subjects to be described here are those whose resting LH values are high. Some of these occurred naturally, as in menopause, Turner's syndrome, and Klinefelter's syndrome, while others were induced artificially by administration of clomiphene, as was done in 6 men.

An example of the results obtained from each type of subject in these three groups is illustrated below. This section discussing clinical studies with LH-RH concludes with some results obtained with human LH-RH and synthetic LH-RH.

In almost every subject to whom porcine LH-RH was administered, a concomitant rise in circulating FSH accompanied the increase in LH levels. At first it was not clear whether this was due to contamination of the LH-RH preparation with FSH-RH or whether LH-RH possessed inherent FSH-releasing activity. Studies with synthetic LH-RH now indicate that LH-RH releases both LH and FSH (Matsuo *et al.*, 1971; Schally *et al.*, 1971c,d).

The natural porcine LH-RH used in the studies described below was purified from porcine hypothalamic tissue, as described previously (Schally *et al.*, 1967). This involved extraction with acetic acid, gel filtration on Sephadex G-25 columns, concentration with phenol, chromatography and rechromatography on carboxymethyl cellulose columns, free-flow electrophoresis, and countercurrent distribution. This preparation was one-tenth as potent in the rat as the most highly purified LH-RH preparations or the synthetic material (Schally *et al.*, 1971a).

The LH-RH was injected intravenously; blood samples were taken 8, 16, 32, 64, and 128 minutes later as well as immediately before the injection. LH and FSH levels were determined in duplicate by radioimmunoassay. The LH-RH preparation used had no detectable LH or FSH activity in the assays. Informed, written consent was obtained from all subjects, none of whom experienced any side effects.

B. Subjects with Normal Resting Levels of LH and FSH

1. Normal, Untreated Men

The salient finding in this subject (Fig. 1) was that porcine LH-RH could elevate LH and FSH levels in a normal human being. This subject also served to answer another question. At one time polyamines such as putrescine were considered to have FSH-RH activity. This concept was based on work performed in the rat which showed that nanogram doses of putrescine could stimulate FSH release *in vivo* (Schally *et al.*, 1968; White *et al.*, 1968). Later, *in vitro* studies using the rat pituitary demonstrated that these polyamine substances do not represent FSH-RH (Schally *et al.*, 1970a). Figure 1 shows that administration of porcine LH-RH resulted in increased release of LH, which was not augmented by the simultaneous administration of 0.1 mg putrescine. The same amount of putrescine by itself was without effect upon LH and FSH release. In this respect, putrescine served as a control, much as did vasopressin in some other tests (Kastin *et al.*, 1969, 1970a).

2. Normal, Untreated Women

Figure 2 illustrates that porcine LH-RH can release LH and FSH in normal women, as well as men. In the small number of cases studied, there have been no statistical differences between the sexes in LH and

FSH release after administration of LH-RH. Results in this group also show the absence of a significant effect of vasopressin on serum LH and FSH levels. The doses of vasopressin administered in these studies ranged from 0.1 to 1.0 unit. These usually corresponded to the pressor

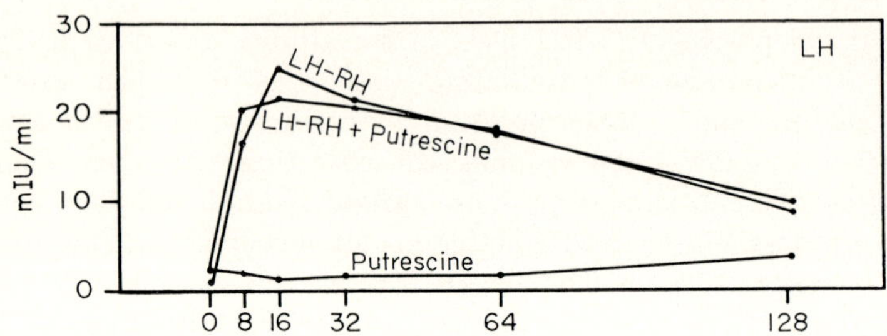

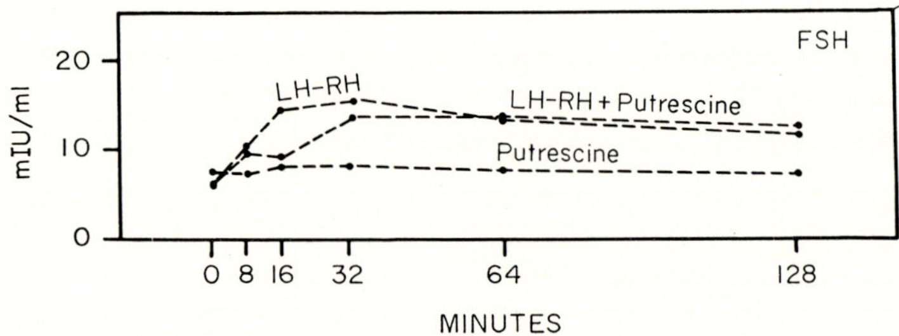

FIG. 1. Serum luteinizing hormone (LH) (●——●) and follicle-stimulating hormone (FSH) (●---●) levels in a normal, untreated man after administration of porcine LH-releasing hormone (LH-RH) with or without putrescine. From Kastin *et al.* (1970a). *Amer. J. Obstet. Gynecol.* **108,** 177.

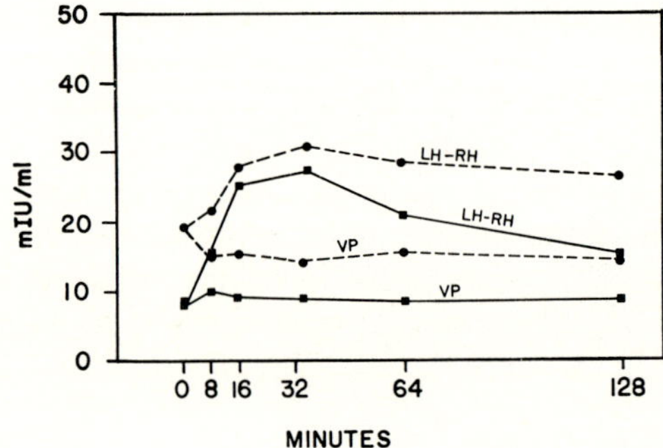

FIG. 2. Serum luteinizing hormone (LH) ■——■ and (FSH) follicle-stimulating hormone (●---●) levels in a normal, untreated woman after administration of porcine LH-releasing hormone (LH-RH). From Kastin *et al.* (1970a). *Amer. J. Obstet. Gynecol.* **108,** 177.

activity in the dose of LH-RH injected, although 1.0 unit represented more vasopressin than was present in LH-RH. It was hoped to be able to induce in the subject of this study ovulation early (day 9) in her menstrual cycle, but this did not occur (Kastin *et al.*, 1969, 1970a). The precise timing of ovulation by administration of LH-RH during the normal menstrual cycle might have importance for the control of fertility.

3. Normal, Untreated Prepubertal Boys and Girls

The children responded to LH-RH by increases in plasma LH (Fig. 3) and FSH similar to those seen after administration of LH-RH to adults (Kastin *et al.*, 1972a). In addition to demonstrating another group of subjects in whom porcine LH-RH is effective, this study served to decrease the possibility that puberty represents the first time that the pituitary is able to respond to LH-RH. Since gonads also respond to LH and FSH in children (Harris, 1955), neither the pituitary nor the gonads appear to be the organs which suddenly change sensitivity at puberty. Rather, it must be the hypothalamus or a higher central

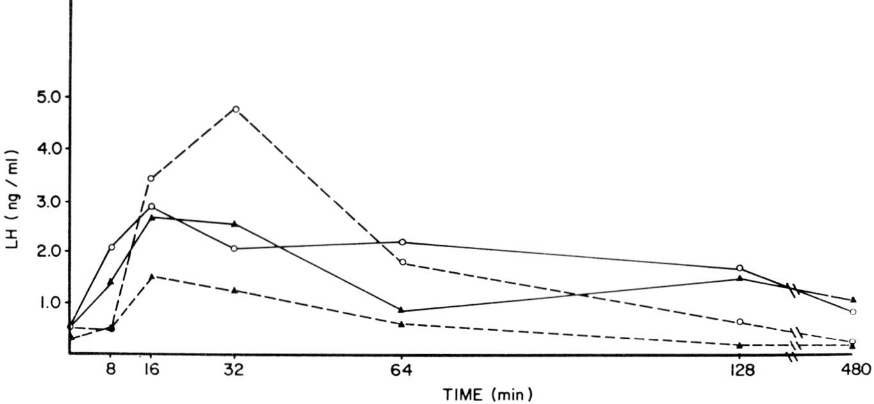

Fig. 3. Serum luteinizing hormone (LH) levels in normal children and adults after administration of porcine LH-releasing hormone (LH-RH). ▲——▲, adult males; ○——○, adult females; ▲---▲, prepubertal males; ●---●, prepubertal females. From Kastin *et al.* (1972a) with permission.

nervous system (CNS) center which controls the onset of puberty, as might be expected.

Only 2 boys, 2 girls, 2 men, and 2 women were compared in this study. Perhaps because of the small number of cases and the wide range of responses, no statistically significant difference was found between males and females or adults and children for either LH and FSH. The question whether porcine LH-RH released growth hormone (GH), thyrotropin (TSH), or cortisol was also answered in this study, in which none of these hormones was significantly changed by the administration of LH-RH. Furthermore, the men and prepubertal boys demonstrated a statistically significant increase in estradiol after injection of LH-RH, showing for the first time that the LH released by LH-RH was able to stimulate a secondary endocrine organ.

4. Ovulation

Another way of demonstrating hormonal response to LH-RH from a target organ would be to induce ovulation. The rapid intravenous method of injection of LH-RH used in the earlier study had failed to cause ovulation in the 4 women in whom it had been previously attempted (Kastin *et al.*, 1970a). Since plasma LH is increased for at least 24 hours in the normal ovulatory menstrual cycle, it was assumed after these first attempts to produce ovulation that a more prolonged stimulation would be required. Therefore, a 34-year-old woman with secondary amenorrhea was infused continuously for 24 hours with a dose of porcine LH-RH similar to that used in other studies (Kastin *et al.*, 1970a), except that 2 additional quick injections of LH-RH (300 μg each) were given at 8 and 24 hours in an attempt to stimulate some of the sudden surges of LH release that appear to occur naturally (Midgley and Jaffe, 1971). In an earlier control cycle, human menopausal gonadotropin (HMG-Pergonal) caused an estrogenic effect but no ovulation. In the cycle in which LH-RH was infused, urinary pregnanediol values and basal body temperature rose to levels indicative of ovulation (Fig. 4). Pregnancy, as detected by an immunological test and subsequently confirmed, constituted final proof of the ovulation, although a causal relationship between LH-RH and ovulation was not proved (Kastin *et al.*, 1971c). A normal boy was delivered at term.

5. Pituitary Tumors

It can be presumed that patients with pituitary tumors may exhibit normal, low, or high levels of LH and FSH and respond to administration of LH-RH with normal, supernormal, or negligible increases in gonadotropins. It would be important to establish the amount of pituitary gonadotropin "reserve" in such cases. Absent as well as normal responses to LH-RH have been found in patients with pituitary tumors (Gual *et al.*, 1971; Kastin *et al.*, 1972b). Moreover, the pituitary response to one hypothalamic hormone does not necessarily correspond to the

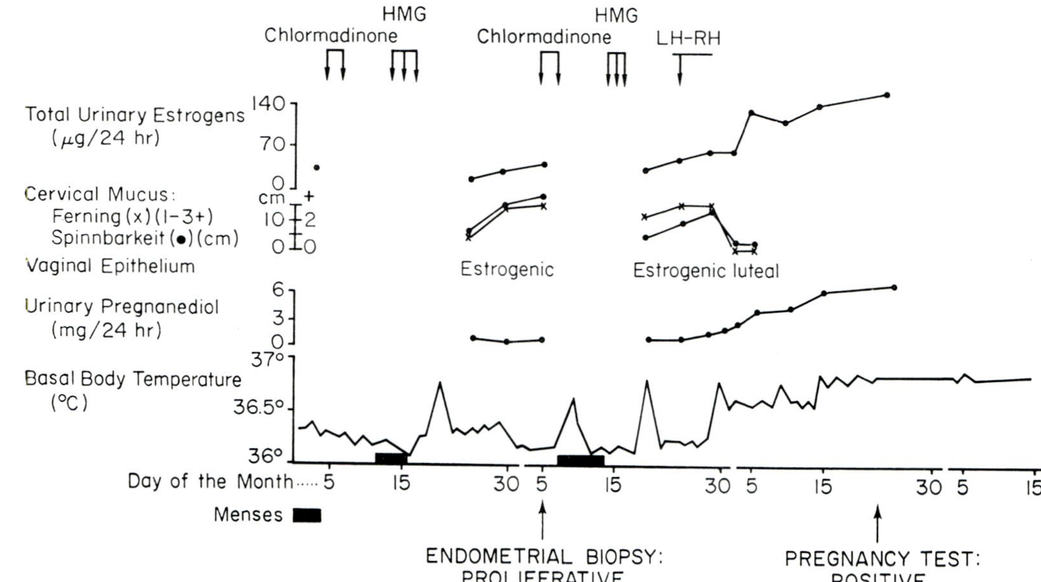

FIG. 4. Basal body temperature and laboratory results in a 34-year-old woman with secondary amenorrhea who was pretreated with human menopausal gonadotopin (HMG; Pergonal) before receiving a 24-hour infusion of porcine luteinizing hormone-releasing hormone (LH-RH). From Kastin *et al.* (1971c). *J. Clin. Endocrinol. Metab.* **33**, 980.

response to another. For example, 3 acromegalies have been tested with both LH-RH and TRH and found to give 3 different sets of responses. One patient responded to TRH but not LH-RH, another to LH-RH but not TRH, and a third responded to both hypothalamic hormones with a corresponding release of the appropriate pituitary hormone (Kastin *et al.*, 1972b).

6. Dose-Response and Minimum Effective Dose

In the studies discussed in this review, almost all the subjects received 300–1500 μg of purified porcine LH-RH. It was determined in 3 normal men, however, that as little as 10 μg of this material could cause statis-

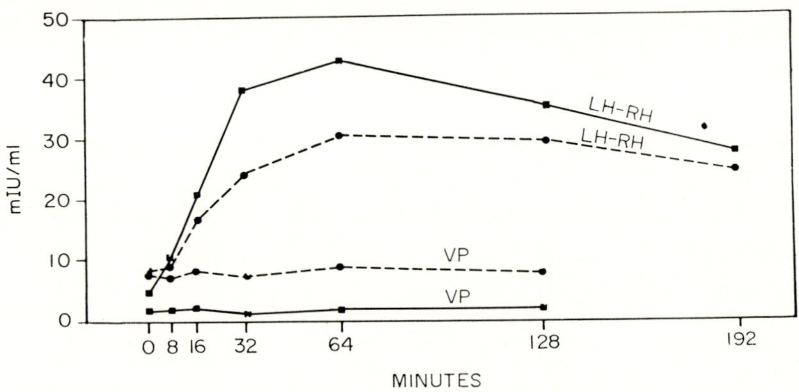

Fig. 5. Serum luteinizing hormone (LH) (■——■) and follicle-stimulating hormone (FSH) (●---●) levels in a man pretreated with 1.5 mg of ethynylestradiol for 3 days before subcutaneous injection of porcine LH-releasing hormone (LH-RH) and lysine vasopressin (VP). From Kastin *et al.* (1970a). *Amer. J. Obstet. Gynecol.* **108**, 177.

tically significant increase in LH release and only 30 μg could significantly increase FSH release. A linear dose-response curve was found (Kastin *et al.*, 1971b). Since natural and synthetic preparations have been made which are ten times more potent than the preparations used in our clinical studies, the minimum effective dose of LH release should be approximately 1 μg.

C. Subjects with Low Resting Levels of LH and FSH

1. Men Pretreated with Ethynylestradiol

The LH and FSH responses of the man represented in Fig. 5 are among the greatest recorded (Kastin *et al.*, 1970a). He is one of only

2 men who have been injected subcutaneously rather than intravenously. LH-RH increased serum LH approximately 900% and FSH 260%. It is not known whether this is because of the subcutaneous route of administration or not, since this subject never received LH-RH intravenously. It will be necessary to inject the same individual at appropriate intervals in several ways and compare the responses. The results obtained with this man do suggest, however, that routes of administration other than intravenous may be feasible for LH-RH.

2. Women Pretreated with Lyndiol

Like the group mentioned above, women pretreated with an oral contraceptive are able to release LH and FSH in response to LH-RH (Fig.

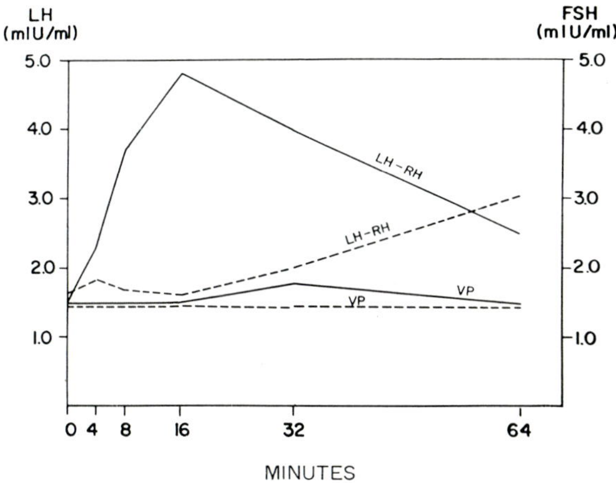

FIG. 6. Serum luteinizing hormone (LH) (———) and follicle-stimulating hormone (FSH) (---) levels in a woman pretreated with the oral contraceptive Lyndiol before administration of porcine LH-releasing hormone (LH-RH) and lysine vasopressin (VP). From Kastin *et al.* (1969). *J. Clin. Endocrinol. Metab.* **29,** 1046.

6). That pretreatment with sex steroids did not prevent the release of LH and FSH by LH-RH in either group of subjects, indicated that most of the action of these compounds in suppressing gonadotropin levels must occur at the hypothalamus or higher CNS center (Kastin *et al.*, 1969; Schally *et al.*, 1968, 1970b; Schally and Kastin, 1970). This differs from the situation discussed earlier in this article in which large amounts of circulating thyroid hormones prevent the release of TSH after injection of TRH. Thus, thyroid hormones unlike the sex steroids, appear to exert their principle negative feedback action at the level of the pituitary gland in man.

3. Subjects with Hypogonadotropic Hypogonadism and Anosmia

Three siblings with isolated low gonadotropin excretion, two of whom also had decreased ability to smell (Kallman's syndrome) have been injected with porcine LH-RH. Although the increase in plasma LH and FSH levels tended to be small, they were statistically significant, indicating that in this family at least part of the difficulty may be at a level higher than the pituitary (Zarate *et al.*, 1972).

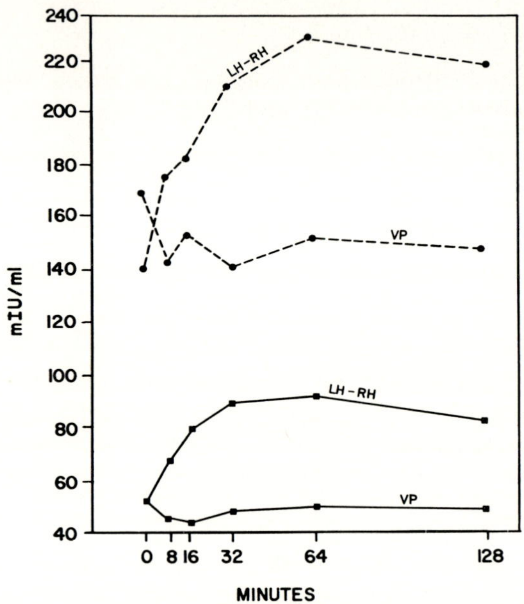

Fig. 7. Serum luteinizing hormone (LH) (■——■) and follicle-stimulating hormone (FSH) (●---●) levels in a postmenopausal woman before administration of porcine LH-releasing hormone (LH-RH) and lysine vasopressin (VP). From Kastin *et al.* (1970a). *Amer J. Obstet. Gynecol.* **108**, 177.

D. SUBJECTS WITH ELEVATED RESTING LEVELS OF LH AND FSH

1. Postmenopausal Women

It was possible that postmenopausal women, having already increased baseline gonadotropin levels, would not be able to release additional LH or FSH. Figure 7 shows that this was not the case. Although the percent increase was not so great, in terms of the 41.6 units released, it was just as great as in most other subjects. This suggested at that time that perhaps LH-RH could release additional LH and FSH from subjects in whom high gonadotropin levels had been induced by treatment with drugs such as clomiphene (Kastin *et al.*, 1970a).

2. Men Pretreated with Clomiphene

Clomiphene is a compound known to release LH as well as FSH. Accordingly, three men were pretreated with 200 mg/day of clomiphene for 8 days, and 3 men for 16 days. The LH response in the group of men who received clomiphene for 8 days is shown in Fig. 8. The increase in LH after 2 weeks of clomiphene was not significantly greater than that seen after 1 week. Administration of porcine LH-RH to these 3

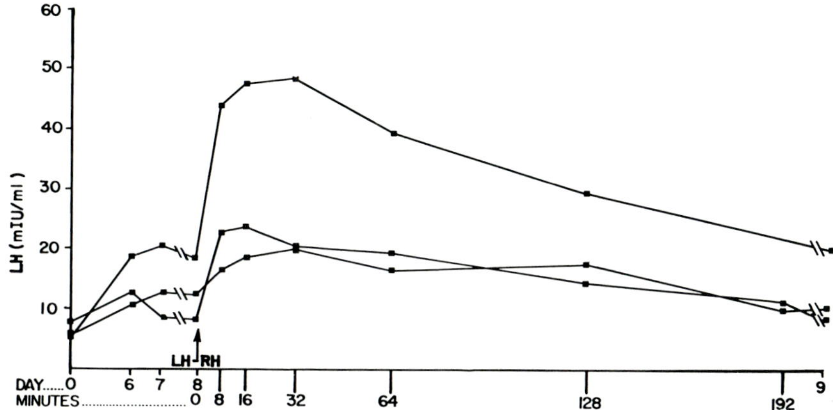

FIG. 8. Serum luteinizing hormone (LH) levels in 3 men pretreated with 200 mg of clomiphene for 8 days before administration of porcine LH-releasing hormone (LH-RH). From Kastin *et al.* (1970b). *J. Clin. Endocrinol. Metab.* **31**, 689.

healthy volunteers resulted in a further sharp mean maximum increase of LH of 153%. This additional increase in LH was statistically significant, as was the increase in serum FSH levels (Kastin *et al.*, 1970b). Since clomiphene has been used clinically to induce ovulation in some infertile women, the findings in this study reinforced the possibility that LH-RH might eventually have a similar function.

3. Turner's Syndrome

The resting LH and FSH levels of the 2 subjects with Turner's syndrome (45, X0 chromosomal pattern) were already elevated, presumably because of the lack of a negative feedback from the "streak" gonads. Both patients responded to LH-RH, as shown in Fig. 9 (Gual *et al.*, 1971).

4. Klinefelter's Syndrome

The baseline values of LH and FSH in this patient with Klinefelter's syndrome were also elevated (Fig. 10). The increase in gonadotropin

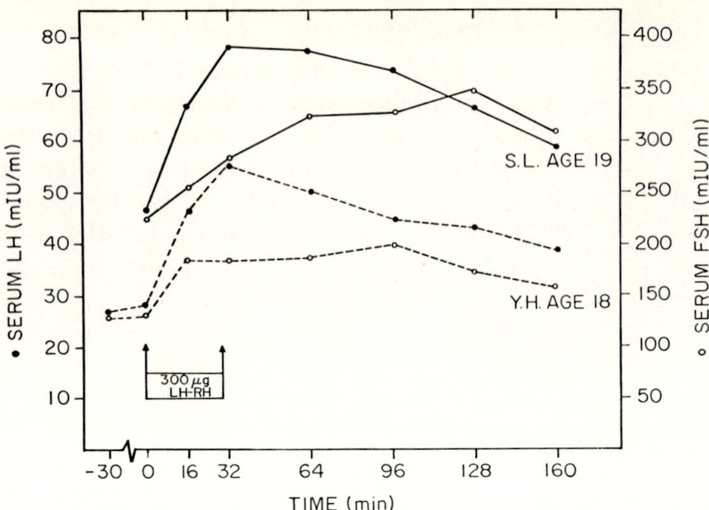

Fig. 9. Serum luteinizing hormone (LH) (●) and follicle-stimulating (FSH) (○) levels in 2 patients with Turner's syndrome before administration of porcine LH-releasing hormone (LH-RH).

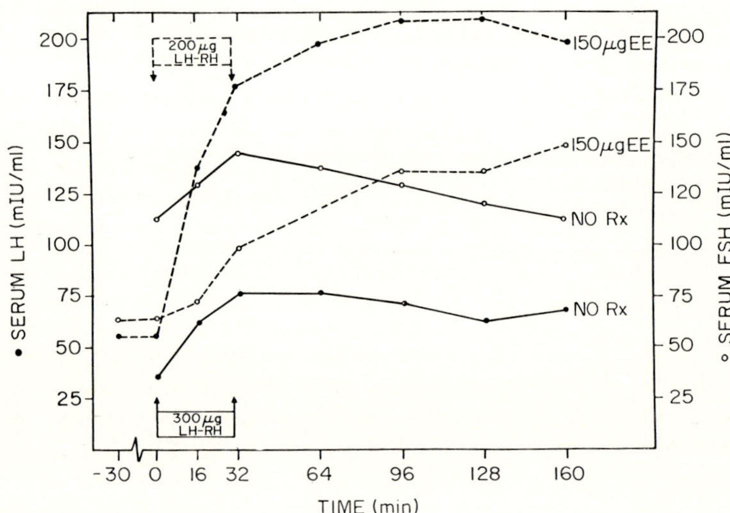

Fig. 10. Serum luteinizing hormone (LH) (●) and follicle-stimulating hormone (FSH) (○) levels after administration of porcine LH-releasing hormone (LH-RH) to a patient (R.A., age 29) with Klinefelter's syndrome before and after pretreatment with 150 μg of ethynylestradiol for 3 days.

release after administration of LH-RH was even greater when he had been pretreated with ethynylestradiol (Gual *et al.*, 1971). This is similar to the normal men who showed a trend (not statistically significant in a small group) toward increased sensitivity to LH-RH after pretreatment with sex steroids (Kastin *et al.*, 1969, 1972c).

E. Effects of LH-RH of Human Origin

Two subjects received LH-RH prepared from human hypothalamic tissue. The response of one of these normal men is shown in Fig. 11, where a significant (290%) increase in serum LH was observed, but

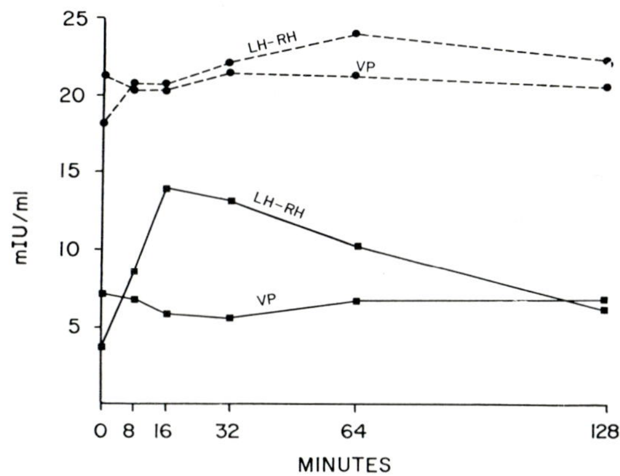

Fig. 11. Serum luteinizing hormone (LH) (■——■) and follicle-stimulating hormone (FSH) (●---●) levels in an untreated man after administration of human LH-releasing hormone (LH-RH) and arginine vasopressin (VP). From Kastin *et al.* (1971a). *J. Clin. Endocrinol. Metab.* **32**, 287.

only a 32% increase in serum FSH. The reason for the lack of significant FSH release is not known, but it may be due to the low dose of LH-RH administered. In contrast to the other subjects who were injected with lysine vasopressin as a control for the porcine LH-RH, this man received arginine vasopressin as a control for the human LH-RH. Again, the vasopressin was without effect (Kastin *et al.*, 1971a).

F. Effects of Synthetic LH-RH

The recent elucidation of the structure and synthesis of LH-RH (Matsuo *et al.*, 1971; Baba *et al.*, 1971; Schally *et al.*, 1971c,d) made it possible to compare the responses of men and women with and without pretreatment with sex steroids to synthetic LH-RH. As in our first study

(Kastin *et al.*, 1969), there were no statistically significant differences in the responses to LH-RH among the groups. A significant release of LH and, to a lesser extent FSH after infusion of synthetic LH-RH did occur (Kastin *et al.*, 1972c). The combined responses (3 per group) are illustrated in Figs. 12 and 13. A large number of clinical studies with the synthetic material are now possible, and it is expected that these will show LH-RH to have both diagnostic and therapeutic value.

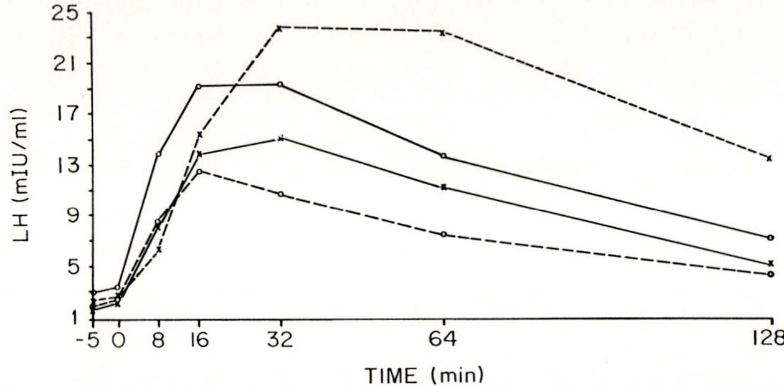

FIG. 12. Plasma luteinizing hormone (LH) levels after administration of synthetic LH-releasing hormone (LH-RH) to normal subjects, some of whom were pretreated with ethynylestradiol (men) or an oral contraceptive (women). X——X, Untreated men; O——O, untreated women; X---X, treated men; O---O, treated women. From Kastin *et al.* (1972c).

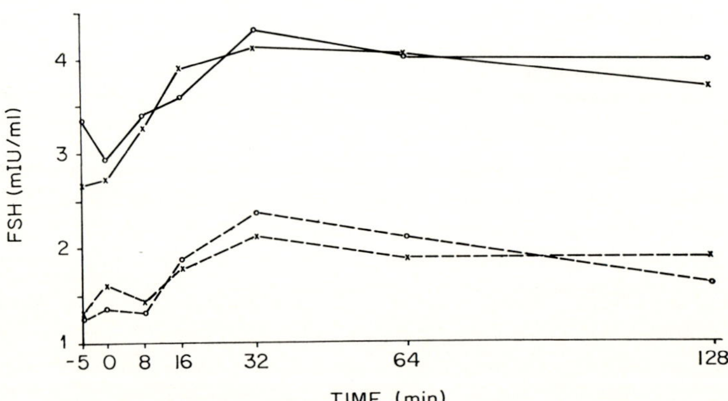

FIG. 13. Plasma follicle-stimulating hormone (FSH) levels after administration of synthetic LH-releasing hormone (LH-RH) to normal subjects, some of whom were pretreated with ethynylestradiol (men) or an oral contraceptive (women). X——X, Untreated men; O——O, untreated women; X---X, treated men; O---O, treated women. From Kastin *et al.* (1972c).

II. Growth Hormone-Releasing Hormone

A substance has been purified from porcine hypothalamic fragments which is capable of releasing growth hormone (GH) from the rat pituitary, as determined by bioassay. Its structure recently has been reported (Schally *et al.*, 1971b). No radioimmunoassay system has detected increased GH release after administration of highly purified or synthetic porcine GH-RH to several species of animals. Accordingly, it was without much hope of observing a significant effect that porcine GH-RH was administered to 6 normal men who had previously been shown to be capable of releasing GH in response to arginine, insulin, or vaso-

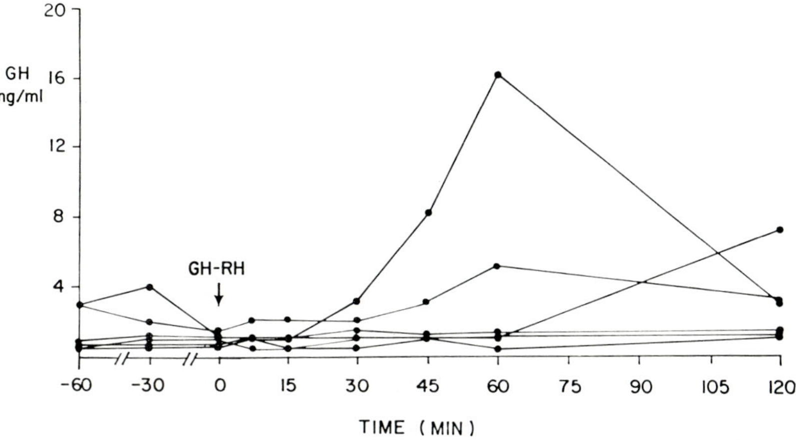

Fig. 14. Plasma growth hormone levels after administration of porcine growth hormone-releasing hormone (GH-RH).

pressin. The results are presented in Fig. 14. In only one subject was there a suggestion of GH release, but he also showed some GH elevation after a control injection of physiological saline (Kastin *et al.*, 1972d).

III. Melanocyte-Stimulating Hormone (MSH)-Release Inhibiting Factor

The isolation of MSH release inhibiting factor (MIF) from bovine hypothalamic tissue and elucidation of its structure as prolylleucylglycinamide recently has been accomplished (Nair *et al.*, 1971). This agrees with the work of Celis *et al.* (1971) that this synthetic substance, which forms the C-terminal tripeptide of oxytocin, has MIF activity. Our concept that MIF exerts the predominant control of MSH release in mammals, as in amphibians, now appears to be generally accepted (Kastin and Schally, 1966, 1971a,b).

A single, acute injection of synthetic MIF into a dozen subjects did not result in a consistent decrease of plasma MSH levels (Kastin *et al.*, 1972e). The subjects included 5 Addisonians off replacement therapy, 3 untreated acromegalics, 3 normals pretreated with metyrapone, a patient with Nelson's syndrome, and a woman with pigmentation of unknown etiology.

It was predicted (Kastin, 1967) that MIF might find use in the therapy of Parkinson's disease. Therefore, several subjects with this disease were given a single intravenous injection of MIF. No significant improvement in the parkinsonism was noted in these acute studies (Kastin *et al.*, 1972e). The results of the chronic studies are not complete. It is possible that if Pro-Leu-Gly-amide were effective, its action(s) might not be mediated by MSH.

IV. Summary

Neuroendocrinology has advanced to the stage where 4 hypothalamic hormones, the structures of all of which are known, have been tested in man. It is reasonable to expect that most of them will find increasing clinical use in diagnosis and treatment.

ACKNOWLEDGMENTS

The authors are grateful to a large number of collaborators, especially Drs. Akira Arimura, David Gonzalez-Barcena, A. Rees Midgley, Jr., Don S. Schalch, and Arturo Zarate. Dr. Edward B. Ferguson, Jr., provided editorial assistance.

REFERENCES

Baba, Y., Matsuo, H., and Schally, A. V. (1971). *Biochem. Biophys. Res. Commun.* **44**, 459.

Celis, M. E., Taleisnik, S., Schwartz, I. L., and Walter, R. (1971). *Biophys. Soc. Abstra.* p. 98a.

Gual, C., Kastin, A. J., Midgley, A. R., and Flores, F. (1971). *Endocrinology* **88**, A-128.

Harris, G. W. (1955). "Neural Control of the Pituitary Gland," p. 93. Arnold, London.

Igarashi, M., Yokota, N., Ehara, Y., Mayuzumi, R., Hirano, T., Matsumoto, S., and Yamasaki, M. (1968). *Amer. J. Obstet. Gynecol.* **100**, 867.

Kastin, A. J. (1967). *New Engl. J. Med.* **276**, 1041.

Kastin, A. J., and Schally, A. V. (1966). *Gen. Comp. Endocrinol.* **7**, 452.

Kastin, A. J., and Schally, A. V. (1971a). *In* "Pigmentation: Its Genesis and Biologic Control" (V. Riley, ed.), p. 215. Appleton, New York.

Kastin, A. J., and Schally, A. V. (1971b). *Proc. 7th Pan-Amer. Congr. Endocrinol. Sao Paulo, 1970*, No. 238, p. 311. *Excerpta Med. Found. Int. Cong. Ser.*

Kastin, A. J., Schally, A. V., Gual, C., Midgley, A. R., Bowers, C. Y., and Diaz-Infante, A. (1969). *J. Clin. Endocrinol. Metab.* **29**, 1046.

Kastin, A. J., Schally, A. V., Gual, C., Midgley, A. R., Bowers, C. Y., and Gomez-Perez, F. (1970a). *Amer. J. Obstet. Gynecol.* **108**, 177.

Kastin, A. J., Schally, A. V., Gual, C., Midgely, A. R., Miller, M. C., and Flores, F. (1970b). *J. Clin. Endocrinol. Metab.* **31**, 689.

Kastin, A. J., Schally, A. V., Gual, C., Midgely, A. R., Arimura, A., Miller, M. C., and Cabeza, A. (1971a). *J. Clin. Endocrinol. Metab.* **32**, 287.

Kastin, A. J., Schally, A. V., Gual, C., Midgley, A. R., Miller, M. C., and Cabeza, A. (1971b). *J. Clin. Invest.* **50**, 1551.

Kastin, A. J., Zarate, A., Midgley, A. R., Canales, E. S., and Schally, A. V. (1971c). *J. Clin. Endocrinol. Metab.* **33**, 980.

Kastin, A. J., Schally, A. V., Schalch, D. S., Korenman, S. G., Miller, M. C., Gual, C., and Perez-Pasten, E. (1972a). *Pediat. Res.* (in press).

Kastin, A. J., Schally, A. V., Gonzalez-Barcena, D., Schalch, D. S., and Lee, L. (1972b). *Arch. Intern. Med.* (in press).

Kastin, A. J., Schally, A. V., Gual, C., and Arimura, A. (1972c). *J. Clin. Endocrinol. Metab.* (in press).

Kastin, A. J., Schally, A. V., Gual, C., Glick, S., and Arimura, A. (1972d). Submitted.

Kastin, A. J., Gonzalez-Barcena, D., Velasco, F., and Schally, A. V. (1972e). In preparation.

Matsuo, H., Baba, Y., Nair, R. M. G., Arimura, A., and Schally, A. V. (1971). *Biochem. Biophys. Res. Commun.* **43**, 1334.

Midgely, A. R., and Jaffe, R. (1971). *J. Clin. Endocrinol. Metab.* **33**, 962.

Nair, R. M. G., Kastin, A. J., and Schally, A. V. (1971). *Biochem. Biophys. Res. Commun.* **43**, 1376.

Root, A. W., Smith, G. P., Dhariwal, A. P. S., and McCann, S. M. (1969). *Nature (London)* **221**, 570.

Schally, A. V., and Kastin, A. J. (1970). *In* "Advances in Steroid Biochemistry and Pharmacology" (M. H. Briggs, ed.), Vol. 2, pp. 41–69. Academic Press, New York.

Schally, A. V., Bowers, C. Y., White, W. F., and Cohen, A. I. (1967). *Endocrinology* **81**, 77.

Schally, A. V., Arimura, A., Bowers, C. Y., Kastin, A. J., Sawano, S., and Redding, T. W. (1968). *Recent Progr. Horm. Res.* **24**, 497.

Schally, A. V., Mittler, J. C., and White, W. F. (1970a). *Endocrinology* **86**, 803.

Schally, A. V., Arimura, A., Kastin, A. J., Reeves, J., Bowers, C. Y., Baba, Y., and White, W. F. (1970b). *In* "Mammalian Reproduction" (H. Gibian and E. J. Plotz, eds.), p. 45. Springer-Verlag, Berlin and New York.

Schally, A. V., Arimura, A., Baba, Y., Nair, R. M. G., Matsuo, H., Redding, T. W., Debeljuk, L, and White, W. F. (1971a). *Biochem. Biophys. Res. Commun.* **43**, 393.

Schally, A. V., Baba, Y., Nair, R. M. G., and Bennett, C. D. (1971b). *J. Biol. Chem.* **246**, 6647.

Schally, A. V., Arimura, A., Kastin, A. J., Matsuo, H., Baba, Y., Redding, T. W., Nair, R. M. G., Debeljuk, L., and White, W. F. (1971c). *Science* **173**, 1036.

Schally, A. V., Kastin, A. J., and Arimura, A. (1971d). *Fert. Steril.* **22**, 703.

White, W. F., Cohen, A. I., Rippel, R. H., Story, J. C., and Schally, A. V. (1968). *Endocrinology* **82**, 742.

Zarate, A., Kastin, A. J., Canales, E., and Schally, A. V. (1972). *Clin. Res.* (in press).

DISCUSSION

A. White: Should the nomenclature of hypothalamic releasing factors or hormones be altered to hypothalamic regulatory hormones, since not all release pituitary hormones but some are inhibitory? Perhaps this might make the terminology a little simpler. As indicated by the speaker, the regulatory hormone for thyrotropin is active by mouth and appears to have a very short duration of action. Thyrotropin-releasing hormone (TRH) is also degraded, e.g., in blood, by an enzyme that catalyzes removal of the pyroglutaminyl residue. Is the short duration of action due to this type of inactivation and/or rapid clearance by the kidney? Also, are these regulatory substances stored in the hypothalamus as a large molecule, e.g., as in the case of neurophysin, and are they liberated by selective proteolysis as requirements arise? In this connection, since TRH is active by mouth and apparently therefore resistant to the proteolytic enzymes of the gastrointestinal tract, I wonder whether anyone has subjected hypothalamic tissue to autolysis and studied the subsequent yield of these regulatory hormones. This might provide an indication as to whether there was a larger neurophysin-like material present in the hypothalamus.

R. Guillemin: It has been a surprise to us in a series of studies recently completed (Braudo *et al.*) to observe that labeled ^{3}H-TRF injected into rats in relatively large doses (1–5 μg/animal) can be recovered in the urine (as much as 50%) in spite of the rapid plasma inactivation which everyone recognizes. The TRF recovered in the urine is biologically active. The labeled TRF is filtered just like inulin as studied by concomitant clearance with inulin-^{14}C.

I. L. Schwartz: This is in response to Dr. White's question about whether there may be precursor or storage forms for hypothalamic releasing or release-inhibiting hormones and whether the hypothalamic hormones may appear after enzymatic action in hypothalamic tissue. The answer is "yes" to both of these questions, at least in the case of the melanocyte-stimulating hormone-releasing factor (MSH–RF) and the melanocyte-stimulating hormone release-inhibiting factor (MSH–RIF). In a recent collaboration between our laboratory and that of Talesnick in Argentina, Celis and Walter found that oxytocin, in addition to its role as a hormone per se, functions as a prohormone for MSH–RIF and possibly also for MSH–RF. Probably an appreciable portion of the oxytocin which is produced in the hypothalamus is cleaved enzymatically to yield its tripeptide tail, L-prolylleucylglycinamide, to serve as the MSH-RIF and a pentapeptide, derived from its ring component to serve as the MSH–RF. Furthermore, there is reason to believe that oxytocin itself may derive from a still larger molecule; thus, in all probability there are prohormones for prohormones, i.e., pro-prohormones.

N. A. Samaan: We have been interested in the use of TRH for treatment of follicular and mixed papillary follicular thyroid carcinoma with metastasis. We hoped that with the increase of TSH release by the pituitary gland after TRH administration we would induce the increase of ^{131}I uptake in the remaining thyroid and metastatic tissues. We noticed that after withdrawal of thyroid therapy for 6 weeks there was a rise of basal TSH levels, as expected. After intravenous administration of TRH in a dose of 500 mg every 8 hours for 2.5 days, there was at least a 100% rise in the serum TSH level 20–30 minutes after the injection of TRH. However, in spite of the rise of the circulating TSH, there was no increase in ^{131}I uptake in the metastatic or remaining thyroid tissues. We also noticed that if we gave the TRH for 2.5 days, the basal level of serum TSH on the second and third days of TRH administration was less than that seen on the first day before the administration of TRH. For treatment of thyroid cancer and metastasis with TRH, maybe we should use it for a longer time. Or did we miss the proper time to give ^{131}I, or were these patients under maximum stimulation before TRH therapy?

C. Gual: Our experience with TRH does not include therapeutic procedures for thyroid cancer with metastasis. Nevertheless, it has been reported that repeated TRH stimulation causes a significant rise in thyroid hormones in blood levels, which in turn induces a blockade to pituitary TSH release to TRH stimulation. This mechanism may be explanation for your decreased TSH levels on the second day of TRH treatment and could in part explain the diminished ^{131}I uptake by metastatic tissue.

J. F. Wilber: Initially thyrotropin-releasing hormone (TRH) was postulated to possess monotropic specificity in the human as the unique stimulus to TSH secretion, since there was no activation of LH or FSH release by TRH, and serum cortisol was demonstrated to fall after TRH administration. However, our recent studies of HGH secretion after TRH in 32 subjects have revealed dramatic excursions in plasma growth hormone concentration in seven subjects, in three of whom the HGH rises were coincident with maximum serum TSH changes. Four other patients manifested later HGH elevations, between 60 and 180 minutes. These HGH responses may not be directly attributable to TRH per se, but dramatize the difficulty in interpreting the rise in HGH in only one patient after receiving "GH-RH." At the present time, I believe the evidence is not sufficient to classify the decapeptide "GH-RH" as a hormone in man.

There is a more remarkable finding in regard to the effect of TRH upon plasma prolactin concentration. Utilizing a heterologous radioimmunoassay for human prolactin [L. Jacobs, P. Schneider, J. Wilber, W. Daughaday, and R. Utiger, *J. Clin. Endocrinol. Metab.* **33**, 996 (1971)], we have identified dramatic (10-fold) increments in plasma HPr, and peak responses corresponded with plasma TSH maxima. These results could not be attributable to TSH or TRH cross-reactivity in the prolactin immunoassay. This phenomenon raises intriguing new questions in neuroendocrine regulation. Conceivably, the prolactin response to TRH may be related to galactorrhea seen in some myxedema subjects.

We have also confirmed in the rat that the female is more responsive to TRH *in vivo* than the male, and we are currently exploring TRH responsivity of isolated adenohypophyses derived from "sex steroid"-treated animals. The enhancement by estrogens and progesterone of anterior pituitary secretory potential seems to be a generalized phenomenon, since not only is TSH release augmented, but also GH release to GRH, LH and FSH to the gonadotropin-releasing decapeptide, and prolactin secretion also are stimulated by estrogens.

R. Guillemin: We have also confirmed that TRF will stimulate the secretion of prolactin *in vitro* with normal pituitary cells or better still with pituitary cells obtained from thyroidectomized rats. We can inhibit the secretion of prolactin due to TRF by thyroxine (T_4) or triiodothyronine (T_3). T_3 inhibits the spontaneous secretion of prolactin *in vitro* much as though it were behaving as a PIF.

J. C. Beck: A further comment on the lack of monotropic specificity for TRH or TRF might be made using the isolated pituitary cell preparation prepared by the technique of Rodbell. Dr. Kudo in our laboratory, and subsequently Dr. Salem, showed that TRF added in this particular *in vitro* system leads to release of growth hormone and also prolactin, using procine anterior pituitaries as the starting preparation.

J. M. Hershman: My colleagues and I have concentrated on defining the dose-response curve in normal subjects given TRH [Hershman *et al., J. Clin. Endocrinol. Metab.* **33**, (1971).]

Fifty-six normal subjects, 4 men and 4 women in each dosage category, were given 15.6, 31.2, 62.5, 125, 250, 500, or 1000 μg of TRH. We found that women responded more sensitively than men in 5 of the 7 dosage categories. All normal subjects have responded with definite increments in serum TSH to doses as small as 15.6 μg. There was a log-linear dose-response curve with marked individual variation; the curve seems to plateau beyond 250 μg. We thought that we might eliminate some of the large individual variation by testing the same normal subjects giving them repeated doses once or twice a week. In 4 subjects tested with each of the 7 doses, 15.6–1000 μg, there tended to be a dose-related response, but the response again was variable. I think this variability makes it difficult to set up strict criteria for a normal response. We found that hypothyroid subjects responded to doses as low as 10 μg of TRH. We have recently begun some studies to define the response to oral doses of TRH. Studies in mice and rats by Dr. Guillemin and his associates have shown that the response ratio, oral: intravenous, was about 20:1 or 40:1. In our preliminary studies, doses of 1 or 2 mg of TRH per os gave only very small responses. I think it will be necessary to use doses of 10 or 20 mg to get consistent responses when TRH is given orally.

P. L. Munson: Dr. Kastin, you showed the structure of growth hormone-releasing factor and then data indicating that it had no detectable effect in man. Would you please comment further on the negative data in man. It also might be well to recall the biological test that established the growth hormone-releasing activity of this material in animals.

A. J. Kastin: The material called GH-RH in the presentation is of porcine origin and has been followed in its purification by bioassay, primarily the tibia test of Greenspan. I expect that, when we test synthetic material of the same structure in man, it will be inactive. The basis for my pessimistic prediction is that GH-RH has not been shown to be active by radioimmunoassay in any species, and in the human apparently only radioimmunoassay can detect growth hormone levels. Another substance with a different structure may be found that will be active by radioimmunoassay. Alternatively, a cofactor may be necessary for the action of the material discussed here.

R. Guillemin: I think that what is at stake is not the tibia test as a test for growth hormone activity itself, but most likely the alleged pituitary depletion part of the assay system, which is further followed by the bioassay.

H. G. Friesen: We have studied the effect of TRF and TRH on prolactin release in man, and it appears from the data that not only is prolaction release accelerated, but prolactin synthesis is also enhanced (Table A). We have carried out these studies in collaboration with Dr. C. Y. Bowers and Dr. K. Folkers, New Orleans, Louisiana, [C. Y. Bowers, *et al., Biochem. Biophys. Res. Commun.* **40** (1971)].

As little at 25 μg of TRH caused a striking release of human prolactin as well as TSH by 15 minutes in 6 normal males and 6 females. The response in females is considerably greater than that in the males. Six females who were hyperthyroid failed to respond to as much as 800 μg of TRF, whereas the response in two hypothyroid females was greatly exaggerated. One male who was hypothyroid has the same response as the 5 normal males. In the two hypothyroid females who were treated with thyroxine (data not shown) there was a marked reduction in both the TSH and the prolactin response to TRH. This is apparently what Dr. Guillemin has found *in vitro* as well; release both of the prolactin and TSH is responsive to the level of thyroxine in the media. One can calculate that the single injection of TRH must release at least four times the total pituitary content of prolactin to produce the increased values in serum that are seen in the normal female. Hence, there must be an immediate, rapid, and profound increase in both the synthesis and release of prolactin. If 20% of the pituitary content of TSH were released, this would be sufficient to account for the increase in serum TSH concentration observed. Hence, much more of the pituitary content of prolactin is released by TRH. These findings support Dr. White's comment: Should the nomenclature of hypothalamic releasing factors or hormones be altered to hypothalamic regulatory hormones since not all release pituitary hormones, but some are inhibitory?

J. Meites: We, too, have tested synthetic TRH *in vitro* on normal rat pituitary tissue (pituitary halves) and found no effect whatsoever on prolactin release. Also, in the intact rat given a single injection of synthetic TRH, we observed no effect on serum prolactin, but there may be a time factor involved here. It is well known that thyroxine stimulates growth hormone and prolactin secretion [see J. Meites and C. S. Nicoll, *Annu. Rev. Physiol.* **28**, 57 (1966).]

TABLE A
Release of Prolactin and TSH by TRH in man[a]

TRH (μg)	Category	Serum HPr (ng/ml)			Serum TSH (μU/ml)		
		0	15 Min	30 Min	0	15 Min	30 Min
25	5 Males	10	28	21	3	13	13
25	6 Females	20	69	58	4	26	22
800	6 Females	13	80	59	4	47	49
800	6 Female hyper- thyroids	12	15	15	3	3	3
800	3 Hypothyroids						
	1 Male	9	27	30	115	165	300
	2 Females	15	225	190	375	1215	2800

[a] TRH was administered as a single bolus intravenously. Prolactin (HPr) and TSH concentrations were measured by radioimmunoassay.

We heard from Dr. Gual that administration of TRH increases TSH levels in human subjects in 3 minutes. It is quite possible, therefore, that the TRH that is administered to human subjects acts first on the pituitary to release TSH and then on the thyroid to release thyroxine, which in turn stimulates pituitary prolactin and growth hormone release into the serum. We have observed that thyroxine can act directly on the pituitary to stimulate prolactin and growth hormone release. Unless the possible intervention of the thyroid is eliminated, it cannot be concluded that the stimulatory effects of TRH on prolactin and growth hormone release in the human are exerted via a direct action on the pituitary. Finally, it should be mentioned that in the original paper by Tashjian and associates (1971), they reported that TRH stimulated prolactin release but *inhibited* growth hormone release.

E. Bogdanove: I would like to add a word or two regarding the *amounts* of hormone thought to be in the pituitary or released by it. Dr. Friesen just stated— on the basis of evidence that a pituitary gland had released more hormone than it originally contained—that new hormone had been synthesized. The logic of this statement is impeccable, but the evidence on which it rests may not be, since the amounts of hormone, either in the gland or released by it, cannot actually be defined at the present time. What we have really been measuring, and talking about, are not *amounts of hormone* at all, but merely relative immunological and/or biological *activities*. These activities are not necessarily quantitative measurements of amounts of hormone. If there were even a constant relationship between the amount of hormone in a tissue or fluid and the biological and immunological activities of that tissue or fluid, the discrepancies between bioassay and immuno-assay findings that have been discussed here would not exist.

I would like to propose that these discrepancies must now be looked at carefully in the light of recent evidence suggesting that some (perhaps all) of these hormones can exist in the body in more than one molecular form. It is certainly possible that the processes of hormone release, as well as those involved in binding and metabolism of circulating molecules, may involve transmutations that influence the biological or immunological activities of these hormones. I believe these points must be borne in mind in any serious attempts to quantitate the transfer of hormone from one bodily compartment to another.

R. Walter: Dr. Kastin, in your publication in which you describe the isolation of MSH-RIF you suggest the existence of two compounds that are capable of inhibiting the release of MSH [*Biochem. Biophys. Res. Commun.* **43**, 1376 (1971).]

A. J. Kastin: Yes, this is based on two thin-layer chromatographic systems of our purified MIF.

R. Walter: May I speculate that the second compound has the structure prolylarginylglycinamide.

A. J. Kastin: It is a good suggestion, but does not agree with our findings. We have tested the compound you mention and have found it to be much less active in our assay system, although I assume that you have found it active in yours.

R. Walter: In preliminary experiments, Dr. Celis found that prolylarginylglycinamide exhibited MSH-RIF activity, as did prolyllysylglycinamide (Fig. A). These data are in accord with our earlier finding that oxytocin is not the only neurohypophysial peptide which yields MSH-RIF activity upon incubation with a hypothalamic microsomal preparation, but that a series of analogs with amino acid replacements in position 8 exhibited this phenomenon [M. E. Celis, S. Taleis-nik, and R. Walter, *Proc. Nat. Acad. Sci. U.S.* **68**, 1428 (1971)]. However, in the series studied, arginine vasopressin gave the lowest MSH-RIF activity values.

The fact that the penultimate leucine residue in the natural MSH-RIF can be substituted with retention of activity has its close analogy with TRH, in which the histidine residue has been successfully replaced [K. Hofmann and C. Y. Bowers, *J. Med. Chem.* **13**, 1099 (1970); D. Gillessen, F. Piva, H. Steiner, and R. O. Studer, *Helv. Chim. Acta* **54**, 1335 (1971)].

F. C. Bartter: In some studies we have been doing in collaboration with Dr. Guillemin, we have found TRF invaluable in separating "true" hypopituitarism from what I hope will *not* be called hypo-hypothalamism. Dr. Gual, are you now ready to reclassify your patients with hypogonadotropic hypogonadism who did not response to TRF as perhaps representing hypopituitarism, or do you have other data on them?

I was disturbed from an academic point of view by Dr. Kastin's suggestion that estrogen inhibits at the hypothalamic level whereas thyroid inhibits at the pituitary level. I wonder whether either speaker would accept the possible alternative suggestion that, with the hypothalamus suppressed, the pituitary loses its TSH whereas it does not lose its FSH or LH.

C. Gual: The illustrated cases of hypogonadotropic hypogonadism responded well to TRH and LH-RH stimulation in a fashion similar to that seen in cases of hypothalamic amenorrhea. On this basis, we feel sure that the pituitary gland was not impaired.

A. J. Kastin: The question of feedback control of any hormone on the pituitary-hypothalamic axis is extremely complicated. I think that there is evidence for cortisol, thyroxine, estrogen, and progesterone acting at both levels: the hypothalamus and the pituitary. There is also some evidence that estrogen may sensitize the pituitary to the action of LH-RH and that progesterone may indeed have some

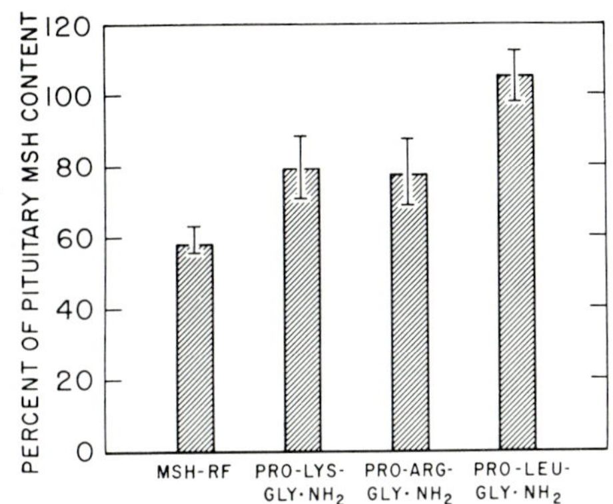

FIG. A. Preliminary experiments of *in vivo* MSH-release-inhibiting activity of synthetic C-terminal tripeptides of lysine vasopressin, arginine vasopressin, and oxytocin. The first bar on the left represents the residual pituitary content of a recipient male rat 20 minutes after the injection of an aliquot of stalk median eminence (SME) homogenate from male rats equivalent to 1 hypothalamus. The other bars represent results with rats which received 40 ng of L-prolyl-L-lysylglycinamide, L-prolyl-L-arginylglycinamide or MSH-R-IF, together with SME homogenate. Controls received phosphate buffer. [S. Hase, R. Walter, M. E. Celis, and S. Taleisnik, unpublished observations.].

blocking action. I think your suggestion is a good one, but most of the evidence points to the fact that the sex steroids exert their principal suppressive negative feedback effects at the hypothalamic level, and thyroxine at the pituitary level.

E. Bogdanove: Since the question has been raised, it may be judicious for me to extend somewhat Dr. Kastin's comments on the suspected sites of negative feedback control of anterior pituitary activity. I want to reemphasize my belief that it is still too early to draw definitive conclusions on this subject. There is certainly very fine evidence that thyroid hormone [E. Bogdanove and Crabill, *Endocrinology* **69**, 581 (1961)], estrogen [E. Bogdanove, *Endocrinology* **73**, 696 (1963)], and testosterone [Kingsley and E. Bogdanove, *Fed. Proc. Fed. Amer. Soc. Exp. Biol.* **30**, 253 (1971)] all can act at the pituitary level. There is also evidence that these same hormones may exert "feedback" effects by acting at (one or more) neural levels.

Such evidence simply *cannot* define the site of *physiological* feedback control. With regard to neural, or indirect, feedback, no one has yet demonstrated the appropriate change in brain function (i.e., a change in the releasing factor secretion rate) in response to physiological fluctuations in target hormone titers. Similarly, even though we have demonstrated that several feedback agents can have effects at the pituitary level, we have no evidence that similar direct effects actually do occur under physiological conditions. Consequently, this question is, at least in my opinion, still completely open.

H. E. Kulin: Although direct evidence for the existence of a feedback relationship between pituitary and hypothalamus in the human is lacking, one recent study does deserve mention. Abrams and co-workers [Abrams *et al., J. Clin. Invest.* **50,** 940 (1971)] have shown that when normal adult males are pretreated with growth hormone their subsequent responsiveness to provocative stimuli for HGH is reduced.

Since purified LRH invariably produces an FSH rise, we have been interested to see whether the reverse phenomenon also is existent. That is, will exogenously administered HCG cause suppression of FSH? In 5 adult men who received 4000 IU of HCG per day for 4 days both serum and urine FSH were significantly suppressed. Two castrate women, however, showed no FSH depression when a similar dose of HCG was administered. The FSH change thus appears to be mediated via the gonad and not as a result of a "short-loop" type effect.

J. R. Goding: You said that when you administered LRH to the normal female the time of the cycle was important. Have you compared the results of administration of LRH in the luteal phase with those in the follicular phase?

A. J. Kastin: We did not have that in mind in our initial studies. We now have a study planned to test that specifically, and I would rather defer answering your question.

J. R. Goding: I would like to make a comment about the effect of circulating progesterone on the release of LH. As Dr. Niswender has shown, in the ewe, FSH and LH are released simultaneously at estrus. This could be a result caused by your LRH-FRH. Estradiol will evoke the release of LH in the anestrous and in the castrate ewe. We have shown that progesterone administered some 20 hours before the estradiol will block this release of LH by estradiol [I. A. Cumming, J. M. Brown, M. A. deB. Blockey, and J. R. Goding, *J. Reprod. Fert.* **24,** 148 (1971)]. If estrogen works by releasing LRH/FRH, it would be expected that administered progestrone would block the action of the releasing hormone in causing LH release by the pituitary. This should be easy to test.

Figure B shows a possible mechanism of control of the estrous cycle in the ewe. The diagram indicates that there is considerable follicular activity during the luteal phase of the ovine cycle. We have seen several waves of excretion of FSH in the urine per luteal phase, and I believe that Dr. Geschwind has observed similar changes in plasma FSH concentration. T. Smeaton and H. A. Robinson [*J. Reprod. Fert.* **25,** 243 (1971)] and others have shown that there are waves of follicular development during the luteal phase, and the Prospect group have reported several large peaks of estradiol secretion then also.

The simplest explanation for these phenomena was to suggest that estradiol caused the release of FSH during the luteal phase, but was prevented from causing simultaneous release of LH by the level of circulating progesterone. I had drawn up this model before I learned of the proposal that LRH and FRH were one and the same thing. This new fact would lead to an obvious modification of

Fig. B; namely that the "permit" step resulting from luteolysis should operate between LRH and the pituitary, rather than as shown. It would be of obvious interest to look at the interaction between LRH–FRH and progesterone in the ewe.

M. R. Henzl: In cases where you have administered LH–FSH releasing hormones to normally ovulating women on day 9, have you been able to induce premature ovulation? If not, what are your thoughts on the possibility of "timing" ovulation in normally ovulating women?

A. J. Kastin: If we could make it possible to predict precisely the day on which ovulation would occur, it would be helpful. Indeed, this is something that will be studied soon.

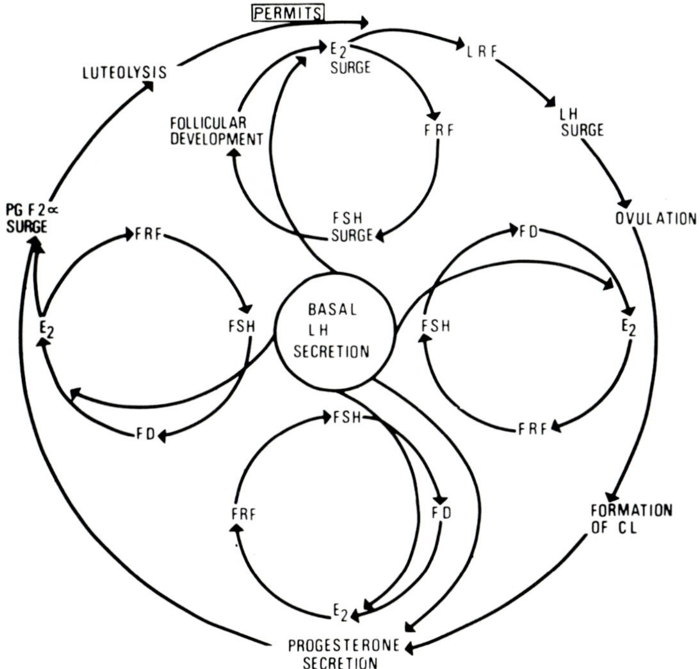

FIG. B. Possible mechanism of control of the estrous cycle in the ewe.

To return to the patient we studied on day 9 of her cycle, we tested a single, acute intravenous injection which gives an elevation of LH lasting at the most only an hour or so. It is our feeling that such a method of administration may not be satisfactory, at least in the early part of the cycle. There are a number of possibilities we hope to test. One would be to prepare the follicle by injecting LH–RH early in the cycle, with perhaps a 24-hour infusion on day 6 and then again on day 13, so that the first injection will be functioning as an FSH–RH, the second injection as an LH–RH. Alternatively, it may be that just a single intramuscular or subcutaneous injection of LH–RH/FSH–RH on each of these days, or even on one day, will be sufficient.

J. Rudinger: Have you any comparison of the ratio in which LH and FSH are released in corresponding situations by the partially purified natural material and by the synthetic material? Is there a possibility that there are in fact two structurally related factors; one for LH and one for FSH, and that they "cross-react" the way that oxytocin and vasopressin do—oxytocin having some antidiuretic effect and vasopressin some uterotonic effect? If this were the situation, then you might get a different ratio of LH and FSH released by an extract containing both the hypothetical factors and the synthetic material having a major and a minor activity. Do the results so far available allow any comment on this possibility?

A. J. Kastin: That is an excellent point. Dr. Arimura is looking at this in animals without, so far, finding a marked difference. I have been concerned in the few studies we have done with synthetic material in man that there is the appearance of a different ratio. There is little question in our minds, however, that LH–RH does release FSH, but this does not exclude the possibility you mentioned.

R. Guillemin: We have recently completed the determination of the structure of ovine LRF of which we had announced the isolation [Amoss *et al., Biochem. Biophys. Res. Commun.* **44**, 205 (1971)]. The amino acid sequence of ovine LRF is identical to that proposed earlier by Matsuo *et al.* for porcine LRF [*Biochem: Biophys. Res. Commun.* **43**, 393 (1971)]. Thus the structure of ovine LRF is pGlu-His-Trp-Ser-Tyr-Gly-Leu-Arg-Pro-Gly-NH_2. We have synthesized a relatively large quantity of that decapeptide by solid phase [Monahan *et al., C. R. Acad. Sci., Ser. D* **273**, 508 (1971)].

The synthetic product has the same physical chemical characteristics as the pure native ovine LRF. Furthermore, in keeping with what we have reported for ovine LRF, and Schally's group for porcine LRF, the synthetic decapetide stimulates concomitantly the secretion of LH and FSH *in vivo* or *in vitro*. However, we do find differences in the ratios of LH to FSH being secreted either *in vivo* or *in vitro* with increasing doses of the synthetic peptide, as in the case of native ovine LRF.

I. MacIntyre: Dr. Kastin, what name would you propose for the structure you described as GH-RH? As far as I can see, there is no evidence that it is a hormone or that it is even active in releasing growth hormone, whatever effect it has on depleting the pituitary.

A. J. Kastin: I prefer to leave matters of terminology to others. The GH-RH which I tested in the human with negative results does deplete pituitary growth hormone as well as elevate plasma growth hormone in the rat as measured by bioassay.

F. C. Greenwood: If an individual does not respond to RH, can one get a response by retesting at a different time or in a different physiological state? Could the one out of 32 growth hormone responses to TRH, noted by Dr. Wilber, be a reflection of one integrative function of the hypothalamus?

R. Guillemin: A point pertinent here is whether occasional effects observed on growth hormone plasma levels following injection of TRF have any relation to the amounts or doses of TRF injected. In our experience with Fleischer in the Medical Center in Houston, we found that there was no relation between the dose of TRF injected and the appearance of "clinical side effects," the occasional elevation of plasma growth hormone levels, or plasma cortisol.

J. Robbins: I would like to return to the comment of Dr. Samaan. Dr. Richard Sachson and I have studied two patients with metastatic functioning thyroid carcinoma and their responses to TRH. Each of the patients received 1500 μg per day in a continuous infusion for 3 days. One of the patients was responsive to TSH, the other was not. The latter patient did not respond to the TRH, as we expected, but the first patient did make a definite and brisk response to TRH in terms of increased iodine accumulation by the tumor. In this dosage range it may be possible to stimulate functioning thyroid carcinoma with this hormone. However, there was no measurable TSH in the blood either before or after TRF, so we cannot be certain that the effect was due to TRF-related increase in TSH.

R. B. Greenblatt: Reference has been made to the fact that ethynylestradiol (EE) acts at the hypothalamic level to block LH-RH. Well known is the inhibition of ovulation in the human by 100 μg of EE from days 5 to 25 of the cycle. However, if clomiphene citrate is simultaneously administered, the antiovulatory effect of EE is negated. This phenomenon cannot wholly be explained on the basis of the antiestrogenic effect of clomiphene, for hypoestrogenic women with Chiari-Frommel syndrome frequently ovulate and conceive after a course of clomiphene. Does clomiphene behave like a hypothalamic-releasing factor, or does it stimulate the release of LH-RH?

C. Gual: When clomiphene citrate is administered to an amenorrheic woman with an intact pituitary–ovarian axis, two types of responses may be obtained: clomiphene negative and clomiphene positive. The first case responds to LH-RH stimulation by a slow gradual increase in plasma LH, and the clomiphene positive responds by a significant surge of LH. It may be conjectured that there are several types of hypothalamic amenorrheas characterized by different responses to clomiphene and LH-RH stimulation.

The Hypothalamus in Pituitary–Thyroid Regulation[1]

Seymour Reichlin, Joseph B. Martin,[2] MaryAnn Mitnick,
Rita L. Boshans,[3] Yvonne Grimm, Judy Bollinger,
Jeffrey Gordon, and Juan Malacara[4]

*Department of Physiology, University of Connecticut, School of Medicine,
Farmington, Connecticut*

I. Introduction

Of mechanisms controlling the anterior pituitary gland, those involved
in the regulation of TSH secretion have been the best defined. This
system, in fact, will serve us in this paper as the paradigm of neuroendocrine control of pituitary function because it includes the classical
elements of both neurogenic and target gland feedback control, and,
in addition, the chemical mediators of control, thyroid hormones, thyroid-stimulating hormone (TSH), and the thyrotropin-releasing hormone
(TRH), have all been well characterized.

Innumerable experiments have led to the wide acceptance of the model
system of hypothalamic–pituitary–thyroid regulation outlined in Fig. 1
(for reviews, see Brown-Grant, 1960; D'Angelo, 1963; Reichlin, 1966,
1971). According to this model, the secretion of TSH is controlled
directly by two factors—a negative feedback signal which is a function
of plasma thyroid hormone level, and a stimulating factor, TRH, derived
from the hypothalamus. Several less direct, adventitious, but nonetheless
important phenomena also control the rate of TSH secretion; these include the peripheral degradation of TSH, of thyroid hormone and of
TRH, and the physiochemical state of the circulating thyroid hormone.
The peculiar metabolism of thyroxine by the anterior pituitary gland
may also be an important though adventitious regulatory factor
(Reichlin *et al.*, 1966b). Ordinarily, the latter factors are relatively
constant, and the interplay between plasma thyroid hormone and the
secretion of TSH are of primary importance in regulation.

The apparent simplicity of this model only partially conceals a number

---

[1] Supported in part by U.S. Public Health Service Grant No. 13695 and a grant
from The Veterans Administration.

[2] Present address: Division of Neurology, Department of Medicine, McGill University and the Montreal General Hospital, Montreal, Canada. Work carried out
during the tenure of a Centennial Fellowship, Medical Research Council, Canada.

[3] Present address: Department of Pharmacology, Washington University School
of Medicine, St. Louis, Missouri.

[4] Present address: Escuela de Medicina de León, Universidad de Guanajuato,
León, Gto, México.

of difficult physiological problems; among these problems are the mechanism by which TSH stimulates the thyroid gland, the mechanism by which the "set point" of plasma thyroid hormone level is determined, the site of feedback control by circulating thyroid hormone, the mechanisms of control of TRH secretion, the mechanism by which ambient temperature controls pituitary–thyroid function, and the mechanism by which neural and hormonal factors interact in the regulation of TSH

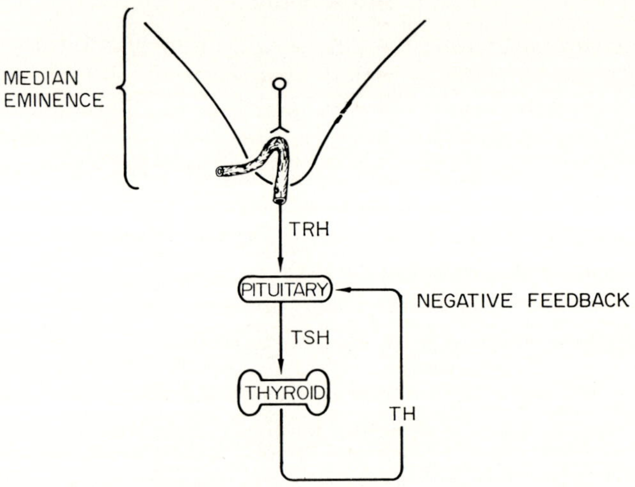

FIG. 1. Model of hypothalamic–pituitary–thyroid axis. In this figure, the classical negative feedback relationship between pituitary and thyroid gland is modified to show that the pituitary is influenced by two opposing factors: inhibition by thyroid hormone and stimulation by the hypothalamic hormone, thyrotropin-releasing hormone (TRH).

secretion. This presentation deals with these questions save for the mechanism of TSH action on the thyroid gland.

II. Nature of Feedback Control of the Pituitary–Thyroid Axis: the Pituitary as a Thyrostat

In his description of the pituitary–thyroid axis as a servo system, Hoskins (1949) pointed out that TSH secretion was decreased when plasma TH was increased, and that when plasma TH was increased, TSH secretion was inhibited. This general descriptive relationship has been defined in quantitative terms under relatively steady-state conditions by means of radioimmunoassay in both the human (Reichlin and Utiger, 1967) and the rat (Reichlin et al., 1970). In experiments with humans, hypothyroid patients were treated with increasing doses of thyroxine (T_4) in stepwise increments that were increased at approximately 10-day intervals. Plasma TSH levels were assayed for TSH, for plasma protein-bound iodine (PBI), and for free T_4 concentration at appropriate intervals during treatment. As might have been anticipated, plasma TSH levels fell as plasma T_4 levels rose. When data relating TSH to T_4 concentration were plotted to show TSH as a function of plasma T_4, a curvilinear relationship between these two variables was evident (Fig. 2). This curve defines a specific TSH plasma level determined by a specific T_4 level. It is of interest that the region of inflection of this curve is close to the normal level of circulating thyroid hormone and that decreases in thyroid hormone level bring about a relatively more striking change in TSH secretion than does an increase in thyroid hormone level.

It is now recognized that plasma triiodothyronine (T_3) makes up a significant fraction of the circulating hormone, so that these derived curves do not reflect precisely the relationship between plasma T_4 and TSH in a normal individual; it is likely nevertheless that the relationship is qualitatively similar. Although we have measured free T_4 levels as well as total T_4 in these patients, our data do not permit any conclusion as to whether it is the free or the bound fraction which is of greater regulatory importance.

This quantitative technique for defining the precise relationship between plasma T_4 and plasma TSH as developed in man offered certain advantages for the study of the role of the hypothalamus in control of the pituitary–thyroid axis if it could be adapted to the experimental animal. Because of the extensive earlier work on hypothalamic control of thyroid function in the rat, it appeared to be of value to develop a radioimmunoassay system for measurement of rat TSH. Lacking pure rat TSH, advantage was taken of the immunological cross reactivity of bovine and rat TSH (Reichlin and Boshans, 1964) to develop a heterologous immunoassay system (Reichlin *et al.*, 1966a, 1970), analogous to the system reported in the human by Lemarchand-Beraud and Vannotti (1965). Wilber and Utiger (1967) and Panda and Turner (1967a) have also reported bovine immunoassay systems suitable for the rat. By use of radioimmunoassay it was possible to analyze the effects of lesions of the thyrotropic area of the hypothalamus on the pituitary–thyroid interaction.

As in the human, under steady-state conditions plasma TSH in the rat is a function of plasma thyroxine. This was shown by treating thyroidectomized rats daily for 2 weeks with doses of thyroxine ranging from 0 to 8 μg/100 gm body weight. Plasma PBI levels rose progressively with increasing doses of T_4, the "normal" PBI level being reached at

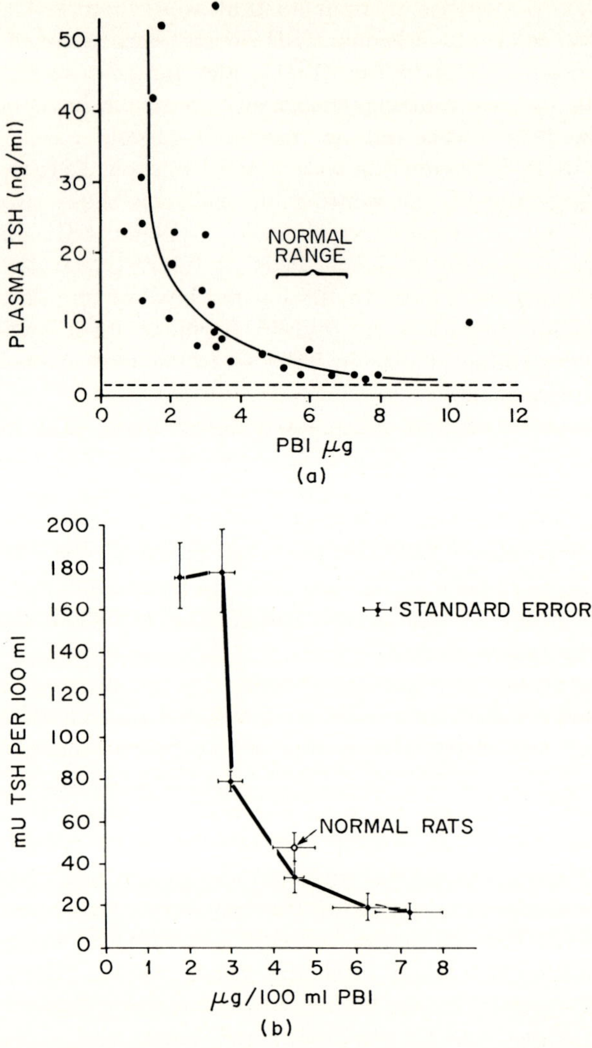

FIG. 2. The curves illustrate in the human (a) and in the rat (b) that plasma thyroid-stimulating hormone (TSH) levels are a curvilinear function of plasma thyroid hormone level. The human studies were carried out by replacing in myxedematous patients successive increments of thyroxine (T_4) at approximately 10-day intervals. Each point represents simultaneous measurement of plasma T_4 and plasma TSH at various times in the six patients studied. The rat studies were done by treating thyroidectomized animals with various doses of thyroxine for 2 weeks prior to assay of plasma TSH and plasma protein-bound iodine (PBI). Top panel from Reichlin and Utiger (1967); bottom panel from Reichlin *et al.* (1970).

a T_4 dose of 2.0 μg/100 gm body weight per day. Plasma TSH levels were unaffected by a T_4 dose of 0.5 μg but fell progressively thereafter. When plotted to show TSH levels as a function of plasma PBI, a curvilinear relationship closely resembling that of man is observed (Fig. 2).

Similar studies were then carried out in rats with hypothalamic lesions. Lesions of the hypothalamus have been shown by many investigators to lead to a decrease in the level of function of the pituitary–thyroid axis (Greer, 1951; Bogdanove and Halmi, 1953; Reichlin, 1957; Florsheim, 1958; Averill *et al.*, 1961; van der Werff ten Bosch and Swanson, 1963; D'Angelo and Traum, 1958; Panda and Turner, 1967a, b; van Rees and Moll, 1968). In such animals, as judged from changes in thyroid function, it has been shown that destruction of the "thyrotropic area" of the hypothalamus, although producing low baseline values for plasma TSH and plasma T_4 concentration, does not prevent activation of pituitary–thyroid function following thyroidectomy or goitrogen administration or the inhibition of the thyroid, which ordinarily follows thyroid hormone treatment.

These observations taken together with earlier studies of effects of direct implantation of thyroxine into the pituitary led to the hypothesis that the pituitary was the principal target of feedback inhibition by thyroid hormone (thus acting as a "thyrostat"), and that the hypothalamus determined the "set point" of control of the pituitary (Reichlin, 1963, 1964, 1966; Knigge, 1964). If this hypothesis were correct, it could be predicted that, in rats with hypothalamic lesions, the curve describing the interaction between plasma thyroxine and plasma TSH would be displaced to the left, indicating that in the absence of hypothalamic input there would be a greater degree of suppressibility of TSH secretion.

To accomplish this study, male rats were subjected to bilateral electrolytic lesions of the hypothalamus (Martin *et al.*, 1970) in the region of the paraventricular nuclei since this area and that between the paraventricular region and the anterior lip of the median eminence have been shown by many workers to be involved in the goitrogenic response to propylthiouracil. Preliminary experiments showed that the lesions blocked goiter production, and thus corresponded to the classical thyrotropic area. In such animals, there were striking changes in resting pituitary–thyroid function. Plasma PBI fell from 3.0 ± 1.0 to 1.7 ± 0.2 μg/100 ml, and plasma TSH levels became unmeasurable. Pituitary TSH content also fell markedly. These observations indicated a profound decrease in TSH secretion and synthesis rate. The response of the lesioned animals to further lowering of plasma thyroid hormone levels by thyroidectomy was next determined (Fig. 3). In normal animals, plasma TSH levels rose after thyroidectomy; in contrast, at 10 days, animals with lesions of the thyrotropic area showed no detectable change in plasma TSH. However, 40 days after thyroidectomy, plasma TSH had risen in both lesioned and control animals although the height of the response was significantly less in thyroidectomized lesioned animals. These findings indicate that lesions of the thyrotropic area lower baseline TSH secretion but do not prevent a delayed, though less marked, TSH response than in controls.

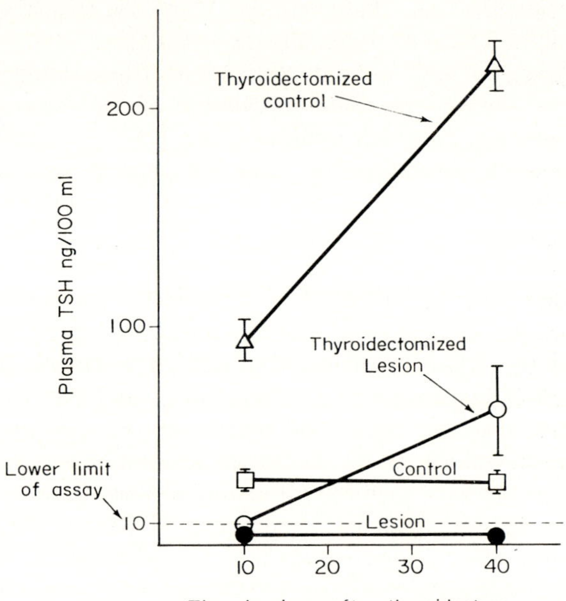

Fig. 3. Plasma TSH levels in rats with "thyrotropic" area lesions and response to thyroidectomy. Lesions reduced plasma thyroid-stimulating hormone (TSH) to unmeasurable levels. By 10 days after thyroidectomy, normals showed a significant rise, whereas animals with lesions had persistently unmeasurable levels. Forty days after thyroidectomy, a small increase was observed in the lesioned group, but much less than in thyroidectomized controls.

The threshold of feedback sensitivity to thyroxine was then tested in animals with lesions. In this experiment, thyroidectomized normal rats and thyroidectomized rats with hypothalamic lesions were treated with replacement doses of 0, 0.5, and 2.0 μg of T_4 per 100 gm body weight per day, a dosage regimen previously shown to be of the order required to reduce the high TSH levels of the thyroidectomized animals into the normal range of resting thyroid hormone levels. Plasma PBI and free T_4 concentrations were determined at each treatment dose to assure that comparable replacement levels of T_4 in plasma were achieved in each group (Fig. 4). This precaution was necessary because of the theoretical possibility that the lesions might alter the peripheral metabolism of T_4, and because of the report by Averill and Newman (1969) that T_4 was cleared more slowly from the circulation into the bile by rats with lesions of the hypothalamus. It was found in these experiments that plasma thyroxine levels were identical in normal and

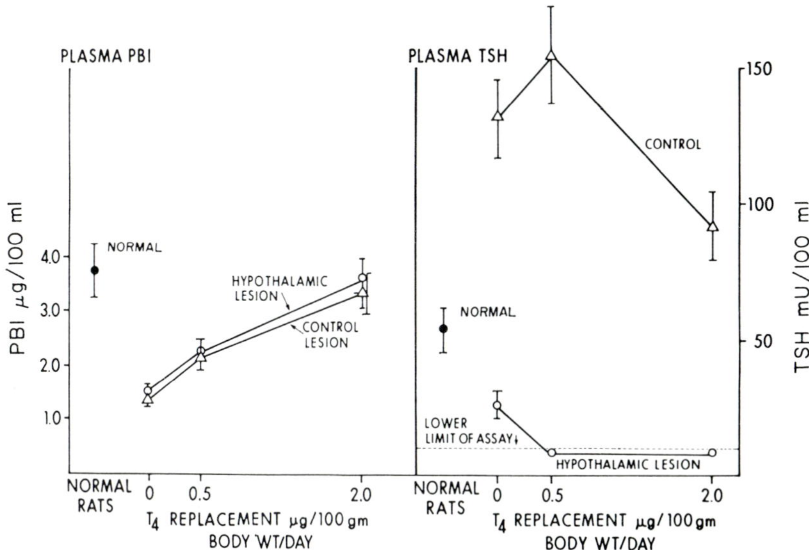

FIG. 4. Effects of T_4 replacement on plasma protein-bound iodine (PBI) and thyroid-stimulating hormone (TSH) levels in thyroidectomized rats with hypothalamic lesions compared with thyroidectomized normals. Equivalent T_4 doses restored plasma T_4 levels to the same degree in both groups of animals. On the other hand, initial TSH levels were much lower in rats with lesions and these fell to unmeasurable levels with T_4, 0.5 μg/ml, a dose which had no effect in normals.

lesioned animals, indicating that peripheral degradation of T_4 is not affected by lesions of the hypothalamus. Nor was the diffusible fraction of T_4 altered by hypothalamic damage. When 0.5 μg of thyroxine was given, no effect on TSH was observed in the normal rats. In contrast, this dose lowered TSH in plasma to undetectable levels in the lesioned animals, thus demonstrating an increased sensitivity in these animals to feedback suppression by thyroid hormone. As predicted, the curve describing TSH as a function of plasma T_4 is shifted markedly to the left (Fig. 5).

These studies indicate, by direct measurement of plasma TSH that hypothalamic lesions lower baseline TSH secretion and modify, but do not prevent appropriate responses to altered thyroid hormone levels. The hypothalamus therefore appears to act by determining the sensitivity of the pituitary to feedback inhibition.

These physiological observations accord well with the demonstrations by Guillemin and collaborators (1963), Reichlin (1965), Bowers *et al.*

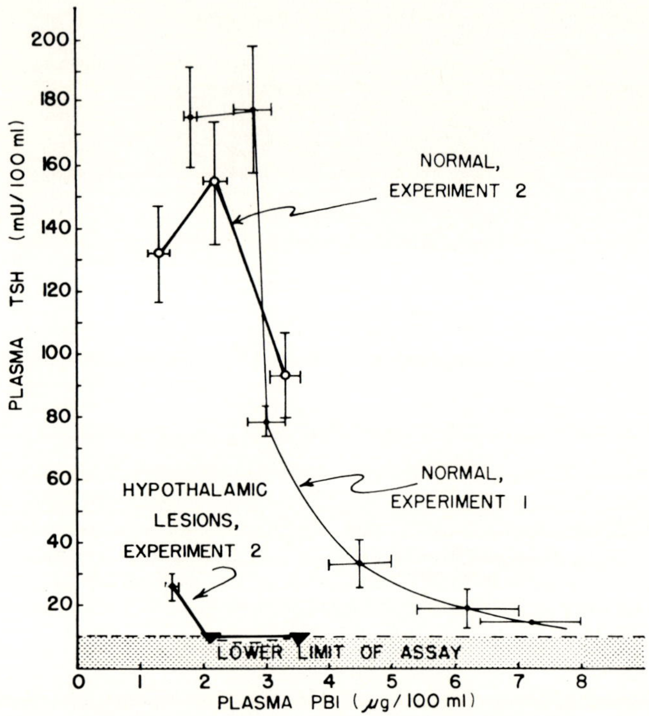

Fig. 5. TSH plotted as a function of plasma protein-bound iodine (PBI). Shown is the curve in normal rats (repeated from Fig. 2). Also shown is the function of a second normal group, replaced with 0, 0.5, and 2.0 µg T_4/100 gm body weight, which resembles the first control group. In contrast, the curve describing the function in animals with lesions is markedly displaced to the left.

(1967), and Wilber and Utiger (1968) that the response of the pituitary to thyroxine is determined by, or is a function of, TRH secretion. High thyroxine dosage inhibits TSH response to TRH, and low thyroxine levels sensitize the pituitary to TRH. The careful quantitative analysis of this interaction by Vale and collaborators (1967), Bowers and collaborators (1967), and Averill (1969) further indicate that these interactions are graded, and that they occur within the physiological level of thyroid hormone in plasma. As an aside, it is of considerable practical importance to note that thyroxine-altered pituitary sensitivity to TRH is probably the best available measure of "effective" tissue thyroid hormone effect and is being widely exploited in tests of pituitary thyroid function in the human (cf. Fleischer *et al.*, 1970).

With respect to the interaction between neural and hormonal feedback effects at the pituitary level it appears reasonable to conclude that the thyrotrope cell is exposed to two opposing influences, hypothalamic drive by TRH and inhibition by thyroid hormone. The demonstration of the importance of new protein synthesis in mediating the inhibitory response to thyroxine (Bakke and Lawrence, 1965; Bowers *et al.*, 1968; Vale *et al.*, 1968) is important because it relates the essential elements of thyroxine regulation of the pituitary to the more general effects of the thyroid hormone in regulating protein synthesis.

III. Hypothalamic Activation of TSH Secretion

ELECTRICAL STIMULATION OF THE HYPOTHALAMUS

Development of an immunoassay system capable of detecting TSH in rat plasma has made it possible to define precisely the effects of electrical stimulation of the hypothalamus on TSH secretion and to identify the anatomical distribution of the "thyrotropic area" (Martin and Reichlin, 1970, 1971a,c). Bipolar or double 0.032 gauge Nichrome electrodes were inserted under stereotactic control into various areas of the hypothalamus and fastened to connectors attached to the skull with dental cement. The animals were stimulated 10–14 days later under pentobarbital anesthesia using constant current biphasic square waves (0.5–1.0 mA) delivered in trains of 4 seconds on and 4–10 seconds off at a frequency of 60 cps and pulse duration of 1–2 msec.

Rapid, marked increases in plasma TSH levels were elicited by electrical stimulation of a number of sites in the medial-basal hypothalamus (Fig. 6). Levels were significantly elevated within 10 minutes after electrical stimulation, and reached peaks between 10 and 25 minutes after the onset of electrical stimulation. They had returned to baseline levels in most animals by 30–60 minutes. In sham-stimulated animals, plasma TSH levels continued to fall during the period of observation, a response observed repeatedly and most likely attributable to the progressive decline in tonic hypothalamic influences on TSH secretion as proposed by Averill and Kennedy (1967). The anatomical distribution of hypothalamic stimulation point effective in activating TSH secretion were widely distributed (Fig. 7). Positive responses were found within the medial-basal hypothalamus extending from the suprachiasmatic and preoptic area to the posterior arcuate area, a region corresponding to the the "hypophysiotropic area" described by Halasz *et al.* (1962). The relative diffuseness of the thyrotropic control area is confirmed also by Guillemin's observation (1970) that TRH is found throughout the hypothalamus (as measured by bioassay), and that the entire hypothalamus has been shown (see below) to synthesize TRH *in vitro*. The two greatest responses occurred following stimulation of the paraventricular area,

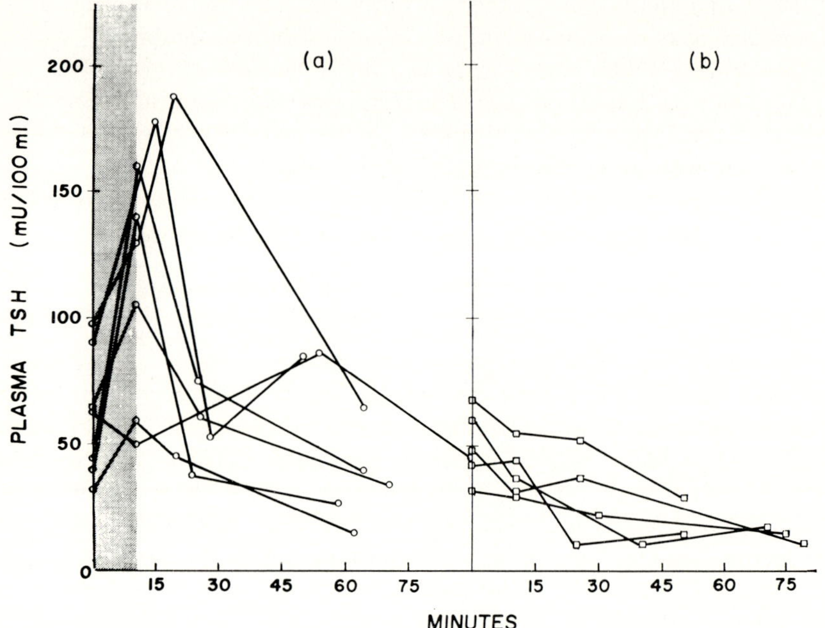

FIG. 6. Plasma thyroid-stimulating hormone (TSH) response in individual rats following electrical (a) or sham (b) stimulation. The gray bar represents the period of stimulation. A prompt rise in plasma TSH is observed; while the anesthetized controls show a gradual fall over the period of observation. From Martin and Reichlin (1970).

the region postulated by Averill *et al.* (1961) to be the chief "thyrotropic area." Positive responses also were noted after preoptic stimulations in regions not generally thought of as being the primary thyrotropic area. It is likely that neurons here are part of the system which mediates cold exposure responses, since this region contains thermosensitive neurons capable of activating body temperature control reflexes, and the thyroid response to local cooling (see below). Lesions in this region were shown by D'Angelo (1960) to have no effect on baseline TSH secretion function in rats, but to prevent the response to acute cold exposure. Only two negative responses were observed in the hypothalamic area, one in an animal with a placement in the mammillary body and the other in the dorsomedial nucleus. Cerebral cortex stimulation was without effect on TSH secretion. Not plotted on this figure are further ineffective stimulation sites in mammillary bodies, lateral hypothalamus and ventral thalamus (Martin and Reichlin, 1972).

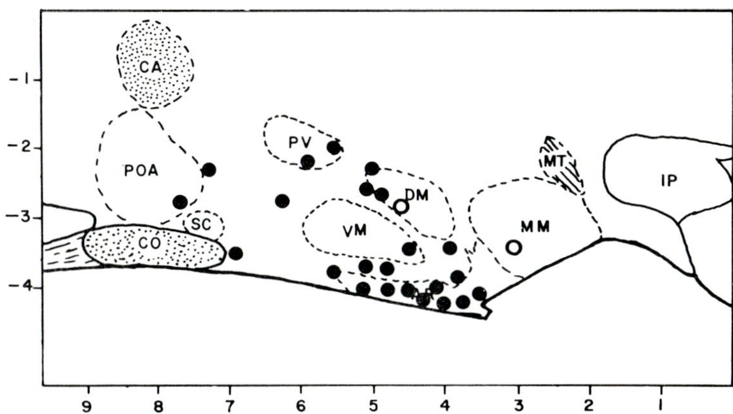

FIG. 7. Diagram of a sagittal section of the rat hypothalamus to show the distribution of areas giving positive thyroid-stimulating hormone (TSH) responses following electrical stimulation (filled circles, 23 animals); negative responses (open circles, 2 animals) are also shown. Negative responses from cerebral cortex stimulation are not shown. Outline and coordinates are from the atlas of de Groot (1959). Abbreviations: *CA*, anterior commissure: *POA*, preoptic area; *SC*, suprachiasmatic nucleus; *AR*, arcuate nucleus; *MM*, mamillary nucleus; *MT*, mammillothalamic tract; *IP*, interpenduncular nucleus.

A number of workers have reported that TSH secretion is increased after electrical stimulation of the hypothalamus; most studies have used thyroid function as an indirect criterion of pituitary activation (Harris and Woods, 1958; Campbell *et al.*, 1960; Shizume *et al.*, 1962; Boop and Story, 1966; Vertes *et al.*, 1965; Averill and Kennedy, 1966). These studies, however, may be criticized on the basis of questionable specificity because in the dog (Ackerman and Arons, 1962) epinephrine, and in the rabbit (Harris *et al.*, 1964) vasopressin, both of which can be released by hypothalamic electrical stimulation, have direct effects on thyroid hormone release from the thyroid gland. The time course studies here correspond most closely to those reported by Averill and Salaman (1967) in the rabbit using a TSH bioassay; the effect is most likely due to the enhanced release of TRH into the portal circulation, since

Wilber and Porter (1969) have recently shown that bioassayable TRH activity appears in hypophysial portal plasma after electrical stimulation of the medial-basal hypothalamus. The time course of response to electrical stimulation is brief with rapid onset of effect, observations which do not correspond to the earlier studies of D'Angelo and collaborators (1964), who had noted a delayed onset of effect followed by a prolonged elevation in bioassayable TSH. The effects seen by D'Angelo (who used stainless steel electrodes) may be due to electrochemical stimulation from locally deposited metallic salts analogous to the electrochemical stimulation of gonadotropin release reported by Everett and Radford (1961).

The possibility has been considered that the localizations were relatively nonspecific because of the high current used in the stimulation. This seems unlikely, however, because parallel studies were made of plasma growth hormone in the same animals, and it was apparent that GH stimulatory responses were strictly limited to the ventromedial nucleus of the hypothalamus and the region between the ventromedial nucleus and the median eminence (Martin and Reichlin, 1971b). We have not, as yet, studied the effects of incrementally varied plasma thyroxine levels on the TSH response to hypothalamic electrical stimulation, but have observed that a high supraphysiological dose of T_4 (50 μg) administered 2 hours prior to electrical stimulation completely blocks the TSH response (Fig. 8). As discussed below, it is likely that the effect here is mediated at the level of the pituitary, since the same dose also effectively blocks responses to the synthetic TRH (Fig. 8), and Wilber and Porter (1969) have shown (but not in quantitatively rigorous studies) that thyroxine administration does not alter the effects of hypothalamic electrical stimulation in causing TRH release into the hypophysial-portal vessels.

Studies with electrical stimulation have confirmed earlier stimulation work which has utilized secondary changes in thyroid function or plasma TSH determined by bioassay. They indicate that the hypothalamus is the site of a widely distributed population of neurones capable of activating TSH secretion in the presence of normal circulating thyroid hormone levels.

Responses to electrical stimulation approximate closely, but not exactly, the time course of TSH secretory response to the intravenous injection of TRH (Martin and Reichlin, 1970). Following intravenous injection of TRH there is within 5 minutes (the first interval tested) elevation of plasma TSH levels, which reach a peak response between 5 and 10 minutes after injection (Fig. 8). The difference in time course between that following electrical stimulation and TRH injection may be partially accounted for by the fact that TRH was administered in a single rapid injection, whereas electrical stimulation was continued for 5–10 minutes. It is reasonable to suppose that with proper manipulation of dose and time, one could mimic completely the effects of electrical stimulation. An intravenous dose of 10 ng of TRH gave a response approximately equal to that obtained with 5 minutes of electrical stimulation.

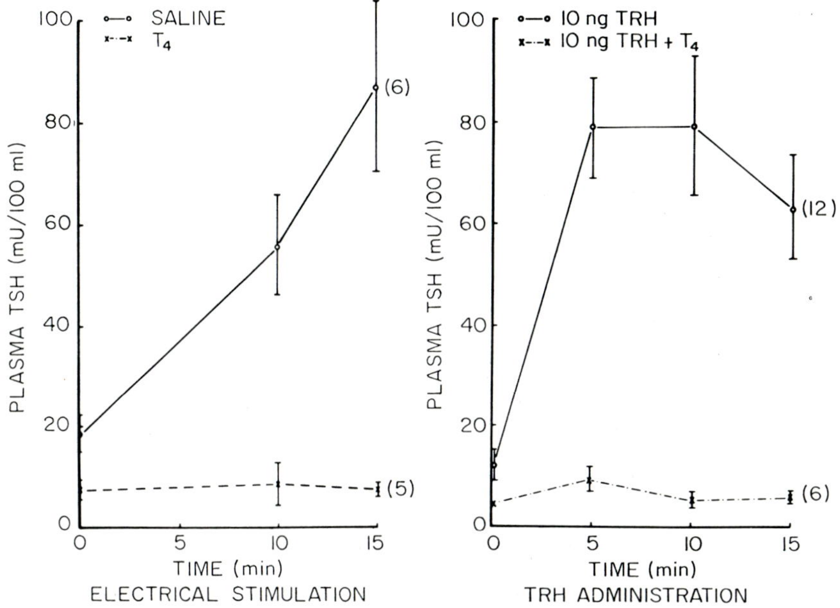

FIG. 8. Effects of thyroxine (T_4) pretreatment on plasma thyroid-stimulating hormone (TSH) response to electrical stimulation or synthetic thyrotropin-releasing hormone (TRH) administration. Injection of 50 μg of T_4 per 100 gm body weight 2 hours prior to each experiment inhibited the TSH response to both stimuli. The number of animals in each group is indicated in parentheses. Vertical bars represent standard errors. Left: O——O, saline; X---X, T_4. Right: O——O, 10 mg TRH; X—·—X, 10 ng TRH + T_4. From Martin and Reichlin (1972).

The rapidity of the response to electrical stimulation suggests that the effect is mediated by release of preformed TRH from the hypothalamus which reached the primary plexus of the hypophysial portal circulation directly.

This point is emphasized here because of the recent proposal most recently championed by Knigge and Scott (1970) that TRH and other releasing factors may reach the anterior pituitary from the third ventricle by way of a specialized ependymal transport system within the median eminence. This route has been suggested because of histological characteristics of the ependyma of the median eminence and because of the relatively well maintained basal thyroid function of animals with surgically isolated median eminences (Voloshin *et al.*, 1968). We have attempted to determine whether TRH can affect TSH secretion when administered by way of the third ventricle (Gordon *et al.*, 1972). Of all routes used (third ventricle, median eminence, anterior pituitary, and intravenous), the third ventricular route was least effective (Fig. 9).

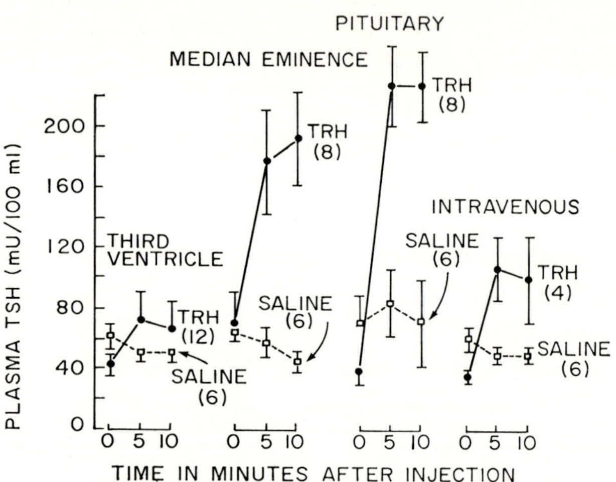

Fig. 9. The effects on plasma thyroid-stimulating hormone (TSH) of the injection of 50 ng of thyrotropin-releasing hormone (TRH) injected into the third ventricle, median eminence, anterior pituitary, and intravenously are shown here. As compared with saline-injected controls, there is a small but significant effect with injection into the third ventricle, but this route is the least effective of those tried. From Gordon *et al.* (1972).

These observations suggest, but do not prove, that the intraventricular route is of relatively little importance in mediating *acute* TSH discharge, although small TRH effects can be seen after TRH injection into the third ventricle.

The TSH immunoassay has permitted the development of a precise bioradioimmunoassay for TRH. Over the dose range of 1–5000 ng of TRH, increment in plasma TSH is a function of the log of the dose administered. In light of the claims of TRH assay by the depletion method (Motta *et al.*, 1970), it is of interest that significant depletion of pituitary TSH concentration is observed only at the highest doses of TRH administered (Table I). This bioradioimmunoassay though highly specific is relatively insensitive. The TRH content of 4 freeze-dried rat hypothalamic fragments (extracted with methanol) was not detectable after intravenous injection whereas 8 hypothalamic fragments gave a response approximately equal to 1.6 ng of TRH, an amount equivalent to approximately 200 pg per hypothalamus.

TABLE I

Plasma and Pituitary TSH Levels Following Intravenous Administration of Synthetic TRH in Pentobarbital-Anesthetized Animals

Dose of TRH (ng)	Number of animals	Plasma TSH peak value (mU/100 ml)	Pituitary TSH (mU/mg)
0	6	5	66.4 ± 5.6
1	6	13.5 ± 2.9[a]	—
10	12	79.7 ± 9.8[b]	71.4 ± 3.8
100	6	118.6 ± 21.3[b]	79.8 ± 10.4
1000	6	146.8 ± 25.8[b]	72.4 ± 11.2
5000	6	168.7 ± 25.1[b]	43.2 ± 1.6[b]

[a] $p < 0.05$ from control.
[b] $p < 0.01$ from control.

IV. Setting the Hypothalamus: Body Temperature Homeostasis

If, as has been inferred from data in animals with lesions, the pituitary acts as the thyrostat, the setting of which is determined by hypothalamic TRH secretion, it is important to determine the factors that the hypothalamus in turn is responsive to.

The possibility that thyroid hormones may act on the hypothalamus as well as the pituitary to regulate TSH secretion must be considered, and will be dealt with below. In anticipation of that discussion it can be stated that the occurrence of negative TH feedback on TRH secretion is subject to conflicting claims.

Other than the thyroid hormones, the most important physiological regulator of hypothalamic–pituitary–thyroid function would appear to be environmental temperature. The literature relating temperature and thyroid responses is vast and has been reviewed by a number of authors (Brown-Grant, 1960; D'Angelo, 1963; Reichlin, 1966; Panda and Turner, 1967b; Bakke and Lawrence, 1971; Galton, 1971. It is fair to conclude from the bulk of this literature that both acute and chronic cold exposure increase TSH secretion, but that only the acute responses can confidently be assumed to be mediated by a direct cold receptor–hypothalamic–pituitary neuroendocrine reflex. Chronic thyroregulatory response may or may not involve a direct hypothalamic system, and they are not always demonstrable; they may be related to such long-term adaptive responses as increased turnover of plasma thyroid hormone, in turn related to increased tissue metabolism or more rapid enterohepatic thyroxine turnover secondary to increased food intake (Freinkel and Lewis, 1957; Galton and Nisola, 1969). Studies of acute cold exposure in this laboratory indicate that temperature transients do alter TSH secretion, but the role of cold receptors in long-term regulation of hypothalamic TRH secretion is less certain.

Physiologists concerned with body temperature homeostasis have recognized that there is a population of thermosensitive neurons in the anterior hypothalamus which serve as a reference index of core temperature (Hammel, 1968), and that the autonomic responses to cold exposure involve an interaction between core temperature and peripheral cold receptors. Our work suggests this to be the case for thyroid regulation as well.

In early studies with McClure (Reichlin, 1964), it was shown that central cooling of the rat hypothalamus, produced by direct application of a water-cooled thermode to the preoptic area, led to thyroid activation as inferred from changes in thyroid ^{131}I release rate. These observations, in confirmation of the work of Andersson and collaborators (1962, 1963) in the goat, indicated that TRH secretion was altered by thermoreceptor neurons in the hypothalamus; from this it was concluded that core temperature is a regulator of TSH secretion.

This work has been extended by Martin and Reichlin (1971a) by measuring TSH rather than thyroid function and by altering core temperature in a less effete manner. Acute lowering of temperature of the core while peripheral thermoreceptors were at constant temperature was accomplished by the administration of the ganglionic blocking agent chlorisondamine (Fig. 10). Core body temperature was measured either by a rectal thermistor probe or by stereotactically implanted hypothalamic thermistors. At room temperature (22°C) sympathetic blockade led to a prompt fall in hypothalamic temperature of approximately 1°C. Plasma TSH responses measured 40 minutes after administration of chlorisondamine showed a rise, a finding which indicates that an acute lowering of core temperature will actuate TSH secretion under constant ambient temperature; from this it may be concluded that TSH secretion is controlled by central thermoreceptor neurons. But peripheral thermoreceptors also control TSH secretion in a situation in which core temperature is either unchanged or slightly elevated.

Plasma TSH levels were measured in a series of individual rats exposed acutely to cold. These animals were prepared with chronic indwelling jugular cannulas. After recovery, they were placed in an isolation chamber and sampled through a remote cannula which extended outside the cage. Acute cold exposure was achieved by the introduction of cold air into the cage; plasma samples were removed at intervals. In five animals so studied, a fall in cage temperature from 26° to 10°C over a period of 10 minutes was accompanied by a rapid increase in plasma TSH, which reached a peak in some animals within 5 minutes (Fig. 11). Plasma TSH returned to baseline levels by 40 minutes despite persistent exposure to cold. The acute changes in TSH secretion following external

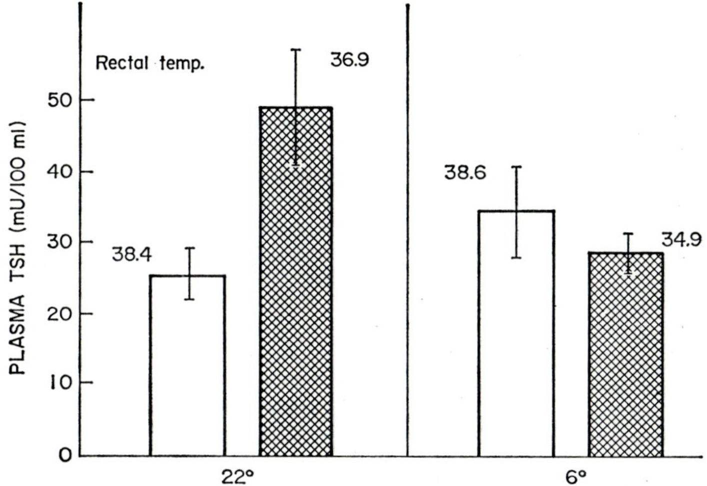

FIG. 10.   Plasma thyroid-stimulating hormone (TSH) levels in rats injected with chlorisondamine (Ecolid) at 22° or 6°C. A significant increase in plasma TSH occurred in rats at 22°C which had a decline of 1.5°C in core temperature (rectal temperatures in each group are indicated beside their respective bars). Ecolid provoked a more marked fall in plasma TSH in cold-exposed animals, but TSH activation did not occur, an effect probably due to "nonspecific" stress. □, Saline; ▨, Ecolid.

cold exposure were not associated with a fall in core temperature. As determined in a separate study, hypothalamic temperature is not lowered when rats are exposed acutely to the cold (Fig. 12).

These findings would indicate that both peripheral and central thermoreceptors initiate reflex TSH release in response to acute cold exposure.

Unlike the situation in the rat, where acute cooling produces relatively easily detected perturbations in thyroid function, change in thyroid secretion of the human after acute cold exposure have been much more difficult to elicit. In our own laboratory, in collaboration with Robert Utiger, Berg et al. (1966) were unable to induce TSH hypersecretion with ice ingestion, a procedure which is documented to lower core temperature by more than 0.5°C while maintaining skin thermal neutrality. On the other hand, pituitary–thyroid activation of the newborn, so elegantly studied by Fisher and Odell (1969) is at least in part analogous to the rat model, and more recently, it has been reported that acute cold exposure in unanesthetized man transiently elevates TSH secretion (Golsteine-Golaire *et al.*, 1970).

On the other hand, there is less certainty about the role of either core or peripheral thermoreceptors in determining the "set point" of hypothalamic–pituitary control over a long time scale. The bulk of ex-

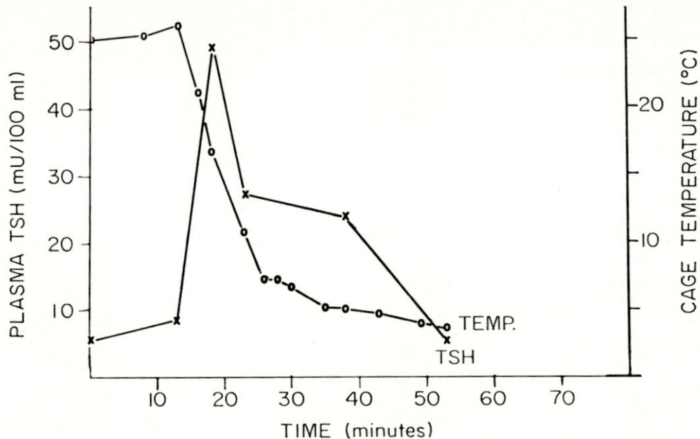

FIG. 11. Plasma thyroid-stimulating hormone (TSH) response to acute cold exposure in an unanesthetized rat. This animal shows a marked early TSH response with a 5-fold rise less than 5 min after onset of cold exposure. Plasma TSH returned to baseline approximately 40 minutes after onset of cold.

perimental work indicates that plasma thyroxine levels are not elevated by prolonged exposure to cold, a finding interpreted to mean that the well established increase in thyroid function of the cold-exposed animal is accompanied by an increased rate of thyroxine disposal—in fact, it has been proposed that the increased thyroid function of the chronically cold-exposed animal is due mainly to increased peripheral thyroid hormone metabolism with secondary activation through hormonal feedback effects. (Galton and Nisola, 1969; Galton, 1971).

In analyzing this problem, we have considered that if the hypothalamus integrates automatic with hormonal heat homeostatic mechanisms, it would seem reasonable to predict that impairment of the autonomic component of temperature homeostatis would lead to a compensatory *increase* in the pituitary–thyroid component, as evidenced by an increase in pituitary–thyroid function accompanied by a higher "set point" of thyroid hormone level in blood. To test this hypothesis, pituitary–thyroid function was studied in rats in which the capacity to maintain normal body temperature in the cold was impaired by immunosympathectomy. Immunosympathectomy is carried out in newborn rats by injecting them with an antibody to the nerve growth factor, a protein substance found in salivary glands of snakes and rodents, and shown by the studies of Levi-Montalcini and Angeletti (1968) to be a potent stimulator of sympathetic nerve development. Rats treated with anti-nerve growth factor fail to develop certain parts of the sympathetic nervous system,

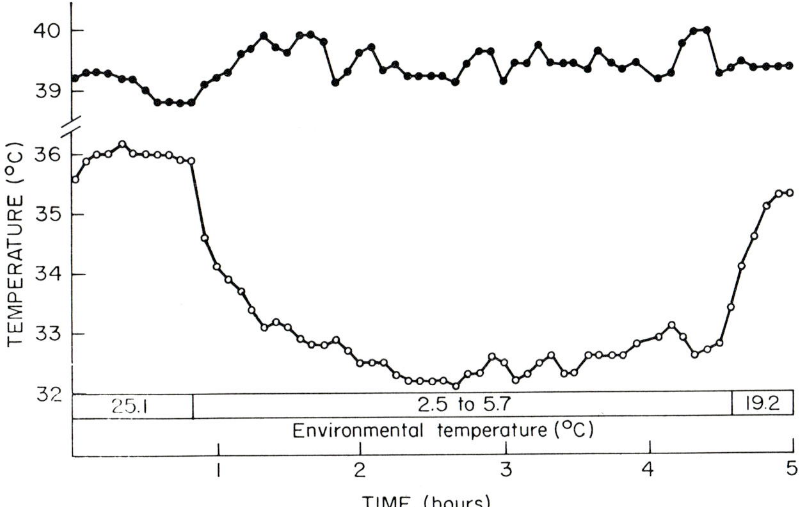

Fig. 12. Acute cold exposure in the rat causes a prompt but transient elevation of hypothalamic temperature with a fall in subcutaneous temperature. Temperature was measured with chronically implanted intrahypothalamic thermistor. ○——○, Body temperature; ●——●, hypothalamic temperature.

evidenced clinically by the appearance of Horner's syndrome. Several important sympathetically innervated areas escape destruction by anti-nerve growth factor, the most important of which are the viscera and the adrenal medulla. More complete loss of sympathetic function can be induced by adrenal demedullation in addition to immunosympathectomy. When animals with combined ablations were exposed to external cold at 6°C for 2 weeks (Table II) it was found that (in contrast to normals similarly exposed) there was a significant degree of hypothermia (36.8 ± 0.1 vs 35.9 ± 0.1°C, $p < 0.01$). The pituitary thyroid system of these cold exposed animals was significantly more activated than cold-exposed controls, as evidenced by increased thyroid weight and plasma TSH, but despite this evidence of enhanced thyroid function, levels of plasma PBI and plasma-free T_4 were not significantly different from control levels.

If, as had been postulated, core temperature or peripheral cold receptors determine the "set point" of plasma T_4 levels, animals with mild chronic hypothermia would have elevated plasma T_4 concentration. The fact that this is not so would indicate either that pituitary–thyroid activation is secondary to enhanced peripheral thyroid hormone degrada-

TABLE II

Pituitary–Thyroid Function in Cold-Exposed Immunosympathectomized-Adrenal Demedullectomized Rats (IS-Adm) and Normals

Statistic	Room temperature (22°C)		Cold exposed (6°C)	
	Normal	IS-Adm	Normal	IS-Adm
Number of animals	16	8	21	10
Final weight (gm)	351 ± 7	319 ± 9	309 ± 8^a	281 ± 8^a
Body temperature (°C)	37.0 ± 0.1	36.8 ± 0.1	36.8 ± 0.1	35.9 ± 0.1^b
Thyroid weight (mg/100 gm)	5.73 ± 0.18	6.17 ± 0.23	6.65 ± 0.21^c	7.83 ± 0.42^d
Plasma PBI (μg/100 ml)	3.3 ± 0.1	3.3 ± 0.1	3.0 ± 0.1^e	3.1 ± 0.2^e
TfT$_4$ (mg/100 ml)	2.08	1.62	1.90	2.19
Plasma TSH mU/100 ml	60.7 ± 8.0	58.0 ± 5.9	83.8 ± 8.1^e	125.6 ± 16.1^d

[a] $p < 0.01$ from each corresponding 22° control.

[b] $p < 0.01$ from each other group.

[c] $p < 0.05$ from corresponding 22° control.

[d] $p < 0.01$ from corresponding 22° control.

[e] Not significantly different from corresponding 22° control.

tion or that, if there is an increased cold-induced change operative through central neural mechanism, it is offset by an independent increase in the rate of thyroid hormone degradation. The possible role of T_3 secretion in cold-adapted rats must also be considered, but data dealing with this question have not been available. From our studies which indicate an enhanced rate of TRH synthesis in chronically cold exposed rats (see below), we believe that there is an important neural component driving TSH secretion during cold exposure and would propose (though without full documentation) that chronic cold exposure does cause an increase in TRH secretion, but that this is compensated for by a change in disappearance rate of thyroxine determined either by increased metabolic rate or intestinal loss. In studies completed since the conference, Nejad and Bollinger working in this laboratory have found, using a specific radioimmunoassay that plasma T3 levels do, in fact, increase in chronic cold-exposed rats while T4 levels do not change. Thus, there may in fact be an elevation in effective thyroid hormone levels in these animals.

V. TRH Biosynthesis by Hypothalamic Incubates

The studies outlined in previous sections have dealt largely with pituitary–thyroid regulation at the level of organized homeostatic responses. The discovery of the chemical structure of TRH (Burgus *et al.*, 1969, 1970; Bøler *et al.*, 1969; Bowers *et al.*, 1970) as the tripeptide amide (pyro)Glu-His-Pro-amide, the extensive information now available about the chemical separation of TRH from other hypothalamic peptides (cf. Meites, 1970; Burgus and Guillemin, 1970), and the development of a highly specific bioassay for TRH in our laboratory has made it possible to study the control of TRH synthesis and release by hormonal and neural factors at a much more fundamental level. Since only a preliminary report of this work has been published (Mitnick and Reichlin, 1971), it will be necessary to describe in some detail the methods used to validate the study of TRH biosynthesis.

That such studies might be feasible was apparent from the work of Sachs and collaborators (1968) (reviewed in this conference in 1967) who had demonstrated biosynthesis of vasopressin and of neurophysin by hypothalamic–neurohypophysial fragments incubated *in vitro*. Morell (1967a,b) has also reported in abstract that dog hypothalamic tissue and subcellular fractions synthesize peptides *in vitro*, although in this work the nature of the newly synthesized products other than oxytocin and vasopressin were not identified. In addition to developing a system which was capable of carrying out TRH biosynthesis, it was of crucial importance that the end product could be identified specifically, an objective which could not have been accomplished until a purified authentic TRH preparation was available.

Hypothalamic tissues were removed from decapitated mature male rats, 250–400 gm body weight in three sections: stalk median eminence, ventral hypothalamus (including arcuate and ventromedial nuclei), and a dorsal hypothalamic fragment which included tissue from the anterior commissure anteriorly, the anterior margin of the mammillary nuclei posteriorly, and the paraventricular nuclei superiorly. Tissue fragments from two rats were added to 1.0 ml of Krebs-Ringer bicarbonate buffer containing 10 or 20 μCi of labeled amino acid, gassed with 95% oxygen and 5% carbon dioxide, and incubated with shaking for 2 hours at 37°C. At the end of the incubation, tissues were homogenized in 90% methanol, a solvent chosen on the basis of the solubility survey of TRH in various organic solvents made by Burgus *et al.* (1967). The supernatant was removed for analysis after centrifugation.

In initial experiments, various chromatographic systems were evaluated to determine the most useful system of identifying TRH. Because it was not certain that *synthetic* TRH was identical with native TRH in the form found in tissue, the distribution of native porcine TRH on Sephadex G-25 columns was first determined by bioassay for later comparison with the biosynthesized material. Crude TRH was first prepared by the method of Schally *et al.* (1969) with modifications from acetone defatted lyophilized pig hypothalamic tissue (Oscar Mayer, Madison, Wisconsin) by extraction with 4 N acetic acid, followed by lyophilization and reextraction with glacial acetic acid. The lyophilized acetic acid extract was reextracted with 90% methanol and subjected to gel filtration (Fig. 13). Column fractions were assayed for TSH by radioimmunoassay and for TRH by bioassay. The unretarded fractions alone had slight amounts of TSH while tubes 40, 41, 42, and 43 alone of those tested contained TRH.

To determine whether newly biosynthesized peptides of rat tissue had the mobility of crude TRH of porcine origin, hypothalamic tissues were incubated with [^{14}C]proline (10 μCi/ml), one of the three probable amino acid precursors of TRH. The extract was added to a carrier consisting of the methanol extract of 100 porcine hypothalami prepared as above and the combined extracts subjected to gel filtration on a column of Sephadex G-25. The peak of incorporated counts (distinct from unreacted proline) was found in tubes 40, 41, 42, and 43, which overlapped the distribution of bioassayable TRH previously determined. However, there was sufficient overlap between the distribution of free (unreacted) and incorporated proline that this chromatographic technique could not be used alone to isolate newly synthesized TRH. The fractions containing biological TRH activity were therefore subjected after lyophilization to thin-layer chromatography on silica gel G sheets. Synthetic TRH, 25 μg, was added as a marker. Radioautographs were prepared by exposure to Kodak medical X-ray film BB-54 for 2 weeks, and the position of the radioactive spot was compared with the Pauly reagent reactive spot. One of the radioactive spots from the peak of biologically active TRH coincided exactly with the Pauly-positive synthetic TRH marker. From this it was concluded (subject to further validation) that native *porcine* TRH and synthetic TRH had identical chromatographic behavior on thin-layer plates, and that a peptide with the chromatographic properties of crude and of synthetic TRH in two separation systems was synthesized by rat hypothalamic fragments *in vitro*.

Thin-layer chromatography permitted excellent separation of free from incorporated proline after preliminary Sephadex separation. It was then

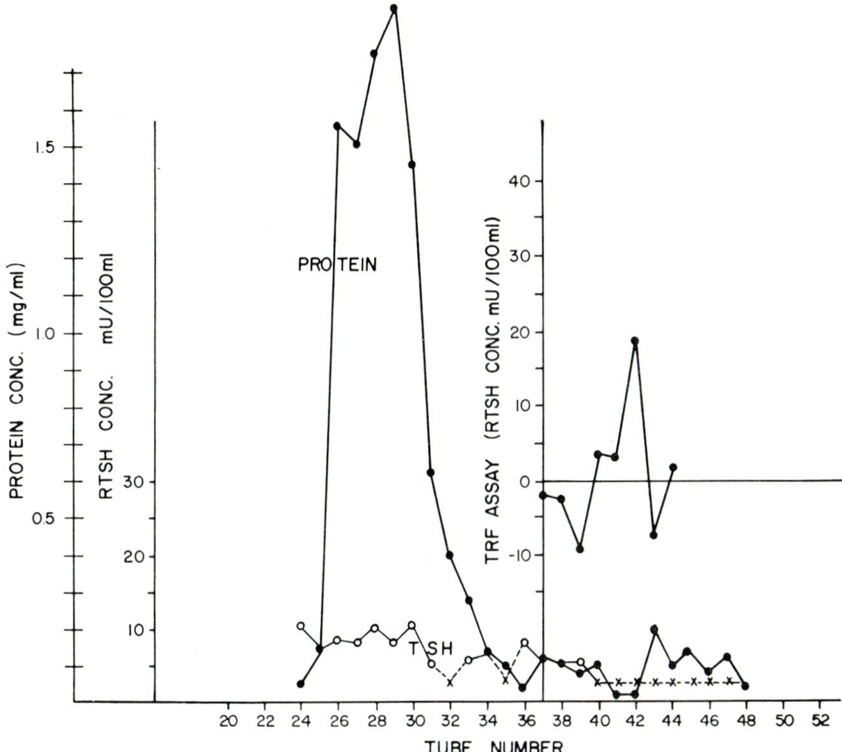

Fig. 13. Distribution of immunoassayable thyroid-stimulating hormone (TSH) and of thyrotropin-releasing hormone (TRH) in acid extract of 100 porcine hypothalamic fragments on a Sephadex G-25 column. Crude TRH was prepared by the method of Schally *et al.* (1969), with modifications, from acetone-defatted lyophilized pig hypothalamic tissue extracted with 4 *N* acetic acid, followed by lyophilization and reextraction with glacial acetic acid. The lyophilized acetic acid extract was reextracted with 90% methanol and filtered on a column of Sephadex G-25, 2 × 66 cm, and eluted in 1.0 *N* acetic acid in 5-ml samples. The protein peak contained small amounts of TSH, while tubes 40, 41, 42, and 43 (tested in 4 animals each) had TRH activity, 42 being the peak.

found that methanol extracts of rat hypothalami could be chromatographed *directly* on thin layer plates. The specificity of synthesis was also tested by use of incubates of rat cerebral cortex and neurohypophysis (Fig. 14). As in the earlier study, the position of TRH was determined using an added marker, and the radioactivity was identified by radioautography. In all the tissues tested, several radioactive areas were identified, indicating the biosynthesis of a variety of proline containing compounds. However, stalk median eminence, ventral hypothalamus, and dorsal hypothalamus alone of the tissues tested included areas with R_f values identical with values for TRH. Cerebral cortex and posterior pituitary incubates had no radioactive spots corresponding to TRH. It was concluded therefore that the latter two tissues did not synthesize this substance. Subsequently, pre-optic area, mammillary body, and anterior pituitary were also studied. None of these tissues incorporate precursor amino acids into TRH.

Although failure to incorporate an amino acid into peptides with the mobility of TRH could be taken as evidence that this tissue did *not* synthesize the material, it was also apparent that incorporation into TRH might not be specific for this substance since other compounds might have identical chromatographic properties in thin-layer separation systems. It was necessary therefore to study the incorporation of all the suspected TRH precursor amino acids and, as controls, other important naturally occurring amino acids. A systematic examination was then made of the pattern of polypeptide incorporation into hypothalamic fragments of glutamic acid, proline, and histidine, the most likely TRH precursors and sixteen other naturally occurring amino acids. They were separately incubated with dissected stalk median eminence, ventral hypothalamus, and dorsal hypothalamus. Each run utilized tissue from two separate rats, and each amino acid was tested in three separate incubations. The three suspected amino acid precursors (proline, histidine, and glutamic acid) and four other amino acids (alanine, glycine, serine, and threonine) were incorporated into polypeptides with R_f values similar to or identical with those for TRH. All the others tested had distributions markedly distinct from that of TRH and could therefore be disregarded from consideration as being precursors of TRH. In an effort to separate further labeled polypeptides with TRH mobility containing the seven precursor amino acids, six other thin-layer systems were studied (Table III). None was capable of separating polypeptides containing the three putative TRH precursors from polypeptides containing the other four amino acids.

Since thin-layer chromatography was ineffective in separating peptides containing the three putative TRH precursors from peptides containing four other amino acids, other chromatographic systems were evaluated in an attempt to identify a TRH-specific area. Methanol extracts of dorsal hypothalami incubated separately or in double-labeling experi-

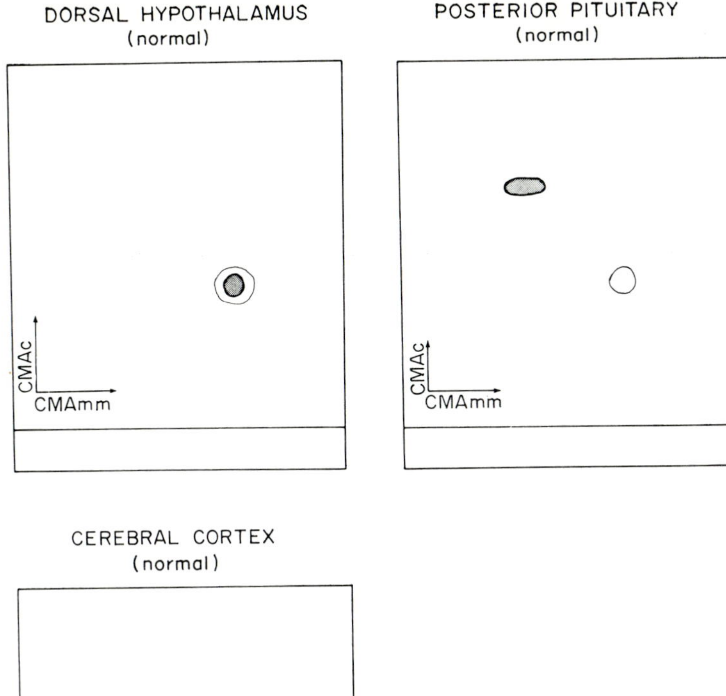

FIG. 14. Composite tracing of thin-layer chromatograms of methanol extracts of dorsal hypothalamic, neurohypophysial, and cerebral cortex fragments which have been incubated in [^{14}C]proline. The open circle area shows the Pauly stain of added thyrotropin-releasing hormone (TRH) marker; the hatched area, the radioactivity of a radioautograph superimposed on the thin-layer plate. CMAc: Chloroform:methanol:acetic acid. 60:40:20; CMAmm: chloroform:methanol:ammonia, 60:45:20. Of the three tissues, only dorsal hypothalamus has an area corresponding to TRH.

TABLE III

R_f Values for Synthetic TRH and for Radioactive Peptides Incorporated by Hypothalamic Tissues in Vitro

Solvent systems[a]	R_f for TRH[b]	R_f for radioactive peptides[c,d]						
		Pro	Glu[e]	His	Ala	Ser	Gly	Threo
(a) Chloroform:methanol:acetic acid (60:40:20)	0.52	0.50	0.51	0.49	0.51	0.51	0.48	0.49
(b) Chloroform:methanol:conc. ammonia (60:45:20)	0.67	0.66	0.67	0.68	0.65	0.63	0.66	0.64
(c) Butanol:acetic acid:water (4:1:5)	0.10	0.10	0.10	0.10	0.11	0.09	0.12	0.11
(d) Butanol:ethyl acetate:acetic acid:water (1:1:1:1)	0.20	0.20	0.20	0.20	0.20	0.20	0.19	0.21
(e) Butanol:ethyl acetate:0.2 N ammonia hydroxide (1:1:2)	0.07	0.07	0.07	0.07	0.08	0.07	0.06	0.07
(f) Ethyl acetate:ethanol:0.2 N ammonia hydroxide (2:1:2)	0.05	0.05	0.05	0.05	0.05	0.05	0.04	0.06
(g) Chloroform:methanol:acetic acid (30:40:20)	0.75	0.75	0.75	0.74	0.74	0.74	0.74	0.74
(h) Chloroform:methanol:acetic acid (60:20:20)	0.25	0.25	0.25	0.25	0.26	0.26	0.24	0.24

[a] Upper phase was used in case of two phases.

[b] Position of TRH spot determined by Pauly reaction using 25 or 50-μg carrier.

[c] Value given for the radioactive spot most closely related to TRH.

[d] Other amino acids tested and found not to correspond to TRH were: arginine, aspartic acid, cystine, cysteine, isoleucine, leucine, lysine, methionine, phenylalanine, tryptophan, tyrosine, and valine. Chloroform:methanol:acetic acid (60:40:20) and chloroform:methanol:concentrated ammonia (60:45:20).

[e] Glutamic acid used for incubation. TRH contains pyroglutamic acid, not glutamic acid.

ments were subjected to paper electrophoresis in barbital buffer, 0.1 N, pH 8.6 (Fig. 15). Radioactivity was localized by scanning, and TRH standards were identified by the Pauly reaction and by the distribution of [^{125}I]TRH prepared by a modification of the method of Hunter and Greenwood (1962). Of the nineteen amino acids tested, proline, histidine,

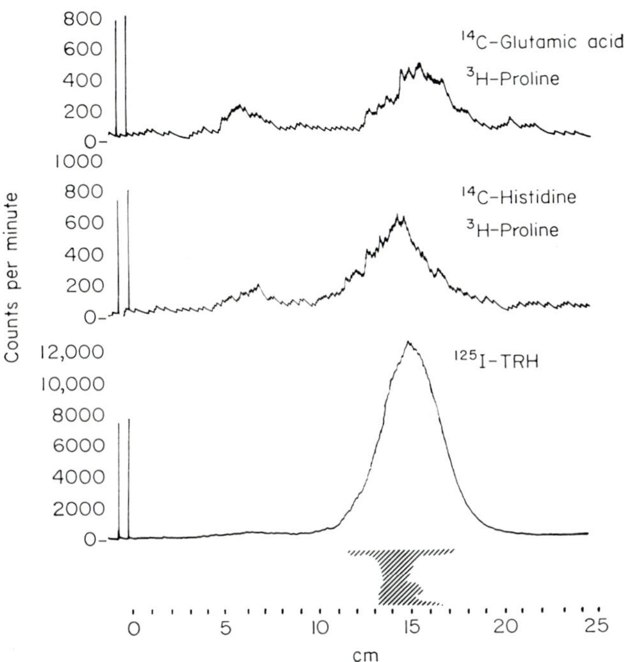

FIG. 15. Paper electrophoresis of synthetic thyrotropin-releasing hormone (TRH) (shown at the bottom as a hatched area corresponding to Pauly-positive material), of ^{125}I-labeled TRH, and of methanol extracts of hypothalamic tissue that had been incubated for 2 hours in buffer solution containing [^{14}C]glutamic acid and [^{3}H]proline or [^{14}C]histidine and [^{3}H]proline. The Pauly-positive material coincides with the ^{125}I-labeled TRH and with the principal peak of the labeled amino acids. Unreacted amino acids migrate to distinctly different regions of the electrophoretogram (not illustrated).

glutamic acid (of the suspected TRH precursor amino acids) and glycine alone coincided with TRH. All the others were separate from TRH.

Further proof of the specificity of proline, histidine, and glutamic acid to TRH was sought by carboxymethyl cellulose (CMC) ion exchange chromatography. This was of particular importance because it had been reported by Schally *et al.* (1969) that discontinuous chromatography on CMC columns achieves complete separation of TRH and LRH.

Methanol extracts of hypothalami incubated in the seven labeled amino acids incorporated into peptides with the mobility of TRH on TLC were subjected to CMC chromatography (Fig. 16). The fractions that contained TRH were shown by ^{125}I-labeled TRH to be located in tubes 42–50. In these same fractions were distributed proline, glutamic acid, and histidine; the other four were located in other fractions.

The experiments described here demonstrate that only glutamic acid, histidine, and proline were consistently found in association with TRH in three different separation systems (TLC, paper electrophoresis, and

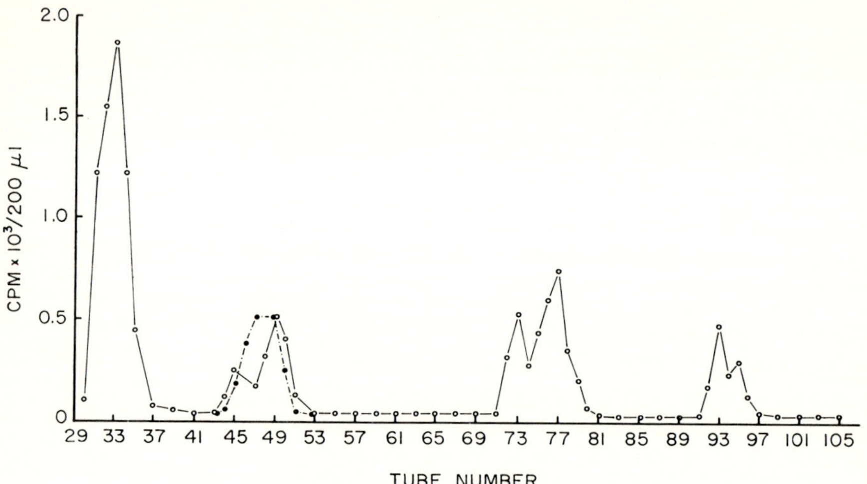

Fig. 16. Hypothalamic incubates exposed to [³H]glutamic acid are chromatographed on a carboxymethyl cellulose (CMC) ion exchange column (2.5 × 45 cm) equilibrated with 0.002 M, pH 4.6 ammonium acetate buffer followed by a gradient to pH 7, 0.1 M ammonium acetate, 10-ml fractions (○——○). The distribution of [125I]TRH from a separate column is plotted on the same coordinates (●---●).

CMC ion exchange chromatography), and they establish with certainty that TRH is synthesized by rat hypothalamic fragments *in vitro*. Since the material has properties identical with synthetic TRH and is extracted by mild procedures, it is reasonable to conclude that native TRH is present in the pyroglutamyl form in tissues.

Of particular note also is the widespread distribution of TRH-synthesizing activity in the hypothalamus, thus confirming Guillemin's observation of the ubiquity of TRH in the hypothalamus. The distribution of TRH synthesizing activity corresponds to the distribution of sites responsive to electrical stimulation (positive in SME, ventral and dorsal hypothalamus, negative in cortex, mammillary bodies) except for the preoptic area which is electrically excitable but does not form TRH. This finding supports the view, stated above, that the pre-optic area is a sensory area which alters hypophysiotropic function through neural control of the TRH—peptidergic neurone.

VI. Mechanism of TRH Biosynthesis

After it was established that TRF biosynthesis could be demonstrated *in vitro*, studies of the mechanisms of biosynthesis were next undertaken. First, we determined whether glutamic acid was cyclized before or after

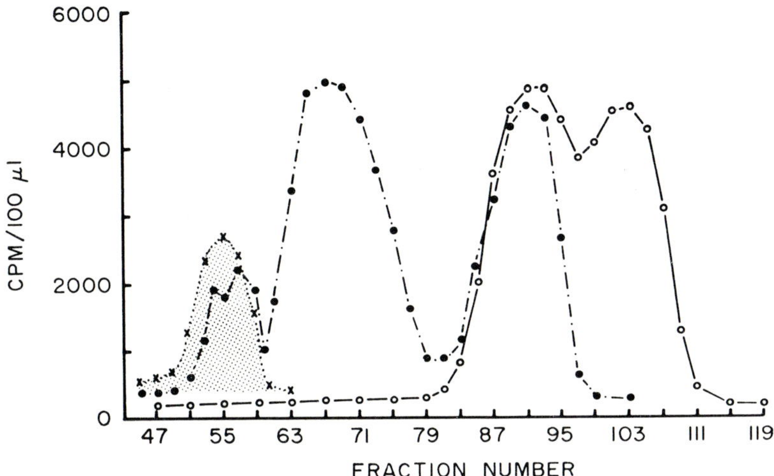

FIG. 17. Sephadex G-10 separation of methanol extracts of rat hypothalamic fragments incubated in either [^{14}C]glutamic acid or [^{14}C]pyroglutamic acid. Peptides with mobility of thyrotropin-releasing hormone (TRH) (distribution shown by [^{125}I]TRH are seen after incubation with labeled glutamic acid, whereas no radioactivity is observed in the TRH area after exposure to labeled pyroglutamic acid, indicating that pyroglutamic acid is not incorporated as such into TRH. Composite of three curves. X···X, [^{125}I]TRH; ●—·—●, [^{14}C]glutamic acid; O——O, [^{14}C]pyroglutamic acid.

incorporation into the tripeptide TRH. This was done by observation of the pattern of polypeptide synthesis by hypothalamic fragments incubated with tracer [^{14}C]glutamic acid or [^{14}C]pyroglutamic acid. Peptides were separated by Sephadex G-10 chromatography (Fig. 17). Glutamic acid, but not pyroglutamic acid, was incorporated into newly synthesized peptides with the mobility of TRH, the TRH region being determined by bioassay and by the distribution of ^{125}I-labeled TRH. It can be concluded therefore that the glutamic acid moiety is cyclized *after* incorporation into the tripeptide (or dipeptide precursor).

Two possible mechanisms of TRH biosynthesis were considered. In view of the small size of this substance, it was reasonable to postulate that synthesis was by a nonribosomal enzymatic mechanism similar to that responsible for the synthesis of the tripeptide glutathione (γ-glutamylcysteinylglycine) which is accomplished by an ATP-dependent synthesis of two peptide bonds by two distinct enzymes (Mooz and Meister, 1967). It was also possible that TRH was synthesized on ribosomes as part of a prohormone, as shown for insulin by Steiner and co-workers (1969) and postulated for vasopressin by Sachs and collaborators (1968). The possibility of a ribosomal prohormone synthesis is further suggested by the recent demonstration that MSH-IF (melanocyte-stimulating hormone inhibitory factor) (prolylleucylglycinamide) was formed from the terminal three amino acids of oxytocin by a stalk median eminence microsomal exopeptidase (Celis *et al.*, 1971).

Ribosomal prohormone type of synthesis of TRH was excluded in two ways: (1) by studies using the antibiotic inhibitor puromycin in incubation systems; and (2) in a more positive way by studies of RNase and puromycin-treated hypothalamic extracts. In the tissue incubation experiments, hypothalamic fragments from two rats, separately dissected into stalk-median eminence, ventral, and dorsal hypothalamic portions, were incubated in concentrations of puromycin of 1, 5, 50, and 100 μg/ml. Only the highest concentration of puromycin significantly reduced proline incorporation into 5% trichloroacetic acid-precipitable counts. Thin-layer chromatography and radioautography for 7 days showed incorporation of proline into the TRH area—these regions were scraped off and counted. In SME, acid-precipitable counts were reduced by 88% by incubation in 100 μg/ml of puromycin while TRH counts were reduced by only 4%. Additional studies, using 200 μg/ml of puromycin gave 98% inhibition of protein synthesis by hypothalamic fragments, but did not interfere with TRH synthesis.

In a more direct way, the enzymatic synthesis of TRH was demonstrated with a soluble fraction of freeze-dried porcine hypothalamic tissue from which particulates had been removed by centrifugation for 13 hours at 100,000 g. For convenience in the repetitive separation of samples for determination of the rate of incorporation of amino acids into TRH over time, an electrophoretic method using cellulose polyacetate strips (Sepraphore III) was chosen because of its greater specificity and rapidity as compared with silica gel thin-layer chromatography and Sephadex chromatography. It is important to recognize in studies of this type that, when histidine is used as a precursor of TRH, separation must be made between TRH and histamine, the decarboxylation product of histidine. Specificity of incorporation of counts into TRH comes also from the demonstration that each of the precursor amino acids gives comparable results in the same system.

An important preliminary step in this experiment was to identify the necessary cofactors and the optimum conditions required for enzymatic activity. The substances were initially tried on the basis of their importance in other well characterized enzymatic reactions. TRH synthesis was not detectable if any one of the three precursor amino acids, proline, histidine, and glutamic acid had been omitted, and the reaction did not occur in the absence of ATP. Optimum conditions for assay were: proline, histidine, and glutamic acid, 10^{-3} M; ATP, 2×10^{-5} M; $MgSO_4$,

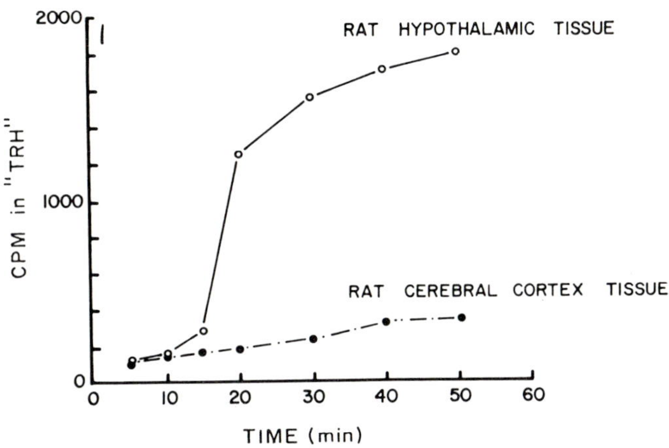

FIG. 18. Progressive incorporation with time of [^{14}C]proline into electrophoretically separated thyrotropin-releasing hormone (TRH) by particulate-free extract of rat hypothalamic tissue in the presence of optimum concentrations of substrates and cofactors. For comparison, the relative inactivity of cerebral cortex extract is shown.

2×10^{-5} M; and at pH 7.4. A crude tissue extract, 5% in the final incubation was found to possess adequate synthesizing capacity (Fig. 18). This was true for fresh rat hypothalamic tissue, and freeze-dried porcine and rat hypothalamic tissue. Cerebral cortex extract had barely detectable effects on the incorporation of precursor amino acids into the "TRH" area (Fig. 18).

In the presence of optimum concentrations of cofactors and precursors, the rate of formation of TRH was a function of the concentration of any one of the amino acids. Lineweaver-Burke plots of the reciprocal of the initial velocity of incorporation of amino acids into TRH as a function of the reciprocal of substrate for each of the three precursor amino acids gave identical slopes within the limit of experimental error (Fig. 19).

As further proof of the nonribosomal nature of the TRH synthetase reaction a preparation of fresh rat hypothalamic tissue was extracted in 0.01 M phosphate buffer and incubated in the usual TRH synthesis system with 1 and 5 μg/ml of pancreatic RNase, a procedure suggested by Dr. Robert Roskoski, The Rockefeller Institute. This material has been used to demonstrate the nonribosomal synthesis of Gramicidin S (Gevers *et al.*, 1968). No inhibition of TRH synthesis was detected in RNase-treated extracts. Puromycin, 0.3 mM, also failed to alter

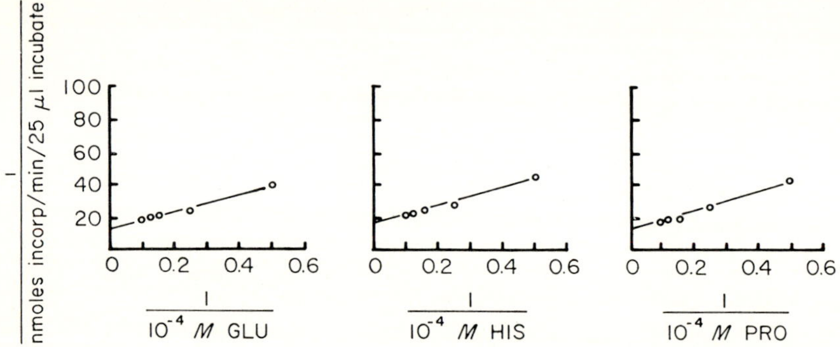

Fig. 19. Lineweaver-Burke plots of rate of incorporation of labeled amino acids into thyrotropin-releasing hormone (TRH) as a function of substrate concentration. In each panel, one of the three precursor amino acids is illustrated. Incubation media contained optimum concentrations of substrates and cofactors (see text) except with the modification of substrate concentration of the amino acid under study. Left panel: glutamic acid; middle panel: histidine; right panel: proline.

the incorporation of proline into TRH by fresh rat hypothalamic homogenates, centrifuged at 5000 rpm for 30 minutes.

These experiments indicate that the synthesis of TRH is an enzymatic one. In view of the fact that in the formation of this molecule two peptide bonds are formed and that there is a cyclization reaction and an amidation step, it is conceivable that at least four separate enzyme reactions may be involved in synthesis although it is possible though less likely only one enzyme is involved. Since the reaction is ATP dependent and involves peptide bond formation, the overall activity of this system can properly be termed "TRH synthetase." Efforts to purify, isolate, and separate the several components of the TRH synthetase system are now underway in our laboratory using various protein separation techniques and labeled and unlabeled precursor molecules.

VII. Regulation of the Concentration of Hypothalamic TRH Synthetase

A. Thyroid Hormone and Cold Exposure

Although it is recognized that TRH synthetase may consist of several enzyme reactions, the quantitative kinetic methods of the enzymologist provide a tool for the analysis in further detail of some of the physiological questions raised in the initial part of this presentation. As was to be expected, the rate of synthesis of TRH was a function of the concen-

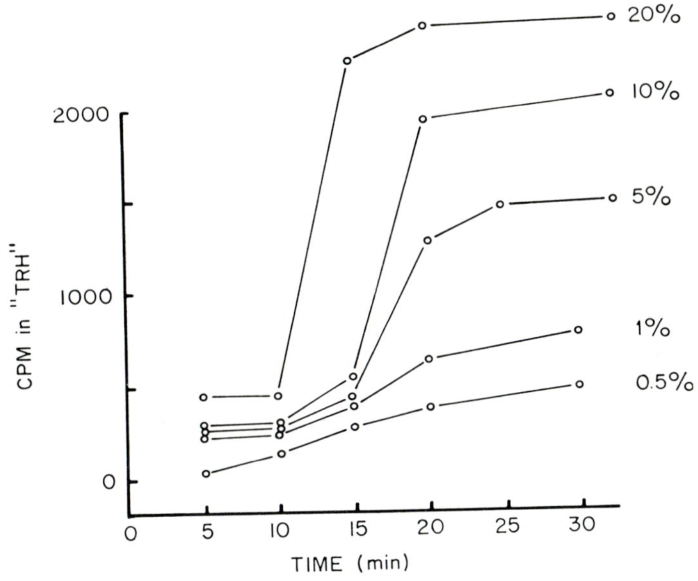

Fig. 20. Curves of incorporation of [^{14}C]proline into thyrotropin-releasing hormone (TRH) by particulate-free hypothalamic extract. The rate of incorporation was a function of extract concentration.

tration of the enzyme(s) as shown by the increasing effectiveness of increasing concentrations of a crude extract of hypothalamic tissue (Fig. 20).

It seemed apparent that the rate of TRH synthesis could be used as a quantitative measure of the effect of various procedures thought to alter pituitary–thyroid function. In particular, it was important to study the possible effects of thyroid hormone deficiency and excess and of cold exposure as regulators of TRH synthesis. Wurtman (1971) among others pointed out the importance (in the case of norepinephrine) of determining the concentration of the synthesizing enzymes as a measure of noradrenergic function rather than the concentration of the neurotransmitter itself. Sinha and Meites (1965) studied the effects of thyroid status on hypothalamic TRH content, and reported that TRH levels were increased in hypothyroid but were unaltered in T_4-treated rats. Since content is a reflection of both synthesis and release, no conclusions can be drawn as to the functional significance of a given hypothalamic TRH content. It remains to be proved that the concentration of TRH synthetase reflects TRH secretion rate into the portal vessels.

Groups of 9 rats each were thyroidectomized (care being taken to preserve at least one parathyroid gland) or were treated with thyroxine, 4 μg/100 gm body weight per day (a dose chosen as the minimum completely suppressing TSH secretion), or placed in the cold at 4–6°C. After 4 weeks, the animals were decapitated, the trunk blood was heparinized and frozen for subsequent TSH immunoassay, and the hypothalami were removed for measurement of TRH synthetase activity. Cold-exposed and thyroidectomized rats had higher plasma TSH levels than normal (47.5 $\pm$ 7.9 SE and 63.8 $\pm$ 9.8 vs 26.1 $\pm$ 6.7 mU/100 ml, respectively), and T_4-treated rats had significantly lowered plasma TSH levels (8.8 $\pm$ 1.0 mU/100 ml).

Enzyme assays were carried out on pools of three hypothalami each, concentration 5% w/v homogenized in 1.5 ml of 0.32 M sucrose in 0.01 phosphate buffer and centrifuged at 100,000 g for 1 hour. The sucrose and smaller molecules were removed by diafiltration (Amicon micro-ultrafiltration system, Model 8 MC) UM 10 membrane. The extracts were then made up in 0.5 ml, and the incorporation of histidine and of proline into TRH was measured in the presence of optimum substrates and cofactors (Fig. 21). We found that enzyme concentration were markedly different in these preparations. Almost no TRH was formed by hypothalamic extracts from hypothyroid animals, while in contrast, T_4-treated animals had somewhat higher than normal synthetase concentrations. From these observations it was concluded that thyroid hormone actually stimulates hypothalamic TRH synthesis.

In terms of the classical question about the site of negative feedback control of TSH secretion, these observations suggest that if thyroxine has an effect on the hypothalamus in the regulation of TSH secretion, it is a positive feedback, rather than a negative, effect. We would suggest that the effects of thyroxine on hypothalamic TRH synthetase synthesis may be a reflection of its stimulating effects on protein synthesis in general, and that this phenomenon may be noticed in certain of the alterations in hypothalamic–pituitary control seen in hypothyroid states, such as growth hormone deficiency, anovulation, and galactorrhea.

Of special interest in terms of physiological regulation of the pituitary-thyroid axis was the observation that TRH synthetase of cold-exposed rats was the highest of all the groups tested (Fig. 21). Since thyroidectomized rats (with high rates of TSH secretion) have low hypothalamic TRH synthetase concentration, it is obvious that one cannot attribute the high hypothalamic synthetase of cold-exposed rats to TSH secretion compensatory in turn to an increased rate of peripheral thyroxine metabolism, one of the several explanations offered to account for thyroid hypersecretion in the cold. Rather, this observation suggests that chronic cold exposure leads to the stimulation of TRH synthesis through a central neurogenic signal. This observation is compatable with

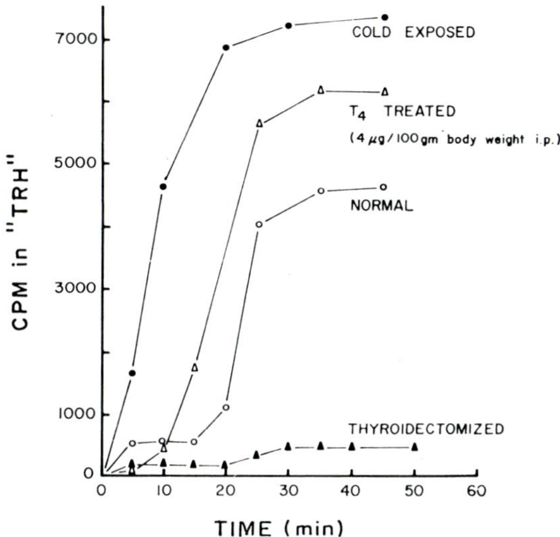

FIG. 21. Curves of incorporation of [^{14}C]proline into thyrotropin-releasing hormone (TRH) by hypothalamic extracts from rats treated for 4 weeks with thyroxine (4 µg/100 gm body weight per day), thyroidectomy, or cold exposure at 4–6°C.

the report by Sakoda and Nakabayashi (1970) that acute cold exposure in the rat increases concentration of hypothalamic TRF.

As mentioned above, a possible explanation of the normal T_4 levels of cold-exposed animals (in the face of increased hypothalamic-pituitary activity) may be that there is an independently determined increase in peripheral T_4 metabolic loss.

The finding that thyroxine concentration apparently stimulates the synthesis of hypothalamic TRH synthetase in the rat immediately brought to mind the studies of tadpole metamorphosis by Etkin (1963) which have led him to postulate that the metamorphic crisis in the tadpole is brought about by progressive increase in pituitary-thyroid secretion, a response in turn dependent upon the development under the influence of thyroid hormone of hypothalamic hypophysiotropic function. The hypothalamic component, according to Etkin's view, is under positive feedback control by thyroxine whereas the pituitary is under negative feedback control. This formulation has not been universally accepted (Frieden and Just, 1970) because TRH has not been identified in tadpole hypothalamus, tadpoles apparently do not metamorphose after injection of native mammalian TRH; appropriate changes in plasma thyroxine have not as yet been demonstrated, and in one species. *Xenopus laevis,* hypothalamic destruction does not prevent metamorphosis.

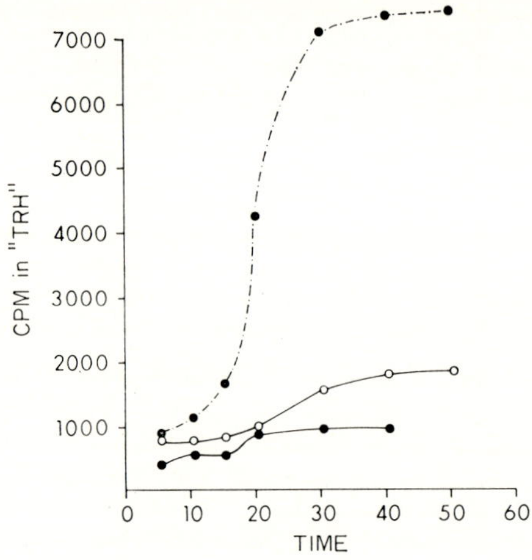

FIG. 22. The rate of incorporation of [¹⁴C]proline into "TRH" by particulate-free extract of whole bullfrog tadpole brain extract used in a 5% concentration (w/v). The preclimax tadpoles (●——●) used in this experiment were so classified on the basis of having no forelimb or opercular development (Frieden and Just, 1970); postclimax tadpoles (●—·—●), the result of spontaneous metamorphosis, had full forelimb development and beginning tail resorption. Adults were leopard frogs (○——○).

Despite these seemingly crucial criticisms of the Etkin hypothesis, we felt that alternative explanations for each of these objections could be found, and that it would still be of value to examine TRH biosynthesis by tadpole brain extract taken from preclimax (hindlimb only) and postclimax (hindlimb and forelimb) animals. These studies done in collaboration with Mr. Dan Keller demonstrated that whole brain extracts from postclimax animals were extremely rich in TRH synthetase (Fig. 22) as measured by incorporation of [¹⁴C]proline counts into TRH. In contrast, preclimax brains had little or no synthetase activity. In adult frog brain, synthetase is present but in lower concentration, presumably because hypothalamus makes up a relatively smaller fraction of adult brain. Present studies are underway to localize the anatomical site of the activity and to measure the time course of response of thyroxine-induced (rather than natural) metamorphosis and compare these changes with the time course of other brain, liver, and tail enzymes known to be activated by thyroxine in the tadpole (Eaton and Frieden, 1969; Cohen, 1970; Frieden and Just, 1970). Subsequent to the conference we have found that T_3 administration induces TRH synthesis in the tadpole, and that all of the TRH synthesizing activity of the adult frog brain is confined to the hypothalamus.

The fact that both frog and rat brain TRH synthetase are thyroxine dependent suggests a fundamental biological similarity in the regulation of TRH secretion, and suggests a molecular biological basis by which the "proper" level of circulating thyroid hormone can be determined. If the rate of TSH secretion is governed by an interaction of the inhibitory effects of thyroxine acting at the pituitary level and the stimulatory effects of TRH acting at the hypothalamic level, both activities being a function of thyroid hormone dependent protein synthetic mechanisms, it is reasonable to propose that the set point of circulating thyroid hormone is genetically determined by the response characteristics of thyroid hormone inducible enzymes in the hypothalamic–pituitary unit.

B. NEUROTRANSMITTER REGULATION OF TRH SYNTHESIS

According to current views of hypothalamic–pituitary control as reviewed recently in a symposium edited by Wurtman (1971), the peptidergic neurones responsible for the secretion of the releasing hormones are in turn regulated by neurotransmitters of which the catecholamines, dopamine, serotonin, and norepinephrine may be the most important. For example, the introduction of dopamine into the third ventricle of the hypothalamus brings about the release of FSH (Kamberi et al., 1971a) and the inhibition of prolactin secretion (Kamberi et al., 1971b); and serotonin, similarly infused, stimulates prolactin secretion and inhibits FSH secretion (Kamberi et al., 1971c).

To determine whether catecholamines regulate the synthesis of TRH synthetase, we have measured enzyme activity in hypothalami from a group of 8 rats treated over a period of 17 days with five doses of reserpine, 100 μg/100 gm body weight. This dosage lowers plasma TSH significantly (19.1 $\pm$ 3.5 vs 10.7 $\pm$ 1.5 mU/100 ml). Assay of a pool of hypothalamic tissue from groups of four to these animals in a concentration of 5% indicates a reduction in TRH synthetase concentration (Fig. 23). These observations support the hypothesis that a monamine neurotransmitter neuron regulates the synthesizing activity of the peptidergic neuron responsible for TRH synthesis. This system may be analogous to that observed in the pineal gland, sympathetic denervation of which is followed by a marked decline in the activity of hydroxy-*o*-methyl transferase, the enzyme responsible for melatonin synthesis (Wurtman *et al.*, 1964) and pineal adenyl cyclase (Weiss and Costa, 1967).

Some caution about this experiment must be voiced because reserpinized rats treated at this dosage level appear chronically ill, and it is known that starvation or stress can inhibit TSH secretion. If in fact the observations are due to nonspecific stress rather than to catechol-

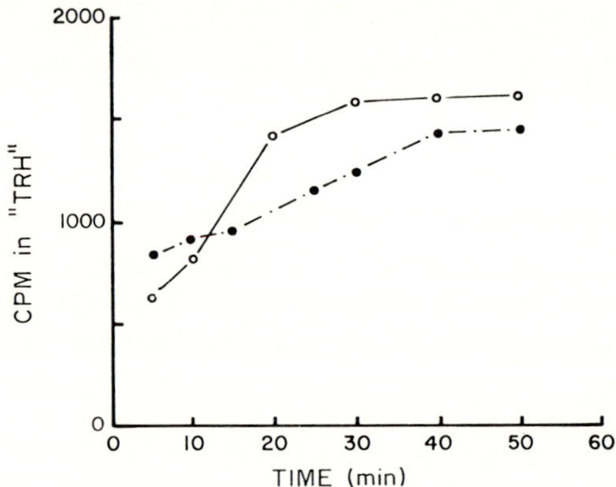

FIG. 23. Incorporation of [¹⁴C]proline into "TRH" by particulate-free hypothalamic extract from rats treated with 5 doses of reserpine, 1 mg/kg administered over a 17-day period. Reserpine-treated (●—·—●) animals had lower synthetase concentrations than normal ones (○——○).

amine depletion, it would then appear likely that stress inhibits TSH secretion through effects exerted at the hypothalamic level.

VIII. TRH Secretion by Isolated Hypothalamic Tissue

As part of the analysis of the physiological regulation of TRH secretion, an *in vitro* system for study of control of TRH secretion has been developed (Grimm and Reichlin, 1972). This system is based largely on the experimental design introduced by Douglas and collaborators (Douglas, 1966) for the study of "stimulus-secretion" coupling in isolated neurohypophysis and adrenal medulla. Instead of measuring hormone discharged into the incubation media (as has been done for vasopressin and epinephrine), the discharge of pulse-labeled TRH has been followed as a measure of TRH secretion.

Mouse hypothalamic fragments (this species chosen so as to provide small, uniform tissue blocks) were exposed to radioactive histidine or proline for 40 minutes in an incubation medium of Locke's solution containing glucose (10 mM) gassed with 5% CO_2–95% O_2 and shaken at 37°C. The tissue fragments, in small nylon mesh baskets, were trans-

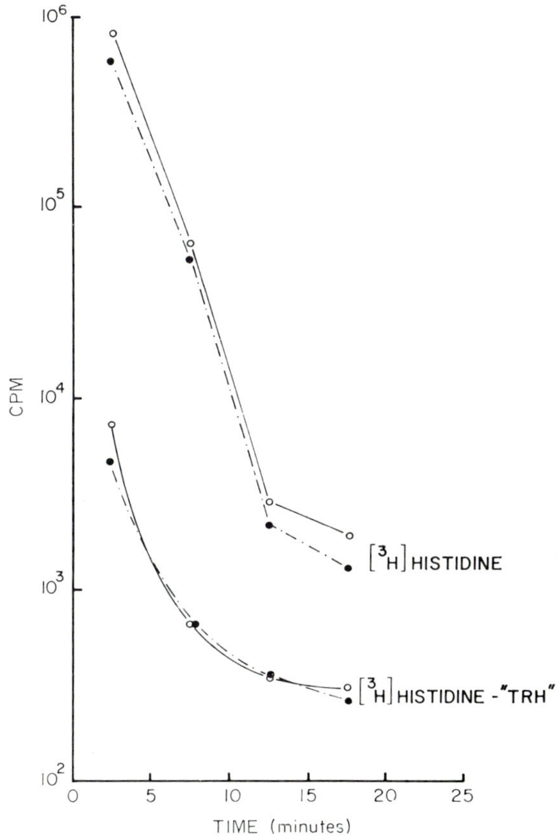

Fig. 24. Rate of appearance of [³H]histidine and of [³H]TRH in incubation medium after pulse labeling of 3 mouse hypothalami in [³H] histidine in glucose-enriched Locke's solution for 40 minutes.

ferred at 5-minute intervals into 0.5 ml of fresh buffer solutions. Methanol (2 ml) was added to each tube which was then air dried, extracted in 90% methanol and subjected to thin-layer chromatography on carboxymethyl cellulose plates (Caltech Inc., Newark, Delaware) using added synthetic TRH stained with the Pauly reagent as a marker. Counts in unincorporated amino acid and in TRH were used as a measure of the TRH released at each time interval (Fig. 24).

Following pulse labeling for 40 minutes, 2–5% of the radioactivity is found in the TRH fraction of the hypothalamus. In successive samples of the media, the concentration of washout counts of unincorporated amino acids falls rapidly in a complex function presumably reflecting in part losses from a freely diffusible tissue pool, from counts adherent to the nylon mesh basket, and in part losses from a sequestered tissue pool. Over the first 15 minutes, rate of appearance of incubate counts follows a first-order curve with a rate of approximately 85% per minute. The counts in labeled TRH secreted into the media also fall rapidly

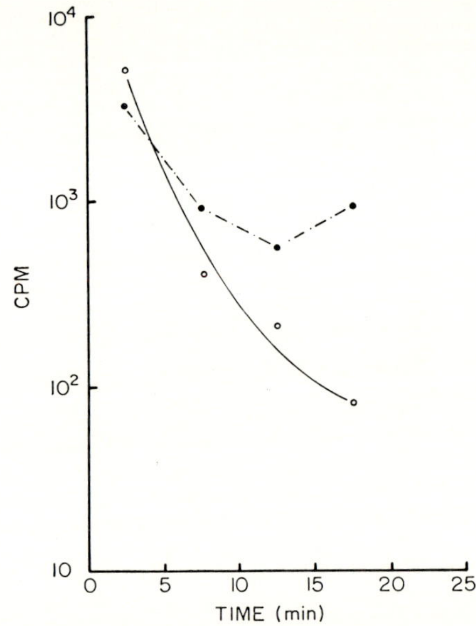

FIG. 25. Effect of norepinephrine, 10^{-4} M, on the discharge of [³H]TRH from pulse-labeled mouse hypothalamic tissue. ●—·—●, norepinephrine; ○——○, control.

over the first 15 minutes roughly paralleling the histidine curve for the first 5–10 minutes. Thereafter, the rate of decline in washout TRH counts falls gradually so that the curve is clearly a more complex function suggesting the participation of several labeled compartments. At the end of 20 minutes of incubation net counts of hypothalamic TRH are within 90% of the original value before incubation despite the loss of more than 20% of initial TRH counts into the media. These observations suggest that there has been continuing synthesis of TRH during incubation, presumably from the large, unincorporated hypothalamic histidine precursor pool.

The mathematical analysis of this complex pool of labeled precursors has not yet been accomplished, but it is likely that passage in and out of the histidine pool is governed by active transport mechanisms.

Although the form of the 20-minute washout curve of labeled TRH does not follow first-order kinetics, change in slope of the line can be used as a measure of relative rates of hormone discharge if it is assumed that the factors causing the change does not alter pool size ,or precursor–product relationships. The effects of putative neurotransmitters were tested at 10^{-4} M concentration. These included acetylcholine,

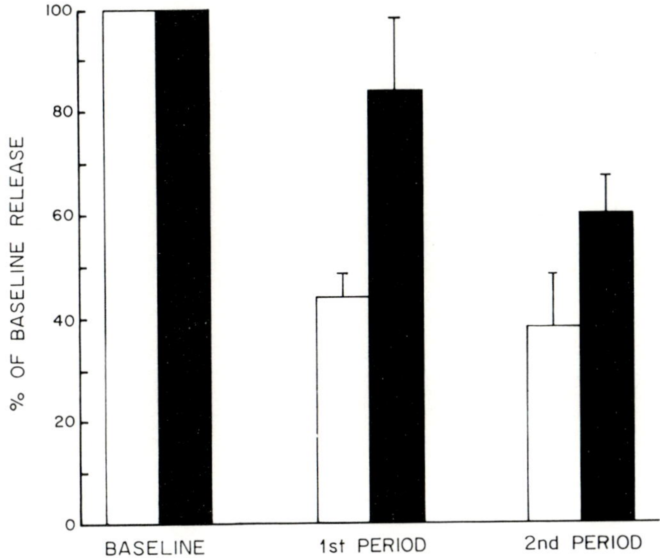

FIG. 26. Summary of studies of effect of norepinephrine on labeled thyrotropin-releasing hormone (TRH) discharge comprising data for 5 separate runs of 3 hypothalami each. Significance of difference in 2 successive 10-minute periods (after an initial 10-minute leach) are shown, comparing discharge in treated (■) with untreated (□) tissue.

γ-amino butyric acid, dopamine, and norepinephrine. Of these, two had a consistent stimulating effect on TRH discharge. Norepinephrine, in 5 separate experiments using pools of three hypothalami each, enhanced the rate of discharge of TRH (Figs. 25 and 26) and dopamine, in two experiments, also caused the release of TRH. Since dopamine may be converted to norepinephrine, the latter transmitter may be the effective one in TRH regulation. This problem is under study. Serotonin consistently inhibited TRH discharge (not shown).

The effects of prior catecholamine depletion on spontaneous TRH release was also examined because of the finding that norepinephrine and dopamine increased TRH discharge and that reserpine inhibited the synthesis of TRH (see above). Hypothalami from reserpine-treated rats (used in the enzyme concentration studies) were pulse labeled, and TRH washout was observed (Fig. 27). Spontaneous release of labeled TRH from the isolated fragment was much slower (5% per minute *vs* 20%

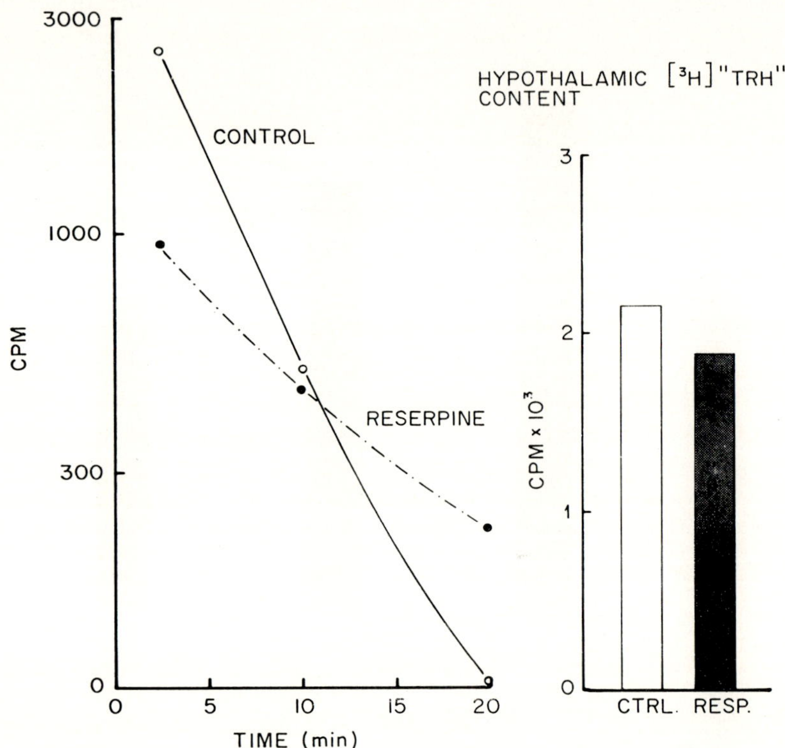

FIG. 27. Effect of pretreatment with reserpine on the rate of discharge of [³H]TRH into the media from pulse-labeled rat hypothalamic fragments. The rate was less than control (rate of decline of radioactivity in media 5% per minute *vs* 20% per minute).

per minute) than from the normal hypothalamic fragment. At the end of the experiment, approximately equal amounts of labeled TRH counts were found in the hypothalamic fragments.

If one considers that hypothalami from reserpine-treated animals have a decrease in TRH synthetase activity, and a decreased rate of discharge of labeled hormone, it would appear reasonable to conclude that synthesis and turnover of TRH are both reduced in catecholamine-depleted hypothalami. Since norepinephrine is capable of increasing TRH release, it may be postulated that the total function of the TRH peptidergic neuron is subject to control by norepinephrine. The function of the hypothalamic neurotransmitter link may thus be analogous to the function of the pineal gland, enzyme activity of which is controlled by norepinephrine. In studies completed since the conference, dibutyryl cyclic AMP has been shown to discharge TRH; adenylcyclase activation may thus be the mechanism by which norepinephrine controls TRH release. It has also been found that pharmacologic blockade of dopamine conversion to norepinephrine blocks dopamine-induced TRH discharge.

IX. Neural Control of TSH Secretion: Summary and Formulation

Studies of hypothalamic TRH regulation summarized in this paper permit us to amplify and provide some generalizing and perhaps speculative insight at a basic level of the mechanisms governing the regulation of the pituitary–thyroid axis (Fig. 28).

It is probable that the thyrotrope cell is regulated by thyroid hormone by a thyroxine-dependent inhibitory substance in the pituitary gland the concentration of which is dependent upon protein synthesis. The inhibitory actions of thyroxine at the pituitary level are opposed by the stimulatory effects of TRH, which determines the sensitivity of the anterior pituitary to negative feedback inhibition.

Our findings indicate that TRH is synthesized enzymatically, presumably by peptidergic neurons, in a widely distributed area of the hypothalamus, the anatomical distribution of which corresponds generally to the same region which is electrically excitable to release TSH. The synthesis of TRH synthetase is apparently stimulated by thyroid hormone, unlike the situation in the pituitary in which the overall effects are inhibitory, and in contrast to previous suggestions by many workers that thyroxine exerts a negative feedback action on the thyrotropic area. We have not as yet been able to show directly that TRH secretion is enhanced by thyroxine administration, but the finding that the synthesizing enzyme is increased in amount suggests that this may be so. If enzyme concentrations are an index of TRH secretion (as is true for the enzymes regulating norepinephrine and melatonin secretion), it would appear reasonable to suggest that the rate of TSH secretion is the resultant of negative feedback at the pituitary level and positive feedback at the hypothalamic level. Since both effects depend upon changes in thyroxine-dependent protein synthesis it would seem reasonable to conclude that in the hypothalamic–pituitary unit, as in the remainder of the body, the fundamental action of thyroxine is in regulating some aspect of protein synthesis. According to this view, the "proper" level of thyroid hormone maintained at the physiological set point is determined by the precise nature of the dose-response characteristics of the protein synthesis regulating systems in pituitary and hypothalamus, which presumably balance at the physiological set point. Such a mechanism would explain the genetic control of a complex regulatory system involving the interaction of several structures.

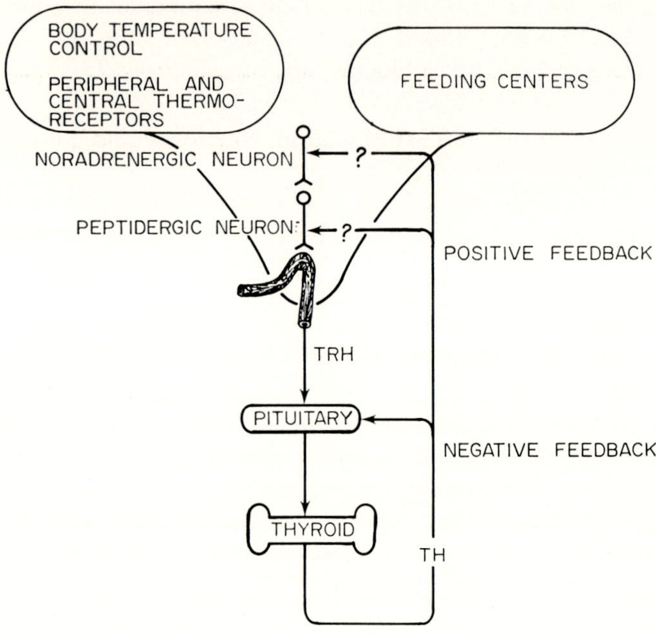

FIG. 28. Hypothalamic–pituitary–thyroid system for control of thyroid (TH) secretion. As in Fig. 1, negative feedback effects of thyroxine on pituitary thyroid-stimulating hormone (TSH) secretion and the interaction between TH and thyrotropin-releasing hormone (TRH) at the pituitary level are illustrated. It is proposed that thyroxine exerts a *positive* feedback effect on TRH secretion and that the set point of control is determined by the interaction between positive stimulatory effects of T_4 on TRH secretion and of inhibitory effects on the thyrotrope, both functions depending upon T_4-inducible protein changes. Peptidergic TRH neuron is shown here as being under control of a noradrenergic neuron which forms part of a noradrenergic thermal homeostatic system involving body temperature regulation and determination of food drive. The level of TRH secretion is thus dependent upon plasma thyroxine *and* is responsive to neurally related homeostatic functions.

In addition to regulation by thyroxine, the TRH peptidergic neuron appears to be dependent upon catecholaminergic and possibly monoaminergic neurons for the regulation of TRH synthesis and release. Through such neurons, cold response reflexes are probably mediated.

Acute TSH responses can be brought about by either central cooling with normal skin temperature or peripheral cooling with normal central temperature—thus, peripheral and central thermoreceptors appear to be involved in these reactions which are superimposed acutely over the long-term balance of hypothalamic–pituitary function maintained by thyroid hormone levels.

We have not as yet arrived at a satisfactory resolution of the question of the significance of thermoregulatory reflexes in the maintenance over the long term of thyroid hormone levels. In our experiments, evidence of enhanced pituitary–thyroid function is unaccompanied by significant changes in circulating thyroid hormone levels which must mean that peripheral T_4 clearance is enhanced in one way or another as proposed by several workers. On the other hand, TRH synthetase is also enhanced by chronic cold exposure. One might propose that the enzyme effects are secondary to peripheral losses of thyroxine, but the studies of thyroxine control of TRH synthetase suggest that low thyroxine levels inhibit hypothalamic function. We would propose that the chronic cold adapted rat has *both* neurogenically stimulated increase in TRH secretion (through a catecholaminergic mechanism) and an increased rate of thyroxine disappearance from the blood secondary to physiologically adaptive mechanisms, the most important of which, as Galton and Nisola have shown, is increased fecal loss secondary to increased food ingestion.

The possible role of central catecholaminergic transmitters in the regulation of TRH secretion permit an additional speculative proposal. In earlier papers (Reichlin, 1963, 1964, 1966) attention was drawn to the close anatomical proximity of the central thermoreceptor area to the thyrotropic area and to the appetite regulating areas of the hypothalamus, and to the observation that central cooling of the hypothalamus activated autonomic temperature regulatory reflexes and, in addition, TSH secretion and increased appetite. It was proposed that the hypothalamus served to integrate thermal homeostasis which involved autonomic control, metabolic control and the provision, through behavioral homeostatic responses of adequate intake of food to provide substrate for energy production. To this view we may now add the additional data which suggests that hypothalamic catecholaminergic, presumably noradrenergic neurons may provide the functional system by which these reactions are mediated. The work of Feldberg and collaborators (Feldberg, 1970) suggests that hypothalamic temperature regulation is mediated by catecholamines, and that, in certain species, including the rat, third ventricular administration of norepinephrine causes a rise in body temperature. When injected into the lateral hypothalamic area (Grossman, 1960) or perifornical area (Leibowitz and Miller, 1969) norepinephrine stimulates food intake, and our own studies suggest that TRH secretion is norepinephrine mediated. It may well be that the neural and hormonal interactions responsible for thermal homeostasis are controlled at the hypothalamic level by a monoaminergic neural system.

ACKNOWLEDGMENT

We wish to acknowledge with thanks our collaborative efforts with Drs. Robert Utiger, Don S. Schalch, John Pierce, Marilyn Kleinman, Andrew Kaufman, and Daniel Keller. Mrs. Marcia Van Camp has given invaluable help with bioassay and histological study. Supplies of TRH used were furnished by Abbott Laboratories, through the courtesy of Dr. Michael Anderson, or were synthesized from the methyl ester of pyroglutamylhistidylproline generously furnished by Hoffmann-La Roche through the courtesy of Dr. W. E. Scott.

The secretarial assistance of Mrs. Frances DiBattista and of Miss Julie Sorensen is gratefully acknowledged.

REFERENCES

Ackerman, N. B., and Arons, W. L. (1962). *Endocrinology* **71,** 7.

Andersson, B., Ekman, L., Gale, C. C., and Sundsten, J. W. (1962). *Acta Physiol. Scand.* **54,** 191.

Andersson, B., Ekman, L., Gale, C. C., and Sundsten, J. W. (1963). *Acta Physiol. Scand.* **59,** 12.

Averill, R. L. W. (1969). *Endocrinology* **85,** 67.

Averill, R. L. W., and Kennedy, T. H. (1966). *N.Z. Med. J.* **65,** 398.

Averill, R. L. W., and Kennedy, T. H. (1967). *Endocrinology* **81,** 113.

Averill, R. L. W., and Newman, W. C. (1969). *Endocrinology* **84,** 950.

Averill, R. L. W., and Salaman, D. F. (1967). *Endocrinology* **81,** 173.

Averill, R. L. W., Purves, H. D., and Sirrett, N. E. (1961). *Endocrinology* **69,** 735.

Bakke, J. L., and Lawrence, N. L. (1965). *Proc. 47th Meet., Endocrine Soc. New York* p. 78.

Bakke, J. L., and Lawrence, N. L. (1971). *Endocrinology* **89,** 204.

Berg, G. R., Utiger, R. D., Schalch, D. S., and Reichlin, S. (1966). *J. Appl. Physiol.* **21,** 1791.

Bøler, J., Enzmann, F., Folkers, K., Bowers, C. Y., and Schally, A. V. (1969). *Biochem. Biophys. Res. Commun.* **37,** 705.

Bogdanove, E. M., and Halmi, N. S. (1953). *Endocrinology* **53,** 274.

Boop, W. C., and Story, J. (1966). *Neurology* **16,** 1167.

Bowers, C. Y., Schally, A. V., Reynolds, G. A., and Hawley, W. D. (1967). *Endocrinology* **81,** 741.

Bowers, C. Y., Lee, K. L., and Schally, A. V. (1968). *Endocrinology* **82,** 75.

Bowers, C. Y., Schally, A. V., Enzmann, F., Bøler, J., and Folkers, K. (1970). *Endocrinology* **86,** 1143.

Brown-Grant, K. (1960). *Brit. Med. Bull.* **16,** 165.

Burgus, R., and Guillemin, R. (1970). *Annu. Rev. Biochem.* **39,** 499.

Burgus, R., Amoss, M., and Guillemin, R. (1967). *Experientia* **23,** 417.

Burgus, R., Dunn, T. F., Desiderio, D., Vale, W., and Guillemin, R. (1969). *C. R. Acad. Sci., Ser. D* **269,** 1870.

Burgus, R., Dunn, T. F., Desiderio, D. M., Ward, D. N., Vale, W., Guillemin,

R., Felu, A. M., Gillessen, D., and Studer, R. O. (1970). *Endocrinology* **86,** 573.

Campbell, H. J., George, R., and Harris, G. W. (1960). *J. Physiol. (London)* **152,** 527.

Celis, M. E., Taleisnik, S., Schwartz, I. L., and Walter, R. (1971). *Biophys. J.* **11,** 98a. Abstr.

Cohen, P. P. (1970). *Science* **168,** 533.

D'Angelo, S. A. (1960). *Amer. J. Physiol.* **199,** 701.

D'Angelo, S. A. (1963). *Advan. Neuroendocrinol., Proc. Symp., Miami, 1961* pp. 158–205.

D'Angelo, S. A., and Traum, R. E. (1958). *Ann. N.Y. Acad. Sci.* **72,** 241.

D'Angelo, S. A., Snyder, J., and Grodin, J. M. (1964). *Endrocrinology* **75,** 417.

de Groot, J. (1959). *J. Comp. Neurol.* **113,** 389.

Douglas, W. W. (1966). *In* "Mechanisms of Release of Biogenic Amines" (U. S. von Euler, ed.), pp. 267–289. Pergamon, Oxford.

Eaton, J. E., Jr., and Frieden, E. (1969). *Gen. Comp. Endocrinol. Suppl.* **2,** 398.

Etkin, W. (1963). *Science* **139,** 810.

Everett, J. W., and Radford, H. M. (1961). *Proc. Soc. Exp. Biol. Med.* **108,** 604.

Feldberg, W. (1970). *In* "The Hypothalamus" (L. Martini, M. Motta, and F. Faschini, eds.), pp. 213–232. Academic Press, New York.

Fisher, D. A., and Odell, W. D. (1969). *J. Clin. Invest.* **48,** 1670.

Fleischer, N., Burgus, R., Vale, W., Dunn, T., and Guillemin, R. (1970). *J. Clin. Endocrinol.* **31,** 109.

Florsheim, W. H. (1958). *Endocrinology* **62,** 783.

Freinkel, N., and Lewis, D. (1957). *J. Physiol. (London)* **135,** 288.

Frieden, E., and Just, J. J. (1970). In "Biochemical Actions of Hormones" (G. Litwack, ed.), pp. 1–52. Academic Press, New York.

Galton, V. A. "Environmental Effects in the Thyroid." 3rd ed. (S. C. Warner and S. H. Ingbar, eds.), 3rd ed., pp. 153–158. Harper (Hoeber), New York, 1971.

Galton, V. A., and Nisola, B. C. (1969). *Endocrinology* **85,** 79.

Gevers, W., Kleinkauf, H., and Lipmann, F. (1968). *Proc. Nat. Acad. Sci.* **60,** 269.

Golsteine-Golaire, J., Vanhaelst, L., Bruno, O. D., Leclercq, R., and Copinschi, G. (1970). *J. Appl. Physiol.* **29,** 622.

Gordon, J., Reichlin, S., and Bollinger, J. (1972). *Endocrinology.* in press.

Greer, M. A. (1951). *Proc. Soc. Exp. Biol. Med.* **77,** 603.

Grimm, Y., and Reichlin, S. (1972). *Science* (submitted for publication).

Grossman, S. P. (1960). *Science* 132, 301.

Guillemin, R. (1970). *In* "Hypophysiotropic Hormones of the Hypothalamus" (J. Meites, ed.), p. 14. Williams & Wilkins, Baltimore, Maryland.

Guillemin, R., Yamazaki, E., Gard, D. A., Jutisz, M., and Sakiz, E. (1963). *Endrocrinology* **73,** 564.

Halasz, B., Pupp, L., and Uhlarik, S. (1962). *J. Endrocrinol.* **25,** 147.

Hammel, H. T. (1968). *Annu. Rev. Physiol.* **30,** 641.

Harris, G. W., and Woods, J. W. (1958). *J. Physiol. (London)* **143,** 246.

Harris, G. W., Levine, S., and Schindler, W. J. (1964). *J. Physiol. (London)* **170,** 516.

Hoskins, R. G. (1949). *J. Clin. Endocrinol. Metab.* **4,** 1429.

Hunter, W. M., and Greenwood, F. C. (1962). *Nature (London)* **194,** 495.

Kamberi, I. A., Mical, R. S., and Porter, J. C. (1971a). *Endocrinology* **88,** 1003.

Kamberi, I. A., Mical, R. S., and Porter, J. C. (1971b). *Endocrinology* **88,** 1012.

Kamberi, I. A., Mical, R. S., and Porter, J. C. (1971c). *Endocrinology* **88,** 1288.

Knigge, K. M. (1964). In "Major Problems in Neuroendocrinology" (E. Bajusz and G. Jasmin, eds.), pp. 261–285. Karger, Basel.

Knigge, K. M., and Scott, D. E. (1970). *Amer. J. Anat.* **129,** 223.

Leibowitz, S. F., and Miller, N. E. (1969). *Science* **165,** 609.

Lemarchand-Beraud, T. H., and Vannotti, A. (1965). *Schweiz. Med. Wochenschr.* **95,** 772.

Levi-Montalcini, R., and Angeletti, P. U. (1968). *Physiol. Rev.* **48,** 534.

Martin, J. B., and Reichlin, S. (1970). *Science* **168,** 1366.

Martin, J. B., and Reichlin, S. (1971a). *Proc. 6th Midwest Thyroid Meet., Columbia, Mo.* pp. 1–24.

Martin, J. B., and Reichlin, S. (1971b). *Neurology* **21,** 436.

Martin, J. B., and Reichlin, S. (1972). *Endocrinology* **90,** 1079.

Martin, J. B., Boshans, R., and Reichlin, S. (1970). *Endocrinology* **87,** 1032.

Meites, J., ed. (1970). "Hypophysiotropic Hormones of the Hypothalamus." Williams & Wilkins, Baltimore, Maryland.

Mitnick, M., and Reichlin, S. (1971). *Science* **172,** 1241.

Mooz, E. D., and Meister, A. (1967). *Biochemistry* **6,** 1722.

Morrell, R. M. (1967a). *Clin. Res.* **15,** 32.

Morrell, R. M. (1967b). *Clin. Res.* **15,** 263.

Motta, M., Piva, F., Fraschini, F., and Martini, L. (1970). *In* "Hypophysiotropic Hormones of the Hypothalamus" (J. Meites, ed.), pp. 44–59. Williams & Wilkins, Baltimore, Maryland.

Panda, J. N., and Turner, C. W. (1967a). *Proc. Soc. Exp. Biol. Med.* **124,** 711.

Panda, J. N., and Turner, C. W. (1967b). *J. Physiol. (London)* **192,** 1.

Reichlin, S. (1957). *Endocrinology* **60,** 567.

Reichlin, S. (1963). *New Eng. J. Med.* **269,** 1182, 1246, 1296.

Reichlin, S. (1964). *Ciba Found. Study Group [Pap.]* **18,** 17–32.

Reichlin, S. (1965). *Excerpta Med. Found. Int. Congr. Ser.* **83,** 499.

Reichlin, S. (1966). *In* "Neuroendocrinology" (L. Martini and W. F. Ganong, eds.), Vol. 1, pp. 445–536. Academic Press, New York.

Reichlin, S. (1971). *In* "The Thyroid" (S. C. Werner and S. H. Ingbar, eds.), pp. 95–111. Harper, New York.

Reichlin, S., and Boshans, R. L. (1964). *Endocrinology,* **75,** 571.

Reichlin, S., and Utiger, R. D. (1967). *J. Clin. Endocrinol. Metab.* **27,** 251.

Reichlin, S., Schalch, D. S., Boshans, R. L., and Pierce, J. (1966a). *Proc. 48th Meet, Endocrine Soc.,* p. 90.

Reichlin, S., Volpert, E. M., and Werner, S. C. (1966b). *Endocrinology* **78,** 302.

Reichlin, S., Martin, J. B., Boshans, R. L., Schalch, D. S., Pierce, J. G., and Bollinger, J. (1970). *Endocrinology* **87,** 1022.

Sachs, H., Fawcett, P., Takabatake, Y., and Portanova, R. (1968). *Recent Progr. Horm. Res.* **25,** 447.

Sakoda, M., and Nakabayashi, H. (1970). *Kobe J. Med. Sci.* **16,** 95.

Schally, A. V., Redding, T. W., Bowers, C. Y., and Barrett, J. F. (1969). *J. Biol. Chem.* **244,** 4077.

Shizume, K., Matsuda, K., Irie, M., Iino, S., Ishii, J., Nagataki, S., Matsuzaki, F., and Okinaka, S. (1962). *Endocrinology* **70,** 298.

Sinha, D., and Meites, J. (1965). *Neuroendocrinology* **1,** 4.

Steiner, D. F., Clark, J. L., Nolan, C., Rubenstein, A. H., Margoliash, E., Aten, B., and Oyer, P. E. (1969). *Recent Progr. Horm. Res.* **25,** 207.

Vale, W., Burgus, R., and Guillemin, R. (1967). *Proc. Soc. Exp. Biol. Med.* **125,** 210.

Vale, W., Burgus, R., and Guillemin, R. (1968). *Neuroendocrinology* **3,** 34.

van der Werff ten Bosch, J. J., and Swanson, H. E. (1963). *Acta Endocrinol. (Copenhagen)* **42,** 254.

van Rees, G. P., and Moll, J. (1968). *Neuroendocrinology* **3,** 115.

Vertes, M., Vertes, Z., and Kovacs, S. (1965). *Acta Physiol.* **27,** 229.

Voloshin, L., Joseph, S. A., and Knigge, K. M. (1968). *Neuroendocrinology* **3,** 387.

Weiss, B., and Costa, E. (1967). *Science* **156,** 1750.

Wilber, J. F., and Porter, J. C. (1969). *Proc. 51st Meet. Endocrine Soc., New York* p. 93.

Wilber, J. F., and Utiger, R. D. (1967). *Endocrinology* **81,** 145.

Wilber, J. F., and Utiger, R. D. (1968). *Proc. Soc. Exp. Biol. Med.* **127,** 488.

Wurtman, R. J. (1971). *Neurosci. Res. Program Bull.* **9,** 172.

Wurtman, R. J., Axelrod, J., and Fischer, J. E. (1964). *Science* **143,** 1329.

DISCUSSION

A. S. Goldman: Being interested in embryonic imprinting of enzymes involved in differentiation, I find your observation of the positive feedback control of the synthetase fascinating. As your experiments with the frog tadpole and with the thyroidectomized rats seem to suggest, it may be that the synthetase must be first "turned on" by thyroid hormone in order for it to play a regulatory role in thyroid gland activity. Two kinds of experiments suggest themselves to me which you may have done. The first would be to determine whether thyroidectomy of the frog tadpole prevents the climax response of the appearance of the synthetase and whether there is a critical period of this inductive effect. The second would be to determine whether thyroxine replacement of the thyroidectomized rat can restore synthetase activity.

S. Reichlin: We have not studied this question.

A. White: In the cell-free system in which you are studying the synthesis of the tripeptide, do you have an indication that the likely two-step reaction is occurring, as in the synthesis of glutathione? Is a dipeptide formed if proline is deleted from the incubation medium, and what is the nature of the reaction in which amidation of the carboxyl group of proline occurs? In relation to the positive feedback effect of thyroid hormone at the neuronal level, perhaps in the hypothyroid animal, with diminished rate of general protein synthesis and lowered metabolic activity, there is a limiting rate of synthesis of TRH in the hypothalamus. In the metamorphosis of the tadpole, for example, Cohen and his colleagues at the University of Wisconsin have shown in some publications that at a specific stage of metamorphosis the synthesis of liver transcarbamylase is initiated, with resultant beginning of urea formation. The synthesis of this key enzyme is induced by thyroxine. Perhaps a depressed rate of protein synthesis, in the hypothalamus, rather than a neuronal locus of thyroxine action, may explain your data.

S. Reichlin: There is at least a two-step reaction, an impression gained from the initial delay. Since the amidation reaction is ubiquitous in the brain, it is not clear whether hypothalamic amidation is specific or nonspecifically related to TRH.

Transamidation reactions probably involve glutamic acid so that it would be feasible to study double-labeled glutamic acid. There may be at least four enzymes involved, a dipeptide synthetase, a tripeptide synthetase, a glutamyl or glutamine cyclase, and a transamidase.

I agree with you about the effect of thyroxine generally on protein metabolism. I think one of the features of this model which intrigues me the most is that it provides a generalizing view of thyroxine action such that wherever thyroxine works, it works by changing protein metabolism and this is the way it works in controlling TSH secretion through effects on TRH and on the regulator protein of the pituitary. Whether thyroxine effects in the brain are specific to the TRH regulatory system is not known.

R. Guillemin: In a series of experiments over the last few months, I have obtained conclusions similar to those presented by Dr. Reichlin today. We can indeed demonstrate biosynthesis of TRF with fragments of rat stalk median eminence or ventral hypothalamus with techniques somewhat different from those used by Dr. Reichlin (incorporation of a single amino acid, [^{3}H]proline with high specific activity of 45 Ci/mmole). Similarly, we observed the synthesis of the tripeptide in the presence of very large doses of puromycin; thus our conclusions are also that we are dealing with a nonribosomal synthesis, probably involving some hypothalamic enzyme system.

D. N. Orth: It appears to me that you have here a unique system in which to look at the question of whether there is really a short-loop feedback mechanism. You are postulating that, in the thyroidectomized animal, the decrease in the hypothalamic TRF synthetase activity is due to the lack of thyroxine which normally exerts a long-loop positive feedback influence. However, it is also possible that it might be due to short-loop negative feedback by increased levels of TSH. Have you looked at the effect of combined thyroidectomy and hypophysectomy on hypothalamic synthetase activity?

S. Reichlin: We have not done the critical experiments to answer your question. We have added TSH to the pulse-labeled system and the results have so far been equivocal. I think the physiological data against the short feedback loop in the case of TSH are relatively overwhelming, but I agree with you that this must be worked out again in terms of our system.

J. Kowal: Would your theory suggest that TSH release from the pituitary may be independent of TRH since it is markedly elevated in hypothyroidism?

S. Reichlin: Yes, I think so. What you see in this system in the hypothyroid is that there is some TRH and an enormous pituitary sensitivity to TRH, as has been demonstrated in humans now by Dr. Gual and others.

J. Kowal: In your *in vitro* system there appears to be a 10-minute lag before there is incorporation of radioactivity into TRH. Do you have an explanation for this?

S. Reichlin: There are some theoretical explanations in terms of enzyme kinetics which are involved. Miss Mitnick has suggestive evidence that there are at least two reactions occurring in that period before one starts to see TRH.

R. Walter: Thus far we have elucidated two general modes for the biosynthesis of releasing and inhibiting factors: (a) the enzyme-catalyzed synthesis of the factor, as we just have heard so lucidly presented by Dr. S. Reichlin for TRH; (b) the enzyme-catalyzed breakdown of an "inactive" precursor molecule. Our studies of the MSH-RIF and MSH-RF derived from oxytocin are a case in point, MSH-RIF being the C-terminal tripeptide of oxytocin, H · Pro-Leu-Gly-NH$_2$

[M. E. Celis, S. Taleisnik, and R. Walter, *Proc. Nat. Acad. Sci. U.S.* **68**, 1428 (1971); R. M. G. Nair, A. J. Kastin, and A. Schally, *Biochem. Biophys. Res. Commun.* **43**, 1376 (1971)] and MSH-RF being the N-terminal pentapeptide of oxytocin, H · Cys-Tyr-Ile-Gln-AsnOH, or a structurally related peptide [M. E. Celis, S. Taleisnik, and R. Walter, *Biochem. Biophys. Res. Commun.* **45**, 564 (1971)].

Two different enzymes compete for the oxytocin pool. While the enzyme giving rise to MSH-RIF exhibits the same activity throughout the various stages of the estrous cycle in rat (determined during proestrus, estrus, and second day of diestrus), the enzymatic activity responsible for the formation of MSH-RF fluctuates as a function of the hormonal cycle; the enzymatic activity is highest during estrus (first reference above). In line with this enzymatic activity profile are the findings that in male rats the pituitary content of MSH is constant, whereas it changes in female rats and is lowest at the estrous stage. In addition, ovariectomy gives rise to an increase of pituitary MSH content, whereas treatment with estrogen reduces the level of MSH in the pituitary [M. E. Tomatis and S. Taleisnik, *Acta Physiol. Lat. Amer.* **18**, 96 (1968); S. Taleisnik and M. E. Tomatis, *Neuro-endocrinology* **5**, 24 (1969)].

On the basis of the above findings we may suggest that estrogen enhances the synthesis of the enzyme responsible for the breakdown of the "prohormone" oxytocin to yield MSH-RF. Presently we are purifying the two enzymes and thus far have made considerable progress in collaboration with Dr. N. Marks with the enzyme responsible for the formation of MSH-RIF. It would be interesting to compare the substrate specificities of these enzymes with some of those from the hypothalamus investigated by others [K. C. Hooper, *Biochem. J.* **83**, 511 (1962); K. C. Hooper, *Biochem. J.* **99**, 128 (1966); X. Arai and T. Kuzama, *Proc. Jap. Acad.* **41**, 734 (1965)]. We are thinking particularly of those enzymes affected by steroid hormones [C. R. Hopkins and K. C. Hooper, *Biochem. J.* **117**, 36p (1970)].

As regards the location of these enzymes capable of forming MSH-RF and MSH-RIF, one may hypothesize that they are present in specific oxytocinergic neurons which are different in their enzymatic content from those transporting oxytocin along the hypothalamoneurohypophysial tract for storage in the neurohypophysis [E. Scharrer and B. Scharrer, *in* "Handbuch der mikroskopischen Anatomie des Menschen" (W. Möllendorf and W. Bargmann, eds.), Vol. 6, p. 953. Springer-Verlag, Berlin and New York, 1954; H. Heller, *in* "Oxytocin" (R. Caldeyro-Barcia and H. Heller, eds.), p. 3. Pergamon, Oxford, 1960]. The enzymes could be located in the perikaryon and, to some degree, in the endoplasmic reticulum of the axons of these neurons. The nonmyelinated nerve fibers which come from the hypothalamus and invade the pars intermedia [W. Etkin, *In* "Neuroendocrinology" (L. Martini and W. F. Ganong, eds.), Vol. 2, p. 261, Academic Press, New York, 1967], where MSH is synthesized, have to be considered as possible candidates. The fact that we have been unable thus far to detect any MSH-RIF in the neurohypophysis is in line with a dual oxytocinergic neuron hypothesis.

S. Reichlin: What Dr. Walter points out is very important in that it may be possible through the use of a study of quantitative enzymology of the releasing factor synthesis systems to get a clue as to what is going on in terms of feedback. In this case, you are studying gonadal feedback on enzyme control. You are talking about hormonal control of hypothalamic enzymes which regulate releasing factor synthesis.

J. F. Wilber: I do not agree that the secretion of TSH is regulated reciprocally by thyroid hormones independent of TRH interaction. For example, if you examine TSH secretion in a pituitary derived from a primary hypothyroid rat *in vitro,* whose TSH secretion rate may be 20-fold normal *in vivo,* in isolation of hypothalamic influences, the TSH secretory rate rapidly declines to a euthyroid rate or lower. Moreover, this level of TSH secretion cannot be lowered further by introduction of thyroid hormones into the medium, nor will TSH release increase in the absence of medium T_4. In addition, the TSH secretory rate *in vitro* cannot be reduced by anaerobiasis, oligomycin, or other known modifiers of TSH release unless TRH stimulation is operative. In the presence of TRH activation, TSH secretion then become subject to inhibition by T_4, T_3, dinitrophenol, etc. Therefore, I think even in the *in vivo* anterior hypothalamic lesion experiments there must be sufficient neurohumor (TRH) to permit thyroid hormone blockade to be manifest.

S. Reichlin: I would agree. There has to be some TRH around to get reasonable TSH secretion.

J. F. Wilber: The hypothalamic hormone most difficult to authenticate has been the "growth hormone releaser." Several years ago, in examining the properties of semipurified TRH preparations *in vitro* (Abbott Laboratories), we observed that this material also could stimulate growth hormone secretion. Initially, we wondered if this were an intrinsic biological property of TRH itself. However, subsequently when synthetic TRH did not release growth hormone in our *in vitro* system, we then decided to reexamine the properties of hypothalamic growth hormone-releasing activity using immunoassay and the *in vitro* rat hemipituitary. After Sephadex G-25 gel filtration of an acetic acid extract derived from 200 rat hypothalami, on a column previously calibrated with $[^{14}C]TRH$ (V_e/V_t 0.8), the "GH-RH" described by Dr. Kastin is reported to elute at a V_e/V_t of 0.5. In contrast, our "growth hormone-releasing activity" emerged in a more retarded position, overlapping $[^{14}C]TRH$ [J. F. Wilber, T. Nagel, and W. White, *Endocrinology* **89,** 1419 (1971)]. No significant activity, capable of releasing *immunoassayable* RGH, was detected in earlier fractions. Similarly, in porcine extracts, we have located the "GH-releasing activity" in a similar elution position. We are confident that our GH stimulatory activity is not due to gonadotropin-releasing hormone (Gn-RH) or vasopressin. Thus, our data support the notion that there can be isolated a hypothalamic hormone that is capable of releasing GH, identified by radioimmunoassay. Similar results have been reported recently by Malacara and Reichlin, and Frohman and associates.

S. Reichlin: Our data are in agreement with those of Dr. Wilber. His group, that of Frohman and Malacara, and I, have found fractions that release growth hormone as judged by elevations of plasma immunoassayable growth hormone.

I. A. Kamberi: What doses of dopamine and norepinephrine have you used in your system?

S. Reichlin: We have used the 10^{-4} M dose.

I. A. Kamberi: We also have been studying the effects of these catecholamines on the release of LH, FSH, and prolactin. When small doses of dopamine hydrochloride are injected into the third ventricle of the brain, within a few minutes there is an increase in plasma concentration of LH and FSH, and a decrease in prolactin levels in the peripheral circulation. Twenty minutes after the intraventricular injection of 1.25 μg of dopamine hydrochloride, the concentrations of LH and FSH are 7.5- and 6.5-fold greater, respectively, than the control levels of saline-injected animals, whereas the plasma prolactin levels were only one-third those of the controls (Fig. A). With an increase in the dose of dopamine, the effect decreases; namely, the extent of the increase of LH and FSH release and inhibition of prolactin release is inversely related to the dose of intraventricularly injected dopamine. In *in vitro* experiments, high dose levels of dopamine (50–100 μg) have been without effect, or have apparently interacted directly with (changed the molecular structure of) FSH, LH, or prolactin. We (Kamberi, Porter, and McCann) have since found by radioimmunoassay that the LH, FSH, and prolactin content of rat pituitary extract decrease following incubation *in vitro* with dopamine. The decrease in the amount of immunologically active LH, FSH, or prolactin after dopamine treatment would seem to result from an alteration (destruction)

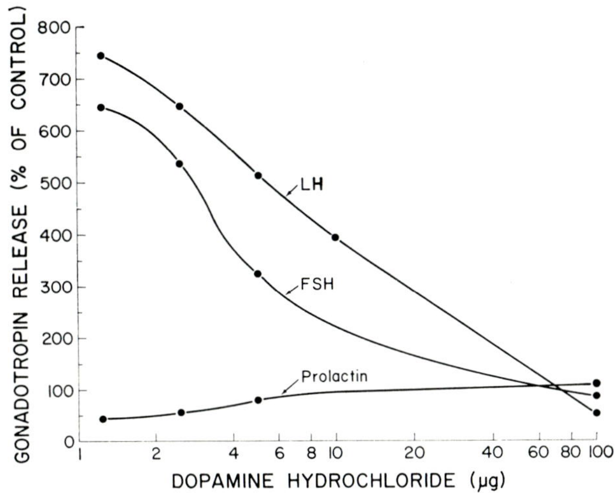

Fig. A. The concentration of LH, FSH, and prolactin in plasma of male rats 20 minutes after the injection of saline or dopamine hydrochloride into the third ventricle of the brain. The values are expressed in terms of the response of those animals given saline alone, which was taken as 100%. Each point represents a mean value of 8–10 animals per group. [Data fom I. A. Kamberi, R. S. Mical, and J. C. Porter, *Endocrinology* **87**, 1 (1970); **88**, 1003 (1971); **88**, 1012 (1971).]

in the structures of these hormones. In contrast, we have observed an increase in FSH and LH concentrations and a decrease in prolactin after the intraventricular injection of high doses of norepinephrine [I. A. Kamberi, R. S. Mical, and J. S. Porter, *Endocrinology* **87**, 1 (1970); **88**, 1003, 1012 (1971)].

The effects of intraventricularly injected dopamine or norepinephrine cannot be seen when these catecholamines are infused directly into the anterior pituitary via a cannulated pituitary stalk portal vessel. This suggests that the effect of these amines of the release of LH, FSH, and prolactin is secondary to the discharge of LRF, FRF, and PIF from neurosecretory elements of the hypothalamus which, in turn, is reflected in the increased release of LH and FSH and the decreased release of prolactin. These conclusions are supported by our *in vivo* and *in vitro* findings.

In *in vivo* experiments, we have collected simultaneously femoral arterial blood and portal blood from pituitary stalks of male rats (donors), untreated (saline injected) or dopamine-treated (dopamine injected into the third ventricle). Plasma from femoral artery blood or stalk blood was then infused back into the anterior pituitary via a cannulated portal vessel of the pituitary stalk of another group of untreated, recipient, male rats at the rate of 2 μl per minute for 30 minutes.

Infusion of stalk plasma from dopamine-treated donors caused an increase in the plasma concentration of LH and FSH, and a decrease in the prolactin levels in peripheral circulation of the recipient animals (Fig. B). On cessation of infusion,

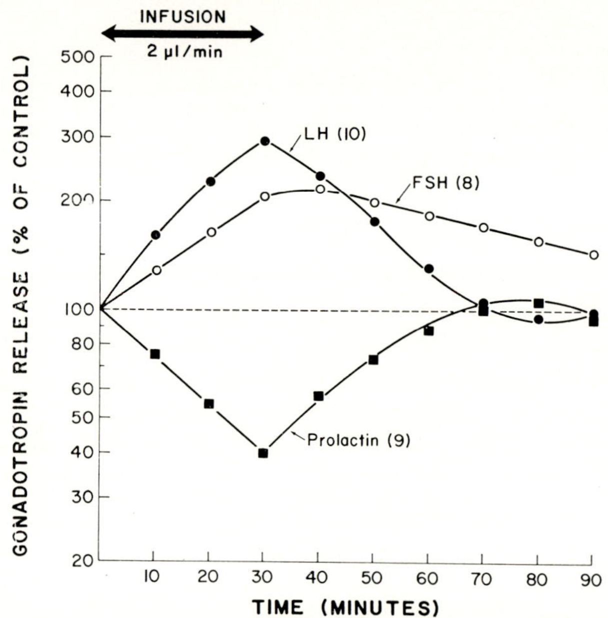

Fig. B. Effect of anterior pituitary infusion, via a cannulated portal vessel, of femoral artery plasma or pituitary stalk plasma from dopamine-treated donors and the release of LH, FSH, or prolactin. The symbols represent mean responses. The values are expressed in terms of the response observed in recipient rats infused with plasma from femoral artery blood from dopamine-treated donors, which was taken as 100%. The number of rats (recipients) is shown in parentheses. [From I. A. Kamberi, R. S. Mical, and J. C. Porter, *Endocrinology* **89**, 1042 (1971).]

the release of LH and FSH and inhibition of prolactin stopped. We have observed the corresponding activities, FSH and LH releasing (FRF, LRF) and prolactin inhibiting (PIF) in the pituitary stalk plasma of untreated donor rats, but they were less pronounced. Infusion of femoral artery plasma from dopamine-treated or untreated donors was without effect.

We have also observed LRF, FRF, and PIF activities under *in vitro* conditions. Pituitary halves of intact male rats were incubated in the presence of femoral artery or pituitary stalk plasma obtained from untreated or dopamine-treated male rats. Pituitary halves incubated in pituitary stalk plasma from untreated donor rats released significantly more LH and FSH, and less prolactin than did the contralateral halves incubated in femoral artery plasma (Fig. C). Furthermore, pituitary halves incubated in pituitary stalk plasma from rats receiving 2.5 μg dopamine hydrochloride via the third ventricle released over four times as much LH and over five times as much FSH as the controls, which were incubated in femoral artery plasma. Prolactin release by hemipituitaries incubated in pituitary stalk plasma from dopamine-treated rats was four times lesss than the controls. In addition, this effect of dopamine seems to be mediated via α-adrenergic receptors. Simultaneous injection into the third ventricle of dopamine and α-blocking agents

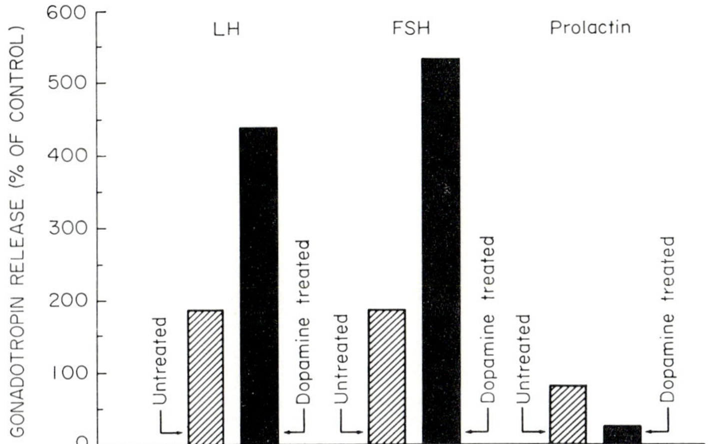

Fig. C. Release of LH, FSH, and prolactin from hemipituitaries incubated *in vitro* in plasma from femoral artery, which served as a control, and from the contralateral halves incubated in stalk plasma. The values are expressed in terms of the response of the hemipituitaries incubated in femoral artery plasma, which was taken as 100%. Each bar represents a mean value of 10 animals per group. [From I. A. Kamberi, R. S. Mical, and J. C. Porter, *Science* **166**, 388 (1969); *Nature* **227**, 714 (1970); *Experientia* **26**, 1150 (1970).]

(either phenoxybenzamine or phentolamine) completely inhibited the dopamine response, whereas pronethalol, a β-blocking agent, was without effect.

G. D. Aurbach: You have introduced again the important concept of a higher control center for thyroid-pituitary regulation; that is, at the hypothalamus with norepinephrine. I presume that this is a β-adrenergic function. I recall that years ago at this conference Sawyer introduced the possibility that catecholamines influenced control of LH secretion. Are the regulatory systems for LRF and TRF interrelated through the influence of catecholamines?

S. Reichlin: Dr. Kamberi and Dr. Porter have shown the role of catecholamines in the regulation of the releasing factors, and there is also work on growth hormone, TSH, and CRF. What I would like to propose in this model is that the median eminence contains a bundle of peptidergic neurons making the releasing factors, and these are in turn played upon by a variety of neurons, the principal neurotransmitters being catecholamines. There may be different catecholamines for different releasing factors. The model is analogous to the pineal gland. In the case of the pineal gland, a sympathetic nerve supply coursing outside of the brain innervates the pineal. In this case, noradrenergic fibers regulate the concentration of HIOMT and adenyl cyclase while in the median eminence, the catecholamines, we propose, regulate releasing factor synthetases.

C. Kordon: I was interested to hear your statement that biosynthesis, and not only release of releasing factors, may be affected by monoaminergic neurons. Some of our data suggest that this is also the case in LH release regulation. Norepinephrine levels in the preoptic area of the hypothalamus have been shown to be related to the estrous cycle. Donoso, *et al., Neuroendocrinology* **4,** 12 (1969)]; microinjections of drugs affecting the metabolism of catecholamines into this part of the hypothalamus are capable of interfering with LH release. However, this effect occurs after a rather long latency only (for instance, *l*-DOPA blocks the ovulatory LH surge when microinjected on diestrous day 1, i.e., 2 days prior to the critical period [C. Kordon, *in* "Neuroendocrinologie" (J. Benoit and C. Kordon, eds.), p. 78. CNRS, Paris, 1971; J. Benoit and C. Kordon, unpublished observations]. This latency is in sharp contrast to the very rapid effect of mono-amines in the median eminence; it may reflect an interference of the drug with processes involved in the biosynthesis of LRF, whereas most authors agree that aminergic terminals in the median eminence affect the release of already synthesized releasing factors.

I. Geschwind: Have you tried glutamine in your system as a precursor?

S. Reichlin: No. [Since the conference, labeled glutamine has been found to serve as a TRH precursor.]

I. Geschwind: There is very little evidence that N-terminal glutamic acid cyclizes readily to pyrrolidone carboxylic acid, whereas N-terminal glutamine does. I think that Dr. Walter has made a most extraordinary finding, and that MRF is the very first hormonal peptide or protein that has a free sulfhydryl group. We always felt that the presence of a free sulfhydryl group should be incompatible with activity because it would be very rapidly oxidized, leading to the disappearance of hormonal activity while the hormone was in the circulation. In this case there is only one cysteine, that would be the N-terminal cysteine, and there must be a free sulfhydryl group in the peptide.

Taleisnik has previously suggested that MRF and MIF are produced in different regions of the hypothalamus [S. Taleisnik, *Endocrinology* **81,** 819 (1967)]. The available evidence suggests that the paraventricular nucleus is the site of formation of oxytocin. The question, then, is how oxytocin gets to the different sites where MRF and MIF are produced. I wonder whether Dr. Walter has any evidence about the anatomical sites of formation of the two peptides that he and his colleagues have described. Does he know anything about the anatomic localization, in terms of the nuclei in the hypothalamus where MRF and MIF are formed? What is the location of the enzyme activities which split between Pro and Leu to form MIF, and between Lys and Pro to form MRF?

R. Walter: The N-terminal pentapeptide of oxytocin, cysteinyl-tyrosyl-isoleucyl-glutaminyl-asparagine, at nanogram amounts, evokes a reduction of the pituitary MSH content and an increase in MSH plasma levels. The pentapeptide exerts these effects by acting on the gland directly, not through the central nervous system, as may be deduced from Fig. D. Although all these properties are consistent with those expected for MSH-RF, we left the door open for the possibility that the natural MSH-RF is not this particular pentapeptide but one of related structure [M. E. Celis, S. Taleisnik, and R. Walter, *Biochem. Biophys. Res. Commun.* in press (1971)]. I should mention, though, that the disulfide dimer of this particular pentapeptide is inactive as MSH-RF.

J. Hershman: I wonder whether your system would explain some puzzling clinical observations. I think that your studies in myxedematous patients treated with thyroxine and in thyroidectomized rats given thyroxine, showing a curvilinear relationship between the blood levels of thyroid hormone and TSH, imply that elevated TSH levels occur only at low levels of thyroid hormone; this is tantamount to deficient secretion of thyroid hormone. One assumes that there is an upper limit to the normal range of serum TSH. In some juveniles with chronic thyroiditis who are considered euthyroid clinically and who have a normal serum thyroxine,

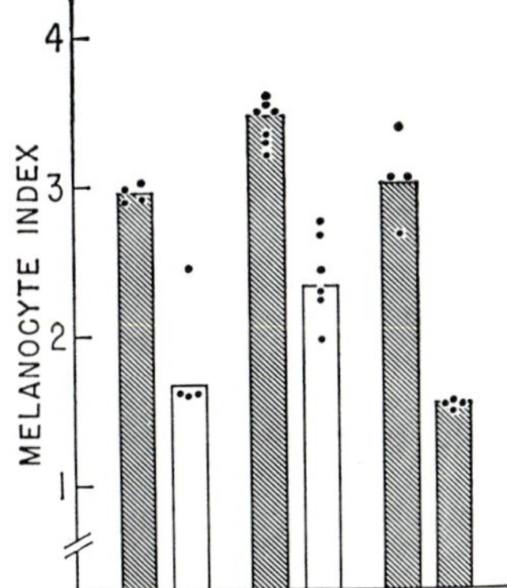

Fig. D. Effect of H-Cys-Ty-Ile-Gln-Asn-OH on plasma MSH activity in intact rats (first set of bars), in rats with median eminence lesions (middle set of bars) and in hypophysectomized rats (last bar; the penultimate bar represents intact animal which served as control). The animals were injected with a constant amount of 40 ng of pentapeptide. Each point represents one experiment, and the bars are the mean values. ▨, Pentapeptide; ☐, saline solution. [From M. E. Celis, S. Taleisnik, and R. Walter, *Biochem. Biophys. Res. Commun.* **45**, 564 (1971).]

one finds distinctly elevated levels of TSH. Also in patients from endemic goiter regions with low iodine intake who are believed to be euthyroid though goitrous, there may be markedly elevated TSH levels. Can your system account for these elevated TSH levels if thyroid hormone secretion is presumed to be in the normal range, or does the elevated serum TSH signify that the secretion of thyroid hormone is deficient?

S. Reichlin: We are looking at two different systems. In the one I present here, we have taken out the thyroid altogether, and replaced with exogenous thyroxine—the loop is open but you are controlling the thyroxine level. In the human with thyroid disease, the plasma thyroxine level rises to the proper level, but only at the expense of high TSH secretion.

K. Sterling: I am especially grateful to Dr. Reichlin for saying that many of these problems are less simple than they appear. On a previous occasion I believe you stated that injected thyroxine had been to a considerable degree converted to T_3 in the hypothalamus. Since you and Dr. Utiger reported a reciprocal relation between free thyroxine and TSH concentrations in sera, can you now add more information along these lines as to the free *vs* total T_4, and T_3 *vs* T_4, in inhibition of TSH secretion at the hypothalamic and hypophysial levels?

S. Reichlin: It is known that there is some conversion of T_4 to T_3 in the pituitary. The role this conversion plays in the regulatory system is not known. We have studied dinitrophenol-treated animals and find that pituitary T_4 concentration is the regulator of the feedback control, even in animals with highly dissociated plasma T_4.

S. L. Cohen: With the positive and the negative feedback mechanisms, what prevents these from neutralizing each other? What determines which one shall take effect?

S. Reichlin: I am proposing that the interaction of a positive and a negative feedback system enables the system to reach a point where they reach stability. Under ordinary circumstances that is what corresponds to the physiological "set point" of plasma thyroid hormone. This can be modified by neurotransmitter activation and by hormonal manipulation. The resulting changes of hypothalamus and pituitary depend upon a genetically programmed change in susceptibility of these respective tissues to the protein-stimulating effect of thyroxine. This system simplifies control and, in servo-systems terms, makes it an inherently more stable system.

18-Hydroxy-11-deoxycorticosterone (18-OH-DOC) Secretion in Experimental and Human Hypertension

James C. Melby, Sidney L. Dale, Roger J. Grekin, Robert Gaunt,* and Thomas E. Wilson

Section of Endocrinology and Metabolism, Robert Dawson Evans Department of Clinical Research, Department of Medicine, Boston University School of Medicine, Boston, Massachusetts

I. Introduction

18-Hydroxy-11-deoxycorticosterone (18-OH-DOC),[1] which has the structure shown in Fig. 1, was isolated and identified as a naturally

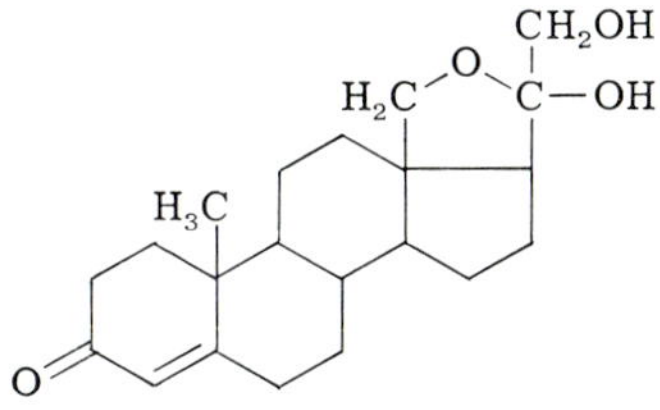

Fig. 1. Structural formula of 18-hydroxy-11-deoxycorticosterone (18-OH-DOC) demonstrates the 18 → 20 hemiketal configuration. The systematic name for 18-OH-DOC is: 20,21-dihydroxy-18,20-epoxypregn-4-en-3-one.

occurring steroid from the medium bathing incubated sectioned rat adrenals by Birmingham and Ward (1961) and Péron (1961). Endogenous secretion of 18-OH-DOC was demonstrated *in vivo* by the sampling

---

* Ciba Pharmaceutical Company.

[1] Steroid nomenclature: 18-hydroxydeoxycorticosterone (18-OH-DOC), 18,21-dihydroxypregn-4-ene-3,20-dione; deoxycorticosterone (DOC), 21-hydroxypregn-4-ene-3,20-dione; aldosterone (aldo), 11β,21-dihydroxy-18-oxo-pregn-4-ene-3,20-dione; corticosterone (B), 11β,21-dihydroxy-4-ene-3,20-dione; 18-hydroxytetrahydrodeoxycorticosterone (18-OH-TH-DOC), 3α,18,21-trihydroxy-5β-pregnan-20-one; tetrahydrodeoxycorticosterone (TH-DOC), 3α,21-dihydroxy-5β-pregnan-20-one; aldosterone lactone, 3-oxo-11β,18-epoxy-4-androstene-17β,18-carbolactone; 18-hydroxydeoxycorticosterone lactone, 3-oxo-4-androstene-17β,18-carbolactone; 18-hydroxycorticosterone (18-OH-B), 11β,18,21-trihydroxypregn-4-ene-3,20-dione; cortisol, 11β,17α,21-trihydroxypregn-4-ene-3,20-dione; tetrahydrocortisone (THE), 3α,17α,21-trihydroxy-5β-pregnan-11,20-dione; 11-deoxycortisol, 17α,21-dihydroxypregn-4-ene-3,20-dione; 6β-hydroxycortisone (6β-OH-E), 6β,17α,21-trihydroxypregn-4-ene-3,11,20-trione; 6β-hydroxycortisol (6β-OH-F), 6β,11β,17α,21-tetrahydroxypregn-4-ene-3,20-dione.

of adrenal venous effluent of the rat by Cortés *et al.* (1963). 18-OH-DOC rivals corticosterone (B) in abundance as a secretory product of the rat adrenal cortex. Because of its predominance it has been anticipated that 18-OH-DOC must have biological significance for this species, although to the present, no unique role for this steroid has been established. In the rat adrenal cortex, 18-OH-DOC has not been shown to be an important precursor of aldosterone (aldo) and is regarded as a final product of steroid biosynthesis (Vecsei *et al.*, 1968). 18-OH-DOC appears to be largely derived from the zona fasciculata and to a very small extent from the zona glomerulosa of the rat adrenal cortex, as evidenced by measurement of the steroid products of incubated decapsulated adrenal gland (zona fasciculata) versus capsular adrenal glands

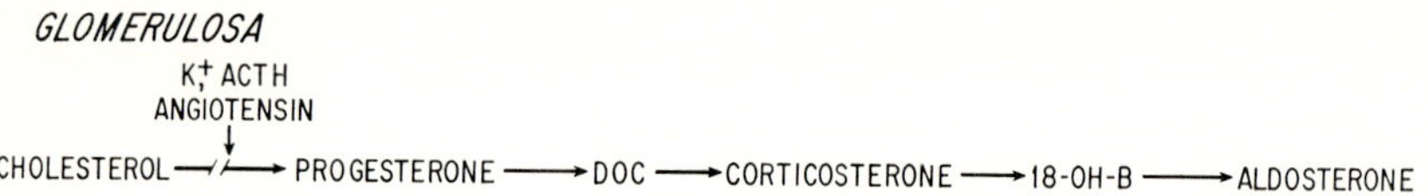

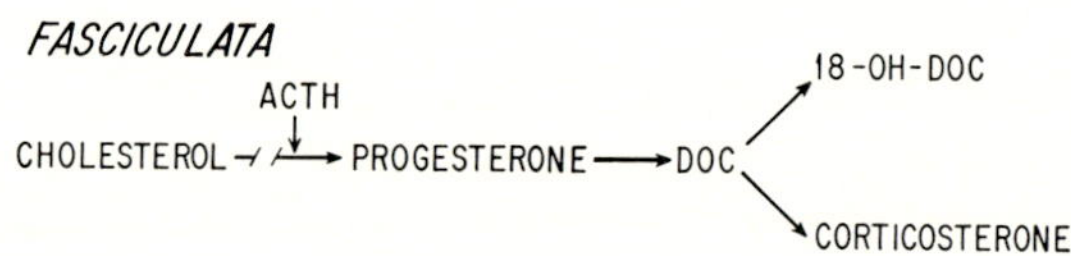

FIG. 2. Steroidogenesis in the zona glomerulosa and zona fasciculata of the rat adrenal cortex.

(glomerulosa) in the studies of Sheppard *et al.* (1963) and of Baniukiewicz *et al.* (1968). Since the rat adrenal cortex is devoid of 17α-hydroxylase activity (Bush, 1951), it is understandable that 18-OH-DOC would be a major secretory product if the biosynthetic pathway is correctly mapped in Fig. 2. Conversion of deoxycorticosterone (DOC) to 18-OH-DOC has been shown to be significant in other mammalian species and in fish (Arai and Tamaoki, 1967).

Conversion of radiolabeled exogenous precursors to 18-OH-DOC when incubated with sectioned human adrenals was shown to occur by Carballeira and Venning (1964) and by de Nicola and Birmingham (1968). Endogenous secretion of 18-OH-DOC has been observed in the camel (Race and Wu, 1964) and in man in preliminary studies (Melby *et al.*, 1970, 1971). In contrast to the rat, it is probable that 18-OH-DOC is quantitatively of less importance in these species.

The nature and extent of physiological activity possessed by 18-OH-DOC have not been well studied. Kagawa and Pappo (1962) found only minor activity of 18-OH-DOC in adrenalectomized rats. However, observations by Birmingham *et al.* (1968) suggest that 18-OH-DOC behaves as a mineralocorticoid, inducing sodium retention and antidiuresis in the adrenalectomized rat. Quantitatively they found 18-OH-DOC activity was similar to deoxycorticosterone acetate (DOCA). 18-OH-DOC exhibits a stimulatory effect on active sodium transport in the isolated urinary bladder of the toad (Porter, 1971). At equimolar concentrations of 7×10^{-7}, 18-OH-DOC was 50% as effective as DOC on transepithelial sodium transport in the toad urinary bladder. Mineralocorticoid activity of 18-OH-DOC by these indices is probably slightly less than that of DOC. In the rat, 18-OH-DOC secretion is severalfold that of DOC and several hundredfold more than aldosterone. Thus, the secretion of 18-OH-DOC in this species may provide the predominant mineralocorticoid function.

18-OH-DOC has been implicated in the genesis of two forms of experimental hypertension in rats. Birmingham *et al.*, (1968) found that 18-OH-DOC was the principal steroid secreted by the enucleate adrenal in response to ACTH in the course of the development of adrenal regeneration hypertension. Rapp and Dahl (1971a,b) have reported that 18-OH-DOC secretion was uniquely increased in rats genetically susceptible to salt-induced hypertension as compared with rats genetically resistant to the hypertensive effects of salt.

In this laboratory, 18-OH-DOC was isolated and identified in human adrenal venous plasma which was obtained by sampling the adrenal venous effluent by percutaneous adrenal vein catheterization (Melby *et al.*, 1967). In the course of studying patients suspected of having mineralocorticoid hypertension, some samples of adrenal venous effluent were found to contain high concentrations of 18-OH-DOC. These latter considerations led to a series of studies in experimental and clinical hypertension which are the subject of this report.

II. 18-OH-DOC Secretion in Animals: Experimental Studies

A. General Methods and Procedures

1. Sampling of Adrenal Venous Effluent of the Rat and Dog

In rats, after a fasting period of 18 hours, a left adrenal vein catheter was placed in animals under sodium pentobarbital anethesia. The animals were heparinized and the blood was drawn from the left adrenal vein via a pocket in the renal vein for periods of approximately 40 minutes by the method of Knigge (1961) devised for hamsters. The blood was obtained between 9:00 AM and 12:00 noon. This same procedure was used in intact and adrenal enucleate animals. Adrenal venous effluent was obtained from the dog by the placing of a permanent lumbo-adrenal vein catheter with a choker for intermittent sampling. The catheter was kept free with heparin sodium chloride solution.

2. Isolation and Quantification of Steroids in Adrenal Venous Effluent

All solvents were spectroanalyzed grade or were distilled prior to use. Methylene chloride was distilled after washing twice with an equal volume of water and filtering through charcoal. Radioactive steroids (New England Nuclear) were checked by chromatography, and purity was found to be greater than 97%. Radioactivity was asssayed in a Packard TriCarb liquid scintillation spectrometer (3000 series) using as the cocktail 42 ml of Liquifluor (New England Nuclear) diluted to 1 liter with toluene. The steroid residue was dissolved in 0.2 ml of methanol prior to adding the counting solution. Counting time was sufficient to ensure a counting error of less than 2%. Thin-layer chromatographic plates were prepared in the usual manner to 375 μ thickness. Silica gel G 254 (Merck, A.G.) plates were activated by heating at 100°C for 1 hour and stored in the desiccator. Kieselguhr G (Merck, A.G.) plates were impregnated with stationary phase prior to sample application by developing in a solution of 5% propylene glycol in acetone. Steroid standards were obtained from Merck, Ciba, or Mann Research Chemicals.

Plasma samples were diluted to 5 ml with water, and 5×10^3 dpm each of tritiated tracers of aldosterone, corticosterone, deoxycorticosterone (specific activity 30–50 Ci/mmole), and 18-OH-DOC (see Section III, B, 1) (specific activity 12–15 μCi/μg) were added. The plasma was extracted twice with five volumes of methylene chloride, and the extract was washed with $\frac{1}{10}$ volume of 0.1 N sodium hydroxide and $\frac{1}{10}$ volume of water and dried in vacuo. The residue was applied over a 1-cm line on a Kieselguhr thin-layer plate impregnated with propylene glycol. Standards of aldosterone, corticosterone, and deoxycorticosterone were applied as a spot on each edge of the plate. The plate was developed to 15 cm using toluene which had been saturated with propylene glycol as the mobile phase. The standards were located by spraying with alkaline blue tetrazolium solution, and the corresponding areas in the samples were eluted. 18-OH-DOC migrates between aldosterone and corticosterone ($R_{\text{corticosterone}}$ 0.65) in the toluene:propylene glycol system. Aliquots of the eluate were removed for tritium recovery assay and quantitative determination. Aldosterone was quantitated by radioimmunoassay utilizing antibodies to the albumin conjugate of D-aldosterone-18,21-dihemisuccinate and a single thin-layer chromatographic separation by the method of St. Cyr *et al.* (1972). Aldosterone assays were kindly performed by Dr. M. St. Cyr. Corticosterone was determined by the competitive protein-binding (CPB) technique of Murphy (1967). Deoxycorticosterone was chromatographed on silica gel in the thin-layer system cyclohexane: ethyl acetate (1:1) prior to quantitation by the CPB method. Further chromatography of deoxycorticosterone on Kieselguhr G impregnated with formamide using cyclohexane:benzene (2:1) saturated with formamide as the mobile phase did not alter the results. 18-OH-DOC was oxidized to the γ-lactone with periodic acid and quantitated by the soda fluorescence method of Tait *et al.* (1967) following paper chromatography on Whatman No. 2 in the system cyclohexane:benzene:methanol:water (5:3:5:1). Recovery of deoxycorticosterone averaged 50%; corticosterone, 70%; 18-OH-DOC, 70%; and aldosterone, 85%. Blank values of an equivalent area of the plate were less than 2 ng for the CPB assay of corticosterone and deoxycorticosterone when working in the range of 20–40 ng of the standard curve. It was usually necessary to quantitate an aliquot of 18-OH-DOC except in the early phases after enucleation, when 95% of the eluate was quantitated. Usually 0.5–1.0 μg of 18-OH-DOC was quantitated with an accuracy of $\pm 10\%$. Values are expressed as micrograms per hour for DOC, corticosterone, and 18-OH-DOC. Aldosterone is expressed as nanograms per hour.

B. REGULATION OF 18-OH-DOC SECRETION

1. Effect of Variations in Dietary Sodium in Rat

Adrenocorticotropic hormone (ACTH) and 3',5'-cyclic AMP both markedly stimulate 18-OH-DOC secretion by rat adrenals *in vivo* and *in vitro* (Cortés *et al.*, 1963; Birmingham *et al.*, 1968). It has been shown that the renin–angiotensin system may not be involved in the activation of aldosterone secretion in the rat adrenal cortex (Eilers and Peterson, 1964). Hypophysectomized rats exhibit a marked reduction in aldosterone secretion as compared to intact controls. Since sodium depletion in the rat is a potent stimulus to the secretion of aldosterone and the stimulatory effect on aldosterone secretion of sodium depletion is probably not mediated by the renin–angiotensin system, it was decided to study the effect of sodium depletion on 18-OH-DOC secretion in intact rats.

2. Procedure

Female Carworth Farms rats weighing between 180 and 200 gm were used for these studies. Twenty-three animals were maintained on Wayne Lab Blox as a standard diet for a period of 3 weeks. Four rats were given a sodium-deficient diet obtained from Nutritional Biochemicals in pellet form for 3 weeks, and six rats were fed 1% sodium chloride solution instead of drinking water for 3 weeks. After a fasting period of 18 hours, left adrenal venous sampling was undertaken.

3. Results

Adrenal venous steroid levels from normal, salt-restricted, and salt-loaded animals appear graphically in Fig. 3. The fall in aldosterone secretion in salt loaded rats is significant ($P < 0.001$). The rise in aldosterone secretion with sodium depletion is not significant although the mean values are nearly twice the controls ($P < 0.1$, >0.05). The lack of significance of the aldosterone response to sodium restriction

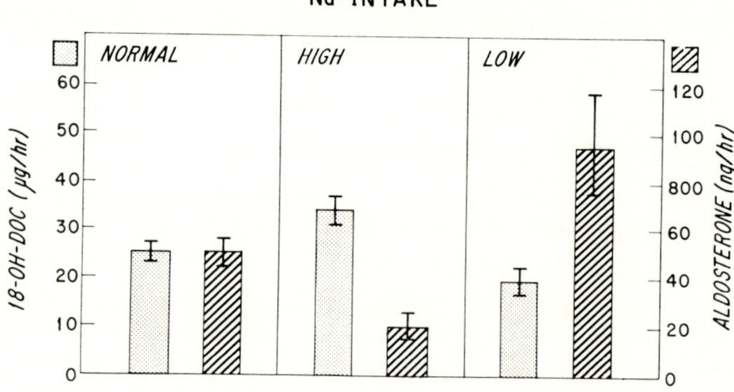

Fig. 3. Adrenal venous steroid levels in normal, salt-loaded, and salt-restricted rats.

is presumably due to the small number of observations made. 18-OH-DOC secretion with sodium depletion and sodium excess are not significantly different from levels in the normal rat. In the rat, at least, sodium loading does not depress, nor does sodium depletion stimulate, secretion of 18-OH-DOC. Since the effects of sodium depletion are manifest only in the zona glomerulosa of the rat, it is not surprising that 18-OH-DOC as a primarily zona fasciculata steroid is not affected by changes in sodium metabolism. Baniukiewicz et al. (1968) demonstrated that sodium loading in rats did not influence in vitro 18-OH-DOC secretion from either the zona glomerulosa (capsular glands) or the zona fasciculata (decapsulated glands).

In summary, rat adrenocortical elaboration of 18-OH-DOC appears not to be affected by sodium intake, and in this respect it behaves predominantly as a product of the zona fasciculata.

4. Effect of Selective Inhibition of 18-Hydroxylase Activity on 18-OH-DOC Secretion from the Canine Adrenal Cortex (Galaburda et al., 1972)

The availability of a pharmacological agent which appears to inhibit selectively the biosynthesis of aldosterone at some point proximal to 18-hydroxycorticosterone (18-OH-B) formation without interfering with cortisol production has provided further insight into the site and regulation of 18-OH-DOC synthesis. 3-Methyl-2-(3 pyridyl) indole methone sulfate (GPA 2282; kindly provided by Dr. Edward Socolow of the Geigy Laboratories), when administered to male mon‸‸‸gs with permanent adrenal vein catheters, produced a mar‸‸‸d fall in aldosterone

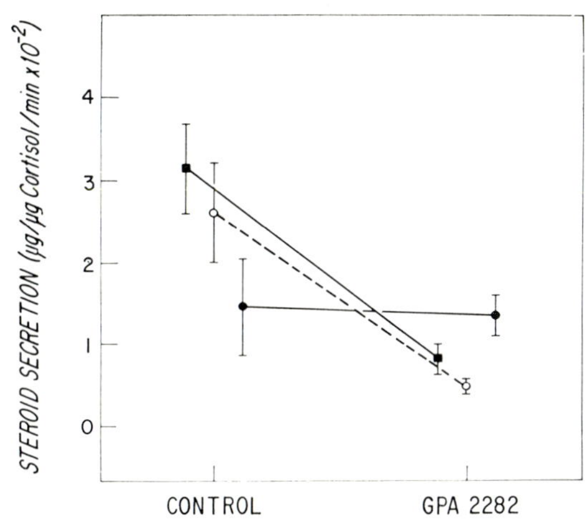

FIG. 4. Effects of inhibition of aldosterone biosynthesis of 3-methyl-2-(3-pyridyl) indole methone sulfate (GPA 2282) on steroid secretion in adrenal venous effluent of the dog. ■——■, 18-OH-corticosterone; ●——●, 18-OH-DOC; ○---○, aldosterone.

(>70% of control) and 18-OH-B (>70% of control) secretion into the adrenal venous effluent without affecting 18-OH-DOC secretion. In the same specimens, DOC and corticosterone secretion increased significantly.

In Fig. 4 are depicted the effects of GPA 2282 on the secretion of 18-OH-DOC, 18-OH-B, and aldosterone in adrenal venous effluent of the dog 1 hour after a dose of 50 mg/kg of body weight. These data are expressed in relation to cortisol secretion which increased slightly after the administration of GPA 2282. The secretion of 18-OH-B and aldosterone dropped significantly to about 30% of the control level, whereas 18-OH-DOC secretion was unchanged in six experiments.

At least two interpretations of these results are possible. The first is that GPA 2282 specifically inhibits the enzymatic conversion of corticosterone to 18-hydroxycorticosterone and does not affect 18-hydroxylation of DOC to 18-OH-DOC in the zona glomerulosa. The second is that GPA 2282 affects all 18-hydroxylase activity of the zona glomerulosa but does not affect 18-hydroxylase activity of the zona fasciculata. These observations provide further support for the idea that 18-OH-DOC is primarily a product of the zona fasciculata.

C. Experimental Hypertension in Rats

1. Adrenal Regeneration Hypertension

The development of severe hypertensive disease which occurs in the rat following unilateral adrenalectomy and nephrectomy and contralateral adrenal enucleation was first described by Skelton (1955). The mechanism by which the regenerating adrenal initiates the hypertension is not known; however, it seems likely that the secretion of one or more adrenocortical steroids is essential to the initiation and perpetuation of the hypertension. Hypophysectomy (Skelton, 1956) and suppression of endogenous corticotropin secretion by exogenous corticosterone administration (Skelton, 1958) block the development of hypertension. Thus, it would appear that steroidogenesis by the regenerating adrenal is ACTH dependent. Mineralocorticoid antagonists also prevent the hypertension sequence (Rapp, 1964; Sturtevant, 1959).

The offending steroid is not aldosterone since secretion is markedly diminished during the development of hypertension (Brogi and Pellegrino, 1959). Rapp (1969) found peripheral blood DOC levels elevated after the establishment of hypertension and failed to find elevations of peripheral blood 18-OH-DOC concentrations (Rapp, 1970). Birmingham et al. (1968) observed that regenerating adrenal slices incubated in vitro elaborated more 18-OH-DOC with ACTH stimulation than did slices of adrenal glands from normal rats. Because of these seemingly conflicting observations, adrenal regeneration hypertension was studied by sampling adrenal venous effluent at various intervals after the enucleation procedure.

Of considerable interest are the observations of Gaunt et al. (1967) on the early phase of adrenal regeneration hypertension. During the first 72 hours following bilateral enucleation of the adrenals, there ensues a marked inability to excrete a light sodium load with some attendant antidiuresis. This effect on sodium metabolism diminishes after 72 hours and disappears in 14 days. The phenomenon of acute sodium retention has characteristics in common with the hypertensive sequence occurring later. It is abolished by hypophysectomy and restored by injections of corticotropin (Gaunt *et al.*, 1969). Exogenous corticosterone inhibits the sodium retention, further emphasizing its dependence on increased ACTH secretory activity. Spironolactone also inhibits the sodium retention after adrenal enucleation suggesting that a mineralocorticoid is involved (Gaunt *et al.*, 1968). However, inhibition of aldosterone secretion by the polysulfated polysaccharide, RO-1-8307, metyrapone, and aminoglutethimide did not abolish the salt retention in response to bilateral enucleation in previous studies (Gaunt *et al.*, 1967, 1968). In this report we have examined the steroid products in the adrenal effluent during the early sodium retaining phase and during the development of adrenal regeneration hypertension.

a. Methods. (1) Male rats were used for observation in the "early" or sodium retaining period (24 hours to 2 weeks). The male rats underwent bilateral adrenal enucleation and adrenal vein blood was obtained at 24 and 72 hours after enucleation corresponding with maximum sodium retention. Sampling was also carried out at 1 and 2 weeks after enucleation.

(2) Female rats were maintained as described in the preceding section. Twenty-three normal rats had normal diets and were fasted 18 hours before adrenal vein catheterization. Four groups of six female rats were designated "enucleate." These animals underwent right adrenalectomy and nephrectomy, and the left adrenal was enucleated by slitting the capsule and extruding the medulla and most of the cortex with gentle forceps pressure. Thereafter they were given 1% NaCl solution as drinking water.

Four groups of six female rats designated "control" underwent right adrenalectomy and nephrectomy, while the left adrenal remained intact. They also were given 1% NaCl solution in place of drinking water. After 24 hours, 1 week, 3 weeks, and 6 weeks, left adrenal vein catheterization was performed as described above.

Mean blood pressures were read 24 hours before catheterization. A cuff on the tail of warmed unanesthetized animals was used with an E and M physiograph.

b. Results. (1) *Sodium retention in male rats.* Studies in bilaterally adrenal enucleated male rats for the "early" phase steroid production are graphically represented in Fig. 5. 18-OH-DOC, corticosterone, and DOC secretion was negligible at 24 and 72 hours and 1 week. At 2 weeks, 18-OH-DOC secretion began to increase significantly from its nadir. Aldosterone secretion declined markedly over the 2-week period, although there was a steep increment at 1 week. The significance of elevated secretion at 1 week is questionable, particularly since it was not seen in the female hypertensive group. It is evident from our observations that none of the steroids studied could have induced severe sodium retention.

(2) *Adrenal regeneration hypertension in female rats.* The results of this study are tabulated in Table I. The secretion of 18-OH-DOC, corticosterone, aldosterone, and DOC in intact normal rats, control and enucleate female rats has been quantified at several intervals during the development of hypertension.

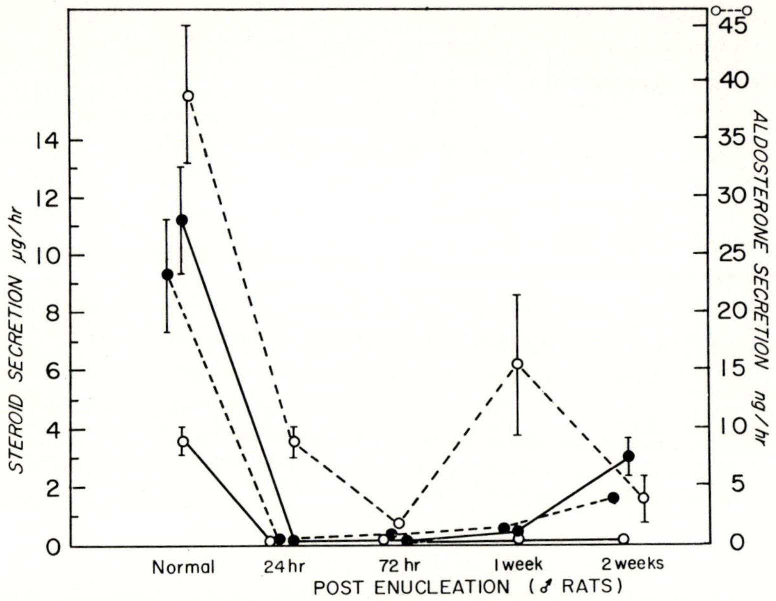

FIG. 5. Steroid secretion in adrenal venous effluent in male rats following bilateral adrenal enucleation. ●——●, 18-OH-DOC; ●---●, corticosterone; ○——○, DOC; ○---○, aldosterone.

Despite the high salt intake, uninephrectomy, and a suggestion of hypertension, secretion of 18-OH-DOC in control rats was not significantly different from intact normal rats. In enucleate animals, 18-OH-DOC secretion was extremely low, less than one-tenth normal, at 24 hours, 1 week, and 3 weeks. Six weeks after enucleation, 18-OH-DOC levels were significantly elevated from controls. As can be seen in Table I, hypertension started to develop by 3 weeks, but was not fully developed until 6 weeks after enucleation. The relation between 18-OH-DOC levels and blood pressure is shown in Fig. 6. The elevated blood pressure was present at 3 weeks and occurred before the rise in 18-OH-DOC secretion.

Corticosterone secretion is low through the first 3 weeks post enucleation, and is restored to normal level, but not elevated, at 6 weeks. 18-OH-DOC to corticosterone ratios were calculated for 6-week control versus enucleate rats, and the ratios were reversed. For control group, 18-OH-DOC/corticosterone = 0.866, while for enucleate animals 18-

TABLE I

*Adrenal Venous Steroid Levels in Enucleate and Control Rats
during the Development of Hypertension*

Group	No. of rats	Systolic blood pressure (mm Hg)	18-OH-DOC (μg/hr)	Corti-costerone (μg/hr)	DOC (μg/hr)	Aldo-sterone (ng/hr)
Normal	23	135 (6 rats) (± 3.7)	25.637 (± 2.004)	26.446 (± 1.803)	4.386 (± 0.642)	50.11 (± 5.66)
Enucleate						
24-Hr	6		0.130 (± 0.088)	0.721 (± 0.170)	0.280 (± 0.052)	9.40 (± 0.84)
1 Week	6		0.958 (± 0.280)	1.239 (± 0.230)	1.520 (± 0.139)	2.00 (± 0.49)
3 Week	6	151 (± 2.6) ($P < 0.05$)	2.529 (± 0.910)	3.291 (± 0.896)	3.268 (± 0.538)	1.51 (± 0.11)
6 Week	9	189 (± 6.3) ($P < 0.05$)	35.779 (± 2.706) ($P < 0.005$)	31.408 (± 4.021) ($P > 0.5$)	4.827 (± 1.717) ($P > 0.5$)	7.26 (± 0.49)
Control						
24 Hr	6		23.250 (± 5.442)	20.625 (± 1.311)	5.565 (± 0.932)	27.06 (± 5.97)
1 Week	6		19.393 (± 4.949)	16.252 (± 3.968)	5.292 (± 0.655)	6.34 (± 1.60)
3 Week	6	132 (± 6.7)	22.800 (± 3.163)	25.458 (± 2.847)	6.248 (± 0.961)	5.31 (± 1.40)
6 Week	12	136 (± 4.0)	23.697 (± 1.769)	29.042 (± 2.193)	4.470 (± 0.983)	4.30 (± 0.57)

OH-DOC/corticosterone = 1.411 ($P < 0.05$). Measurements of DOC were very low 24 hours after enucleation, but were already returning to normal at 1 week. By 3 weeks, adrenal vein DOC secretion was almost back to normal. There was no elevation above normal of DOC secretion at 6 weeks. Aldosterone secretion in enucleate animals was markedly decreased throughout the 6-week period, as has been found by others (Brogi and Pellegrino, 1959). After the first 24 hours, control animals had aldosterone levels as low as those found in enucleate ani-

mals. This was no doubt due to the high sodium intake. These levels
stayed low throughout the entire 6-week period. Corticosterone to aldosterone ratios at 1 and 3 weeks post enucleation were not different
from normal.

   c. Discussion.   It is clear that 18-OH-DOC secretion in hypertensive
enucleate rats at 6 weeks is significantly increased above comparable
controls and normal intact rats. A 40% increment in 18-OH-DOC secretion probably has biological significance if sustained. If 18-OH-DOC
possesses 50% of the mineralocorticoid activity of DOC, the elevation
at 6 weeks is equivalent to a doubling of normal DOC secretion. It

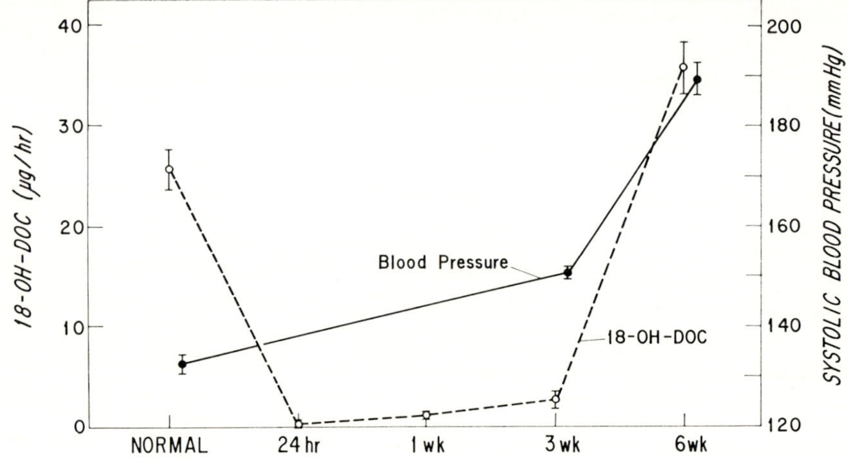

FIG. 6.   18-OH-DOC secretion in adrenal venous effluent of adrenal enucleate
rats during the development of adrenal regeneration hypertension.

is possible that excessive secretion of 18-OH-DOC perpetuates and
worsens adrenal regeneration hypertension. However, it is difficult if
not impossible to explain the major increment in blood pressure observed
at 3 weeks in adrenal enucleate rats on the basis of 18-OH-DOC excess,
as 18-OH-DOC secretion at 3 weeks averages approximately 10% of
the intact normal rat secretory level for this steroid. Excessive secretion
of DOC was not observed in our studies at any stage of the development
of adrenal regeneration hypertension, as enucleate rats did not secrete
more DOC than intact normal rats or control rats. We found DOC
to be the first steroid measured to return to control levels, and acting
relatively alone could be a factor in initiating hypertension in animals
sensitized by uninephrectomy and high sodium intake. It must be remembered that the catheterization procedure represents marked stress. Our
results would be expected to differ quantitatively from those existing

in a basal, unstressed state. This difference is, however, minimized by the fact that ACTH secretion is apparently high after adrenal enucleation (Fortier and de Groot, 1959).

In summary, the rat adrenal steroid product which behaves as a mineralocorticoid, is ACTH dependent and is somehow involved in the development of adrenal regeneration hypertension, has not been positively identified as yet. The sequential actions of steroids secreted in different relative amounts at different times, e.g., early production of DOC in normal amounts followed by the hypersecretion of 18-OH-DOC, could well be involved.

2. Spontaneously Hypertensive Rats (SHR)

A strain of Wistar rats has been developed by Okamoto and Aoki (1963), in which 100% of the rats develop hypertension. Hypertension is always present by 15 weeks of age, although it is not usually severe enough to cause nephrosclerosis or periarteritis nodosa (Okamoto *et al.*, 1964). These rats have been found to have low and unresponsive plasma renin levels (Sokabe, 1965, 1966). Adrenal hyperplasia and hypokalemia have also been observed (Aoki *et al.*, 1963; Baer *et al.*, 1971), and the possibility of a form of mineralocorticoid hypertension is clearly suggested. We have measured adrenal venous effluent steroids in SHR as well as in normotensive rats.

a. Methods and Results. Seven Wistar rats of the Okamoto strain, and 23 normotensive control rats from Carworth Farms weighing 180–200 gm were used. Rats were fed a normal diet, and left adrenal vein blood was obtained as described, except that animals were not fasted prior to catheterization. Blood pressures were measured 24 hours prior to catheterization.

18-OH-DOC, aldosterone, corticosterone, and DOC were quantitated in adrenal vein blood as described. Steroid levels are dipicted in Fig. 7. Values for 18-OH-DOC, corticosterone, and aldosterone are not significantly different in hypertensive rats than in controls. Values for DOC are strikingly low in hypertensive rats ($P < 0.001$).

b. Discussion. Aoki (1963) demonstrated increased weight of anterior pituitary and adrenal cortex in SHR and suggested these organs have a significant role in the development of hypertension. He then demonstrated that removal of these organs prevents the development of hypertension (Aoki *et al.*, 1963). Sokabe (1965, 1966) found low and unresponsive plasma renin activity in SHR which also suggested the possibility of an adrenal role in the etiology of hypertension. Louis *et al.* (1969) showed that the hypertension was relatively unaffected by changes in salt in the diet. He found, however, significantly higher blood pressure with 4% NaCl diet, and lower blood pressure in SHR on entirely salt-free diets.

In a recent report, Baer *et al.* (1971) found that exogenous corticosterone therapy delays the onset of hypertension in young SHR. They showed, however, that by the eighth week of treatment, hypertension in the corticosterone-treated rats was no different than in controls. After adrenalectomy, they found blood pressure continued to rise in SHR, but not as quickly as in control animals.

The importance of markedly decreased DOC secretion in SHR can only be speculated upon. The normal levels of aldosterone and corti-

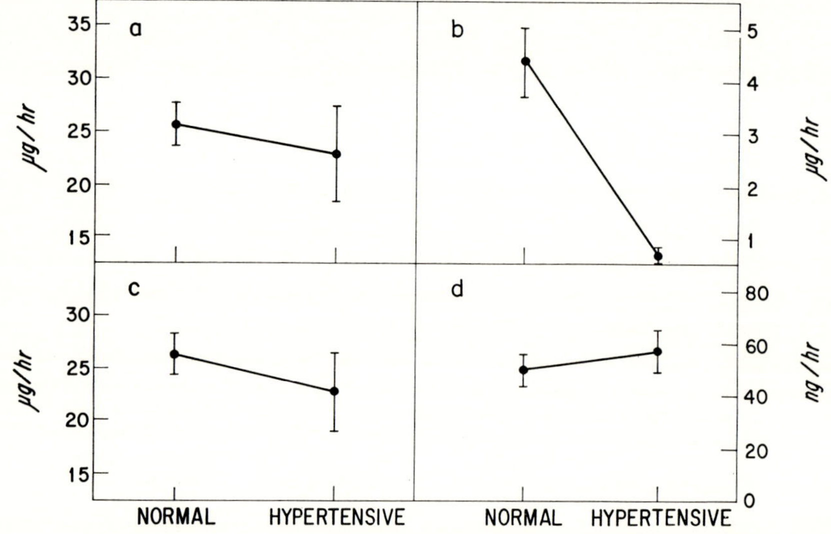

FIG. 7. Steroid secretion in adrenal venous effluent in normal and spontaneously hypertensive rats. (a) 18-OH-DOC; (b) DOC; (c) corticosterone; (d) aldosterone.

costerone suggest there is no 21-hydroxylase deficiency in either the zona glomerulosa or fasciculata. A possibility remains that there is increased turnover of DOC as it is metabolized to an unknown steroid.

The role of 18-OH-DOC in the development of hypertension in these rats seems unlikely to be of importance. The normal levels of this steroid along with corticosterone suggest a normal pituitary adrenal axis.

III. 18-OH-DOC Secretion in Man

Percutaneous bilateral catheterization of the adrenal veins and sampling of the adrenal venous effluent have been carried out in this laboratory for several years as part of the diagnostic evaluation of patients suspected of having primary aldosteronism (Melby *et al.*, 1967). During the analysis of the adrenal vein blood, a soda fluorescent material was noted which migrated with the mobility of 18-OH-DOC γ-lactone in a Bush type chromatographic system cyclohexane:benzene:methanol:water (5:3:5:1) used for the quantitation of aldosterone γ-lactone. 18-OH-DOC had not been known to be an endogenous secretory product of the human adrenal cortex, although it had been shown to be formed from exogenous radioactive steroid precursors when incubated with sectioned human adrenals *in vitro* (Carballeira and Venning, 1964; de Nicola and Birmingham, 1968).

In this report the identification of 18-OH-DOC isolated from human adrenal effluent was established by demonstrating: (1) physical and chemical properties of isolated 18-OH-DOC and its derivatives identical to reference 18-OH-DOC, (2) radiochemical homogeneity of labeled 18-OH-DOC as parent and derivatives, and (3) mass spectral analysis identical to the published spectrum of this compound.

Preparation of labeled 18-OH-DOC with specific activity sufficient to ensure quantitative radioactive significance as a tracer dose has permitted the study of the secretion and metabolism of this compound. Using the technique for aldosterone developed by Tait *et al.* (1961) to determine the metabolic clearance rate (MCR), volume of distribution, and plasma biological half-life, the behavior of this steroid was further characterized in man.

A. Methods-General

1. Solvents and Chemicals

All solvents used were of spectroanalyzed quality or distilled prior to use. Methylene chloride was purified by washing twice with an equal volume of water and filtering through charcoal followed by distillation.

Silica gel GF-254 and Kieselguhr G (Celite) (Merck, A.G.) thin-layer plates were prepared in the usual manner to 375 μ thickness. Impregnation of Kieselguhr G plates with stationary phase was accomplished by predevelopment in 15% formamide in acetone or 6%, 10%, or 15% propylene glycol in acetone as indicated. Silica gel was impregnated by adding 2% w/v of sodium borate to the slurry used in preparing the plates. These plates were only used when acetone:benzene:water (90:10:8) was the mobile phase. Radioactivity was assayed using a Packard TriCarb liquid scintillation spectrometer-3000 series. Counting solution consisted of 42 ml of Liquifluor (New England Nuclear) diluted to 1 liter with toluene. Tritium efficiency was 35%, and carbon 14 efficiency 45%, at the settings used when both isotopes were assayed simultaneously. Scanning of paper chromatograms was carried out using a Tracer Lab 4π scanner.

Gas chromatography was carried out on a Packard Model 7401 gas chromatograph. Combined gas chromatography–mass spectrometry was performed through the courtesy of Dr. Lewis Engel, Massachusetts General Hospital, Boston, Massachusetts, on an LKB-9000 instrument. Steroid standards were obtained from Merck or Mann Research Laboratories. Standard 18-OH-DOC was generously supplied by Dr. R. Pappo, G. D. Searle and Co. Pure 4-^{14}C-labeled 18-OH-DOC was kindly supplied by Dr. M. K. Birmingham. Labeled steroids were purchased from New England Nuclear Corporation.

2. Animals

Male Sprague-Dawley rats (150–200 gm) were used in the incubation experiments and for the preparation of tritiated 18-OH-DOC. All incubations were carried out in Krebs-Ringer buffer containing 200 mg of glucose per 100 ml according to the method of Saffran and Schally (1955).

3. Extraction Procedure

Samples were extracted three times with five volumes of methylene chloride (unless otherwise noted). The extract was washed with $\frac{1}{10}$ volume of 0.1 N sodium hydroxide and $\frac{1}{10}$ volume of distilled water and reduced to dryness on a rotating evaporator *in vacuo* with the temperature of the bath not exceeding 40°C. The dried residue was partitioned between 70% aqueous ethanol and cyclohexane, and the aqueous ethanol layer was reduced to dryness.

4. Chromatography

Some of the thin-layer and paper chromatographic systems used in the isolation and identification of 18-OH-DOC are listed in Table II.

B. Isolation and Identification

1. Preparation of 1,2-^{3}H-Labeled 18-OH-DOC

The adrenal glands of twelve rats were homogenized in 15 ml of Krebs-Ringer bicarbonate buffer containing 200 mg of glucose per 100 ml. The homogenate was incubated with 250 μCi of 1,2-^{3}H-labeled DOC (specific activity = 30–50 Ci/mmole) for 6 hours. The tritiated DOC was purified prior to use by thin-layer chromatography on Celite (15% impregnation) using cyclohexane:benzene (2:1) saturated with formamide.

The extract of the medium was purified by thin-layer chromatography successively on silica gel in chloroform:acetone:water (80:20:2), on

Celite (15% impregnation) in system 3 (Mattox and Lewbart, 1958) and on Celite (6% impregnation) in system 5 (Zaffaroni *et al.*, 1950). 18-OH-DOC has the following mobilities in these systems respectively: $R_{corticosterone}$ 0.55, $R_{cortisol}$ 0.88, and $R_{corticosterone}$ 0.65. A 10–15% conversion of DOC to 18-OH-DOC was generally achieved. The purity of ^{3}H-labeled 18-OH-DOC was checked by addition of carrier, oxidation to the γ-lactone with periodic acid, and chromatography on Whatman No. 2 paper in system 10 (Tait *et al.*, 1967). Upon scanning for tritium, only one peak of radioactivity was present which corresponded to the UV-absorbing and soda fluorescent material with the mobility relative to F-11

TABLE II

Thin-Layer and Paper Chromatographic Systems Used for Isolation
and Identification of 18-OH-DOC

System No.	Support	System
1	Silica gel	Benzene:acetone:water (75:50:2)
2	Silica gel and 2% sodium borate	Acetone:benzene:water (90:10:8)
3	Kieselguhr and formamide	Butyl acetate:formamide:water (100:5:5)
4	Kieselguhr and formamide	Chloroform:formamide (100:sat.)
5	Kieselguhr and propylene glycol	Toluene:propylene glycol (100:sat.)
6	Kieselguhr and formamide	Benzene:chloroform:formamide (1:1:sat.)
7	Paper W No. 2	Benzene:methanol:water (100:50:50)
8	Paper W No. 2	Petroleum ether:benzene:methanol:water (33:17:40:10)
9	Silica gel	Benzene:acetone:water (80:25:0.5)
10	Paper W No. 2	Cyclohexane:benzene:methanol:water (5:3:5:1)
11	Silica gel	Benzene:ethanol (92:8)

dye of 1.03. Acetylation of the original material in pyridine:acetic anhydride (1:2) for 12 hours followed by chromatography in the above paper system demonstrated a major peak (95%) of ^{3}H with $R_{corticosterone\ acetate}$ 1.01, but a tailing of the scan toward the solvent front was noted. In addition, a small portion (5%) of ^{3}H was demonstrated at the origin. The tailing effect and origin ^{3}H may be due to alteration of 18-OH-DOC in the conditions of the acetylation procedure. Pure ^{14}C-labeled 18-OH-DOC was added to the ^{3}H-labeled 18-OH-DOC and the ^{3}H:^{14}C ratio checked of the parent compound and lactone derivative. The ^{3}H:^{14}C ratios remained constant following chromatography of the parent compound in the toluene:propylene glycol system and the lactone in the paper system described above.

The specific activity of the purified 1,2-³H-labeled 18-OH-DOC determined by the soda fluorescence method (Tait *et al.*, 1967) was 10–15 μCi/μg.

2. Isolation and Identification of 18-OH-DOC from Human Adrenal Venous Effluent

Pooled adrenal vein plasma (220 ml) obtained by the percutaneous catheterization technique (Melby *et al.*, 1967) consisting of unstimulated and ACTH-stimulated samples, was extracted with dichloromethane after addition of 1.5×10^5 cpm (100 ng) of 1,2-³H-labeled 18-OH-DOC (prepared as described). The extract was chromatographed successively in systems 1 (Quesenberry and Ungar, 1964), 3, 4 (Zaffaroni *et al.*, 1950), 5, and 2 (Melby *et al.*, 1967) (Table II). 18-OH-DOC has the following relative mobilities in these systems, respectively: R_{cortisol} 0.89; R_{cortisol} 0.88; R_{cortisol} 2.4; $R_{\text{corticosterone}}$ 0.65, and R_{cortisol} 0.45. An aliquot of the eluate from the last system was further chromatographed succesively in systems 7 and 8 with 18-OH-DOC having the following mobilities: system, 7, R_f 0.65; system 8, R_{DOC} 0.59.

Aliquots of the eluate from system 8 were oxidized to the lactone and acetylated and chromatographed in system 10. Soda fluorescence was developed by spraying with alkaline blue tetrazolium, and the papers were dried. Soda fluorescent material was observed migrating with standard 18-OH-DOC lactone and acetate with the respective relative mobilities $R_{\text{F-11 dye}}$ 1.03 and $R_{\text{corticosterone 21-acetate}}$ 0.99. The lactone of 18-OH-DOC isolated from plasma following chromatography in system 8 migrated at R_{DOC} 0.90, identical to that reported by Peron (1961). Aliquots of the remaining residue of system 2 were removed for UV absorption analysis in concentrated sulfuric acid and phenylhydrazine reaction (Silber and Porter, 1954) product spectrum.

The spectrum obtained in concentrated sulfuric acid (du Pont) at a concentration of 6 μg/ml after 2 hours at 22°C from 220 to 400 nm demonstrated absorption maximum at 290 nm with a small shoulder at 340 nm in both standard and plasma 18-OH-DOC. The phenylhydrazine reaction product, developed for 12–16 hours, produced a yellow color with maximal absorption at 410 nm and a minor peak at 300 nm in standard and plasma 18-OH-DOC. The remaining residue of system 2 was oxidized to the γ-lactone. After purification in system 9 ($R_{\text{corticosterone}}$ 0.72), mass spectra of the plasma lactone and standard 18-OH-DOC lactone were obtained on a Perkin-Elmer Hitachi RMV-6E mass spectrometer at an electron energy of 70 eV and an electron current of 60 μA by direct probe at a temperature of 155°C with the ion at 195°C. The principal fragmentation ions of both the standard and plasma compound

are shown in Table III. The spectrum agrees with respect to ion peaks
and relative intensity with the published spectrum of Genard *et al.*
(1968). The base peak is at *m/e* 272 (M-42) with a strong M$^+$ at 314
(46% of base). The ion at *m/e* 229 (56% of base) probably represents
the combined loss of ketene and C_3H_7 from ring A. The intense peak
at *m/e* 124 (83% of base) is more abundant than that reported by
Genard *et al.* and is characteristic of a Δ^4-3-ketone resulting from B ring

TABLE III

Principal Fragment Ions in the Mass Spectra of Standard
and Human Plasma 18-OH-DOC Lactones

Mass/electron	Relative intensity		Cleavage or loss
	Standard	Plasma	
314	56	46	M$^+$
299	5	9	—CH$_3$
296	6	7	—H$_2$O
286	11	14	—CO
272	100	100	—CH$_2$CO
229	60	56	—CH$_2$CO and C$_3$H$_7$
256	5	9	
214	11	15	γ-Lactone ring
171	18	24	
124	81	83	B-ring cleavage

cleavage (Budzikiewicz *et al.*, 1964; Genard *et al.*, 1968). Other peaks
at *m/e* 229, 296, 256, 214, and 171 are in agreement with the published
spectrum.

3. Quantitative Method for Measurement of 18-OH-DOC in Adrenal Vein Plasma

Heparanized adrenal vein blood, obtained by bilateral adrenal vein
catheterization, was centrifuged and the plasma frozen until analysis.
After addition of 1×10^4 cpm of 1,2-^{3}H-labeled 18-OH-DOC (10 ng)
to 10 ml or more of plasma for recovery purposes, the plasma was extracted. The extract was chromatographed in systems 3 and 5 (Table
II). The eluate from system 5 was oxidized with periodic acid, and
the procedure of Tait *et al.* (1967) was used for purification and quantitation of the γ-lactone. The sensitivity of the method is approximately
0.05 μg with an average recovery of 65%. The minimal detectable level
of 18-OH-DOC in adrenal venous effluent if 10 ml of plasma is extracted
is about 0.75 μg/100 ml. The basal 18-OH-DOC level was 1.5 ± 0.6

μg/100 ml plasma in thirty-seven subjects with hypertensive disease of all causes (Melby *et al.*, 1971).

C. METABOLISM OF 18-OH-DOC

1. Metabolic Clearance Rate

Two normal human males and two normal human females were injected with 1 μCi of 1,2-^{3}H-labeled 18-OH-DOC in 1 ml of ethanol and 20 ml of saline. Heparanized blood samples were obtained at 5, 7.5, 15, 22.5, 30, 45, 60, 90, and 120 minutes after the injection. The plasma was separated immediately and frozen until analysis. Ten milliliters of plasma was used for analysis in samples through 45 minutes, and 15 ml after 45 minutes. 4-^{14}C-labeled 18-OH-DOC (225 cpm) plus

TABLE IV

18-OH-DOC Metabolism in Healthy Subjects

Patient	Secretion rate (μg/24 hr)	MCR (L/24 hr)	$\bar{V}$ Apparent volume of distribution	Plasma $T_{1/2}$ (min)	Calculated plasma concentration (ng/100 ml)
J. V. ♀ 28	121	1699	69.3	34	7.0
D. R. ♀ 23	115	1137	75.6	52	10.1
N. E. ♂ 24	94	1242	60.3	35	7.5
J. D. ♂ 23	100	857	34.5	36	11.6

10 μg of 18-OH-DOC was added to all samples prior to extraction with methylene chloride to correct for losses incurred during purification. The extract was chromatographed on system 3, followed by chromatography on silica gel in chloroform:acetone:water (80:20:2). The eluate was counted in a Packard TriCarb liquid scintillation spectrometer. Three 100-minute counts were obtained for each sample, and the points were plotted using least squares analysis (Snedecor, 1956). Oxidation to the γ-lactone and chromatography in system 9 ($R_{\text{aldo lactone}}$ 1.11) and system 11 ($R_{\text{aldo lactone}}$ 1.02) did not change the ^{3}H:^{14}C ratio. The metabolic clearance rate (MCR), plasma half-life, apparent volume of distribution, secretion rate, and calculated average plasma concentrations are shown in Table IV. The metabolism of 18-OH-DOC is very similar to that described by Tait *et al.* (1961) for aldosterone using the two-compartment model. These data for 18-OH-DOC, presumably produced primarily in the zona fasciculata (Baniukiewicz *et al.*, 1968), are not consistent with those for the other fasciculata steroids, cortisol and corti-

costerone, which have a much smaller volume of distribution and MCR and longer plasma half-life. The calculated plasma concentration of 7–11.6 ng/100 ml represents an average concentration over a 24-hour period. Since 18-OH-DOC secretion is ACTH sensitive, the daily circadian rhythm of ACTH secretion probably influences 18-OH-DOC blood levels similar to those of cortisol.

2. 18-OH-DOC Secretion Rate and Urinary Excretion of 18-Hydroxytetrahydrodeoxycorticosterone (18-OH-TH-DOC)

a. Preparation of 18-OH-TH-DOC. An acetone powder of rat liver was incubated according to the method of August (1961) at 37°C for 6 hours with cold or tritiated 18-OH-DOC in shaking flasks containing 200 mg of powder, 3 ml of Krebs-Ringer bicarbonate buffer, 5 mg of glucose 6-phosphate, 0.3 mg of NADP, and 0.2 mg of glucose-6-phosphate dehydrogenase. 18-OH-TH-DOC was isolated by thin-layer chromatography as follows.

1. Chloroform:benzene:formamide (1:1:saturation) on Celite (15% impregnation) developed twice. R_{cortisol} 2.1.

2. Acetone:benzene:water (90:10:8) on silica gel (2% sodium borate impregnation). R_{cortisol} 0.58.

3. Toluene:ethyl acetate:propylene glycol (80:20:saturation) on Celite (15% impregnation) developed twice. The specific activity of the free alcohol was determined by the Porter-Silber reaction. An aliquot of the eluate was acetylated in acetic anhydride:pyridine (2:1) at 60°C for 30 minutes, and the acetate was chromatographed as follows: benzene:ethyl acetate (3:1) on silica gel (R_f 0.34), and heptane:propylene glycol on Celite (10% impregnation) developed twice (R_f 0.33).

The specific activity of the acetate was determined as for the free alcohol. The isolated material was Porter-Silber positive, did not reduce blue tetrazolium, and did not absorb at 240 nm in the UV spectrum. Reaction in methanolic 0.1 N acetic acid produced a compound migrating approximately twice as fast as 18-OH-TH-DOC. A similar reaction occurred with 18-OH-DOC. This is believed to be the product (L-form) described by Dominguez (1965). The specific activity of the derivative was identical to the free alcohol and acetate when measured by the Porter-Silber reagent using cortisol as the reference standard. The conversion of 18-OH-DOC to 18-OH-TH-DOC was 85% as estimated by conversion of 1,2-³H-labeled 18-OH-DOC to 1,2-³H-labeled 18-OH-TH-DOC. The absorption spectrum in sulfuric acid:ethanol (7:3) demonstrated a maximum at 325 nm (Fig. 8).

b. Isolation of 1,2-³H-Labeled 18-OH-TH-DOC from Urine after Injection of 1,2-³H-Labeled 18-OH-DOC and Secretion Rate Determina-

tion. One fifth of a 48-hour urine collection after injection of 2 μCi of 1,2-³H-labeled 18-OH-DOC was hydrolyzed for 18 hours at 37°C in acetate buffer at pH 4.5 with 500 U/ml of β-glucuronidase (Glusulase, Endo. Labs.). Urinary free 18-OH-TH-DOC was isolated by the procedure described above for the preparation of 18-OH-TH-DOC. The acetate was formed, and, after chromatography in benzene:ethyl acetate (1:1) on silica gel, aliquots were removed for quantitation by Porter-

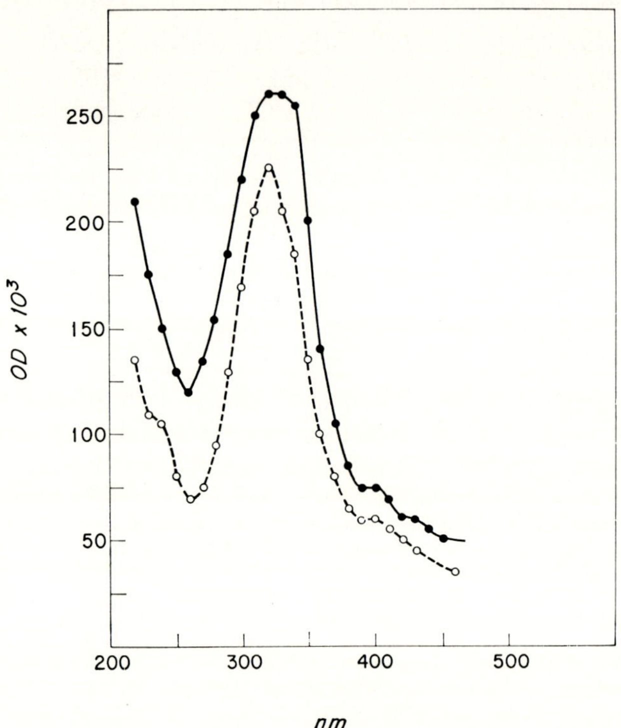

FIG. 8. Absorption spectra of standard (●——●) and urinary 18-OH-TH-DOC (○---○).

Silber reaction and radioactivity measurement. The secretion rate was calculated by dividing the injected dose (in cpm) by the specific activity of the isolated 18-OH-TH-DOC (in cpm/μg). Recovery of 1,2-³H-labeled 18-OH-TH-DOC, added to urine following β-glucuronidase hydrolysis, averaged 58 ± 7% in other samples studied.

Urinary 18-OH-TH-DOC was shown to be identical in chromatographic mobility to 18-OH-TH-DOC prepared by the reduction of 18-OH-DOC by rat liver acetone powder in the above systems both as the free alcohol and as the acetate derivative. The UV spectrum in absolute ethanol demonstrated the lack of absorption at 240 nm. The spectra of urinary 18-OH-TH-DOC and 18-OH-TH-DOC prepared from the rat liver acetone powder gave similar absorption maxima at a concentration of 5 μg/ml in sulfuric acid:ethanol (7:3) with major absorption at 325 nm and a small shoulder at 400 nm (Fig. 8). The specific activity of the free alcohol, acetate, and L-form product produced by reaction in methanolic 0.1 N acetic acid agreed within 5%. Secretion rate determination from 18-OH-TH-DOC and 18-OH-DOC isolated from the urinary neutral fraction and quantitated by the procedure for adrenal vein blood analysis agreed within 15%. The mean 18-OH-DOC secretion rate in fourteen healthy adults on a normal uncontrolled diet was 83.4 $\pm$ 8.9 μg/24 hours. Urinary 18-OH-TH-DOC measurement in twenty-seven healthy adults on a normal uncontrolled diet was found to have a mean of 15.3 $\pm$ 1.0 μg/24 hours.

3. ^{3}H-Distribution after Injection of 1,2-^{3}H-Labeled 18-OH-DOC in Man

Five healthy, adult males received intravenous injections of labeled 18-OH-DOC in this study. Radioactivity in whole urine was determined in a Packard liquid scintillation counter using 0.5 ml of urine in 10 ml of counting cocktail containing 100 ml of Scintisol (Isolabs), 42 ml Liquifluor (New England Nuclear) and 858 ml of toluene. Correction for quenching in all samples was performed using an internal ^{3}H standard. An aliquot of urine from a 48-hour collection after injection of 2 μCi 1,2-^{3}H-labeled 18-OH-DOC was extracted with methylene chloride for the neutral fraction. The extracted urine was hydrolyzed with β-glucuronidase at pH 4.5 for 18 hours and reextracted with methylene chloride (glucuronide fraction). The spent urine was adjusted to pH 1 with hydrochloric acid, hydrolyzed at room temperature for 18 hours, and extracted with methylene chloride (acid fraction).

A patient with an incomplete biliary fistula was injected with 2 μCi ^{3}H-labeled 18-OH-DOC. Bile was collected for a 24-hour period, and the amount of radioactivity was determined after bleaching with peroxide. A total volume of 414 ml of bile was collected in the 24-hour period. The results of the urinary distribution of radioactivity in five normal males are shown in Table V.

The urinary excretion of tritium was practically complete in 48 hours, accounting for 47 $\pm$ 4% of the injected dose. About 6% was excreted in the second 24-hour collection. The neutral fraction accounted for 2.7 $\pm$ 0.3% of the injected dose: glucuronidase hydrolysis liberated 37 $\pm$ 5% of the injected dose or 78% of the tritium excreted into urine. Pure 18-OH-TH-DOC account for about 40% of the secretion rate.

After acid hydrolysis, an additional $2.1 \pm 0.3\%$ of the injected dose was extracted with methylene chloride. A considerable portion of the injected radioactivity was excreted into bile. Actual measurement of tritium accounted for 16.2% of the injected dose in the 24-hour collection. This figure is corrected to 24–39% based on the data given for normal bile flow of 600–1000 ml per day (Bockus, 1965) since the subject had an incomplete biliary fistula (Table V). This mode of excretion does not occur for aldosterone or cortisol since greater than 95% of injected radioactive doses of these compounds can be accounted for in a 48-hour urine collection in normal subjects. 18-OH-DOC has been isolated and measured by the method used for isolating it from adrenal vein blood in the neutral fraction. Efforts to isolate 18-OH-DOC from the acid hydrolyzable fraction failed proabably because of the acid-labile nature of the compound.

TABLE V

^{3}H-Distribution in Man after Intravenous Injection 1,2-^{3}H-Labeled 18-OH-DOC

	Percent of injected dose/48 hours
Urine	46.85 ± 5.98
Glucuronidase hydrolysis	37.34 ± 6.24
18-OH-TH-DOC	38.90 ± 11.00
Acid hydrolysis	2.12 ± 0.32
Neutral extract	2.70 ± 0.37
Bile (est)	24–39
Total	71–86

In summary, 18-OH-DOC is a natural secretory product of the human adrenal cortex. Under basal conditions, the secretion rate of 18-OH-DOC is somewhat less than that of aldosterone (approximately 80 μg/24 hours). The MCR, volume of distribution, plasma half-life, and estimated plasma concentration are roughly similar to those of aldosterone. The physiological disposal of 18-OH-DOC and its metabolites is very different from most human steroid products in that less than 50% is excreted in urine and a significant portion is excreted in bile. The rate of excretion of 18-OH-TH-DOC glucuronide in urine is a fairly reliable index of 18-OH-DOC secretion, reflecting nearly 20% of the secretion rate uncorrected for recovery.

IV. Regulation of 18-OH-DOC Secretion in Man

In the rat, 18-OH-DOC is by and large a product of the zona fasciculata and its secretion is under the control of the anterior pituitary, specifically corticotropin. Changes in sodium intake did not appear to influence 18-OH-DOC secretion in the rat. Man and rat differ not only in the nature of the corticosteroids they produce, but in the regulation of secretion of corticosteroids they have in common. A partial assessment of factors influencing 18-OH-DOC secretion in man follows.

A. Methods

1. Subjects

Patients suspected of having mineralocorticoid hypertension were hospitalized in the Metabolic Research Unit of the University Hospital. Unless stated to the contrary, patients and healthy adult control subjects received a diet containing 128 meq of sodium and 50 meq of potassium daily. Twenty-seven healthy adults of both sexes participated in these studies.

2. Adrenal Vein Catherization

Catheterization in hypertensive patients was carried out as described by Melby *et al.* (1967). In the course of adrenal vein catheterization and after collection of 25 ml of blood from the adrenal veins, angiotensin was infused through a peripheral vein at a rate sufficient to maintain the diastolic pressure 10 mm Hg above baseline. Doses averaged 200 ng/min to 500 ng/min. After 10 minutes additional blood samples were obtained. Immediately after, 0.25 mg of β-1,24-corticotropin was injected over 30 seconds into a peripheral vein; 5 minutes after this injection adrenal venous blood samples were obtained. Thirty-six patients were studied. Adrenal venous plasma 18-OH-DOC concentrations, 18-OH-DOC secretion rate, and urinary 18-OH-TH-DOC measurements were performed as described in Section III. Aldosterone secretion rate (ASR) was measured by the method of Melby *et al.* (1967).

B. Effect of ACTH Stimulation and Pituitary ACTH Suppression by Dexamethasone

ACTH or β-1,24-corticotropin, within 10 minutes after injection intravenously into patients during adrenal vein catheterization, evoked an enormous increment in adrenal venous plasma level of 18-OH-DOC. In Fig. 9 this response is depicted. The increment in 18-OH-DOC level was greater than that of aldosterone (not shown).

Intramuscular injections of ACTH gel (80 units b.i.d.) also stimulated a nearly 20-fold rise in the 18-OH-DOC secretion rate in four healthy subjects as represented in Fig. 10.

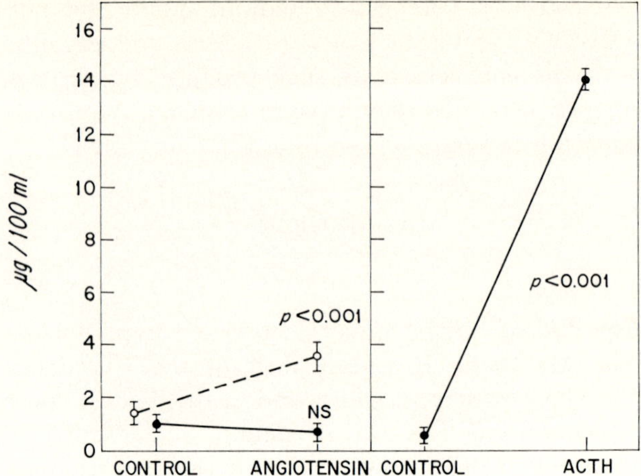

Fig. 9. Human adrenal venous plasma level of aldosterone (○---○) and 18-OH-DOC (●——●) after angiotensin and 18-OH-DOC (●——●) after ACTH stimulation. NS, not significant.

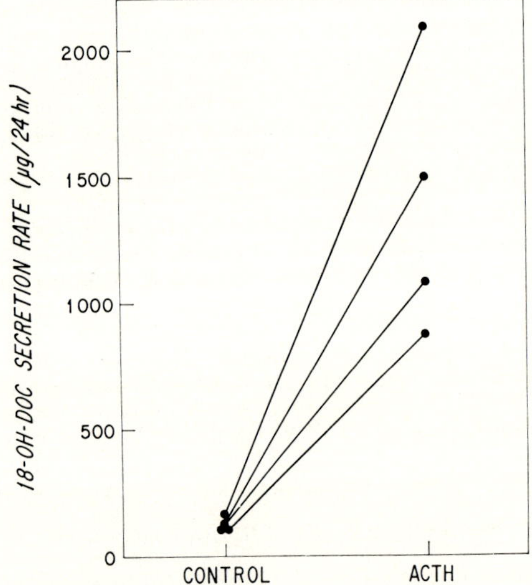

Fig. 10. 18-OH-DOC secretion rates after intramuscular ACTH injections in four healthy human subjects.

When dexamethasone (0.5 mg every 6 hours) was given to healthy subjects for 2 days, urinary 18-OH-TH-DOC excretion declined significantly (50–60% reduction from control). The response to dexamethasone is shown in Fig. 11.

It is apparent from these observations that ACTH is a potent stimulus to 18-OH-DOC seecretion and that suppression of pituitary ACTH discharge in the basal state markedly diminishes 18-OH-DOC secretion.

C. Effect of Exogenous Angiotensin on 18-OH-DOC-Secretion

In Fig. 9 the effect of brief infusions of subpressor doses of angiotensin on levels of aldosterone and 18-OH-DOC in the adrenal venous effluent of patients is demonstrated. If a unilateral aldosterone-producing adenoma was present, the venous effluent of the contralateral adrenal

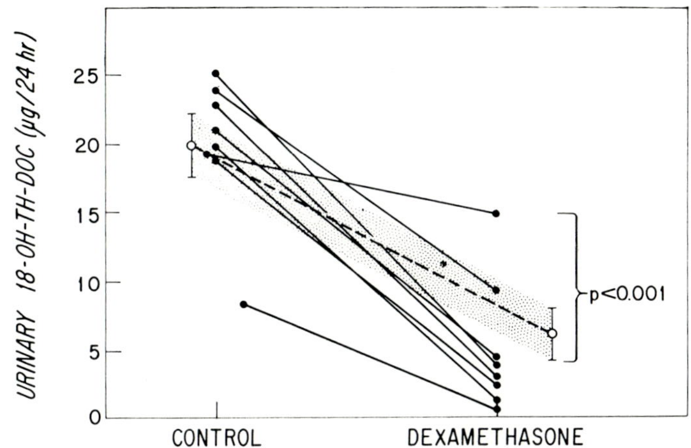

Fig. 11. Effect of dexamethasone suppression on urinary 18-OH-TH-DOC excretion in healthy human subjects.

was sampled. Angiotensin evoked a 3-fold increment in mean aldosterone concentration which was highly significant, whereas 18-OH-DOC concentration was unaffected by angiotensin.

D. Effect of Alterations in Dietary Sodium on Secretion of 18-OH-DOC

Seven healthy adult males received a diet containing more than 150 meq of sodium daily for 4 days. In the last 2 days of moderately high salt intake, an 18-OH-DOC secretion rate was performed. These same subjects were then given a diet containing less than 10 meq of sodium for four more days. During the last 2 days of salt restriction, 18-OH-DOC secretion rates were repeated. The results of this study appear in Fig. 12. 18-OH-DOC secretion was lower on the average during sodium loading than in the basal state (57 μg/24 hours versus 83 μg/24 hours). During sodium restriction 18-OH-DOC secretion increased in all but two subjects. These few observations suggest an influence of the sodium ion on the regulation of 18-OH-DOC secretion. In some of these same subjects, plasma aldosterone concentrations were depressed to unmeasurable levels during sodium loading and stimulated 4- to 5-fold by sodium restriction (St. Cyr *et al.*, 1972). Aldosterone secretion rates (ASR) varied proportionately. As expected, plasma renin activity (PRA) varied directly with ASR and sodium intake. Küchel (1971) in the dis-

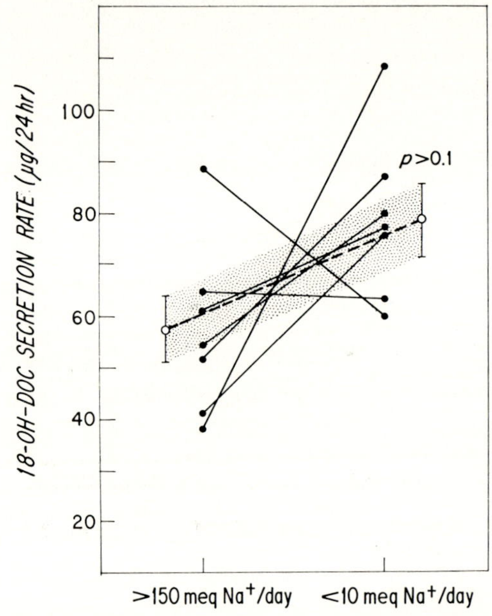

Fig. 12. Effect of high and low sodium diet on 18-OH-DOC secretion in seven healthy human subjects.

cussion of a preliminary report from this laboratory (Melby *et al.*, 1971) observed a 2- to 3-fold rise in 18-OH-DOC secretion with sodium deprivation. It seems possible that sodium deprivation evoked a modest rise in 18-OH-DOC secretion which is not commensurate with the changes observed in plasma aldosterone levels and PRA. Observations in one control subject suggested an interaction of pituitary ACTH and an activated renin–angiotensin system (sodium deprivation). In this individual, sodium deprivation was accompanied by a modest increment in 18-OH-DOC secretion. After an appropriate interval of sodium loading, sodium restriction was again imposed. During sodium deprivation dexamethasone (dose of 0.5 mg every 6 hours) was given. Dexamethasone greatly diminished 18-OH-DOC secretion despite the marked sodium restriction in the diet.

A few explanations suggest themselves, the most appealing being that a small but measurable portion of secreted 18-OH-DOC arises from the zona glomerulosa and as such is regulated as is aldosterone. It is possible that the modest increments in 18-OH-DOC secretion observed during sodium restriction are largely accounted for by increased secretion by the glomerulosa. When the pituitary dependent portion of the total

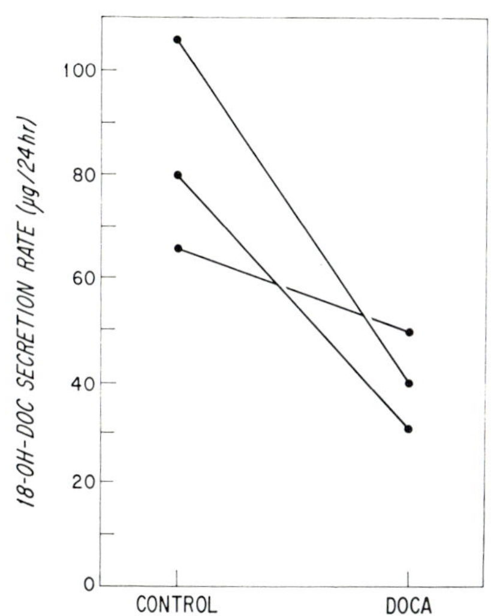

FIG. 13. Effect of DOCA injections on 18-OH-DOC secretion in three healthy human subjects.

18-OH-DOC secreted is suppressed, 18-OH-DOC secretion from the glomerulosa is relatively insignificant.

E. EFFECT OF DEOXYCORTICOSTERONE ACETATE (DOCA) ON 18-OH-DOC SECRETION

In a limited number of control subjects 18-OH-DOC secretion rates were obtained on a control diet (125 meq sodium) and during 2 days of DOCA intramuscular injections (10 mg every 12 hours). The results are depicted individually in Fig. 13. The reduction in 18-OH-DOC secretion observed after DOCA administration was unexpected. DOCA administered in this way also induces a variable but significant decrement in aldosterone secretion. It is conceivable that exogenous mineralocorticoid suppresses any contribution to 18-OH-DOC secretion by the zona

glomerulosa, however, the magnitude of the decrements in 18-OH-DOC
secretion is rather more than expected if this were the case.

V. 18-OH-DOC Secretion in Human Hypertension

Since Conn (1955) first described the clinical and laboratory manifestations of aldosterone hypersecretion by an adenoma of the adrenal cort-
tex, a concept of the physiological consequence of human mineralocorticoid excess has gradually evolved. Hypertension and varying degrees
of potassium depletion (Conn *et al.*, 1966) accompanied by excessive
secretion of DOC, aldosterone, cortisol (Biglieri *et al.*, 1968), and possibly other unknown adrenal steroids were thought to be the principal
features of the mineralocorticoid hypertensive syndrome. Another characteristic of this syndrome is that its manifestations may be abolished by the
administration of sufficient doses of specific competitive antagonists of
mineralocorticoids such as spironolactone (Spark and Melby, 1968).

A most consistent observation in patients with the mineralocorticoid
hypertensive syndrome is that plasma renin activity (PRA) is usually
decreased, and it responds not at all or minimally to potent physiological stimuli such as sodium deprivation or assumption of the erect posture
(Conn *et al.*, 1964). Hyporesponsiveness of PRA, or suppressed PRA,
is a reliable indicator of mineralocorticoid excess. However, a significant
proportion (perhaps 20%) of patients with "essential" hypertension exhibits suppressed PRA (Küchel *et al.*, 1967; José *et al.*, 1970; Channick
et al., 1969). The majority of these patients have no identifiable cause
for the presence of suppressed PRA.

The possibility that as yet unidentified mineralocorticoids may be
responsible for suppressed PRA in these hypertensive patients is supported by studies of Woods *et al.* (1969) in which inhibition of adrenal
steroidogenesis by aminoglutethimide restored normal physiological responsiveness of plasma renin and reduced blood pressure significantly.
Spark and Melby (1971) have demonstrated a similar restoration of
PRA responsiveness and reversal of hypertension with large doses of
spironolactone in their patients with suppressed PRA and hypertension.
Mineralocorticoid antagonism by spironolactone in high dose did not
affect blood pressure in patients with hypertension and normally responsive PRA.

The isolation and identification of 18-OH-DOC in human adrenal
venous effluent coincided with the growing indirect evidence that suppressed PRA hypertension might be due to a mineralocorticoid the structure of which was unknown. The possibility that 18-OH-DOC may somehow be involved in suppressed PRA hypertension has been examined
in several patients.

18-OH-DOC secretion was evaluated in patients with mineralocorticoid hypertension in which the offending steroid(s) was known and in hypertensive patients with normally responsive and suppressed PRA.

A. METHODS AND PROCEDURES

In patients studied prior to 1967, measurements of PRA were not made consistently. PRA was determined by the method of Gunnels *et al.* (1967). Aldosterone secretion rate or urinary aldosterone metabolite excretion and urinary 18-OH-TH-DOC excretion were carried out in all patients. 18-OH-DOC secretion rates, urinary TH-DOC, and cortisol secretion rates (where indicated) were measured in fewer patients.

Methodology for the measurement of urinary 18-OH-TH-DOC, 18-OH-DOC secretion rate, and aldosterone secretion rate have been described. Tetrahydrodeoxycorticosterone (TH-DOC) excretion in urine was measured as follows:

One fifth of a 24-hour urine was hydrolyzed with β-glucuronidase. A tracer of 1,2-^{3}H-labeled TH-DOC, prepared by reduction of 1,2-^{3}H-labeled DOC with rat liver acetone powder, was added and urine was extracted with methylene chloride. The extract was purified by thin-layer chromatography in the following systems:

1. Benzene:hexane:formamide (1:1:saturation) on kieselguhr impregnated with 20% formamide.

2. Toluene:hexane:propylene glycol (80:20:saturation) on kieselguhr impregnated with 15% propylene glycol.

3. Butyl acetate:acetone:water (85:15:3) on silica gel impregnated with 2% sodium borate.

4. Benzene:ethyl acetate (1:1) on silica gel. The eluate off the last system was aliquoted for quantitation of TH-DOC by the modified Porter-Silber reaction (Lewbart and Mattox, 1961) and counting for recovery of 1,2-^{3}H-labeled TH-DOC. Recovery of added ^{3}H-labeled TH-DOC averaged 60%, with normal TH-DOC levels ranging from 5 to 20 μg/24 hours.

Cortisol secretion rates were determined by intravenous injection of 1×10^6 cpm of ^{3}H-labeled cortisol followed by 24-hour urine collection. After β-glucuronidase hydrolysis, urine was extracted with methylene chloride, and chromatographed in benzene:ethyl acetate:ethylene glycol (80:20:saturation) on Celite predeveloped in 15% ethylene glycol in acetone. Tetrahydrocortisone (THE) was quantitated by the Porter-Silber method (Silber and Porter, 1954), and secretion rate was calculated by dividing the amount of ^{3}H injected by the specific activity of urinary THE.

Hypertensive patients in whom data on 18-OH-DOC secretion and/or

18-OH-TH-DOC excretion was obtained had the disorders discussed below.

B. 18-OH-DOC Secretion in Cushing's Syndrome

Approximately 75% of patients with Cushing's syndrome have adrenocortical hyperplasia due to excessive and unvarying stimulation by ACTH. Ectopic ACTH production from nonendocrine tumors accounts for about 20% of patients with ACTH excess. In 25% of the patients, Cushing's syndrome is produced by primary adrenocortical neoplasms, and because of autonomous cortisol production, the cybernetic mechanism for ACTH secretion becomes nil. The physiological consequences of mineralocorticoid excess are observed in a few patients with pituitary Cushing's (Cushing's disease) and nearly all patients with ectopic ACTH production from nonendocrine tumors. Frequency of appearance of the manifestations of mineralocorticoid excess is proportionate to the quantitative production of cortisol by the adrenals. Aldosterone secretion is not increased, and in fact is often quite low. Suppressed plasma renin activity is observed in a significant proportion of patients. In patients with ACTH excess (pituitary Cushing's and ectopic ACTH production) increased secretion of 18-OH-DOC may be expected. Biglieri *et al.* (1968) found increased DOC secretion in patients with ectopic ACTH production and severe pituitary Cushing's. In Table VI, 18-OH-DOC secretion and/or 18-OH-TH-DOC excretion in urine are compared with the cortisol secretion rate and, in a few instances, urinary tetrahydro-DOC excretion. It can be seen that with pituitary Cushing's, 18-OH-TH-DOC excretion in urine is moderately elevated in some patients and markedly so in one. In patients with ectopic ACTH production from nonendocrine tumors, 18-OH-DOC secretion and/or 18-OH-TH-DOC excretion are markedly elevated. Finally, in the patient with a primary adrenocortical neoplasm, the excretion of 18-OH-TH-DOC is limited while the excretion of TH-DOC is increased. We have no explanation for this observation. In some of the patients in whom milligram quantities of 18-OH-DOC were secreted, it is reasonable to assume that a significant mineralocorticoid activity is present. If DOC secretion parallels 18-OH-DOC secretion, as it is reasonable to assume, then both hypertension and hypokalemic alkalosis can be explained on the basis of these steroids acting in concert.

Many patients with Cushing's syndrome are hypertensive without accompanying electrolyte disturbances. Hypertension is present in 6 of the 8 patients listed with pituitary Cushing's syndrome. A 3-fold increment in secretion of 18-OH-DOC and DOC and unvarying secretion of these steroids in this syndrome might provoke arterial hypertension.

TABLE VI

18-OH-DOC Secretion and/or Excretion in Patients with Cushing's Syndrome

Patient	Age	Sex	Race	Type of Cushing's syndrome	18-OH-DOC CSR (mg/24 hr)	18-OH-DOC SR (μg/24 hr)	18-OH-TH-DOC excretion in urine (μg/24 hr) Control	18-OH-TH-DOC excretion in urine (μg/24 hr) Dexamethasone	TH-DOC excretion in urine (μg/24 hr)
1	16	F	W	Pituitary	53	194	34	—	32
2	20	F	W	Pituitary	44	—	38	14	—
3	46	M	W	Pituitary	110	2492	353	63	4 (dexamethasone)
4	19	F	W	Pituitary	101	—	58	12	16
5	19	M	W	Pituitary	98	—	89	—	—
6	20	F	W	Pituitary	102	—	68	—	—
7	32	F	W	Pituitary	63	—	28	—	65
8	38	F	W	Pituitary	38	—	43	—	—
9	36	F	W	Pituitary	47	—	45	—	103
10	49	M	W	Ectopic ACTH	934	3030	706	—	—
11	57	F	W	Ectopic ACTH	475	486	178	—	732
12	67	F	W	Ectopic ACTH	235	—	425	—	537
13	67	M	W	Ectopic ACTH	371	—	324	—	258
14	58	M	W	Adrenocortical carcinoma	144	—	18	—	370

Patients without endocrine disease, treated with therapeutic doses of cortisol in amounts similar to secretion rates observed in pituitary Cushing's, ordinarily would not develop hypertension. The major contributor to excessive mineralocorticoid activity in patients with Cushing's syndrome appears, however, to be DOC.

C. 18-OH-DOC Secretion in Hypertension with Disorders of Adrenal Steroid Biogenesis

1. 11β-Hydroxylase Deficiency

Three patients with this disorder were studied. All these patients exhibited hypertension, increased secretion of 11-deoxycortisol and DOC, and diminished secretion of cortisol. 18-OH-TH-DOC excretion contrasted with urinary TH-DOC excretion as shown in the first three patients in Table VII. Basal excretion of 18-OH-TH-DOC is not different from that observed in normal subjects. In a single subject, ACTH failed to evoke as great an increment in 18-OH-TH-DOC excretion as is seen in normal subjects. On the other hand, basal TH-DOC excretion in urine was 10- to 20-fold basal excretion in normal subjects.

2. 17α-Hydroxylase Deficiency

Urine specimens from two patients with this interesting disorder were analyzed through the courtesy of Dr. E. G. Biglieri. Absent secondary sex characteristics in these genotypic females along with hypertension, hypokalemia, and the characteristic steroid pattern in urine (low 17-ketosteroids, low 17-hydroxycorticosteroids, elevated TH-DOC and corticosterone metabolites) met the diagnostic criteria for this syndrome described by Biglieri *et al.* (1966). Urinary 18-OH-TH-DOC and TH-DOC excretion in these women are listed in Table VII as cases 4 and 5. In both patients, 18-OH-TH-DOC excretion was markedly increased, and TH-DOC excretion was increased to an even greater degree. In one patient, the 18-OH-TH-DOC excretion could be diminished by the administration of dexamethasone.

In 11β-hydroxylase deficiency, the rise in DOC secretion results from inhibition of its conversion to corticosterone both in the zona glomerulosa and the zona fasciculata. This block is exaggerated by the heightened ACTH activity which is a reciprocal response to decreased cortisol production. 18-OH-DOC secretion is ACTH dependent and stimulatable, yet 18-OH-DOC excretion was in the normal range. Large doses of exogenous ACTH increased 18-OH-DOC secretion only to a small extent compared to normal subjects. Kraulis and Birmingham (1965) demonstrated that the 11β-hydroxylase inhibitor, metyrapone, inhibited con-

TABLE VII

Urinary 18-OH-TH-DOC Excretion in Hypertensive Patients with Disorders of Adrenal Steroid Biogenesis

Patient	Age	Sex	Race	Diagnosis	18-OH-TH-DOC excretion (μg/24 hr)			Basal TH-DOC excretion (μg/24 hr)
					Control	ACTH	Dexamethasone	
1	16	F	N	11β-Hydroxylase deficiency	12	43	8	588
2	38	M	W	11β-Hydroxylase deficiency	22		6	309
3	37	F	W	11β-Hydroxylase deficiency	28		0	287
4[a]	18	F	W	17α-Hydroxylase deficiency	110			488
5[a]	36	F	W	17α-Hydroxylase deficiency	211		31	620

[a] Patients studied by Dr. E. G. Biglieri, San Francisco, California.

version of DOC to 18-OH-DOC in sectioned rat adrenals. Since both the 11β-hydroxylase and 18-hydroxylase enzymes are located in the mitochondria, it is possible that these enzyme systems are related. It is known that metyrapone probably inhibits both 11β-hydroxylation and 18-hydroxylation by competitive inhibition of the binding of DOC to the cytochrome P-450 necessary for both 11β- and 18-hydroxylation (Simpson *et al.*, 1969). On the other hand, 17α-hydroxylase deficiency does not interfere with the production of progesterone, DOC, 18-OH-DOC, and corticosterone. It is probable that 17α-hydroxylation of pregnenolone is diminished. It is of interest that in 17α-hydroxylase deficiency, hypokalemic alkalosis is usual, whereas in 11β-hydroxylase deficiency, severe electrolyte disturbances are usually not observed. Aldosterone secretion is usually reduced in 17α-hydroxylase deficiency so that the manifestations of mineralocorticoid excess must be due entirely to excessive secretion of DOC and to a lesser extent 18-OH-DOC. Reduced aldosterone secretion may be the consequence of suppressed plasma renin activity which has been observed in this syndrome.

D. 18-OH-DOC-Secretion in Patients with Essential Hypertension and Normally Responsive PRA

Twenty-eight patients with benign essential hypertension were studied. PRA was measured after several hours in recumbency and 4 hours after being up and about. Mean upright PRA averaged 275 ng/100 ml with the range from 184 to 425 ng per 100 ml. In none of these patients did the aldosterone secretion rate exceed 150 μg/24 hr while on a diet of 128 meq sodium and 50 meq potassium per day. Urinary 18-OH-TH-DOC excretion in these patients is plotted on the second panel of Fig. 14. The values in the hypertensive population with normally responsive PRA were not different than those observed in healthy control subjects. TH-DOC excretion was measured in a few of these subjects and was found to be low. Normally responsive PRA virtually excluded the possibility that there was significant mineralocorticoid excess. Additional indirect evidence that mineralocorticoid excess is not a factor in hypertension with responsive PRA is the fact that large doses of spironolactone (400 mg/day) for a period of 3–5 weeks failed to influence the blood pressure in 75% of these patients.

E. 18-OH-DOC Secretion in Hypertensive Patients with Suppressed PRA

In thirty patients with hypertension, PRA did not increase significantly from the recumbent to the erect posture. Mean PRA in the erect posture in these patients was 94 ng/100 ml with a range of less than

50 to 120. Aldosterone secretion was lower in patients with suppressed PRA than in those who were normally responsive. Mean aldosterone secretion rate was 77 μg/24 hr with a range of 30–104. Urinary 18-OH-TH-DOC excretion in these patients averaged more than twice the mean for hypertensive subjects with normally responsive PRA (Fig. 14). The mean 18-OH-TH-DOC excretion in urine was 38.4 $\pm$ 4.3 μg/24 hours. The difference between the mean excretion of 18-OH-TH-DOC in urine in these patients as compared to both healthy adult controls and patients with normally responsive renin activity is highly significant ($P < 0.001$). Spironolactone was administered in high dose to 22 of

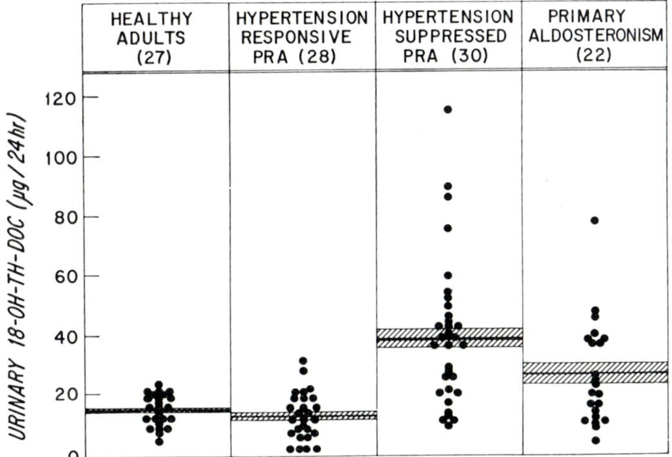

Fig. 14. 18-OH-TH-DOC excretion in healthy subjects, hypertensive patients with normally responsive PRA, hypertensive patients with suppressed PRA, and patients with primary aldosteronism.

the 30 subjects studied. In all these patients a significant reduction in blood pressure was obtained and in 60% spironolactone reduced the blood pressure consistently into the normal range. Urinary TH-DOC excretion was quantitated in more than 50% of these patients, and, with a single exception, excretion was within normal limits. In Table VIII the essential clinical information and steroid patterns are listed for ten patients with hypertension, suppressed plasma renin activity, and elevated 18-OH-DOC secretion. Six of these patients were studied elsewhere. Patients 1 through 4 were evaluated in this clinic. In each of these patients an adrenal vein catheterization was performed because of the near certainty that they were producing excessive amounts of a mineralocorticoid. Adrenal venous plasma 18-OH-DOC levels in

TABLE VIII

Clinical Features in Patients Found to Have Increased 18-OH-TH-DOC Excretion and Suppressed Plasma Renin Activity (PRA)

Patient	Sex	Age	Race	PRA[a] (ng/100 ml)	ASR (μg/24 hr)	Urinary aldo-sterone (μg/24 hr)	18-OH-TH-DOC excretion in urine (μg/24 hr) Con-trol	Dexa-methasone	Blood pressure response to dexa-methasone	Disposition	Results of treatment
1	F	40	W	<50E	50		79			Spironolactone	Normotensive
2	F	54	W	55E	44		42			Spironolactone	Unknown
3	M	43	W	50E	38		48		Significant reduction	Bilateral adrenalectomy, nodular hyperplasia	Normotensive
4	M	37	W	<50E	30		115	3	Normotensive	Spironolactone	Normotensive
5	M	54	W	130$\downarrow$Na$^+$		12.3	86.	9	Normotensive	Spironolactone	Normotensive
6	M	51	N	135$\downarrow$Na$^+$		11.9	52			Spironolactone	Normotensive, recent MI
7	F	8	W			7.3	73		Normotensive	Bilateral adrenalectomy, hyperplasia	Normotensive
8	F	47	W	87E		12.7	79			Right adrenalectomy, (2 adenomas)	Hypertensive
9	M	60	W	142E 195$\downarrow$Na$^+$		14.2	82,94			Left adrenalectomy, (adenoma)	Hypertensive
10	M	44	W	(1.1 ng angiotensin I/ml/hr)		15.2	56		No change	Bilateral adrenalectomy, hyperplasia	Normotensive

[a] E = erect posture; $\downarrow$Na$^+$ = low salt diet.

all 4 patients were greatly elevated (6.1–13.1 μg/100 ml), whereas in most other patients 18-OH-DOC levels did not exceed 1 μg/100 ml. Because venography demonstrated bilateral nodular hyperplasia of the adrenal gland, patient number 3 underwent bilateral total adrenalectomy following a short-lived attempt to reduce his blood pressure with dexamethasone in a dose of 0.75 mg twice daily. Dexamethasone administration was carried out for only 3 weeks. Blood pressure actually increased during the first week of dexamethasone therapy from 180/120 mm Hg to 202/130 mm Hg. After about 2.5 weeks, blood pressure began to fall, and at the discontinuation of therapy it was 146/100 mm Hg. After preparation with spironolactone, a bilateral total adrenalectomy was

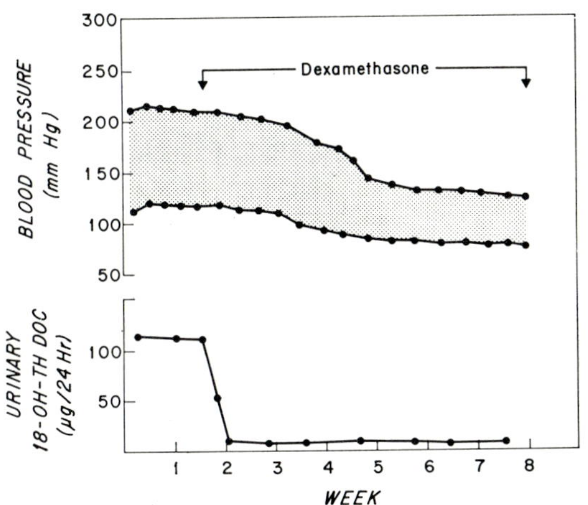

FIG. 15. Effect of dexamethasone therapy in a 37-year-old man with elevated urinary 18-OH-TH-DOC, hypertension, and suppressed PRA.

performed and the patient remains normotensive more than 22 months after the operation.

Patient number 4, a 37-year-old male with hypertension, hypokalemic alkalosis, suppressed PRA, and a low aldosterone secretion, underwent adrenal venography, and the adrenals appeared to be perfectly normal in size. The adrenal venous effluent contained 6.1 μg of 18-OH-DOC per 100 ml of plasma. Urinary 18-OH-TH-DOC was elevated at 115 μg/24 hours. TH-DOC excretion in urine was 23 μg/24 hours. After a series of control observations, dexamethasone 0.5 mg was begun and continued for a period of 8 weeks. In Fig. 15 is plotted the blood pressure response and urinary 18-OH-TH-DOC excretion before and during dexamethasone therapy. 18-OH-TH-DOC excretion was promptly suppressed with dexamethasone, and, within 3 weeks of therapy, blood pressure began to decrease. Therapy was discontinued at the end of the eighth week, and the patient became hypertensive 6 weeks after the cessation of therapy. He was again treated with dexamethasone for a prolonged period during which he remained normotensive. He was then placed on spironolactone for maintenance to the present. After 1.5 weeks of therapy, the patient had lost nearly 2 kg of body weight. Urinary sodium was increased over the control period.

Patient 5, a hypogonadotropic, hypogonadal male (studied by Dr. C. A. Paulsen of Seattle), has been treated with depotestosterone since 1965. He was mildly hypertensive at that time. Hypertension worsened; hypokalemia was induced with a thiazide diuretic and persisted after its discontinuance. He was demonstrated to have hyporesponsive renin activity to sodium depletion. Urinary aldosterone secretion was within the normal range, and urinary 18-OH-TH-DOC excretion was significantly elevated. Spironolactone reduced his blood pressure to normal. He then was allowed to become hypertensive and was placed on dexamethasone. During the first week and one half, blood pressure increased, but by the third week it was significantly reduced and he eventually became normotensive. No other corticosteroid metabolite was found to be altered in his urine. Urinary TH-DOC excretion was 13 μg/24 hours.

Patient 7, an 8-year-old girl (studied by Dr. Gerald Kerrigan of Marquette University), had spontaneous hypertensive episodes and hypertensive paroxysms could be induced by prolonged infusions of ACTH. Spironolactone prevented both the spontaneously occurring and ACTH-induced attacks. Urinary steroid patterns were measured by a number of investigators and no abnormal steroid was demonstrated. Urinary aldosterone excretion was normal, urinary TH-DOC excretion was 20 μg/24 hours. The patient was studied before the advent of PRA as a diagnostic procedure. Glucocorticoids prevented spontaneous hypertensive episodes. Eventually a bilateral total adrenalectomy was performed with the finding of hyperplasia. After adrenalectomy the patient had no further episodes of hypertension. Several years later, frozen specimens of this patient's urine were examined for urinary 18-OH-TH-DOC, which was found to be elevated.

Patient 8 (studied by Dr. Caulie Gunnels, Duke University) underwent bilateral adrenal exploration because of hypertension and suppressed PRA. The right adrenal gland, which contained two clinically significant adenomas, was removed, and the patient remained hypertensive (Gunnels *et al.*, 1970). Increased 18-OH-TH-DOC excretion was found preoperatively. Urinary aldosterone and TH-DOC were in the normal range. The possibility that the remaining adrenal continues to produce a mineralocorticoid must be considered.

Patient 9 (reported by Drs. Rovner, Conn, and Vader), a 60-year-old hypertensive man with hyporesponsive PRA and normal amounts of urinary aldosterone, underwent adrenal venography which demonstrated a left adrenal adenoma. A left adrenalectomy was performed, and the adenoma was sectioned and incubated and found to convert labeled progesterone to 18-OH-DOC at a rate ten times as fast as that observed in several other incubated tumors (Rovner *et al.*, 1971). Urinary 18-OH-TH-DOC excretion measured in this laboratory was elevated. The patient remained hypertensive after the unilateral adrenalectomy.

Patient 10 (studied by Dr. C. Grim), a 44-year-old male with suppressed PRA, normal aldosterone excretion, and elevated urinary 18-OH-TH-DOC excretion, was given dexamethasone for several weeks without any reduction in blood pressure. Bilateral total adrenalectomy was performed, and adrenocortical hyperplasia was demonstrated. The patient is now normotensive.

In nine of these ten patients exhibiting suppressed PRA, normal or low aldosterone secretion, and normal TH-DOC excretion, only one steroid abnormality could be found, excessive excretion of 18-OH-TH-DOC in urine. Dexamethasone suppression of 18-OH-DOC secretion was carried out in five of these patients. Four of the five patients responded, three of these became normotensive.

F. 18-OH-DOC SECRETION IN PRIMARY ALDOSTERONISM

Twenty-two patients were studied. All these patients were found to have significantly increased aldosterone secretion, suppressed PRA, spontaneous or easily inducible hypokalemia, and reversal of these manifestations during the administration of large doses of spironolactone. Diagnosis of primary aldosteronism was confirmed by the presence at operation of solitary adenoma in 16 of the patients and bilateral nodular hyperplasia in six of the patients. 18-OH-TH-DOC excretion in urine is plotted in Fig. 14. Although mean 18-OH-TH-DOC excretion was less than that of patients with suppressed PRA hypertension, it was significantly greater than that in healthy subjects ($P < 0.005$).

A 42-year-old male with hypertension, studied by Dr. David Kem (Tripler General Hospital, Hawaii), was found to have elevated urinary aldosterone and plasma aldosterone levels, hyporesponsive PRA, and evidence for adrenal hyperplasia by venography. Urinary 18-OH-TH-DOC excretion was increased at 52 μg/24 hours. With dexamethasone suppression, the blood pressure fell dramatically as did the urinary

18-OH-TH-DOC (13 μg/24 hours) and plasma aldosterone became un-measurable. Glucocorticoid-suppressible aldosteronism is probably the diagnosis in this instance, but it is of interest that 18-OH-DOC secretion is also increased in this condition.

In summary, excess 18-OH-DOC secretion has been found to occur in four varieties of suppressed PRA hypertension including: Cushing's syndrome with excessive ACTH, 17α-hydroxylase deficiency, primary aldosteronism and suppressed or hyporesponsive PRA hypertension of unknown etiology. In Cushing's syndrome with excessive ACTH and in 17α-hydroxylase deficiency, DOC secretion is also increased, and this probably accounts for the larger share of mineralocorticoid activity. In primary aldosteronism, aldosterone secretion is usually sufficiently increased to account for all the manifestations of mineralocorticoid ex-cess. Of patients with suppressed PRA hypertension of unknown etiol-ogy, approximately 15% excreted markedly abnormal quantities of 18-OH-TH-DOC in urine as the sole alteration from normal steroid pattern. The suppression of ACTH secretion with dexamethasone is as-sociated with a decline in both blood pressure and 18-OH-TH-DOC excre-tion in urine, in four of five patients studied. Correlation of these events immediately suggests that 18-OH-DOC has some role in the development of hypertension. In limited cases, morphological abnormalities of the adrenal producing excessive 18-OH-DOC were found ranging from hy-perplasia to solitary tumor. It is difficult to understand how 18-OH-DOC could have causal importance in suppressed PRA hypertension because 18-OH-DOC is a weak mineralocorticoid, and, although secretion is raised in some patients with suppressed PRA hypertension, the level still is not sufficient to explain the manifestations of mineralocorticoid activity observed. It is possible that 18-OH-DOC serves as a biological marker for some unidentified mineralocorticoid of high potency.

The possibility that 18-OH-DOC secretory excess in patients with suppressed PRA hypertension is some variant of Biglieri's syndrome of 17α-hydroxylase deficiency has to be considered. However, the lack of excessive DOC secretion mitigates against this interpretation. A par-tial 17α-hydroxylase deficiency might exist which would increase the secretion of 18-OH-DOC sufficiently to produce hypertension without overloading the 18-hydroxylase enzyme system of the zona fasciculata, thereby avoiding the buildup of precursor, DOC. Such a scheme is depicted in Fig. 16. The evidence for partial 17α-hydroxylase deficiency is lacking in this study. A single patient with suppressed PRA and exces-sive 18-OH-DOC secretion was also found to have hypogonadism. The possibility that the hypogonadism is related to 17α-hydroxylase deficiency in the testes is unlikely because gonadotropin secretion was also diminished. Appropriate further studies of these patients might include maximal stimulation with ACTH to make more obvious the presence of 17α-hydroxylase deficiency.

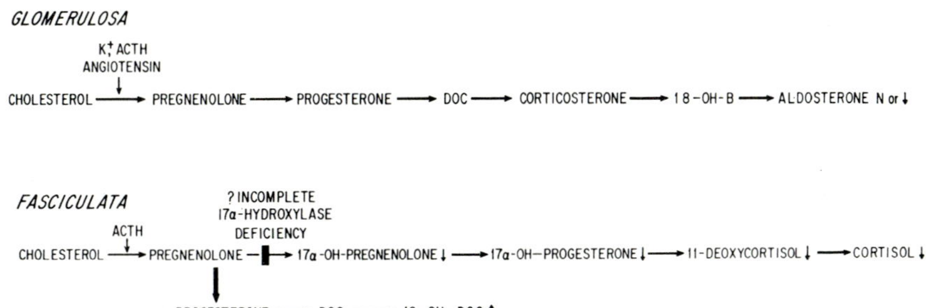

FIG. 16. Theoretical sequence of steroidogenesis in human hypertension in partial 17α-hydroxylase deficiency with elevated 18-OH-DOC secretion and normal DOC secretion.

VI. 18-OH-DOC as a Precursor

A. INTRODUCTION

The possibility that 18-OH-DOC is a precursor of some more active steroid was immediately suggested by analogy to 18-OH-B. Nicolis and Ulick (1965) demonstrated the conversion of 18-OH-B to aldosterone by adrenal tissue, although the percentage of conversion of labeled 18-OH-B was small. However, Pasqualini (1964) used much lower concentrations of 18-OH-B and demonstrated a marked improvement in the yield of label incorporated into aldosterone. Neher and Wettstein (1956) isolated a compound from adrenal extracts in which the 18-hydroxyl of 18-OH-DOC had been oxidized to a carboxyl group [Δ^4-3-keto-pregnene-20,20,21-triol-18-acid lactone $(18 \rightarrow 20)$]. 18-OH-B's conversion to aldosterone in the glomerulosa is analogous to the conversion of 18-OH-DOC to a hypothetical 11-deoxyaldosterone in the fasciculata. Whether this conversion is even theoretically possible is not known. To this end, labeled 18-OH-DOC was prepared and incubated with sectioned rat and human adrenals.

B. INCUBATION—RAT ADRENALS

Male Sprague-Dawley strain rats weighing 150–200 gm were killed by decapitation, the adrenals were removed, immediately quartered or homogenized in Krebs-Ringer bicarbonate buffer, and incubated for 2 hours at 37°C under a 95% oxygen–5% carbon dioxide gas mixture in the presence of 1 μCi of 1,2-^{3}H-labeled 18-OH-DOC. The adrenals from 10 rats were used for each incubation. The methylene chloride extract of the medium was partitioned between aqueous ethanol and cyclohexane and chromatographed in the thin-layer system butyl acetate:formamide:water (100:5:5) with aldosterone and cortisol standards. The area migrating with cortisol was eluted for 18-OH-DOC and assayed for radioactivity. The area from the origin to 0.5 cm above aldosterone was eluted and oxidized with periodic acid and chromatographed on paper in cyclohexane:benzene:methanol:water (5:3:5:1). The areas corresponding to aldosterone and 18-OH-B lactone were eluted and chromatographed on silica gel in the system benzene:ethanol (92:8).

TABLE IX

Mobilities of Some Steroids in Benzene: Acetone: Water (2:1:2) System

Steroid	$R_{6\beta\text{-OH-E}}$
6β-OH-E	1.00
6β-OH-F	0.42
18-OH-B	1.35
Aldosterone	1.87
Cortisol	2.07
18-OH-DOC	2.20
Polar conversion product	0.69

18-OH-B lactone has a mobility relative to aldosterone lactone of 0.65 in this system. The areas were eluted and counted. An aliquot of the origin to aldosterone zone off the butyl acetate system was chromatographed on paper (Whatman No. 2) in a Bush-type system in benzene:acetone:water (2:1:2). Standards of aldosterone, 18-OH-B (isolated from previous rat adrenal incubations) 6β-hydroxycortisone (6β-OH-E) and 6β-hydroxycortisol (6β-OH-F) were applied. The radioactivity scan indicated two peaks of ^{3}H running with aldosterone and 18-OH-B. See Table IX for relative mobilities. In Table X is shown the percentage conversion of ^{3}H-labeled 18-OH-DOC to aldosterone and 18-OH-B. The studies of Baniukiewicz *et al.* (1968) suggest that 18-OH-DOC is produced largely in the fasciculata and could not serve as a major precursor of aldosterone *in vivo*. Our studies confirm that 18-OH-DOC does not appear to be a quantitatively important precursor to aldosterone under normal conditions.

C. HUMAN ADRENAL INCUBATION

Surgically obtained human adrenal slices from patients with breast carcinoma were incubated (as above) with 1,2-^{3}H-labeled 18-OH-DOC and 500 μg of 18-OH-DOC within 1 hour after removal. One gram of tissue per 10 ml of Krebs buffer was used. Incubation of ^{3}H-labeled 18-OH-DOC with only Krebs buffer served as a control to determine whether 18-OH-DOC was altered by the incubation procedure. Aldosterone, 18-OH-B, and 18-OH-DOC were isolated as for the rat adrenal incubations. A band of radioactivity, more polar than 18-OH-B in the paper system benzene:acetone:water (2:1:2) ($R_{6\beta\text{-OH-E}}$ 0.69) was

TABLE X

Conversion of ^{3}H-18-OH-DOC to Other Compounds in Adrenal Incubations

Compound	% Conversion[a]	
	Rat	Human
Aldosterone	0.5–1.0	0.2–0.5
18-OH-corticosterone	2.5–5.0	1.2–3.0
Hydroxylated conversion product of 18-OH-DOC	Not found	5–10
Unchanged 18-OH-DOC	90	65

[a] Values not corrected for procedural losses.

TABLE XI

Distribution of ^{3}H in Extracts of Adrenals Incubated with 1,2-^{3}H-18-OH-DOC

Extract	Percent of amount incubated
Petroleum ether	3.3
CH_2Cl_2 (pH 7.4)	80.0
CH_2Cl_2 (pH 6)	3.6
EtAc of pH 6 after CH_2Cl_2	10.0
Remaining aqueous	3.1

present, and the isolation procedure was changed to enable maximum recovery of this material. The distribution of ^{3}H in extracts of the medium of human adrenal incubations with 1,2-^{3}H-labeled 18-OH-DOC was studied first. Analyses of radioactivity in the medium of incubations accounted for only 65–70% of the total ^{3}H added to the flask. Homogenization of the adrenal slices in water followed by extraction into ethanol resulted in the recovery of the remaining label. Adrenal slices were therefore homogenized in the medium and the resultant homogenate subjected to various extraction procedures. These results are shown in Table XI. Petroleum ether extraction of the homogenate removed 3.3% of the incubated label. The aqueous phase (pH 7.4) was extracted with methylene chloride four times with five volumes. The major amount of ^{3}H was recovered in this extract (80%). The aqueous phase was then adjusted to pH 6 with hydrochloric acid and reextracted with methylene chloride (3.6% of ^{3}H) followed by ethyl acetate. Ten percent of the tritium was found in the ethyl acetate extract with 3.1% remaining in the spent aqueous phase.

D. Isolation and Partial Identification of Polar Conversion Product of 18-OH-DOC

The various extracts of adrenal incubations were pooled and after partitioning between 70% ethanol:petroleum ether were chromatographed on silica gel in benzene:acetone:water (80:25:0.5) with corticosterone as the reference standard. The area migrating from the origin to the top of the corticosterone spot was eluted with ethanol. This system serves to remove the more nonpolar materials. 18-OH-DOC runs just slightly slower ($R_{corticosterone}$ 0.72) than corticosterone. The eluate was chromatographed on Whatman No. 2 paper in benzene:acetone:water (2:1:2) (Table IX). The radioactive scan indicated, in addition to zones of tritium migrating with 18-OH-B, aldosterone, and 18-OH-DOC, a band of ^{3}H running at $R_{6\beta\text{-OH-E}}$ 0.69. A very small amount of tritium was also found migrating with 6β-OH-E. The radioactive material migrating between 6β-OH-E and 6β-OH-F absorbed UV light and did not reduce blue tetrazolium. The material was eluted with ethanol and chromatographed on Whatman No. 2 paper in the system chloroform:acetone:water (2:1:2). In this system, the polar conversion product has an $R_{6\beta\text{-OH-F}}$ 1.68. The eluate was chromatographed on silica gel GF in methylene chloride:methanol: acetone:water (70:5:25:0.7) with aldosterone standard and the UV absorbing material migrating at $R_{aldosterone}$ 0.59 was eluted. The UV spectrum in methanol from 220 to 330 nm demonstrated maximal absorption at 240 nm. An aliquot of the eluate was oxidized with periodic acid and chromatographed in benzene:ethanol (92:8) on a silica gel thin-layer plate with aldosterone lactone standard. The unknown oxidized material had an $R_{aldo\ lactone}$ 0.76 in this system. The lactone was acetylated in acetic acid anhydride:pyridine (2:1) overnight at room temperature and rechromatographed in the benzene:ethanol system. The acetylated product migrated with an $R_{aldo\ lactone}$ 0.99 and $R_{18\text{-OH-DOC 21-acetate}}$ 0.89. The parent material gave the typical yellow color when allowed to react with the Porter-Silber reagent, peaking at 410 nm. When allowed to sit overnight in 0.1 N methanolic acetic acid, a Porter-Silber UV positive compound was formed which migrated approximately twice as fast as the parent compound. This may be similar to the reaction 18-OH-DOC undergoes to form the "L form" (Dominguez, 1965). Gas–liquid chromatography of the trimethylsilyl ethers (TMSE) according to Pinkus *et al.* (1971) on OV-101 on 100/120 mesh supelcoport column 6 feet $\times$ 4 mm with column temperature 265°C at a N_2 flow of 60 ml/minute resulted in the following retention times. 18-OH-DOC TMSE, 3 peaks, 12-minute shoulder, 12.4-minute major and 14-minute minor; conversion product TMSE, 14-minute minor, 16-minute major, 17.5-minute shoulder; conversion product lactone TMSE, 12.7 minutes; conversion product lactone acetate, 16.5 minutes; conversion product acetate, decomposition.

E. Mass Spectral Analysis of Derivatives of Polar Conversion Product

(Gas chromatography–mass spectrometry analysis through the courtesy of Dr. Lewis Engel, Massachusetts General Hospital, Boston, Massachusetts)

Mass spectra were obtained on the material emerging from a 6 foot, 2% OV-1 gas chromatographic column with N_2 flow of 27 ml/min and

TABLE XII

Mass Spectral Studies of the Adrenal Conversion Product of 18-OH-DOC

Product[a]	Mass per electron (m/e)	
	Molecular ion	Base peak[b]
Conversion product TMSE	578	475
Conversion product lactone TMSE	402	387
Conversion product MO TMSE	607	607
Conversion product lactone MO TMSE	431	151
18-OH-DOC lactone MO	343	312

[a] TMSE = trimethylsilyl ether; MO, methoxime.
[b] Base peak arbitrarily chosen as largest ion peak $> m/e$ 150.

flash heater at 260°C. The column was maintained for 5 minutes at 220°C, then programmed at 3°/min to 260°C. Mass spectra were obtained on the following derivatives of the polar conversion product; trimethylsilyl ether (TMSE), TMSE-3-methoxime (MO), lactone MO, and the lactone MO TMSE. The 3-MO of 18-OH-DOC lactone was run as a reference standard. Derivatives were formed in the gas phase according to the method of Engel *et al.* (1970) and Engel (1971). Molecular ions and base peaks of these derivatives are summarized in Table XII.

1. 18-OH-DOC Lactone MO

The mass spectrum of this standard compound gives the correct molecular ion at m/e 343 with the base peak at m/e 312 (M-31) representing loss of the 3-methoxime. This is different from the base peak in the spectrum of the lactone obtained by direct probe in which ketene is lost from the A-ring. The ion at m/e 124 in the lactone spectrum

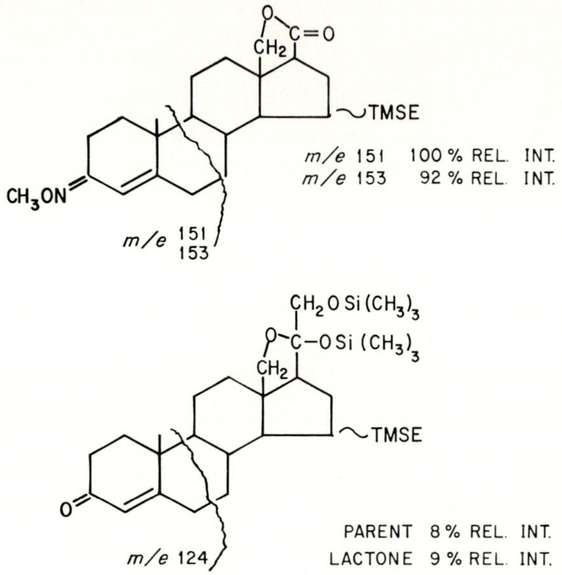

FIG. 17. Mass spectral evidence for hydroxyl group location in 18-OH-DOC derivative.

is comparable to the ion at m/e 153 in the MO derivative (124 + 29) representing B-ring cleavage (Engel, 1971; Dray and Weliky, 1970). This type of cleavage is shown in Fig. 17.

2. Conversion Product TMSE

A very weak molecular ion peak at m/e 578 was present with the base peak at m/e 475 (M-103) in the spectrum of this derivative of the polar conversion product of 18-OH-DOC. The base peak is indicative of bond cleavage between C-20 and C-21 with loss of $CH_2 = OSi(CH_3)_3$. Intense ions were observed at m/e 129, 103, 91, 75, and 73 and probably are characteristic of trimethylsilyl ethers. The 124 ion was not very intense, but present at 8% relative abundance.

3. Conversion Product MO TMSE

The spectrum of this derivative gave an intense M^+ at 607 which was also the base peak. The ion at m/e 155 was of equal intensity to the base peak. The interpretation of this fragment ion is uncertain.

4. Conversion Product Lactone TMSE

The molecular ion at m/e 402 was of small intensity (1% of base) in this spectrum. The base peak at m/e 387 (M-15) represents loss of CH_3. The rest of the spectrum was rather unimpressive with the exception of ions (approximately 50% of base or greater) at m/e 129, 105, 93, 91, 75, and 73 similar to the spectrum of the conversion product TMSE. The ion at m/e 124 was 9% of the base.

5. Conversion Product Lactone MO TMSE

The spectrum of this derivative gave a good M^+ at 431 (22% of base) with a base peak at m/e 151. Probably the most significant features of this spectrum are the abundant ions at m/e 153 (92% of base) and at 151 (base) representing cleavage of the B-ring.

F. Discussion

18-OH-DOC is converted to a more polar material in addition to aldosterone and 18-OH-corticosterone in incubations with human adrenal glands. We were unable to isolate this material from rat adrenal gland incubations. This material absorbs UV light at 240 nm (possesses a Δ^4-3-ketone), gives a positive Porter-Silber and negative blue tetrazolium reaction, undergoes reaction in methanolic 0.1 N acetic acid similar to 18-OH-DOC and reacts with periodic acid to yield a more nonpolar material (presumably the lactone). The lactone readily acetylates and forms a TMSE to give a material which migrates differently from the unacetylated lactone in thin-layer systems and gas-liquid chromatography. This evidence, together with the mass spectra molecular ions of the various derivatives, indicates the compound is a hydroxylated derivative of 18-OH-DOC. In the mass spectra of the TMSE and lactone TMSE, the typical B-ring cleavage is evident by the 124 peak. This is further substantiated by the intense 153 and 151 peaks in the lactone MO TMSE derivative (Fig. 17). On this basis, carbons at positions 1, 2, 6, and 19 may be ruled out as the site of the additional hydroxyl group. Carbons 7, 12, 15, and 16 are the only remaining carbon atoms in the steroid molecule in which a hydroxyl group could exist which is capable of mild acetylation and formation of the TMSE. The actual location and configuration of the additional hydroxyl group will have to await isolation of sufficient amounts of the material for additional chemical and physical analysis. The tentative structure of this compound is shown in Fig. 18.

The stimulatory effect of the hydroxylated derivative of 18-OH-DOC on active sodium transport was assessed by measuring the short-circuit current of the isolated urinary bladder of the toad. Drs. George A. Porter and Goeffrey Sharp independently tested this compound for effect and found no activity in stimulating active sodium transport. It is possible, though not very likely, that the hydroxylated derivative of 18-OH-DOC behaves differently in man and animals than in the isolated toad bladder.

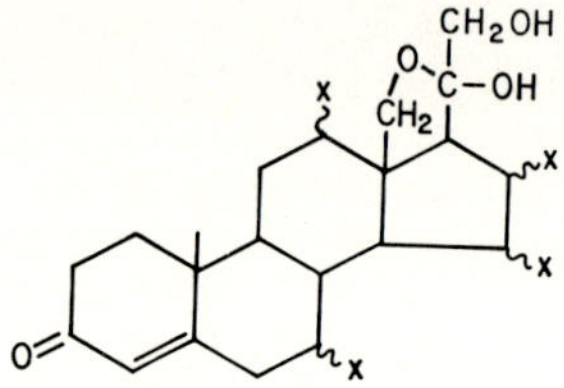

FIG. 18. Tentative structure of hydroxylated derivative of 18-OH-DOC. Additional OH on carbons at positions 7, 12, 15, or 16.

VII. Concluding Remarks

18-OH-DOC, the most abundant mineralocorticoid secreted by the rat adrenal cortex, was found to be secreted excessively after the establishment of adrenal regeneration hypertension in rats. Secretion was negligible at the onset of hypertension and therefore cannot be implicated in the development of adrenal regeneration hypertension. On the other hand, DOC secretion was restored to normal at 3 weeks, the time of onset of adrenal regeneration hypertension. It is possible that hypersecretion of 18-OH-DOC may be involved in the perpetuation of hypertension in the enucleate animals.

18-OH-DOC was isolated from human adrenal vein blood and its identification established. In healthy, adult human subjects the secretion of 18-OH-DOC is quantitatively similar to the secretion rates of aldosterone and DOC. Indirect evidence was provided in man, dog, and rat that 18-OH-DOC is mainly a secretory product of the zona fasciculata of the adrenal cortex. The secretion of 18-OH-DOC appears to be primarily ACTH dependent, but small increments in 18-OH-DOC secretion may possibly be associated with activation of the zona glomerulosa by sodium depletion. In animals and man, angiotensin evokes no increase in 18-OH-DOC secretion.

Excessive secretion of 18-OH-DOC was demonstrated in four forms of hypertension with suppressed PRA. In 17α-hydroxylase deficiency,

Cushing's syndrome, primary aldosteronism, and, in a small number of patients with essential hypertension with suppressed PRA, 18-OH-DOC secretory excess or excessive excretion of urinary 18-OH-TH-DOC was found. In some of the patients with essential hypertension and suppressed PRA with hypersecretion of 18-OH-DOC, dexamethasone administration resulted in a significant reduction in blood pressure associated with a fall in the urinary excretion of 18-OH-TH-DOC. 18-OH-DOC possesses relatively modest mineralocorticoid activity, and it is difficult to accept 18-OH-DOC as the cause of mineralocorticoid hypertension in those individuals studied.

Studies on the precursor role of 18-OH-DOC showed it to be converted to a more highly hydroxylated steroid which exhibits less mineralocorticoid activity. The possibility that another derivative of 18-OH-DOC possesses significant activity still exists.

ACKNOWLEDGMENTS

We are very grateful to Miss Elvira Gisoldi and Miss Nancy Smith for their expert technical assistance in studies in adrenal regeneration, and to Messrs. Barry Portnoy and Nelson Evans, and Dr. Marlys St. Cyr for their excellent work in steroid analysis.

This work was supported in part by Grants-in-Aid AM-12027-04, TO1-AM-05446-06, and 2-PO2-AM-08657-07 from the National Institutes of Health.

REFERENCES

Aoki, K. (1963). *Jap. Heart J.* **4**, 443.

Aoki, K., Tankawa, H., Fujinami, T., Miyazaki, A., and Hashimoto, Y. (1963). *Jap. Heart J.* **4**, 426.

Arai, R., and Tamaoki, B. (1967). *J. Endocrinol.* **39**, 453.

August, J. T. (1961). *Biochim. Biophys. Acta* **48**, 203.

Baer, L., Knowlton, A. I., Kirshman, J. D., and Laragh, J. H. (1971). *Endocrinology* **88**, Suppl., Abstr. No. 143.

Baniukiewicz, S., Brodie, A., Flood, C., Motta, M., Okamoto, M., Tait, J. F., Tait, S. A. S., Blair-West, J. R., Coghlan, J. P., Denton, D. A., Goding, J. R., Scoggins, B. A., Wintour, E. M., and Wright, R. D. (1968). *In* "Functions of the Adrenal Cortex" (K. W. McKerns, ed.), Vol. 1, pp. 153–221. Appleton, New York.

Biglieri, E. G., Herron, M. A., and Brust, N. (1966). *J. Clin. Invest.* **45**, 1946.

Biglieri, E. G., Slaton, P. E., Schambelan, M., and Kronfield, S. J. (1968). *Amer. J. Med.* **45** ,170.

Birmingham, M. K., and Ward, P. J. (1961). *J. Biol. Chem.* **236**, 1661.

Birmingham, M. K., MacDonald, M. L., and Rochefort, J. G. (1968). *In* "Functions of the Adrenal Cortex" (K. W. McKerns, ed.), Vol. 2, pp. 647–689. Appleton, New York.

Bockus, H. L. (1965). "Gastroenterology," Vol. 3, p. 109. Saunders, Philadelphia, Pennsylvania.

Brogi, M. P., and Pellegrino, C. (1959). *J. Physiol. (London)* **146**, 165.

Budzikiewicz, H., Djerassi, C., and Williams, D. H. (1964). "Structure Elucidation of Natural Products by Mass Spectrometry," Vol. 2, p. 87. Holden-Day, San Francisco, California.

Bush, I. E.   (1951).   *J. Physiol. (London)* **115**, 12P.

Carballeira, A., and Venning, E. H.   (1964).   *Steroids* **4**, 329.

Channick B. J. Adlin, E. V., and Marks, A. D.   (1969).   *Arch. Intern. Med.* **123**, 131.

Conn, J. W.   (1955).   *J. Lab. Clin. Med.* **45**, 3.

Conn, J. W., Cohen, E. L., and Rovner, D. R.   (1964).   *J. Amer. Med. Ass.* **190**, 213.

Conn, J. W., Rovner, D. R., Cohen, E. L., and Nesbit, R. M.   (1966).   *J. Amer. Med. Ass.* **195**, 21.

Cortés, J. M., Peron, F. G., and Dorfman, R. I.   (1963).   *Endocrinology* **73**, 713.

de Nicola, A. F., and Birmingham, M. K.   (1968).   *J. Clin. Endocrinol. Metab.* **28**, 1380.

Dominguez, O. V.   (1965).   *Steroids Suppl.* **2**, 29.

Dray, F., and Weliky, J.   (1970).   *Anal. Biochem.* **34**, 387.

Eilers, E. A., and Peterson, R. E.   (1964).   *In* "Aldosterone—A Symposium" (E. E. Baulieu and P. Robel, eds.), pp. 251–264. Blackwell, Oxford.

Engel, L. L.   (1971).   Personal communication.

Engel, L. L., Neville, A. M., Orr, J. C., and Raggatt, P. R.   (1970).   *Steroids* **16**, 377.

Fortier, C., and de Groot, J.   (1959).   *Amer. J. Physiol.* **196**, 589.

Galaburda, A., Dale, S. L., and Melby, J. C.   (1972).   In preparation.

Gaunt, R., Renzi, A. A., Gisoldi, E., and Howie, N. C.   (1967).   *Endocrinology* **81**, 1331.

Gaunt, R., Gisoldi, E., Herkner, J., Howie, N., and Renzi, A. A.   (1968).   *Endocrinology* **83**, 927.

Gaunt, R., Gisoldi, E., Smith, N., and Giannina, T.   (1969).   *Endocrinology* **84**, 1193.

Genard, P., Palem-Vliers, M., Coninx, P., and Margoulies, M.   (1968).   *Steroids* **12**, 763.

Gunnells, J. C., Grim, C. E., Robinson, R. R., and Wildermann, N. M.   (1967).   *Arch. Intern. Med.* **119**, 232.

Gunnells, J. C., McGuffin, W. L., Robinson, R. R., Grim, C. E., Wells, S., Silver, D., and Glenn, J. F.   (1970).   *Ann. Intern. Med.* **73**, 901.

José, A., Crout, J. R., and Kaplan, N. M.   (1970).   *Ann. Intern. Med.* **72**, 9.

Kagawa, C. M., and Pappo, R.   (1962).   *Proc. Soc. Exp. Biol. Med.* **109**, 982.

Knigge, K. M.   (1961).   *Anat. Rec.* **141**, 145.

Kraulis, I., and Birmingham, M. K.   (1965).   *Can. J. Biochem.* **43**, 1471.

Küchel, O.   (1971).   *Circ. Res.* **28/29**, Suppl. 2, 150.

Küchel, O., Fishman, L. M., Liddle, G. W., and Michelakis, A.   (1967).   *Ann. Intern. Med.* **67**, 791.

Lewbart, M. L., and Mattox, V. R.   (1961).   *Anal. Chem.* **33**, 559.

Louis, W. J., Tabei, R., Spector, S., and Sjoerdsma, A.   (1969).   *Circ. Res.* **24**, Suppl. 1, 93.

Mattox, V. R., and Lewbart, M. L.   (1958).   *Arch. Biochem.* **76**, 362.

Melby, J. C., Spark, R. F., Dale, S. L., Egdahl, R. H., and Kahn, P. C.   (1967).   *New Engl. J. Med.* **277**, 1050.

Melby, J. C., Wilson, T. E., and Dale, S. L.   (1970).   *J. Clin. Invest.* **49**, 64a. Abstr.

Melby, J. C., Dale, S. L., and Wilson, T. E. (1971). *Circ. Res.* **28/29**, Suppl. 2, 143.

Murphy, B. E. P. (1967). *J. Clin. Endocrinol. Metab.* **27**, 973.

Neher, R., and Wettstein, A. (1956). *Helv. Chim. Acta* **39**, 2062.

Nicolis, G. L., and Ulick, S. (1965). *Endocrinology* **76**, 514.

Okamoto, K., and Aoki, K. (1963). *Jap. Circ. J.* **27**, 282.

Okamoto, K., Aoki, K., Nosaka, S., and Fukushima, M. (1964). *Jap. Circ. J.* **28**, 943.

Pasqualini, J. R. (1964). *Nature (London)* **201**, 501.

Péron, F G. (1961). *Endocrinology* **69**, 39.

Pinkus, J. L., Charles D., and Chattoraj, S. C. (1971). *J. Biol. Chem.* **246**, 633.

Porter, G. A., and Kimsey, J. (1971). *Endocrinology* **89**, 353.

Quesenberry, R. O., and Ungar, F. (1964). *Anal. Biochem.* **8**, 192.

Race, G. J., and Wu, H. M. (1964). *Gen. Comp. Endocrinol.* **4**, 199.

Rapp, J. P. (1964). *Endocrinology* **75**, 326.

Rapp, J. P. (1969). *Endocrinology* **84**, 1409.

Rapp, J. P. (1970). *Endocrinology* **86**, 668.

Rapp, J. P., and Dahl, L. K. (1971a). *Endocrinology* **88**, 52.

Rapp, J. P., and Dahl, L. K. (1971b). *Circ. Res.* **28/29**, Suppl. 2, 153.

Rovner, D. R., Conn, J. W., and Vader, S. D. (1971). *Endocrinology* **88**, Suppl., Abstr. No. 104.

Saffran, M., and Schally, A. V. (1955). *Endocrinology* **56**, 523.

St. Cyr, M. J., Sancho, J. M., Dale, S. L., and Melby, J. C. (1972). In preparation.

Sheppard, H., Swenson, R., and Mowles, T. F. (1963). *Endocrinology* **73**, 819.

Silber, R. H., and Porter, C. C. (1954). *J. Biol. Chem.* **210**, 923.

Simpson, E. R., Cooper, D. Y., and Estabrook, R. W. (1969). *Recent Progr. Horm. Res.* **25**, 523.

Skelton, F. R. (1955). *Proc. Soc. Exp. Biol. Med.* **90**, 342.

Skelton, F. R. (1956). *Arch. Intern. Med.* **98**, 449.

Skelton, F. R. (1958). *Endocrinology* **62**, 365.

Snedecor, G. W. (1956). "Statistical Methods," Iowa State College Press, Ames, Iowa.

Sokabe, H. (1965). *Nature (London)* **205**, 90.

Sokabe, H. (1966). *Jap. J. Physiol.* **16**, 380.

Spark, R. F., and Melby, J. C. (1968). *Ann. Intern. Med.* **69**, 685.

Spark, R. F., and Melby, J. C. (1971). *Ann. Intern. Med.* **75**, 831.

Sturtevant, F. M. (1959). *Endocrinology* **64**, 299.

Tait, J. F., Tait, S. A. S., Little, B., and Laumas, K. R. (1961). *J. Clin. Invest.* **40**, 72.

Tait, S. A. S., Tait, J. F., Okamoto, M., and Flood, C. (1967). *Endocrinology* **81**, 1213.

Vecsei, P., Lommer, D., and Wolff, H. P. (1968). *Experientia* **24**, 1199.

Woods, J. W., Liddle, G. W., Stant, E. G., Michelakis, A. M., and Brill, A. B. (1969). *Arch. Intern. Med.* **123**, 366.

Zaffaroni, A., Burton, R. B., and Keutman, E. H. (1950). *Science* **111**, 6.

DISCUSSION

R. Gaunt: Dr. Ronald Brown, Department of Medicine, Vanderbilt University, has been doing analogous studies in rats on DOC levels after adrenal enucleation. His work, like that of Dr. Melby's group, was done on blood samples which we supplied, and for that reason it suffers some of the same limitations. His emphasis has been on peripheral plasma levels of DOC and in that respect his results differ from the secretion rate data presented today. His methods have been published [R. Brown and C. Strott, *J. Clin. Endocrinol.* **32,** 74 (1971)] and involve a competitive binding procedure that can detect as little as 0.2 ng of DOC.

Variability in individual animals and on a day-to-day basis has bothered us and arises, I suspect, largely in our part of the shop. We try to do the same thing to a group of animals, and they end up with disconcertingly variable steroid levels. If, however, we adrenalectomize the rats and put in a pellet of DOC, the plasma DOC levels are then much more constant. This seems to show that rats do, indeed, vary greatly in the amount of steroids they secrete or at least in the amounts that circulate.

Nevertheless, certain general patterns have emerged, and I believe they will turn out to be consistent with those reported here today.

In the first place, and entirely in agreement with the Boston group and everyone else who has studied this problem, because other steroids may be involved the plasma levels of DOC are unmeasurable, i.e., less than 0.6 ng/ml, in the first few days after adrenal enucleation.

This is a question of particular interest to me because during this period enucleate animals cannot excrete a salt load and do not have the increased salt appetite characteristic of adrenalectomized animals. In other words, they behave as though they were under the influence of some potent mineralocorticoid. Whatever is there, it is dependent on ACTH and inhibited by spironolactone. I am getting ready to retire and despair that I will ever know what causes this phenomenon.

The question of broader interest concerns what causes, some weeks after enucleation, the phenomenon that the late Floyd Skelton, its discoverer, called adrenal regeneration hypertension (AHR). Those of you familiar with the subject are aware that there are fairly widespread opinions, and some evidence, supplied paticularly by the Skelton-Brownie group and by Rapp, that some excess of DOC secretion causes AHR.

As I would interpret the evidence on our samples, as read both in Nashville and Boston, that hypothesis is not proved but certainly is not excluded. Under the conditions as originally used by Skelton, in which animals are sensitized by unilateral nephrectomy and a high-salt diet, Dr. Melby's group notes that the secretion of DOC is the first to recover from the near zero levels to which all steroid output drops immediately after enucleation.

Figure A shows that Brown finds in the same type of animals an apparent elevation above normal of DOC plasma levels by 3 weeks after operation; that is the period during which a developing hypertension becomes clearly evident. To date, at least, this elevation is not apparent at 7 weeks despite the progression of the hypertension; the time course of events is being checked further. But at these later stages the Boston group finds that the 18-OH-DOC secretion rate rises.

Having found that plasma DOC levels of rats with AHR were elevated at 3 weeks, we next asked the question: Are these plasma levels of DOC sufficient to cause hypertension? To answer this question we measured plasma DOC in a group of adrenalectomized rats made hypertensive by implantation of either one or two 25-mg DCA pellets.

As can be seen from Table A, hypertension developed approximately 3 weeks after implantation, presumably solely as a result of the exogenous DOC. This table compares blood pressure and plasma DOC levels of rats with ARH to rats with DCA hypertension. Note that hypertension can develop at three weeks with plasma levels of DOC even less than those in ARH. Thus, the amount of DOC present in the plasma of rats with ARH is sufficient to cause the hypertension following adrenal enucleation.

Eric Hall [E. Hall, and S. Ayachio, *Can. J. Physiol. Pharmacol.* **49**, 139 (1971)] has shown that it takes very little DOC to trigger the advent of hypertension in

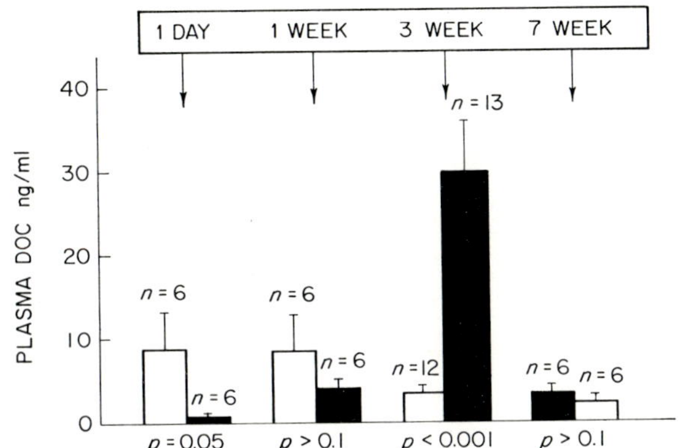

FIG. A. Adrenal regeneration hypertension in rats. Data of Dr. R. Brown and associates.

such sensitized animals. One could well imagine, therefore, whether or not high levels of DOC prevail only transiently or indefinitely, a situation in which a temporary relatively high output of DOC followed by increased 18-OH-DOC could initiate events leading, as in metacorticoid hypertension, to a self-sustaining process of hypertensive disease. This would be a rather complex chain of cause-and-effect events, but if hypertension of other types were due to simple single causes, we probably would not have had to spend so many years identifying them.

J. C. Melby: It would appear from the data you have presented that DOC initiates the hypertension and that if 18-OH-DOC excess has a role in adrenal regeneration hypertension, it only perpetuates it and intensifies it. The time

TABLE A

*Plasma DOC in Adrenal Regeneration Hypertension
(ARH) vs DCA Hypertension at 3 Weeks[a]*

	Blood pressure (mmHG)	Plasma DOC (ng/ml)
ARH	176	29.9 ± 6.1
DCA (25 mg)	149	7.8 ± 2.2
DCA (50 mg)	177	11.8 ± 1.4

[a] Data of Dr. R. Brown *et al.*

course of appearance of qualitative alterations in steroidogenesis appears to be crucial.

N. Varsano-Aharon: How often do you find overproduction of 18-hydroxy-deoxycorticosterone among the hypertensive patients with suppressed plasma renin activity? How many such patients have been studied, and how many of those studied have shown evidence of overproduction? We have, in Dr. Ulick's laboratory, thus far looked at a small number (5) of those patients, all of which were known to have responded very favorably to spironolactone therapy, but who did not show raised aldosterone production rates. The secretory rates of 18-OH-DOC were measured at least 4 weeks after discontinuation of the spironolactone, and all were below 140 μg/24 hours, which is within the normal range of our method based on the isolation of the tetrahydro derivative of the tritiated steroid injected intravenously and its quantitation as a ^{14}C-labeled etiolactone monoacetate.

J. C. Melby: Approximately 20% of patients with suppressed plasma renin activity and hypertension responsive to spironolactone exhibit increased excretion of 18-OH-TH-DOC in urine. It is possible that had you examined a larger number of patients 18-OH-DOC excess might have been found. On the other hand, we have studied such a highly selective population that it is conceivable that our experience is not representative.

R. Horton: In order to look directly at cause and effect in this problem, have you administered an equivalent amount of 18-OH-DOC secreted in essential hypertensives (low PRA) following renin activity in normal subjects? This would not only test the direct effect of the compound, but also cover the possibility of other active metabolites of 18-OH-DOC *in vivo*.

J. C. Melby: We have not administered 18-OH-DOC to normal subjects over a prolonged period to observe its effect on plasma renin activity because of an inadequate supply of material. We have injected 500 μg of 18-OH-DOC intravenously over a 24-hour period and found that it produced no significant alteration in electrolyte metabolism or plasma renin activity in a single human subject. The infusion of 18-OH-DOC was continued for only 8 hours of the 24-hour period of observation, and we did not feel that this was sufficient to permit conclusions as to physiological activity of 18-OH-DOC in the human.

R. Horton: Since angiotensin does not alter 18-OH-DOC secretion, the increased production of this compound in some hypertensives suggests that the steroid is derived from the fasciculata and is presumably under the influence of ACTH. What then is the correlation between dexamethasone effect on blood pressure in these patients and 18-OH-DOC secretion or excretion?

J. C. Melby: In four of five patients given suppressive doses of dexamethasone daily for a period of 3–5 weeks, a significant reduction in diastolic blood pressure was achieved. The reduction in diastolic blood pressure was correlated with diminution of excretion of 18-OH-TH-DOC in the urine of those patients tested. Plasma renin activity is no longer suppressed, and normal physiological responsiveness to the erect posture is observed during dexamethasone suppression.

F. G. Péron: Would you clarify for me again the potency of 18-OH-DOC and its mineralocorticoid activity with respect to aldosterone and DOC. Recently, G. A. Porter and J. Kimsey [*Endocrinology* **89**, 353 (1971)], using the short-circuit current (SCC) technique in the toad bladder to assess sodium transport, found the SCC pattern for 18-OH-DOC to be quite similar to that of the mineralocorticoids aldosterone, DOC, and 9α-flurocortisol. However, by the use of dissociation constants derived for 18-OH-DOC and DOC to assess relative mineralocorticoid potencies, it was found that 18-OH-DOC had only about $\frac{1}{7}$ the potency of aldosterone and $\frac{1}{5}$ that of DOC but was equal to that found for the glucocorticoid cortisol. It might be interesting to point out that when I had isolated crystalline 18-OH-DOC from biological material [*Endocrinology* **69**, 39 (1961)] and submitted this material for bioassay, one sample to Dr. Eugenia Rosemberg and another to a commercial testing laboratory, the 18-OH-DOC showed about 1 to $\frac{1}{40}$ the mineralocorticoid activity of DOC and aldosterone.

Finally, I would like to mention some results of Dr. Bernard Shapiro, a Fellow on my Steroid Training Program who has been working on the problem of hypertension in the adrenal enucleated rat. Briefly his approach has been to bring about adrenal regeneration hypertension in a group of unilaterally nephrectomized-adrenalectomized and adrenal-enucleated rats kept on a 1% salt solution regimen for 60 days. Rats in the control group, which were not enucleated, did not develop the hypertension. By preparing adrenal cell suspensions by trypsin and collagenase digestion of the adrenals [see I. D. K. Halkerston *et al., Fed. Proc. Fed. Amer. Soc. Exp. Biol.* **27**, 626 (1968); Sayers *et al. Proc. Soc. Exp. Biol. Med.* **136**, 619 (1971]] of the control as well as the hypertensive group, he was able to show differences in steroidogenic response to ACTH between these groups when the cells were incubated *in vitro* in the presence of this tropic hormone. More specifically, there was a marked depression in corticosterone synthesis in the cells derived from the hypertensive group as compared to those derived from the control group. This finding does not exactly agree with yours where corticosterone levels in the 40-day hypertensive rat were not different from those in the controls. It does, on the other hand, support the interpretation of the Buffalo group [*Endocrinology* **86**, 1085 (1970)] that there is a failure of the adrenal steroid enzyme 11β-hydroxylase which brings about a lowering of circulating corticosterone levels but an increase in DOC production in the hypertensive rats.

J. C. Melby: The mineralocorticoid potency of 18-OH-DOC when compared to DOC depends upon the concentrations at which comparisons are made. At higher concentrations 18-OH-DOC appeared to possess about half of the activity of DOC in stimulating active sodium transport.

In the Porter studies at equimolar concentrations 7×10^{-7} the effect of 18 hydroxy DOC was half of what DOC is and that may be $\frac{1}{7}$ of what aldosterone is at that point.

F. C. Bartter: According to the standard dog assay or some standard rat assays the activity of aldosterone would be 30 times DOC.

J. C. Melby: They are not the same in the toad bladder.

N. Nowaczynski: We have some recent results of determination of 18-OH-DOC secretion rates in normotensive and essentially hypertensive subjects. The work was done in collaboration with Drs. O. Küchel, J. Genest, C. Sasaki, K. Seth, and E. Mancheno Rico.

A double isotopic procedure was used, involving: (a) injection of 2 μCi of ^{3}H-labeled 18-OH-DOC and 2 consecutive 24-hour urine colllections, (b) 18-OH-TH-DOC isolation from urine (benzene–hexane–methanol–water system), (c) periodate oxidation of 18-OH-TH-DOC to γ-lactone, (d) acetylation with ^{14}C-labeled acetic anhydride and purification in 5 chromatographic systems to isolate 18-OH-TH-DOC lactone acetate.

18-OH-DOC secretion rates were determined in 11 healthy control subjects and 32 hypertensive patients (Fig. B). All subjects were studied on the fourth day of a diet containing 135 meq of sodium and 90 meq of potassium. Hypertensive patients underwent a complete clinical checkup before the investigation to eliminate all known causes of hypertension and were divided into two groups. The first group of 17 patients was defined as suffering from labile hyperkinetic hypertension with a return of blood pressure to normal with rest and reassurance, but with a "hyperkinetic" response to posture, cold, and emotion (excessive increase in blood pressure and/or heart rate) and a favorable response to treatment by beta-blockers.

The second group of 15 patients met all the known criteria of stable, benign essential hypertension with persistent blood pressure readings above 140/90 mmHg.

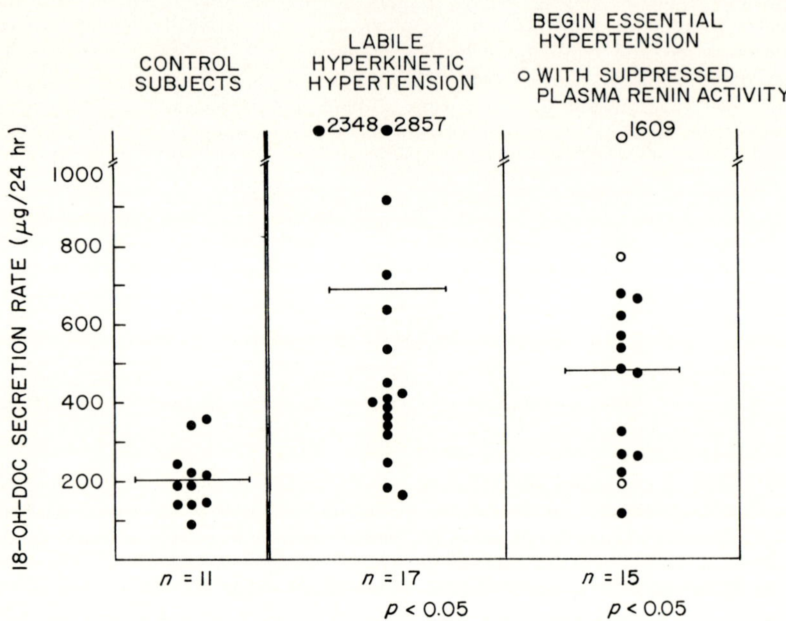

Fig. B. Secretion rates for 18-hydroxy-11-deoxycorticosterone (18-OH-DOC) control subjects and hypertensive patients receiving daily diet of 135 meg of sodium and 90 meg of potassium.

Three of them had suppressed plasma renin activity, and normal aldosterone secretion rates between 60 and 90 μg/day.

Our results show a mean 18-OH-DOC secretion rate of 209 μg/24 hours in the control group, and in labile hyperkinetic hypertension a significantly above-normal mean of 680 μg/day. In stable benign essential hypertension, there is a significant increase with a mean of 482 μg/day. Patients with suppressed renin are marked in Fig. B by open circles, and it is evident that the two highest values in this group are those of two patients with suppressed PRA.

In addition, sodium depletion induced an 18-OH-DOC secretion rate increase in all 3 control subjects from a mean of 123 to 310 μg/day (Fig. C).

Two hypertensive patients had the same response and two others a decreased response to sodium depletion, probably due to a higher baseline 18-OH-DOC secretion rate, pointing to an acceleration of this biosynthetic pathway.

This pattern, when compared to the changes in plasma progesterone (protein-binding method) is very similar.

Sodium depletion stimulates the biosynthetic pathway very early, i.e., before progesterone and 18-OH-DOC, leading to an increase of both steroids. This stimulation is probably not due to ACTH, since plasma cortisol (protein binding method),

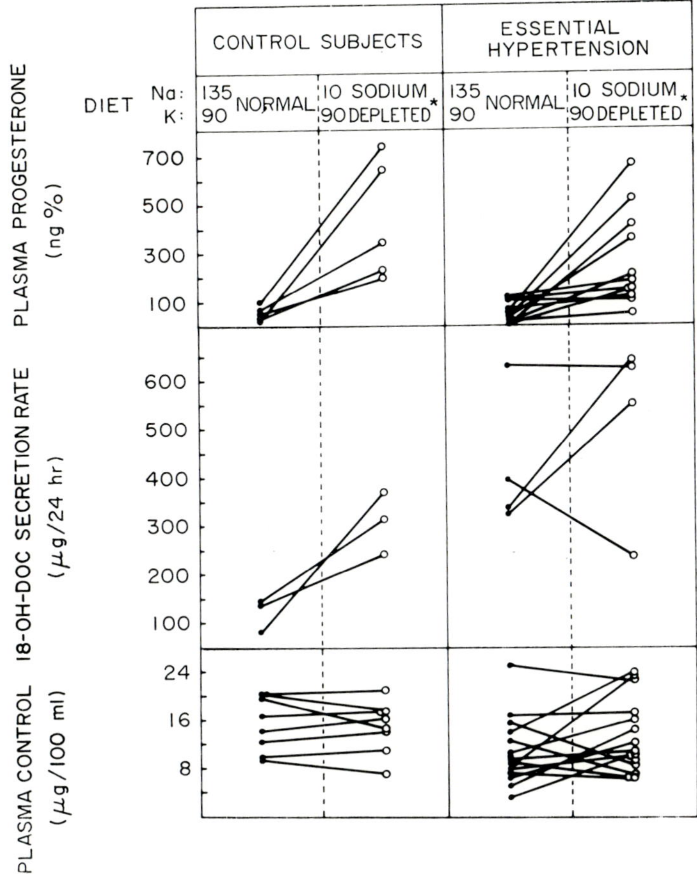

FIG. C. Effect of sodium depletion on 18-hydroxy-11-deoxycorticosterone (18-OH-DOC) in control subjects and in patients having essential hypertension. Asterisk in column 4 indicates mean weight loss of 26 ± 1 kg.

the main ACTH-dependent steroid, as shown on Fig. C, did not change under the same conditions.

In addition, metabolic clearance rates of 18-OH-DOC determined by a continuous infusion procedure [N. Nowaczynski *et al. J. Clin. Invest.* **50**, 2184 (1971)], in 4 normal subjects gave values of 2002, 1430, 945, and 1625 liters/24 hours (mean 1500 liters/24 hours) and in 3 patients with stable BEH and normal PRA values of 1450, 2220, and 1910 liters/24 hours (mean 1860 liters/24 hours) showing excellent agreement with values reported by Dr. Melby.

We have also determined 18-OH-DOC secretion rates and plasma concentrations and secretion rates of some other steroids in a patient of Dr. E. McGarry from Royal Victoria Hospital, Montreal (Table B). This female patient, with a very marked 17α-hydroxylation deficiency with feminizing testes (removed about 1 year prior to this study) and dexamethasone-suppressible hypertension and hypokalemia, showed a very marked acceleration of the mineralocorticoid biosynthetic pathway up to 18-OH-DOC. Aldosterone secretion rates were near the lower limit of the normal range and plasma cortisol concentration was very low.

Our data indicate that the 18-OH-DOC secretion rate is elevated in the majority of patients with BEH, and even in early labile hypertension are compatible with a hypothesis that 18-OH-DOC, another ACTH-dependent steroid, when secreted

TABLE B
Patient with 17α-Hydroxylation Deficiency and Dexamethasone-Suppressible Hypertension and Hypokalemia

	Before dexamethasone	After dexamethasone
Plasma steroid ($\mu g/100\ ml$)		
Cortisol	0.41(15 ± 2)[a]	0
Corticosterone	10.4(0.27 ± 0.07)	7.2
Progesterone	1.3(0.035 ± 0.014)	0.14
Secretion rate ($\mu g/day$)		
DOC	21,600(174 ± 14)	
Corticosterone	34,700(2,770 ± 232)	
18-OH-DOC	8,460(209 ± 18)	
	7,880	1420

[a] In parentheses: normal mean ± SE.

in excess may lead to sodium retention, even in the absence of hyperaldosteronism. An excess of ACTH may be due to a neurogenic stimulation via hypothalamic neurohormones and adenohypophysis, or to a relative deficiency of 17α-hydroxylation with subsequent decreased feedback similar to the very exaggerated situation existing in Dr. McGarry's patient. This may be supported by the fact that 13 out of 15 plasma cortisol determinations shown in Fig. C in benign hypertensive patients on normal sodium intake, were below the normal mean.

Even though the salt-retaining activity of 18-OH-DOC is weak, it can be argued that its excessive secretion for prolonged periods of time, and with high sodium intake, may be sufficient to increase the blood pressure to hypertensive levels.

J. C. Melby: We have not examined patients with labile hypertension, specifically.

A. S. Goldman: The thought occurs to me that in adrenal regeneration hypertension after the initial adrenal enucleation the residual tisssue must be mostly glomerulosa zone and thus contain relatively high amounts of 18-hydroxylase. If during regeneration, activity of 11β-hydroxylase is relatively slower in reappearing than 21-hydroxylase, the primary adrenal products would be 18-hydroxy DOC and DOC. This may explain the origin of the syndrome, especially if 18-hydroxy DOC potentiated the peripheral angiotensin responses, as has been suggested elsewhere in this discussion. Another possible explanation is that 18-OH-DOC may be an inhibitor of 11β-hydroxylase. Has anyone looked at 11β-hydroxylase activity in adrenal regeneration hypertension or whether 18-OH-DOC may inhibit 11β-hydroxylase.

J. C. Melby: Brownie and Skelton [*In* "Functions of the Adrenal Cortex" (K. W. McKerns, ed.), Vol. 2, p. 710. Appleton, New York, 1968], have shown impairment of 11β-hydroxylation in regenerating adrenals.

A. White: In connection with the metabolism of 18-OH-DOC you showed a slide in which you indicated the distribution and urinary products of labeled 18-OH-DOC given intravenously in normal man. Would you kindly give those values again. Also, have you done the same sort of metabolism studies of 18-OH-DOC in individuals with hypertension?

J. C. Melby: No we have not. Regarding the urinary distribution in normal subjects, 47% of the radioactivity was excreted in the urine in 2 days. A negligible amount was excreted thereafter. Thirty-seven percent of the administered dose was in the glucuronide fraction and was practically all 18-OH-TH-DOC. About 3% was excreted as unchanged 18-OH-DOC in the neutral fraction and we cannot account for the rest. Anywhere from 20 to 40% was excreted in the bile. This could only be estimated because we had a partial collection. We based our calculations on a collection of 50% of the bile.

A. H. Vagnucci: I wish to congratulate Dr. Melby for the extensive and the complete work he has done in this particular area. One of my questions concerns the regulation of the secretion of 18-OH-DOC. Apparently the 18-OH-DOC depends on ACTH. If this dependence is true, why should the 18-OH-DOC be high and the cortisol secretion be normal in patients with idiopathic hypertension and suppressed PRA and in patients with primary aldosteronism? Why would secretion of 18-OH-DOC be increased selectively without an augmentation of other secretory products by the fasciculata? Furthermore, on the basis of other forms of endocrine hypertension (e.g., primary aldosteronism and Cushing's) we are used to think in terms of autonomous secretion. In the described condition, however, autonomy was not present, because the patients always showed suppression of 18-OH-DOC by dexamethasone.

The second question concerns the mineralocorticoid effect of 18-OH-DOC. You stated that the mineralocorticoid effect is a weak one. Do you have any evidence (for instance, expansion of the extracellular fluid volume) that sodium retention is present in these patients with idiopathic hypertension and PRA suppression? Do these patients have any hypokalemia? How do you visualize the mechanism of action of the 18-OH-DOC in causing a sodium retention without other electrolyte disturbance?

J. C. Melby: Most of our patients who have had this disorder have had hypokalemia. Two patients studied when given dexamethasone within 7 days lost between 2 and 3.6 kg of body weight. Sodium balance was negative during dexamethasone administration within the first week. The first question is the one that bothers us. We can conceptualize an increase in 18-OH-DOC secretion secondary to a small reduction in cortisol secretion and a resulting mild increase in ACTH output. There is not, however, any decrease in urinary 17-hydroxycorticosteroids in our patients. The possibility of a mild 17α-hydroxylase deficiency exists, but further investigation is necessary.

E. D. Bransome: In respect to the comment made about maturation of steroid secretion, have you been able to examine the metabolic fate of exogenous 18-hydroxy-DOC in hypertensive rats? The question really relates to the activity of 18-01-dehydrogenase.

It may be naive on my part, but I still think that the basic diathesis involved in hypertension is mysterious. An example that comes to mind is Bartter's syndrome, where, despite considerable overproduction of aldosterone, patients may be normotensive. Apropos of this question: Have you been able to separate your patients from the garden variety of essential hypertensives, e.g., in terms of measurements of volume expansion?

J. C. Melby: We have not examined the metabolic fate of exogenous 18-OH-DOC in hypertensive rats, although this question is pertinent. While 18-OH-DOC secretion is definitely markedly increased at 6 weeks after adrenal enucleation, studies by Rapp using gas chromatography failed to show increased peripheral levels of 18-OH-DOC. We have not made observations in these patients with low renin hypertension and 18-OH-DOC excess in comparison with essential hypertensive patients and normally responsive plasma renin activity of the kind you suggest. Some of the patients have been hypokalemic and others have not. All have responded to spironolactone with a response to dexamethasone. In most instances these patients more nearly resemble those with the syndrome of primary aldosteronism.

F. C. Bartter: We believe there is no good evidence for volume expansion in primary aldosteronism.

O. Kuchel: Did you do your infusion of 18-OH-DOC in man? Did you measure urinary electrolyte excretion? Could you find that there is really a qualitative difference in the mineralocorticoid effect of 18-OH-DOC when compared to other mineralocorticoids? Does 18-OH-DOC cause less potassium loss, as shown by Birmingham? On which basis did you express the mineralocorticoid activity as 50% of DOC? Is it calculated in sodium retention, or as a global mineralocorticoid effect on sodium and potassium?

J. C. Melby: We know little as to the *in vivo* effects of 18-OH-DOC on sodium and potassium metabolism except for Birmingham's experiments in rats. 18-OH-DOC does stimulate active sodium transport to an extent of from 20% to 50% of DOC [G. A. Porter and J. Kinsey, *Endocrinology* **89,** 353 (1971)].

G. T. Bryan: What is the familial incidence of hypertension in the patients studied by you? I think there is growing evidence that some types of hypertension have a high familial incidence [S. Londe, J. J. Bourgoinie, A. M. Robson, and D. Goldring, *J. Pediat.* **78,** 569 (1971)]. The data you have presented would be easy to use for genetic analysis.

Do you have evidence that urinary excretion of the infused, radioactive 18-OH-DOC is complete at the end of 48 hours? This condition must be met for valid secretion rate determinations, and should be assured in all patients with impaired renal function.

J. C. Melby: It is my impression that many of our patients have a family history of hypertension. Essentially all radioactivity is excreted in urine 48 hours after injection of 1,2-³H-labeled 18-OH-DOC.

L. L. Engel: There have been several references this evening to an 18-dehydrogenase. There are two other situations in the biosynthesis of steroid hormones in which an angular methyl group is oxidized. One is the 14α-methyl group of lanosterol. The first step is hydroxylation to the hydroxy methyl after which there occurs dehydrogenation catalyzed by an oxidized pyridine nucleotide-linked dehydrogenase. The second instance is in estrogen biosynthesis where C-19 is attacked—first by a hydroxylase to give the 19-hydroxy compound and then by a second hydroxylase to give the *gem*-diol or the 19-oxo compound. Is there any direct evidence that the second step in the formation of the aldehydic group of aldosterone is a dehydrogenation, not a hydroxylation with a requirement for reduced NADP and molecular oxygen?

J. C. Melby: I know of no direct evidence that this is so.

R. Gaunt: Several of the pertinent questions raised in this discussion concerned the biological properties of 18-OH-DOC. What are the potencies and properties of this steroid in standard corticosteroid tests? Will it cause hypertension when administered chronically and at what dosage? These questions could be and would have been simply answered if anyone had ever had available sufficient compound to make the tests. When one approaches synthetic chemists and asks that 18-OH-DOC be made, one produces signs of central nervous agitation but does not get any compound. Until enough of this substance is available for adequate biological characterization, we are going to have to continue to work in half shadows.

S. Cohen: Is there perhaps a general increased activity in 18-hydroxylation mechanisms here? Have you examined for the presence of any other 18-hydroxy compounds such as 18-hydroxycorticosterone.

J. C. Melby: There is no increase in 18-hydroxycorticosterone secretion in either the patients or the animals studied.

B. F. Rice: In essence you have postulated a 17-hydroxylase defect. Could you explain why you chose an unnatural corticoid such as dexamethasone for replacement or suppression rather than cortisol or cortisone acetate? Dexamethasone has differing pharmacologic properties from cortisol which could complicate the interpretation of the beneficial effects obtained. For example, was the improvement due to suppression of endogenous adrenal activity via the pituitary or was it due to some direct pharmacologic effect of dexamethasone on the kidney?

J. C. Melby: The half-life of dexamethasone is so long that we can assure suppression by giving a single dose in 24 hours. This is not true for cortisol; the suppression is not as pronounced even when divided doses are given daily.

B. F. Rice: Have you measured 18-OH-DOC in any women with ovarian tumors causing virilization and hypertension? Drs. Huseby and Dominguez have reported that a transplantable mouse luteoma possesses 21-hydroxylase activity and formed DOC from labeled progesterone and pregnenolone used as precursors. We have studied this same luteoma in tissue culture and found *de novo* synthesis of DOC from acetate-1-^{14}C as well as synthesis of microgram amounts of DOC. I would suspect that one might find this compound produced by some human ovarian tumors in the future if one looks for it.

J. C. Melby: We have not.

K. Savard: In my own simplistic approach to the adrenal and its enzymes, I have tended to view, and to teach, that the biochemical differences between cells of the zona glomerulosa and cells of the fasciculata are primarily associated with the presence or the absence of certain steroid-hydroxylating enzymes. One of the basic rules would be that in the glomerulosa there is an absence of 17-hydroxylase, and conversely in the fasciculata there is an absence of 18-hydroxylase. There would be, of course, common enzymes in both. These distributions and localizations reflect the transcription or translation of highly specific genetic information during the fetal or developmental stages of these cells. I am a bit dismayed, and even behind in my awareness of the literature, to find my simplistic approach upset by your evidence for the presence of 18-hydroxylase in the zona fasciculata. Some years ago a most interesting study was done in Montreal by Dr. Andres Carballeira working with Dr. Eleanor Venning, who studied a pheochromocytoma, which I think was totally devoid of adrenalcortical tissue, but contained all the steroid hydroxylases for the synthesis of cortisol, with the exception of the cholesterol side-chain cleavage system. How they got there, I do not know. My question is, do you feel that there may be some—I do not know what it might be called—"leakage" of this enzyme or its genetic determinant from glomerulosa into the fasciculata cells?

J. C. Melby: The idea that 18-OH-DOC is largely a steroid secreted by the zona fasciculata is supported by the work of Sheppard and of the Tait group, who have reported experiments in which as much of the glomerulosa of rat adrenals was peeled off as possible. They found that most of the 18-OH-DOC was secreted by the decapsulated or glomerulosa-free gland. A little 18-OH-DOC was secreted by the capsule. Evidence that 18-hydroxylation occurs in the zona fasciculata of the rat adrenal cortex is overwhelming. When we use an inhibitor of aldosterone biosynthesis, GPA 2282, a reduction of 75% in the secretion of aldosterone and 18-hydroxycorticosterone is observed while no alteration in 18-OH-DOC secretion or cortisol secretion is found. This suggests to us that inhibition of 18-hydroxylation in the glomerulosa does not materially influence the total secretion of 18-OH-DOC emanating from the fasciculata.

O. V. Dominguez: I have a technical question related to a very peculiar characteristic of the 18-hydroxydeoxycorticosterone and its behavior. This is the only compound reported until now that has two different interconvertible forms that can be separated by paper chromatography in a ratio more polar to less polar form of about 70 to 30. You mentioned that 65% of your 18-hydroxy-DOC was recovered. Have you taken into account in your final results the correction due to the existence of the other form of the compound, which represents 30% of the total 18-OH-DOC?

J. C. Melby: If the use of methanol is avoided, one can reduce the formation of other forms of 18-OH-DOC. Birmingham has called the more polar form produced in methanol a dimer. We do not include the transformation to this other compound when we are calculating recovery. The tritiated 18-OH-DOC we add to plasma for recovery purposes is purified to remove the nonpolar form. The method used in purifying the plasma extract effectively separates these two forms. We recover 65% of the added tritiated 18-OH-DOC.

J. C. Orr: You did not specify the stereochemistry of position 20 in 18-hydroxy-DOC. Do you have equilibration, as you presumably could, of the $20R$- and $20S$-isomers? Presumably the ether ring could open up easily enough to give equilibration of the two forms. Can you get an equilibrium of this type with acid or base? It is possible that one 20-isomer could be considerably more polar than the other, but "dimer" would suggest much higher molecular weight. Have you evidence for molecular weight?

J. C. Melby: The answer to the first two questions is, "I don't know." We do not have evidence for molecular weight of the so-called "dimer."

O. V. Dominguez: In relation to the two forms of 18-hydroxy-DOC, the regular form of 18-hydroxy DOC is slightly more polar than corticosterone. The other form, which 18-hydroxy-DOC adopts, occurs when it is eluated off the paper with methanol, and it is less polar than the first one. Actually, I called M the more polar form and L the less polar form of 18-hydroxy-DOC.

J. C. Melby: It is less polar—I stand corrected.

O. V. Dominguez: Correct—so this is different from the dimer that Dr. Birmingham suggested.

This comment is concerned with Dr. Savard's question. Incubations in which we used 11-deoxycorticosterone as substrate, in the presence of rat adrenals which were previously decapsulated, we found that, depending on the substrate concentration, in some instances the 18-OH-DOC became the major product formed, which indicates that the 18-hydroxylase is present in various layers in the adrenal gland, rather than present only in the external zona of the adrenal. The 18-hydroxylase in the rat adrenal has high activity and high concentration, and perhaps that enzyme reaction is not a limiting factor in aldosterone biosynthesis.

In relation to Dr. Engel's comment indicating that so far no one has demonstrated definite evidence of the existence of 18-hydroxylase, enzyme required prior to the conversion to aldosterone, there seems to be a question as to how the 18-hydroxylated compounds are converted to 18-al in aldosterone biosynthesis. I have been thinking for some time about the possibility of an alternative pathway of aldosterone biosynthesis in which corticosterone is converted to 11-dehydrocorticosterone (cmpd. A). A limited amount is then hydroxylated at the 18 position in a next step to form 18-hydroxy-A (18,21-dihydroxy-Δ^4-pregnene-3,11,20-trione). The rearrangement between the 11-keto and 18-hydroxy occurs in similar fashion as it has been accepted for the rearrangement between th 11β-hydroxy and 18-aldo groups of aldosterone to adopt the hemiketalic configuration.

J. C. Melby: We have consistently observed the formation of a more nonpolar form of 18-OH-DOC when it is allowed to stand in methanol. The rate of conversion of 18-OH-DOC to this compound is enhanced by the presence of a trace of acetic acid. The "dimer" as suggested by Birmingham and the "L form" you describe may well be one in the same. We have no evidence that it is or is not a dimer.

Structure, Synthesis, and Mechanism of Action of Parathyroid Hormone

G. D. AURBACH, H. T. KEUTMANN, H. D. NIALL, G. W. TREGEAR, J. L. H. O'RIORDAN, R. MARCUS, S. J. MARX, AND J. T. POTTS, JR.

Section on Mineral Metabolism, Metabolic Diseases Branch, National Institute of Arthritis and Metabolic Diseases, National Institutes of Health, Bethesda, Maryland; Endocrine Unit, Massachusetts General Hospital, Boston, Massachusetts; Middlesex Hospital, London, England

I. Introduction

In the six years since our earlier discussion of this topic (Potts *et al.*, 1966), knowledge concerning the chemistry, physiology, and mechanism of action of parathyroid hormone has advanced remarkably. The bovine and porcine hormones have been isolated in completely purified form, the amino acid sequences of these molecules have been deciphered, and the biologically important region of the molecule at the amino terminus identified. These advances have led to synthesis of the amino-terminal tetratriacontapeptide of bovine parathyroid hormone; the product shows qualitatively all the biological properties of the native hormone. The chemical studies also allowed the identification of immunologically reactive regions of the molecule. Within the molecule there are several segments that react immunologically but are insufficient for expression of biological activity. These sites, possibly including those near the carboxyl terminus, are probably represented in the circulation as immunologically reactive but biologically inert fragments that are formed from the molecule after secretion.

Important new information has been gathered indicating that the mechanism of action of the hormone is mediated by cyclic 3',5'-adenosine monophosphate (3',5'-AMP). These studies combined with biological testing of synthetic peptides show that the hormone acts on bone and kidney through the same biochemical mechanism. Thus, it is implied that receptor sites in both renal and skeletal tissue must be virtually identical.

This report presents a discussion of the new information that has been accumulated so rapidly in this field.

II. Isolation and Chemical Properties of Parathyroid Hormone

Efficient means for extracting and purifying parathyroid hormone (Aurbach, 1959a,b) had been developed more than ten years ago and highly purified preparations of the hormone were described during our presentation at this conference in 1965 (Potts *et al.*, 1966). In general, the procedure involved extraction of crude gland powder with phenol, fractionation with salt and acetic acid and precipitation with trichloracetic acid. This material was then readily purified with the use of Sephadex G-100 columns and chromatography on carboxymethyl cellulose (Potts and Aurbach, 1965). These procedures seemed to allow isolation of pure material in sufficient quantity to undertake sequence determination. The original procedure (Aurbach, 1959b) effectively utilized countercurrent distribution to purify the hormone. This method, however, seemed to be too cumbersome and time consuming to be applied to preparation of the large quantities of material classically needed for sequence work.[1] At the 1965 conference (Potts *et al.*, 1966) we reported that the purest preparations of parathyroid hormone obtained by ion exchange chromatography on carboxymethyl cellulose contained two readily identifiable peptides on disc gel electrophoresis. Each of the bands of peptide material were isolated and shown to be biologically active. One accounted for the bulk (approximately 75%) of the peptide material present and the other minor component represented a genetic variant of the major hormonal polypeptide. End group analysis with dinitrobenzyl fluoride indicated only a single end group, alanine, for either component. Later, however, application of the extremely sensitive phenylisothiocyanate reaction of Edman for N-terminal amino acids with identification by gas chromatography revealed at least two additional peptides in the best hormone preparations purified by chromatography on carboxymethyl cellulose. Sequential analysis for several cycles indicated that these additional peptides must be contaminants unrelated to the parathyroid hormone peptides. The search for a further purification step to allow complete removal of these contaminating peptides as well as separation of the major form of the hormone and genetic variants led to the use of chromatography on carboxymethyl cellulose columns using buffers containing 8 M urea (Keutmann et al., 1971a). This procedure allowed isolation of the hormone in completely pure form and revealed the presence of two minor hormonal variants (Fig. 1). The fractions of interest were recovered by adsorption to and elution from a second column of carboxymethyl cellulose (CMC) using buffers without urea. Results of end-group analyses emphasize the differences in purity of the hormone between the Sephadex

---

[1] At the time this project was begun, the primary structure of but a few proteins had been deciphered. In each instance, large quantities of purified polypeptide or protein had been required for analysis of amino acid sequence. With the highly sophisticated and sensitive methods currently available, only milligram quantities of polypeptide hormone are required. Had these methods been available a decade earlier, the small amounts of purified hormone obtained from countercurrent distribution systems would have doubtless been sufficient for sequence determination.

G-100 stage and the CMC-urea column stage (Fig. 2). The latter (CMC) preparation showed undetectable (less than 1%) contaminants with N-terminal valine or leucine. In contrast to this result the hormone prepared by gel filtration on G-100 alone showed N-terminal valine- and leucine-peptides representing contamination up to 25 or 30%. Two other polypeptides elute near the major bovine hormone fraction during chromatography on the CMC-urea columns (Fig. 1). These fractions (II and IV representing the forms BPTH-II and -III) are variants of the major form of bovine parathyroid hormone (BPTH-I). Both these variants show N-terminal alanine on end group analysis. Amino acid analyses show that BPTH-II contains one threonine substituted for one

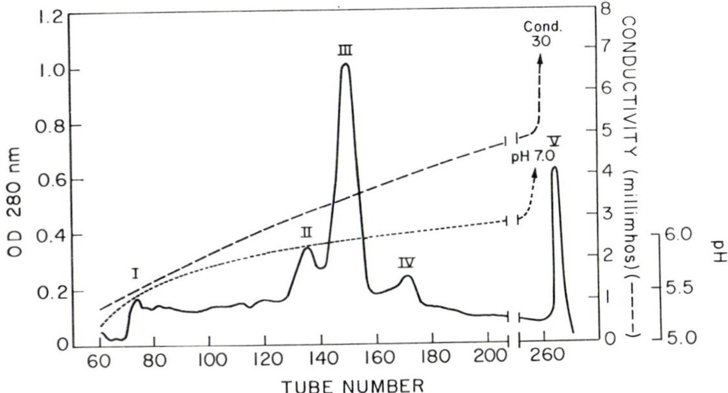

Fig. 1. Chromatography of 150 mg of bovine parathyroid hormone on a 2 × 40 cm column of carboxymethyl cellulose using a linear gradient of ammonium acetate in 8 *M* urea. The major form of bovine parathyroid hormone (BPTH-I) constitutes peak III. Peaks II and IV represent BPTH-II and BPTH-III, that represent minor genetic variants of the hormone. From Keutmann *et al.* (1971a).

valine (Table I). Results to date show similar amino acid compositions for BPTH-II and BPTH-III. The amino acid sequence is the same for the first fifteen residues of each form of the hormone, BPTH-I, -II, and -III. Further characterization of these variants, in order to explain their distinct chromatographic behavior, depends upon isolation of further quantities and analyses for amino acid squence.

Drs. Woodhead, O'Riordan, and collaborators (Woodhead *et al.*, 1971) have recently carried out the isolation of porcine parathyroid hormone utilizing methods very similar to those applied to the bovine hormone. The porcine hormone was obtained free of nonhormonal contaminants using chromatography on carboxymethyl cellulose; in this instance, it was not necessary to add 8 *M* urea to the buffer systems as was required for final purification of the bovine hormone. The product was homogeneous by disc gel electrophoresis (Fig. 3). It contains the same total number of amino acids and shows a composition very much analogous to that of the bovine hormone (Table I). There are, however, notable differences. The amino-terminal residue of the porcine hormone is serine instead of alanine and there is one less methionine, one less phenyl-

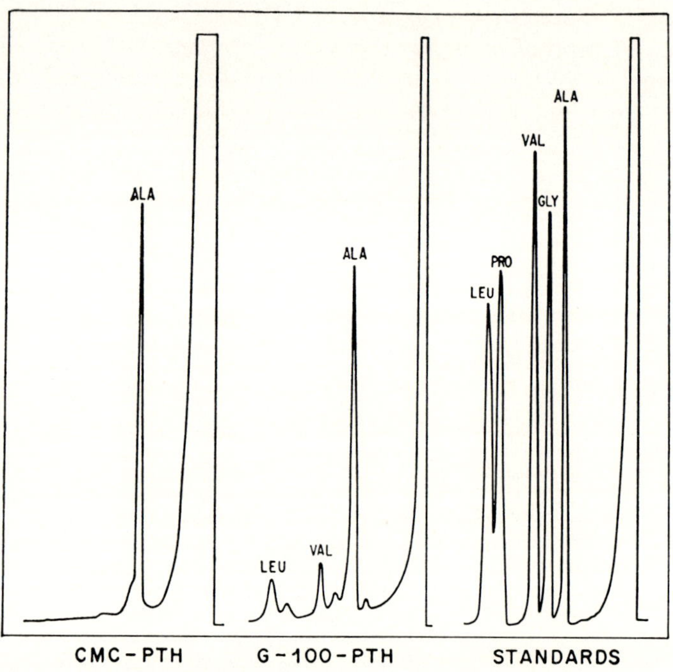

FIG. 2. Gas–liquid chromatography of phenylthiohydantoin derivatives representing amino-terminal groups obtained from bovine parathyroid preparations after gel filtration on Sephadex G-100 and chromatography on carboxymethyl cellulose in 8 M urea. The alanine end group represents that of the bovine parathyroid hormone molecule; valine and leucine derivatives represent contaminants. From Keutmann *et al.* (1971a).

alanine, and no tyrosine. As in the major form of the bovine hormone, there is no threonine. In the bovine hormone there are two methionines at positions 8 and 18. Oxidation of the methionines to the sulfoxide form causes biological inactivation of the hormone. The porcine hormone contains only one methionine; nevertheless, oxidation of the single methionine at position 8 causes similar inactivation of the porcine hormone. The lack of tyrosine in the porcine molecule is readily discernible from the ultraviolet spectrum (Fig. 4). The porcine hormone shows the spectrum characteristic of tryptophan alone whereas the spectrum of the bovine hormone represents the sum of absorption for 1 mole of tryptophan plus 1 mole of tyrosine.

Small quantities, in the microgram range, of human parathyroid hormone have been isolated in highly purified form (O'Riordan *et al.*, 1971). There has not been sufficient material to carry out exhaustive determinations on the purity of the preparation, but by several criteria it

TABLE I

Amino Acid Analyses of Parathyroid Hormones[a]

Amino acid	BPTH I	BPTH II	PPTH	HPTH
Aspartic acid[b]	9	—[c]	8	—
Threonine	0	1	—	3
Serine	8	—	—	7
Glutamic acid[d]	11	—	—	10
Proline	2	—	—	3
Glycine	4	—	5	5
Alanine	7	—	6	8
Valine	8	7	9	6
Methionine	2	—	1	—
Isoleucine	3	—	—	2
Leucine	8	—	10	9
Tyrosine	1	—	0	—
Phenylalanine	2	—	1	—
Tryptophan	1	—	—	ND[e]
Lysine	9	—	—	—
Histidine	4	—	—	—
Arginine	5	—	—	—

[a] BPTH, bovine parathyroid hormone; PPTH, porcine parathyroid hormone; HPTH, human parathyroid hormone.

[b] Three of the aspartics exist as asparagine in BPTH and PPTH.

[c] Dash (—) space indicates composition same as for BPTH I.

[d] Five of the glutamics exist as glutamine in BPTH and PPTH.

[e] Not determined.

seemed to be highly purified. The physical behavior of the human hormone was similar by ultracentrifugation, gel filtration, and ion exchange chromatography to the properties of bovine and porcine hormone. It is likely, then, that the human hormone is similar in size and charge distribution to the bovine and porcine molecules. Amino acid analyses of the human preparation are generally similar to those for porcine and bovine hormones, but show several significant differences. Like the bovine variants BPTH-II and BPTH-III, the human hormone contains

threonine. A comparison of amino acid analyses for the bovine parathyroid isohormones and porcine and human hormones, is presented in
Table I.

III. Biological and Immunological Activities of the Hormones

A number of tests were used to evaluate the biological properties
of the several parathyroid hormones. *In vivo* bioassays were based upon
the procedure of Munson (1955) modified as described earlier (Aurbach,
1959c; Marcus and Aurbach, 1969); the concentration of calcium in
serum was determined 5 hours after injection of preparations into para-

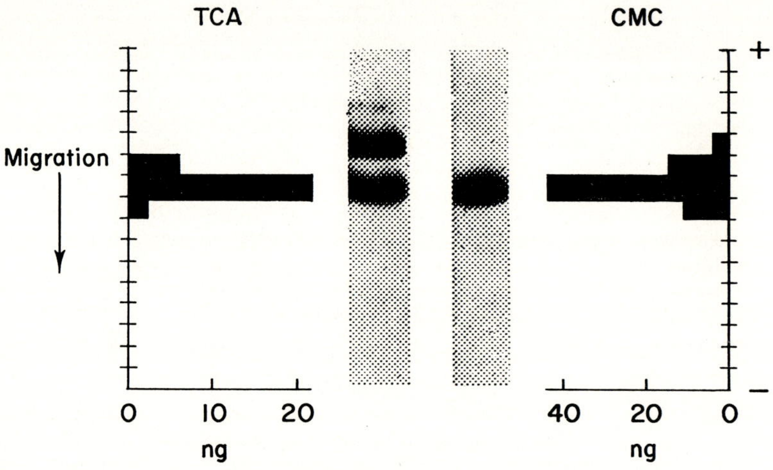

FIG. 3. Disc gel electrophoretic patterns obtained with porcine parathyroid hormone preparations. Histograms show distribution of hormone eluted from the gels
and detected by radioimmunoassay. TCA, preparation at the trichloroacetic acid
precipitate stage; CMC, hormone after purification on carboxymethyl cellulose.
From Woodhead *et al.* (1971).

thyroidectomized rats. Occasionally the procedure of Amer (1968) was
used, in which instance thyroparathyroidectomized rats were the test
animals. Further evaluation of biological properties *in vivo* was based
upon the finding that parathyroid hormone causes an increased excretion
of cyclic 3′,5′-AMP as well as phosphate into the urine of rats and
man (Chase and Aurbach, 1967). *In vitro,* biological properties of the
hormones were evaluated using the bioassay based on activation of renal
adrenyl cyclase (Marcus and Aubach, 1969) and by the effect of the
hormone in causing a rise in endogenous concentration of 3′,5′-AMP
in skeletal tissue (Chase and Aurbach, 1970). Immunological properties
of the preparations were tested by the radioimmunoassay described

earlier (Berson *et al.*, 1963) utilizing an antiserum, GP-1, obtained from one guinea pig. Cyclic 3′,5′-AMP was determined in tissues by the method of Aurbach and Houston (1968) or by radioimmunoassay (Steiner *et al.*, 1969). The Medical Research Council biological standard, MRC 67/342, was used as the biological standard for *in vitro* as well as *in vivo* studies. It was assigned an activity of 180 USP units per vial (Marcus and Aurbach, 1969).

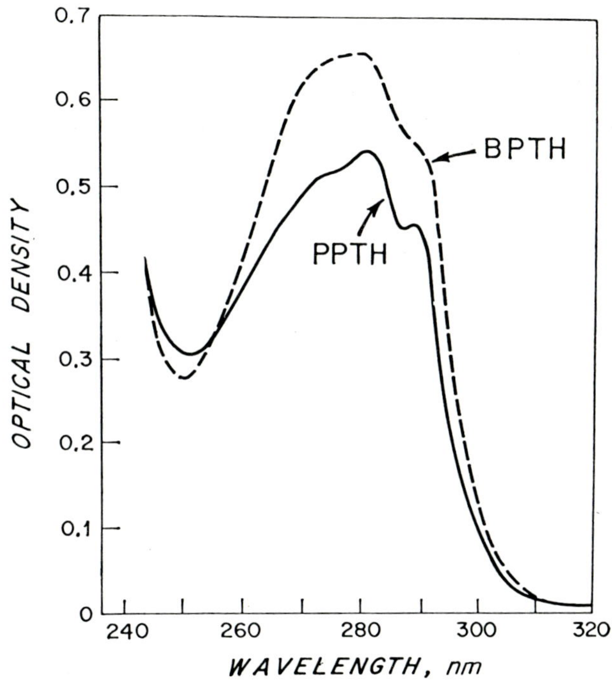

FIG. 4. Absorption spectrum of porcine parathyroid hormone (PPTH) compared to bovine parathyroid hormone (BPTH). Each is in solution at a concentration of 1 mg/ml in 0.1 *M* acetic acid. From Woodhead *et al.* (1971).

Purified BPTH-I and BPTH-II showed similar specific biological activity, 1500 units/mg when tested with the *in vitro* adenyl cyclase method and 2000–3000 units/mg *in vivo*. It is of interest that the purified porcine hormone shows biological activity not significantly different *in vivo* from that for the bovine hormone. However, activity of the porcine hormone *in vitro* with the renal adenyl cyclase system was consistently in the range of 500–600 units/mg as compared to an activity of 1400 units/mg for purified parathyroid hormone-I. The dose-response curves for the two hormone preparations were parallel, though, whether tested *in vitro* or *in vivo*. Membranes from the renal cortex contain an enzyme which destroys parathyroid hormone *in vitro*. This same membrane fraction contains adenyl cyclase which is activated by parathyroid hormone. The two activities, that is, adenyl cyclase and inactivating enzyme, are not separated by fractionating the membranes on a column of Bio-Gel® A-0.5 m. Several experiments indicate that the procine hormone is more susceptible to inactivation by the membrane-bound enzyme than is the bovine hormone. Conversely, there is no significant difference in activities of the two hormones when tested *in vitro* with bone fragments. Thus, it is apparent that the bovine and porcine hormones are of comparable biological activity on bone (which contains no detectable hormone-inactivating enzyme), but the bovine hormone, being more resistant to inactivation by the renal membrane system, shows greater apparent biological activity on the kidney. This problem of inactivation of the hormone *in vitro* as well as *in vivo* is discussed later in greater detail. The biological activity of human parathyroid hormone has not been studied in great detail because of scarcity of supply. The studies carried out so far indicate that the human hormone is considerably less potent (300–500 units/mg) *in vivo* as well as *in vitro*. Further, in each bioassay, either *in vitro* with renal adenyl cyclase or *in vivo* determining serum calcium, the slope of the response was considerably less than that for the bovine hormone.

IV. Amino Acid Sequences of the Parathyroid Hormones

The development of the automated sequencer based on serial one-step degradations with the phenylisothiocyanate reaction as developed by Edman and Begg (1967), proved to be a key tool in deciphering the amino acid sequence of parathyroid hormone. The improved purification techniques discussed above allowed the production of bovine hormone of such high purity that as many as 54 repetitive cycles were possible with one 8-mg sample of pure BPTH-I[2] (Niall *et al.*, 1970). At each cycle, the newly exposed amino terminal residue was allowed to react with phenylisothiocyanate and, after extraction and conversion, was recovered as the phenylthiohydantoin of that amino acid. The phenylthiohydantoins were identified by gas–liquid chromatography (Pisano and Bronzert, 1969; Niall and Potts, 1970). The repetitive yield on parathyroid hormone was 93.6%. Other analyses involved manual and/or automated degradations on cleavage products prepared with cyanogen bro-

---

[2] Brewer and Ronan (1970) reported that it was possible to carry through as many as 66 consecutive cycles in their analysis of the structure of the major form of bovine parathyroid hormone.

mide and the use of selective ϵ-amino blocking of lysine residues coupled with digestion with trypsin and chymotrypsin (Fig. 5). At the fifty-second cycle PTH-arginine was identified which then accounted for all of the five arginine residues known to exist in the hormone molecule. After succinylation of lysine epsilon amino groups and subsequent digestion with trypsin it was possible to recover in high yield a 32-amino acid peptide (residues 53 to 84 of Figure 5) representing the entire C-terminal region of the molecule. The succinylated carboxyl-terminal fragment was then digested with chymotrypsin. Sequential Edman degradation of the three peptides showed them to represent residues

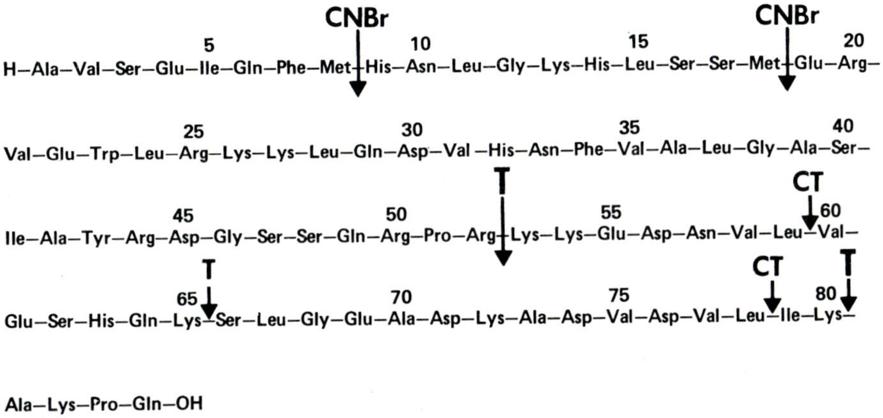

```
            CNBr                                    CNBr
     5            10              15            20
H—Ala—Val—Ser—Glu—Ile—Gln—Phe—Met—His—Asn—Leu—Gly—Lys—His—Leu—Ser—Ser—Met—Glu—Arg—

        25            30            35            40
Val—Glu—Trp—Leu—Arg—Lys—Lys—Leu—Gln—Asp—Val—His—Asn—Phe—Val—Ala—Leu—Gly—Ala—Ser—
                                                                    CT
        45            50            55             60
Ile—Ala—Tyr—Arg—Asp—Gly—Ser—Ser—Gln—Arg—Pro—Arg—Lys—Lys—Glu—Asp—Asn—Val—Leu—Val—
                                                         CT        T
        65            70            75            80
Glu—Ser—His—Gln—Lys—Ser—Leu—Gly—Glu—Ala—Asp—Lys—Ala—Asp—Val—Asp—Val—Leu—Ile—Lys—
```

Ala—Lys—Pro—Gln—OH

FIG. 5. Amino acid sequence of bovine parathyroid hormone illustrating approach in sequence analysis. Specific cleavages produced with cyanogen bromide (CNBR) and trypsin (T) are indicated by the longer arrows. The shorter arrows indicate points of cleavage during subdigestion of the isolated dotriacontapeptide (residues 53 to 84).

53 to 59, 60 to 78, and 79 to 84. These fragments were analyzed by manual degradation with the Edman reaction; the degradation of the hexapeptide was stopped after five cycles and the carboxyl-terminal glutamine detected as the free amino acid on the amino acid analyzer. Another set of fragments constituting the C-terminal region of the molecule was prepared after blocking ϵ-amino groups by reaction with maleic anhydride. After digestion with trypsin the maleoylated dotriacontapeptide (residues 50 to 84) was recovered by gel filtration on Sephadex G-50. The blocking maleolyl groups were then removed by treatment with acid, and the peptide was digested further with trypsin to produce fragments representing residues 53 to 65, 66 to 80, and 81 to 84. These three fragments were separated by ion exchange chromatography on carboxymethyl cellulose, and manual degradations with the Edman reac-

tion allowed confirmation of the sequence at the carboxyl-terminal region (Fig. 5).

The succinylation and/or maleoylation procedures represent highly efficient means for producing the dotriacontapeptide representing the C-terminus of parathyroid hormone molecules. In addition to use in structural analyses, this fragment has been valuable in studies on the

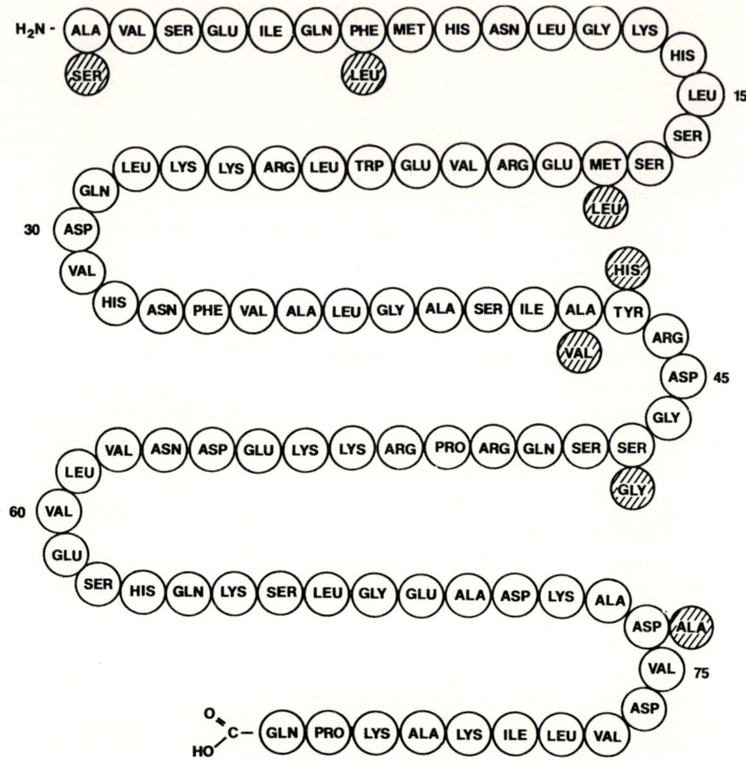

FIG. 6. Structures of bovine parathyroid hormone and porcine parathyroid hormone. The main structure represents the bovine hormone; the shaded residues represent the differences in amino acid sequence found in the porcine hormone.

immunologically reactive sites of hormone molecules. This will be discussed in further detail. The same general procedures have been applied also to produce fragments of the porcine parathyroid hormone. Analyses of these fragments allowed localization of amino acid differences (Table I) between bovine and porcine hormone as distinct changes in sequence. The overall results are shown in Fig. 6, wherein the structures of the two hormones are compared. Note particularly that serine rather than alanine is the amino-terminal residue in the porcine hormone and that

the reduction in methionine content of the latter is accounted for by
its replacement by leucine at position 18. Tyrosine is lacking and is
replaced by histidine at position 43. Several other changes in sequence
are noted. Each of these can be accounted for by single-base substitution
in the genetic coding for biosynthesis of the hormone.

V. Biologically Active Fragments of the Hormone

It has been known for several years that a biologically active fragment
could be produced by hydrolysis of purified parathyroid hormone in

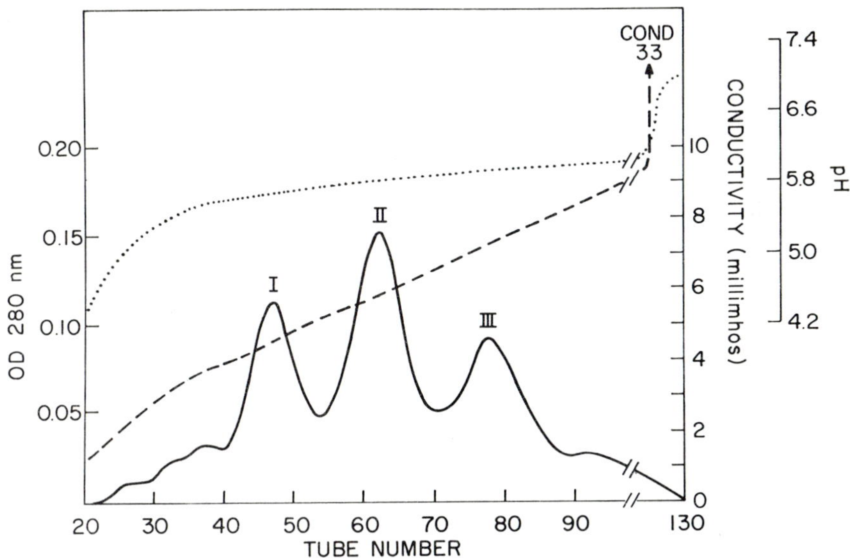

FIG. 7. Chromatography of the nonacosapeptide obtained after hydrolysis of
the native bovine hormone molecule in dilute acid. The central peak (II) contains
the pure nonacosapeptide that was found to be biologically and immunologically
active.

dilute hydrochloric acid (Potts *et al.*, 1966). Indeed, hot dilute hydrochloric acid was used by Collip (1925) as the first effective means of
extracting activity from parathyroid glands. We had reported earlier
(Potts *et al.*, 1968) that a biologically active fragment could be prepared
from the amino terminus of purified bovine parathyroid hormone by
hydrolysis in dilute acid. Recently it has been possible to isolate the
biologically active fragment produced by this type of hydrolysis
(Keutmann *et al.*, 1971b). The fragment was prepared by heating a
sample of pure BPTH-I, 10 mg/ml, at 110° for 4 hours in 0.03 *N* hydrochloric acid containing 1/2000 mercaptoethanol sealed in an evacuated

tube. The product was purified by gel filtration on G-50 Sephadex and then ion exchange chromatography on carboxymethyl cellulose. Three major peptide components were eluted during this ion exchange chromatography (Fig. 7). All three fractions showed some biological activity, but peak II appeared to be the most highly purified on thin-layer chromatography. This fraction (II) was recovered and analyzed for amino acid content as well as biological activity. The preparation was active in the renal adenyl cyclase system *in vitro* at 580 units/mg. The amino acid composition of this fraction corresponded to residues 1 to 29 of the native hormone (see Fig. 5). Characteristically, hydrolysis in dilute hydrochloric acid causes cleavage of peptide bonds at either side of aspartate residues (Tsung and Fraenkel-Conrat, 1965). It has been possible to isolate each of the polypeptides bounded by aspartate residues within the molecule. Thus, in addition to the nonacosapeptide 1–29, fragments have been isolated representing residues 31 to 44, 46 to 55, 57 to 71, and 77 to 84. The aspartic acid liberated from the molecule was identified by direct amino acid analysis as was free valine representing residue 75, which is bounded by aspartic acid at positions 74 and 76. The nonacosapeptide (residues 1 to 29) isolated as a hydrolysis product from the native molecule represents the smallest peptide currently known to be biologically active.

VI. Chemical Synthesis of an Active Parathyroid Hormone Peptide

The lack of aspartate within the amino-terminal one-third of the molecule and the biological importance of this region suggested that a fragment approximately 30 amino acids in length would be biologically active. On that basis, synthesis was begun on the fragment represented by amino acids 1 through 34 (Potts *et al.*, 1971). The peptide chain was assembled by stepwise addition of protected amino acids onto an active site at the surface of a polymer support using the general procedure of Merrifield (1963). A graft copolymer (Tregear, 1969) of poly(trifluorochloroethylene-G-methyl) styrene was used as the solid support phase. Synthesis commensed with the preparation of resin esterified phenylalanine (residue 34). t-BOC-amino acids were coupled successively in the order residues 33, then 32, etc. After each coupling step the protective t-BOC group was removed with 4 M HCl in dioxane (trifluoroacetic acid was used to deblock glutamine). Removal of the t-BOC group exposed the amino group for the next coupling step. Sensitive side chain functions were blocked with appropriate groups (benzyl esters for the side chain carboxyls, benzyl ethers for serine hydroxyls, and the nitro derivative of the guanadino group of arginine). Imidazole nitrogen was blocked with the dinitrophenyl group and the ϵ-amino group of lysine with trifluoroacetyl. At the completion of the synthesis of the tetratriacontapeptide representing residues 1 to **34**, the peptide was cleaved from the resin with hydrogen fluoride. Simultaneously the protecting groups of aspartic, glutamic, serine, and arginine were removed. Further treatment with 1 M piperidine in 8 M urea removed the trifluoroacetyl groups from the lysines. The synthetic peptide was then purified by gel filtration on Sephadex G-50 and then chromatography on carboxymethyl cellulose.

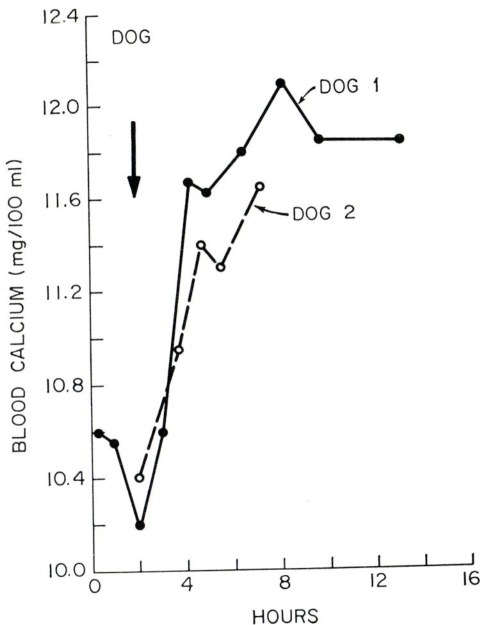

FIG. 8. Biological activity in dogs. At arrow, 1 mg of synthetic hormone was injected intravenously into each of two dogs. From Potts *et al.* (1971).

The synthetic product showed biological activity qualitatively identical to that of the natural hormone isolated from bovine glands. The synthetic peptide caused a rise in serum calcium when injected into dogs or parathyroidectomized rats, and caused increased urinary excretion of phosphate and 3′,5′-AMP. It was highly active in *in vitro* assays utilizing renal adenyl cyclase or the rise in tissue concentration of cyclic AMP in bone. In addition, it showed immunological reactivity virtually identical to that of the active fragment (residues 1 to **29**) hydrolyzed from the native polypeptide hormone with dilute acid. These results are illustrated in Figs. 8 to 12. More recently, several further polypeptide fragments representing analogs or portions of the tetratriacontapeptide

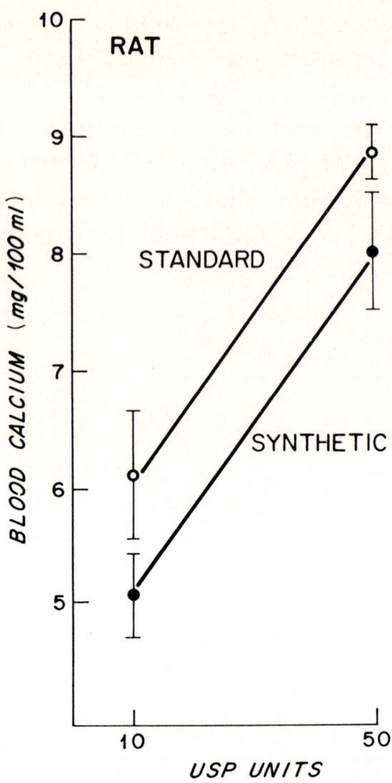

FIG. 9. Bioassay comparing synthetic parathyroid polypeptide to standard hormone preparation. Each substance at two comparable dilutions was injected subcutaneously (in 15% gelatin) into parathyroidectomized rats fed a calcium-deficient diet. From Potts *et al.* (1971).

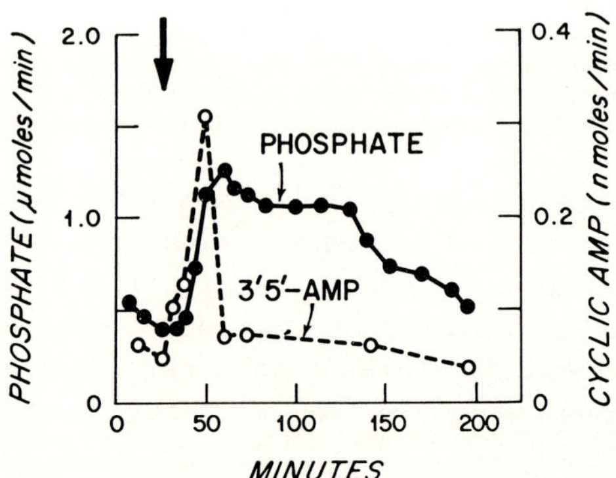

FIG. 10. The effect of synthetic parathyroid hormone on urinary excretion of 3',5'-AMP (○---○) and phosphate (●——●). At the time marked by arrow, the equivalent of 25 μg of synthetic hormone was injected intravenously into a conscious constantly perfused parathyroidectomized rat. From Potts *et al.* (1971).

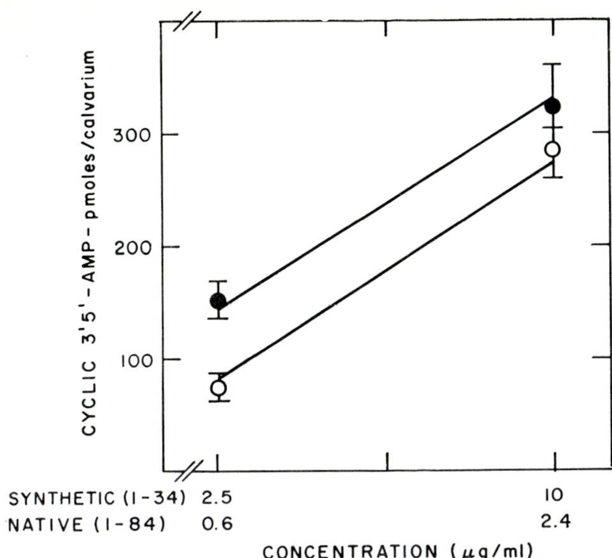

FIG. 11. Parathyroid hormone-induced accumulation of 3′,5′-AMP in fetal rat calvaria. Calvaria were incubated for 10 minutes in Krebs-Ringer phosphate buffer containing synthetic (1 to 34) or native (1 to 84) bovine parathyroid hormone.

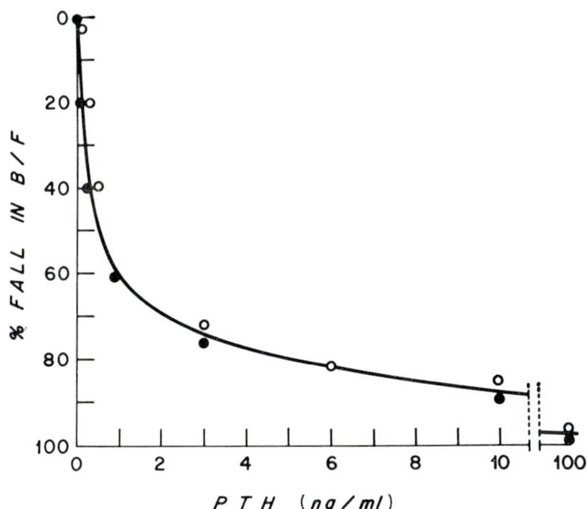

FIG. 12. Comparison of immunological reactivity of synthetic parthyroid hormone (●) (1 to 34) with that of corresponding active fragment (1 to 29) hydrolyzed from native hormone (○) with dilute acid. B/F, ratio of bound ^{125}I-labeled hormone to free (F) hormone. The synthetic and natural fragments are equivalent in immunological activity, but neither showed more than 50% reactivity against the native bovine parathyroid hormone molecule. From Potts *et al.* (1971).

have been synthesized. Results with these synthetic peptides are presented later in connection with the discussion on characteristics and minimum size for a biological activity of parathyroid hormone.

VII. Mechanism of Action of Parathyroid Hormone

There is now considerable evidence that cyclic 3',5'-adenosine monophosphate (3',5'-AMP) is a key biochemical intermediate in the action of parathyroid hormone. In physiological experiments leading to development of this thesis it was observed that parathyroid hormone acts directly on the kidney influencing the urinary excretion of 3',5'-AMP (Chase and Aurbach, 1967). Pharmacological experiments showed that injection of theophylline into parathyroidectomized rats caused a rise in serum calcium similar to that produced by PTH (Wells and Lloyd, 1967). Theophylline is a known inhibitor of cyclic 3',5'-nucleotide phosphodiesterase and causes an increase in concentration of 3',5'-AMP in skeletal tissue (Chase and Aurbach, 1970). These several experiments indicated that parathyroid hormone produces an increase in concentration of 3',5'-AMP in the receptor cells of the kidney and bone. The increase in 3',5'-AMP concentration, then, leads to a series of biochemical events that account for the physiological actions of the hormone. Subsequent to these initial studies there has been considerable further progress in explaining the diverse actions of parathyroid hormone at the biochemical level.

The site of action of parathyroid hormone in the kidney has been localized to the tubules of the cortex (Melson *et al.*, 1970). In the bone, an adenyl cyclase system similar to that of the kidney has been identified which is activated by parathyroid hormone *in vitro* (Chase *et al.*, 1969a). In addition, it has been shown that parathyroid hormone incubated with bone fragments *in vitro* causes a marked increase in the concentration of endogenous cyclic AMP in the tissue (Chase and Aurbach, 1970). Studies from other laboratories show that dibutyryl-cyclic AMP added *in vitro* mimics the action of parathyroid hormone on bone. The dibutyryl analog has been shown to cause release of calcium (Raisz *et al.*, 1969), bone resorption (Herrmann-Erlee, 1972), and increased production of lysozomal enzymes by bone explants (Vaes, 1968). The action of parathyroid hormone on the kidney is reproduced by injecting dibutyryl-cyclic AMP or 3',5'-AMP intravenously (Russell *et al.*, 1968; Rasmussen *et al.*, 1968).

The overall conclusions from these experiments indicate that the series of events involved in the action of parathyroid hormone on specific target tissues include: (1) binding of hormone to the cell membrane; (2) activation of adenyl cyclase; (3) a rise in intracellular concentration of 3',5'-AMP; (4) stimulation of a kinase or phosphorylating enzyme;

(5) phosphorylation of an enzyme or membrane system; (6) enhanced activity of a transport system for calcium.

VIII. Binding to Membranes

So far it has not been possible to separate the phenomenon of binding to receptor for parathyroid hormone from activation of adenyl cyclase. In studies on the action of corticotropin and glucagon, evidence has been obtained that binding of the hormone to membrane receptor is a process closely linked, although separate and distinct, from activation of adenyl cyclase in specific receptor tissues. For example, calcium has been found essential for activation of adenyl cyclase in membranes prepared from adrenal or adipose cells (Rodbell *et al.*, 1970; Lefkowitz *et al.*, 1970) in response to ACTH. But calcium is not required for binding of the hormone to the receptor as shown by Lefkowitz *et al.* (1970), who developed a dispersed micellar fraction containing the receptor from adrenal tumor cells. They were able to determine binding to the receptor directly using radioiodinated ACTH. Rodbell and his collaborators (Rodbell *et al.*, 1970, 1971) found that binding of labeled glucagon could be separated as a phenomenon from activation of adenyl cyclase in liver membranes. More recently, Levey (1971) has succeeded in solubilizing and purifying adenyl cyclase from cat heart. After isolation, the solubilized enzyme no longer responded to glucagon but the remarkable discovery was made that hormonal responsiveness could be restored by adding phosphatidylcholine (Levey, 1971).

It is quite probable, then, that binding of parathyroid hormone to its receptor will ultimately be shown to be distinguishable from activation of the enzyme adenyl cyclase. We have initiated an investigation of binding of radioiodinated parathyroid hormone to membrane receptors with a view toward attempting to resolve binding versus enzyme activity. These studies are at only the most rudimentary stage. We still depend primarily on the indirect index of binding to membrane by determination of adenyl cyclase activity. Addition of hormone *in vitro* to membranes prepared from either kidney or bone show that the enzyme is half maximally activated at hormone concentrations of 4×10^{-7} M (Fig. 13). One can conclude, then, that the association constant for binding of hormone to membrane must be at least as high 2×10^6. In the ACTH receptor system, there appear to be at least two distinct orders of association constants for binding of hormone to membrane. One set shows a very high affinity in the range of 10^{11} to 10^{12}. It may be possible that there is a very high affinity site for parathyroid hormone as well; detection of such a high affinity site requires methods more sensitive than have been developed so far. In particular, a suitable high specific activity

tracer for parathyroid hormone must be prepared without destroying
biological activity.

It is possible that interaction of the hormone with the cell membrane
also influences the entry of calcium into cells. Parsons *et al.* (1971)
have found that within minutes after injection of parathyroid hormone
there is a transient shift of calcium into bone. Borle (1972) has described
a direct effect of the hormone on isolated renal cells causing increased

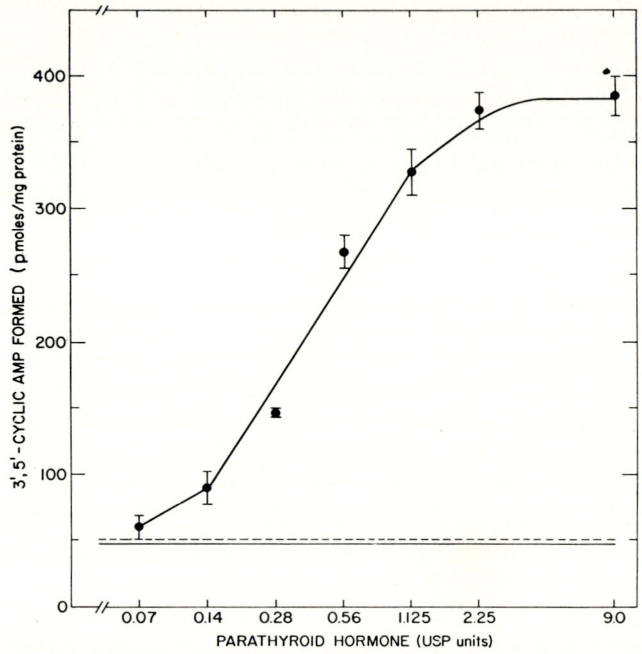

Fig. 13. Increase in adenyl cyclase activity as a function of concentration of
native bovine parathyroid hormone. Adenyl cyclase preparation obtained from
renal cortex of rats and assayed as described in text. Half-maximal activation of
the enzyme occurs with **0.5 USP** unit of parathyroid hormone (equivalent to
$4 \times 10^{-7}\ M$) in the assay. From Marcus and Aurbach (1971).

concentrations of ^{45}Ca within the cells. This effect develops rather slowly,
though. Rasmussen (1970) feels that calcium entry into cells is a specific
obligatory phenomenon, one of primary importance, in the mechanism
of action of many hormones. However, this postulate cannot be properly
evaluated until required studies are carried out on the polypeptide and
hormonal specificity of these effects and their relationship to adenyl
cyclase reaction as well as the intracellular function of cyclic 3',5'-AMP.
There is no question that calcium is required for many cell functions

including several under hormonal control. The question is whether this characterizes calcium as another "intracellular messenger" of hormonal action or whether calcium is simply one more "permissible" factor needed in general cell biology.

IX. Adenyl Cyclase in the Membrane

Assays for adenyl cyclase in isolated kidney tubules have allowed identification of the anatomical region containing the receptor for para-

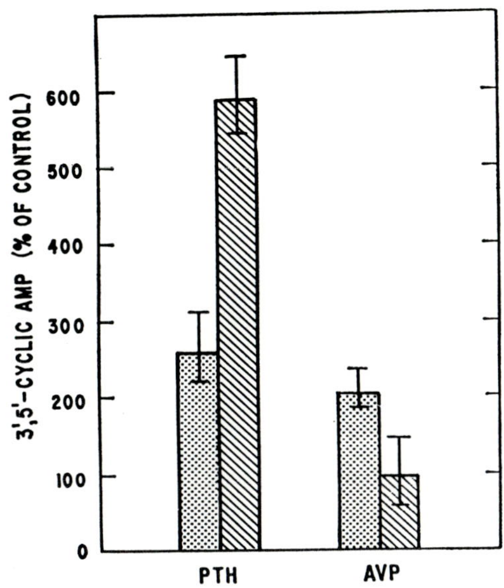

Fig. 14. Differential effect of parathyroid hormone (PTH) and arginine vasopressin (AVP) on adenyl cyclase obtained from renal tubules of the cortex (▨) and medulla (▨). Adenyl cyclase was assayed as described in text. Data are expressed as the percent (mean ±SD) of the control reactions which contained no added hormones. From Melson *et al.* (1970).

thyroid hormone. The receptor for parathyroid hormone lies within renal cortical tubules (Chase and Aurbach, 1968), most likely the proximal tubular cell (Melson *et al.*, 1970). The receptor site for parathyroid hormone is anatomically separate from that for vasopressin, the latter being located within medullary tubular elements (Fig. 14).

Recently studies on renal adenyl cyclase have been extended to learn more of the biochemical nature of the system. Development of a method to stabilize the enzyme to prolonged storage facilitated these studies by permitting repeated experiments with a single preparation of enzyme (Marcus and Aurbach, 1969). Using this preparation it was possible to achieve some fractionation of the particulate enzyme on Bio-Gel A-0.5m (Fig. 15). This procedure removed some contaminating cell particles as well as certain contaminating enzymes, and provided further confirmation of the large particulate (hence likely cell-membrane derived) origin of the receptor for parathyroid hormone. Phosphodiesterase was removed from the preparation by this procedure, but membrane-bound ATPase appeared in the same fraction as adenyl cyclase. The particles containing adenyl cyclase and ATPase appeared at the void

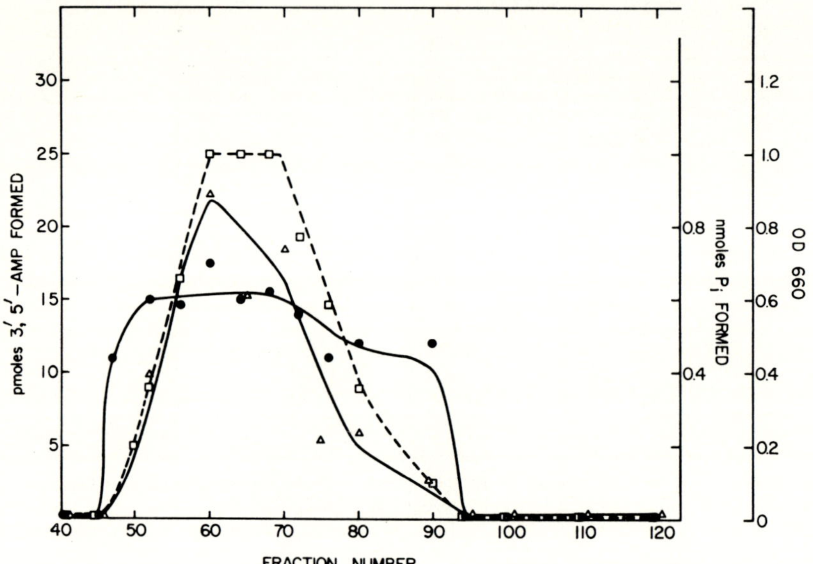

Fig. 15. Distribution of adrenyl cyclase (●——●) and ATPase (▲———△) after fractionation of adenyl cyclase in renal membranes as Bio-Gel A-0.5m. Adenyl cyclase and ATPase appeared in fractions 46 to 90, corresponding to the calculated void volume and the appearance of turbidity (□—·—□). Cyclic nucleotide phosphodiesterase was removed from the membrane-bound adenyl cyclase enzyme by this procedure. From Marcus and Aurbach (1971).

volume of the column (Fig. 15). Marcus and Aurbach (1971) have studied extensively the kinetic properties of the enzyme. Reactions were carried out with varying ATP concentrations and magnesium/ATP ratios to determine whether parathyroid hormone activation causes predominantly a change in K_m or V_{max} As shown in Fig. 16, the increased rate of reaction caused by parathyroid hormone represented effectively an increase in V_{max} without detectable change in K_m for ATP.

The stability of the enzyme at liquid nitrogen temperature makes the renal adenyl cyclase particularly valuable as the basis for a sensitive and precise *in vitro* bioassay for parathyroid hormone. Enzyme activity is determined by the rate of conversion of $AT^{32}P$ to cyclic $3',5'$-$AM^{32}P$ (Marcus and Aurbach, 1969) according to the method of Krishna *et al.* (1968). Enzyme activity is a direct function of the log of parathyroid

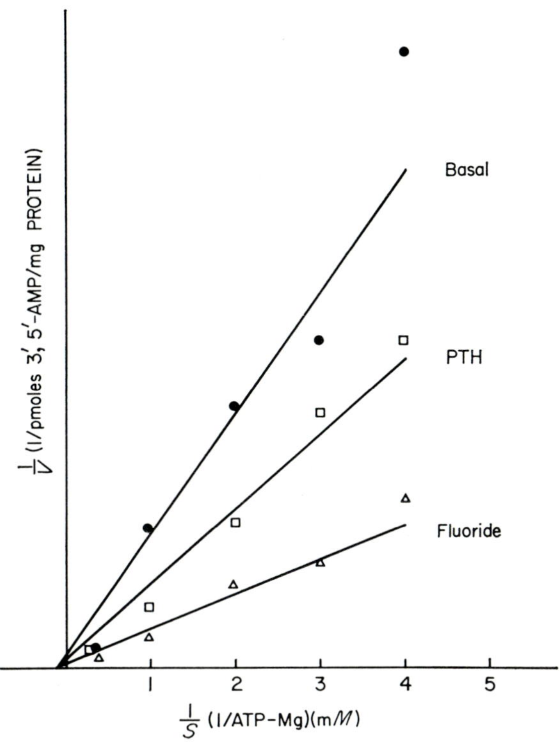

FIG. 16. Double reciprocal plots of enzyme activity against substrate concentration. Magnesium/ATP concentration ratio was maintained at 1.3 over the entire range of concentration tested. Circles represent control activity; squares plus parathyroid hormone (PTH) (5 μg/ml); triangles plus fluoride (7 $\times$ 10^{-3} M). Activation of the enzyme by hormone and fluoride effected a change in the V_{max} of the enzyme, but without significant change in K_m.

hormone concentration (Marcus and Aurbach, 1969). This assay has been utilized effectively to test hormone preparations available only in microgram amounts and as a means to identify biological and metabolic properties of several hormone analogs.

X. Concentration of 3′,5′-AMP in Cells

The rise in intracellular concentration of 3′,5′-AMP in response to parathyroid hormone in each receptor tissue leads to expression of the physiological actions of the hormone. An increase in intracellular concentration of cyclic AMP in kidney cells has been detected after injection of the hormone *in vivo* (Rasmussen and Tenenhouse, 1968; Aurbach *et al.*, 1969) or after addition of the hormone to isolated renal tubules (Melson *et al.*, 1970; Nagata and Rasmussen, 1970). The original observation that urinary cyclic 3′,5′-AMP increases in response to parathyroid hormone (Chase and Aurbach, 1967) has been extended to man (Chase

TABLE II

Effect of Parathyroid Hormone (PTH) on Cyclic Adenosine 3′, 5′-Monophosphate (AMP) in Renal and Skeletal Tissue[a]

	PTH (dose or conc.)	Time after dose (min)	Cyclic AMP	
			−PTH	+PTH
In vivo				
Kidney	10 μg	1	0.76 ± 0.13 (4)	1.68 ± 0.28 (6)
In vitro				
Renal tubules	65 μg/ml	1	0.80 ± 0.09 (4)	1.74 ± 0.22 (4)

[a] For experiments carried out *in vivo*, the left kidney was removed 1 min after injecting parathyroid hormone (PTH) or saline into the dorsal tail vein of anesthetized rats. For experiments carried out *in vitro*, tubules isolated from the renal cortex were incubated with and without PTH. Results for cyclic AMP are the mean ± SE. The number of determinations is in parentheses. Values are expressed as nanomoles per gram (net weight) of tissue.

et al., 1969b; Kaminsky *et al.*, 1970) and can be explained as a consequence of the hormone-induced rise in intracellular concentration of the nucleotide in the tubule cells of the renal cortex (Table II).

Skeletal tissue similarly shows a very sensitive response to parathyroid hormone *in vitro* (Chase and Aurbach, 1970; Aurbach and Chase, 1970) with an increase in intracellular concentration of cyclic AMP (Fig. 17). Studies in other laboratories (Vaes, 1968; Raisz *et al.*, 1969; Herrmann-Erlee, 1972) have established that dibutyryl-cyclic AMP added *in vitro* mimics the action of parathyroid hormone on bone. It is implied, then, that a rise in concentration of cyclic AMP in the skeletal tissue *in vivo* leads to resorption of bone mineral and transfer of calcium from bone to the extracellular fluid. The experiments of Rasmussen *et al.* (1968) and Russell *et al.* (1968) have shown that intravenous injection of dibutyryl-cyclic AMP as well as 3′,5′-AMP itself causes an increase in rate of phosphate excretion, thereby mimicking the effect *in vivo* of parathyroid hormone on the kidney. These several experiments lend further support to the thesis that cyclic 3′,5′-AMP indeed is the intracellular mediator of the actions of parathyroid hormone.

Other recent studies (Heersche *et al.*, 1971) indicate that the effect of dibutyryl-cyclic AMP on bone is effected through inhibition of cyclic nucleotide phosphodiesterase in the tissue, causing thereby an

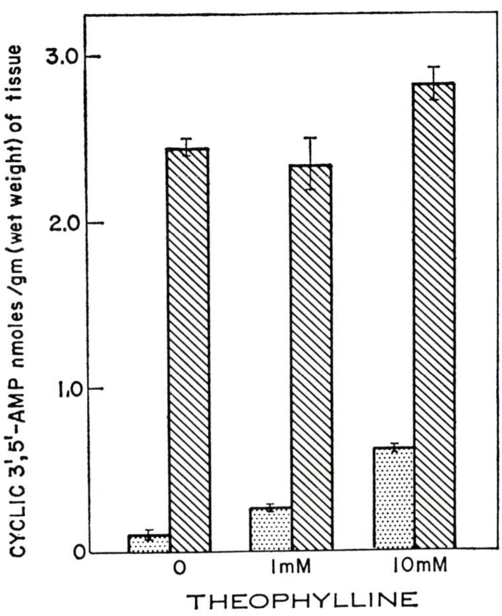

FIG. 17. Parathyroid hormone(PTH)-induced increase in concentration of 3′,5′-AMP in skeletal tissue incubated *in vitro*. Fetal rat calvaria were incubated for 5 minutes at 37°C after adding parathyroid hormone (2.5 µg/ml) with or without theophylline. Vertical bars indicate standard error for triplicate determinations. ▨, without PTH; ▧, with PTH. From Chase and Aurbach (1970).

accumulation of endogenous cyclic 3′,5′-AMP. Experiments with radioactive dibutyryl-cyclic AMP showed that the rise in tissue concentration of 3′,5′-AMP could not be accounted for by degradation of dibutyryl-cyclic 3′,5′-AMP. Further, it was shown that the concentration of dibutyryl-cyclic AMP reached within the cell was sufficient to cause inhibition of phosphodiesterase.

XI. Role of Cyclic 3,′5′-AMP within Cells of Receptor Tissues

As noted above, it is apparent that a rise in concentration of cyclic AMP within the cells of the receptor tissues leads ultimately to physiological expression of the response to parathyroid hormone. The mechanisms important in mediating this effect of heightened tissue concentration of cyclic AMP are under study in several laboratories. Protein kinases (Walsh *et al.*, 1968; Langan, 1968; Kuo and Greengard, 1969; Winickoff and Aurbach, 1970; Gill and Garren, 1970; Tao *et al.*, 1970; Reimann *et al.*, 1971; Erlichman *et al.*, 1971) have been found in a number of tissues whose metabolism is regulated by hormones through the intermediation of cyclic AMP. Studies in Greengard's laboratory

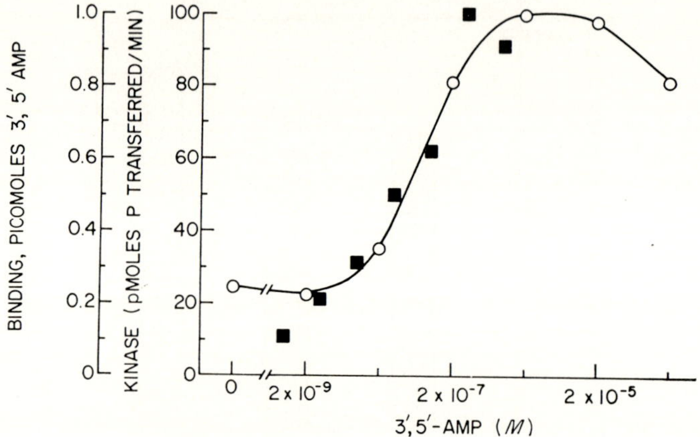

Fig. 18. Cyclic 3',5'-AMP binding (■) to and activation (○) of crude protein kinase enzyme obtained from bovine renal cortex. The K_m for binding of tritiated 3',5'-AMP was the same as the concentration of 3',5'-AMP required for half-maximal activation of the enzyme.

(Kuo and Greengard, 1969) as well as those in our laboratory (Winickoff and Aurbach, 1970) have identified a histone kinase in kidney tissue that is activated by cyclic 3',5'-AMP. It is probable that phosphorylation of a specific protein catalyzed by the protein kinase is part of the chain of events leading to the physiological effects of the hormone. Figure 18 illustrates an increase in kinase activity obtained from bovine kidney under the influence of increasing concentrations of cyclic 3',5'-AMP.

There is evidence from several laboratories that activation of protein kinase is effected through binding of cyclic AMP to a specific binding protein. The binding protein in heart, adrenal, muscle, reticulocytes, and possibly salivary glands, can be detected with tritium-labeled 3',5'-AMP and appears to be distinct from the kinase itself. Certain experiments suggest that this is a controlling mechanism in the kidney. The concentration of 3',5'-AMP required for half-maximal binding to the binding protein was the same (2×10^{-7} M) as the K_m for 3',5'-AMP in activating the crude kinase enzyme. Further, it was found that the kinase was separable from the cyclic 3′,5′-AMP-binding protein by gel filtration on Bio-Gel P-200 (Fig. 19). Thus, one would conclude that in the resting state the binding protein is bound to the kinase enzyme which then shows relatively low catalytic activity. When the concentration of cyclic AMP rises, the nucleotide becomes bound to the inhibitor protein which then dissociates from the kinase accounting for the apparent activation of the enzyme by cyclic 3′,5′-AMP.

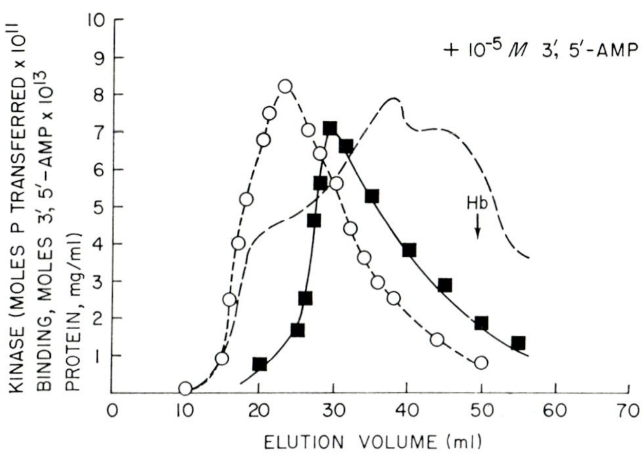

Fig. 19. Fractionation of crude protein kinase from renal cortex on Bio-Gel P-200. The protein binding tritiated cyclic AMP (○) was clearly separable from the kinase enzyme (■). Total protein (---).

XII. Physiological Substrate for Phosphorylation

Several functions mediated in receptor tissues by cyclic AMP, particularly secretory processes, appear to involve microtubules of cells. Melanocyte-stimulating hormone, whose action is mediated by cyclic AMP, causes dispersion of melanin granules apparently along lines demarcated by microtubules (Bickle *et al.*, 1966) within the melanophores. Microtubular systems bind colchicine, and in several instances colchicine has been shown to block secretory processes in cells in which microtubules have been identified. For example, it has been shown that colchicine blocks release of iodinated compounds from the thyroid in response to TSH (Williams and Wolff, 1970) as well as insulin secretion (Lacy *et al.*, 1968) from pancreatic islets in response to glucose. In melanocytes, colchicine interferes with aggregation of dispersed melanin granules (Malawista, 1965). More recently, it has been reported (Goodman *et al.*, 1970) that microtubular protein may be the substrate for phosphorylation catalyzed by protein kinases which in turn are activated as noted above by cyclic AMP. It is possible that bone resorption under the influence of parathyroid hormone is another process involving microtubules. Such a microtubular system might be involved in transfer of calcium across cells and membranes from bone or renal tubular fluid

TABLE III
Colchicine Inhibition of Response to Parathyroid Hormone in Thyroparathyroidectomized Rats

	Colchicine (mg)	Serum calcium (mg/100 ml)
Control	0	4.9 ± 0.35
	0.2, i.p.	5.1 ± 0.24
PTH, 100 USP units	0	7.6 ± 0.77
	0.2	4.9 ± 0.37

to the extracellular fluid. Some recent experiments suggest the possibility that this type of mechanism is involved in the action of parathyroid hormone. Administration of colchicine to rats causes hypocalcemia and, as illustrated by the experimental results presented in Table III, colchicine prevents the rise in serum calcium of thyroparathyroidectomized

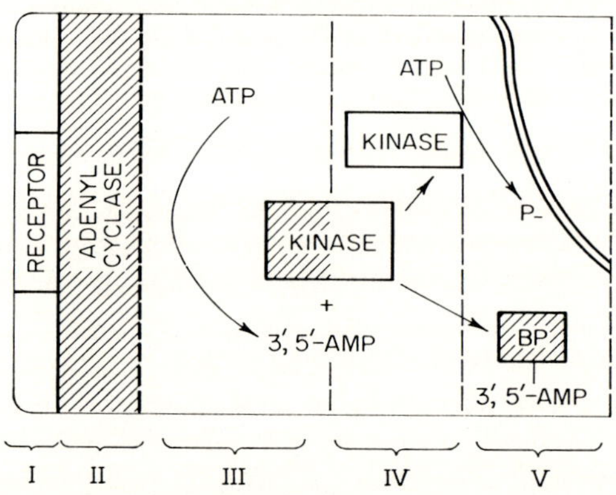

FIG. 20. Scheme proposed for sequence of metabolic events in the mechanism of action of parathyroid hormone on bone and kidney. I, binding of the hormone to receptor site; II, activation of adenyl cyclase; III, increased intracellular concentration of 3',5'-AMP; IV, binding of 3',5'-AMP to receptor protein causing its dissociation from kinase and consequent activation of kinase enzyme; V, increased phosphorylation of substrate protein. It is possible that the substrate protein is part of a transport system for calcium (perhaps involving microtubules).

rats given parathyroid hormone. It is of particular interest in this regard that Holtrup (1972) has found microtubules in osteocytes and osteoblasts of neonatal bone cells.

A scheme representing our current understanding of the mechanism of action of parathyroid hormone is presented in Fig. 20. The first several steps are clearly established. The exact mechanism remains to be elucidated whereby increased intracellular concentration of cyclic 3',5'-AMP leads to transfer of calcium to the extracellular fluid. The question whether this parathyroid hormone-mediated response in bone and kidney indeed involves microtubules must be resolved by further investigation.

XIII. Biological Inactivation and Metabolic Fate of Parathyroid Hormone

Kinetic studies with the renal adenyl cyclase system and parathyroid hormone indicate that the membrane system contains not only the receptor for parathyroid hormone and a link between the receptor and activation of adenyl cyclase, but also a system for inactivating the hormone. Biological inactivation of the hormone is readily detected by incubating the hormone with renal cell membranes for periods of time and then determining loss of hormonal activity utilizing renal adenyl cyclase as the test system. Incubation of purified bovine parathyroid hormone for 20 minutes at 37° with plasma membrane fractions of rat kidney causes biological inactivation of 90% or more of the hormone added (Fig. 21). Incubation of the hormone with a boiled membrane preparation under identical conditions causes no significant loss of biological activity. There is little or no biological inactivation when the hormone is incubated with plasma membrane fractions at 4°; inactivation at 37° proceeds at a rate proportional to time (Fig. 22). These observations make it extremely likely that biological degradation of the hormone with kidney membranes is due to an enzyme associated with the same plasma membrane fraction containing the adenyl cyclase that is activated by the hormone. Inactivation of parathyroid hormone at the receptor site does not appear to be directly related to nor a necessary condition for activation of adenyl cyclase by the hormone. Activation of adenyl cyclase in skeletal tissue by parathyroid hormone occurs without significant inactivation (Marcus, Heersche, and Aurbach, unpublished).

Experiments to date indicate that biological inactivation of parathyroid hormone is specific for membranes prepared from renal tissue. This is illustrated by experiments shown in Fig. 23. The bovine hormone remained biologically active after incubation with membranes from calvaria, heart, or liver. The rate of inactivation of parathyroid hormone preparations seems to vary with chemical structure. Since the enzyme

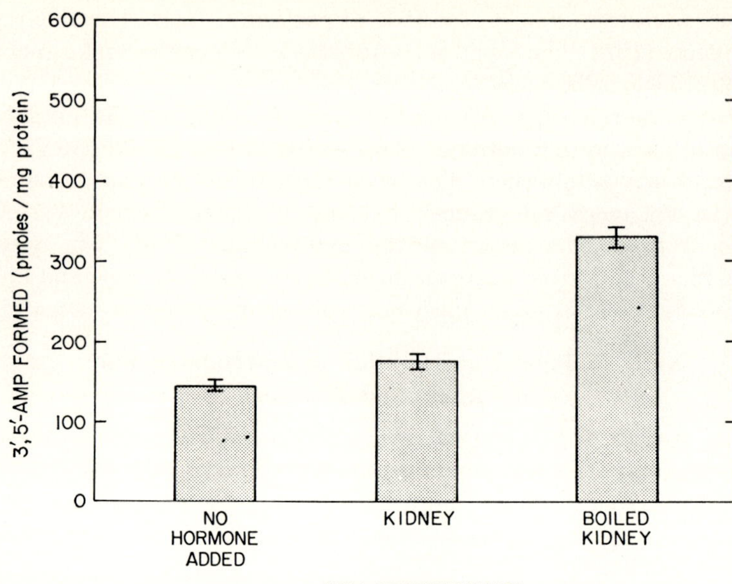

FIG. 21. Inactivation of bovine parathyroid hormone (PTH) during incubation with membrane fraction from renal cortex. Incubation of hormone with kidney membranes causes almost complete inactivation within 20 minutes. Incubation with boiled kidney membrane does not influence the activity of the hormone (determined by bioassay with the renal adenyl cyclase system).

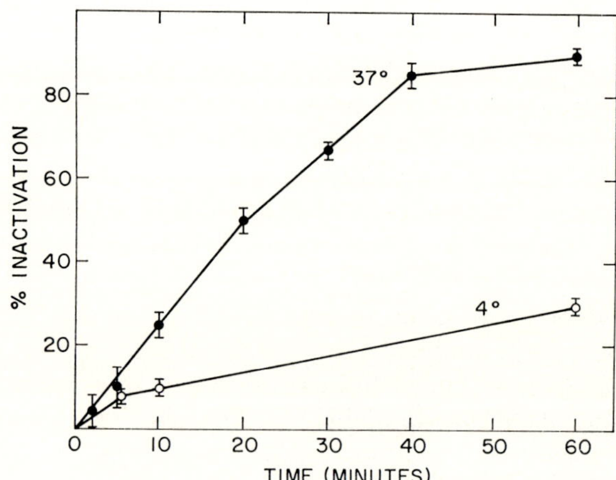

FIG. 22. Rate of inactivation of parathyroid hormone by renal membrane at 37° or 4°C.

causing inactivation of the hormone appears to be located primarily in the kidney with little or no detectable enzyme in bone, this allows for discrepant ratios of apparent activity on renal adenyl cyclase *in vitro* versus skeletal adenyl cyclase. Hormone structures with high resistance to enzymatic destruction will thus have high apparent ratios of activity on kidney relative to bone. This problem is discussed further in connection with the apparent biological activity of the different forms of natural and synthetic parathyroid polypeptides.

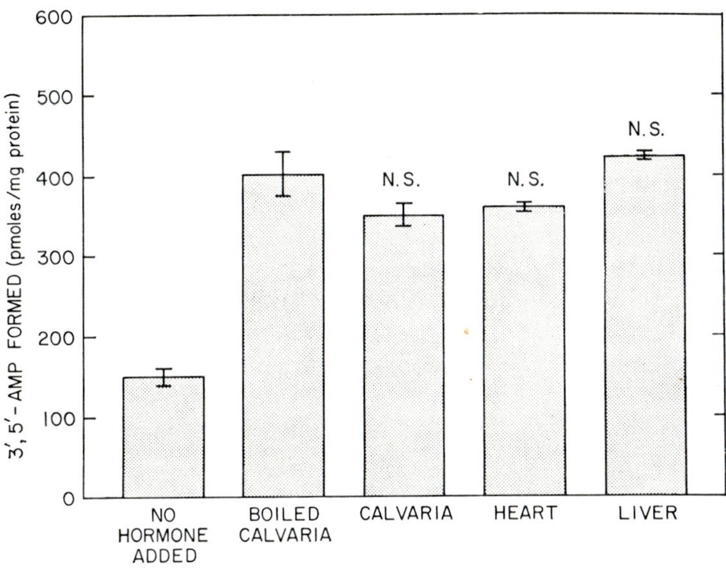

Fig. 23. There is no significant inactivation of parathyroid hormone (PTH) by membranes prepared from calvaria, heart, or liver.

The suggestion that the kidney may be important in the biological inactivation of parathyroid hormone is not new. Orimo *et al.* (1965) reported that crude parathyroid extracts were inactivated biologically upon incubation with kidney slices. They also found some apparent inactivation with slices from other tissues of the rat. Martin *et al.*, (1969) have approached the problem of metabolism of parathyroid hormone utilizing ^{125}I-labeled hormone. They found that microsomes obtained from rat kidney caused very rapid degradation of iodine-labeled parathyroid hormone *in vitro* (Martin *et al.*, 1969). This system seemed to be specific for parathyroid hormone and for microsomes obtained from the kidney. The exact relationship between the two systems (degradation of radioiodine-labeled hormone versus biological inactivation) remains to be clarified. The possibility has not been excluded that the microsomal system represents the same enzyme found on the plasma membrane; partial disruption of the system from the plasma membrane during homogenization could account for its appearance in the "microsomal fraction." In any event, the studies with *in vitro* systems as well as the high concentration and prolonged half-life of parathyroid hormone (by radioimmunoassay) in chronic renal disease, clearly imply that the kidney is important in the peripheral metabolism of the hormone.

XIV. Primary Structure and Immunological Reactivity

NATURE OF THE HORMONE IN THE CIRCULATION

At the 1965 conference (Potts *et al.*, 1966), we presented physiological studies on cows indicating that the hormone in the circulation was immunologically identical to that purified from parathyroid glands. This information suggested that there was no chemical change in the hormone after it was biosynthesized in the gland, and secreted into the circulation to reach specific receptor sites at the periphery. The first suggestion that this thesis might be incorrect was reported by Berson and Yalow (1968) concerning the hormone circulating in man. They had developed two different types of antisera to bovine parathyroid hormone. With one type of antiserum, the apparent half-life of the hormone in the circulation appeared to be inordinately long. The other type of antiserum detected a relatively rapid rate of disappearance for the hormone from plasma after parathyroidectomy or induction of hypercalcemia. These results indicated that there might be more than one form of parathyroid hormone in the circulation of human beings. Further, they found that the half-life of either form of the hormone (i.e., detected by either of the two antisera types) was prolonged markedly in uremic subjects. One might conclude, then, that after secretion human parathyroid hormone is converted into a form that disappears more slowly from the circulation.

Arnaud *et al.* (1970) as well as Sherwood *et al.* (1970) have also proposed that the hormone in the circulation of man is different immunologically from that in the gland. The latter group utilized organ cultures for parathyroid glands to study regulation of synthesis as well as secretion into the medium. The hormone in the culture medium was immunologically different from that purified from bovine glands. A similar discrepancy was found when utilizing human parathyroid tissue *in vitro*. Examination of the media from the tissue cultures revealed a form of the hormone lower in molecular weight than that extracted directly from the gland. It is to be emphasized that all these studies utilized radioimmunoassay rather than biological determinations to detect the parent hormone as well as apparent metabolites. Potts *et al.* (1972) have also found evidence for circulating metabolites of parathyroid hormone that differ by size as well as by immunological activity from the hormone in the gland. Gel filtration studies indicate that there is native hormone (immunologically reactive material eluting in the position of intact parathyroid hormone) as well as metabolites of lower molecular weight in samples of peripheral plasma. These studies were extended to determine the nature of the hormone in plasma obtained directly from the inferior thyroid veins. These samples, obtained primarily as a diagnostic aid in localizing abnormal parathyroid tissue in patients with primary hyperparathyroidism, contained hormone representing that just secreted from the gland. Hormone in these samples was equivalent in size to the native 1 to 84 polypeptide hormone. The implication, then, is that the hormone immediately elaborated into the circulation does not differ from that in the gland, but after secretion into the peripheral circulation, smaller hormonal metabolites are rapidly produced. Whether any of these fragments of low molecular weight are of biological significance remains to be determined. The studies on the fragment (1–29) of the native hormone and of the synthetic parathyroid polypeptide (1–34) clearly indicate that the biologically significant area of the hormone molecule lies at the amino terminus. On the other hand, only part of the immunological reactivity of the native hormone molecule can be ascribed to this segment of the molecule. There are clearly areas peripheral to this region that react significantly with antibodies developed against the native 1 to 84 polypeptide.

Studies carried out recently elucidate some of the marked differences in immunological reactivity between the amino-terminal portions of the molecule and the peripheral or C-terminal portion (Keutmann *et al.*, 1971b). In the discussion on the chemistry of parathyroid hormone it was noted that the entire C-terminal fragment, residues 53 to 84, could be prepared from native hormone after acylation of the ϵ-amino groups and cleavage catalyzed by trypsin. A comparison could then be made between the C-terminal dotriacontapeptide, the amino terminal nonacosapeptide (residues 1 to 29), and native hormone (1 to 84) reacting with antiserum developed toward the entire molecule. Neither of the polypeptide fragments alone (i.e., the amino terminus or the carboxyl terminus) caused complete displacement of ^{125}I-labeled native hormone from antibody (Fig. 24). However, when the two fragments were added together in equimolar concentrations, they caused almost complete displacement of native hormone from antibody. These experiments indicate that some of the peripheral immunologically important sites lie within the C-terminal dotriacontapeptide. Further experiments confirming this thesis are illustrated in Fig. 25. In this instance, the antiserum was adsorbed with the fragment representing the carboxyl terminus (residues 53 to 84) or with a fragment (residues 1 to 29) representing the biologically important amino terminus. By this means, antisera can be selected out which react predominantly with one end of the molecule or the other. Note that the antisera thus selectedly preadsorbed continue to

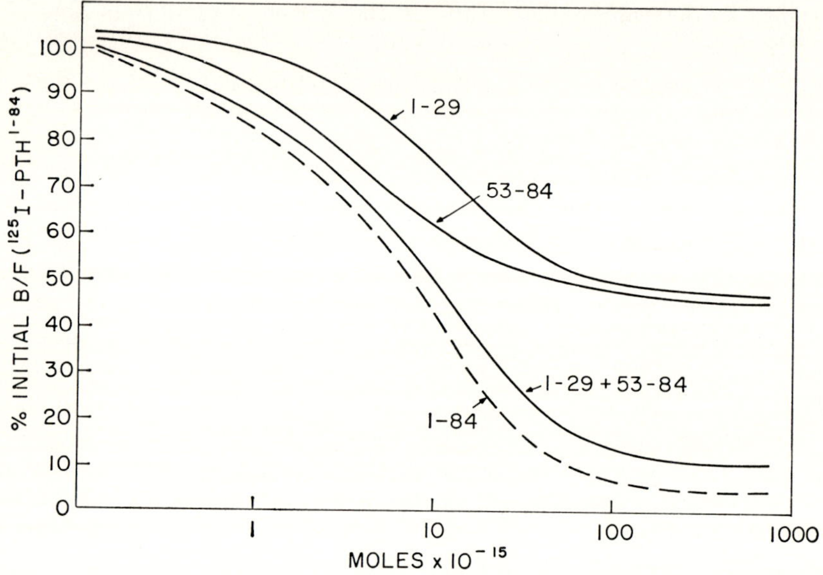

Fig. 24. Immunoreactivity of fragments of parathyroid hormone molecule compared to native bovine hormone (1 to 84). Neither the C-terminal fragment (53 to 84) nor the amino-terminal biologically active fragment (1 to 29) caused complete displacement of ^{125}I-PTH$_{1-84}$ from antibody. A mixture of the two fragments, however, produced almost complete immunological reactivity with the native hormone molecule. B/F = ratio of bound to free.

react with the native hormone molecule (residues 1 to 84). These experiments substantiate further that at least two different types of immunological determinants exist in the hormone molecule. One set is clearly outside the biologically important amino-terminal region of the molecule. This, then, affords a possible explanation for persistence in the circulation of biologically inert fragments of the native hormone molecule that are immunologically reactive. We cannot prove that none of the lower molecular weight species is biologically active; some might represent biologically effective fragments from the amino terminus. On the other hand, it is more probable that most, if not all, of the low molecular weight immunologically reactive fragments of the molecule in the circulation represent biologically inert polypeptides.[3] Use of the selectedly preadsorbed antisera may help resolve this issue. Even more important would be the development of a sensitive assay, perhaps a radioreceptor assay, that would permit identification of biologically active fragments of the hormone in the circulation.

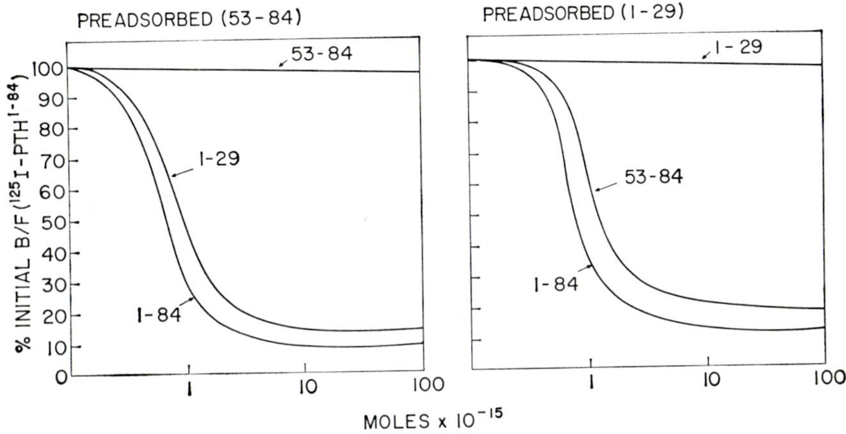

Fig. 25. Species of antibody selected for reactivity against the amino terminal (1 to 29) or carboxyl-terminal (53 to 84) fragments of bovine parathyroid hormone. By selective adsorption, one can obtain antisera specific for one region of the molecule alone. B/F = ratio of bound to free.

XV. Primary Structure and Biological Activity

Chemical synthesis of an active parathyroid polypeptide provides for the possibility of preparing and studying analogs of known composition. Synthesis of the tetratriacontapeptide representing residues 1 to 34 of bovine parathyroid hormone proved that the same limited sequence at the amino terminus was sufficient for action of the hormone on both kidney and bone. More recently, the shorter polypeptide representing residues 1 to 29 has been isolated in pure form from hydrolyzates of the native molecule and similarly has been found active on both renal and skeletal tissue.

---

[3] The native molecule (i.e., obtained from the gland) is considerably more active *in vivo* as well as *in vitro*. It would not seem efficient to synthesize and secrete a highly active molecule just to convert it to a considerably *less active* fragment before it reaches its specific peripheral receptor sites.

These studies have been extended in an effort to define the minimum length polypeptide sufficient for biological activity. Tests for biological activity have involved almost exclusively the *in vitro* renal adenyl cyclase system which requires only microgram quantities of material for bioassay. Fragments tested include those produced from the native hormone by enzymatic or chemical means as well as those synthesized. The fragment representing residues 2 to 34, obtained by chemical synthesis, was biologically inactive.[4] Mixtures of fragments representing residues 1 to 20, 26 to 44, and 53 to 84, also were inactive biologically as were the fragments 1 to 20 (obtained by digestion of the native molecule with trypsin) or 1 to 13 (obtained by chemical synthesis). Thus, to date, the minimal fragment obtained that is biologically active consists of the 1 to 29 peptide hydrolyzed from the native molecule by dilute acid. The analog Tyr_1-34 showed reduced but significant biological activity. This analog may prove to be useful for studies requiring radioiodinated hormonal peptides. Extension of the Ala_1-34 by one tyrosine at the amino terminus also yielded an active analog, but extension by three additional residues gave an inactive polypeptide. Information on activity is summarized for these several synthetic peptide fragments in Fig. 26.

Some of the biological properties of porcine parathyroid hormone were discussed in connection with the presentation of its chemical structure. It was noted that the porcine molecule shows biological potency equal to that of the bovine molecule when assayed *in vivo* using serum calcium as the parameter. Bioassays using the renal adenyl cyclase *in vitro*, however, showed the porcine molecule to be only one-third as active as the bovine molecule. Some recent experiments seem to explain at least part of this discrepancy.

We have shown that the membrane-bound adenyl cyclase obtained from rat kidney also contains an enzyme that inactivates parathyroid hormone (Figs. 20–22). This enzyme is lacking or undetectable in membranes prepared from bone (Fig. 22). This would allow for discrepant activity ratios for various parathyroid hormone structures on skeletal versus renal tissue. The bovine hormone is relatively resistant to inactivation by the enzyme in renal membrane preparations and shows a relatively high ratio of potency on renal versus skeletal tissue. Con-

---

[4] Dr. Martin Rodbell and his associates (Rodbell *et al.*, 1971) have found that des-His-glucagon (lacking the amino-terminal residue of the native hormone) is a competitive inhibitor of binding of ^{125}I-labeled glucagon to liver receptors. This observation prompted us to test extensively for the possibility that this parathyroid hormone analog (residues 2–34) might compete for parathyroid hormone binding sites. No inhibition was found even at very high concentrations of analog.

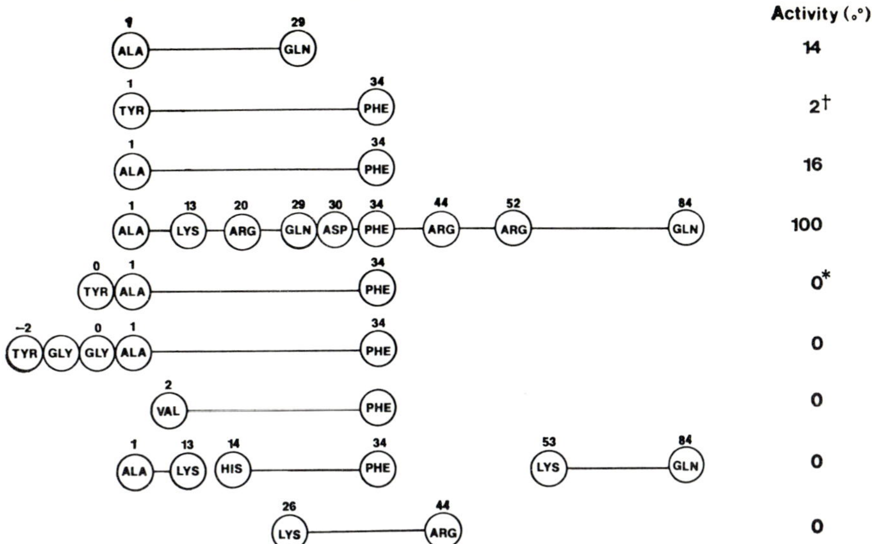

Fig. 26. Biological activity of native bovine parathyroid hormone and fragments of the molecule. Data shown represent percent activity on a molar basis using the *in vitro* bioassay with renal adenyl cyclase system. †Maximal response less than with native hormone. *Less than 0.5%.

versely, hormonal species like the porcine molecule, more sensitive to the renal enzyme, will show discrepantly low activity when tested with the renal adenyl cyclase assay. The porcine hormone shows an activity of only 500 to 600 USP units/mg with the *in vitro* adenyl cyclase bioassay. When tested *in vivo*, the porcine hormone showed a potency comparable to the bovine hormone (2400 units/mg). It is presumed that the latter bioassay is more an index of the action of the hormone on bone. Thus, when the porcine hormone was compared to bovine hormone in assays with skeletal tissue *in vitro*, the two molecules were equally effective (Fig. 27). In the latter instance, adenyl cyclase from fetal rat calvaria was used as the test system. The interpretation of these several results, then, is that the porcine hormone is more susceptible to (or is a better substrate for) the renal enzyme that inactivates the hormone. It is not certain that the inactivating enzyme is a significant factor *in vivo*. The porcine hormone was tested *in vivo* by determining the renal responsiveness using urinary 3'5'-AMP as the parameter. In this instance the porcine hormone was virtually as active as the bovine standard (Fig. 28).

In contrast to the relatively low activity of porcine parathyroid hormone on renal adenyl cyclase *in vitro*, the smaller active framents (1–29 and 1–34) of the bovine hormone show relatively greater activity *in*

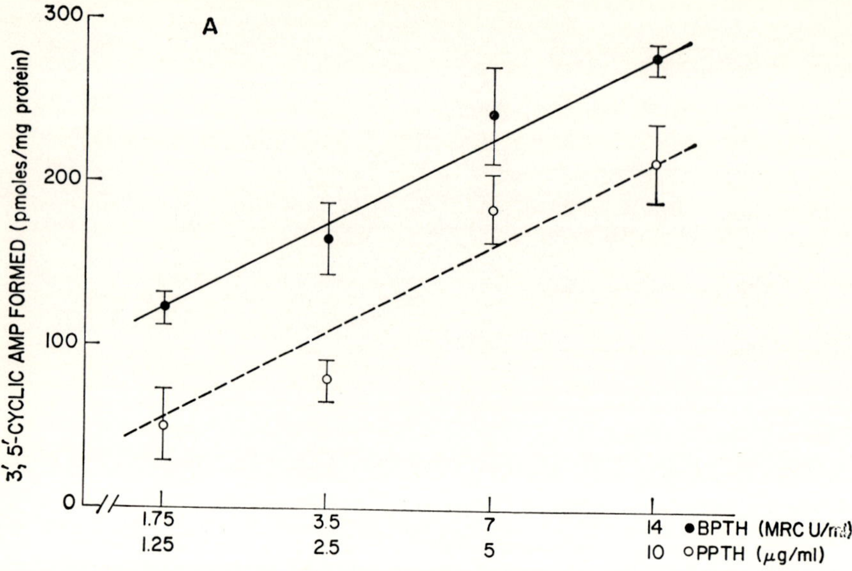

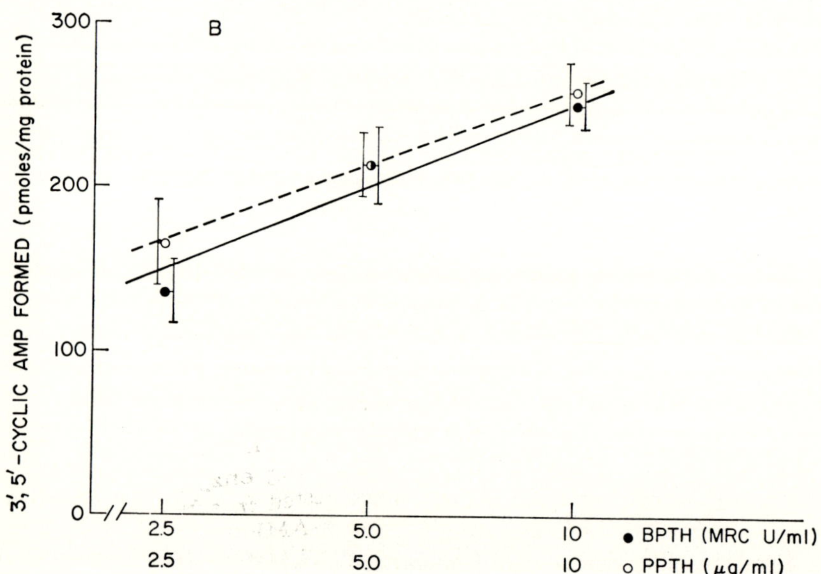

FIG. 27. Comparative bioassays of porcine hormone (PPTH) against bovine parathyroid hormone (BPTH) *in vitro* using adenyl cyclase from kidney (panel A) or bone (panel B). The porcine hormone is only one-third as active as the bovine hormone in assay A; it is equal in potency to the bovine hormone when tested with skeletal adenyl cyclase (B) or *in vivo*. MRC, Medical Research Council.

vitro than *in vivo* (Potts *et al.*, 1971). For example, the synthetic tetria-contapeptide was active at 400–600 USP units/mg with the renal cyclase system; *in vivo* it showed a potency of only 150 units/mg. This discrepancy presumably reflects a shorter half-life *in vivo* for the smaller peptide fragments of the bovine hormone. It will be important to determine the half-lives for the fragments as well as for the porcine hormone and compare results with the bovine hormone. Such a comparison would

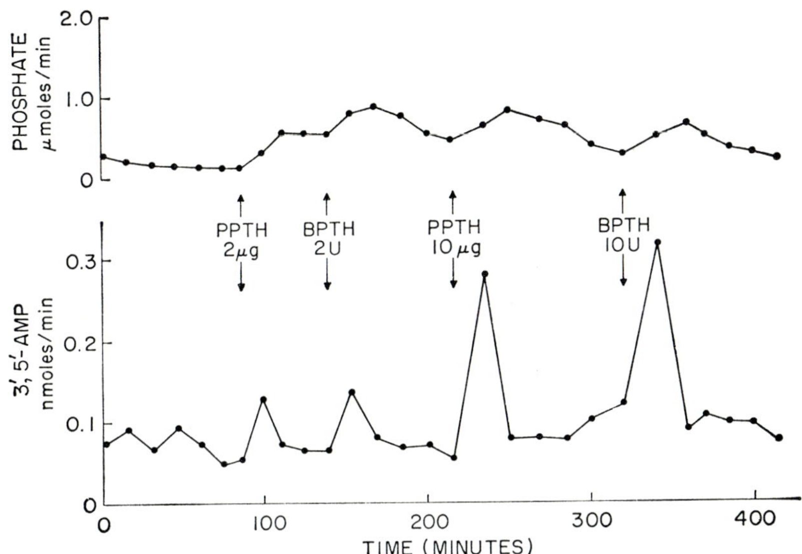

Fig. 28. Comparative responses of porcine and bovine parathyroid hormones *in vivo* using urinary 3′,5′-AMP excretion as the biological parameter. The arrows indicate points at which purified porcine hormone (PPTH) or bovine parathyroid hormone (BPTH, Medical Research Council standard) was injected intravenously into a conscious, constantly infused parathyroidectomized rat.

help determine the physiological significance of the renal inactivating enzyme and also delineate any function of C-terminus in protecting the native hormone against inactivation *in vivo*.

ACKNOWLEDGMENT

The authors wish to acknowledge the important contributions made to this work by Drs. L. J. Deftos, J. S. Woodhead, C. J. Robinson, J. N. M. Heersche, D. M. Heath, and R. N. Winickoff. We are indebted to B. F. Dawson, M. L. Hogan, R. Sauer, M. L. Stoltz, S. A. Fedak, J. Palmer, W. Cheatham, and C. Woodard for expert technical assistance. We wish to thank Mrs. L. Perry for secretarial assistance in preparing and typing the manuscript.

REFERENCES

Amer, M. S. (1968). *Endocrinology* **82,** 166.

Arnaud, C. D., Tsao, H. S., and Oldham, S. B. (1970). *Proc. Nat. Acad. Sci. U.S.* **67,** 415.

Aurbach, G. D. (1959a). *Arch. Biochem. Biophys.* **80,** 467.

Aurbach, G. D. (1959b). *J. Biol. Chem.* **234,** 3179.

Aurbach, G. D. (1959c). *Endocrinology* **64,** 296.

Aurbach, G. D., and Chase, L. R. (1970). *Fed. Proc. Fed. Amer. Soc. Exp. Biol.* **29,** 1179.

Aurbach, G. D., and Houston, B. A. (1968). *J. Biol. Chem.* **243,** 5935.

Aurbach, G. D., Potts, J. T., Jr., Chase, L. R., and Melson, G. L. (1969). *Ann. Intern. Med.* **70,** 1243.

Berson, S. A., and Yalow, R. S. (1968). *J. Clin. Endocrinol. Metab.* **28,** 1037.

Berson, S. A., Yalow, R. S., Aurbach, G. D., and Potts, J. T., Jr. (1963). *Proc. Nat. Acad. Sci. U.S.* **49,** 613.

Bikle, D., Tilney, L. G., and Porter, K. R. (1966). *Protoplasma* **61,** 322.

Borle, A. B. (1972). *In* "Parathyroid Hormone and The Thyrocalcitonins" (R. V. Talmage and P. L. Munson, eds.). Excerpta Med. Found., Amsterdam. In press.

Brewer, H. B., Jr., and Ronan, R. (1970). *Proc. Nat. Acad. Sci. U.S.* **67,** 1862.

Chase, L. R., and Aurbach, G. D. (1967). *Proc. Nat. Acad. Sci. U.S.* **58,** 518.

Chase, L. R., and Aurbach, G. D. (1968). *Science* **159,** 545.

Chase, L. R., and Aurbach, G. D. (1970). *J. Biol. Chem.* **245,** 1520.

Chase, L. R., Fedak, S. A., and Aurbach, G. D. (1969a). *Endocrinology* **84,** 761.

Chase, L. R., Melson, G. L., and Aurbach, G. D. (1969b). *J. Clin. Invest.* **48,** 1832.

Collip, J. B. (1925). *J. Biol. Chem.* **63,** 395.

Edman, P., and Begg, G. (1967). *Eur. J. Biochem.* **1,** 80.

Erlichman, J., Hirsch, A. H., and Rosen, O. M. (1971). *Proc. Nat. Acad. Sci. U.S.* **68,** 731.

Gill, G. N., and Garren, L. D. (1970). *Biochem. Biophys. Res. Commun.* **39,** 335.

Goodman, D. B. P., Rasmussen, H., DiBella, F., and Guthrow, C. E., Jr. (1970). *Proc. Nat. Acad. Sci. U.S.* **67,** 652.

Heersche, J. N. M., Fedak, S. A., and Aurbach, G. D. (1971). *J. Biol. Chem.* **246,** 6770.

Herrmann-Erlee, M. P. M. (1972). *In* "Parathyroid Hormone and The Thyrocalcitonins" (R. V. Talmage and P. L. Munson, eds.). Excerpta Med. Found., Amsterdam. In press.

Holtrup, M. (1972). *In* "Parathyroid Hormone and The Thyrocalcitonins" (R. V. Talmage and P. L. Munson, eds.). Excerpta Med. Found., Amsterdam. In press.

Kaminsky, N. I., Broadus, A. E., Hardman, J. G., Jones, D. J., Jr., Ball, J. H., Sutherland, E. W., and Liddle, G. W. (1970). *J. Clin. Invest.* **49,** 2387.

Keutmann, H. T., Aurbach, G. D., Dawson, B. F., Niall, H., Deftos, L. J., and Potts, J. T., Jr. (1971a). *Biochemistry* **10,** 2779.

Keutmann, H. T., Niall, H. D., Tregear, G. W., Murray, T. M., O'Riordan, J. L. H., Aurbach, G. D., and Potts, J. T., Jr. (1971b). *Proc. 53rd Meet. Endocrine Soc., San Francisco* Abstr No. 41.

Krishna, G., Weiss, B., and Brodie, B. B. (1968). *J. Pharmacol. Exp. Ther.* **163,** 379.

Kuo, J. F., and Greengard, P. (1969). *Proc. Nat. Acad. Sci. U.S.* **64**, 1349.

Lacy, P. E., Howell, S. L., Young, D. A., and Fink, C. J. (1968). *Nature (London)* **219**, 1177.

Langan, T. A. (1968). *Science* **162**, 579.

Lefkowitz, R. J., Roth, J., Pricer, W., and Pastan, I. (1970). *Proc. Nat. Acad. Sci. U.S.* **65**, 745.

Levey, G. S. (1971). *Biochem. Biophys. Res. Commun.* **43**, 108.

Malawista, S. E. (1965). *J. Exp. Med.* **122**, 361.

Marcus, R., and Aurbach, G. D. (1969). *Endocrinology* **85**, 801.

Marcus, R., and Aurbach, G. D. (1971). *Biochim. Biophys. Acta* **242**, 410.

Martin, T. J., Melick, R. A., and deLuise, M. (1969). *Biochem. J.* **111**, 509.

Melson, G. L., Chase, L. R., and Aurbach, G. D. (1970). *Endocrinology* **86**, 511.

Merrifield, R. B. (1969). *Advan. Enzymol. Relat. Areas Mol. Biol.* **32**, 221.

Munson, P. L. (1955). *Ann N.Y. Acad. Sci.* **60**, 776.

Nagata, N., and Rasmussen, H. (1970). *Proc. Nat. Acad. Sci. U.S.* **65**, 368.

Niall, H., and Potts, J. T., Jr. (1970). *Proc. Biochem. Symp. 1st., New York,* p. 215. Dekker, New York.

Niall, H., Keutmann, H., Sauer, R., Hogan, M., Dawson, B., Aurbach, G. D., and Potts, J. T., Jr. (1970). *Hoppe-Seyler's Z. Physiol. Chem.* **351**, 1586.

Orimo, H., Fujita, T., Morii, H., and Nakao, K. (1965). *Endocrinology* **76**, 255.

O'Riordan, J. L. H., Potts, J. T., Jr., and Aurbach, G. D. (1971). *Endocrinology* **89**, 234.

Parsons, J. A., Neer, R. M., and Potts, J. T., Jr. (1971). *Endocrinology* **89**, 735.

Pisano, J. J., and Bronzert, T. J. (1969). *J. Biol. Chem.* **244**, 5597.

Potts, J. T., Jr., and Aurbach, G. D. (1965). *In* "The Parathyroid Glands: Ultrastructure, Secretion and Function" (P. J. Gaillard, R. V. Talmage, and A. M. Budy, eds.), pp. 53–67. Univ. of Chicago Press, Chicago, Illinois.

Potts, J. T., Jr., Aurbach, G. D., and Sherwood, L. M. (1966). *Recent Progr. Horm. Res.* **22**, 101.

Potts, J. T., Jr., Keutmann, H. T., Niall, H. D., Deftos, L. J., Brewer, H. B., and Aurbach, G. D. (1968). *In* "Parthyroid Hormone and Thyrocalcitonin (Calcitonin)" (R. V. Talmage and L. F. Belanger, eds.), pp. 44–53. Excerpta Med. Found., New York.

Potts, J. T., Jr., Niall, H. D., Keutmann, H. T., Tregear, G. W., Deftos, L. J., Habener, J., Murray, T. M., O'Riordan, J. L. H., and Aurbach, G. D. (1972). *In* "Parathyroid Hormone and the Thyrocalcitonins" (R. V. Talmage and P. L. Munson, eds.). Excerpta Med. Found., Amsterdam. In press.

Potts, J. T., Jr., Tregear, G. W., Keutmann, H. T., Niall, H. D., Sauer, R., Deftos, L. J., Dawson, B. F., Hogan, M. L., and Aurbach, G. D. (1971). *Proc. Nat. Acad. Sci. U.S.* **68**, 63.

Raisz, L. G., Brand, J. S., Klein, D. C., and Au, W. Y. W. (1969). *In* "Progress in Endocrinology" (C. Gaul, ed.), pp. 696–703. Excerpta Med. Found., Amsterdam.

Rasmussen, H. (1970). *Science* **170**, 404.

Rasmussen, H., and Tenenhouse, A. (1968). *Proc. Nat. Acad. Sci. U.S.* **59**, 1364.

Rasmussen, H., Pechet, M., and Fast, D. (1968). *J. Clin. Invest.* **47**, 1843.

Reimann, E. M., Brostrom, C. O., Corbin, J. D., King, C. A., and Krebs, E. G. (1971) *Biochem Biophys. Res. Commun.* **42**, 187.

Rodbell, M., Birnbaumer, L., Pohl, S. L., and Krans, H. M. J. (1970). *Acta Diabet. Lat.* **7**, Suppl. 1, 9.

Rodbell, M., Birnbaumer, L., Pohl, S. L., and Sundby, F. (1971). *Proc. Nat. Acad. Sci. U.S.* **68**, 909.

Russell, R. G. G., Casey, P. A., and Fleisch, H. (1968). *Calcif. Tissue Res.* **2**, Suppl., 54.

Sherwood, L. M., Rodman, J. S., and Lundberg, W. B. (1970). *Proc. Nat. Acad. Sci. U.S.* **67**, 1631.

Steiner, A. L., Kipnis, D. M., Utiger, R., and Parker, C. (1969). *Proc. Nat. Acad. Sci. U.S.* **64**, 367.

Tao, M., Salas, M. L., and Lipmann, F. (1970). *Proc. Nat. Acad. Sci. U.S.* **67**, 408.

Tregear G. W. (1969). Ph.D. Thesis, Monash Univ., Australia.

Tsung, C. M., and Fraenkel-Conrat, H. (1965). *Biochemistry* **4**, 793.

Vaes, G. (1968). *Nature (London)* **219**, 939.

Walsh, D. A., Perkins, J. P., and Krebs, E. G. (1968). *J. Biol. Chem.* **243**, 3763.

Wells, H., and Lloyd, W. (1967). *Endocrinology* **81**, 139.

Williams, J. A., and Wolff, J. (1970). *Proc. Nat. Acad. Sci. U.S.* **67**, 1901.

Winickoff, R., and Aurbach, G. D. (1970). *Proc. 52nd Meet. Endocrine Soc.* Abstr. No. 17, p. 45.

Woodhead, J. S., O'Riordan, J. L. H., Keutmann, H. T., Stoltz, M. L., Dawson, B. F., Niall, H. D., Robinson, J., and Potts, J. T., Jr. (1971). *Biochemistry* **10**, 2787.

DISCUSSION

A. White: Do the receptors in bone and kidney for the parathyroid hormone also bind calcitonin, and is there any degree of competition between the two hormones?

Inasmuch as you have demonstrated success in separating the kinase from the cyclic 3′,5′-AMP binding protein, have you been able to detect the release of any moiety from the inhibited enzyme when it is activated by binding cyclic AMP *in vitro?* The data of Dr. Rudinger on vasotocin indicated that oxytocinase is an amino peptidase which inactivates the molecule by scission of the first peptide bond from the amino terminal end of the molecule. Does inactivation of parathyroid hormone occur by enzymatic degradation beginning at the carboxyl or amino-terminal end of the molecule? Is parathyroid hormone attacked *in vitro* by amino peptidases?

G. D. Aurbach: We have found, as have Murad *et al.*, that bone as well as kidney contain receptors for calcitonin as well as parathyroid hormone. In each tissue calcitonin activates adenyl cyclase and causes increased concentrations of cyclic 3′,5′-AMP. However, there is no competitive inhibition by one hormone on the effect of the other. Thus, the apparent physiological antagonism of the hormones must be accounted for at some level other than adenyl cyclase or generation of 3′,5′-AMP. We propose that the receptors for parathyroid hormone and calcitonin must be located on distinct cells. We are very interested in attempting to design experiments to show which particular cells respond to each hormone.

We do not know the detailed mechanism whereby the hormone is inactivated by tissues *in vitro* or *in vivo*. We know that fragments of the hormone are produced *in vivo*, but the enzymes responsible are unknown. The hormone is a substrate for amino peptidase *in vitro*, but the native hormone is relatively resistant to carboxypeptidase A.

We have found evidence that the 3′,5′-AMP-binding protein can be separated from the kinase, but have no information on concurrent degradation of the enzyme.

B. F. Rice: I would like to ask a question about the phosphaturic effect of parathyroid hormone. Renal physiologists would characterize this as due to an increased clearance of inorganic phosphorus. This same phenomenon occurs in thyroparathyroidectomized animals given vitamin D. Could you explain these similar phenomena on a molecular basis? Do you believe the increase in phosphorus clearance is occurring through the same mechanism?

G. D. Aurbach: Experiments at the gross physiological level as well as those with *in vitro* systems indicate that the release of phosphate in the urine is probably a secondary phenomenon. The phosphaturic effect has been considered for many years a primary and specific action of the hormone. In fact, until recently it was known as the most rapid physiological action of the hormone. The experiment with determination of cyclic AMP excretion, however, showed that the latter increases considerably before phosphate excretion. Perhaps one can take as an analogy the work of Williamson with the isolated, perfused rat heart. Subsequent to the epinephrine-produced inotropic action on the heart, there is an immediate increase in cyclic AMP within the cell which parallels the immediate rise in force of contraction. Moments later there is elaboration of phosphate into the perfusing medium. Williamson showed that this phosphate comes from the cardiac cells at the expense of creatine phosphate. Thus phosphate elaborated into the medium simply reflects utilization of intracellular high-energy phosphate compounds. An analogous situation in the kidney would represent a reasonable explanation for the phosphaturic action of parathyroid hormone. In other words, hormone-induced phosphaturia represents the end result of hormonal activation of the cell with consequent increased utilization of high-energy phosphate for metabolic activity.

We have no information concerning the phosphaturic action of vitamin D. We presume that its action is mediated by mechanisms different from those for parathyroid hormone.

J. M. McKenzie: Would you confirm for me, what I assume is the case, that once you separate the kinase from the inhibitor that is binding cyclic AMP the nucleotide then has no effect on kinase activity.

Since you quoted related evidence from work on the thyroid, I take the liberty of referring to that gland again. It is rather difficult to see in the thyroid that all the actions that cyclic AMP is thought to mediate could be explained by a kinase causing phosphorylation of a microtubular membrane, although this could possibly explain some of the actions. Is it likely, therefore, that there are perhaps several substrates for the cyclic-AMP-dependent kinases in cells or, perhaps even more likely, that there are several kinases dependent on cyclic AMP?

G. D. Aurbach: Your question is most appropriate and I want to reemphasize the point that I attempted to make during the talk: We really do not know what is the true intracellular physiological substrate for phosphorylation. Your question leads into the area of what other things cyclic AMP can be affecting in the cell, and protein synthesis is certainly one of them. Many actions of hormones are absolutely dependent upon new protein synthesis. Garren's group has reported within the past year that ribosomal protein may be a substrate for phosphorylation, and this might be a mechanism through which cyclic AMP could influence new protein synthesis. Also, as you indicate, there appears to be more than one protein kinase in each of several tissues. Some may not be influenced by cyclic AMP. Finally, after separating the kinase from the binding protein, the kinase no longer responds to cyclic AMP.

J. C. Orr: Molecular weights are known for these compounds, and you know exactly the concentrations with which you are dealing. In the dose response comparison of the synthetic and the natural polypeptides, you used "hormone dilution" as the abcissa; this, at least to me, does not mean nearly as much as a molar concentration would have. Now that molecular weights are known, perhaps expressing quantities of hormones in moles and molarities could signal that one is confident of the identity and quantity of material.

G. D. Aurbach: You will find in the legends to figures in the text the necessary information for converting dilution or weight to molarity. Notice in the figure comparing relative activities of the several analogs that percent activity is expressed on a molar basis. I agree that molar concentrations are appropriate when dealing with substances of known molecular weight. In general, concentrations are thus represented in the text.

E. D. Bransome: I am concerned with the apparent lack of discrimination in the inside of the cell compared to the rather finicky aspect of the plasma cell membrane. Let me describe what we found in the adrenal cortex and then ask you whether you found that a similar situation pertains to the renal cell. Prompted by Garren's work, we have looked for relations between cyclic AMP binding and kinase activity. Unfortunately, there was a multiplicity of kinases and of the substrates for kinase activity in adrenal cortex. Everything was phosphorylated. Is your system any cleaner? Have you determined whether the parathyroid hormone itself can prompt kinase activation *in vitro?*

G. D. Aurbach: It is quite possible that more than one kinase exists in the kidney. So far we have not found evidence for such by the usual methods of gradient ultracentrifugation, gel filtration, or column chromatography. However, other tissues, e.g., liver, muscle and reticulocytes, have multiple forms and you seem to have several in the adrenal. We have not, as stated above, discovered the true intracellular substrate. We have only worked with the enzyme using histone as substrate. There is some background phosphorylation as in every system, but the major fraction is histone-dependent in our particular assay.

I. MacIntyre: If the phosphaturia produced by parathyroid hormone is merely a secondary consequence of a more important primary cellular event, the phosphaturia should be produced by other hormones acting on the kidney. I think this is not the case, although phosphaturia of some degree is produced by many hormones—for example, calcitonin. But as far as I know the phosphaturia produced by parathyroid hormone is very much more striking.

Do you have any comment on how Parsons' findings relate to your work [J. A. Parsons and C. J. Robinson, *Nature* **230**, 581–582 (1971)]?

G. D. Aurbach: Parsons' very interesting experiment showed that injection of parathyroid hormone causes acute hypocalcemia by increasing the entry of calcium into bone, possibly bone cells. This acute, transient hypocalcemic action of the hormone was discussed during Dr. Copp's presentation concerning calcitonin some years ago at this conference [*Rec. Progr. Hormone Res.* **24**, 589 (1968); see p. 637 of discussion]. This is a true property of purified parathyroid hormone. Borle has shown that in prolonged *in vitro* incubations with kidney cells there is a gradual increase in intracellular calcium. Malaise's group has shown that calcium enters islet cells when stimulated by glucose, Kraicer has shown entry of calcium into pituitary cells when secretion of hormone is stimulated by potassium, and Rudman's group has found calcium uptake into adipocytes under conditions of hormone-induced lipolysis. Thus this phenomenon may be general in the endocrine world and reflect simply an interruption or change in configuration of the cell membrane brought about by interaction with hormone. This is a topic that we discussed at some length in the text. Its overall importance remains to be established.

Concerning the specificity of phosphaturia, in the course of our investigations Dr. Chase and I found that vasopressin causes a definite phosphaturic response in rats. One can predict that all substances causing an increase in cyclic AMP in the kidney, particularly in the proximal tubule, will cause phosphaturia. This would include catecholamines, calcitonin, dibutyryl-cyclic AMP, and perhaps glucagon. The effects of glucagon and calcitonin are slight, presumably because only a few cells are capable of responding. A rise in cyclic AMP in the proximal tubule seems to inhibit reabsorption of sodium; water and phosphate go with sodium to the distal tubule where most of the latter is reabsorbed, leaving water and phosphate (and some other cation, H^+ or K^+) to be excreted. This mechanism is nicely discussed in recent papers in the *Journal of Clinical Investigation* by Gill and Casper, as well as by Agus *et al.*

O. H. Pearson: Is colchicine an effective agent in lowering the serum calcium in a patient with hyperparathyroidism?

G. D. Aurbach: I do not know the answer to that question. Perhaps a hypercalcemic patient undergoing an acute gouty attack would offer an opportunity to find out. Hypocalcemic tetany is a significant part (not necessarily the sole cause) of the toxicity of colchicine in rats.

M. A. Kirschner: In view of the occasional clinical difficulties in establishing the diagnosis of hyperparathyroidism, what are your thoughts concerning the measurement of cyclic AMP in the urine of patients whom we suspect have this condition?

G. D. Aurbach: We as well as several other groups have looked at this. The rapid effect of parathyroid hormone on urinary cyclic AMP was initially discovered with rats and it was found also that calcium suppresses excretion of cyclic AMP by normal rats through inhibition of secretion of parathyroid hormone. The postulate was immediately made that this should be a useful diagnostic test for hyperparathyroidism [L. R. Chase and G. D. Aurbach, *Proc. Nat. Acad. Sci. U.S.* **58,** 518 (1967)]. It turned out, however, that rats are different from man or dogs. In a rat the half-life of cyclic AMP in the plasma is 2 minutes or less. Therefore, most, approximately 80%, of the cyclic AMP excreted into the urine by the rat is under the control of parathyroid hormone through its influence on the kidney. In dog and in man, it is only about 40–50%. The latter data come from the elegant clearance studies of Broadus *et al.* [*J. Clin. Invest.* **49,** 2222 (1970)]. Consequently, calcium infusion in man does not cause such great inhibition of excretion of cyclic AMP. Thus, the potential clinical utility of this type of suppression test is not as great as was originally projected. A very high rate of 3′,5′-AMP excretion associated with hypercalcemia would indicate hyperparathyroidism. In general the 24-hour excretion of cyclic AMP is significantly higher in hyperparathyroidism than in normal subjects, and invariably following parathyroidectomy in human beings there is a sharp fall in urinary excretion of cyclic AMP. However, there is too much overlap between the hyperparathyroid and normal subjects to be of absolute diagnostic value in any one case.

K. Sterling: Regarding your comment on the purely surface cell membrane action of parathyroid hormone, would you consider doing, or have you done, the type of study that Pedro Cuatrecasas did of attaching insulin to Sepharose and demonstrating that the hormone still works when attached to large particles—much too large to enter the cell.

G. D. Aurbach: Dr. O'Riordan has successfully coupled parathyroid hormone to a solid phase support, but I do not believe that it has been tested for biological activity.

K. Sterling: Could the diminution of urinary cyclic AMP conceivably be the best index of successful resection of parathyroid adenoma, even better than the transient fall in serum calcium we are always looking for after resectional surgery?

G. D. Aurbach: In every patient we have tested so far we have invariably shown a dramatic 60–70% fall in the cyclic 3′,5′-AMP excretion after parathyroidectomy. I do not remember how many cases have been studied; probably only a half-dozen or so. We have tested more rats than we have patients.

C. Y. C. Pak: We have followed 10 patients with hyperparathyroidism after surgery in collaboration with Murad. In 8 of them the urinary cyclic AMP decreased to the normal range. In one in whom it did not decrease, there was some question whether all the abnormal parathyroid tissue was removed.

Have you tested the effect of the parathyroid hormone fragments on urinary excretion of calcium? It has been postulated, as you know, that the secretion of an abnormal parathyroid hormone may be responsible for the development of hypercalciuria in normocalcemic primary hyperparathyroidism. Could this hormone be one of the peptide fragments?

G. D. Aurbach: As discussed in replies to questions of Dr. Sterling and Dr. Kirshner, we have found the same result after surgery for hyperparathyroidism.

We have not tested many analogs *in vivo*. Even with the native hormone in the rat, we find little change in calcium excretion under our test conditions. We are interested in the results from Dr. Munson's laboratory showing that the golden hamster responds acutely to parathyroid hormone almost exclusively through a renal mechanism. Nordin and his associates also have refocused our attention on the importance of the kidney in regulation of calcium in acute responses to parathyroid hormone. But in the rat we do not see a very rapid or significant effect under the conditions we utilize for measuring urinary phosphate and cyclic AMP. Therefore our test preparation would offer little promise as a system to test for the altered hormones you postulate.

J. Kowal: We have found that ACTH stimulates the phosphorylation of a myriad of substances in the cell comprising many phospholipids and proteins as well as RNA. Recently we have found a variety of proteins in cell extracts which act as substrate for the cyclic AMP-stimulated protein kinase. As long as histone, which has not been implicated in ACTH action, is such a good substrate for the reaction, are we missing an essential aspect of this reaction that confers specificity?

G. D. Aurbach: I agree that merely identifying an enzyme in a cell-free extract does not prove that the enzyme takes part in the expression of hormone activity *in vivo*. The phenomenon is most clear-cut in the action of epinephrine or glucagon in the liver on activating phosphorylase kinase. I think that is the best clear-cut evidence we have that a hormone does influence the activity of the kinase in expressing its activity. Much further work needs to be done to evaluate the role of phosphorylating systems in the actions of other hormones, including parathyroid hormone.

M. Manning: I was intrigued to note that the amino acid differences between the porcine and bovine hormones can all be related by single point mutations in their respective RNA codons. This suggests that these hormones may have arisen from a single ancestral precursor.

You remarked about the relative activities of the fragment of parathyroid hormone and parathyroid hormone itself. I am puzzled by your statement to the effect that on a molar basis the fragment is only 16% as active as parathyroid hormone itself, yet on a weight basis the fragment would be more active than the complete sequence. I cannot see how this could be since the fragment comprises only about 40% by weight of the total PTH sequence. Have you any information on the biosynthesis of PTH? Might there be a prohormone involved?

G. D. Aurbach: The synthetic molecule is only 40% the size of the native molecule. Therefore, even equipotency on a weight basis would represent only 40% activity on a molar basis. The hormone fragment 1–34 showed *in vitro* an activity of 400–600 units/mg, compared to 1600 mU/mg for native hormone (1–84).

We have also concluded that a single base change would account for each of the changes in amino acid sequence in the porcine molecule.

J. M. McKenzie: Could you elaborate on the difference between bovine and porcine parathormone and their inhibition by the two tissues? You referred to this inhibition as being due to enzyme activity but did not tell us why the action was enzymatic and not a reflection of binding or other mode of inhibition.

G. D. Aurbach: There is an activity associated with kidney membrane which causes loss of biological activity on incubation with parathyroid hormone. This loss of activity can be determined by simply measuring the residual activity (biological activity) using the adenyl cyclase system. This activity is destroyed by boiling. The activity is much reduced at 4°C, fully expressed at 37°C and is inhibited to some extent by adding a high concentration of exogenous protein. We have shown in addition that this enzyme activity remains associated with the cell membrane fraction on gel filtration. That represents the information suggesting that it is an enzyme. The hormone is bound by membranes from bone as well as kidney, but it is only the kidney membrane fraction that shows significant amounts of the inactivating enzyme.

J. Rudinger: You said that 1–29 is the minimal active sequence. On the evidence you had, can you really say that? Would you not rather say it is the upper limit of the minimal required sequence?

G. D. Aurbach: Thank you for your comments. Drs. Tregear and Keutmann have some results that bear on that question.

G. W. Tregear: The smallest fragment we have synthesized so far with biological activity is the 1–30 sequence. Dr. Keutmann has, however, found that the 1–29 natural fragment has activity. We have not yet made 1–28, 1–27, or 1–26.

H. T. Keutmann: The shortest fragment which we have found to possess biological activity is 1–29. We have tested synthetic fragment 1–13 and natural fragment 1–20 and found neither to be active in the renal adenyl cyclase system. Therefore, we can say that the minimum fragment required for biological activity occupies a continuous sequence of the molecule running from residue 1 certainly through some point between residue 21 and residue 29. We are now attempting to shorten the 1–29 fragment, by use of exopeptidases, to 27 or 24 residues. These fragments appear to contain little, if any, activity but this information is only preliminary. We have also tested the possibility that inactive fragment 2–34 could act as an inhibitor to the activity of 1–34 or 1–84. Presence of excess 2–34 does not diminish the biological activity of either 1–34 or 1–84, suggesting that 2–34 binds to the receptor site very weakly or not at all.

F. C. Bartter: It sounds as though you are suggesting that tubular secretion of phosphate could account for the phosphaturic action of parathyroid hormone. Many investigators, as you know, have failed to show secretion by standard measures of kidney function. I think that is not so disturbing if you postulate that the phosphate that is released in renal cells blocks transport, and that it is the filtered phosphorus that you see in the urine. At least, it is not possible to see *more* than the filtered phosphorus in the urine.

G. D. Aurbach: One can question whether phosphate is "transported" at all. It may simply follow sodium, as suggested by the work of Gill and Casper as well as the work of Goldberg's group. The classical methods of renal physiology are not really adequate to determine precisely what happens to phosphate throughout the nephron. One cannot exclude the possibility that phosphate "secretion" as well as reabsorption goes on at one or more loci in the nephron between the grossly observed phenomena of filtration and excretion. In this connection we do not imply that release of phosphate from the cell is an "active" process such as transport. It may be simple diffusion from the cell of the phosphate hydrolyzed from high energy 'intermediates utilized in metabolism. That is what I was trying to imply in the analogy to Williamson's experiment with the perfused heart.

F. C. Bartter: I had hoped Dr. Pak would enlarge a bit further on the use of cyclic AMP in the urine in the diagnosis of hyperparathyroidism. We find this is the most useful single test—except the measurement of parathyroid hormone itself. This is so when the measurement is combined with calcium infusion; in hyperparathyroidism, urinary cyclic is not depressed.

G. D. Aurbach: Earlier in the discussion we referred to the original experiments showing suppression of urinary cyclic AMP excretion in response to calcium infusion [L. R. Chase and G. D. Aurbach, *Proc. Nat. Acad. Sci. U.S.* **58**, 518 (1967)]. We pointed out that this has not been uniformly benefiical in proving the diagnosis of hyperparathyroidism and suggested the reason why this is so. Certainly there is a significant difference between the normal population and that with hyperparathyroidism. The question is whether the test provides 95% diagnostic confidence when confronted with the individual patient. Particularly critical will be the results with borderline cases. As you know well, the several "diagnostic" tests for hyperparathyroidism score well in obvious cases. It is in the borderline or complex case that so many of these tests fail and those alone represent the real challenges for the clinician.

Studies in Familial (Medullary) Thyroid Carcinoma

Kenneth E. W. Melvin, Armen H. Tashjian, Jr., and
Harry H. Miller

*New England Medical Center Hospitals and Tufts University School of Medicine,
Departments of Medicine and Surgery; Harvard School of Dental Medicine
and Harvard Medical School, Department of Pharmacology,
Boston, Massachusetts*

I. Introduction

It is little more than a decade since the emergence of medullary carcinoma of the thyroid gland (MCT) as a distinct clinical entity. With the recognition by Hazard *et al.*, (1959), of the characteristic histological and clinical features of this solid tumor, distinguished by its amyloid stroma and relatively indolent clinical course, began an appreciation of the remarkable individuality of this particular variety of thyroid cancer. In the course of that decade, more than 250 cases have been reported (Hazard *et al.*, 1959; Woolner *et al.*, 1961; Freeman and Lindsay, 1965; Williams, 1965; Williams *et al.*, 1966; Ibanez *et al.*, 1967; Ljungberg *et al.*, 1967; Block *et al.*, 1967; Sarosi and Doe, 1968; Steiner *et al.*, 1968; Williams and Brewer, 1969; Tashjian *et al.*, 1970), and a strong familial incidence has been noted (Schimke and Hartmann, 1965; Ljungberg *et al.*, 1967; Block *et al.*, 1967; Steiner *et al.*, 1968; Melvin *et al.*, 1971). A characteristic syndrome has been described, associating this carcinoma with pheochromocytomas, which, when present, are bilateral in 75% of cases (Sipple, 1961; Cushman, 1962; Manning *et al.*, 1963; Finegold and Haddad, 1963; Nourok, 1964; Williams, 1965; Schimke and Hartmann, 1965; Steiner *et al.*, 1968; Melvin *et al.*, 1971), multiple mucosal neuromas, Marfanoid habitus (Mielke *et al.*, 1965; Williams and Pollock, 1966; Gorlin *et al.*, 1968), and hyperparathyroidism (for a review, see Steiner *et al.*, 1968; Moertel *et al.*, 1965; Urbanski, 1967; Gonzales-Licea *et al.*, 1968; Sarosi and Doe, 1968; Paloyan *et al.*, 1970; Mandelstam *et al.*, 1970; Melvin *et al.*, 1971). More recently, much has been learned of the extraordinary endocrine and biochemical propensities of this tumor, the study of which will constitute a major part of this report.

Although MCT is a rare tumor, probably comprising between 5 and 10% of all thyroid carcinomas (for reviews, see Steiner *et al.*, 1968; Robbins, 1969), its frequently familial incidence creates important epidemiological considerations among the relatives of affected individuals. Our studies have shown that the unusual endocrine function of this tumor provides a means of establishing an early diagnosis in individuals at risk by virtue of their heredity, improving their chance of effective surgical cure, and providing a sound basis for genetic counseling.

To be presented in this paper is an account of studies performed in sporadic and familial cases of medullary thyroid carcinoma which established the calcitonin-secreting nature of the tumor, and the application of this observation to the early diagnosis of clinically occult MCT in 12 members of a single large kindred.

II. The Syndrome of Medullary Thyroid Carcinoma, Multiple Mucosal Neuromas, and Pheochromocytoma

A. CLINICAL FEATURES

In its fullest expression, patients suffering from this syndrome exhibit the features shown in Fig. 1. Only a minority of patients with MCT, of the order of 5%, possess the full spectrum of physical features demon-

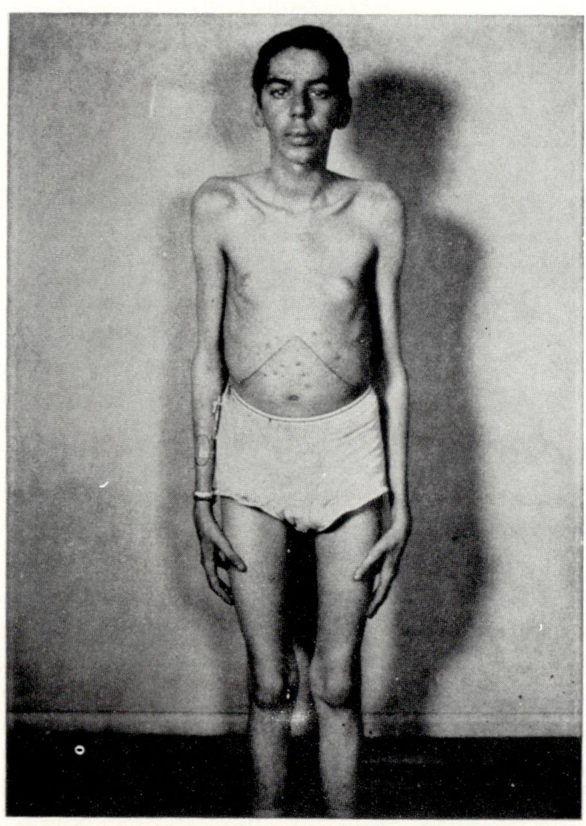

FIG. 1. Patient with medullary thyroid carcinoma (sporadic), bilateral pheochromocytomas, Marfanoid habitus, multiple mucosal neuromas, and gynecomastia.

strated by this patient, who died of disseminated carcinoma at the age of 21 years, one year after diagnosis. Studies in this patient were the subject of an earlier report (Melvin *et al.*, 1970a). He presented in left ventricular failure precipitated by bilateral pheochromocytomas; during adrenalectomy he was noted to have metastatic MCT in the liver and in one of the three pheochromocytomas. Body habitus was Marfanoid in type, with arachnodactyly and poorly developed musculature. Pectus excavatus, and pes cavus were additional features. The thyroid gland was enlarged bilaterally. There was no known family history of thyroid carcinoma or of pheochromocytoma, and investigation of his immediate family revealed no evidence of either tumor. The gynecomastia evident in this patient has been described in only one other case (Aach and Kissane, 1969). The ectopic production of gonadotropins has not been demonstrated, but the possibility should be considered. The

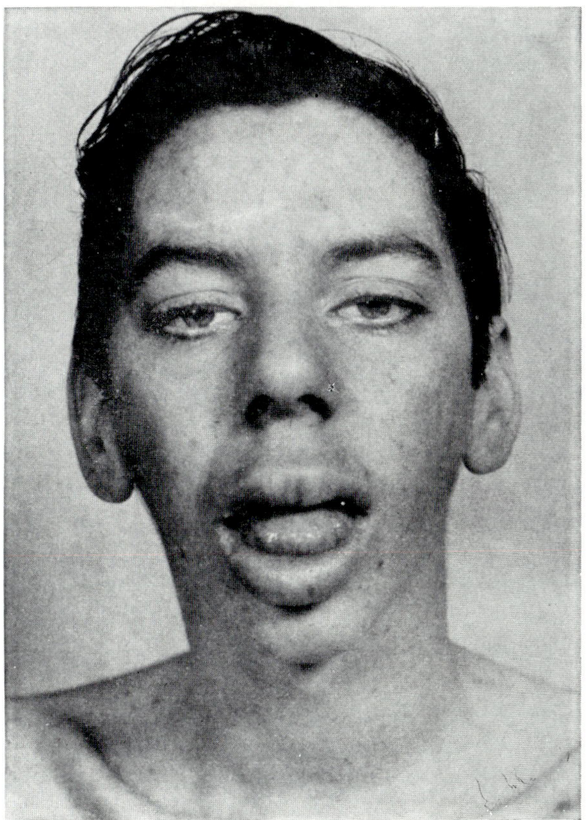

FIG. 2. Typical facies associated with the syndrome of medullary thyroid carcinoma, pheochromocytoma, and multiple mucosal neuromas. Neuromas evident on the tongue, and producing hypertrophy of the lips and tarsal margins.

facial appearance (Fig. 2) was characteristic and may be so similar from one patient to another as to suggest blood relationship. Multiple mucosal neuromas in the lips and tarsal margins give the appearance of hypertrophy of these structures. The tongue may also be involved in a similar fashion, and in this patient a dentist had removed the tip of the tongue in early childhood on account of its "barnacle encrusted appearance." Medullated corneal nerves were readily demonstrated by the slit lamp, their presence in some patients with this syndrome being in some way related to the associated pheochromocytoma (for review, see Gorlin *et al.*, 1968). There were no neurofibromas or café au lait spots, such as have been reported in a small number of patients (Williams and Pollock, 1966). As a child this patient had suffered from megacolon, a not uncommon feature in reported cases, and shown to be due to ganglioneuromatosis (Williams and Pollock, 1966), which may involve the entire intestinal tract. Barium studies showed the colon to be the site of multiple diverticuli, another frequent finding in this syndrome (Gorlin *et al.*, 1968). The patient suffered from intractable diarrhea, a not uncommon symptom in patients with advanced metastatic disease, and for which a humoral cause seems likely (Williams *et al.*, 1968). In this patient as in every case which we have tested, calcitonin (CT) was present in high concentrations in both peripheral plasma and tumor tissue.

B. HUMORAL ACTIVITY OF MEDULLARY CARCINOMA OF THE THYROID

1. Calcitonin

It was Williams (1966), whose careful studies constitute a major contribution to our understanding of the medullary carcinoma syndrome, who first suggested that the tumor might arise from the parafollicular or C cells of the thyroid gland. Studies carried out in the first patient seen by us, and the subject of previous reports (Melvin and Tashjian, 1968; Tashjian and Melvin, 1968) established the fact of synthesis and secretion of CT by MCT, providing functional evidence in support of Williams' hypothesis. At the same time, Meyer and Abdel-Bari (1968) reported close ultrastructural similarities between the cells of MCT and parafollicular cells, also finding elevated levels of hypocalcemic activity in the medullary carcinoma tissue. These findings have since been confirmed by numerous investigators (Cunliffe *et al.*, 1968; Milhaud *et al.*, 1968; Dube *et al.*, 1969; Woodhouse *et al.*, 1969; Johnston *et al.*, 1970; Deftos *et al.*, 1971a).

In addition to the secretion of CT, which appears to be a consistent feature of MCT, the tumor has on occasion been shown to be capable of a variety of other biosynthetic activities.

2. *ACTH*

Shown in Fig. 3 are data from a case of Cushing's syndrome caused by an ACTH-secreting MCT, reported by us a year ago, and bringing the total recorded cases to 17 (for review, see Melvin *et al.*, 1970b). Very high levels of plasma cortisol and ACTH fell to normal levels

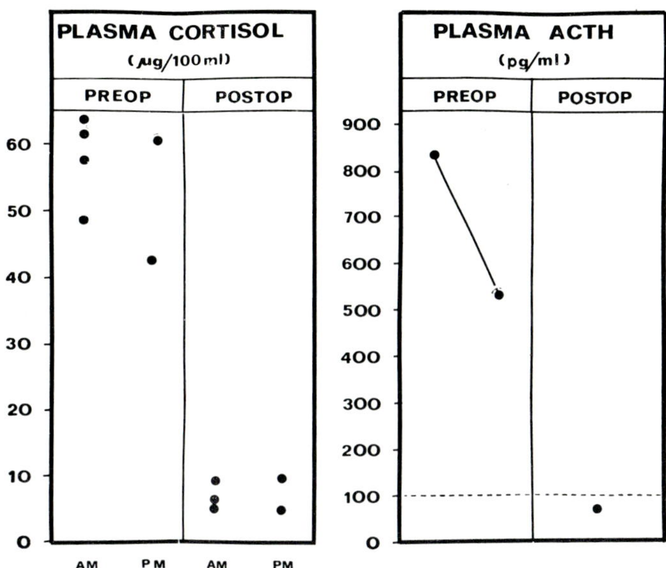

Fig. 3. Plasma cortisol and ACTH values before and after excision of ACTH-secreting medullary thyroid carcinoma. Dots surrounded by circles give values obtained while the patient was receiving suppressive doses of Dexamethasone. Broken line indicates upper limit of normal for plasma ACTH. From Melvin *et al., Metab. Clin. Exp.* **19,** 831 (1970b) by permission.

after thyroidectomy (Fig. 3). The thyroid contained a unilateral medullary carcinoma in which there were large quantities of both ACTH and CT. Removal of the MCT resulted in a prompt remission of her Cushing's syndrome. High serum CT levels fell postoperatively but still remained above the upper limit of the normal range, suggesting metastatic disease (Fig. 4).

3. *Histaminase*

Elevated levels of histaminase activity have been demonstrated in the plasma of several patients suffering from MCT with metastases

(Baylin *et al.*, 1970). The physiological and pathological significance of this finding remains obscure. Increased activity of this enzyme was measured in many of our patients, details of which will be presented later.

4. Prostaglandins

High levels of prostaglandins E_2 and $F_{2\alpha}$ were found by Williams *et al.* (1968) in the plasma of 2 patients suffering from MCT and in

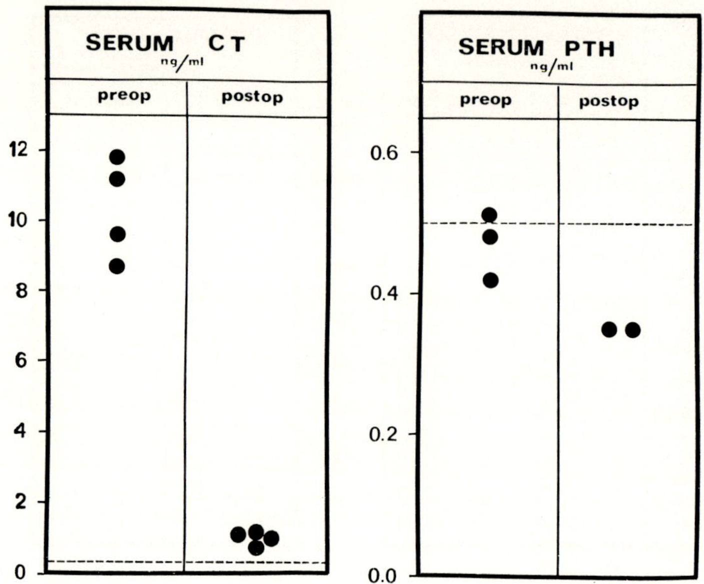

FIG. 4. Levels of serum calcitonin (CT) and parathyroid hormone (PTH) as measured by radioimmunoassays before and after excision of medullary thyroid carcinoma (MCT) in a patient with Cushing's syndrome from ACTH-secreting MCT. Preoperative values for both hormones were obtained during a 4-hour infusion of calcium. Upper limits of normal ranges are indicated by broken lines. From Melvin *et al. Metab. Clin. Exp.* **19**, 831 (1970) by permission.

the tumor tissue of 4. He believed that the levels present in the plasma were adequate to produce intestinal hypermotility and might in some patients account for the symptom of diarrhea frequently encountered in association with this tumor.

With the assistance of Dr. Lawrence Levine at Brandeis University, who has developed an immunoassay for prostaglandins (Levine and Van Vunakis, 1970), we have looked for prostaglandins in the serum of 6 patients with familial MCT, none of whom suffered from diarrhea, and 6 patients with sporadic MCT, in 3 of whom diarrhea was severe (Table

I). On the left are shown the findings in 10 samples from 6 subjects with familial carcinoma, none of which contained high concentrations of prostaglandins of the ABE series. Prostaglandin $F_{2\alpha}$ was sought in the serum from 3 of these subjects, with negative results. On the right are the findings of similar studies on 12 samples from 6 patients with sporadic MCT. Of the 3 subjects in this group with diarrhea, only 1 had elevated levels of prostaglandins (8–12 ng/ml), and this occurred in the ABE series. The concentration of prostaglandin $F_{2\alpha}$ was not elevated in that subject or in 2 other patients without diarrhea.

TABLE I

Measurements of Prostaglandins (PG) in Serum of Patients with Medullary Carcinoma[a]

	J-Kindred with MCT:		Sporadic MCT:	
	Prostaglandins		Prostaglandins	
	ABE series	$F_{2\alpha}$	ABE series	$F_{2\alpha}$
Number of subjects	6	3	6	3
Number of samples	10	3	12	5
Subjects with no PG detected	6[b]	3[c]	5[b]	3[c]
Subjects with high PG detected	0	0	1[d]	0

[a] From Melvin *et al.*, *New Engl. J. Med.* **285**, 1115(1971) by permission.
[b] Less than 1.6 ng/ml.
[c] Less than 5 ng/ml.
[d] Four samples from one subject ranged from 8 to 12 ng/ml.

Thus, while prostaglandin production by MCT may explain the diarrhea in some of the patients affected, it may not be the explanation in all.

5. Serotonin

Argentaffinity and chromaffinity have been demonstrated in MCT (Ljungberg *et al.*, 1967), a finding which we confirm. In addition, the rare patient has shown an increased urinary excretion of 5-hydroxyindoleacetic acid (Moertel *et al.*, 1965; Ibanez *et al.*, 1967; Williams *et al.*, 1968). The possible relevance of this finding will be the subject of later discussion.

C. EFFECTS OF CALCIUM AND GLUCAGON ON THE SECRETION OF CALCITONIN BY MCT

Our initial studies of calcitonin in patients with MCT employed the bioassay as described by Cooper *et al.* (1967). Subsequent studies were by means of a radioimmunossay (Tashjian *et al.*, 1970) which has the following characteristics:

1. Labeled synthetic human CT-M; guinea pig antihuman CT
2. Sensitivity: 0.002 ng CT (about 1000 times more sensitive than the bioassay)
3. Human serum, thyroid gland extracts, and MCT extracts superimposable on the standard curve
4. Sensitivity for human serum: about 0.10 ng/ml (inhibition greater than 2 SD beyond "no cold")
5. At least 50% of normal basal scrum CT levels <0.10 ng/ml; all of 140 normal basal values <0.38 ng/ml
6. Physiological observations with this assay:
   a. No cross-reactions with nonprimate CT's
   b. Calcium infusion in 63 normal subjects resulted in a 2- to 3-fold rise in serum CT in some but not all tests; the maximum level reached never exceeded 0.55 ng/ml
   c. All of 33 subjects with clinically overt MCT had basal serum CT levels of >1.0 ng/ml
   d. Immunoassay and bioassay results of the same MCT serum samples in excellent agreement
   e. Calcium infusion in 27 patients with overt MCT resulted in 2- to 60-fold increase in serum CT to 4–2200 ng/ml; in occult MCT, levels of >2 ng/ml have led to no false positive predictions in 22 cases
   f. CT was not detected in serum of 22 of 25 totally thyroidectomized patients.

On the assumption that medullary thyroid carcinoma is a tumor of C cells, which in their nonneoplastic state respond physiologically to an increase in serum calcium concentration by further increasing their rate of secretion of CT, we studied the responsiveness of the carcinoma to induced hypercalcemia. This was effected by means of a standard calcium infusion test, in which 15 mg of calcium per kilogram body weight was infused intravenously in 400 ml of 0.9% saline over 4 hours.

Glucagon-induced hypocalcemia has been described in the dog (Paloyan *et al.*, 1967a), rabbit (Paloyan *et al.*, 1967b), man (Paloyan, 1967; Birge and Avioli, 1969), and rat (Morain and Aliapoulios, 1968; Williams *et al.*, 1969; Tanzer *et al.*, 1970). In the pig, perfusion of the thyroid gland with glucagon results in the release of calcitonin from the gland, and hypocalcemia (Care *et al.*, 1969). If glucagon-induced hypocalcemia in man occurs as a result of calcitonin release, it might be expected to demonstrate this phenomenon readily in patients with calcitonin-secreting thyroid carcinoma. This was tested by administering glucagon as an initial intravenous loading dose of 0.01 mg per kilogram body weight, followed immediately by a 4-hour infusion of 0.01 mg/kg per hour in a total volume of 400 ml of 0.9% saline. The results of both calcium and glucagon infusion tests are shown in Fig. 5. Plotted are the peak values obtained for serum CT during the 4-hour infusions,

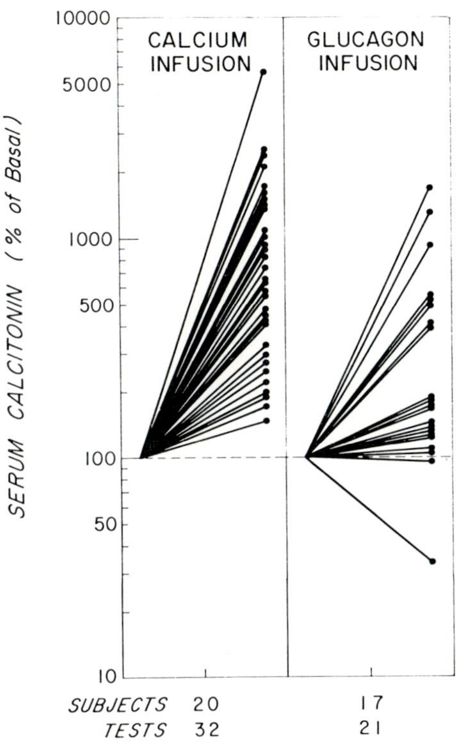

Fig. 5. Response of serum calcitonin (CT) to intravenous infusions of calcium and glucagon in patients with medullary thyroid carcinoma. Basal serum CT values were plotted as 100%; values at the end of the 4-hour infusion were plotted to the right of the connecting line in each panel, as percentage of basal value. Values below the broken line represent a decrease in serum CT in response to the infusion; values above the line, an increase.

expressed as a percentage of basal value. CT was measured by biological assay in 3 patients and by radioimmunoassay in the remainder. In 32 infusions of calcium in 20 patients with MCT, an increase in serum CT was noted in all patients, values ranging from 145 to 5700% of basal. In 21 infusions of glucagon in 17 patients, the rise in serum CT was generally much less than observed in response to calcium, and in one patient the infusion resulted in a significant decrease in serum CT.

These results demonstrated that MCT responds to induced hypercalcemia, and to glucagon, in a fashion similar to that demonstrated for normal C cells in a variety of other mammalian species. Similar findings were subsequently reported by Deftos *et al.* (1971a), who noted in addition that the rate of secretion of CT by MCT diminished in response to EDTA-induced hypocalcemia.

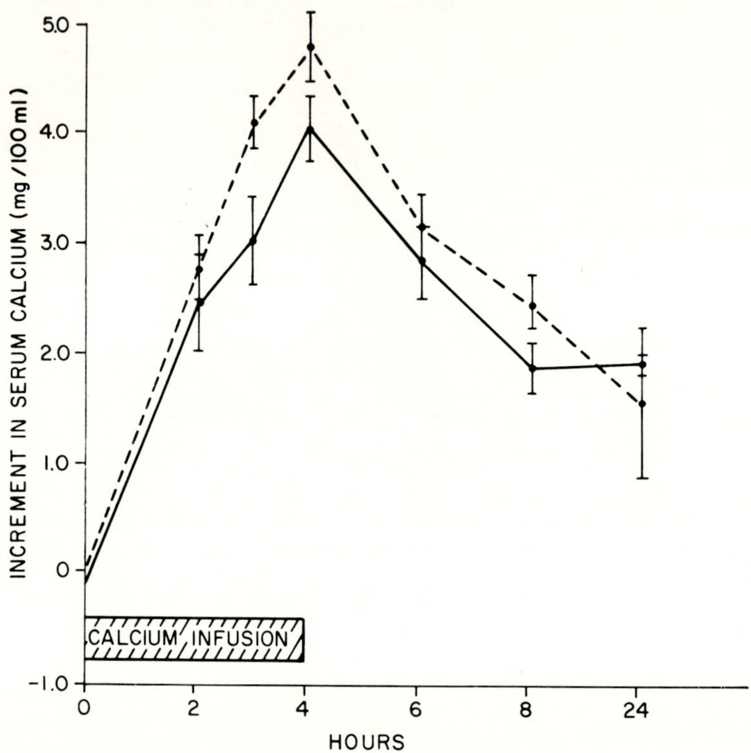

Fig. 6. Increment in serum calcium concentration in response to calcium infusion in 12 patients with familial medullary thyroid carcinoma (MCT) (—), and 12 age-matched relatives with normal serum calcitonin (CT) (---). Plotted are the mean values and standard error limits. During the infusion serum CT rose to high levels (>2 ng/ml) in all patients with MCT (see Fig. 5) but not in the normal relatives.

D. COMPARISON OF SERUM CALCIUM LEVELS DURING AND AFTER CALCIUM INFUSIONS IN NORMAL SUBJECTS AND PATIENTS WITH MCT

If CT plays an important role in the human, protecting against an acute increase in serum calcium concentration, it might be expected that patients with MCT would show a smaller rise in serum Ca in response to calcium infusion than normal subjects. In Fig. 6 are shown the increments in serum calcium observed during the standard calcium infusion in 12 subjects with MCT, and 12 age-matched controls. The mean and standard error at each time interval are plotted for both groups. Although slight differences were observed, the rise in serum calcium being greater in the normal subjects than in those with MCT, the differences were not statistically significant.

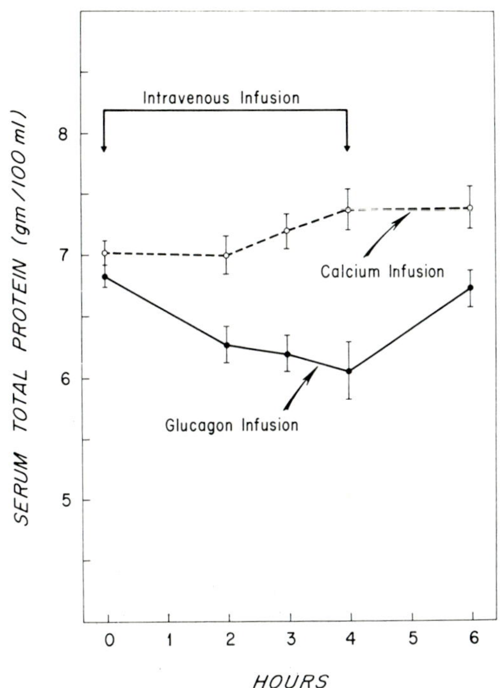

Fig. 7. Changes in serum total protein concentration in response to infusions of calcium (○---○) and of glucagon (●——●) in 15 patients with medullary thyroid carcinoma. Plotted are the mean values and standard error limits. A dilutional effect was apparent during the infusion of glucagon, but not with the calcium infusion.

E. Effects of Glucagon Infusion on Serum Calcium, Inorganic Phosphate, and Calcitonin

A small reduction in serum calcium concentration was observed in most patients in response to infusion of glucagon. This was accompanied however by a reduction in plasma protein concentration, suggesting that the apparent fall in serum calcium was due to hemodilution rather than to a true hypocalcemic effect of glucagon. Shown in Fig. 7 are the mean values and standard error limits for serum total protein concentration in 15 patients with MCT, in response to infusions of calcium and of glucagon. Although both calcium and glucagon were infused in identical volumes of saline at the same rate of 100 ml per hour, a fall in plasma protein concentration occurred only in response to glucagon, not in response to calcium. The reason for this is not clear, but it is possible that the documented rise in blood glucose which accompanied the infusion of glucagon may have altered plasma osmolality and distribution

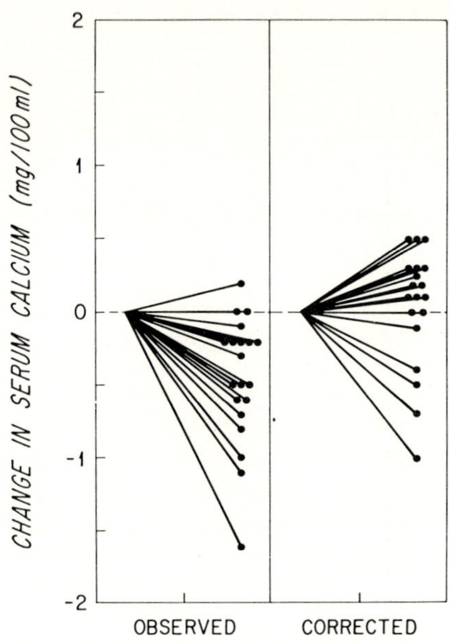

FIG. 8. Changes in serum calcium concentration in response to intravenous glucagon infusions in patients with medullary thyroid carcinoma. In the left panel are shown the observed changes in serum calcium, and in the right panel, the same changes after correction for the decrease in the serum protein concentration which occurred during the infusions (see Fig. 7).

of body water sufficient to produce a dilutional change in plasma protein concentration that was not evident upon the infusion of a similar volume of calcium and saline. Correction (Dent, 1962) of the serum calcium values for the observed changes in plasma protein concentration confirmed this, as shown in Fig. 8. Shown in the left panel are the observed changes in serum calcium in response to glucagon infusions, and in the right panel the same values after correction for changes in plasma proteins. These corrected data fail to confirm a convincing or consistent hypocalcemic effect of glucagon in the normocalcemic human subject, despite an increase in serum CT. It is concluded therefore that if glucagon-induced hypocalcemia occurs at all in normocalcemic man, it is unlikely to be on the basis of CT action.

The infusion of glucagon resulted in moderate hypophosphatemia in both normal subjects and patients with MCT (Fig. 9). During the 4-hour infusion, serum phosphate fell by an average of 1.25 mg/100 ml in 15 patients with MCT, a similar fall occurring in 60 normal relatives. The mean basal serum phosphate was lower in the MCT group than in the

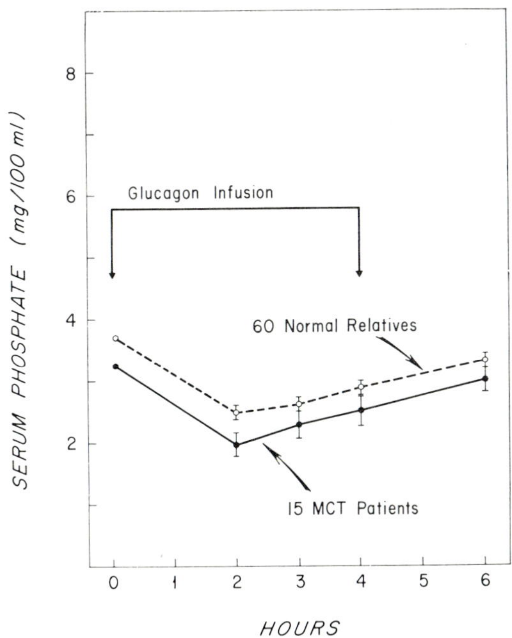

Fig. 9. Fall in serum phosphate levels in response to the intravenous infusion of glucagon in 15 patients with medullary thyroid carcinoma (MCT) (●———●) and 60 relatives with normal serum calcitonin (CT) (○---○). Serum CT rose during the infusion in most of the patients with MCT (see Fig. 5), but not in the normal relatives.

normal subjects, but the latter group included many young children in whom the serum phosphate tends to be higher than in the adult.

It is interesting to note that Tanzer *et al.* (1970), who compared the effects of CT and glucagon on plasma calcium and phosphate in rats, found that the hypocalcemic and hypophosphatemic effects of glucagon occurred independently of the thyroid gland and were probably mediated by a direct effect of glucagon on bone. A similar conclusion was reached by Williams *et al.* (1969). The effects of glucagon in hypercalcemic states, as described by Paloyan (1967) in hyperparathyroidism, need further study.

There is evidence that many of the metabolic effects of glucagon are mediated through activation of the adenyl cyclase–cyclic AMP system (Sutherland and Robison, 1966). In a variety of tissues, the concentration of cyclic AMP is influenced by sympathetic nervous activity, being increased by β-adrenergic stimulation and decreased by α-adrenergic

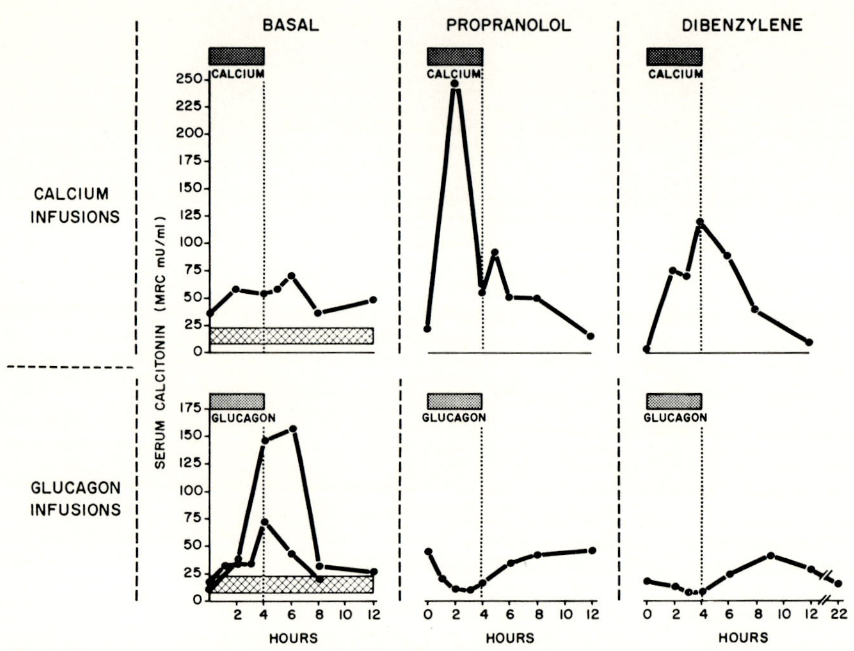

FIG. 10. Serum calcitonin (CT) response to infused calcium and glucagon basally, and after α- and β-adrenergic blockade, in a patient (Fig. 1) with MCT. The range of untreated, resting values of serum CT is shown in the shaded bar at bottom of the basal infusion graphs. From Melvin *et al.* (1970a).

stimulation (Murad *et al.*, 1962; Sutherland and Robison, 1966; Turtle and Kipnis, 1967).

In one patient with MCT, a series of experiments was performed, aimed at determining the extent to which the α- and β-adrenergic systems, and indirectly therefore, the adenyl cyclase-cycle AMP mechanism, are involved in the release of CT by calcium or glucagon. Accordingly, calcium and glucagon were each infused under basal conditions, and again after β-adrenergic blackade with propranolol (120 mg daily for 5 days), and after α-adrenergic blockade with dibenzyline (30 mg daily for 7 days). The results, which have been reported previously (Melvin *et al.*, 1970a), are shown in Fig. 10 and were obtained from study of the patient portrayed in Figs. 1 and 2. Measurements of serum CT in these studies, were by biological assay.

Initially, calcium produced a modest rise in serum CT to 70 Medical Research Council (MRC) mU/ml. Glucagon caused a 13-fold increase in serum CT from 12 to 158 MRC mU/ml.

β-Adrenergic blockade was accompanied by a greatly augmented response to a calcium load, serum CT values increasing 13-fold from 22 to 251 MRC mU/ml, but propranolol abolished the response to glucagon—possibly even leading to a fall in serum CT.

Following α-adrenergic blockade, the infusion of calcium produced a rise in serum CT that was much greater than that seen in the untreated state, but in absolute values less than that occurring after β-blockade. Again, the response to glucagon was abolished.

The patient subsequently received an intravenous infusion of dibutyryl cyclic AMP (0.25 mg/kg per minute for 30 minutes, with aminophylline 10 mg per minute), which failed to produce a significant rise in serum CT, while producing testicular pain, severe tachycardia and hyperhydrosis, and symptoms described by the patient as resembling exactly those experienced during paroxysmal activity of his pheochromocytomas prior to adrenalectomy. No firm conclusions can be drawn from such limited data, other than to suggest that calcium and glucagon differ in their mode of action in stimulating CT secretion by MCT. Similar studies carried out in the pig showed that the secretion of CT increased in response to direct perfusion of the thyroid gland with dibutyryl cyclic AMP (Care and Gitelman, 1968), and upon stimulation of β-adrenergic receptors in the thyroid (Bates *et al.*, 1970).

With the availability of the radioimmunoassay, and in the knowledge of the CT-secretory activity of MCT and the responsiveness of this tumor to provocative testing with calcium and glucagon, an opportunity arose to apply these observations to the screening of a large family (J-kindred) afflicted by MCT.

A 29-year-old man (DJ), the primary propositus, sought advice because of what he felt to be more than a coincidental occurrence of thyroid cancer in his sister (LG), the secondary propositus, and first cousin (DF). His father and a paternal uncle and aunt had died in their 50's of cancer, site unspecified, and another surviving paternal uncle (JJ) and aunt (PT) had undergone thyroidectomy for thyroid cancer 18 and 2 years ago, respectively. Although DJ was found to be free from disease, the family history was highly suggestive of MCT. A review of surgical and autopsy records, and tissue sections, confirmed this diagnosis in each of the 7 relatives, 3 of whom had died of their disease. Autopsy in one (EJ) had revealed, in addition, bilateral pheochromocytomas.

Accordingly, it was proposed to study the entire kindred by means of provocative testing and CT assay. All 89 members were apprised of the fact that, by virtue of their inheritance, many were at risk of harboring an occult thyroid carcinoma, which might be detected by means of serum and urinary CT measurements.

All gave informed consent to the study, and were admitted to the Clinical Study Unit of the New England Medical Center Hospitals, where the results of clinical examination, and thyroid scans, proved negative in all but one case. In that subject (DT), tumor was easily palpable in each lobe of the thyroid gland, and clearly evident upon scanning of the gland with ^{99}Tc and ^{131}I. In the remainder, the thyroid was entirely normal to palpation and radioisotope scanning. None showed any of the physical features associated with the syndrome of MCT, but 4 gave a history of renal calculi. Pertinent clinical details for each member of the kindred are contained in Appendix I, and for each of the 12 subjects diagnosed in the course of this study, in Table II.

With the exception of children less than 4 years old, who were subjected only to a basal venipuncture, all subjects underwent provocative testing with 4-hour intravenous infusions of calcium and glucagon.

Serum obtained at intervals (usually hourly) during the infusions was separated and frozen immediately for assay of CT and parathyroid hormone (PTH), which were both measured by means of radioimmunoassay techniques, details of which have been published previously (Tashjian *et al.*, 1966, 1970), and described above. Characteristics of the radioimmunoassay for PTH will be defined further under the section on serum PTH.

III. Results of Studies

A. CALCITONIN

1. Serum Calcitonin

The validity of the CT assay and the finding of elevated serum CT levels in cases of MCT were subjects of earlier reports (Tashjian *et al.*, 1970, 1972; Melvin *et al.*, 1971). Summarized in Fig. 11 are the serum CT values obtained in 60 patients with proven MCT. As described earlier, CT was not detected in the sera from most of 140 normal subjects, and in all was less than 0.38 ng/ml. In contrast, 33 patients with clinically evident MCT were found regularly to have levels of CT in

TABLE II

*Summary of Clinical and Hormone-Assay Data in 13 Members of the J Family
with Medullary Carcinoma of the Thyroid, Pheochromocytoma, or Both[a]*

Patient	Age	Sex	Medullary carcinoma		Pheochromocytoma		Renal stone	Serum calcitonin[b] (ng/ml)		Local node metastases	Calcitonin content of tumor[c] (MRC U/gm)		Parathyroid hyperplasia	Serum parathyroid hormone[d] (ng/ml)	Postoperative calcium infusion test[e]
			R	L	R	L		Basal	4 Hr		R	L			
RH	29	M	+	+				0.60	3.8	−	3200	930	+	0.41–0.51	−
HJ Sr.	61	M	+	+	+	+	+	0.97	6.3	−	1000	255	+	0.16–0.75	−
PJ	29	M	+	+				0.65	5.9	−	2150	500	−	1.3 –1.8	−
			+									800			
TJ	64	M	+	+			+	0.40	7.0	+	2155	1055	+	0.26–0.38	−
JP	37	F	+	+	+			1.8	29.0	+	1560	1155	+	0.19–0.20	+
				+								255			
RR	34	F	+	+				0.45	6.4	+	600	840	+	1.2– 1.3	−
CT	24	F	+	+				0.99	2.2	+	355	130	+	0.29–0.50	−
HJ Jr.	30	M	+	+			+	0.55	13.0	−	[f]	2120	+	0.36–0.64	−
RJ	36	M	+	+			+	0.79	6.5	−	1070	1475	−	1.4 –2.7	−
JS	27	F	+	+	+	+		0.40	4.1	+	2800	2400	+	0.10–0.85	−
BDeY	39	F	+	+	+	+		0.30	2.9	+	290	1200	+	0.72–1.1	−
DT	25	M	+	+				1.6	8.0	+			+	0.49–0.60	−
WT	26	M	NO[g]	NO[g]		+		<0.3	<0.3					0.28–0.59	

[a] From Melvin *et al., New Engl. J. Med.* **285,** 1115(1971) by permission.

[b] Normal basal levels in control subjects = < 0.38 ng/ml. The 4-hour column designates values at the end of a 4-hour calcium infusion test.

[c] Calcitonin content of normal adult human thyroid = 0.05–0.25 MRC U per gram fresh weight. The results given were determined by biological assay; the mean standard error limits of the estimates were ± 15%.

[d] The values given are the range of 3–8 separate samples. The range in normal control subjects is from undetectable levels to 0.60 ng/ml.

[e] A minus sign (−) indicates normal or undetectable serum calcitonin levels postoperatively which did not rise abnormally during a 4-hour calcium infusion test. A plus sign (+) indicates high basal serum calcitonin or an abnormal rise with calcium infusion.

[f] Tumor less than 1 mm in diameter; entire specimen was taken for histology.

[g] Not operated.

the serum >1.0 ng/ml. In addition, in 27 patients in whom the diagnosis had been confirmed surgically, but in whom, following surgery, residual tumor was not clinically evident, 10 showed high basal CT levels, probably reflecting residual occult tumor tissue. Not included in these data are the results obtained in the study of the J-kindred.

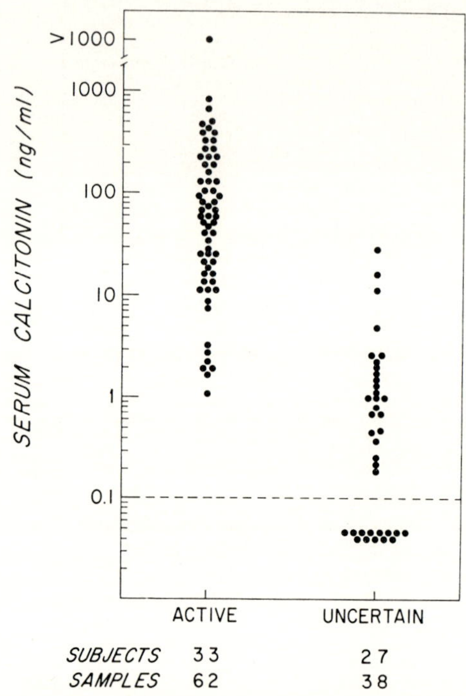

FIG. 11. Basal serum calcitonin (CT) values in 60 patients with proven medullary thyroid carcinoma (MCT). In the left column are plotted values from patients with clinically obvious tumor; values in right column were obtained from patients in whom, as a result of surgery, tumor tissue was not clinically evident. High serum CT values in 10 of the latter probably reflect undetected, residual tumor. Broken line represents the limit of sensitivity of the assay. Not included in these data are the results from members of the J-kindred. Note log scale on the ordinate.

In Fig. 12 are shown the results of calcium infusion tests in the J-kindred. In 55 individuals the response of the serum CT to calcium infusion did not exceed that seen in normal subjects, but in 12 subjects (15 tests) the response was abnormal, the serum CT rising to levels seen only in patients with MCT. In normal control subjects the basal concentration of CT in serum is often undetectable (< 0.10 ng/ml) and always less than 0.38 ng/ml, and values of >0.55 ng/ml have not been detected in 63 normal subjects during or after a calcium infusion test. In the J-kindred, elevated basal CT levels were present in 9 subjects, the range being 0.45–1.8 ng/ml. In 3 subjects, the basal levels were borderline (Table II), but in all 12, a 4-hour infusion of calcium produced a rise in serum CT to levels greater than 2.0 ng/ml, well beyond those values

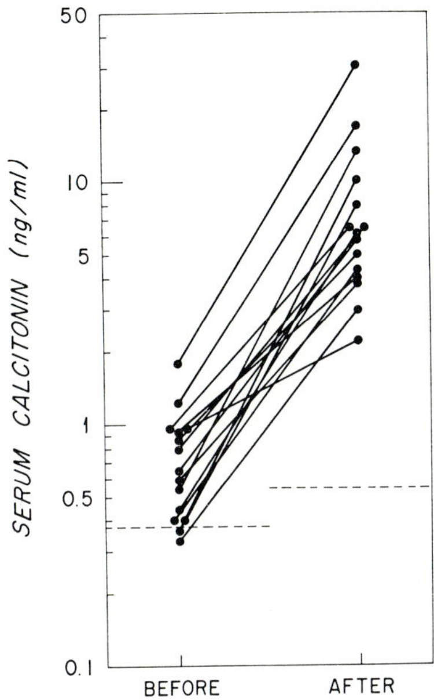

Fig. 12. Preoperative serum calcitonin values, basally (left column) and in response to calcium infusion (right column) in the J-kindred. Broken lines indicate, on the left, the maximum basal value obtained in 140 normal subjects, and on the right, the maximum stimulated values from 63 normal subjects. In 55 members of the J-kindred, low or undetectable values were obtained both basally and in response to the calcium infusion. Plotted are the abnormal results obtained in 12 members of the kindred, 9 of whom had elevated basal values, and all of whom showed an abnormal response to the calcium infusion. Note log scale on the ordinate.

observed in normal control subjects or in 55 other members of the kindred who had low or undetectable basal levels of CT. On the basis of these results, MCT was suspected in these 12 subjects, 11 of whom showed no clinical evidence of disease.

It has recently been reported by MacIntyre et al. (1972) and Deftos et al. (1971) that in measuring the very low levels of CT that obtain in normal subjects, the assay may be subject to nonspecific interference. No attempt is made, therefore, in these data to define the range of basal serum CT values in normal subjects, and values of < 0.10 ng/ml are reported as undetectable. As stated previously, stimulated values of > 0.55 ng/ml have not been detected in normal subjects. By adopting what is almost certainly an overestimate of the upper limit of normal for basal serum CT, it is likely that, without recourse to the calcium infusion test, some cases of MCT would be missed on basal criteria as presently defined. These considerations do not, however, apply in evaluating the exaggerated responses of serum CT to calcium infusion observed in patients with MCT.

2. Urinary Calcitonin

Immunoreactive CT was assayed in unconcentrated urine, as described in earlier reports (Tashjian *et al.*, 1970, 1972).

Evidence for a calcitonin-like material in human urine is as follows:
- I. Biological assay (Voelkel and Tashjian, 1971)
    - A. Hypocalcemic activity (0.2–80 MRC mU/ml) was measured in 65 of 80 urine samples from 19 patients with MCT.
    - B. Hypocalcemic activity in urine of patients with MCT is labile at acid pH, and inactivation is prevented by boiling.
    - C. Concentrated CT-like material had the following properties:
        1. Was inactivated by proteolytic enzymes.
        2. Gave log dose-response lines which were linear and parallel to that of the MRC CT standard.
        3. Was eluted from Sephadex in the same position as tumor CT and synthetic CT-M.
        4. Gave rise to antibodies in guinea pigs which bound labeled human CT (Tashjian *et al.*, unpublished data).
        5. Was detected at very low levels in some but not all normal human urine samples.
- II. Radioimmunoassay
    - A. Immunoreactive material is present (>1 ng/ml) in urine of all patients with overt MCT, often in large quantities requiring 100- to 1000-fold dilution.
    - B. There is no reaction of this material in the porcine or salmon CT assays, or bovine PTH assay.
    - C. The material is not regularly detected (<1 ng/ml) in unextracted normal urine, but is usually present in concentrated normal urine.
    - D. The amount present in urine from both MCT patients and normal subjects increases in response to calcium infusion.
    - E. The material decreases markedly, or disappears from the urine following removal of MCT.

F. Immunological activity in urine and urine concentrates is regularly much greater (2–100 times) than the biological activity; yet elution from Sephadex is in the same position as the biological activity; hence, "CT-like."

Timed 4- or 8-hour urine collections were made basally before calcium infusion and during and after the infusion test. All urine samples were assayed at 25 μl or less per 500 μl reaction mixture. Although nonspecific effects of urine at high concentrations (dilutions 1:15) have been observed in many radioimmunoassays, no nonspecific effects were observed in the CT assay when the final dilution of urine was 1:20 or greater.

In normal subjects immunoreactive CT in urine is present at concentrations of less than 1 ng/ml, and values of greater than 2 ng/ml have not been detected in 20 normal subjects during or after calcium infusion. In contrast, the 12 members of the J-kindred with exaggerated serum CT responses to calcium infusion, each had basal urine CT concentrations of >1.0 ng/ml and levels after calcium infusion of >2.0 ng/ml (Fig. 13).

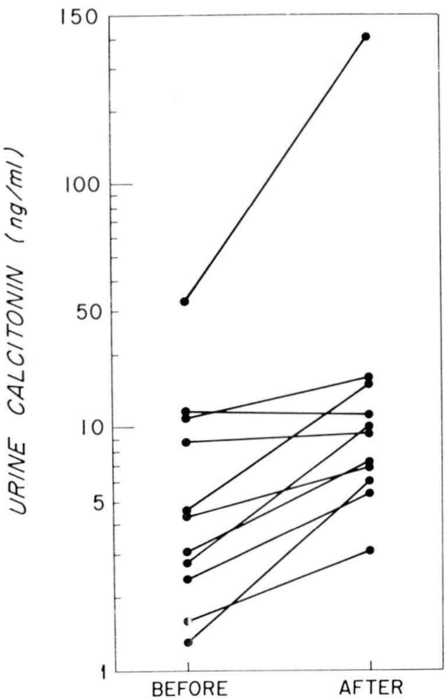

FIG. 13. Urine calcitonin (CT) values, basally (left), and in response to calcium infusion (right), obtained preoperatively in 11 of the 12 members of the J-kindred with elevated serum CT values. In normal subjects, urine CT has not exceeded 1 ng/ml basally, or 2 ng/ml during or after calcium infusion. In the 55 members of the J-kindred with normal serum CT, urine CT values were likewise low or undetectable. Note log scale on the ordinate.

In these 12 patients a good correlation was observed between basal serum and urine CT concentrations (Fig. 14). Correlation coefficient equaled 0.9.

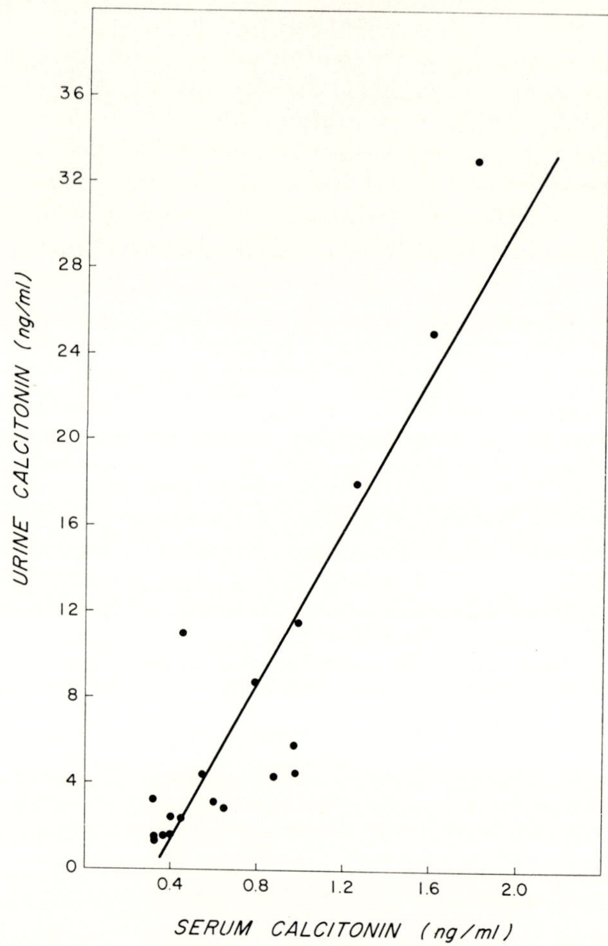

Fig. 14. Correlation between serum and urine calcitonin in 12 subjects with medullary thyroid carcinoma (MCT). $r = 0.9$; $n = 19$; $y = 18.1x - 5.7$.

These results strengthened further the suspicion that the 12 subjects with abnormal serum CT values suffered from MCT.

3. Thyroidectomy

When finally courage matched convictions, total thyroidectomy with clearance of the lymph nodes of the central zone of the neck (Crile,

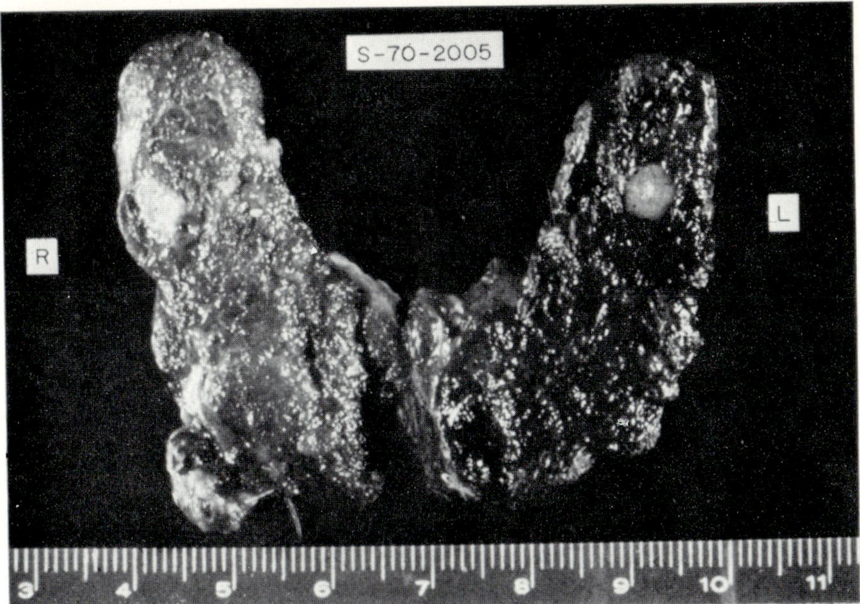

FIG. 15. Sagittal section through the thyroid gland from one of the J-kindred, showing bilateral medullary carcinoma.

1956) was performed in the 12 subjects with abnormal serum and urine CT values.

Bilateral MCT was present in every case. In each instance, the tumors were situated at the junction of the upper and middle thirds of each lobe of the thyroid gland (Fig. 15). Tumor size varied from microscopic to 14 mm in diameter. In two cases, the tumor appeared to be multifocal.

Lymph node metastases were present locally in 7 patients, there being no evidence of metastases in the remaining five.

Histologically, the tumors showed the typical features of medullary carcinoma, with closely packed polygonal or spindle-shaped cells, with eosinophilic cytoplasmic granules, and amyloid stroma (Fig. 16). In the three instances in which tissue was fixed in Zenker's solution, chromaffinity was demonstrated.

An unusual feature of each of the 12 thyroid glands was moderate lymphocytic infiltration.

4. Calcitonin Content of Tumors

Tumor tissue was extracted and assayed for CT by methods described previously (Melvin and Tashjian, 1968; Tashjian and Melvin, 1968). CT was present in very high concentrations in each of the thyroid tumors

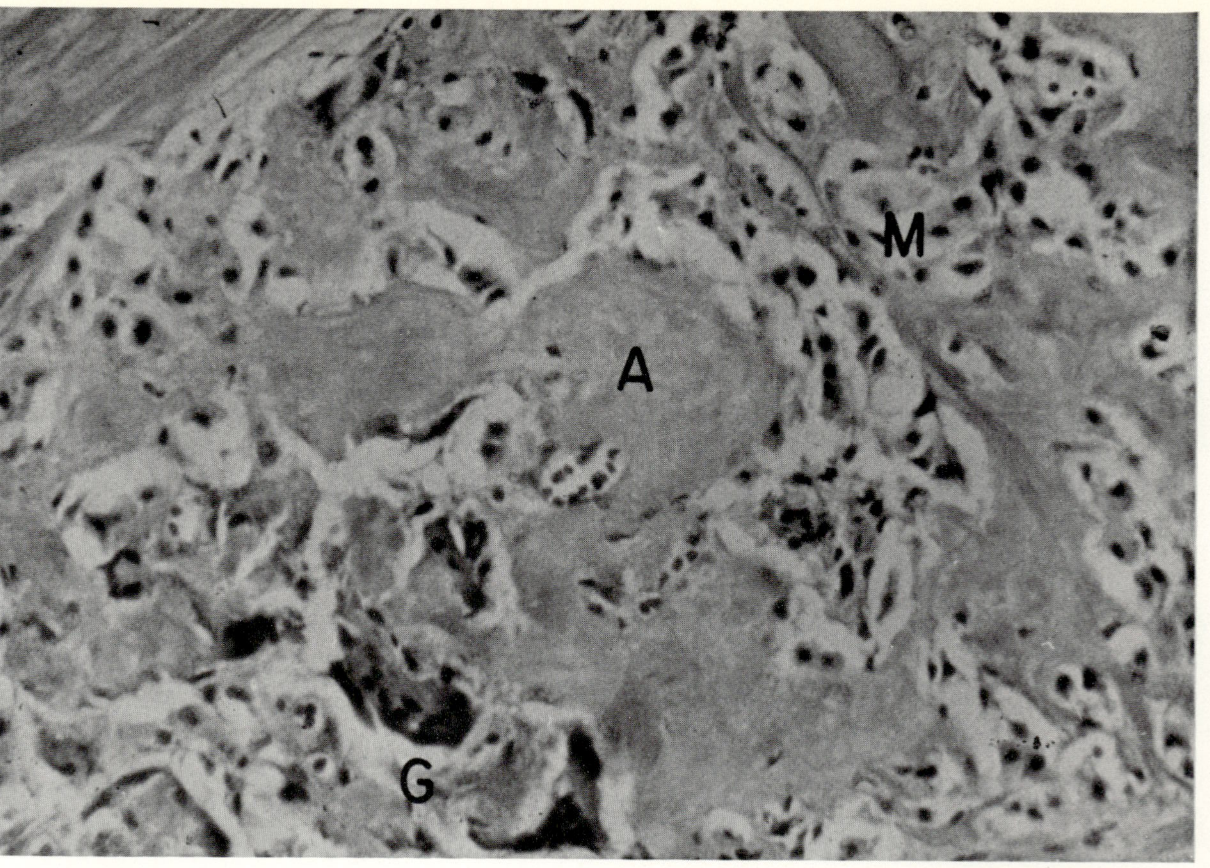

Fig. 16. Histological preparation from one of the tumors shown in Fig. 15. The fibrous stroma with amyloid deposition is typical of medullary thyroid carcinoma. A, amyloid; M, medullary carcinoma cells; G, foreign body giant cells associated with amyloid.

(Fig. 17). The range was 130 to 3200 MRC U/gm fresh weight. These values represent a 650- to 16,000-fold increase over the average concentration of the peptide occurring in normal human thyroid tissue. This is

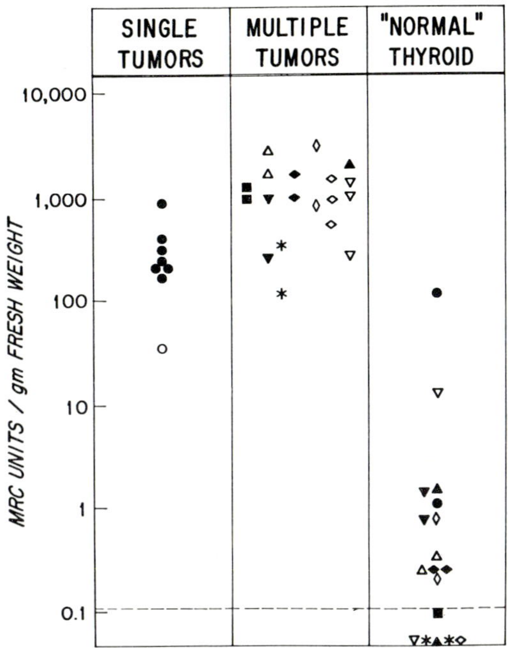

Fig. 17. Calcitonin (CT) content of medullary thyroid carcinoma (MCT) tissue and of "normal" thyroid tissue from areas of the gland remote from the site of gross tumor. All results shown were determined by biological assay. "Single tumors" were obtained from patients with the sporadic form of MCT. The single value given by the open circle (○) was from a patient in whom the MCT was producing ACTH (see Figs. 3 and 4) in addition to CT. Results given for "multiple tumors" are those obtained from 9 members of the J-kindred. Each symbol gives the value for a separate tumor. The last column gives values for thyroid tissue from patients with MCT, but taken from sites distant from identifiable tumor. The same symbol is used consistently to identify each patient. The dashed horizontal line gives the mean value for CT content of normal human thyroid tissue. Symbols on the baseline indicate that CT was not detected in those specimens. Note log scale on the ordinate. From Tashjian *et al.* (1972).

in accordance with earlier reports (Melvin and Tashjian, 1968). The CT content of thyroid tissue in areas of the thyroid gland remote from overt tumor sites was undetectable or normal in 12 specimens, and clearly elevated in 9 others, ranging from 0.3 to 23 MRC U/gm fresh weight (CT content of normal adult human thyroid ranges from 0.05

to 0.25 MRC U/gm fresh weight). It is notable in this respect that Ljungberg (1970a) has reported a greatly increased population of C-cells scattered throughout the substance of a thyroid gland which contained in addition a medullary carcinoma. It is of interest to speculate whether or not our findings indicate an inherited tendency to generalized C-cell

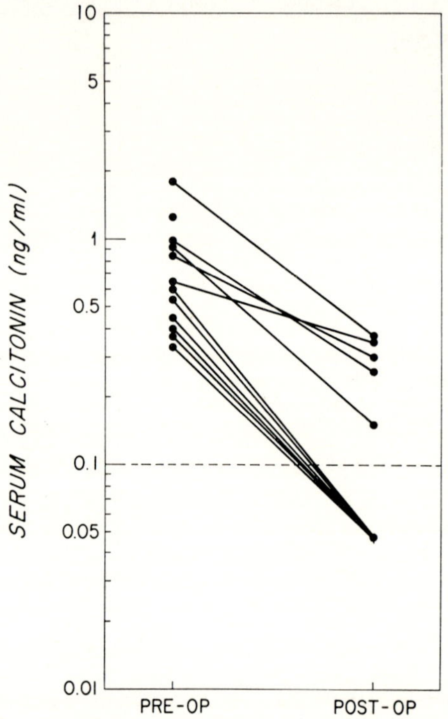

FIG. 18. Fall in serum calcitonin (CT) observed in 11 patients after excision of medullary thyroid carcinoma. Preoperative basal serum CT values shown on the left; postoperative values on the right. Limit of sensitivity of the assay is denoted by the broken line. Postoperatively, serum CT fell to undetectable levels in 6 patients. Result in a twelfth patient has yet to be determined. Note log scale on the ordinate.

hyperplasia, or whether the findings merely reflect the frequently multi-focal nature of the tumors. One of our patients (JP) had in addition to the main focus of tumor in each thyroid lobe, many minute foci visible grossly throughout the substance of the gland. The uniformly consistent anatomical situation of the gross tumor in each lobe of the thyroid glands of our patients may have explanation in the embryological development and subsequent migration of the lateral thyroid (Tashjian *et al.*, 1972). The demonstration of chromaffinity in medullary carcinoma cells is also in accordance with the observations of Ljungberg (1970b), who has described this syndrome as a familial chromaffinomatosis.

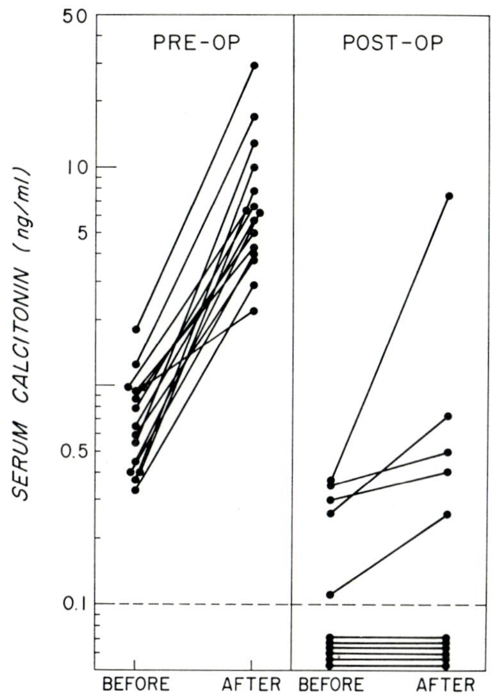

FIG. 19. Change in the response of serum calcitonin (CT) to calcium infusion following excision of medullary thyroid carcinoma in 11 members of the J-kindred. Shown in the left panel are the preoperative responses; postoperative responses are in the right panel. The latter fell in all but one patient (JP), who had the highest serum CT values preoperatively, and who at operation was found to have multiple cervical node metastases. In 6 patients, serum CT remained undetectable in response to the postoperative calcium infusion. Note log scale on the ordinate.

5. *Postoperative Changes in Serum and Urine Calcitonin*

A. Serum Calcitonin. Postoperatively, basal serum CT values fell markedly in 11 patients, being undetectable in 6. The result in the twelfth patient, who only recently underwent surgery, remains to be determined (Fig. 18). Similarly the response to calcium infusion postoperatively was diminished markedly in all patients, falling to within normal limits in all but one patient (Fig. 19). This patient (JP) had shown the highest values of serum CT preoperatively, at operation was found to have many lymph nodes involved, and was the only patient in whom abnormally high CT levels persisted in the urine postoperatively. In 6 of the patients, serum CT remained undetectable both basally and in response to the calcium infusion.

B. Urine Calcitonin. After thyroidectomy, urine CT fell to undetectable levels in all but one patient (Fig. 20). This patient alone had shown abnormal serum CT values postoperatively. Urine CT remained

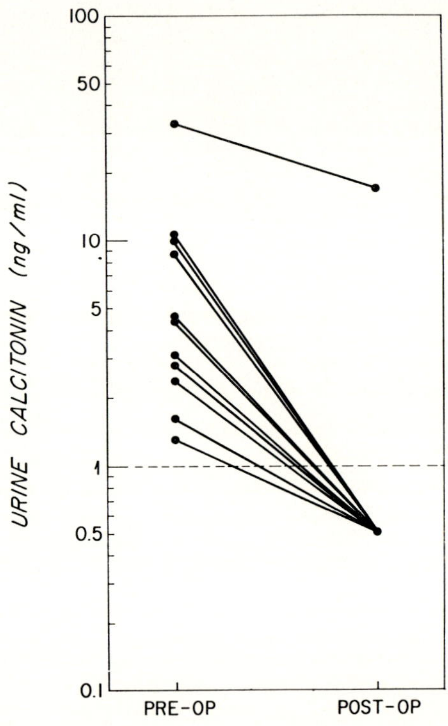

FIG. 20. Fall in urine calcitonin (CT) observed in 11 patients after excision of medullary thyroid carcinoma. Preoperative basal urine CT values shown on the left; postoperative values on the right. Limit of sensitivity of the assay is denoted by the broken line. Basal urine CT fell to undetectable levels in all but one patient (JP), who had shown an abnormal serum CT response to calcium infusion postoperatively (see Fig. 19). Note log scale on the ordinate.

undetectable or not elevated in all but this patient in response to calcium infusion (Fig. 21).

B. PHEOCHROMOCYTOMA

All members of the kindred were screened for pheochromocytoma by means of urinary vanilmandelic acid (VMA) measurement (Pisano *et al.,* 1962). Bilateral adrenal angiography was performed in those cases with an increased urinary VMA.

Previously unsuspected pheochromocytomas were demonstrated in 5 patients, being bilateral in 2 (Fig. 22).

In 4, mild symptoms of palpitations and sweating had been present for several years, and in one patient the tumor was completely asymp-

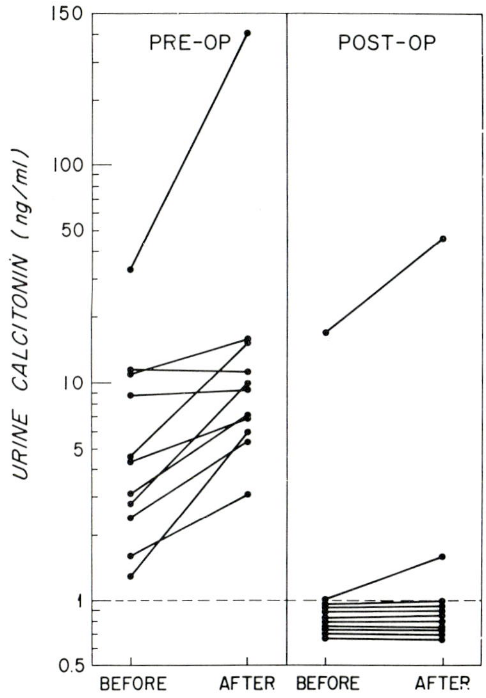

Fig. 21. Change in the response of urine calcitonin (CT) to calcium infusion following excision of medullary thyroid carcinoma in 11 members of the J-kindred. Preoperative responses are shown in the left panel; postoperative responses in the right panel. The response remained abnormal postoperatively in only one patient (JP), who had shown abnormal serum CT values postoperatively. Note log scale on the ordinate.

tomatic. None were hypertensive, and only 2 showed a hypertensive response to intravenous glucagon.

The diagnosis was confirmed surgically in 4 patients, there being no present indication for surgical removal in the asymptomatic patient (HJ, Sr). In 4 of the 5 patients, the pheochromocytomas occurred in association with MCT; in the remaining patient (WT) the thyroid was not explored because of repeatedly low serum CT values both basally and

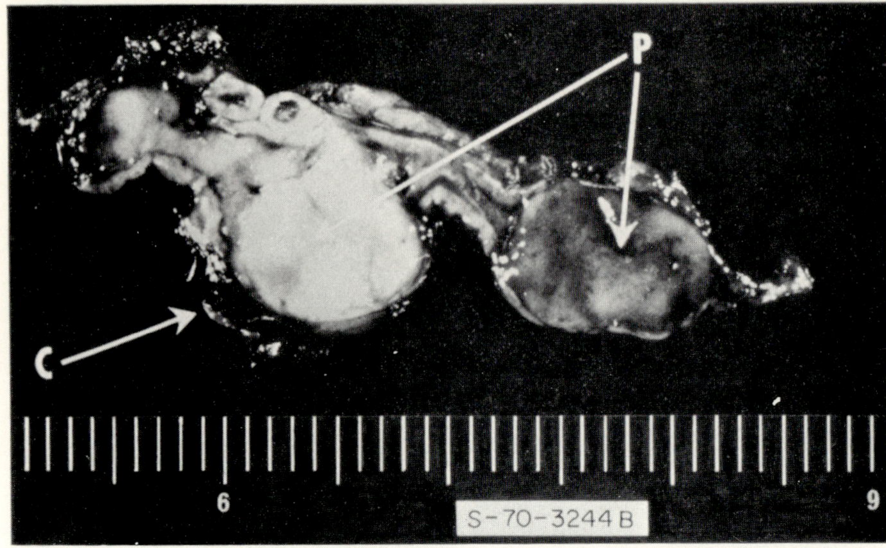

Fig. 22. Section through left adrenal gland from patient (JS) with medullary thyroid carcinoma and bilateral pheochromocytomas. Shown are two pheochromocytomas (*P*) within the left adrenal medulla. Attenuated rim of adrenal cortex (*C*) is seen on lower left. Centimeter scale.

in response to provocative testing. A sixth patient (EJ) was known from autopsy records to have had bilateral pheochromocytomas in association with bilateral MCT.

In view of the report of CT-like material in the adrenal medulla of pigs (Kaplan *et al.*, 1970), it is noteworthy that extracts of the pheochromocytomas from our patients contained no detectable hypocalcemic activity by biological assay, and no more CT than could be accounted for by entrapped plasma when the extracts were assayed immunologically.

C. Parathyroid Function and Histology

1. Serum Calcium

Serum calcium was measured by atomic absorption spectrophotometry, the total range in normal subjects being 8.2 to 10.6 mg/100 ml, with a mean ±2SD of 9.56 ± 0.74 (Robert *et al.*, 1968). In all adult patients, serum calcium was measured under fasting basal conditions on at least 3, and often 4, occasions. Shown in Fig. 23 are basal fasting serum calcium values in 14 chosen normal subjects, 56 relatives of patients with familial MCT, in whom serum CT values were not elevated, and *preoperatively* in 14 patients with proven MCT. Open circles denote those subjects shown to have parathyroid hyperplasia. Three or 4 separate samples were obtained from each patient. All determinations fell within the normal range, and no differences were observed between the three groups.

To our knowledge, there is reference in the literature (Melvin and Tashjian, 1968; Woodhouse *et al.*, 1969; McDermott and Hart, 1970; Aach and Kissane, 1969), to only 3 cases, in addition to our own, of MCT with hypocalcemia. Although in our case high levels of both im-

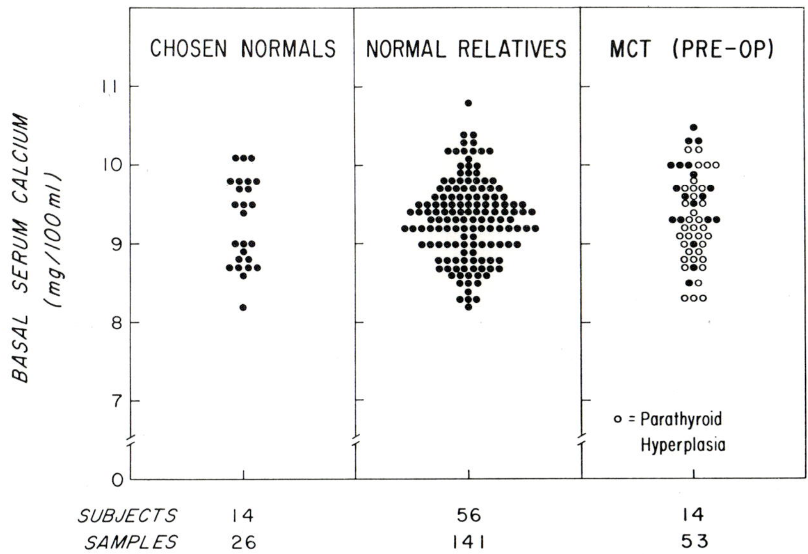

FIG. 23. Serum calcium levels as measured by atomic absorption spectrophotometry in 14 chosen normal subjects, 56 normal relatives of patients with familial medullary thyroid carcinoma (MCT) [normal serum calcitonin (CT)], and *preoperatively* in 14 patients with proven MCT (high serum CT). Open circles (○) denote values obtained in patients with histological evidence of parathyroid hyperplasia.

munoreactive PTH and CT were measured in the serum, the patient was not included in Fig. 23 since he had previously undergone thyroidectomy. In the case described by McDermott, mild hypocalcemia (serum calcium 8.2 mg/100 ml) was corrected by total thyroidectomy, postoperative serum levels being 9.0–10.0 mg/100 ml. The authors concluded that the tumor was producing CT and that the preoperative hypocalcemia was on this basis. In a third case (Aach and Kissane, 1969), hypocalcemia occurred after thyroidectomy, and at autopsy no parathyroid tissue was found.

The rarity with which hypocalcemia is seen in patients with MCT raises serious doubts as to the relevance of increased CT secretion to this phenomenon. In our hypocalcemic patient, prolonged treatment with calciferol, 500,000 units daily, failed completely to modify his severe hypocalcemia. It might be argued therefore that in rare cases MCT secretes a substance which either singly or in conjunction with CT, inhibits the action of vitamin D, thereby producing hypocalcemia.

2. Serum Parathyroid Hormone

Serum PTH was assayed by means of a radioimmunoassay having the following characteristics:

i. Employed labeled bovine PTH, and guinea pig antibovine PTH.

ii. Sensitivity was 0.01 ng bovine PTH.

iii. PTH in human serum or parathyroid gland extracts is *not* completely superimposable on the standard curve for bovine hormone.

iv. Sensitivity for human serum was about 0.15–0.20 ng (bovine equivalents) per milliliter.

v. Values for 35% of 177 normal human sera were ‘<0.20 ng/ml; upper limit of normal range 0.60 ng/ml.

vi. Physiological observations with this assay:

a. In the cow, EDTA infusion produced a rapid increase in serum PTH, and calcium infusion a rapid decrease.

b. Parathyroid adenoma or hyperplasia has been diagnosed correctly in 37 human subjects, with serum PTH > 0.60 ng/ml.

c. Parathyroid adenomas were correctly localized preoperatively by selective venous catheterization in 7 of 7 patients, and in 1 patient with hyperplasia.

d. PTH was not detected (< 0.20 ng/ml), in the serum of 12 patients with hypoparathyroidism.

e. Serum PTH > 1.0 ng/ml in patients with chronic renal failure.

f. In man, serum immunoreactive PTH is not suppressed rapidly by calcium infusion, as measured by our assay method.

The present assay detects about 0.20 ng of bovine PTH equivalents per milliliter of human serum. The range of fasting morning levels measured in the sera of 177 normal subjects was from undetectable (<0.20 ng/ml) in 35% of subjects to 0.60 ng/ml (Fig. 24). In none of the members of the J-kindred with MCT were the fasting levels undetectable, the values ranging up to 2.7 ng/ml. Of considerable interest is the finding that PTH levels in excess of 0.60 ng/ml were present in at least one, and often 2 or 3 serum samples from approximately 40% of members of the J-kindred who do not at present show clinical or laboratory evidence of MCT. Although half of the patients in this group are children, for which we do not have entirely adequate age-matched control data, the results suggest at least sporadic parathyroid hyperfunction in a number of these subjects in whom serum CT levels are not yet elevated. Similarly, elevated serum PTH levels were found in several "normal" relatives in other kindreds with familial MCT, currently being studied by Jackson, Block, and Tashjian. If confirmed, these results may have important prospective significance,

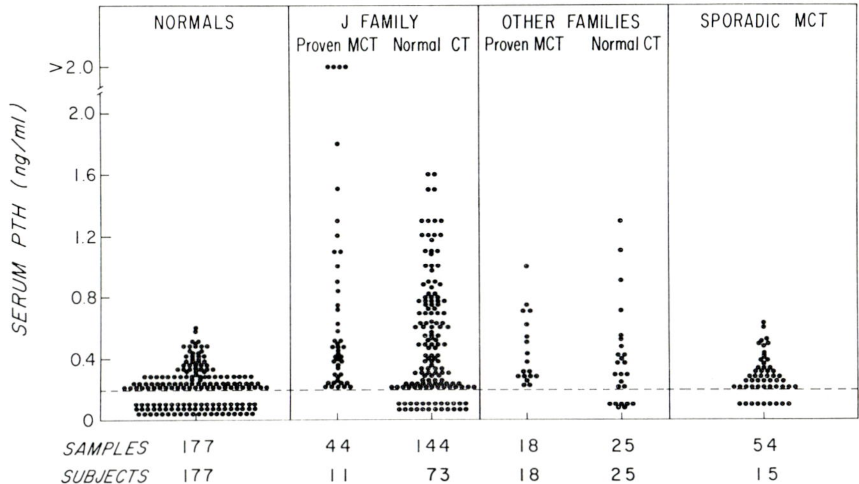

Fig. 24. Serum parathyroid hormone (PTH) levels, as measured by radioimmunoassay, in 177 fasting, normal control subjects, 29 patients with familial medullary thyroid carcinoma (MCT) and 98 of their relatives with normal serum calcitonin (CT) values, and 15 cases of sporadic MCT. As denoted by the broken line, the assay detects 0.20 ng of bovine PTH equivalents per milliliter of human serum. Normal values range from undetectable (<0.20 ng/ml), up to 0.60 ng/ml. In none of the subjects with proven familial MCT were serum PTH values undetectable, these ranging up to 2.7 ng/ml. Of considerable interest is the finding of serum PTH levels in excess of 0.6 ng/ml in approximately 40% of the J-kindred with presently normal serum CT values (i.e., no evidence of MCT) and among similarly normal relatives of other familial cases of MCT. Serum PTH levels were consistently normal in all the patients with sporadic (nonfamilial) MCT.

for on the basis of Mendelian dominant inheritance, at least 25 individuals in the J-kindred, can be expected to develop MCT at some time in the future. If it could be shown in this and other affected kindreds that hyperparathyroidism may antecede hypercalcitoninemia, the association between familial MCT and hyperparathyroidism would appear to be on a genetic rather than a reactive basis. Consistent with this hypothesis are the findings (Fig. 24) that there was no increase in serum PTH observed in 15 cases of nonfamilial (sporadic) MCT.

Abnormally high serum parathyroid hormone levels were present in at least 2 samples from each of 6 patients from the J-kindred with MCT (Table II). Of these 6, 4 had histological evidence of parathyroid hyperplasia and 2 had previous history of renal stones.

In Fig. 25 are shown the response of serum PTH to hypercalcemia induced by calcium infusion in 6 patients with MCT. Following the infusion, serum PTH levels fell slightly in all but 1 patient, in whom

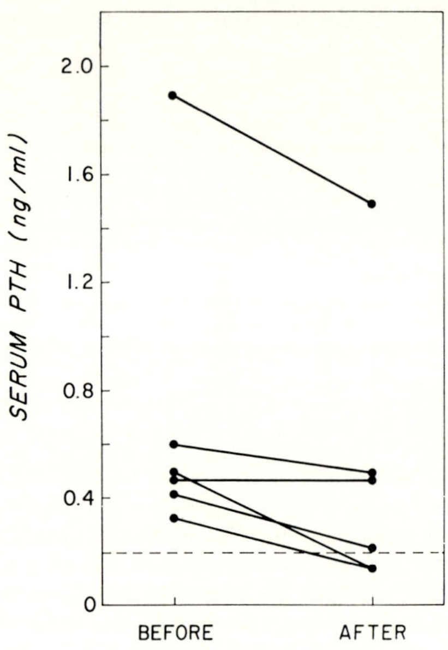

Fig. 25. Response of serum parathyroid hormone (PTH) to hypercalcemia induced by calcium infusion in 6 patients with medullary thyroid carcinoma (MCT). Serum calcium rose to 13–14 mg/100 ml in each patient, and was accompanied by no increase or a fall in serum PTH levels in 5 of the 6 cases by the end of the 4-hour infusion.

no difference was noted. Our results differ from those reported recently by Deftos *et al.* (1972b), who observed an increase in serum PTH in response to induced hypercalcemia in patients with MCT and suggested that CT might act directly to release PTH.

3. Phosphate Excretion Index

The phosphate excretion index (Nordin and Fraser, 1960) was measured in only 6 patients. An abnormally high value was obtained in 3 (HJ, Sr., TJ, JP) two of whom gave a past history of renal stones, and all showing parathyroid hyperplasia histologically. Of these three, only one (HJ, Sr.) was found to have high circulating levels of PTH.

4. Parathyroid Histology

The superior parathyroid glands were preserved intact in each case. In order to ensure total lymph node clearance, it was necessary to sacrifice the inferior parathyroid glands. The diagnosis of parathyroid hyperplasia was based upon overall measurements of the glands, and upon an increase in the parenchymal cell to fat cell ratio as seen in the light microscope (Castleman, 1952; Roth, 1962) (Fig. 26). Most of the glands were only minimally enlarged grossly, being 7–10 mm in greatest dimensions (normal 4–6 mm), but in 2 cases larger glands up to 1.5 cm in diameter were observed (Fig. 27). All histological sections were reviewed by at least 2 pathologists, and in some instances, a third, all of whom concurred in the diagnosis. The sections were later reviewed "blind" with confirmation of the original diagnosis in each instance. Based upon these criteria, parathyroid glands in 10 of the 12 members of the J-kindred who underwent thyroidectomy in the course of this study were considered to be hyperplastic. Of these 10 subjects, 4 had elevated levels of serum PTH (Table II), 2 had a previous history of renal stones, in one of whom increased phosphaturia was present. An additional family member (DH) had undergone neck exploration in 1962, for hypercalcemia. A parathyroid adenoma was removed, and in the course of this procedure the patient was found to have bilateral MCT.

5. Renal Stones

Four members of the J-kindred suffered from renal stones. Each of the 4 was shown to have MCT, one having in addition a unilateral pheochromocytoma. In 3 of the 4 cases with renal stones, parathyroid hyperplasia was evident, serum PTH levels being elevated in 2.

6. Discussion of Parathyroid Studies

From these data it is clear that as measures of parathyroid hyperfunction, serum PTH levels and the histological features of the glands as seen by light microscopy did not always correspond. The reasons for this may be several. Light microscopy is of limited value in assessing the functional status of a tissue, and there is evidence (Birge *et al.*, 1970) that electron microscopy may reveal evidence of hyperfunction in parathyroid glands considered to be normal by light microscopy. It is recognized also that at the present time many questions remain to be answered concerning the relationship between immunoreactive and biologically active PTH

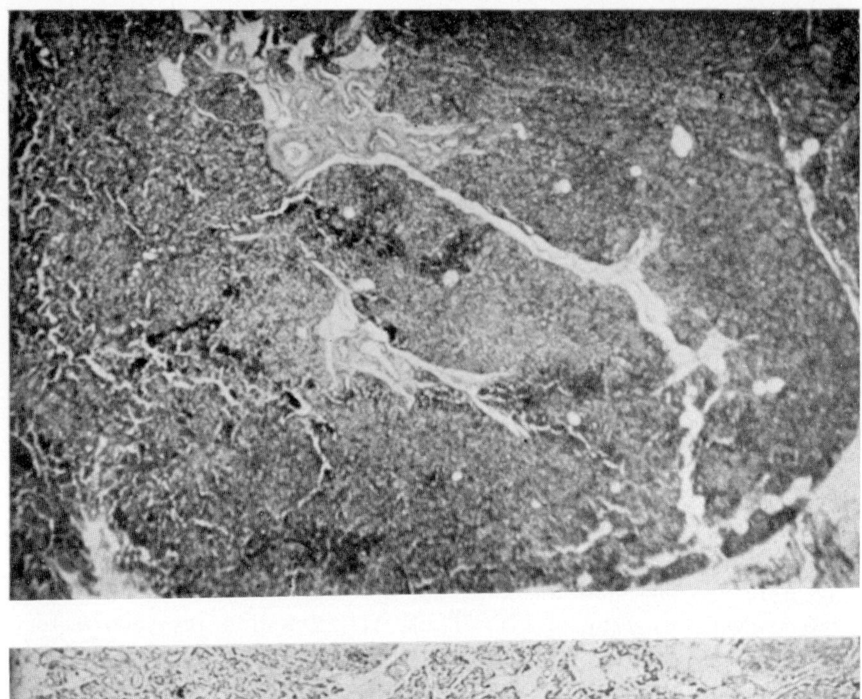

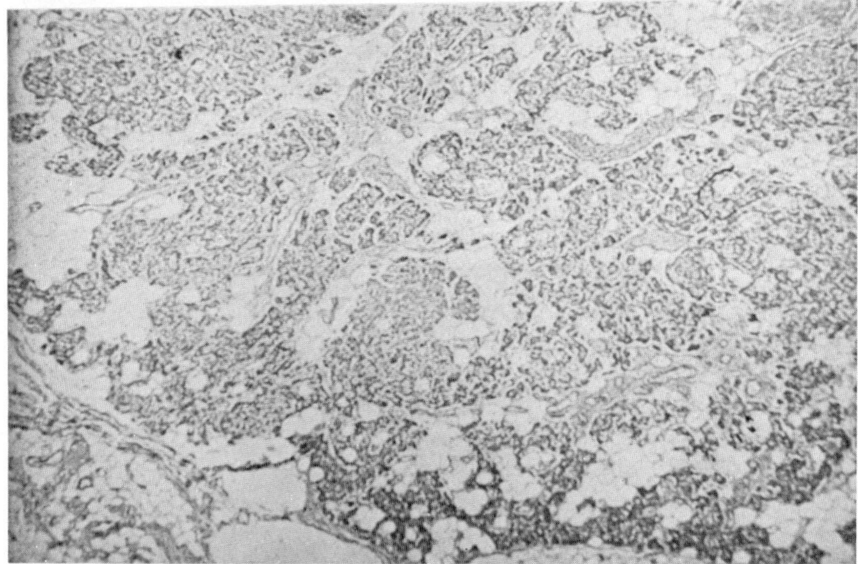

Fig. 26. Composite of histological section from (*top*) hyperplastic parathyroid gland from a member of the J-kindred with medullary thyroid carcinoma (MCT), and (*bottom*) a normal parathyroid gland from a normal subject matched for age and sex. Clearly demonstrated is the increased parenchymal cell-to-fat cell ratio in the gland from the patient with MCT.

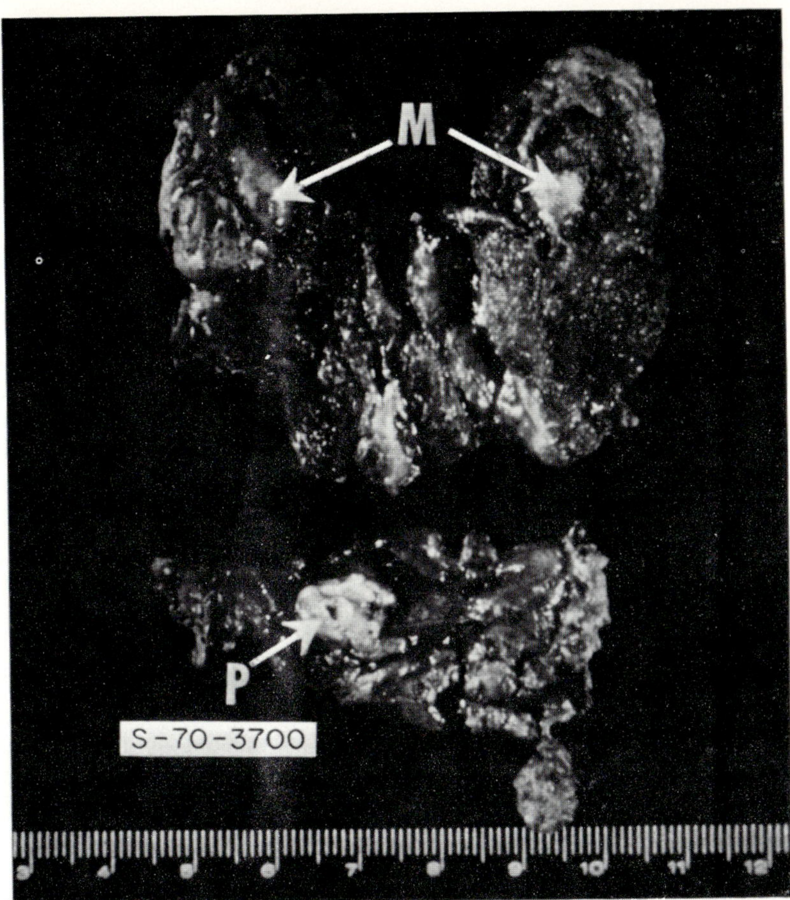

FIG. 27. Sagittal section through the thyroid gland (above) and thymus (below) from a member of the J-kindred, showing bilateral medullary thyroid carcinoma (*M*) and gross enlargement of a parathyroid gland (*P*) which lay in the thymus and which histologically showed chief cell hyperplasia.

as measured in the peripheral plasma (Arnaud *et al.*, 1971; Potts *et al.*, 1971). Nonetheless, the findings of structural abnormalities in the parathyroid glands of 10 of the 12 members of this kindred with MCT, of increased serum PTH levels in 6 of the 12, and of the occurrence of renal calculi in 4, reveal that, in this kindred, parathyroid hyperfunction occurred in the majority of those with MCT. Whether such hyperfunction occurs as a compensatory response to the hypocalcemic influence of chronic CT excess, or as a genetically determined event along with MCT and pheochromocytoma, remains to be determined. In our patients, the unusually high incidence of renal calculi favors primary rather than secondary hyperparathyroidism, as does the finding—already discussed—of increased serum PTH levels in a number of relatives with normal serum CT.

In a comprehensive review of the literature, Steiner *et al.* (1968) found reference to 11 patients with adenoma or hyperplasia of the parathyroid glands in association with MCT, and added 2 further cases of their own. To this review must be added 8 additional patients with MCT and hyperparathyroidism (Moertel *et al.*, 1965; Urbanski, 1967; Gonzales-Licea *et al.*, 1968; Sarosi and Doe, 1968; Paloyan *et al.*, 1970; Mandelstam *et al.*, 1970). Excluding our own patients, the total number of cases described having both MCT and hyperparathyroidism is 21, 6 with chief-cell hyperplasia, and 15 with adenoma, 4 of which were multiple. Among the 21 cases described, a family history of MCT and/or pheochromocytoma was present in 12, was not mentioned in 6, and was described as negative in the remaining 3. Hypercalcemia was noted in 8, with no reference to the level of serum calcium in the other 13 cases.

It is evident from the observations in our own patients, and from a review of the literature, that the incidence of hyperparathyroidism in association with MCT is too high to be accounted for by chance, and that involvement of more than one parathyroid gland occurs in the majority of cases. The differentiation between adenomatous changes in the parathyroids, and hyperplasia is often difficult, and it is possible that some of those described as having multiple adenomas may in fact have shown chief cell hyperplasia. The incidence of multiple parathyroid adenomas in primary hyperparathyroidism unrelated to thyroid carcinoma is reported as 4.3 to 7.3% (Woolner *et al.*, 1961; Cope, 1966). In nonfamilial hyperparathyroidism, 80% of cases have a single parathyroid adenoma as the etiology of the disease, and chief cell hyperplasia occurs in approximately 10% of cases (Cope, 1966). In familial hyperparathyroidism, however, in the absence of pheochromocytoma, involvement of more than one gland is usual, and primary chief cell hyperplasia is the most common pathologic entity (Steiner *et al.*, 1968). Thus, the pattern of pathologic change seen in the parathyroid glands in hyperparathyroidism associated with MCT and/or pheochromocytoma, is similar to that seen in familial hyperparathyroidism occurring alone. Whether the relationship between hyperparathyroidism and MCT is genetic or reactive in nature must await the answer whether in affected kindreds hyperparathyroidism may occur in association with pheochromocytoma, but in the absence of the thyroid tumor; or whether hyperparathyroidism may antecede hypercalcitoninemia. Two patients have been reported (O'Brien, 1963; Crout, 1966) in which parathyroid adenoma occurred in association with pheochromocytoma, but in neither instance was MCT positively excluded.

7. Postoperative Changes in Serum PTH

Postoperatively, transient hypocalcemia occurred in 10 of the 12 patients. The serum calcium had returned to normal in all but one subject (HJ, Sr.) by the end of the third postoperative month, and in this patient alone (serum calcium 8.2 mg/100 ml) has treatment with calciferol been necessary.

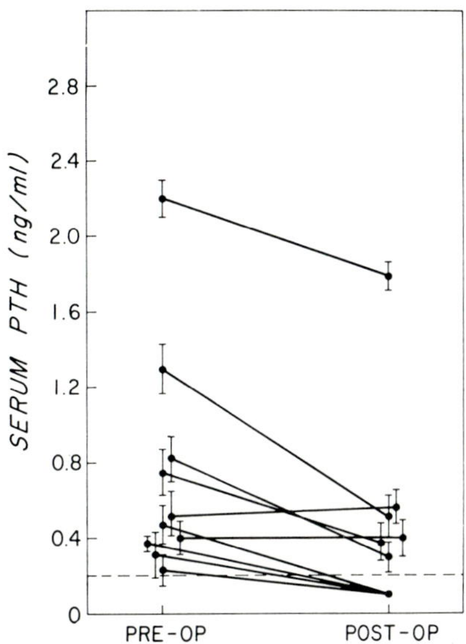

Fig. 28. Serum parathyroid hormone (PTH) levels in 10 members of the J-kindred before and after excision of medullary thyroid carcinoma (MCT) and partial parathyroidectomy. Adequate resection of the thyroid tumors and adjacent lymph nodes necessitated removal of the inferior parathyroid glands. The broken line denotes the limit of sensitivity of the assay. Serum PTH fell to undetectable levels in 4 subjects, fell significantly in 4 others, and remained essentially unchanged in the remaining two. Each point gives a mean value for one subject and the bars give the standard errors.

During the first 2 months after operation, serum PTH levels were unchanged or less than 25% lower than preoperative values in four cases (Fig. 28). In 7 of the 11 subjects for whom data are presently available, three to five determinations in each were all less than 0.60 ng/ml, and about 25% of the samples contained less than 0.20 ng/ml. Thus, in approximately 65% of the subjects, serum PTH fell to levels one-half to one-eighth of those observed before thyroidectomy. More prolonged

study of these patients will be needed to determine whether this fall
is due merely to the partial parathyroidectomy or to removal of the
thyroid gland with its source of CT.

D. Effects of Oral Calcium Load
on Serum Calcitonin Levels

Gray and Munson (1969) have shown that, in the rat, the thyroid
gland protects the animal against hypercalcemia following calcium ingestion. Cooper and Deftos (1970) and Cooper *et al.* (1971) have shown
that in the pig CT secretion increased after oral administration of calcium even in the absence of a detectable hypercalcemia. These authors
suggested that the CT-secreting cells in the thyroid gland might be sufficiently sensitive to the blood calcium concentration as to respond to
small, analytically undetectable increases in blood calcium resulting from
intestinal absorption of the ion. They also suggested, as did Care (1970)
independently, an alternative hypothesis that the presence of calcium
in the gastrointestinal tract might have caused secretion of a gastrointestinal hormone (such as gastrin) which in turn might act as a CT
secretagogue. Care (1970) observed that the addition of porcine pan-
creozymin to blood perfusing the isolated pig thyroid gland produced
a 2-fold increase in CT concentration in the venous effluent. Subsequently, Cooper *et al.* (1971), reported that a 40-fold increase in CT
secretion occurred in the pig following a small intravenous dose of pentagastrin, a synthetic pentapeptide containing the active carboxyterminal
tetrapeptide amide of gastrin, the hormone secreted by the mucosa of
the gastric antrum. Injection or infusion of pentagastrin produced hypocalcemia and hypophosphatemia before, but not after, functional thyroidectomy, accompanied by marked elevations of CT in the thyroid
venous effluent blood.

We have examined the effects of a large oral load (1300 mg) of calcium
on serum CT levels in patients with MCT, and in addition report preliminary observations on the effect of infused pentagastrin in these patients. The results of an oral calcium load given without pentagastrin
to 4 patients with MCT are shown in Fig. 29. After 2 basal fasting
venous blood samples, the patients ingested 60 ml of calcium glucono-
galactogluconate (Neocalglucon®), containing approximately 1300 mg
of elemental calcium. Venous blood samples were then taken at 15-min-
ute intervals for the first hour, and again at 2 hours. Little or no response,
either in serum calcium or serum CT, was seen in three of the patients,
but in one (EC) the serum calcium rose promptly from 8.8 to 10 mg/100
ml, this being accompanied by a rapid increase of CT levels from 200
to 550 ng/ml. Note that the patient CC, the subject of an earlier report

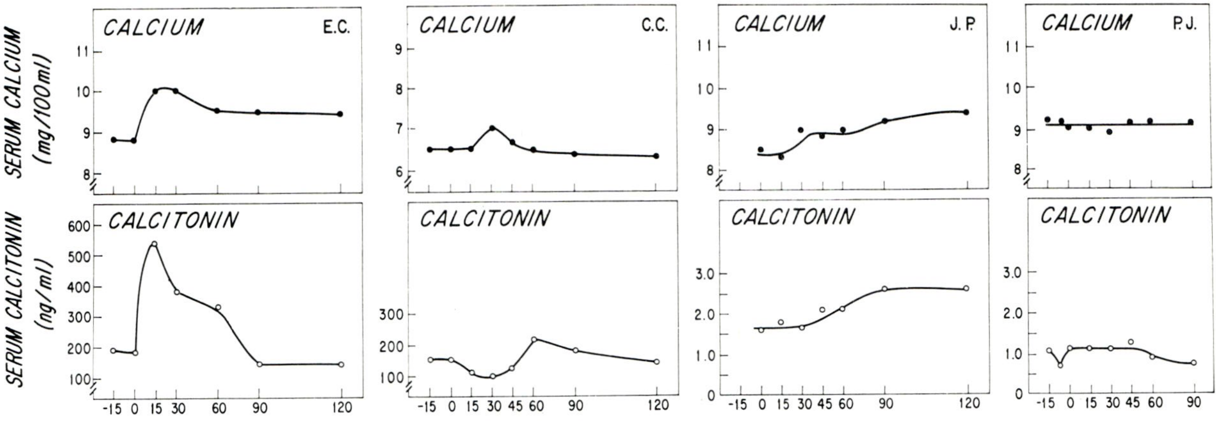

FIG. 29. Serum calcium and calcitonin (CT) responses to an oral calcium load, in 4 patients with medullary thyroid carcinoma. Patients ingested 1300 mg of elemental calcium at time 0. In one (EC), despite greatly increased levels of serum CT, the serum calcium concentration increased by 1.2 mg/100 ml at 15 minutes. This was accompanied by a further increase in serum CT from 200 to 500 ng/ml. In a second patient (JP) a slight increase in serum calcium concentration was accompanied by a small increase in serum CT from 1.6 to 2.6 ng/ml. From Tashjian *et al.* (1972).

(Melvin and Tashjian, 1968), was severely hypocalcemic. Patient JJ showed a rise in serum calcium concentration from 8.4 to 9.4 mg/100 ml accompanied by a small increase in serum CT. Both patients CC and JJ were restudied at a later date, receiving both an oral calcium load and intravenous pentagastrin.

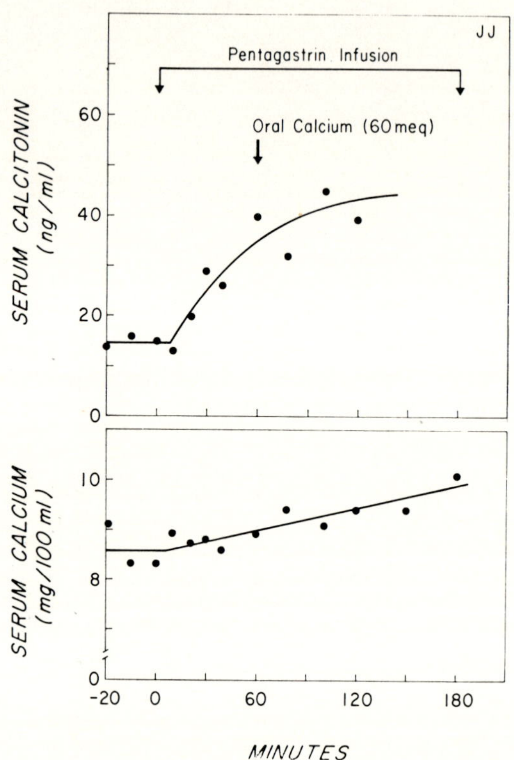

Fig. 30. Effects of intravenous pentagastrin and an oral calcium load on serum calcium and calcitonin (CT) in a patient with medullary thyroid carcinoma. Pentagastrin was infused for 3 hours from time 0. A 3-fold increase in serum CT levels occurred, commencing with the start of the pentagastrin infusion. A rise in serum calcium concentration became evident after the administration of 1300 mg of calcium orally at 60 minutes, without further increase in serum CT.

The effects of intravenous pentagastrin on serum calcium and CT in patient JJ, with advanced metastatic MCT, are shown in Fig. 30. Following 3 basal fasting venous samples, an intravenous infusion of pentagastrin 0.5 μg/kg per hour was commenced. Venous blood samples were then obtained at 10-minute intervals for 40 minutes, and thereafter every 20 minutes to 2 hours, and thereafter at 0.5-hour intervals until conclusion of the test at 3 hours. One hour after commencement of the infusion, an oral calcium load containing 1300 mg of elemental calcium was administered. During the first hour, in response to intravenous pentagastrin alone, no change was observed in the serum calcium concentration. Serum CT, however, increased 3-fold from 14 to 40 ng/ml. Subsequently, following ingestion of the oral calcium load, the serum calcium rose approximately 1 mg/100 ml without further increase in the level of serum CT.

In the second patient, CC, neither serum calcium nor serum CT showed any change in response to either pentagastrin or oral calcium.

These data show that in patients with MCT no consistent increase in either serum calcium or CT appears in response to a large oral dose of calcium, and in the one patient in whom a response was observed, the increased secretion of CT did not prevent a rise in serum calcium concentration. The results of pentagastrin infusion are at present inconclusive and require further study.

E. Serum Histaminase Activity in Patients with Medullary Thyroid Carcinoma

Baylin and co-workers (1970) have reported elevated levels of histaminase activity in the serum of 4 patients with metastatic MCT, and found high activity of this enzyme in the tumor tissue. Administration of the drug aminoguanidine sulfate, a specific inhibitor of histaminase, resulted in total inhibition of serum histaminase activity in 2 patients with metastatic MCT, this being accompanied in one patient by temporary clinical improvement. These authors suggested that measurement of histaminase activity in the serum might be of value in the diagnosis of MCT, and that aminoguanidine might prove useful in the treatment of this condition.

Subsequently, in collaboration with Baylin, we have carried out parallel studies of serum histaminase activity and serum CT in patients with MCT. Shown in Fig. 31 are corresponding measurements of serum CT and histaminase activity in 16 patients with MCT, 5 of whom were without evidence of metastases. Serum CT is plotted against the ordinate, using a log scale; serum histaminase against the abscissa.

Twelve of the 16 subjects (75%) had a clearly abnormal basal serum CT value, but in only 5 of these (30%) was there an increase in serum histaminase activity. Serum histaminase was increased in a total of 8 subjects, 3 of whom had normal or marginally abnormal serum CT. Otherwise stated, in 16 patients with MCT, normal basal values of serum CT occurred in 4, and of histaminase, in 8. As an indicator of disease, therefore, measurement of serum CT would appear to be the more sensitive of the two. And as already discussed, induced hypercalcemia produced an abnormal serum CT level in all subjects with MCT.

The effects of aminoguanidine were studied in 4 of our patients. Total inhibition of serum histaminase activity occurred in all, without change in the levels of serum CT basally, or in the rise in the serum CT occurring in response to calcium infusion. The failure of aminoguanidine to influence CT secretion by the tumor, suggests that the effects of this

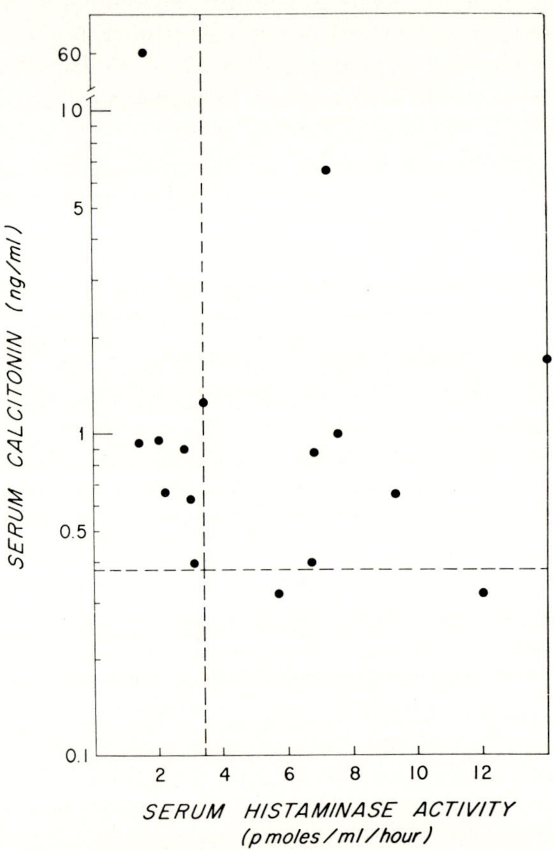

Fig. 31. Corresponding basal serum calcitonin (CT) and serum histaminase levels in 16 patients with medullary thyroid carcinoma. Broken lines indicate, respectively, the maximum value of serum CT observed in 140 normal subjects, and 2 SD above the mean value for serum histaminase as measured in 62 normal subjects. Elevation of serum CT was evident in 12 patients, and of histaminase in 8. Of the 4 patients with marginal or normal serum CT values basally, all subsequently showed an abnormal rise in serum CT in response to calcium infusion. Note log scale used for serum CT.

drug are limited to a specific inhibition of the enzyme histaminase, that a more generalized inhibition of tumor-cell activity is unlikely, and that histaminase is not intimately involved in the calcium-mediated stimulation of CT secretion from the tumor. These results do not support a likely role for aminoguanidine in the treatment of this condition.

IV. Genetic Aspects

In its familial form, MCT appears to be transmitted as a Mendelian dominant trait with a high degree of penetrance. Analysis of the genealogical data in the J-family confirms this mode of inheritance (Fig.

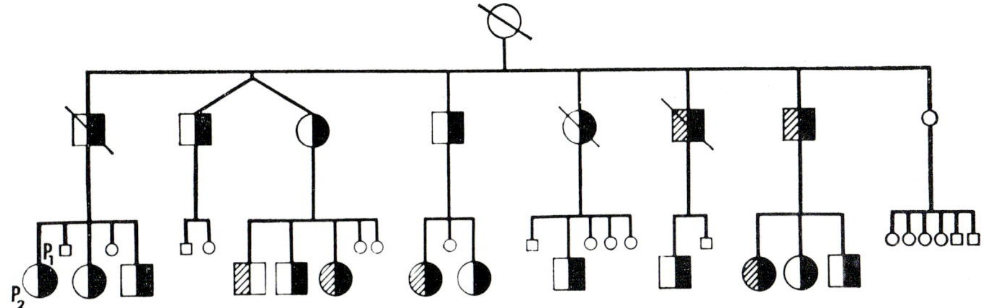

FIG. 32. Genealogy of the J-kindred. Deceased subjects are indicated by an oblique stroke. The fourth generation, comprising 54 children, has been omitted for the sake of clarity, since none presently have clinical or laboratory evidence of MCT or pheochromocytoma. From Melvin *et al.* (1971). Normal: □, male; ○, female. Medullary carcinoma: ▉, male; ◑, female. Pheochromocytoma: ▨, male; ◮. Medullary carcinoma plus pheochromocytoma: ▰, male; ◕, female. P_1 is the primary propositus, and P_2 is the secondary propositus. The presence or absence of MCT in the original female member of this family remains uncertain.

32). The total number of proven cases of MCT is 19, 5 in association with pheochromocytoma. A sixth subject with pheochromocytoma has no evidence of MCT. The fourth generation, comprising 54 children, has been omitted, since none have yet shown evidence of disease.

A. GENEALOGICAL DATA

1. Generation I

This lady died of thromboembolism in 1947, at the age of 65 years. Autopsy showed no disseminated carcinoma, and the thyroid and adrenal glands were grossly normal. The thyroid was not, however, sectioned, and in view of the findings in other members of this family, could have contained occult tumor. The patient had immigrated to the United States in 1901 from Sweden leaving many relatives who have been the subject of a study by Dr. O. Ljungberg of Malmo. Among these, 13 have been found to have MCT and/or pheochromocytoma (Ljungberg, O., personal communication).

2. Generation II

In this generation comprising 8 sibs, 7 were found to have MCT, 2 in association with pheochromocytoma. This prevalence appears greater than would be expected for a disease due to a Mendelian dominant gene. In clinical data, however, the small size of the average family, which necessarily includes at least one affected, often biases the estimate of the Mendelian segregation ratio. The bias is in the direction of over-estimating the true proportion of affected. Unfortunately, the methods for correcting for the bias of ascertainment are not precisely applicable to large pedigrees. As a first approximation, however, data from generation II were corrected by the sib method (Neel and Schull, 1954):

$$P_{II} = \frac{R - N}{T - N}$$

where P = estimated proportion of affected; R = observed number of affected; N = number of propositi; T = total number of children.

$$P = \frac{7 - 1}{8 - 1} = \frac{6}{7} = 0.86 \pm 0.13 \text{ (SE)}$$

Expected $P = 0.50$. The difference betweeen estimated and expected values of $P = 0.36$, which by the t test is just significant: $0.05 > p > 0.025$ ($t = 2.77$ for 6 degrees of freedom).

3. Generation III

Similar analysis of the data of the third generation comprising 31 individuals, 12 with MCT, 3 with associated pheochromocytoma, and 1 with pheochromocytoma alone leads to a corrected proportion of affected individuals of 0.37. In this case, generation III was considered to contain a single propositus. The calculations are shown below:

$$P = \frac{R - N}{T - N} = \frac{12 - 1}{31 - 1} = \frac{11}{30} = 0.37$$

$$\text{Variance } P_{III} \text{ or } \sigma_{III}{}^2 = \frac{(T - R)(R - N)}{(T - N)^3}$$

$$= \frac{(31 - 12)(12 - 1)}{(31 - 1)^3} = 0.0077$$

$$\text{Standard error} = \sqrt{\sigma_{III}} = 0.088$$

Thus $P_{III} = 0.37 \pm 0.09$, which is not significantly different from expectation, 0.50.

4. Pooled Data from Generations II and III

Correction of the pooled data for bias of ascertainment results in an estimated prevalence of 0.47 ± 0.088, which is in agreement with the expected segregation ratio if the condition were transmitted by a Mendelian dominant gene.

Other features of the J-family which support the hypothesis that this syndrome is due to a rare autosomal dominant gene are: (1) all affected individuals have one (and only one) affected—or presumably affected—parent. (2) Those who are affected are capable of producing two types of offspring—affected and nonaffected. As noted earlier, the corrected proportion of affected individuals agrees with what one would expect from Mendelian theory. (3) The disease can be transmitted by a parent of either sex to offspring of either sex. (4) The overall sex ratio is essentially 1:1 (10 male:9 female).

It is known, from a study of marriage records, that from at least 1794, the family has been free of consanguinity (Ljungberg, personal communication). In a preliminary account of studies of the Swedish branch of this family, Ljungberg *et al.* (1967) also noted an autosomal dominant mode of inheritance for the thyroid carcinoma.

B. Pheochromocytoma

The genetic aspects of pheochromocytoma as it occurred in this family are much less clearly defined. In only one instance was the parent of an affected offspring shown to have the adrenal tumor. Neither of two offspring of another individual known to have had both pheochromocytoma and medullary thyroid carcinoma, have any evidence of the former. In one instance (WT), pheochromocytoma was present in the absence of evidence of MCT. Familial pheochromocytoma has been reported in at least 85 instances in 29 families (Steiner *et al.*, 1968). In several of these families the inheritance pattern was that of an autosomal dominant gene with high penetrance. It might be argued that the association between pheochromocytoma and MCT in the J-family is fortuitous, and represents the chance inheritance of two independent genetically determined diseases. However, we consider such a chance association to be unlikely for the following reasons:

1. The incidence of MCT in patients with familial pheochromocytoma is much greater than that seen in the general population (Sipple, 1961; Steiner *et al.*, 1968). In their exhaustive review, Steiner *et al.* (1968)

notes that of 85 patients with familial pheochromocytoma, 20 had associated MCT.

2. In sporadic cases of pheochromocytoma alone, bilateral tumors occur in only 5% of cases. In patients with both medullary thyroid carcinoma and pheochromocytoma, however, the latter is bilateral in over 70% (Williams, 1966), irrespective of whether the thyroid tumors are sporadic or familial.

It appears much more likely, therefore, that the syndrome of pheochromocytoma and MCT is the result of a single genetic abnormality, which in familial cases, may be variably expressed from one generation to another.

C. Karyotypes

Analysis of the karyotypes as determined in peripheral leukocytes (Bloom *et al.*, 1966) in 6 patients in the second and third generations of the J-kindred with MCT has revealed no abnormality in either number or structure of the chromosomes. We do not therefore confirm the findings of Al-Saadi and Lieberman (1968), who reported a high incidence of aneuploidy in leukocyte preparations from 3 patients with MCT. It must be recognized, however, that the methods used were relatively crude and that newer more delicate methods of analyzing chromosomes may reveal abnormalities not detected by us.

D. Blood Groups

Evidence for linkage of the gene responsible for transmission of this syndrome is being sought in detailed studies of the blood groups in all members of the second and third generations of the J-kindred. Results are as yet incomplete. In the unlikely event that a clear pattern of linkage emerges from such studies, such information might prove useful in identifying those children in the fourth generation at greatest risk of developing the disease in the future.

E. Follow-up Studies

Each member of the J-kindred has now been retested one year after the initial study. Abnormally high levels of CT in both serum and urine, both basally and in response to calcium infusion, have recurred in two of the patients (JP, CT) who underwent thyroidectomy a year ago. Both patients had extensive local lymph node involvement at the time of the initial operation, and in the immediate postoperative period, calcium infusion had resulted in a clearly abnormal rise in serum CT in one (JP) and a borderline response in the other, suggesting residual tumor tissue in both. In these patients, it is planned to carry out selective venous catheterization of the neck, mediastinal and hepatic veins, attempting to localize the metastases by means of CT assay, and to direct further surgical management, possibly with exploration of the mediastinum, on the basis of such information.

Since on the evidence presented at least half of the 54 children comprising the fourth generation of this kindred can be expected to develop the disease at some time, and since we do not know whether presently unaffected adults in the third generation may yet become affected, this kindred poses an open-ended problem in follow-up and genetic counseling. Whether this constitutes an unparalleled opportunity for prospective study of a fascinating disorder, or an investigative millstone, will depend at least in part upon the degree of restraint that can be introduced into this hitherto fecund family.

V. Treatment

At present, the most effective treatment of MCT is surgical. Early diagnosis, i.e., detection of the carcinoma *in situ*, which is now possible by means of CT assay, makes surgical cure a reasonable expectation. Experience with radiotherapy in this condition is limited, but if CT secretion can be considered a measure of tumor mass, radiotherapy has proved ineffective in modifying the disease in each of three of our patients with inoperable metastatic MCT. In view of the frequently multifocal and multiglandular nature of the disease as it affects the thyroid, adrenal medulla, and parathyroid glands, surgical management should probably include total thyroidectomy, identification and careful examination of all 4 parathyroids, and direct examination of both adrenal glands in the event of laboratory evidence of pheochromocytoma. Total thyroidectomy should be accompanied by total clearance of the lymph nodes of the central zone of the neck, and, in the light of our experience, extension of this procedure to include clearance of the mediastinal lymph nodes through a sternal splitting procedure should frozen section reveal local node involvement. As argued by Crile (1956), we do not believe that radical neck dissection offers any advantage over the management just outlined. When pheochromocytoma is present in addition, it is suggested that adrenalectomy be undertaken before thyroidectomy, simplifying operative management and permitting direct examination of the liver for metastases, which would make mediastinal exploration unjustified. The assay of CT in hepatic venous blood obtained by selective catheterization of the hepatic veins may also be useful in determining the extent of the surgical procedures.

A quantitative assessment of the efficacy of surgery (or any other form of treatment) can be obtained by means of serial postoperative (or posttreatment) measurement of serum and urine CT, and of the response to calcium infusion.

Although there is one report (Wahner *et al.*, 1968) of a case of metastatic medullary carcinoma of the thyroid apparently regressing on treatment with thyroid hormone, there is no *a priori* reason for believing that this tumor is TSH dependent. In 2 of our patients with advanced metastatic disease and in 3 normal subjects, the administration of TSH, as a single intramuscular dose of 10 units, produced no rise in serum CT levels measured 2 and 18 hours later. Further, in one patient rendered hypothyroid by radiotherapy to cervical metastases, serum calcitonin levels were not suppressed following treatment with full replacement doses of thyroid hormone. It would appear, therefore, that the role of thyroid hormone in the management of this condition is limited to replacement following thyroidectomy.

It is posssible that the case described with apparent regression of metastases on treatment with thyroid hormone was an example of mixed papillary or follicular and parafollicular cell tumors, such as described in one case by Williams (1966) and seen in one other by us (Melvin *et al.*, unpublished data).

The role of chemotherapy in the treatment of MCT remains to be determined. For the reasons already stated, aminoguanidine does not appear to have an antitumor effect. We are presently evaluating the efficacy of Bleomycin (Shomaoka *et al.*, 1971) in the treatment of MCT.

The results of any treatment must be evaluated in the light of the natural history of the untreated disease. Such evaluation is difficult in the case of MCT, since the age at onset is unknown. In a review of 249 cases of MCT, Fletcher (1970) noted that the age at diagnosis ranged from 10 to 82 years, with a mean of 36 years. The youngest case was reported by Mandelstam *et al.* (1970), who described a "prophylactic" thyroidectomy in a $4\frac{1}{2}$ year old boy with a strong family history of MCT, pheochromocytoma, mucosal neuromas, and hyperparathyroidism. This patient was found at operation to have MCT with cervical metastases. It may be inferred from the slow progression of the tumor in many cases that MCT may be present for several years before it becomes clinically manifest.

The rate of progression of this tumor may, however, be very variable. In our experience of patients other than the J-kindred, 2 died of disseminated MCT at the age of 21 years, and a third patient, also aged 21 years, is declining rapidly on account of disseminated disease. In contrast, 2 of Williams' cases (1966) were alive more than 17 years after the discovery of metastases, and among the J-kindred, one (JJ) is surviving 18 years since metastases were found, and 2 (DH, WJ) died 18 years after thyroidectomy. Two other patients in our series are surviving 13 years after the initial operation, both with metastases. Those of the J-kindred who died of MCT, did so at ages 50, 54, and 58 years.

Thus, it can be concluded, that while MCT is compatible with a prolonged survival, even in the presence of metastases, a normal life-span is unusual, and in some instances, the disease may be rapidly fatal at an early age.

Steiner *et al.* (1968), in reviewing the mortality in 91 of 145 cases (Hazard *et al.*, 1959; Woolner *et al.*, 1961; Williams *et al.*, 1966) for which follow-up data were available, noted that the 5-year survival rate after diagnosis was 50%.

Fletcher (1970), who reviewed the same series, but with the addition of 33 cases reported by Ibanez *et al.* (1967) noted that after definitive surgical treatment, 45% of the patients traced were alive after 5 years, but only 15% of the total 162 patients traced were known to have died from their disease within 5 years. Of these patients, only 23% were surviving 10 years after surgery.

It must be noted, however, that in the 4 series upon which these mortality figures are based, lymph node metastases were present in 48–70% of the patients at the time of initial operation (Robbins, 1969). It is our submission that early diagnosis of this disease, such as can be achieved by means of calcitonin assay before signs or symptoms develop, may be expected to improve markedly the chances of surgical cure in this condition.

VI. Unifying Concept of Histogenesis

It has been suggested, (Ljungberg *et al.*, 1967; Schimke *et al.*, 1968; Weichert, 1970; Paloyan *et al.*, 1970) that the medullary carcinoma syndrome can be attributed to a defect in a single-cell system, originating in the neural crest, and representing a familial chromaffinomatosis. We believe this to be an attractive hypothesis, for which there is considerable supporting evidence.

It is suggested that the thyroid, parathyroid, and ultimobranchial body—all of endodermal derivation, share with the adrenal medulla, which is an ectodermal derivative, a common stem-cell component deriving early in embryonic development from the neural crest; that these neuroectodermal stem-cells migrate into the primitive alimentary tract in which they form the enterochromaffin system, and contribute to the developing endocrine glands which bud off from the foregut (Weichert, 1970). In addition, they form the chromaffin system, including the adrenal medulla. Endocrine glands deriving from the foregut have in common the production of amino acid and peptide hormones, whereas those of mesodermal origin—ovary, testis, and adrenal cortex—produce steroid hormones.

Weichert (1970) has reviewed the evidence that neuroendocrine cells are present in the mucosa of the alimentary tract, that they are carried by the various endocrine glands in their embryonic migration, and that they have the potential to secrete a variety of peptide hormones.

Carcinoid tumors of the foregut, and nonendocrine tumors of foregut derivatives—the lung in particular—have a remarkable potential for the ectopic production of peptide hormones. Milhaud *et al.* (1970) have reported an example of CT secreting carcinoid tumor of the bronchus.

Argentaffin and chromaffin cells are found in MCT (Ljungberg *et al.*, 1967), together with serotonin and catecholamines, and these tumors have been known to secret serotonin (Ibanez *et al.*, 1967; Moertel *et al.*, 1965; Williams *et al.*, 1968). Chromaffinity was demonstrated in the tumors from 2 of our patients in whom it was sought.

Thus it appears possible that C-cells, whether situated in the ultimobranchial body, thyroid or parathyroid glands, or thymus, originate in the neural crest and share a common embryonic origin with the neurochromaffin cells of the adrenal medulla.

The hypothesis seems reasonable, therefore, that the syndrome of MCT can be explained on the basis of a genetically determined dysplasia of neural ectodermal cells.

On the basis of this argument, it would be expected that other tumors of neuroectodermal origin—including pheochromocytoma, paraganglioneuroma and chemodectoma, together with tumors of the thymus and parathyroid glands—might be capable of the ectopic production of CT. The secretion of CT by such tumors should be looked for in the future.

VII. Summary

1. The syndrome of MCT has been reviewed, with reference to the clinical features, and the variety of biosynthetic activities of the tumor.

2. Studies have been described which first established the CT-secreting nature of MCT and demonstrated that despite the autonomy of neoplasia, these tumors remain responsive to the stimulus of an increase in serum calcium concentration.

3. The effects on serum CT, calcium, and phosphate of infusions of calcium and glucagon are reported in 23 patients with active MCT studied personally. Calcium infusion, more effectively and consistently than did glucagon, results in a rise in the serum CT level. No convincing hypocalcemic effect of glucagon was observed, although hypophosphatemia was marked. Patients with MCT and increased circulating CT

appeared equally susceptible, as did normal subjects, to hypercalcemia induced by calcium infusion. With one exception, all subjects were normocalcemic basally. (One was hypocalcemic as discussed.) None had gross changes in radiological bone density as seen in standard skeletal radiographs. It is concluded that, if CT plays an important role in calcium and bone metabolism in the human, effects of chronic hypercalcitoninism were not evident in this population. Indeed, it is our belief that the physiological importance of CT in the adult human has yet to be convincingly demonstrated.

In view of our observations of high CT levels in early human fetal thyroid tissue (Tashjian *et al.*, 1972), the possibility remains that CT may play a role in early development, possibly of the skeleton or even of some other tissue.

4. We have shown the value of immunoassay measurements of CT in blood and urine in the early diagnosis of MCT. Twelve previously unsuspected thyroid carcinomas, 11 being clinically occult, were detected in a single large family (J-kindred) as a result of such studies.

5. On the basis of finding an increased concentration of CT in areas of the thyroid gland remote from the principal site of tumor, it was proposed that patients with familial MCT inherit, in addition, a tendency to generalized hyperplasia of C-cells.

6. An unexpectedly high incidence of parathyroid hyperplasia was observed in the J-kindred with MCT. It remains to be determined whether the parathyroid abnormalities found in this syndrome are reactive, or genetic in origin. Our findings favor the latter.

7. Preliminary studies were reported in which patients with MCT failed to show regularly a significant stimulation of CT secretion by the ingestion of a large calcium load, or an important role for the gut-hormone, gastrin, as a secretagogue of calcitonin.

8. Studies in which parallel measurements were made of both basal serum CT and histaminase activity in patients with MCT, showed the former to be elevated more frequently than the latter. In these patients, provocative testing with intravenous calcium regularly produced a rise in serum CT to abnormal levels.

The histaminase inhibitor, aminoguanidine, while producing complete inhibition of serum histaminase activity in patients with MCT, failed to modify serum CT levels either basally or in response to calcium infusion.

9. The incidence of MCT in the kindred studied was consistent with an autosomal Mendelian dominant inheritance. A total of 19 members were found to have MCT, the tumor being definitely bilateral in 18, and probably bilateral in the other. Pheochromocytoma was present in

5 with MCT, being bilateral in 3, and occurred in a sixth patient in whom there is at present no evidence of MCT. Analysis of karyotypes revealed no gross chromosomal abnormality.

10. Evidence was reviewed favoring a unified concept of histogenesis in which the medullary carcinoma syndrome is considered to be a genetically determined dysplasia of neuroectodermal tissue.

Appendix I

Clinical Details of Kindred J

Individuals are identified according to generation, and individual number, allocated in sequence according to their position from left to right in the genealogical chart (Fig. 32).

Initials of those with MCT are italicized, e.g., *EJ*: with pheochromocytoma, underlined e.g., WT; with MCT plus pheochromocytoma, italicized and underlined, e.g., *JS*.

Generation I

1. AG: Female. Died aged 65, of thromboembolism. Autopsy showed no disseminated carcinoma, and thyroid and adrenal glands were grossly normal. Thyroid gland was not, however, sectioned, and it may have contained occult tumor.

Patient had immigrated to the United States in 1901, from Sweden. Many relatives remaining in Sweden were studied by Ljungberg *et al.* (1967), 13 with MCT and/or pheochromocytoma.

Eight children, comprising generation II.

Generation II

1. *EJ:* Male. Died 1969, aged 58, "carcinomatosis,—? bronchial carcinoid." Metastases in spine and ribs. Diagnosis made from mediastinal lymph node biopsy. No autopsy. Sections reviewed by us and found to be typical MCT.

Five children (*LG*, DJ, *CT*, And. J, *RJ*).

2. *JJ:* Male. Living, aged 50. Partial thyroidectomy 1952, aged 32, for "undifferentiated thyroid carcinoma," found to be inoperable. Treated with radiation. Sections reviewed by us, and found to be typical MCT. Now has cervical and hepatic metastases. Normal urine VMA.

Two children (JPJ, AT).

3. *RT:* Female. Living, aged 50. Dizygotic twin of JJ. Thyroid nodule noted age 41. Total thyroidectomy 1968, with tumor in left lobe and cervical nodes. Pathologic diagnosis "follicular adenocarcinoma of the thyroid"; sections reviewed by us, typical MCT. Now has hepatic

metastases evident on liver scan, and by CT assay of hepatic venous effluent. Normal urine VMA.

Five children (WT, *DT*, JS, KT, JT).

4. *TJ:* Male. Living, aged 65. Bilateral MCT diagnosed in the course of this study. Thyroid normal to palpation and scan. Local cervical node metastases. Parathyroid hyperplasia; previous history of renal stones. Normal urine VMA.

Three children (*BDeY*, ER, *DF*).

5. *DH:* Female. Died, 1968, aged 54. Disseminated "Hurtle cell carcinoma of the thyroid." Thyroid nodule noted 15 years previously, and bilateral thyroid carcinoma found during total thyroidectomy 1962. Neck exploration performed primarily for hypercalcemia (serum calcium 13.0 mg/100 ml), found to be due to parathyroid adenoma. No autopsy. Sections of thyroid tumor reviewed by us, typical of MCT.

Five children (RoH, *RiH*, AJ, LR, DW).

6. *WJ:* Male. Died, 1958, aged 50. Autopsy report "bilateral mixed follicular and papillary adenocarcinoma of the thyroid, and bilateral pheochromocytomas." A "malignant lump" had been removed from his neck at age 32. Autopsy sections reviewed by us typical MCT.

Two children (*PLJ*, PWJ).

7. *HJ*, Sr: Male. Living, aged 61. Thyroid normal to palpation and scan, but bilateral MCT found in the course of this study; no metastasis evident. Pheochromocytoma on left side by adrenal angiography, but not removed for lack of symptoms. Urine VMA 11.5 mg/24 hours. Renal stone aged 30 years; parathyroid hyperplasia present at thyroidectomy.

Three children (*JP*, RR, *HJ, Jr*).

8. EM: Female. Living, aged 52. No evidence of MCT or pheochromocytoma.

Six children (SM, JT, GM, LM, GrM, JM).

Generation III

1. *LG:* Female. Secondary propositus. Living, aged 28. Thyroid nodule noted 1969. Bilateral radical neck dissection and total thyroidectomy 1969 (prior to this study) confirmed bilateral MCT, with local node metastases. Urine VMA normal. Sections from thyroid tumors reviewed by us, typical MCT.

Four children ($1\frac{1}{2}$–$6\frac{1}{2}$ years), none with evidence of MCT or pheochromocytoma.

2. DJ: Male. Primary propositus. Living, aged 30. No evidence of MCT or pheochromocytoma.

Two children (4, 6 years), none with evidence of MCT or pheochromocytoma.

3. *CT:* Female. Living, aged 24. Bilateral MCT with local node metastases found in course of this study. Thyroid gland was normal to palpation and scan. Normal urine VMA. Parathyroids showed hyperplasia.

Three children (1–4 years), none with evidence of MCT or pheochromocytoma.

4. And. J: Female. Living, aged 19. No evidence of MCT or pheochromocytoma (serum and urine CT normal repeatedly; urine VMA normal). Thyroid gland diffusely enlarged to twice normal, but regressed completely with suppressive doses of thyroid hormone. Thyroid scan normal.

5. *RJ:* Male. Living, aged 35. Bilateral MCT without node metastases found in the course of this study. Thyroid gland normal to palpation and scan. Previous history of renal stones; parathyroids normal. Urine VMA normal.

Five children (4–$11\frac{1}{2}$ yr) without evidence of MCT or pheochromocytoma.

6. JPJ: Male. Living, aged 25. No evidence of MCT or pheochromocytoma.

No children.

7. AT: Female. Living, aged 22. No evidence of MCT or pheochromocytoma.

One child (2 years) without evidence of MCT.

8. WT: Male. Living, aged 26. Left-side pheochromocytoma diagnosed by high VMA, plasma catecholamines, and adrenal angiography. No evidence of MCT; thyroid gland normal to palpation and scan; normal serum and urine CT.

Two children (2, 3 years) without evidence of MCT or pheochromocytoma.

9. *DT:* Male. Living, aged 25. Bilateral MCT with local node metastases found in the course of this study. Thyroid gland diffusely enlarged with bilateral tumors easily palpable preoperatively. No evidence of pheochromocytoma.

No children.

10. *JS:* Female. Living, aged 27. Bilateral MCT and bilateral pheochromocytomas found in the course of this study. Thyroid tumors not palpable preoperatively, and scan normal. Parathyroid hyperplasia present.

Two children (4, 8 years) without evidence of MCT or pheochromocytoma.

11. KT: Female. Living, aged 11. No evidence of MCT or pheochromocytoma.

12. JT: Female. Living, aged 14. No evidence of MCT, but elevated urine VMA currently being investigated.

13. _BDeY:_ Female. Living, aged 39. Bilateral MCT and bilateral pheochromocytomas found in the course of this study. Cervical node metastases present. Not diabetic, but necrobiosis lipoidea present on both lower legs. Thyroid tumors not palpable preoperatively, and scan normal.

Two children (15, 17 years).

14. ER: Female. Living, aged 36. No evidence of MCT or pheochromocytoma.

Four children (7–15 years) with no evidence of MCT or pheochromocytoma.

15. _DF:_ Female. Living, aged 34. Nodule palpated in right lobe of thyroid 1969. Evident on scan as "cold" area. Right thyroid lobectomy 1969 (prior to this study)—"compact small cell carcinoma of the thyroid with amyloid stroma." No record of lymph node status. Subsequently given cobalt irradiation. No evidence of pheochromocytoma; urine VMA normal; adrenal vein catecholamines normal. Suffers from diarrhea.

Five children (5–15 years) with no evidence of MCT or pheochromocytoma.

16. RoH: Male. Living, aged 28. No evidence of MCT or pheochromocytoma. Idiopathic epilepsy.

Two children (5, 7 years) without evidence of MCT or pheochromocytoma.

17. _RiH:_ Male. Living, aged 29. Bilateral MCT found in course of this study. Tumors not palpable preoperatively. Negative thyroid scan. No metastases evident. Parathyroid hyperplasia present. No evidence of pheochromocytoma.

Three children (4–8 years) without evidence of MCT or pheochromocytoma.

18. AJ: Female. Living, aged 24. No evidence of MCT or pheochromocytoma.

Two children (5, 8 years) without evidence of MCT or pheochromocytoma.

19. LR: Female. Living, aged 21. No evidence of MCT or pheochromocytoma.

One child (9 months) without evidence of MCT.

20. DW: Female. Aged 27 years. No evidence of MCT or pheochromocytoma. No children.

21. _PLJ:_ Male. Living, aged 29. Bilateral MCT found in the course of this study, without evidence of metastases. No evidence of pheochromocytoma. Normal parathyroids.

Three children (2–8 years) without evidence of MCT or pheochromocytoma.

22. PWJ: Male. Living, aged 35. No evidence of MCT or pheochromocytoma.

Two children (6, $7\frac{1}{2}$ years) without evidence of MCT or pheochromocytoma.

23. _JP_: Female. Living, aged 37. Bilateral MCT with local node metastases found in the course of this study. Thyroid gland normal to palpation and scan. Pheochromocytoma on right side. Parathyroid hyperplasia present.

Three children (12–16 years) without evidence of MCT or pheochromocytoma.

24. RR: Female. Living, aged 34. Bilateral MCT with cervical node metastases found in the course of this study. Thyroid gland normal to palpation and scan. No evidence of pheochromocytoma.

Two children (3, 6 years) without evidence of MCT or pheochromocytoma.

25. HJ, Jr: Male. Living, aged 30. Bilateral MCT found in the course of this study. No metastases evident. No evidence of pheochromocytoma. Previous history of renal stones; parathyroid hyperplasia present.

Two children (4, 6 years) without evidence of MCT or pheochromocytoma.

26. SM: Female. Living, aged 27. No evidence of MCT or pheochromocytoma.

One child ($3\frac{1}{2}$ years) normal thyroid and serum CT.

27. JT: Female. Living, aged 25. No evidence of MCT or pheochromocytoma.

Three children (1–4 years), no evidence of MCT or pheochromocytoma.

28. GM: Female. Living, aged 19. No evidence of MCT or pheochromocytoma.

One child ($1\frac{1}{2}$ years), normal thyroid and serum CT.

29. LM: Female. Living, aged 18. No evidence of MCT or pheochromocytoma.

No children.

30. GrM: Male. Living, aged 16. No evidence of MCT or pheochromocytoma.

31. JM: Male. Living, aged 15. No evidence of MCT or pheochromocytoma.

ACKNOWLEDGMENTS

This study was possible only through the joint efforts of a team of co-workers, to each of whom grateful acknowledgment is made: Misses Barbara Howland, Norma Barowsky, Diane Jensen, Yolanda Santo, and Mr. Edward Voelkel;

Mrs. Marilyn Melvin and Miss Zoila Torres; the nursing staff of the Clinical Study Unit; Mr. Andrew Macaulay and his staff of the Core Laboratory; Drs. Lewis Williams, Stephen Martyak, Luis Reyes; Dr. Paul Kahn and staff; and Drs. James Gibson and Hubert Wolfe, Department of Pathology; Dr. Stephen Baylin and colleagues of the National Institutes of Health for serum histaminase measurements; Dr. Lawrence Levine of Brandeis University for the prostaglandin assays; Dr. Robert Krooth, College of Physicians and Surgeons, Columbia University, for expert advice on the genetic aspects of this study. Particular thanks are due Dr. E. B. Astwood whose invaluable help and encouragement imparted great momentum to this investigation. We are also grateful to the Ayerst Company and Dr. Thomas Robitscher for the gift of pentagastrin (Peptavlon®).

This investigation was supported in part by research grants (AM 612, AM 10206) from the National Institute of Arthritis and Metabolic Diseases; and General Clinical Research Centers Branch (FR 0054).

References

Aach, R., and Kissane, J. (1969). *Amer. J. Med.* **46**, 961.

Al-Saadi, A., and Lieberman, L. (1968). *Proc. Amer. Thyroid Ass. Annu. Meet., Abstr.*

Arnaud, C. D., Sizemore, G. W., Oldham, S. B., Fischer, J. A., Tsao, H. S., and Littledike, E. T. (1971). *Amer. J. Med.* **50**, 630.

Bates, R. F., Bruce, J. B., and Care, A. D. (1970). *J. Endocrinol.* **46**, 11.

Baylin, S. B., Beaven, M. A., Engelman, K., and Sjoerdsma, A. (1970). *New Engl. J. Med.* **283**, 1239.

Birge, S. J., and Avioli, L. V. (1969). *J. Clin. Endocrinol. Metab.* **29**, 213.

Birge, S. J., Haddad, J., Teitelbaum, S., and Avioli, L. (1970). *Proc. 52nd Meet. Endocrine Soc., New York* p. 119.

Block, M. A., Horn, R. C., Jr., and Miller, J. M. (1967). *Ann. Surg.* **166**, 403.

Bloom, G. E., Warner, S., Park, S. G., and Diamond, L. K (1966). *New Engl. J. Med.* **274**, 8.

Care, A. D. (1970). *Fed. Proc. Fed. Amer. Soc. Exp. Biol.* **29**, 253.

Care, A. D., and Gitelman, H. J. (1968). *J. Endocrinol.* **41**, XXI.

Care, A. D., Bates, R. F., and Gitelman, H. J. (1969). *J. Endocrinol.* **43**, LV.

Castleman, B. (1952). *In* "Atlas of Tumor Pathology" Sect. IV, Fasc. 15, p. 1. Armed Forces Inst. Pathol., Washington, D.C.

Cooper, C. W., and Deftos, L. J. (1970). *Fed. Proc. Fed. Amer. Soc. Exp. Biol.* **29, 253**.

Cooper, C. W., Hirsh, P. F., Toverud, S. U., and Munson, P. L. (1967). *Endocrinology* **81**, 610.

Cooper, C. W., Schwesinger, W. H., Mahgoub, A. M., and Ontjes, D. A. (1971). *Science* **172**, 1238.

Cope, O. (1966). *New Engl. J. Med.* **294**, 1174.

Crile, G., Jr. (1956). *Ann. Surg.* **145**, 317.

Crout, J. R. (1966). *Pharmacol. Rev.* **18**, 651.

Cunliffe, W. J., Black, M. M., Hall, R., Johnston, I. D. A., Hudgson, P., Shuster, S., Gudmundsson, T. V., Joplin, G. F., Williams, E. D., Woodhouse, N. J. Y., Galante, L., and MacIntyre, I. (1968). *Lancet* **ii**, 63.

Cushman, P., Jr. (1962). *Amer. J. Med.* **32**, 352.

Deftos, L. J., Goodman, A. D., Engelman, K., and Potts, J. T., Jr. (1971). *Metab. Clin. Exp.* **20**, 428.

Deftos, L. J., Murray, T. M., Powell, D., Habener, V. F., Singer, F. R., Meyer, G. P., and Potts, J T., Jr. (1972). *Proc. 4th Parathyroid Conf., Chapel Hill, N.C., Excerpta Med. Found. Int. Congr. Ser.* **243**, in press.

Dent, C. C. (1962). *Brit. Med. J.* **ii**, 1419.

Dube, W. J., Bell, G. O., and Aliapoulios, M. A. (1969). *Arch. Intern. Med.* **123**, 423.

Finegold, M. J., and Haddad, J. R. (1963). *Arch. Pathol.* **76**, 449.

Fletcher, J. R. (1970). *Arch. Surg. (Chicago)* **100**, 257.

Freeman, D., and Lindsay, S. (1965). *Arch. Pathol.* **80**, 575.

Gonzales-Licea, A., Hartmann, W. H., and Yardley, J. H. (1968). *Amer. J. Clin. Pathol.* **49**, 512.

Gorlin, R. J., Sedano, H. O., and Vickers, R. A. (1968). *Cancer* **22**, 293.

Gray, T. K., and Munson, P. L. (1969). *Science* **166**, 512.

Hazard, J. B., Hawk, W. A., and Crile, G., Jr. (1959). *J. Clin. Endocrinol. Metab.* **19**, 152.

Ibanez, M. L., Cole, V. W., Russell, W. O., and Clark, R. L. (1967). *Cancer* **20**, 706.

Johnston, C. I., Martin, T. J., and Riddell, J. (1970). *Australas. Ann. Med.* **19**, 50.

Kaplan, E. L., Arnaud, C. D., Hill, B. J., and Peskin, G. W. (1970). *Surgery* **68**, 146.

Levine, L., and Van Vunakis, H. (1970). *Biochem. Biophys. Res. Commun.* **41**, 1171.

Ljungberg, O. (1970a). *Acta Pathol. Microbiol. Scand.* **78**, 364.

Ljungberg, O. (1970b). *Acta Pathol. Microbiol. Scand.* **78**, 621.

Ljungberg, O., Cederquist, E.. and von Studnitz, W. (1967). *Brit. Med. J.* **1**, 279.

McDermott, F. T., and Hart, J. A. L. (1970). *Brit. J. Surg.* **57**, 657.

MacIntyre, I., Foster, G. V., Woodhouse, N. J. Y., Bordier, P., Gudmundsson, T. V., and Joplin, G. F. (1972). *Proc. 4th Parathyroid Conf., Chapel Hill, N.C., Excerpta Med. Found. Int. Congr. Ser.* **243**, in press.

Mandelstam, P., Rush, B. F., Mabry, C. C., and Bartlett, R. C. (1970). *J. Lab. Clin. Med.* **76**, 867.

Manning, P. C., Jr., Molnar, G. D., Black, B. M., Priestley, J. T., and Woolner, L. B. (1963). *New Engl. J. Med.* **268**, 68.

Melvin, K. E. W., and Tashjian, A. H., Jr. (1968). *Proc. Nat. Acad. Sci. U.S.* **59**, 1216.

Melvin, K. E. W., Voelkel, E. F., and Tashjian, A. H., Jr. (1970a). *In* "Calcitonin 1969 Proceedings" (S. Taylor, ed.), p. 487. Heinemann, London.

Melvin, K. E. W., Tashjian, A. H., Jr., Cassidy, C. E., and Givens, J. R. (1970b). *Metab. Clin. Exp.* **19**, 831.

Melvin, K. E. W., Miller, H. H., and Tashjian, A. H., Jr. (1971). *New Engl. J. Med.* **285**, 1115.

Meyer, J. S., and Abdel-Bari, W. (1968). *New Engl. J. Med.* **278**, 523.

Mielke, J. E., Becker, K. L., and Gross, J. B. (1965). *Gastroenterology* **48**, 379.

Milhaud, G., Tubiana, M., Parmentier, C., and Coutris, G. (1968). *C. R. Acad. Sci., Ser. D* **266**, 608.

Milhaud, G., Calmettes, C., Raymond, J.-P., Bignon, J., and Moukhtar, M. S. (1970). *C. R. Acad. Sci., Ser. D* **270**, 2159.

Moertel, C. G., Beahrs, O. H., Woolner, L. B., and Tyce, G. M. (1965). *New Engl. J. Med.* **273**, 244.

Morain, W. D., and Aliapoulios, M. A. (1968). Presented at 3rd International Congress of Endocrinology, Mexico City.

Murad, F., Chi, Y.-M., Rall, T. W., and Sutherland, E. W. (1962). *J. Biol. Chem.* **237**, 1233.

Neel, J. V., and Schull, W. J. (1954). "Human Heredity," p. 211. Univ. of Chicago Press, Chicago, Illinois.

Nordin, B. E. C., and Faser, R. (1960). *Lancet* **1**, 947.

Nourok, D. S. (1964). *Ann. Intern. Med.* **60**, 1028.

O'Brien, D. (1963). *New Engl. J. Med.* **268**, 1365.

Paloyan, E. (1967). *Surg. Clin. N. Amer.* **47**, 61.

Paloyan, E., Paloyan, D., and Harper, P. V. (1967a). *Surgery* **62**, 167.

Paloyan, E., Paloyan, D., and Harper, P. V. (1967b). *Metab. Clin. Exp.* **16**, 35.

Paloyan, E., Scanu, A., Straus, F., Pickleman, J. R., and Paloyan, D. (1970). *J. Amer. Med. Ass.* **215**, 1443.

Pisano, J. J., Crout, J. R., and Abraham, D. (1962). *Clin. Chim. Acta* **7**, 285.

Potts, J. T., Jr., Murray, T. M., Peacock, M., Niall, H. D., Tregear, G. W., Keutmann, H. T., Powell, D., and Deftos, L. J. (1971). *Amer. J. Med.* **50**, 639.

Robbins, J. (1969). *Med. Ann. D.C.* **38**, 7.

Robert, J., Johnson, K., and Riechmann, G. C. (1968). *Clin. Chem.* **14**, 1218.

Roth, S. I. (1962). *Arch. Pathol.* **73**, 495.

Sarosi, G., and Doe, R. P. (1968). *Ann. Intern. Med.* **68**, 1305.

Schimke, R. N., and Hartmann, W. H. (1965). *Ann. Intern. Med.* **63**, 1027.

Schimke, R. N., Hartmann, W. H., Prout, T. E., and Rimoin, D. L. (1968). *New Engl. J. Med.* **279**, 1.

Shomaoka, K., Ohnuma, M. D., and Bakshi, S. (1971). Presented at American Thyroid Association Meeting, Birmingham, Alabama. *Abstr.*

Sipple, J. H. (1961). *Amer. J. Med.* **31**, 163.

Steiner, A. L., Goodman, A. D., and Powers, S. R. (1968). *Medicine (Baltimore)* **47**, 371.

Sutherland, E. W., and Robison, G. A. (1966). *Pharmacol. Rev.* **18**, 145.

Tanzer, F. S., Kennedy, J. W., 3rd, and Talmage, R. V. (1970). *Proc. Soc. Exp. Biol. Med.* **133**, 500.

Tashjian, A. H., Jr., and Melvin, K. E. W. (1968). *New Engl. J. Med.* **279**, 279.

Tashjian, A. H., Jr., Frantz, A. G., and Lee, J. B. (1966). *Proc. Nat. Acad. Sci. U.S.* **56**, 1138.

Tashjian, A. H., Jr., Howland, B. G., Melvin, K. E. W., and Hill, C. S., Jr. (1970). *New Engl. J. Med.* **283**, 890.

Tashjian, A. H., Jr., Melvin, K. E. W., Voelkel, E. F., Howland, B. G., Zuckerman, J. E., and Minkin, C. (1972). *Proc. 4th Parathyroid Conf., Chapel Hill, N.C., Excerpta Med. Found. Int. Congr. Ser.* **243**, in press.

Turtle, J. R., and Kipnis, D. M. (1967). *Biochem. Biophys. Res. Commun.* **28**, 797.

Urbanski, F. X. (1967). *J. Chronic Dis.* **20**, 627.

Voelkel, E. F., and Tashjian, A. H. Jr. (1971). *J. Clin. Endocrinol. Metab.* **32**, 102.

Wahner, H. W., Cuello, C., and Aljure, F. (1968). *Amer. J. Med.* **45**, 789.
Weichert, R. F. (1970). *Amer. J. Med.* **49**, 232.
Williams, C. R., and Brewer, D. B. (1969). *Brit. J. Surg.* **56**, 437.
Williams, E. D. (1965). *J. Clin. Pathol.* **18**, 288.
Williams, E. D. (1966). *J. Clin. Pathol.* **19**, 114.
Williams, E. D., and Pollock, D. H. (1966). *J. Pathol. Bacteriol.* **91**, 71.
Williams, E. D., Brown, C. L., and Doniach, I. (1966). *J. Clin. Pathol.* **19**, 103.
Williams, E. D., Karim, S. M. M., and Sandler, M. (1968). *Lancet* **1**, 22.
Williams, G. A., Bowser, E. N., and Henderson, W. J. (1969). *Endocrinology*
      85, 537.
Woodhouse, N. J. Y., Gudmundsson, T. V., and Galante, L. (1969). *J. Endocrinol.*
      45, Suppl. xvi.
Woolner, L. B., Beahrs, O. H., McConahey, W. M., and Keating, F. R., Jr. (1961).
      Amer. J. Surg. **162**, 354.

DISCUSSION

G. A. Bray: The anatomical localization of medullary carcinoma in the center of the thyroid leads me to ask whether you had done just a simple removal of this tumor or whether you removed the whole thyroid. What approach would you recommend for those of us who may see such a patient in the future? Did any of your patients experience postoperative hypoparathyroidism or injury of the recurrent laryngeal nerve?

K. E. W. Melvin: We believe that because of the diffuse nature of the distribution of the C-cells, and on the assumption that this is in fact a tumor of C-cells, the present approach to the surgical treatment of this tumor should include total thyroidectomy. In addition, one would aim at a clearance of the lymph nodes of the central zone of the neck. We agree with the argument developed by Crile that it is the central zone which should be cleared and that more extensive and radical procedures are unlikely to offer better chances of surgical cure. If frozen sections confirm cervical node involvement, a sternal splitting procedure and removal of mediastinal nodes might be included in such a central zone clearance. None of the patients suffered recurrent laryngeal nerve damage, and only one was left with hypoparathyroidism beyond the immediate postoperative period.

J. R. Gill, Jr.:   What was the occurrence of hypophosphatemia or elevated alkaline phosphatase in your patients with medullary carcinoma who had parathyroid hyperplasia or an adenoma. Did any of them show evidence of bone disease by X-ray?

K. E. W. Melvin: There was no evidence of bone resorption as sought in gross radiological procedures, and in particular there was no evidence of subperiosteal erosion. In the entire series, we were unable to detect by gross methods any alteration in overall bone mineralization. The mean serum phosphate value for the group of patients with medullary carcinoma and high serum calcitonin values was 0.5 mg/100 ml less than in the corresponding group of 55 normal relatives. But once again we must remember that the normal relatives were much younger; in fact half of that group were children, and it is well recognized that their serum phosphate levels tend to run higher than in the adult. Therefore, we cannot interpret the lower serum phosphate levels in the MCT group as necessarily reflecting the effects of either parathyroid hormone or calcitonin.

F. C. Bartter: Dr. Gill is thinking of a patient we studied some time ago who had this mysterious combination of von Recklinghausen's neurofibromatosis on the one hand, and severe osteomalacia on the other. This patient had very marked hypophosphatemia, her plasma calciums were in general normal, and her alkaline phosphatase was very high. Her urinary calciums were normal, but I hasten to add that they were thus very high for any form of osteomalacia that is nonrenal in origin. Because of evidence for autonomous parathyroid overactivity, she was subjected to exploration, and had massive hyperplasia of all parathyroid tissue. Unfortunately, in the postoperative period she suffered a cardiac arrest. On autopsy she was found to have a medullary carcinoma of the thyroid, and a pheochromocytoma. Thus she appears to represent a syndrome quite similar to that of the families that Dr. Melvin has just discussed. We believe the medullary carcinoma was "primary," the parathyroid hyperplasia secondary. Have you seen any evidence of osteomalacia in your patients, Dr. Melvin, and have you seen any association of classical neurofibromatosis with the syndrome?

K. E. W. Melvin: We have not seen evidence of osteomalacia in these patients. Serum alkaline phosphatase has been normal in all instances, and in the children appropriate to their ages. Urine calcium values have been normal, and the patients have been uniformly without any symptoms of bone disease. None of our patients has shown café-au-lait patches or neurofibromas. In his review of this subject, E. D. Williams [*J. Clin. Pathol.* **19**, 114 (1966)] refers to one patient with MCT who appeared to have, in addition, typical von Recklinghausen's neurofibromatosis.

J. T. Potts, Jr.: As you mentioned, there is uncertainty about the actual concentration of calcitonin circulating in normal individuals. Our view is that in most normal people one cannot clearly measure the hormone in the circulation. Your view apparently is that it is very low and near the limits of detection. In either case, this would mean that, when man is compared with other animal species that Dr. Deftos and others in our group and workers elsewhere have examined, the concentration of calcitonin in humans is much lower than in other species, such as cows, pigs, rabbits, sheep, and even salmon. This is unusual in itself; furthermore, the situation complicates our attempts to study abnormal calcitonin production in man. What is the defect in the patients with medullary carcinoma? Since we cannot look at normals very easily, such studies as a definition of the threshold of serum calcium at which medullary carcinoma turns on and turns off cannot be interpreted because we have so much difficulty in looking at control of secretion in normals. One wonders whether it is just the mass of tissue in medullary carcinoma that is responsible for the abnormality or whether there is a set point error.

I would like to comment about the phenomenon of the rather paradoxical parathyroid response. You mentioned that Deftos has reported seeing a paradoxical increase in parathyroid hormone after calcium infusion. My initial reaction to this was that it was such an apparently crazy result that I did not like it; it disturbed my previous concept. However, Deftos has definitely seen this. I believe, however, that the observation was made only in several patients in whom the concentration of calcitonin during the calcium infusion was extremely high, in the range of 3000–5000 pg/ml. We have just reported [R. E. Reitz, G. P. Mayer, L. J. Deftos, and J. T. Potts, Jr., *Endocrinology* **89**, 932 (1971)] an animal study in which infusions of calcitonin in moderate doses clearly had no direct effect in parathyroid hormone production. So if there is an effect of calcitonin, direct or indirect, it must occur only at very high concentrations. Nevertheless, the story is unusual and interesting.

K. E. W. Melvin: With reference to your comment about the possible relevance of the concentration of calcitonin to any effect, direct or indirect, on the secretion of parathyroid hormone, the patients we studied, and in whom no paradoxical response to calcium infusion was evident, had relatively small increases in serum CT concentration.

I wonder whether the tumors in these patients represent neoplastic change in what may be a persistence of fetal tissue into adult life. We have been impressed by the absence of any convincing evidence that calcitonin is metabolically active in these patients, and this raises the question whether calcitonin is important physiologically in the adult human. One might have expected that patients with MCT, subjected to chronic calcitonin excess often for many years, might show evidence of increased bone density, and that hypocalcemia would be observed more frequently. Further, our preliminary studies have failed to demonstrate an important role for the calcitonin mechanism in controlling any postprandial rise in serum calcium concentration in patients with MCT. The response of one of our patients to infused pentagastrin, although of a much lower order than that observed by Cooper *et al.* in the pig, was consistent with the view that calcitonin secretion following a calcium meal may be mediated by a gut hormone [C. W. Cooper, W. H. Schwesinger, A. M. Mahgoub, and D. A. Ontjes, *Science* **172,** 1238 (1971)]. In that patient, however, the acute rise in serum calcitonin did not prevent a rise in serum calcium following ingestion of the calcium load.

A. White: You referred to metastases of the lymphoid structures in some of your patients and you also listed several endocrine glands, namely the thyroid, the parathyroids, the ultimobranchial body, and, in the pig, the adrenal medulla, which might produce calcitonin. I would like to add to the list another endocrine gland, namely the thymus. Galante and his colleagues [L. Galante, T. V. Gudmundsson, E. W. Mathews, A. Tse, E. D. Williams, N. J. Y. Goodhouse, and I. MacIntyre, *Lancet* **ii,** 537 (1968)] reported the presence in human thymomas of significant concentrations of calcitonin, and at the International Congress of Immunology last August in Washington, A. Mizutani of Japan reported the extraction and purification of calcitonin activity from calf thymus. Dr. T. F. Dougherty, in a personal communication, has informed me that a preparation from calf thymosin, which we provided him and which we have termed thymosin, had calcitonin activity when tested in normal mice. However, this was a relatively crude preparation. Subsequently, Dr. A. L. Goldstein in our laboratory examined pure calcitonin, kindly provided by Dr. O. K. Behrens of the Eli Lilly Co., for thymosinlike activity in two systems which assess the ability of thymosin to augment cell-mediated immunological competence. Calcitonin was inactive in these two assay systems at the doses tested.

I. MacIntyre: Pearse in London applied his immunofluorescent technique with human calcitonin antiserum to a child's thymus. The C-cells do exist sometimes in "normal" human thymus. Of course, in this instance the thymus was not removed in the belief that it was normal. One question arises from this. Should one think of thymectomy when one is operating on patients with medullary carcinoma?

There is now good experimental evidence to explain the coexistence of pheochromocytoma with medullary carcinoma. The evidence is that the calcitonin is not an ultimobranchial hormone after all [A. G. E. Pearse and J. M. Polak, *Histochemie* **27,** 96 (1971); N. LeDouarin and C. LeLièvre, *C. R. Acad. Sci. Ser. D.* **270,** 2857 (1970)]. The most direct evidence was supplied by Professor LeDouarin. She excised the neural crest from a quail embryo and transplanted this to a chick embryo. The quail neural crest cells are recognizable. She found that the C-cells in the chick ultimobranchial body came from the transplanted neural crest.

Further, the quail embryo from which the neural crest had been excised developed ultimobranchial bodies free of C-cells. We can therefore regard the syndrome of medullary carcinoma with phechromocytoma as a disorder of neuroectodermal tissues.

K. E. W. Melvin: Present knowledge of the embryological development of the C-cells, and their subsequent migration, makes an extrathyroidal site for MCT—such as the thymus—eminently possible. We have made a practice of removing the thymus in our patients with MCT, not because the organ may contain C-cells, primarily, but rather to facilitate complete clearance of lymph nodes. I have no evidence that medullary carcinoma has occurred outside of the thyroid gland, although an example of chemodectoma occurring in association with thyroid carcinoma has been described.

Dr. Astwood has been interested in the possibility that the lateral aberrant thyroid might represent a hang-up in migration of the lateral thyroid, together with its C-cell component. Perhaps examples of this tumor should be reviewed. They may prove to be of the medullary type histologically. I think that the concept of a thymic medullary carcinoma is a theoretical possibility.

H. G. Friesen: Are there enough data available to suggest a prognosis in any given patient? Also, with the histological evidence of amyloid deposition and lymphocytic infiltration, are there any immunological abnormalities documented in these patients?

K. E. W. Melvin: I believe it is not possible to assess the prognosis on the basis of any particular clinical or histological characteristic. Williams has made certain observations about subtleties of histology in assessing the likely prognosis in a given instance, and relevant to this was the amount of amyloid present. I am not sure that this has really withstood the test of time. The problem in assessing prognosis in this condition generally is that there is a wide variation in survival from one patient to another. The tumor has been shown to be compatible with a survival of 30 years following the discovery of metastases. In our experience, we have had two patients die at the age of 21 from a very rapid dissemination of their disease. We have two patients alive 13 years after the demonstration of metastases, and one for 18 years. How one would differentiate in advance or prognosticate from one case to another I think cannot be defined with any clarity.

Concerning possible immunological abnormalities, we have noted a marked lymphocyte infiltration in most of the thyroid glands removed, and a significantly elevated titer of antithyroid antibodies in a few of these patients. More generally, there has been no quantitative abnormality in immunoglobulins, and skin testing against mumps, candida, and tuberculin antigens proved normal in all cases.

J. M. McKenzie: Would you care to say a little more about the expression of peripheral features that were so prominent in the first patient that you showed us. In the family photograph we saw, there was no suggestion of these gross changes, but what about their occurrence in other sporadic examples of the syndrome?

For those of us who cannot readily measure calcitonin, I gathered from what you said that there is not much to be done in routine examining of the thyroid in the hope of finding a medullary carcinoma; is there any question about a selenomethionine scan being useful, however? Presumably the carcinoma tissue is very active in protein synthesis, although, since the lesion tends to be rather small, it might well be beyond the capacity of such a scanning technique.

Finally, in the family tree that you showed, I presume you were not implicating parthenogenesis in the first generation. Why did you offer us only the first mother? Is there no possibility of a contribution to this syndrome from the father's side?

K. E. W. Melvin: I should point out that the family of the progenitor has been traced back by Dr. Ljungberg to 1794, and he has clearly established that there is no consanguinity in the family at least to that time. Many of her relatives

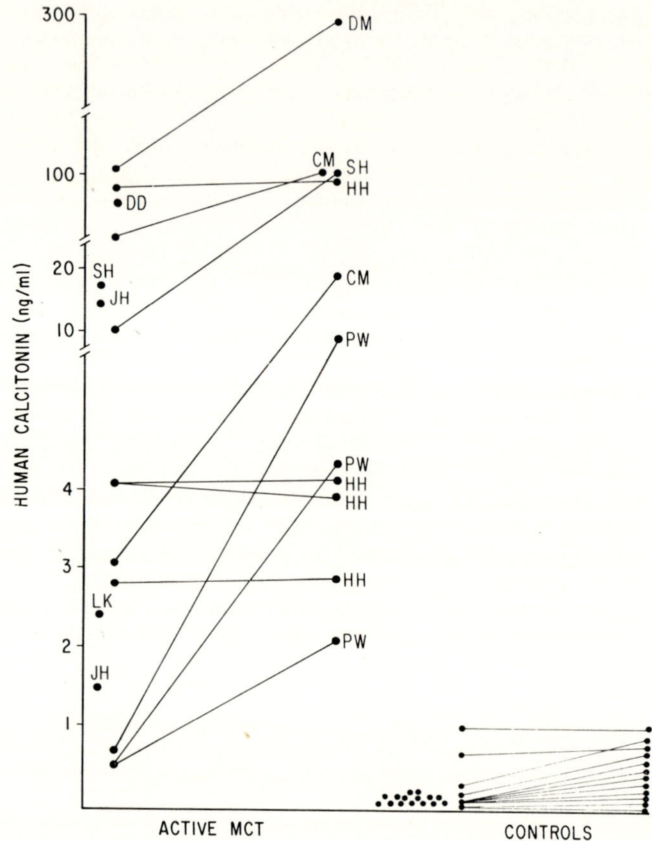

Fig. A. Measurement of basal levels of calcitonin and of levels during calcium infusion in patients with medullary thyroid carcinoma and in controls.

suffer from medullary carcinoma. Concerning her husband, there is no incidence whatsoever of any feature of this syndrome in any of the spouses or their families, and that is why no mention was made of the spouses in the discussion of genealogy.

We have done technitium, selenium, and radioiodine scans in all those with high calcitonin levels and are about to evaluate gallium in future cases as being possibly more tumor-specific. However, in only one patient where tumor was clearly evident upon palpation was there any abnormality evident in the scan—even after suppression of the thyroid with triiodothyronine for a week before scanning with selenium.

N. A. Samaan: We have 34 patients living with clinically active medullary carcinoma of the thyroid, and all these patients except one have high basal levels of calcitonin. The patient who had a normal basal level of calcitonin in spite of extensive metastasis in bone showed high levels of calcitonin during calcium infusion. If medullary carcinoma is suspected, one should measure basal levels of calcitonin as well as during calcium infusion. With progress of the disease and deterioration of a patient's condition, the calcitonin level rises either at the basal level or during calcium infusion (see Fig. A).

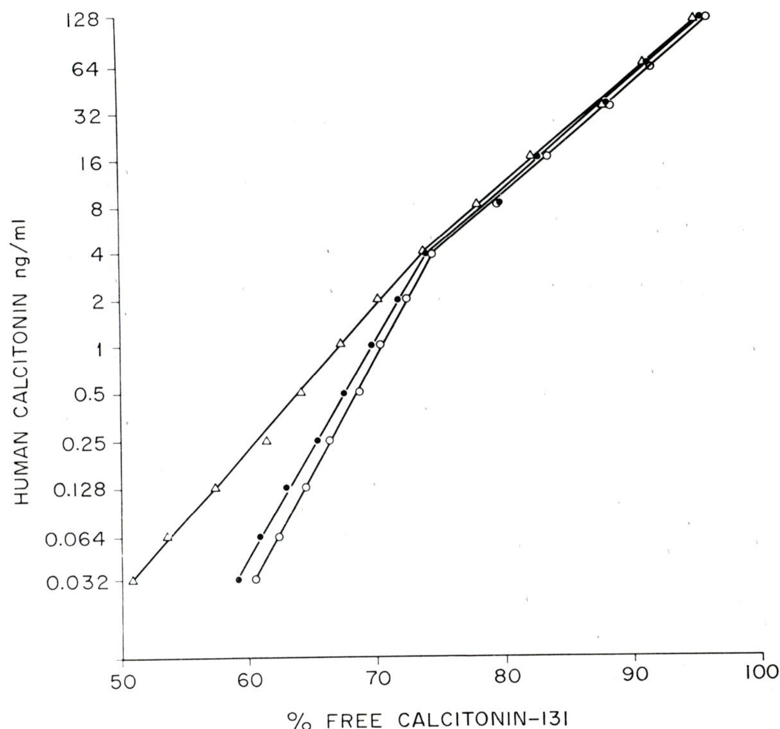

Fig. B. Standard dose response curves of calcitonin for pooled human serum (O——O), serum from an athyroid patient (●——●), and normal guinea pig serum (△——△).

What is the normal basal level in man? It depends on the assay you are running. One of the factors to be considered in the radioimmunoassay is compensation for the protein fraction in the unknown serum. To demonstrate this point, there are shown in Fig. B standard dose response curves of calcitonin using pooled human serum, serum from an athyroid patient, and a third curve using normal guinea pig serum. There is no significant difference in the two curves using pooled human serum or serum from an athyroid patient, but there is a difference in the curve when guinea pig serum was used. For example, a value of 0.25 ng/ml read on a standard curve using serum from an athyroid patient will read 1 ng/ml on the standard curve when guinea pig serum is used. In the radioimmunoassay, we have to compensate for the protein in the unknown serum and we have to standardize our assays. If we use serum from an athyroid patient, in a standard dose response curve we get similar results to those Dr. Melvin showed us. I would expect variation in the normal basal level according to the method used.

The key to one family we studied was a patient who died during surgery for gastric ulcer. At postmortem both medullary carcinoma of the thyroid and pheochromocytoma were found. In the second generation of this family (see Fig. C), we found patients with medullary carcinoma or pheochromocytoma, or both. One of these

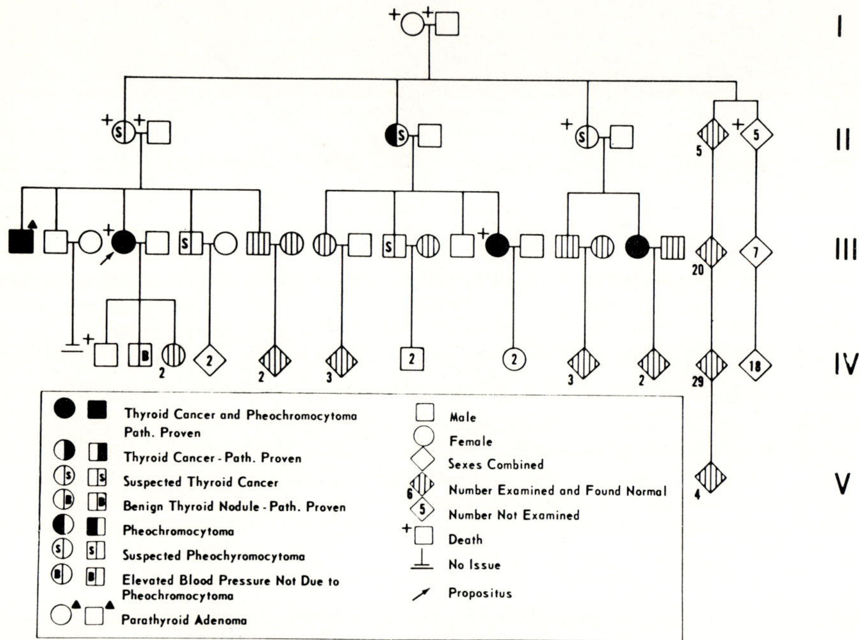

FIG. C. Genealogical study of family with high incidence of medullary thyroid carcinoma and pheochromocytoma.

patients also had an adenoma of the parathyroid. We found that some of these patients with medullary carcinoma had normal basal calcitonin levels, but abnormally high levels during calcium infusion. In one of the patients who belonged to the families with medullary thyroid carcinoma, we found a tumor in the thyroid gland but her calcitonin level was normal at the basal level and during calcium infusion. At surgery, the tumor was found to be a papillary follicular carcinoma.

J. H. Dussault: We had the opportunity to study a young female patient who had a pheochromocytoma, two parathyroid adenomas, and medullary carcinoma situated exactly in the same position as in Dr. Melvin's patients. The diagnosis was suspected on the basis of neonatal hypocalcemia in her last two infants. Her family history revealed that her mother had a pheochromocytoma and a medullary carcinoma. The brother of her mother also had both diseases, and his son had surgery for pheochromocytoma when he was 14 years old. We are now investigating the other members of this family.

What was the youngest patient in whom you could detect elevated calcitonin levels, and at what age should investigation of members of such a family be started?

K. E. W. Melvin: The youngest member of this kindred in whom we diagnosed MCT was 24 years of age. There is no way, however, of determining how long the tumors had been present prior to our investigation. We have commenced investigation of the infants in this kindred from the age of one year, and by the end of 20 years from now we may have the answer to your question. As I mentioned earlier, there has been described a "prophylactic" thyroidectomy in a $4\frac{1}{2}$-year-old child with a strong family history of this syndrome. Even at that age, the child was found to have bilateral MCT with cervical lymph node metastases. It is even possible, I suppose, that some individuals may be born with the tumors. We have to start investigating any future members of this family *in utero,* by means of amniocentesis.

B. M. Dobyns: I am interested in the histology in your cases because pathologists seem to have varied in their interpretations of some of these tumors. There are thyroid neoplasms that are considered by some to be medullary carcinomas but without amyloid production in them. Did you find some of the high calcitonin levels in patients with neoplasms which contained no evidence of amyloid formation? How many cases of medullary carcinomas of the thyroid with high calcitonin levels might we expect to see sporadically, compared to the numbers seen related to a family with the trait?

K. E. W. Melvin: In each affected member of this kindred, the carcinoma occurred at an identical position, toward the upper pole in each lobe of the thyroid. But, in addition, in two cases there were very minute foci of tumor grossly visible throughout the substance of the thyroid gland. While there was gross tumor at each of the normal sites bilaterally, in two cases there was diffuse occurrence of tumor as well. I suspect that what was observed was the multifocal nature of the tumor. Ljungberg, using specific staining techniques, has shown greatly increased numbers of C-cells distributed throughout both lobes of a thyroid gland which contained, in addition, bilateral MCT. All our cases did have amyloid demonstrable in the tumor to a varying degree. In some instances, it may be necessary to use a battery of stains to be quire confident of the presence of amyloid.

I cannot give you a factual answer as to the ratio of sporadic to familial cases. Certainly sporadic cases have been described more frequently than familial, but the facts relating to inheritance are not always known. I believe that the percentage of familial cases will ultimately be shown to exceed 50%.

B. F. Rice: I believe a question has been raised in your presentation because of the high levels of circulating thyrocalcitonin with little or no effect in serum calcium levels in most patients. The answer to the question of a role for the thyrocalcitonin may be answered when you do a total thyroidectomy in a patient with hypercalcemia but leave behind a parathyroid adenoma in the mediastinum. Such an unplanned study as this was performed at New Orleans Charity Hospital in a patient with medullary carcinoma and a mediastinal parathyroid adenoma. Following thyroidectomy the patient's hypercalcemia rose to the lethal range and he died with severe hypercalcemia. We believe that removal of his source of thyrocalcitonin without removal of his parathyroid adenoma was the cause of his severe postoperative hypercalcemia and certainly suggests a role for the thyrocalcitonin in the situation of coexisting parathyroid adenoma.

A. H. Tashjian, Jr.: One comment relating to the location of the major tumors in the thyroid gland and to the possibility of multiple tumor foci as these questions relate to human embryology. If one recalls the studies that were done in the 1930's, the *Contributions to Embryology* of the Carnegie Institution, it becomes evident that in man the fourth and possibly the fifth branchial pouches give rise to the structure which is later known as the "lateral thyroid body." This structure merges with the developing medial thyroid body between the 1.4 and 2.2 crown–rump stage of human embryonic development. The lateral structures persist as somewhat discrete masses; however, until about the 6.5 cm stage of fetal development when they are engulfed in the rapidly growing medial portion of the gland. Based on that sort of embryological knowledge, Dr. James Zuckerman and I have begun to correlate early human fetal thyroid development with calcitonin concentration. It is difficult anatomically to dissect the lateral thyroid from very early human embryos, and this study is still in its infancy. We have begun by examining about 40 embryos, and in three which were less than 6 cm in crown–rump length we have found concentrations of calcitonin in total thyroid tissue about 3 to 15 times greater than that found in late fetal or adult thyroid glands. If one notes where the bilateral medullary carcinomas appear in the adult human thyroid, it is right where the lateral thyroid cell masses were last seen as discrete structures. I believe that may be an embryological explanation for the anatomical localization of the tumors.

P. L. Munson: Would you please discuss this question, which appears to be controversial: Is there really any circulating thyrocalcitonin in normal man?

A. H. Tashjian, Jr.: I do believe that calcitonin circulates in normal man although I would state clearly that it has not yet been proved. Probably the most compelling evidence that we have for such a dogmatic statement is the finding that Edward Voelkel can concentrate material from normal human urine which is hypocalcemic in the bioassay and reacts in the immunoassay in a manner similar to that of calcitonin from normal thyroid glands or medullary carcinoma tissue and from the plasma of patients with medullary carcinoma. It is possible, but unlikely, that this material in urine is of nephrogenous origin and totally unrelated to anything being produced by C-cells in the thymus, parathyroid, or thyroid glands.

I also believe that we do not yet know what the circulating concentration of calcitonin is in normal man. In nearly all of over 150 normal control subjects, the levels were less than 200 pg/ml, and in over 50% the value is below 100 pg/ml, the lower limit of sensitivity of the assay for serum. The problem in quantitating the low levels present lies both in the sensitivity of the assay method and in the knowledge that certain polypeptide hormone assays are more or less sensitive to nonspecific interference of the assay by large amounts of serum, particularly as related to phase separation procedures. With greater experience with the assay, I now think the earlier data presented from our laboratory which led us to conclude that we could measure calcitonin concentrations in most normal serum samples was erroneous. We do, however, occasionally see normal subjects in whom the levels are in the range of 0.20–0.30 ng/ml in whom, using the methods suggested by L. Deftos, this activity can be removed by charcoal adsorption of the plasma. Furthermore, over a very small portion of the standard curve, this immunoreactive material does dilute as if it were calcitonin. In addition, in some individuals it also increases with calcium infusion, as does the material in normal urine. Thus, I believe the problem is one of sensitivity of the assay both in terms of absolute quantities of calcitonin and in sensitivity of the phase separation method to nonspecific interference with large quantities of plasma. But I also believe that once these problems are solved, if they can be, the peptide will be found to circulate. That finding alone, of course, does not say that calcitonin has physiological significance.

I. MacIntyre: I agree that calcitonin almost certainly circulates in normal man, but our immunoassay is insufficiently sensitive for us to have any real confidence in the normal levels apparently present. I believe that the normal level of calcitonin M as determined by immunoassay is below 100 pg/ml, or perhaps even below 50 pg/ml. Our bioassay figures for calcitonin-like activity in normal plasma are much higher [T. G. Gudmundsson, N. J. Y. Woodhouse, L. Galante, E. W. Matthews, T. D. Osafo, A. D. Kenny, R. C. Wiggins, and I. MacIntyre, *Lancet* **i**, 443 (1969)]. This is doubtless due in part to the interference by noncalcitonin substances in our bioassay. But there may be other reasons for the discrepancy, such as the circulation of calcitonin in a form in which it does not react in the immunoassay system. A great deal more work is necessary.

J. M. Hershman: One might expect to find large amounts of urinary cyclic AMP in these subjects. Have you measured urinary cyclic AMP in patients with medullary carcinoma?

A. H. Tashjian, Jr.: I cannot state what the absolute values are yet as we have just set up an assay for cyclic AMP in urine. So far we have examined urine from too few patients with medullary carcinoma and hyperparathyroidism to draw any conclusions. I believe we want also to look at the effect of calcitonin itself on urinary cyclic AMP before one can interpret urinary cyclic AMP levels in the urine of patients with medullary carcinoma.

I should like to add one comment about the urinary calcitonin-like material. This urinary peptide which reacts exactly like the serum peptide or gland peptide in the immunoassay exhibits one important additional property. It has a ratio of biological to immunological activity which is unlike that of serum or gland calcitonin. It is about $\frac{1}{10}$ to $\frac{1}{100}$ as active biologically as the serum or gland form of the hormone, and yet it appears to be unchanged immunologically or in size as measured by gel filtration. What kind of change in the molecule might give rise to this discrepancy? Dr. Keutmann's suggestion was, I believe, the loss of the amide group on the C-terminal proline residue.

K. Sterling: May I mention some rather early antecedent information regarding the existence of medullary carcinoma of the thyroid. When I was a fourth-year medical student at Johns Hopkins I was shown slides of medullary carcinoma in the surgical pathology course of the late Dr. Sam S. Blackman. Histologically the sections looked like those you have shown, and were so designated, although without any clinical or genetic information concerning other associated disorders, but only the thyroid histology. May I ask whether in comparative vertebrate studies there is any information that calcitonin might be secreted in greater quantities by carnivores that may chew bones.

K. E. W. Melvin: There have been sporadic reports of this thyroid cancer dating back many years. In 1951 it was first pointed out that there was a group of tumors rather different from the more usual epithelial carcinomas of the thyroid. More specifically, Hazard, Hawk, and Crile collected 23 cases and in 1959 described the clinical as well as the characteristic features of this carcinoma, proposing the term "medullary." I am not aware of comparative studies of calcitonin concentrations in carnivores.

A. H. Tashjian, Jr.: There is one clear example of the fact that medullary carcinoma develops in an animal that does not usually eat meat. In a herd of bulls which are the major sperm donors for artificial insemination in the northeastern United States there occurs a disease which looks similar to medullary carcinoma of the thyroid in man and the tumors contain high concentrations of calcitonin. This disorder may possibly be related to the unusually high calcium intake which these bulls have over many years rather than to bone or meat consumption.

C. Y. C. Pak: Calcitonin does not appear to have a direct effect on the urinary excretion of cyclic AMP in man. We administered both porcine and salmon calcitonin to patients with hypoparathyroidism and found no change in the urinary excretion of cyclic AMP.

Dr. Melvin, I noted with interest the high incidence of renal stones in your patients. Would you comment on the pathogenesis for stone formation in them; in particular, was urinary calcium elevated in any of them? Certainly, hypercalcemia is one of the major determinants for the state of supersaturation of urine with respect to brushite (nidus for calcium-containing stones) [C. Y. C. Pak, *J. Clin. Invest.* **48**, 1914 (1969); C. Y. C. Pak, J. W. Cox, J. W. Powell, and F. C. Bartter, *Amer. J. Med.* **50**, 67 (1971)].

K. E. W. Melvin: As I recall the results of the urinary calcium measurements, none was hypercalcuric. I do not know the mechanism whereby patients with hyperparathyroidism may develop stones. It is not necessarily related to the concentration of calcium in urine.

F. C. Bartter: If the concentration of calcitonin in the urine were in fact 1 ng/ml, and if none of the filtered calcitonin were reabsorbed, you would need a plasma value of 0.01 ng/ml, which is well below anything that has been measured.

Recent Studies on Functions and Control of Prolactin Secretion in Rats

JOSEPH MEITES,[1] K. H. LU, W. WUTTKE,[2] C. W. WELSCH,[3]
H. NAGASAWA,[4] AND S. K. QUADRI

*Department of Physiology, Michigan State University,
East Lansing, Michigan*

I. Introduction

Several reviews on the functions and regulation of prolactin secretion have appeared in recent years (Meites, 1966; Meites and Nicoll, 1966; Baldwin, 1969; Nicoll, 1971). No attempt will be made here to deal with all aspects of prolactin physiology. The functions and control of prolactin secretion in nonmammalian species were thoroughly analyzed by Bern and Nicoll (1968), and information on the biochemical effects of prolactin were reported by Baldwin (1969). These will not be reviewed here. Recent studies on the chemistry and physiology of primate prolactins were reported in May 1971 at a Ciba Foundation (London) symposium, "Lactogenic Hormones," (G. E. W. Wolstenholme and J. Knight, eds., Churchill Livingstone, Edinburgh and London, 1972).

The development of radioimmunoassays for prolactin, as for other hormones, has revolutionized the scope and depth of investigations on prolactin, and has resulted in significant new information. However, prolactin radioimmunoassays have usually confirmed and extended many earlier studies based on bioassays. Thus investigations and conclusions drawn on the basis of pigeon crop sac assays of pituitary prolactin levels during the estrous cycle, pregnancy, lactation, and as influenced by estrogens, other hormones and several drugs, appear to be essentially confirmed by radioimmunoassays. Radioimmunoassays have made it possible to measure prolactin in the blood for the first time, and most of the work reported here will emphasize investigations in which radioimmunoassays were used.

Prolactin is best known for its effects on mammary growth and lactation. Since these actions of prolactin have been reviewed extensively

---

[1] Aided in part by NIH research grants AM 04784 and CA 10771.

[2] Postdoctoral Research Fellow of the Max Kade Foundation, New York, New York. Present address: Max Planck Institute for Biophysical Chemistry, Department of Neurobiology, Göttingen-Nikolausberg, German Federal Republic.

[3] Department of Anatomy, Michigan State University, East Lansing, Michigan.

[4] Postdoctoral Research Fellow of the International Agency for Research on Cancer, World Health Organization. Present address: Pharmacology Division, National Cancer Center Research Institute, Tokyo, Japan.

in the past, they will be considered only briefly here and major emphasis here will be placed on the secretion of prolactin during the life-span of the rat, on its luteolytic and luteotropic actions, and on its relation to development and growth of mammary tumors. The discussion of the regulation of prolactin secretion will emphasize the inhibitory influence of the hypothalamus, the role of biogenic amines, the inhibitory action of prolactin on its own secretion, and the direct effects of some hormones and drugs on the pituitary.

II. Secretion of Prolactin during Life-Span of Rats

A. Prolactin Secretion prior to and at the Onset of Puberty

Early work by Meites and Turner (1948a) showed that pituitary prolactin content in sexually immature female rats, guinea pigs, and rabbits was significantly lower than in mature females of the same species. More recent studies in our laboratory by Minaguchi *et al.* (1968) using the pigeon crop assay, and by Voogt *et al.* (1970) using a prolactin radioimmunoassay, demonstrated that pituitary prolactin content and concentration were uniformly low from 21 to about 37 days of age, and then rose dramatically with the onset of puberty on about day 38 of age and during estrus. Serum prolactin was also low from 21 to 36 days of age, and then showed a sharp rise on the day of vaginal opening, day 37. Significant increases in serum prolactin were also observed after cycling began on the day of estrus. Daily injections of 0.1, 0.3, or 0.5 μg of estradiol benzoate from 26 to 30 days of age significantly increased pituitary and serum prolactin concentrations, suggesting that estrogen is responsible for the increase in prolactin secretion at the onset of puberty.

B. Prolactin Secretion during Recurrent Estrous Cycles

Early studies of the estrous cycles of rats and guinea pigs indicated that pituitary prolactin content was higher during proestrus and estrus than during diestrus (Reece and Turner, 1937; Reece, 1939). This was confirmed later by Sar and Meites (1967), who also observed that pituitaries removed from proestrous and estrous rats released more prolactin upon incubation than pituitaries from diestrous rats. Hypothalamic PIF activity was lower in proestrous and estrous than in diestrous rats.

More recent work on serum prolactin concentrations during the estrous cycle of the rat, as measured by radioimmunoassay, has confirmed the earlier pituitary studies, but in addition has shown that peak levels of prolactin are reached on the afternoon of proestrus (Fig. 1) (Kwa and Verhofstad, 1967; Niswender *et al.*, 1969; Amenomori *et al.*, 1970; Wuttke and Meites, 1970; Gay *et al.*, 1970). Peak levels in serum LH and FSH also are seen on the afternoon of proestrus in the rat (Wuttke and Meites, 1970; Gay *et al.*, 1970), and plasma GH levels are elevated on the day of estrus (Fig. 2) (Dickerman and Meites, 1971). Serum prolactin values are elevated in the goat prior to ovulation (Bryant and Greenwood, 1968). Raud *et al.* (1971), but not Karg and Schams (1970),

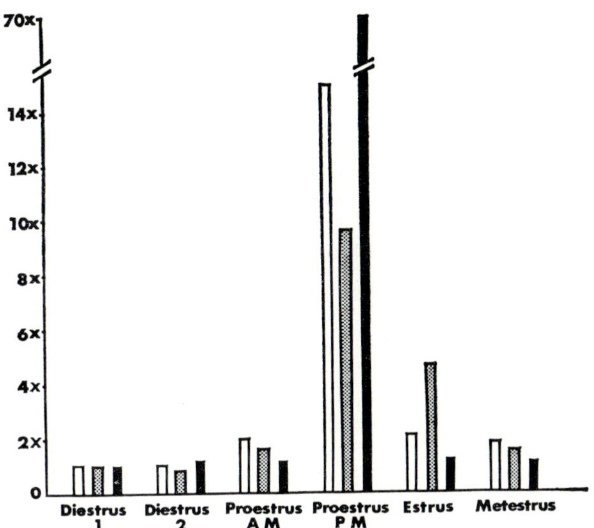

FIG. 1. Changes in serum levels of prolactin (□), follicle-stimulating hormone (FSH) (▦), and luteinizing hormone (LH) (■) during the estrous cycle in rats with 4-day cycles. Note peak rise in all 3 hormones on afternoon of proestrus, with lowest serum prolactin on days 1 and 2 of diestrus. LH shows a rise only on the afternoon of proestrus and is many fold greater than the increases in prolactin and FSH. From Wuttke, Lu, and Meites (unpublished).

noted a rise in serum prolactin in cows during proestrus or estrus. No changes in blood prolactin were seen in women during the menstrual cycle (Jaffe and Midgley, 1971). The rise in serum prolactin on the days of proestrus and estrus in the rat is believed to be due to estrogen, since ovariectomy on the day before proestrus prevented any rise in serum prolactin (Clark and Meites, unpublished). Neill *et al.* (1971) also observed that administration of an antiestrogen on the day before proestrus prevented any rise in serum prolactin on the following day.

Estrogen is believed to be the most important factor regulating prolactin secretion in the female rat, and many early studies showed that it increased pituitary prolactin levels and initiated lactation in a variety of species (Meites, 1966). Serum prolactin concentration is also markedly increased by estrogen administration, and even high doses of estrogen do not inhibit prolactin release (Fig. 3) (Chen and Meites, 1970). Progesterone has relatively little ability to increase serum prolactin values

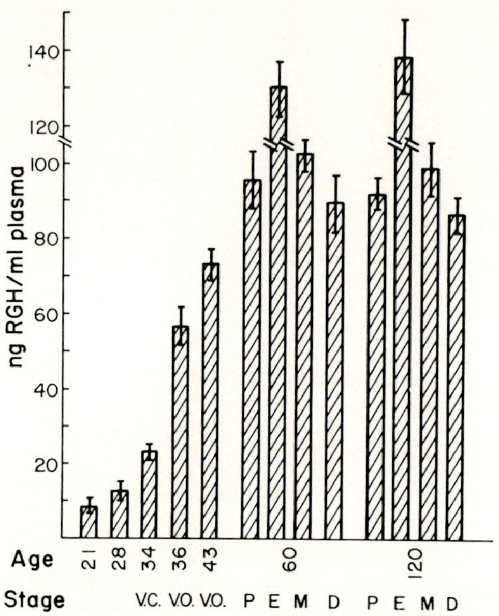

Fig. 2. Changes in plasma growth hormone (GH) during pre- and postpubertal stages and during the estrous cycle in the rat. V.C. = vagina closed; V.O. = vagina open; P = proestrus; E = estrus; M = metestrus; D = diestrus. Note significant increase in plasma GH during the cycle only on the day of estrus. Since plasma samples were collected only once each morning, a possible rise in plasma GH on the afternoon of proestrus cannot be excluded. After Dickerman and Meites (1971).

but can partially counteract the stimulatory effects of small doses of estrogen.

C. PROLACTIN SECRETION DURING PSEUDOPREGNANCY AND PREGNANCY

It is generally agreed that prolactin is essential for the maintenance of pseudopregnancy in the rat, although LH also is believed to have an important role (Greenwald and Rothchild, 1968). However, serum prolactin values have been found to be elevated for only the first 2 to 3 days after induction of pseudopregnancy by cervical stimulation and then decline to very low levels (Fig. 4) (Kwa and Verhofstad, 1967; Amenomori, Dickerman and Meites, unpublished). Spies and Niswender (1971) observed a peak rise in serum prolactin by 8 hours after mating, which is in agreement with our observation of a maximum rise in serum prolactin between 5.5 and 7.5 hours after cervical stimulation by glass rod (Wuttke and Meites, 1972). We also observed that

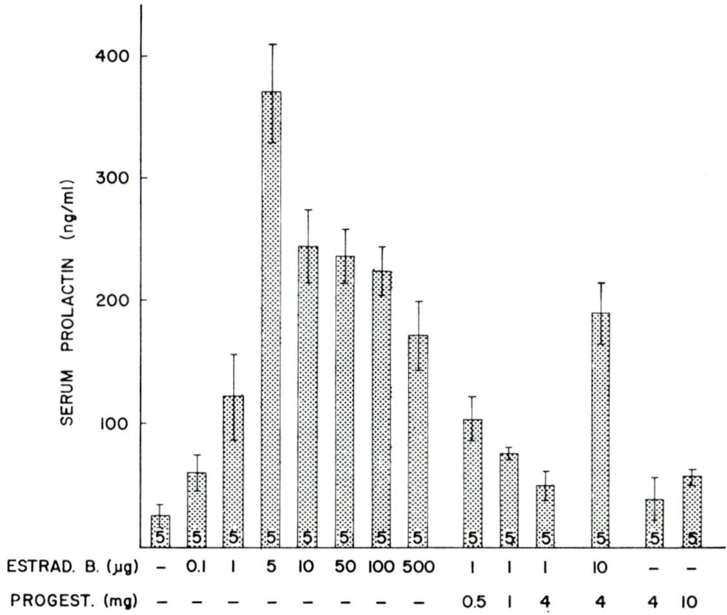

FIG. 3. Effects of graded doses of estradiol benzoate injected for 6 days into mature ovariectomized rats on serum prolactin concentrations. Note that even the highest dose of estrogen (500 μg) did not depress prolactin release. Only the 10-mg dose of progesterone produced a small rise in serum prolactin, but 4 mg of progesterone was able to partially prevent 1 μg of estradiol from elevating serum prolactin. From Chen and Meites (1970).

pseudopregnancy can be induced in rats even if the initial rise in serum prolactin is completely prevented by injecting the drug ergocornine.

During pregnancy in the rat, serum prolactin was elevated for about the first 3 days and then declined to very low levels, until the last day of gestation (Fig. 5) (Amenomori et al., 1970). Pituitary ACTH and blood corticosterone levels also are low during gestation in the rat (Voogt et al., 1969b), and an insufficiency of adrenal cortical hormones or prolactin or both are believed to account for the absence of lactation during pregnancy (Meites, 1966). At the time of parturition there is

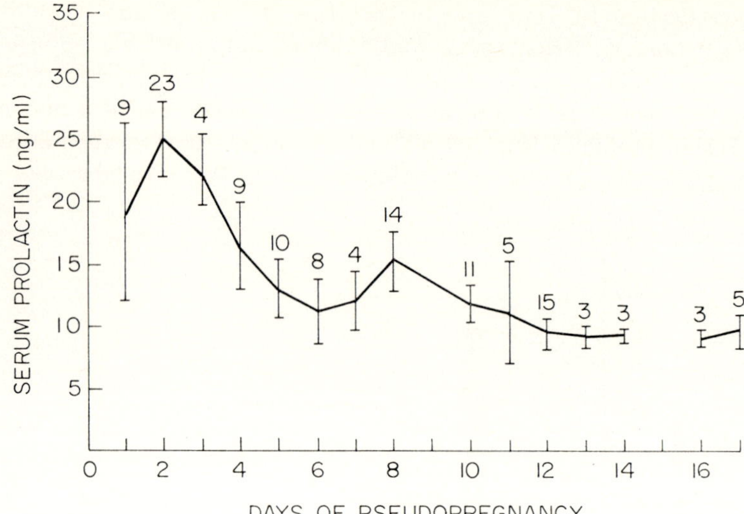

FIG. 4. Serum prolactin levels during pseudopregnancy in rats. Blood samples were collected beginning 24 hours after cervical stimulation of the uterine cervix with a glass rod on the morning of estrus. The reference standard used was HIV-8-C (supplied by Dr. S. Ellis) expressed in terms of NIH-P-Bl. Note that serum prolactin is elevated only for the first 3 days of pseudopregnancy. The two values on days 16 and 17 are from rats that have resumed cycling and are in the diestrous phase. Bars indicate standard errors of mean, and numbers on top indicate number of rats from which blood samples were collected. From Amenomori, Dickerman, and Meites (unpublished).

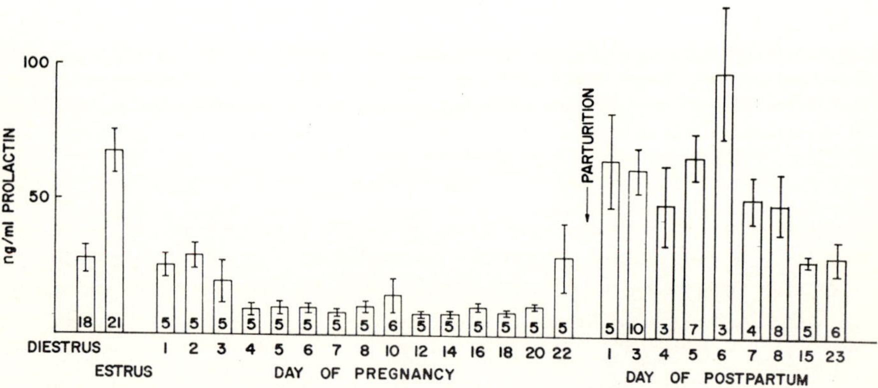

FIG. 5. Serum prolactin values during pregnancy in the rat. Serum prolactin declines after first 3 days and rises on last day of pregnancy. High serum prolactin levels are maintained postpartum by the frequent suckling stimulus. From Amenomori et al. (1970).

a pronounced rise in blood prolactin, ACTH, and corticosterone in the rat, and these are believed to initiate lactation.

The low serum prolactin values observed during most of pregnancy probably are associated with the low estrogen secretion reported during gestation in the rat (Yoshinaga *et al.*, 1969), and perhaps also with the ability of appropriate levels of progesterone to override the stimulating action of estrogen on prolactin secretion (Meites and Turner, 1948a; Meites, 1966). The rise in prolactin at about the time of parturition is probably associated with the increase in estrogen secretion, and perhaps also with the stress of parturition. The latter also may account for the increase in ACTH and corticosterone secretion at this time. Increases in pituitary prolactin content at the time of parturition were observed in the mouse, guinea pig, rabbit, goat, and cow (Meites and Turner, 1948a; Meites, 1966), and in the serum of the goat (Bryant and Greenwood, 1968), cow (Karg and Schams, 1970) and human (Frantz and Kleinberg, 1970).

D. PROLACTIN SECRETION DURING POSTPARTUM LACTATION

The principal stimulus for high prolactin secretion during postpartum lactation is suckling or milking. Stimulation of the numerous sensory nerves in the nipples and surrounding skin evokes a reflex release of prolactin from the pituitary via a hypothalamic mechanism, by depressing PIF activity (Ratner and Meites, 1964). Suckling elicits a decrease in pituitary prolactin content (Reece and Turner, 1937; Grosvenor and Turner, 1958; Sar and Meites, 1969) and an increase in serum prolactin concentration (Fig. 6) (Amenomori *et al.*, 1970). The milking stimulus also has been observed to raise blood prolactin values in the goat (Bryant and Greenwood, 1968), cow (Karg and Schams, 1970), and human (Frantz and Kleinberg, 1970). Milking also evokes release of ACTH and corticosterone in the rat (Voogt *et al.*, 1969b); of oxytocin, necessary for milk ejection, and of GH (Grosvenor *et al.*, 1968; Sar and Meites, 1969), but inhibits release of LH and FSH (Meites, 1966; Minaguchi and Meites, 1967a,b). There is some evidence that exteroceptive stimuli can induce release of prolactin from the pituitary of rats (Grosvenor, 1965) and earlier work from our laboratory (Nicoll *et al.*, 1960) suggested that stresses induced prolactin release.

E. PROLACTIN SECRETION DURING POSTREPRODUCTIVE LIFE

Normal estrous cycles rarely occur in old female rats (Mandl and Shelton, 1959; Aschheim, 1961; Clemens and Meites, 1971). Vaginal smears from old rats commonly show a pattern of constant estrus, repeated pseudopregnancies of irregular length or no definite pattern. Old rats frequently have enlarged pituitaries and occasional pituitary tumors, enlarged mammary glands and frequent mammary tumors. The incidence of mammary tumors in Sprague-Dawley female rats, 24 months or older, in our laboratory may be as high as 50% or greater.

Earlier work indicated that anterior pituitaries removed from old female rats released more prolactin when cultured *in vitro* than pitu-

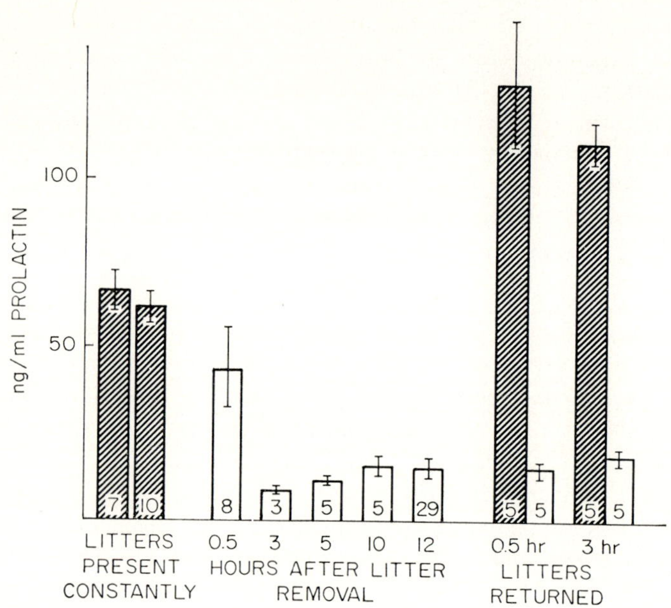

FIG. 6. Effects of suckling on serum prolactin values. Rats with litters constantly had elevated serum prolactin levels, since litters suckle frequently. Removal of litters for 3 or more hours resulted in a fall in serum prolactin to basal levels. Suckling for 30 minutes or 3 hours resulted in marked elevations of serum prolactin concentration. From Amenomori *et al.* (1970).

itaries removed from young female rats (Meites *et al.*, 1961a). Old constant-estrous rats, approximately 21 months of age, were found to have about twice as much prolactin in their pituitaries as 3-month-old female rats on the day of estrus, and their pituitaries were significantly heavier than in the younger rats (Clemens and Meites, 1971). These old rats also contained about twice as much pituitary FSH and hypothalamic FRF and about half as much pituitary LH as young 3-month-old rats on the day of estrus. We have postulated that a fundamental change occurs in hypothalamic control of anterior pituitary function in old rats, and have reported that electrical stimulation of the preoptic area or injections of epinephrine or progesterone can induce ovulation and re-initiate estrous cycles in some rats (Clemens *et al.*, 1970).

F. Prolactin Secretion in Males

Early investigation demonstrated that the pituitaries of male animals usually contained less prolactin than the pituitaries of females. Thus the pituitaries of mature male rats had less than half as much prolactin as the pituitaries of mature female rats, and the pituitaries of mature male rabbits contained only about one-fourth as much prolactin as the pituitaries of mature female rabbits (Meites and Turner, 1948a). Male guinea pigs appear to be an exception, since their pituitaries contain as much prolactin as female guinea pigs. Castration resulted in reduced pituitary prolactin levels, and administration of androgen increased pituitary prolactin although it was much less effective than estrogens (Meites and Turner, 1948b). The pituitaries of male rats and rabbits also appear to be less responsive to the prolactin-stimulating action of administered estrogen. Serum prolactin in male rats shows no cyclic variations, and is about as low as in female rats during diestrus or after ovariectomy (Amenomori *et al.*, 1970). The function(s), if any, of prolactin in male mammals remains to be determined (see p. 518).

III. Relation of Prolactin to Mammary Growth and Lactation

Prolactin is the most important hormone involved in mammary growth and lactation, although it is not the only hormone that influences these processes. Practically the entire neuroendocrine system has some regulatory influences on the mammary gland.

In the classical experiments of Lyons *et al.* (1958), rats were hypophysectomized, ovariectomized, and adrenalectomized to remove all anterior pituitary and steroid hormones. A combination of prolactin, GH, estrogen, and progesterone was shown to produce full lobuloalveolar development comparable to that seen in rats in late pregnancy. Earlier, it had been demonstrated by Turner (1939) that ovarian hormones have little or no effect on mammary growth in the absence of the anterior pituitary. It was not recognized until more recently, as a result of an observation by Clifton and Furth (1960) in adrenalectomized-ovariectomized rats carrying a pituitary tumor graft, that pituitary hormones alone may be capable of stimulating full mammary development. In subsequent experiments in our laboratory (Talwalker and Meites, 1961, 1964), it was demonstrated that injections of prolactin and GH, or prolactin alone, could induce mammary lobuloalveolar growth in ovariectomized-adrenalectomized or ovariectomized-adrenalectomized-hypophysectomized rats. In hypophysectomized rats, implantation of rat anterior pituitary near a single mammary gland, produced localized lobuloalveolar growth (Meites and Kragt, 1964). Subsequently, Dilley and Nandi (1968) reported that alveolar development could be induced in organ cultures of rat mammary tissue by incorporation of prolactin and insulin alone into the medium. Insulin is essential for maintenance of mammary tissue viability during culture. These and related observations suggest that the ovarian hormones act in part via the pituitary, as well as directly on the mammary glands together with pituitary hormones. Evidence was reported recently that estrogen increases plasma GH as well as prolactin levels in rats (Dickerman and Meites, 1971).

Estrogen and progesterone appear to provide the inciting stimulus for mammary growth during recurrent estrous and menstrual cycles and during pregnancy. However, under some conditions, the initiating stimulus to mammary growth may come from the pituitary via the hypothalamus rather than from the ovaries: (1) During postpartum lactation, the suckling stimulus induces release of prolactin and ACTH but inhibits release of FSH and LH. Suckling stimulates mammary growth and lactation but inhibits gonadal function. (2) Certain drugs, such as reserpine and chlorpromazine, can increase prolactin secretion and inhibit gonadotropin secretion, resulting in mammary stimulation and ovarian inhibition (Meites *et al.*, 1963a). (3) Removal of hypothalamic connections to the anterior pituitary by placement of appropriate lesions in the arcuate nucleus-median eminence area, by pituitary stalk section, or pituitary transplantation, result in increased secretion of prolactin and greatly reduced secretion of all other anterior pituitary hormones. (4) Certain abnormal states in human patients (Chiari-Frommel and Forbes-Albright syndromes) are characterized by breast development, persistent lactation, and amenorrhea. It these states, the pituitary rather than the ovaries appears to be responsible for mammary growth and lactation.

Prolactin and adrenal cortical hormones are essential for initiation and maintenance of lactation in most species studied. The rabbit appears to be an exception, since it has been demonstrated that in hypophysectomized (Fredrikson, 1939; Kilpatrick *et al.*, 1964) or adrenalectomized rabbits (Cowie and Watson, 1966), administration of prolactin alone can initiate lactation. It is interesting, however, that cortisol as well as prolactin can initiate lactation in pregnant rabbits (Meites *et al.*, 1963b), and ACTH as well as prolactin can initiate milk secretion in pseudopregnant rabbits (Chadwick and Folley, 1962).

The absence of lactation during most of pregnancy, assuming the animal is not lactating as a result of a previous gestation, has been explained on the basis of an insufficiency of prolactin and/or adrenal cortical hormones or both, and to an action by estrogen and progesterone on the mammary gland which renders it relatively refractory to the lactogenic effects of prolactin and adrenal steroids (Meites, 1966). Administration of adrenal glucocorticoid hormones initiated lactation in the pregnant rat (Talwalker and Meites, 1961), mouse (Nandi and Bern, 1961), and cow (Tucker and Meites, 1965). Injections of either prolactin or hydrocortisone acetate induced lactation in the pregnant rabbit (Meites *et al.*, 1963b). At about the time of parturition, there is an increase in secretion of prolactin and adrenal cortical hormones, permitting the onset of lactation. The milking stimulus together with milk removal from the mammary gland serves to maintain postpartum lactation.

The question has been raised frequently about the mechanism by which large doses of estrogen inhibit lactation. Since even large doses of estrogen do not inhibit prolactin release (Meites, 1966; Chen and Meites, 1970), a different mechanism must be responsible for this phenomenon. We have reported that lactation inhibition by large doses of estrogen or estrogen and progesterone in rabbits can be largely overcome by injecting relatively large doses of prolactin, and suggested that large doses of estrogen interfere with the peripheral action of prolactin on the mammary gland (Meites and Sgouris, 1954). More recent work in our laboratory appears to bear out the correctness of this hypothesis.

IV. Luteolytic Role of Prolactin during the Estrous Cycle of the Rat

Luteolysis of corpora lutea formed during the previous cycle occurs during each estrous cycle in the rat (Greenwald and Rothchild, 1968; Schwartz and Waltz, 1970). Prolactin has been observed to promote luteolysis of nonfunctional corpora lutea present in hypophysectomized rats (Malven and Sawyer, 1966; Piacsek and Meites, 1967). Since serum prolactin values reach peak levels on the late afternoon of proestrus and are still somewhat elevated on the day of estrus of each cycle (Gay *et al.*, 1970; Wuttke and Meites, 1970), it was of interest to determine whether prolactin was responsible for luteolysis of the earlier crop of corpora lutea formed during each cycle.

Early work by Shelesnyak (1954, 1958) indicated that ergot drugs could inhibit prolactin secretion. We have observed that the drug ergocornine can completely block the normal rise in serum prolactin on the late afternoon of proestrus without interfering with ovulation or with regularity of cycling in rats (Wuttke *et al.*, 1971). Nagasawa and Meites (1970) and Heuson *et al.* (1970) also noted that injections of ergot drugs induced enlargement of the ovaries of rats due to the accumulation of corpora lutea. It was of interest therefore to determine whether injections of ergocornine could prevent luteolysis of corpora lutea during the cycle, and whether injection of prolactin in ergocornine-treated rats could produce luteolysis (Wuttke and Meites, 1971).

Vaginal smears were taken from mature female Sprague-Dawley rats, 3–4 months old, and only rats with regular 4- or 5-day estrous cycles were used. These rats were injected with 50 μg of ergocornine per 100 gm body weight during a single cycle, beginning on the afternoon before the expected day of proestrus and continuing for 3 days thereafter. Some of the ergocornine-treated rats were also injected with a single dose of 1 mg of prolactin (NIH ovine prolactin, 28 IU/mg). Control cycling rats were injected with physiological saline. In another experiment, rats were injected with ergocornine for 3 cycles, and some of these rats were

TABLE I

Effect of Ergocornine (EC) and Prolactin during Three Cycles on Ovaries[a]

Treatment and number of rats	Average body weight (gm)	Average ovarian weight (mg)	Average number of corpora lutea[b]
Saline (12)	283.3 ± 6.6	63.2 ± 3.2	6.0 ± 0.3
EC (13)	278.5 ± 4.6	89.4 ± 3.5[c]	13.4 ± 0.9[c]
EC plus prolactin (11)	286.4 ± 6.7	70.0 ± 3.2	5.8 ± 0.4

[a] From Wuttke and Meites (1971).

[b] Per ovarian cross section.

[c] EC vs. saline, and EC vs. EC + prolactin, $P < 0.001$.

given 1 mg of prolactin on the days of proestrus and estrus. These rats were killed during the diestrus after the third cycle, and the ovaries were removed, weighed, fixed, and stained for microscope examination.

The results of the 1- and 3-cycle experiments were similar, and only those from the 3-cycle study are shown in Table I. It can be seen that the ovaries of the rats given ergocornine (EC) alone weighed significantly more and contained about twice as many corpora lutea per ovary as the ovaries of saline-injected controls. Only corpora lutea seen in stained sagittal cross sections of each ovary under the microscope were counted. Rats given prolactin together with ergocornine had ovarian weights and average number of corpora lutea similar to those of the controls (Fig. 7).

These observations suggest that the primary action of prolactin on the ovaries of rats during the estrous cycle is to induce luteolysis of the previous crop of corpora lutea. Luteolysis apparently occurs as a result of the peak rise in serum prolactin on the afternoon of proestrus and perhaps also because of the relatively high prolactin on the day of estrus. Similar conclusions were drawn independently by Billeter and

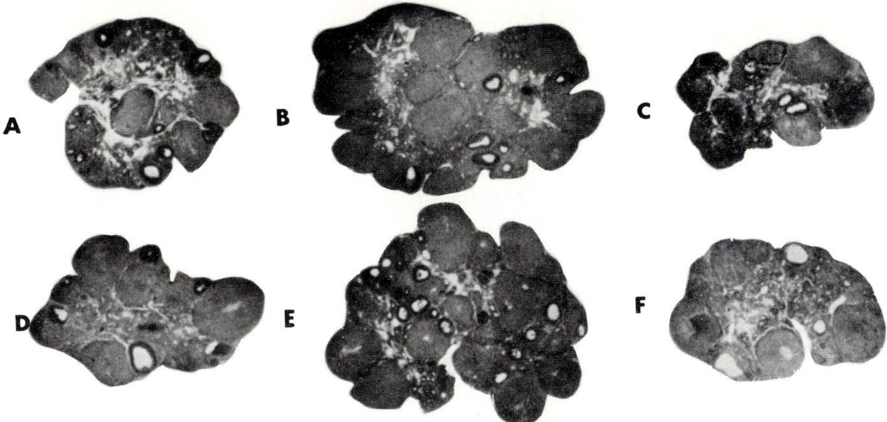

FIG. 7. Ovaries from rats treated during one estrous cycle with physiological saline (A, controls), ergocornine (B), or ergocornine and prolactin (C). Ovaries from rats treated during 3 estrous cycles with physiological saline (D, controls), ergocornine (E), or ergocornine and prolactin (F). Prolactin decreased the number of corpora lutea in ergocornine-treated rats to those of control rats. From Wuttke and Meites (1971).

Fluckiger (1971) on the basis of a study in which ergocryptine (CB154) was given alone or together with prolactin continuously for a period of 3 weeks. They observed both luteolytic and luteotropic effects. We have noted that when ergocornine was injected daily for 30 days into cycling rats, the ovaries weighed about 50% more than in control cycling rats and contained numerous corpora lutea; single injections of 1 mg prolactin on the last 2 days of the 30-day period in ergocornine-treated rats returned the ovaries to normal weight and number of corpora lutea (Lu and Meites, unpublished).

Although Rothchild (1965) has expressed the view that LH is responsible for luteolysis of the older corpora lutea during each cycle in the rat, this seems unlikely since the dose of ergocornine used does not interfere with cycling or ovulation (Wuttke et al., 1971). Whether the peak rise in serum prolactin on the late afternoon of proestrus also has a luteotropic role on the newly formed corpora lutea on the day of estrus is unknown. This appears to be doubtful, however, since coitus or cervical stimulation on the day of estrus produces a rise in serum prolactin and LH in the rat (see Section II,C), and apparently it is this initial rise in these two hormones that is associated with the onset of pseudopregnancy. Thus the major function of prolactin on the ovaries during the estrous cycle appears to be to induce luteolysis of the old corpora lutea.

V. Role of Prolactin in Induction of Pseudopregnancy in the Rat

Prolactin has been shown to be essential for induction and maintenance of pseudopregnancy and early pregnancy in the rat (Greenwald and Rothchild, 1968). LH is believed to synergize with prolactin to promote progesterone secretion (Armstrong and Greep, 1962). Inasmuch as serum prolactin rises for the first few days of pseudopregnancy after initiation by stimulation of the uterine cervix (see Section II, C), it was of interest to determine whether prevention of this rise by ergocornine (EC) would interfere with pseudopregnancy. Previously Shelesnyak (1958) had reported that large doses of ergot drugs could produce termination of existing pseudopregnancy or early pregnancy in rats, and hence only a small dose of EC was used in this study.

Mature female Sprague-Dawley rats, undergoing regular 4 or 5 day estrous cycles, were injected intraperitoneally with 50 μg EC/100 gm body weight beginning at 5:15 PM on the last day of diestrus before the expected day of proestrus, at 12 AM and 5:15 PM on the day of proestrus, and at 9:30 AM on the day of estrus. Control rats were given the injection vehicle only. Except for 9 untreated rats permitted to continue cycling, each of the EC and control rats were stimulated with a glass rod on the uterine cervix for 45 seconds at 9:30 AM on the day of estrus to induce pseudopregnancy. Blood samples from these rats were collected before and after cervical stimulation and measured for prolactin (Niswender *et al.*, 1969), LH (Niswender *et al.*, 1968), and FSH (Parlow *et al.*, 1969) by radioimmunoassays. Daily vaginal smears were collected from each rat to determine the length of pseudopregnancies.

It can be seen (Fig. 8) that EC prevented the rise in serum prolactin on the afternoon of proestrus, after cervical stimulation on the day of estrus and for the following 2 days. By contrast, in the control rats there was a normal rise in serum prolactin on the afternoon of proestrus, and a secondary rise after cervical stimulation on the day of estrus which continued for the next 2 days. Rats not stimulated via the uterine cervix on the morning estrus showed no increase in serum prolactin values.

Figure 9 shows that serum LH values were elevated equally in both EC-treated and control rats on the afternoon of proestrus, and rose again but to a lesser degree after cervical stimulation on the morning of estrus. Thus EC injections did not prevent the normal rise in serum LH in these rats as a result of cervical stimulation. Figure 10 shows that neither EC nor cervical stimulation altered the normal cyclic changes in serum FSH on the days of this experiment. Cervical stimulation in control rats resulted in pseudopregnancies of an average duration of 14.3 ± 0.3 days, and in EC-treated rats of an average length of 10.7 ± 0.2 days. Rats not cervically stimulated continued to cycle normally with an average duration per cycle of 4.3 ± 0.1 days.

The above results indicate that a pseudopregnancy of shortened duration can be induced and maintained after cervical stimulation on the morning of estrus, despite prevention of any rise in serum prolactin

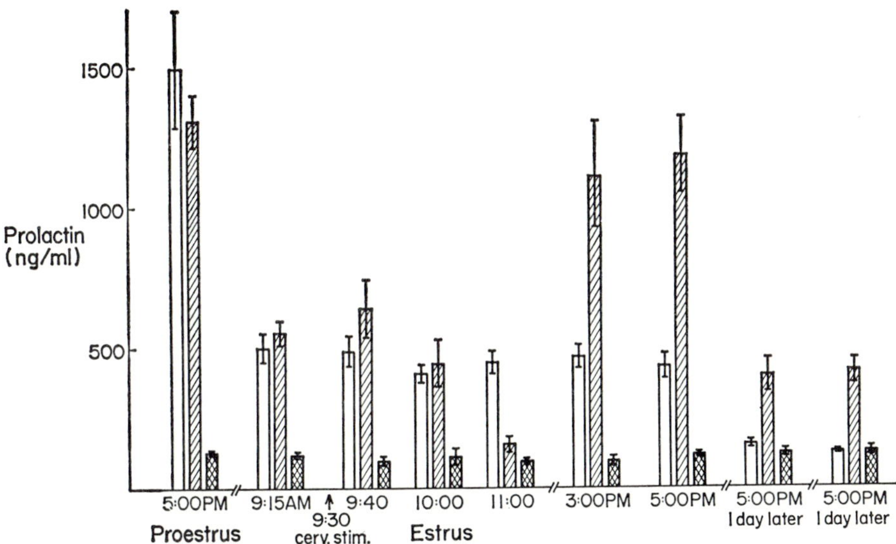

FIG. 8. Effects of ergocornine (EC) injections and cervical stimulation on morning of estrus on serum prolactin values. Note that EC prevented the rise in serum prolactin at 5:00 PM on the day of proestrus, and also after cervical stimulation at 9:30 AM on the day of estrus. By contrast, similar cervical stimulation of rats injected with physiological saline resulted initially in a fall in serum prolactin until 11:00 AM followed by a peak rise at 3:00 and 5:00 PM; serum prolactin remained elevated in these rats for the next 2 days. Average length of pseudopregnancy was 10.7 ± 0.2 days in the EC-treated rats and 14.3 ± 0.3 days in the saline-injected rats. □, Saline only; ▨, saline plus cervical stimulation; ▨, EC plus cervical stimulation. From Wuttke and Meites (unpublished).

by EC and in the presence of a normal increase in serum LH. In a subsequent experiment, we observed that a dose of 0.2 mg EC/100 gm body weight, given only on the morning of estrus prior to cervical stimulation, produced no change in duration of pseudopregnancy, which was the same as in control, cervically stimulated rats. It should be emphasized that the 50-μg dose of EC used in the present study did not completely inhibit prolactin release, and serum prolactin values were more than twice as high as in hypophysectomized rats (H-10-10-B reference standard prolactin).

Induction of pseudopregnancy in the EC-treated rats cannot be attributed solely to the rise in LH after cervical stimulation, since it has not been demonstrated that LH administration alone on the day of estrus can elicit pseudopregnancy (Rothchild, 1965). The removal of any luteolytic action by prolactin as a result of reducing serum prolactin

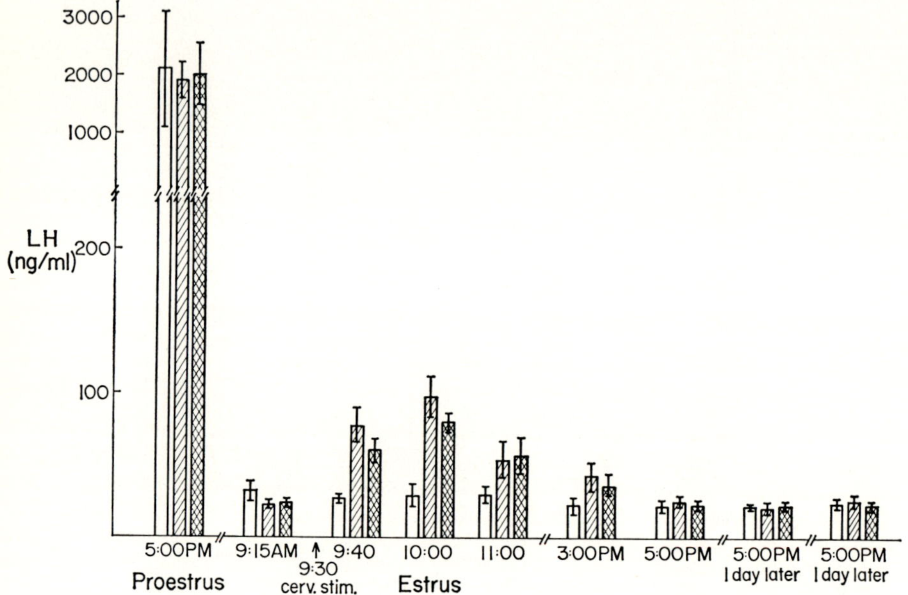

FIG. 9. Effects of ergocornine (EC) injections and cervical stimulation on morning of estrus on serum luteinizing hormone (LH) values. Note that the dose of EC used (50 μg) did not prevent the rise in serum LH after cervical stimulation at 9:30 AM on the day of estrus. Serum LH remained elevated in the cervically stimulated rats until 3:00 PM and then fell to basal levels. □, Saline only; ▨, saline plus cervical stimulation; ▧, EC plus cervical stimulation. From Wuttke and Meites (unpublished).

levels by EC injections may partially account for the pseudopregnancy. Although progesterone secretion was not measured in the EC-treated rats, it does not appear probable that pseudopregnancy could have been maintained in these rats without progesterone secretion. Further work will be required to determine the functional state of the ovaries of the EC-treated rats made pseudopregnant by cervical stimulation.

The ability of large doses of EC to produce termination of an existing pseudopregnancy in rats (Shelesnyak, 1958) may be due not only to a more severe inhibition of prolactin release, but also to marked reduction in LH release. We have reported that large doses of EC can significantly depress serum LH values, although they do not appear to interfere with ovulation (Wuttke *et al.*, 1971). Large doses of ergot drugs can also produce toxic manifestations, including loss of body weight, vasoconstriction, and antiadrenergic effects (Brazeau, 1970).

VI. Relation of Prolactin to Mammary Tumors

Prolactin and estrogen appear to be the most important hormones for mammary tumor development and growth in rats, mice, and perhaps

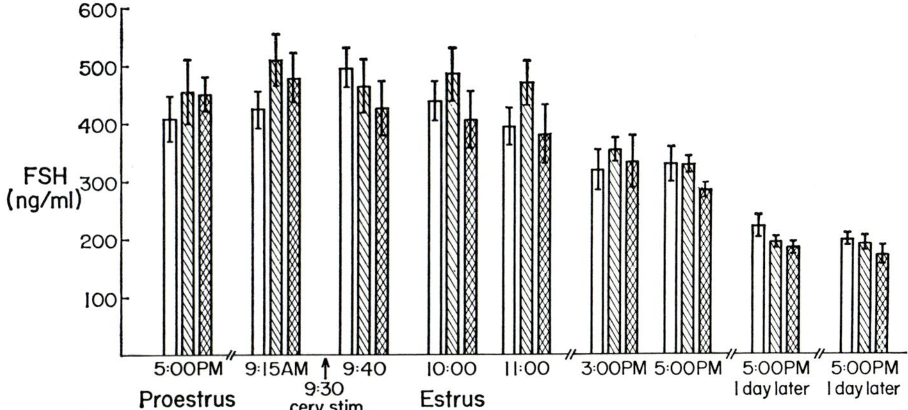

FIG. 10. Neither ergocornine (EC) injections nor cervical stimulation on morning of estrus had any effect on serum follicle-stimulating hormone (FSH) values. Apparently a rise in FSH release is not essential for induction of pseudopregnancy in the rat. □, Saline only; ▨, saline plus cervical stimulation; ▧, EC plus cervical stimulation. From Wuttke and Meites (unpublished).

other species. Whether these two hormones are themselves capable of acting as mammary carcinogens, or whether they merely act as cocarcinogens in the presence of other agents, such as viruses, certain chemicals, or radiation, remains to be elucidated. Carcinogenic agents do not appear to be capable of evoking mammary tumor development in the absence of a proper hormonal environment. On the other hand, there is considerable evidence that prolonged action by estrogen or prolactin or both in intact rats and mice may result in mammary tumorigenesis. The possibility cannot be ruled out that some latent carcinogenic agent may be present and activated by these hormones.

Estrogens do not produce mammary tumors in rats in the absence of the pituitary (Sterental *et al.*, 1963) but act synergistically with pituitary hormones on the mammary gland, and via the pituitary and hypothalamus to increase prolactin (Chen *et al.*, 1970) and GH secretion (Dickerman *et al.*, 1971). We have reported that increases in prolactin secretion result in enhanced growth of spontaneous and carcinogen-induced mammary tumors in rats. Elevated blood levels of prolactin also can inhibit development of mammary tumors prior to treatment with carcinogenic chemicals. Some of these studies will be cited here.

A. Relation of Prolactin to Development of Spontaneous Mammary Tumors

Injection of prolactin for prolonged periods of time or grafting of extra pituitaries into intact mice presumably free of any mammary viral agent, resulted in mammary tumorigenesis in a high percentage of animals (Muhlbock and Boot, 1967; Boot, 1969). Leob and Kirtz (1939) first observed that grafting of extra pituitaries into randomly bred mice increased mammary tumor incidence, but it was not appreciated at that time that the major hormonal contribution from such grafts was persistent prolactin secretion. Spontaneous mammary tumors frequently appear in old female rats, and in our laboratory Sprague-Dawley female rats commonly reach an incidence of 50% or higher by the time the rats are 2 years old. These appear as a single benign adenoma or fibroadenoma and differ in their appearance and growth characteristics from carcinogen-induced mammary adenocarcinomas in rats.

The effect of placing bilateral lesions in the median eminence was tested in intact female Sprague-Dawley rats (Welsch *et al.*, 1971). Such lesions have been shown to result in a significant and prolonged increase for many months in serum prolactin levels and in a marked reduction in secretion of other hormones (Meites *et al.*, 1963a; Chen *et al.*, 1970). The rats were lesioned at 10 months of age and were killed 25 weeks later (16.5 months of age). Control rats received sham lesions. The effects of the lesions are shown in Table II. It can be seen that 4 out of 21 control rats developed a total of 4 tumors, whereas 12 out of 23 lesioned rats developed a total of 20 tumors or an average of 1.6 tumors per rat. Serum prolactin concentration in the lesioned rats was more than 3 times as great as in the control rats. The appearance of more than 1 spontaneous mammary tumor per rat in untreated old female rats is extremely rare, and the increased number of tumors per rat in the lesioned group can be attributed to the high prolactin secretion. In agreement with our results in rats, Bruni and Montemurro (1971) reported that placement of lesions in the anterior and median hypothalamus increased the incidence of spontaneous mammary tumors in mice.

TABLE II

Effects of Median Eminence Lesions on Development of Normal and Neoplastic Mammary Tissue and Serum Prolactin Levels in Female Rats[a]

Treatment[b]	Total number of rats	Final body weight (gm)	Serum prolactin levels (ng/ml)	Average mammary gland ratings	Number and percent of rats with tumors	Total number of tumors
Controls, sham lesion	21	360 ± 7[c]	50.9 ± 9.6[c]	3.1 ± 0.1[c]	$4(19\%)$[c]	4[c]
Median eminence lesions	23	451 ± 19[d]	179.8 ± 23.9[d]	4.2 ± 0.2[d]	$12(52\%)$[d]	20[d]

[a] From Welsch *et al.* (1970).

[b] All rats were sacrificed 25 weeks after placement of median eminence or sham lesions. Final body weight, serum prolactin levels, and average mammary gland ratings are represented as the mean value $\pm$ SE.

[c,d] $p < 0.001$.

An increased incidence of spontaneous mammary tumors in intact female Sprague-Dawley rats also was produced by grafting multiple pituitaries (Welsch *et al.*, 1970). In addition, we have occasionally observed mammary tumors in inbred Wistar-Furth female rats grafted with a "mammosomatotropic" pituitary tumor (MtT.W5 or MtT.W15) (Meites, 1972). These pituitary tumors, originally developed by Dr. Jacob Furth, secrete enormous amounts of prolactin and GH. Approximately 3.5–4.5 months after grafting small amounts of pituitary tumor tissue subcutaneously, mammary tumors were observed in a few rats. These mammary tumors were transplanted to other inbred Wistar-Furth rats, but grew readily only in the presence of a pituitary tumor. They also developed well in ovariectomized—adrenalectomized rats with a pituitary tumor, but the average latency period was somewhat prolonged. This suggests a primary dependency on pituitary hormones for growth of these mammary tumors.

B. Relation of Prolactin to Development and Growth of Carcinogen-Induced Mammary Tumors

Huggins (1965) developed a method for rapid induction of mammary cancers in rats. Female Sprague-Dawley rats, 50–60 days old, are given a single intravenous injection of a lipid emulsion of 5 mg of 7,12-dimethylbenzanthracene (DMBA). Multiple mammary adenocarcinomas appear 1–4 months later, with a mean latency period in our laboratory of about 55 days. These mammary cancers are hormone dependent and respond readily to changes in prolactin or estrogen secretion. In general, when prolactin levels are raised prior to DMBA administration, there is a decreased incidence of mammary cancers. When prolactin levels are raised after the appearance of mammary cancers, development and growth of these cancers are increased. Apparently, stimulation of mammary growth by prolactin and ovarian hormones prior to administration of DMBA, protects the mammary tissue from the action of the carcinogen, whereas prolactin stimulates growth of existing DMBA-induced mammary tumors.

The relative importance of ovarian versus pituitary hormones for development and growth of DMBA-induced tumors was assessed in a preliminary study (Talwalker *et al.*, 1964). Sprague-Dawley rats were given a single dose of DMBA 7 days after ovariectomy at 52 days of age. Some of the rats were then injected with estradiol and others with a combination of prolactin (NIH ovine prolactin, 15 IU/mg) and bovine GH (approximately 1 unit/mg, E. R. Squibb and Son, New Brunswick, New Jersey). The results are shown in Table III. The 18 intact sham-operated controls all developed an average of 3.9

TABLE III

Effects of Hormone Administration on Mammary Tumor Induction in Ovariectomized Rats Treated with 7,12-Dimethyl-1,2-benzanthracene (DMBA)[a]

Group and treatment[b]	Duration of treatment (days)	Total number of rats	Number and percent of rats with tumors	Average number of tumors per tumor-bearing rat	Range and mean latency period (days)
1. Sham-operated controls, saline	14	18	18(100)	3.9	52–110(68)
2. Ovax-controls, saline	14	20	0(0)	—	—
3. Ovax, 1 μg Esd 1 $\times$ daily	14	12	4(33)	1.5	72–130(104)
4. Ovax, 10 μg Esd 1 $\times$ daily	14	13	3(23)	1.0	76–102(89)
5. Ovax, 1 mg STH[c] 1 mg prolactin 2 $\times$ daily	14	9	4(44)	1.8	114–210(146)
6. Ovax, STH[c] prolactin[c] 1 $\times$ daily	75	9	6(66)	2.3	94–207(122)

[a] From Talwalker *et al.* (1963).

[b] Ovax = ovariectomized; STH = porcine growth hormone; Esd = estradiol. DMBA was administered 7 days after ovariectomy.

[c] Dosages of STH and prolactin were increased every 15 days.

cancers per rat. None of the 20 ovariectomized control rats developed mammary cancers. In the ovariectomized rats given 1 or 10 μg of estradiol for 14 days after DMBA treatment, there was an incidence of 33 and 23% mammary cancers, respectively. The ovariectomized rats given prolactin and growth hormone for 14 or 75 days after DMBA treatment, showed an even higher incidence of mammary cancers and more cancers per rat. This experiment suggests that administration of pituitary hormones alone into ovariectomized rats can produce mammary cancers after DMBA treatment. Since the adrenals remained in these rats, the presence of some estrogen and progesterone in these rats cannot be excluded completely. In a subsequent experiment (Quadri and Meites, unpublished), rats were first ovariectomized and adrenalectomized, and then given 1 mg of DMBA as a dry powder placed directly over a single mammary gland in the inguinal region. These rats were injected with prolactin and GH. Single mammary cancers developed only in the mammary glands treated with DMBA. This strongly indicates that mammary cancers can develop after DMBA treatment in the absence of ovarian hormones and in the presence of high levels of prolactin and GH.

In another study, the effects of placing bilateral lesions in the median eminence, before or after intravenous injection of DMBA, were tested in female Sprague-Dawley rats (Clemens *et al.*, 1968). All rats were given a single intravenous injection of 5 mg of DMBA at 56 days of age. Bilateral lesions were placed in the median eminence at 50 days of age, and some of these rats were ovariectomized at 64 days of age. All rats were examined once weekly for 6 months for appearance of tumors. The results (Table IV) show that 21 out of 22 control intact rats developed an average of 3.1 tumors per rat, whereas only 6 out of 20 intact rats lesioned prior to DMBA treatment developed an average of 1.1 tumors per rat. The mean latency period for development of tumors was longer (104 ± 14 days) in the lesioned than in the intact control rats (78 ± 5 days). In the control ovariectomized rats, 7 out of 13 rats developed mammary tumors with a mean latency period of 156 ± 7 days, whereas none of the 20 lesioned, ovariectomized rats developed mammary tumors. Some mammary tumors appeared in the ovariectomized control rats because the ovaries were not removed until 14 days after DMBA administration. The presence of the ovaries has been reported to be most critical for development of mammary tumors during the first 2 weeks after DMBA treatment (Dao, 1962).

The effects of placing bilateral lesions in the median eminence also was tested in rats 75 days after injecting DMBA, when most of the rats already had mammary tumors. Some rats were also ovariectomized at this time. The mammary tumors of all rats were examined at 0, 10, and 25 days after placement of lesions. The results (Table V) show that in the control, sham-lesioned rats (group 1), there was a small increase in average number of tumors per rat by 10 and 25 days later. In the rats lesioned in the median eminence (group 2) there was a pronounced increase in average number of tumors per rat by 10 days and a further large increase by 25 days after lesion placement. The average number of tumors per rat decreased in the control ovariectomized rats by 10 to 25 days after surgery (group 3), whereas in the lesioned,

TABLE IV

Effects of Median Eminence Lesions Placed in Rats before Treatment with DMBA[a,b]

Group and treatment	Total number of rats	Number and percent of rats with tumors	Average number of tumors/tumor-bearing rat	Range and mean latency period (days)
1. Control intact	22	21(95)	3.1 ± 0.11[c]	40–175(78 ± 5)[c]
2. Lesioned intact[f]	20	6(30)	1.16 ± 0.03[e]	65–177(104 ± 14)[e]
3. Control ovariectomized[g]	13	7(54)	2.33 ± 0.42[d]	102–186(156 ± 7)[d]
4. Lesioned ovariectomized	20	0(0)	0	0(0)

[a] From Clemens *et al.* (1968).

[b] DMBA was administered on day 56.

[c,d,e] Groups possessing different superscripts are significantly different from each other ($p < 0.05$). Standard error of the mean is indicated for each group.

[f] ME lesion placed on day 50.

[g] Ovariectomy performed on day 64.

ovariectomized rats (group 4) there was an increase in average number of tumors 10 days later. By 25 days tumor regression was observed in these rats. The latter suggests that sustained estrogen secretion is necessary to permit high serum levels of prolactin to promote mammary tumor growth beyond a relatively short period of time.

The above observations on the effects of median eminence lesions before and after DMBA treatment were confirmed and extended by Welsch *et al.* (1969). They found that lesions placed in the median eminence of rats with DMBA-induced mammary tumors resulted in enhanced tumor growth, but lesions placed in the preoptic area of the hypothalamus or in the amygdaloid area resulted in a reduction in tumor growth. There were indications that the preoptic and amygdaloid lesions were associated with a reduction in estrogen secretion. In related studies,

TABLE V

Effects of Median Eminence Lesions Placed in Rats 75 Days after Treatment with DMBA[a]

Group and treatment[b]	Total number of rats	Average number of palpable tumors per rat				
		Initial	10 days after treatment	Change (%)	25 days after treatment	Change (%)
1. Controls, intact	13	3.2 ± 0.7[c]	3.8 ± 0.9[c]	$+ 19$	4.3 ± 0.8[c]	$+ 33$
2. Lesioned, intact	12	3.5 ± 1.2	7.7 ± 2.3	$+120$	10.1 ± 1.7	$+189$
3. Control, ovariectomized	16	4.0 ± 0.7	2.9 ± 0.5	$- 27$	1.4 ± 0.5	$- 56$
4. Lesioned, ovariectomized	16	4.6 ± 1.1	8.3 ± 1.1	$+ 80$	2.8 ± 0.6	$- 39$

[a] From Clemens *et al.* (1968).

[b] DMBA was administered at 55 days of age. The rats were ovariectomized and/or lesioned approximately 75 days after DMBA treatment.

[c] Standard error of the mean.

it was reported that grafting of multiple pituitaries (Welsch *et al.*, 1968), or administration of a norethynodrel–mestranol combination (Welsch and Meites, 1969), inhibited mammary cancer development when given prior to DMBA injection but promoted growth of existing DMBA-induced cancers. These observations indicate that stimulation of mammary growth by prolactin prior to carcinogen treatment protects the mammary tissue from the action of the carcinogen, whereas increased levels of prolactin stimulate growth of existing mammary tumors.

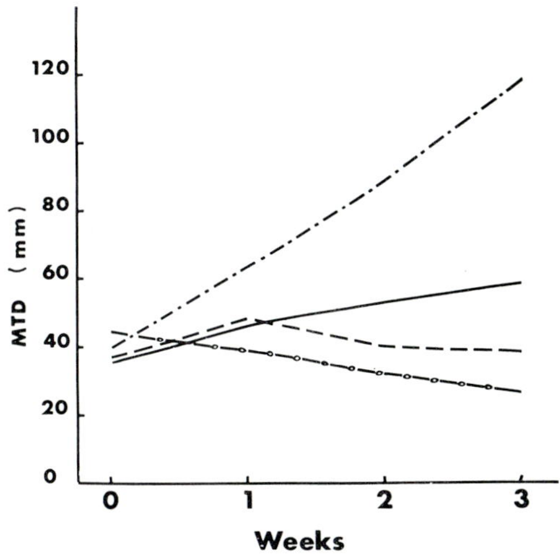

Fig. 11. Effects of daily injections of prolactin (–––) (1 mg NIH ovine prolactin, 28 IU/mg), iproniazid (–––) (2.5 mg/100 gm body weight), or pargyline (—O—O—) (2.5 mg/100 gm body weight) on growth of 7,12-dimethyl-1,2-benzanthracene (DMBA)-induced mammary tumors in Sprague-Dawley female rats. —, Control; MTD = mean tumor diameter. From Quadri, Clark, and Meites (unpublished).

Related studies have shown that administration of drugs that increase prolactin secretion result in enhanced growth whereas drugs that decrease prolactin secretion result in decreased growth of mammary tumors in rats. The drugs that increase mammary tumor growth include reserpine (Welsch *et al.*, 1970) and haloperidol (Quadri, Clark, and Meites, unpublished). The drugs that inhibit mammary tumor growth include several ergot compounds (Nagasawa and Meites, 1970; Cassell *et al.*, 1971), iproniazid (Nagasawa and Meites, 1970) *l*-DOPA and pargyline (Quadri, Clark, and Meites, unpublished) (Figs. 11 and 12). The ergot drugs have been of particular interest since they decrease prolactin secretion via the hypothalamus by increasing hypothalamic PIF activity (Wuttke *et al.*, 1971) and also directly inhibit pituitary prolactin release (Lu and

Meites, 1971). The effects of injecting ergocornine and ergocryptine for 4 weeks to rats with DMBA-induced mammary tumors are shown in Fig. 13. It can be seen that tumor regression in the EC-treated rats was similar to that of the ovariectomized rats. When drug treatment was terminated at the end of 4 weeks, tumor growth promptly resumed. These same 2 ergot drugs similarly inhibited growth of spontaneous mammary tumors in old female rats (Quadri and Meites, 1971). Heuson *et al.* (1970) also observed inhibition of DMBA-induced mam-

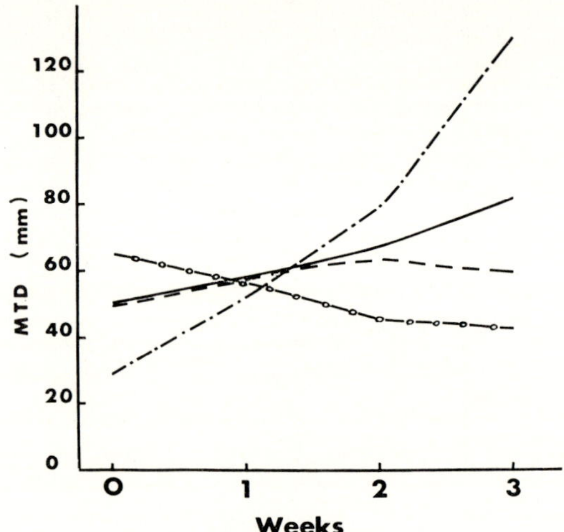

FIG. 12. Effects of daily injections of haloperidol (—·—·) (50 μg/100 gm body weight), LSD (---) (1 μg/100 gm body weight), or pargyline (—○—○—) (2.5 mg/100 gm body weight) on growth of DMBA-induced mammary tumors in Sprague-Dawley female rats. MTD = mean tumor diameter. From Quadri, Clark, and Meites (unpublished).

mary tumors in rats injected with ergocryptine, and Yanai and Nagasawa (1970) reported that ergot drugs suppressed growth of mammary hyperplastic alveolar nodules in mice.

C. How do Large Doses of Estrogen Suppress Mammary Tumor Growth?

Although estrogens can promote mammary tumor development and growth in rats, large doses of estrogens (and androgens) can inhibit growth of established mammary tumors in rats (Huggins *et al.*, 1959a,b; Dorfman, 1965), and also can produce remission of breast cancer in more than 20% of treated human patients (Hayward, 1970). Since there is no evidence that large doses of estrogen can inhibit prolactin secretion in rats (Chen and Meites, 1970) and administration of relatively large doses of prolactin can overcome the inhibitory effects of estrogen on lactation (Meites and Sgouris, 1954), we suggested that large doses of estrogen interfered with the peripheral action of prolactin on the mammary gland. Preliminary evidence indicates that a similar mechanism

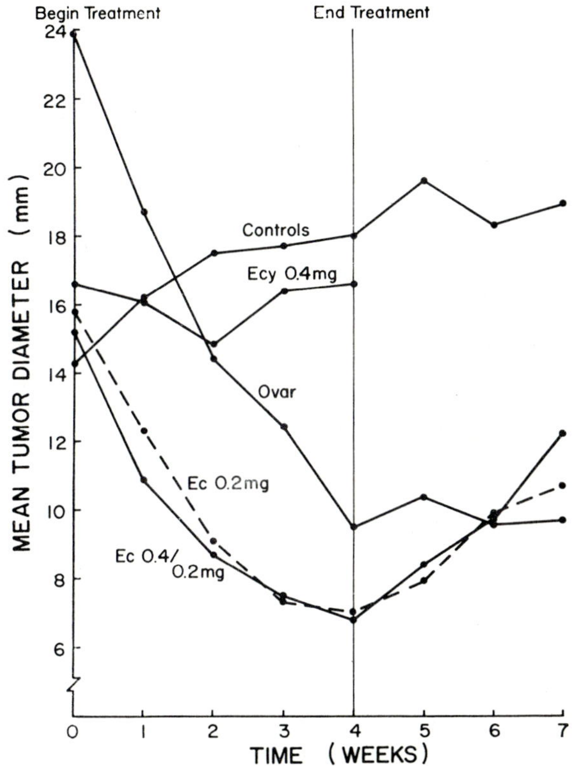

FIG. 13. Effects of daily injections of ergocornine (Ec) or ergocryptine (Ecy) on growth of 7,12-dimethyl-1,2-benzathracene-induced mammary tumors in Sprague-Dawley female rats. Doses are indicated for each drug. Ovar = ovariectomized. Note resumption of tumor growth after ergocornine treatment was terminated. From Cassell *et al.* (1971).

operates with respect to the action of large doses of estrogen on mammary tumor growth in rats (Meites *et al.*, 1971).

Sprague-Dawley female rats with developed DMBA-induced mammary cancers were divided into 4 groups, and injected once daily for 20 days with corn oil (controls); 20 μg of estradiol benzoate (EB); 1 mg of prolactin (NIH-P-S-8 ovine prolactin, 28 IU/mg); 20 μg of

EB and 1 mg of prolactin. Mammary cancer diameter and number of tumors per rat were recorded every 5 days. Figure 14 shows that the mammary cancers in the control rats increased by 50.3% in average diameter, whereas growth of cancers in the EB-treated rats was completely inhibited. Rats given EB together with prolactin showed an average gain in cancer diameter of 49.5% or about the same as in the control

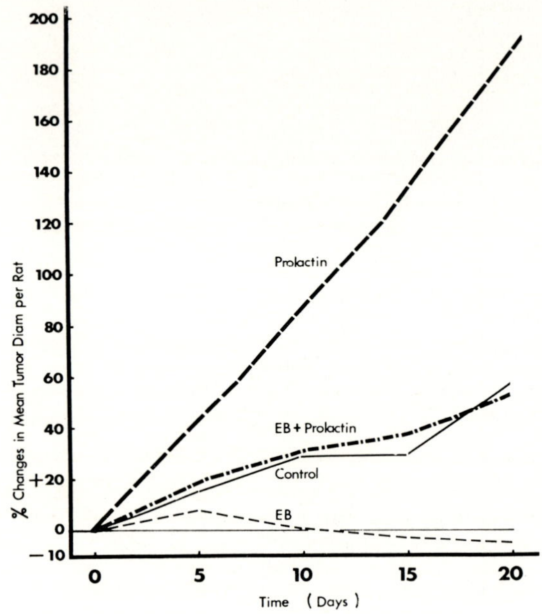

FIG. 14. Effects of daily injection of prolactin (1 mg NIH ovine prolactin, 28 IU/mg), estradiol benzoate (EB) (20 µg), or both on growth of 7,12-dimethyl-1,2-benzanthracene-induced mammary tumors in Sprague-Dawley rats. Note that EB inhibited tumor growth-stimulating action of prolactin and also prevented tumor growth in untreated tumor-bearing rats. The combination of prolactin and EB prevented the estrogen from inhibiting mammary tumor growth. From Meites, et al. (1971).

rats. Rats injected with prolactin alone showed about a 190% increase in average cancer diameter. These results indicate that the large dose of EB used inhibited growth of DMBA-induced mammary cancer by interfering with the peripheral action of prolactin. This is shown by the suppression by EB of the cancer-growth action of the administered prolactin and by the ability of prolactin when given together with EB to counteract its cancer growth-inhibiting effect. Previously we found that daily injections of 20 µg of EB, the same dose as used in the present experiment, increased serum prolactin values in rats with DMBA-induced tumors from 27.3 ± 2.6 to 177.0 ± 21.0 ng/ml (Nagasawa and Meites, unpublished). Thus mammary tumor growth can be inhibited by estrogen in the presence of increased serum prolactin concentration. The mechanism(s) by which large doses of estrogen interfere with the peripheral action of prolactin (and perhaps GH) on the mammary cancer tissue is unknown, but it may block binding of prolactin to receptor sites.

VII. Hypothalamic Regulation of Prolactin Secretion

A. HYPOTHALAMIC INHIBITION

The predominant action of the mammalian hypothalamus on prolactin secretion appears to be inhibitory in nature. When hypothalamic influence is removed or decreased by placement of lesions in the median eminence, by pituitary stalk section, by pituitary transplantation, by incubation or culture of the anterior pituitary or by administration of appropriate drugs, prolactin secretion continues at a high level whereas secretion of all other anterior pituitary hormones is sharply reduced (Meites *et al.*, 1963a; Meites and Nicoll, 1966). Placement of bilateral lesions in the median eminence, to destroy the "final common pathway" to the anterior pituitary, resulted within 30 minutes in a 10-fold increase in serum prolactin levels in rats, and even 6 months later serum prolactin was still 3 to 4 times above control values (Chen *et al.*, 1970; Welsch *et al.*, 1971). Lesions placed in the anterior or posterior hypothalamus also resulted in elevations of serum prolactin (Fig. 15), suggesting that these areas influence prolactin secretion as well. On the other hand, bilateral lesions placed in the amygdaloid nuclei had no effect on serum prolactin. It is of interest that the 3 lesion sites found to be effective in raising serum prolatin levels fall within the "hypophysiotropic region" of Halasz *et al.* (1962), shown to be the principal hypothalamic area regulating anterior pituitary function. Lesions placed in the "hypophysiotropic region" have been observed to initiate lactation in the rabbit (Haun and Sawyer, 1960), cat (Grosz and Rothballer, 1961) and rat (DeVoe *et al.*, 1966).

Crude extracts from the hypothalamus of the rat, sheep, bovine, pig and human have been demonstrated to inhibit prolactin release from the pituitary under both *in vitro* and *in vivo* conditions (Meites *et al.*, 1961b; Pasteels, 1961; Talwalker *et al.*, 1963; Schally *et al.*, 1967). The presumed hypophysiotropic hormone was named "prolactin inhibiting factor" (PIF) by us because hypothalamic extracts appear to inhibit synthesis as well as release of prolactin *in vitro* (Talwalker *et al.*, 1963).

This factor has not yet been characterized chemically, although it appears to be a small molecule (Talwalker *et al.*, 1963). Early the view was expressed that LRF and PIF might be the same (Everett, 1961), but subsequent work has demonstrated that these 2 factors are separate (Schally *et al.*, 1964, 1967). A negative dose-response relationship was demonstrated between the quantity of hypothalamic extract placed into an incubation medium and the amount of prolactin released by male rat pituitary after 4 hours of incubation (Kragt and Meites, 1965). PIF activity also has been assayed *in vivo* in rats on the basis of its

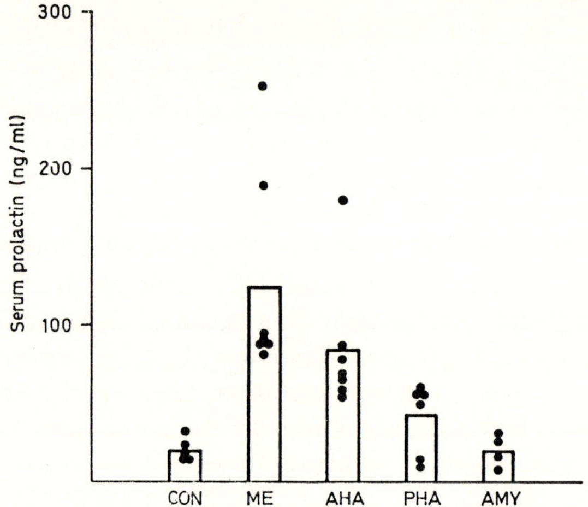

FIG. 15. Serum prolactin values 2 weeks after placement of lesions in brain of ovariectomized rats. CON, sham operated controls; ME, median eminence lesions; AHA, anterior hypothalamic lesions; PHA, posterior hypothalamic lesions; AMY, amygdaloid lesions. From Chen *et al.* (1970).

ability to prevent depletion of pituitary prolactin content by suckling during lactation (Grosvenor *et al.*, 1964) or by stimulation of the uterine cervix during estrus (Schally *et al.*, 1967). The effects of an injection of rat hypothalamic extract on serum prolactin values in female rats are shown in Fig. 16.

Changes in hypothalamic PIF content (activity) by a variety of stimuli were demonstrated early in our laboratory (Ratner and Meites, 1964; Ratner *et al.*, 1965; Meites, 1970a,b). The suckling stimulus, reserpine, perphenazine (Danon *et al.*, 1963), haloperidol (S. Dickerman, unpublished), estrogen, testosterone, progesterone, cortisol, and a norethynodrel-mestranol combination (Enovid) were found to decrease hypothalamic PIF activity, providing an explanation of how these stimuli evoked increases in prolactin release. An anesthetic dose of sodium pentobarbital was found to raise serum prolactin levels for the first 30 minutes after administration, and subsequently to depress prolactin release for a prolonged period (Wuttke and Meites, 1970). Hypothalamic PIF activity was completely depelted by 30 minutes after administration of this drug.

Agents that increase hypothalamic PIF activity and reduce prolactin release include prolactin itself (Chen *et al.*, 1967; Voogt and Meites, 1971), ergot drugs (Wuttke *et al.*, 1971), *l*-DOPA and several monoamine oxidase inhibitors (Lu and Meites, 1971). We also have observed that

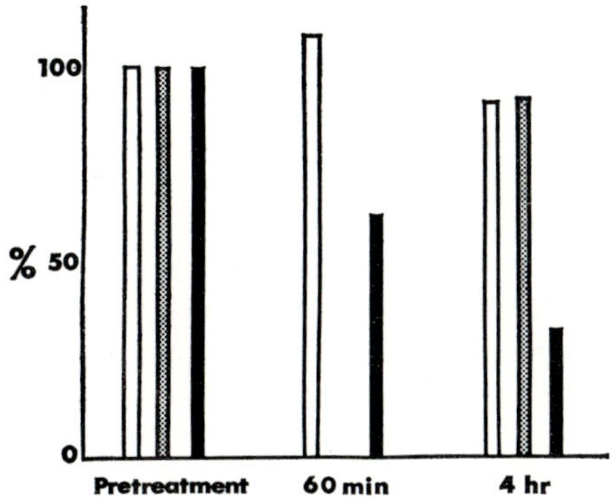

FIG. 16. Effects of a single injection of extract from 8 hypothalami (HE) (■) or equivalent amount of cerebral cortical tissue (□) from rats on serum prolactin levels. Only HE decreased serum prolactin levels. ▨, Saline. After Amenomori *et al.* (1970).

l-DOPA increased release of PIF into the systemic blood of hypophysectomized and intact rats (Lu and Meites, unpublished), and Kamberi *et al.* (1970a,b) reported that a single injection of dopamine into the third ventricle of rats elevated PIF activity in the hypothalamopituitary portal blood. PIF activity is readily detected in the general circulation of hypophysectomized but not in intact rats (Meites *et al.*, 1963a; Meites, 1970a; Lu and Meites, unpublished).

B. Hypothalamic Stimulation

The possibility that the hypothalamus may contain a factor that stimulates prolactin release was considered early. After the demonstration by Bargmann and Scharrer (1951) that the posterior pituitary hormones are produced in hypothalamic nuclei, Benson and Folley (1956) postulated that oxytocin may be responsible for the release of prolactin. This was based on their observation that injections of oxytocin, like prolactin, could retard mammary involution in postpartum lactating rats after litter removal. Prolactin secretion was not actually measured. Subsequent reports by a number of laboratories failed to substantiate the view that oxytocin could release prolactin (see Meites *et al.*, 1963a). More recently, Greenwood (1971) observed that oxytocin injections failed to increase serum prolactin levels significantly in sheep. Although both prolactin and oxytocin can inhibit mammary involution in postpartum lactating rats, their effects on the mammary gland differ markedly and they each appear to act by different mechanisms (see Meites *et al.*, 1963a; Meites, 1966). It appears therefore, that oxytocin is not a significant releasor of prolactin.

Our laboratory attempted to demonstrate prolactin-releasing activity in the rat hypothalamus by injecting small doses of a crude extract into estrogen-primed rats (Meites *et al.*, 1960). Lactation was initiated in these rats although a few animals also responded positively to injections of cerebral cortical extract. Injections of extracts of kidney, liver, or physiological saline had no effect. It was concluded that the hypothalamic extract "induced discharge of prolactin and probably ACTH from the anterior pituitary." Although this was not pursued subsequently, Mishkinsky *et al.* (1968) confirmed our observations, and concluded that this constituted evidence in favor of the existence of a prolactin releasing factor (PRF) in the hypothalamus.

Other studies also suggested that presence of a PRF in the mammalian hypothalamus. Nicoll *et al.* (1970) observed both prolactin inhibiting and prolactin stimulating activity in rat hypothalamic extract when incubated with rat pituitary. For the first 4 hours of incubation prolactin release was inhibited, but for the following 4 hours prolactin release was increased. Our laboratory, using a somewhat different system of incubation, observed that a rat hypothalamic extract only inhibited prolactin release during 8 hours of incubation (Meites, 1970a). Krulich *et al.* (1971) sectioned freshly frozen hypothalamus from normal male rats, and found PIF activity predominantly in the dorsolateral part of the preoptic area, and PRF activity mainly in the median eminence and a narrow basal portion of the preoptic area. However, numerous observations indicate that lesions placed in the median eminence result in enhanced prolactin release (see Section VI, A) indicating that this area mainly contains PIF activity. If PRF activity were predominant here, a fall in serum prolactin might be expected after placing lesions in the median eminence. Lesions in the preoptic area have been reported to influence LH but not prolactin release (Everett and Quinn, 1966; Kordon, 1966), suggesting that this area is not involved in prolactin release.

Valverde and Chieffo (1971) reported finding only PRF activity in an extract of porcine hypothalamus, as measured by its ability to increase serum prolactin levels in male rats pretreated with estrogen and progesterone. However, previous work indicated that porcine hypothalamic extract inhibited prolactin release under both *in vitro* and *in vivo* conditions (Schally *et al.*, 1967). The reason for these contradictory observations is not clear, although the methods used for extraction and assay of the extracts were not the same.

In contrast to mammals, the avian hypothalamus appears to exert a stimulatory rather than an inhibitory influence on prolactin secretion and apparently contains a prolactin-releasing factor (PRF). Early studies indicated that the pigeon pituitary failed to respond to the same stimuli as the mammalian pituitary with prolactin release. Thus pigeons showed no increase in pituitary prolactin levels after injections of estrogen, testosterone, or progesterone (Meites and Turner, 1947), and failed to respond to reserpine (Meites and Turner, unpublished). Pituitaries removed from many mammalian species and cultured *in vitro* continued to release prolactin at a good rate but pigeon pituitary failed to do so (Nicoll and Meites, 1962b). Subsequently it was demonstrated that an extract of pigeon hypothalamus stimulates prolactin release by the pigeon pituitary *in vitro* (Kragt and Meites, 1965). Hypothalamic extract from the chicken and quail (Meites, 1967b), tricolored blackbird (Nicoll, 1965), duck (Gourdji and Tixier-Vidal, 1966), and turkey (Chen *et al.*, 1968a) also induced release of prolactin when incubated with pituitaries from these species. Thus the predominant influence of the avian hypothalamus on prolactin release is stimulatory. The explanation for this interesting difference in control of prolactin release by the mammalian and avian hypothalamus is not yet available.

C. Role of Biogenic Amines

Reports over many years have suggested that biogenic amines may influence release of pituitary hormones (see Harris, 1955). The possible role of biogenic amines took on greater significance when methods were developed by Swedish workers for measuring them in the brain (see review by Hillarp *et al.*, 1966a), and it was demonstrated that norepinephrine and serotonin were highly concentrated in the hypothalamus (Vogt, 1954; Brodie *et al.*, 1959). These amines were shown to be particularly concentrated in nerve terminals, and it was suggested that they serve as neurotransmitters. Since the median eminence is the "final common pathway" between the brain and pituitary, and contains the peptidergic nerve terminals that release the hypophysiotropic hormones (releasing factors) into the adjacent hypophysial portal blood vessels and travel to the anterior pituitary to regulate its function, it was of considerable interest that the median eminence was observed to be rich in dopaminergic nerve terminals (Fuxe, 1964; Carlsson *et al.*, 1965). From this and related work the concept evolved that the catecholamines in the hypothalamus, and perhaps mainly dopamine in the median eminence, act as neurotransmitters to control the release of the hypophysiotropic hormone (Coppola, 1968; Fuxe and Hokfelt, 1969; Fuxe *et al.*, 1970; Wurtman, 1970).

Early work by Markee *et al.* (1946) and Sawyer *et al.* (1949) indicated that adrenergic drugs produced ovulation and therefore presumably elicited LH release in rats and rabbits, whereas antiadrenergic drugs blocked these effects. This has been substantiated and extended in more recent investigations by Kamberi *et al.* (1970a,b) and Schneider and McCann (1970a,b) who observed that catecholamines, particularly dopamine, induced release of LH and FSH by increasing hypothalamic LRF and FRF activities.

In our laboratory, we observed that injections of epinephrine and norepinephrine induced lactation in estrogen-primed rats and rabbits, but lack of specificity was indicated when we found that antiadrenergic and nonspecific agents were also effective in initiating lactation in these animals (Meites, 1958, 1962). The effective drugs included serotonin, acetylcholine, atropine, pilocarpine, amphetamine, morphine, reserpine, chlorpromazine, and meprobamate. Since lactation is not a specific indicator of prolactin release (intact rats and rabbits with developed mammary glands can lactate in response to administration of ACTH or adrenal cortical steroids alone), these early experiments did not prove that biogenic amines could induce prolactin release. By use of a specific radioimmunoassay for prolactin in rats (Niswender *et al.*, 1969), we found that systemic injection of a single dose of epinephrine, norepinephrine, or dopamine had no effect on serum prolactin levels, although epinephrine and dopamine produced a small but significant fall in pituitary prolactin concentration (Lu *et al.*, 1970). Inasmuch as these drugs do not readily pass through the blood-brain barrier (Innes and Nickerson, 1970), no definite conclusions could be made as to their role in prolactin release.

In a more recent study, we employed drugs that readily pass through the blood-brain barrier, and can either increase or decrease hypothalamic catecholamine activity (Lu and Meites, 1971). A single intraperitoneal injection of drugs that increase hypothalamic catecholamines, including *l*-DOPA (the immediate precursor of dopamine), or monoamine oxidase inhibitors (pargyline, iproniazid, or Lilly compound 15641) which inhibit the catabolism of catecholamines, all significantly depressed serum prolactin values in rats (Fig. 17). On the other hand, a single injection of drugs that depress hypothalamic catecholamine activity, including reserpine, chlorpromazine, α-methyl-*p*-tyrosine, α-methyl-*m*-tyrosine, *m*-DOPA and *d*-amphetamine, all greatly elevated serum prolactin values (Fig. 18). Haloperidol, another neuroleptic drug that lowers hypo-

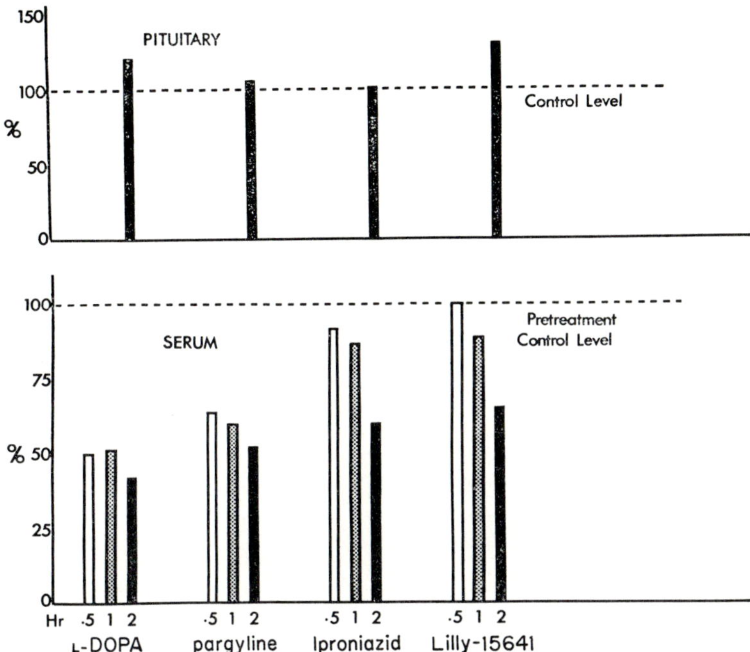

FIG. 17. Depression of serum prolactin levels by drugs that increase hypothalamic catecholamines. Two hours after a single injection of each drug, the rats were killed and their pituitaries were removed and assayed for prolactin. Only *l*-DOPA and Lilly compound 15641 produced a small but significant increase in pituitary prolactin stores, presumably reflecting the reduced prolactin release. All drugs increased hypothalamic activity. After Lu and Meites (1971).

thalamic catecholamine activity, also markedly increased serum prolactin concentration in rats (S. Dickerman and Meites, unpublished).

It is significant that all 4 drugs that increased hypothalamic catecholamine activity (*l*-DOPA and the 3 monoamine oxidase inhibitors) also increased hypothalamic PIF activity (Lu and Meites, 1971). Moreover, *l*-DOPA also elevated PIF activity in the serum of hypophysectomized and intact rats (Lu and Meites, unpublished data). In a related study, Kamberi *et al.* (1970b,c) reported that a single injection of dopamine into the third ventricle of rats depressed serum prolactin concentration and increased PIF activity in the hypophysial portal vessels. Thus *l*-DOPA and dopamine increase both synthesis and release of PIF by the hypothalamus. On the other hand, reserpine, chlorpromazine, and haloperidol significantly reduced hypothalamic PIF activity, and in the case of chlorpromazine, also decreased serum PIF activity. These related studies

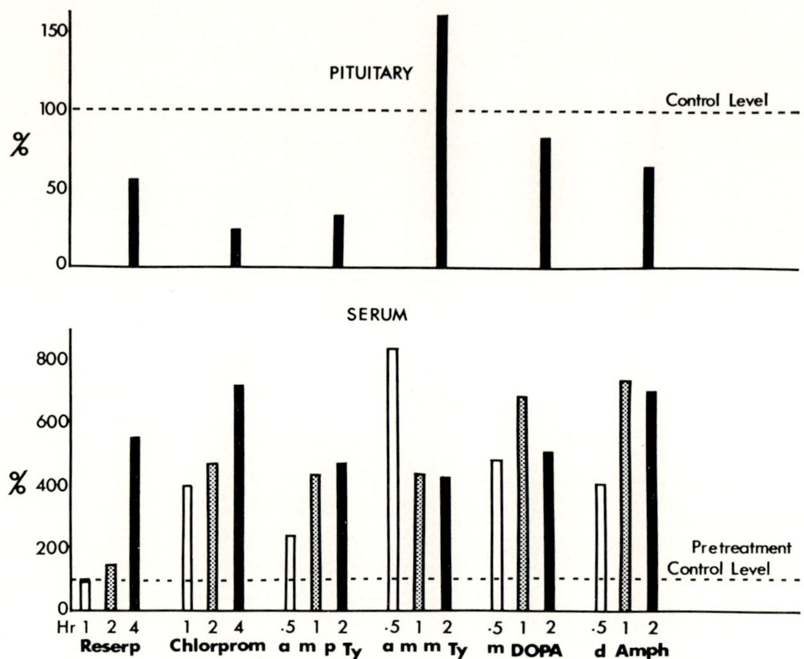

FIG. 18. Elevation of serum prolactin levels by a single injection of drugs that depress hypothalamic catecholamines and decrease PIF activity. All drugs with the exception of α mmTy significantly decreased pituitary prolactin stores, reflecting the increased release of this hormone. Presumably α mmTy produced a rapid increase in synthesis as well as release of prolactin. Reserp = reserpine, chlorprom = chlorpromazine, α mpTy = α-methyl-*p*-tyrosine, α mmTy = α-methyl-*m*-tyrosine, d amph = *d*-amphetamine. After Lu *et al.* (1970) and Lu and Meites (1971).

indicate that hypothalamic catecholamines act as neurotransmitters to increase release of PIF, which in turn depresses release of prolactin from the pituitary. Stimuli that reduce hypothalamic catecholamines produce a decrease in PIF activity, and a resultant increase in prolactin release. It remains to be determined whether all catecholamines or principally dopamine are involved in control of prolactin secretion.

The role of other biogenic amines in the hypothalamus on prolactin secretion is also under investigation. Kamberi *et al.* (1971) reported that an injection of serotonin or melatonin into the third ventricle of rats elevated the release of prolactin into the plasma. We have recently confirmed this observation and noted that intravenous injection of several precursors of serotonin (tryptophane or 5-hydroxytryptophane significantly raised serum prolactin levels in rats (Lu and Meites, unpublished). The role of these biogenic amines has not yet been determined, but they may be involved in the diurnal variation we have observed in serum prolactin in rats. Rats show higher prolactin values on the late afternoon of each day than in the morning (Koch *et al.*, 1971); rats similarly show higher pituitary levels of prolactin on the afternoon than in the morning (Clark and Baker, 1964). These findings suggest that there is a dual control in the hypothalamus over prolactin secretion, with a predominant adrenergic tonus acting to depress release of prolactin under most conditions, and a serotonergic system that may be responsible for the diurnal rise on the late afternoon and perhaps under other conditions in which serotonin is increased.

D. FEEDBACK OF PROLACTIN ON THE HYPOTHALAMUS

Many years ago we postulated that prolactin may act to inhibit its own secretion by the pituitary, since there was no target gland or tissue that produced hormones to inhibit prolactin secretion (Sgouris and Meites, 1953). This hypothesis has been verified by many recent studies, although its physiological significance is not yet clear. Transplants of pituitary "mammosomatotropic" tumors into intact rats were reported to decrease pituitary prolactin content (MacLeod *et al.*, 1966; Chen *et al.*, 1967). In addition the latter workers found increased hypothalamic PIF activity.

The site of the inhibitory action of prolactin on its own secretion appears to be in the hypothalamus, and there is no definite proof that prolactin acts directly on the pituitary to inhibit its secretion (Clemens *et al.*, 1969; Nicoll, 1971). Implantation of a minute dose of prolactin into the median eminence of rats, an amount too small to exert any systemic effect, decreased pituitary and serum prolactin levels, increased hypothalamic PIF activity, inhibited mammary growth and lactation, and resulted in termination of pseudopregnancy or early pregnancy (Clemens and Meites, 1968; Chen *et al.*, 1968b; Welsch *et al.*, 1968; Voogt *et al.*, 1970). Subsequently, Fuxe and Hokfelt (1970) demonstrated that prolactin injections into rats markedly activated the tuberoinfundibular dopaminergic neurons. It appears therefore that prolactin acts via the hypothalamus to increase dopamine activity, resulting in increased release of PIF into the portal vessels and inhibition of prolactin release.

One of the most interesting aspects of the above studies has been the repeated observation that prolactin not cnly can shut off its own secretion, but can stimulate release of FSH and LH. An implant of prolactin in the median eminence during pseudopregnancy, early pregnancy, or postpartum lactation in rats, increased FSH and LH release and resulted in stimulation of follicular growth, ovulation, and resumption of cycling (Clemens and Meites, 1968; Clemens *et al.*, 1968; Voogt and Meites, 1971). Serum LH and FSH were elevated when prolactin was implanted in the median eminence during pseudopregnancy (Voogt and Meites, 1971). Systemic injection or implantation of prolactin into the median eminence of 21-day-old female rats hastened the onset of puberty by an average of 6.6 days (Clemens *et al.*, 1969), and resulted in increased pituitary FSH levels and a probable rise in LH release (Voogt *et al.*, 1969a). The increase in hypothalamic dopamine activity observed as a result of prolactin administration by Fuxe and Hokfelt (1970), could well account for the reduction in prolactin release and the increase in FSH and LH release. Thus, in addition to its direct luteotropic or luteolytic effects on the ovaries of some species, prolactin can also influence the secretion of FSH and LH by the pituitary. These observations deserve further study.

VIII. Direct Effects of Hormones and Drugs on Pituitary Prolactin Secretion

Not all agents necessarily act through the hypothalamus to regulate prolactin secretion. Evidence that estradiol can directly stimulate rat pituitary prolactin release in a 3-day organ culture system was reported by Nicoll and Meites (1962a) and confirmed subsequently (Nicoll and Meites, 1964; Lu *et al.*, 1971). Since estrogen injections *in vivo* can depress hypothalamic PIF activity in the rat (Ratner and Meites, 1964), this indicates that estrogen acts to promote prolactin release both via the hypothalamus and by a direct action on the anterior pituitary. The direct action of estrogen on the pituitary may explain why serum prolactin, LH and FSH all reach peak values on the late afternoon of proestrus in rats. Estrogen administration has been reported to reduce LRF, FRF, and PIF activities in the hypothalamus (see Meites, 1970a), and if this were its only mode of action, it would depress release of FSH and LH but elevate release of prolactin. However, estrogen has been found to directly stimulate pituitary LH release (Piacsek and Meites, 1966), and it also may directly promote pituitary FSH release.

It long has been known that the thyroid can influence prolactin secretion (see Meites, 1960, 1966). In general, hypothyroidism results in reduced prolactin secretion, and thyroid administration can promote prolactin secretion. Incorporation of small amounts of thyroxine and triiodothyronine into a culture system were found to directly increase prolactin release by the rat pituitary (Nicoll and Meites, 1963). Subsequently it was reported that *in vivo* injection of thyroxine does not alter hypothalamic PIF activity in the rat (Chen and Meites, 1969). This suggests that thyroxine promotes prolactin release only by a direct

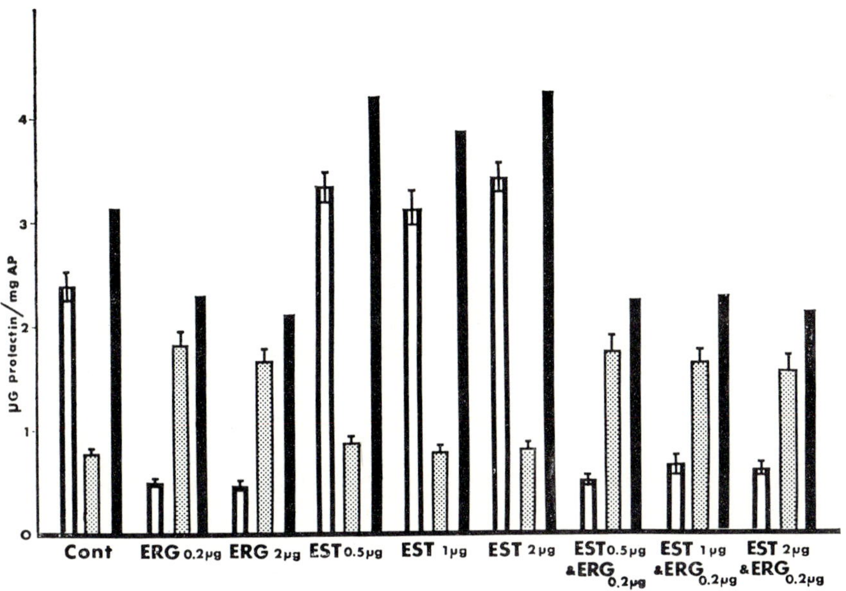

FIG. 19. Effects of ergocornine (ERG), estradiol (EST), or both on release of prolactin *in vitro* by male rat (AP) halves incubated for 12 hours. ERG decreased release and increased pituitary accumulation of prolactin. ERG also prevented increased release of prolactin by EST. □, Medium; ▨, pituitary; ■, total. From Lu *et al.* (1971).

action on the pituitary. Thyroxine has been found to directly inhibit pituitary TSH release (Guillemin *et al.*, 1963) and also may directly stimulate pituitary GH release (Deuben and Meites, unpublished).

Several drugs have been found to directly inhibit or stimulate pituitary prolactin release. Mention has already been made of the ability of ergot drugs to depress prolactin release. Figure 19 shows that *in vitro* incubation of rat pituitary with ergocornine diminishes prolactin release, resulting in an increase in pituitary prolactin stores. It also can be seen that ergocornine prevents estradiol from increasing prolactin release. When ergocornine was injected together with estrogen into rats, it prevented estrogen from producing enlargement of the pituitary (Lu *et al.*, 1971). Incorporation of sodium pentobarbital into an incubation medium inhibited prolactin release by the rat pituitary *in vitro* (Wuttke *et al.*, 1971), apparently accounting for its ability to depress serum prolactin when given *in vivo* (Wuttke and Meites, 1970). On the other hand, when pentobarbital was implanted into the median eminence, it reduced hypothalamic PIF activity and elevated serum prolactin values (Wuttke *et al.*, 1971).

Catecholamines have been found to alter pituitary prolactin release *in vitro*. MacLeod (1969) and Birge *et al.* (1970) reported that norepinephrine, epinephrine, and dopamine inhibited prolactin release by the incubated rat pituitary, and suggested that these might account for hypothalamic PIF activity. Later it was demonstrated that much smaller doses of catecholamines than those used by these workers could stimulate release of prolactin *in vitro* (Koch *et al.*, 1970). Catecholamines were not detected in the pituitary portal circulation (see Wurtman, 1970), and had no effect on prolactin release when given via infusion into a single hypophysial portal vessel (Kamberi *et al.*, 1970d). It is probable, therefore, that the direct effects of catecholamines on pituitary prolactin release observed *in vitro* are mainly of pharmacological significance.

Tashjian *et al.* (1971) reported the interesting observation that addition of thyrotropin-releasing hormone (TRH) to the medium of 2 clonal strains of cells from rat pituitary tumors, stimulated release of prolactin and inhibited release of GH. In our laboratory, several doses of synthetic TRH were added to incubation medium containing anterior pituitary halves from normal male rats and rat pituitary tumor tissue. No effects on prolactin release were observed. A single injection of TRH also failed to raise serum prolactin levels in rats (Quadri and Lu, unpublished). Further work is required to evaluate the significance of these observations.

IX. Summary

The principal functions of prolactin in female rats are to promote mammary development and lactation, to induce luteolysis of the previous crop of corpora lutea during each cycle and to help maintain functional corpora lutea together with LH during pseudopregnancy and early pregnancy. Prolactin also has an important role in development and growth of benign and malignant mammary tumors in rats. The principal stimulator of prolactin secretion in the female rat is estrogen, and apparently this hormone is responsible for the rise in prolactin release at the onset of puberty, for the proestrous surge of prolactin on the afternoon of proestrus during each estrous cycle, and for the increase in prolactin at the time of parturition. The suckling stimulus is the major promotor of prolactin release during postpartum lactation. The function(s) of prolactin in the male rat is unknown.

Prolactin secretion is regulated mainly by the hypothalamus which exerts a predominantly inhibitory influence under most conditions. Disruption of hypothalamic connections to the anterior pituitary by placement of appropriate lesions in the median eminence area results in rapid and prolonged elevations of serum prolactin levels. The presence of a prolactin-inhibiting factor (PIF) has been demonstrated in hypothalamic extracts from many mammalian species. Many stimuli that increase prolactin release, including estrogen, reserpine, and other tranquilizers, suckling and stressful agents, act by depressing hypothalamic PIF activity. On the other hand, agents that reduce prolactin release, including prolactin itself, l-DOPA, monoamine oxidase inhibitors and ergot drugs act by increasing hypothalamic PIF activity. Serotonin and its precursors also can increase prolactin release, but their mode of action presently is unknown. A prolactin-releasing factor (PRF) in the mammalian hypothalamus has not been definitely demonstrated, although such a factor has been found in the hypothalamus of avian species.

Hypothalamic catecholamines have an important role in regulating prolactin secretion and apparently serve as neurotransmitters to stimulate or inhibit release of PIF from nerve endings in the median eminence. Administration of agents that increase hypothalamic catecholamines result in increased hypothalamic PIF activity and in a reduction in prolactin release, whereas agents that lower hypothalamic catecholamines produce a decrease in hypothalamic PIF activity and a rise in prolactin release. High blood levels of prolactin depress prolactin secretion by the *in situ* pituitary by increasing hypothalamic PIF activity, but whether this represents a functional mechanism for controlling prolactin secretion remains to be established. In addition to hypothalamic regulation of prolactin secretion, several agents including estrogen, thyroxine, triiodothyronine, ergocornine, pentobarbital, and several catecholamines, have been reported to act directly on the anterior pituitary to increase or depress prolactin release.

REFERENCES

Amenomori, Y., Chen, C. L., and Meites, J. (1970). *Endocrinology* **86**, 506.
Armstrong, D. T., and Greep, R. O. (1962). *Endocrinology* **70**, 701.
Aschheim, P. (1961). *C. R. Acad. Sci.* **253**, 1968.
Baldwin, R. L. (1969). *In* "Reproduction in Domestic Animals" (H. H. Cole and P. T. Cupps, eds.), 2nd Ed., pp. 441–472. Academic Press, New York.
Bargmann, W., and Scharrer, E. (1951). *Amer. Sci.* **39**, 255.
Benson, G. K., and Folley, S. J. (1956). *Nature (London)* **117**, 700.

Bern, H. A., and Nicoll, C. S. (1968). *Recent Progr. Horm. Res.* **24**, 681.

Billeter, E., and Fluckiger, E. (1971). *Experientia* **27**, 464.

Birge, C. A., Jacobs, L. S., Hammer, C. T., and Daughaday W. H. (1970). *Endocrinology* **86**, 120.

Boot, L. M. (1969). "Induction by Prolactin of Mammary Tumors in Mice." North-Holland Publ., Amsterdam.

Brazeau, P. (1970). *In* "The Pharmacological Basis of Therapeutics" (L. S. Goodman and A. Gilman, eds.), 4th Ed., pp. 897–905. Macmillan, New York.

Brodie, B. B., Spector, S., and Shore, P. A. (1959). *Pharmacol. Rev.* **11**, 548.

Bruni, J. E., and Montemurro, D. G. (1971). *Cancer Res.* **31**, 854.

Bryant, G. D., and Greenwood, F. C. (1968). *Biochem. J.* **109**, 831.

Carlsson, A., Dahlstrom, A., Fuxe, K., and Hillarp, N. A. (1965). *Acta Pharmacol. Toxicol.* **22**, 270.

Cassell, E. E., Meites, J., and Welsch, C. W. (1971). *Cancer Res.* **31**, 1051.

Chadwick, A., and Folley, S. J. (1962). *J. Endocrinol.* **24**, 11.

Chen, C. L., and Meites, J. (1969). *Proc. Soc. Exp. Biol. Med.* **131**, 576.

Chen, C. L., and Meites, J. (1970). *Endocrinology* **86**, 503.

Chen, C. L., Minaguchi, H., and Meites, J. (1967). *Proc. Soc. Exp. Biol. Med.* **126**, 317.

Chen, C. L., Bixler, E. J., Weber, A. I., and Meites, J. (1968a). *Gen. Comp. Endocrinol.* **11**, 489.

Chen, C. L., Voogt, J. L., and Meites, J. (1968b). *Endocrinology* **83**, 1273.

Chen, C. L., Amenomori, Y., Lu, K. H., Voogt, J. L., and Meites, J. (1970). *Neuroendocrinology* **6**, 220.

Clark, R. H., and Baker, B. L. (1964). *Science* **143**, 375.

Clemens, J. A., and Meites, J. (1968). *Endocrinology* **82**, 878.

Clemens, J. A., and Meites, J. (1971). *Neuroendocrinology* **7**, 249.

Clemens, J. A., Welsch, C. W., and Meites, J. (1968). *Proc. Soc. Exp. Biol. Med.* **127**, 969.

Clemens, J. A., Sar, M., and Meites, J. (1969). *Endocrinology* **84**, 868.

Clemens, J. A., Amenomori, Y., Jenkins, T., and Meites, J. (1970). *Proc. Soc. Exp. Biol. Med.* **132**, 561.

Clifton, K. H., and Furth, J. (1960). *Endocrinology* **66**, 893.

Coppola, J. A. (1968). *J. Reprod. Fertil. Suppl.* **4**, 35.

Cowie, A. T., and Watson, S. C. (1966). *J. Endocrinol.* **35**, 213.

Danon, A., Dikstein, S., and Sulman, F. G. (1963). *Proc. Soc. Exp. Biol. Med.* **114**, 366.

Dao, T. L. (1962). *Cancer Res.* **22**, 973.

DeVoe, W. F., Ramirez, V. D., and McCann, S. M. (1966). *Endocrinology* **78**, 158.

Dickerman, E., and Meites, J. (1971). *Excerpta Med. Found. Int. Congr. Ser.* **236**, 11.

Dilley, W. G., and Nandi, S. (1968). *Science* **161**, 59.

Dorfman, R. I. (1965). *Methods Horm. Res.* **4**, 165.

Everett, J. W. (1961). *In* "Sex and Internal Secretions" (W. C. Young, ed.), Vol. 1, pp. 497–555. Williams & Wilkins, Baltimore, Maryland.

Everett, J. W., and Quinn, D. L. (1966). *Endocrinology* **78**, 141.

Frantz, A. G., and Kleinberg, D. L. (1970). *Science* **170**, 745.

Fredrikson, H. (1939). *Acta Obstet. Gynecol. Scand. Suppl.* **1**, 1–167.

Fuxe, K. (1964). *Z. Zellforsch. Mikrosk. Anat.* **61**, 710.

Fuxe, K., and Hokfelt, T. (1969). *In* "Frontiers in Neuroendocrinology, 1969" (W. F. Ganong, and L. Martini, eds.), pp. 47–96. Oxford Univ. Press, London and New York.

Fuxe, K., and Hokfelt, T. (1970). *In* "The Hypothalamus" (L. Martini, M. Motta, and F. Fraschini, eds.), pp. 123–138. Academic Press, New York.

Fuxe, K., Hokfelt T., and Jonsson, G. (1970). *In* "Neurochemical Aspects of Hypothalamic Function" (L. Martini, and J. Meites, eds.), pp. 61–83. Academic Press, New York.

Gay, V. L., Midgley, A. R., Jr., and Niswender, G. D. (1970). *Fed. Proc. Fed. Amer. Soc. Exp. Biol.* **29**, 1880.

Gourdji, D., and Tixier-Vidal, A. (1966). *C. R. Acad. Sci.* **263**, 162.

Greenwald, G. S., and Rothchild, I. (1968). *J. Anim. Sci.* **27**, Suppl., 139.

Greenwood, F. (1972). *Ciba Found. Symp.: Lactogenic Hormones,* pp. 197–206.

Grosvenor, C. E. (1965). *Endocrinology* **77**, 1037.

Grosvenor, C. E., and Turner, C. W. (1958). *Endocrinology* **63**, 535.

Grosvenor, C. E., McCann, S. M., and Nallar, M. D. (1964). *Proc. 46th Meet., Endocrine Soc., San Francisco* p. 96.

Grosvenor, C. E., Krulich, L., and McCann, S. M. (1968). *Endocrinology* **82**, 617.

Grosz, H. J., and Rothballer, A. B. (1961). *Nature (London)* **190**, 349.

Guillemin, R., Yamazaki, E., Gard, D. A., Jutisz, M., and Sakiz, E. (1963). *Endocrinology* **73**, 564.

Halasz, B., Pupp, L., and Uhlarik, S. (1962). *J. Endocrinol.* **25**, 147.

Harris, G. W. (1955). "Neural Control of the Pituitary Gland." Arnold, London.

Haun, C. K., and Sawyer, C. H. (1960). *Endocrinology* **67**, 270.

Hayward, J. (1970). "Recent Results in Cancer Research," p. 69. Springer-Verlag, Berlin and New York.

Heuson, J. C., Gaver, W. V., and Legros, N. (1970). *Eur. J. Cancer* **6**, 353.

Hillarp, N. A., Fuxe, K., and Dahlstrom, A. (1966a). *Pharmacol. Rev.* **18**, 727.

Hillarp, N. A., Fuxe, K., and Dahlstrom, A. (1966b). *In* "Mechanisms of Release of Biogenic Amines" (U. S. von Euler, S. Rosell, and B. Uvnas, eds.), p. 31. Pergamon, London.

Huggins, C. (1965). *Cancer Res.* **25**, 1163.

Huggins, C., Briziarelli, G., and Sutton, H. (1959a). *J. Exp. Med.* **109**, 25.

Huggins, C., Grand, L. C., and Brillantes, F. P. (1959b). *Proc. Nat. Acad. Sci. U.S.* **45**, 1294.

Innes, I. R., and Nickerson, M. (1970). *In* "The Pharmacological Basis of Therapeutics" (L. S. Goodman, and A. Gilman, eds.), pp. 478–523. Macmillan, New York.

Jaffe, R. B., and Midgley, A. R., Jr. (1971). *Proc. 53rd Meet., Endocrine Soc., San Francisco,* XXX.

Kamberi, I. A., Mical, R. S., and Porter, J. C. (1970a). *Experientia* **26**, 1150.

Kamberi, I. A., Mical, R. S., and Porter, J. C. (1970b). *Endocrinology* **87**, 1.

Kamberi, I. A., Schneider, H. P. G., and McCann, S. M. (1970c). *Endocrinology* **86**, 278.

Kamberi, I. A., Mical, R. S., and Porter, J. C. (1970d). *Fed. Proc., Fed. Amer. Soc. Exp. Biol.* **29**, 378.

Kamberi, I. A., Mical, R. S., and Porter, J. C. (1971). *Endocrinology* **88**, 1288.

Karg, H., and Schams, D. (1970). *In* "Lactation" (I. R. Falconer, ed.), pp. 141–143. Butterworth, London.

Kilpatrick, R., Armstrong, D. T., and Greep, R. O. (1964). *Endocrinology* **74,** 453.

Koch, Y., Lu, K. H., and Meites, J. (1970). *Endocrinology* **87,** 673.

Koch, Y., Chow, Y. F., and Meites, J. (1971). *Endocrinology.* **89,** 1303.

Kordon, C. (1966). Thesis, Univ. of Paris, Paris.

Kragt, C. L., and Meites, J. (1965). *Endocrinology* **76,** 1169.

Krulich, L., Quijada, M., and Illner, P. (1971). *Proc. 53rd Meet., Endocrine Soc., San Francisco* p. A-83.

Kwa, H. G., and Verhofstad, F. (1967). *J. Endocrinol.* **39,** 455.

Leob, L., and Kirtz, M. M. (1939). *Amer. J. Cancer* **36,** 56.

Lu, K. H., and Meites, J. (1971). *Proc. Soc. Exp. Biol. Med.* **137,** 480.

Lu, K. H., Amenomori, Y., Chen, C. L., and Meites, J. (1970). *Endocrinology* **87,** 667.

Lu, K. H., Koch, Y., and Meites, J. (1971). *Endocrinology* **89,** 229.

Lyons, W. R., Li, C. H., and Johnson, R. E. (1958). *Recent Progr. Horm. Res.* **14,** 219.

MacLeod, R. M. (1969). *Endocrinology* **85,** 916.

MacLeod, R. M., Smith, M. C., and DeWitt, G. W. (1966). *Endocrinology* **79,** 1149.

Malven, P. V., and Sawyer, C. H. (1966). *Endocrinology* **79,** 268.

Mandl, A. M., and Shelton, M. (1959). *J. Endocrinol.* **12,** 444.

Markee, J. E., Sawyer, C. H., and Hollingshead, W. H. (1946). *Endocrinology* **38,** 345.

Meites, J. (1958). *Proc. Soc. Exp. Biol. Med.* **97,** 742.

Meites, J. (1960). *Proc. Soc. Exp. Biol. Med.* **103,** 208.

Meites, J. (1962). *Proc. Int. Pharmacol. Meet., 1st, Stockholm, 1961* **1,** 151–179.

Meites, J. (1966). *Neuroendocrinology. 1966–1967* **1,** 669.

Meites, J. (1967a). *Arch. Anat. Microsc. Morphol. Exp.* **56,** Suppl. Au. N. 3–4, 401.

Meites, J. (1967b). *Arch. Anat. Microsc. Morphol. Exp.* **56,** Suppl. Au. N. 3–4. 516.

Meites, J. (1970a). *In* "Hypophysiotropic Hormones of the Hypothalamus" (J. Meites, ed.), pp. 261–281. Williams & Wilkins, Baltimore, Maryland.

Meites, J. (1970b). *In* "Neurochemical Aspects of Hypothalamic Function" (L. Martini and J. Meites, eds.), pp. 1–13. Academic Press, New York.

Meites, J. (1972). *In* "Breast Cancer Workshop" (T. Dao, ed.). Univ. of Chicago Press, Chicago, Illinois. In press.

Meites, J., and Kragt, C. L. (1964). *Endocrinology* **75,** 565.

Meites, J., and Nicoll, C. S. (1966). *Annu. Rev. Physiol.* **28,** 57.

Meites, J., and Sgouris, J. T. (1954). *Endocrinology* **55,** 530.

Meites, J., and Turner, C. W. (1947). *Proc. Soc. Exp. Biol. Med.* **64,** 488.

Meites, J., and Turner, C. W. (1948a). *Mo. Agr. Exp. Sta. Res. Bull.* **415.**

Meites, J., and Turner, C. W. (1948b). *Mo. Agr. Exp. Sta. Res. Bull.* **416.**

Meites, J., Talwalker, P. K., and Nicoll, C. S. (1960). *Proc. Soc. Exp. Biol. Med.* **103,** 298.

Meites, J., Kahn, R. H., and Nicoll, C. S. (1961a). *Proc. Soc. Exp. Biol. Med.* **108,** 440.

Meites, J., Nicoll, C. S., Talwalker, P. K. (1961b). *Symp. Neuroendocrinol., Miami, Fla., 1961.*

Meites, J., Nicoll, C. S., and Talwalker, P. K. (1963a). *In* "Advances in Neuroendocrinology" (A. V. Nalbandov, ed.), pp. 238–277. Univ. of Illinois Press, Urbana, Illinois.

Meites, J., Hopkins, T. F., and Talwalker, P. K. (1963b). *Endocrinology* 73, 261.

Meites, J., Cassel, E., and Clark, J. (1971). *Proc. Soc. Exp. Biol. Med.* 137, 1225.

Minaguchi, H., and Meites, J. (1967a). *Endocrinology* 80, 603.

Minaguchi, H., and Meites, J. (1967b). *Endocrinology* 81, 826.

Minaguchi, H., Clemens, J. A., and Meites, J. (1968). *Endocrinology* 82, 555.

Mishkinsky, J., Khazen, K., and Sulman, F. G. (1968). *Endocrinology* 82, 611.

Muhlbock, O., and Boot, L. M. (1967). *Biochem. Pharmacol.* 16, 627.

Nagasawa, H., and Meites, J. (1970). *Proc. Soc. Exp. Biol. Med.* 135, 469.

Nandi, S., and Bern, H. A. (1961). *Gen. Comp. Endocrinol.* 1, 195.

Neill, J. D., Freeman, M. E., and Tillson, S. A. (1971). *Proc. 55th Annu. Meet., Fed. Amer. Soc. Exp. Biol., Chicago* p. 474.

Nicoll, C. S. (1965). *J. Exp. Zool.* 158, 203.

Nicoll, C. S. (1971). *In* "Frontiers in Neuroendocrinology, 1971" (L. Martini and W. F. Ganong, eds.), pp. 291–330. Oxford Univ. Press, London and New York.

Nicoll, C. S., and Meites, J. (1962a). *Endocrinology* 70, 272.

Nicoll, C. S., and Meites, J. (1962b). *Nature (London)* 195, 606.

Nicoll, C. S., and Meites, J. (1963). *Endocrinology* 72, 544.

Nicoll, C. S., and Meites, J. (1964). *Proc. Soc. Exp. Biol. Med.* 117, 579.

Nicoll, C. S., Talwalker, P. K., and Meites, J. (1960). *Amer. J. Physiol.* 198, 1103.

Nicoll, C. S., Fiorindo, R. P., McKennee, C. T., and Parsons, J. A. (1970). *In* "Hypophysiotropic Hormones of the Hypothalamus" (J. Meites, ed.), pp. 115–150. Williams & Wilkins, Baltimore, Maryland.

Niswender, G. D., Midgley, A. R., Jr., Monroe, S. E., and Reichert, L. E. (1968). *Proc. Soc. Exp. Biol. Med.* 128, 807.

Niswender, G. D., Chen, C. L., Midgley, A. R., Jr., Meites, J., and Ellis, S. (1969). *Proc. Soc. Exp. Biol. Med.* 130, 793.

Parlow, A. F., Daane, T. A., and Schally, A. V. (1969). *Proc. 51st Meet., Endocrine Soc.* p. 83.

Pasteels, J. L. (1961). *C. R. Acad. Sci.* 253, 2140.

Piacsek, B. E., and Meites, J. (1966). *Endocrinology* 79, 432.

Piacsek, B. E., and Meites, J. (1967). *Neuroendocrinology* 2, 129.

Quadri, S. K., and Meites, J. (1971). *Proc. Soc. Exp. Biol. Med.* 138, 999.

Ratner, A., and Meites, J. (1964). *Endocrinology* 75, 377.

Ratner, A., Talwalker, P. K., and Meites, J. (1965). *Endocrinology* 77, 315.

Raud, H. R., Kidely, C. A., and Odell, W. D. (1971). *Proc. Soc. Exp. Biol. Med.* 136, 689.

Reece, R. P. (1939). *Proc. Soc. Exp. Biol. Med.* 42, 54.

Reece, R. P., and Turner, C. W. (1937). *Mo. Agr. Exp. Sta. Res. Bull.* 266.

Rothchild, I. (1965). *Vitam. Horm. (New York)* 23, 209.

Sar, M., and Meites, J. (1967) *Proc. Soc. Exp. Biol. Med.* 125, 1018.

Sar, M., and Meites, J. (1969). *Neuroendocrinology* 4, 25.

Sawyer, C. H., Markee, J. E., and Townsend, B. F. (1949). *Endocrinology* 44, 18.

Schally, A. V., Meites, J., Bowers, C. Y., and Ratner, A. (1964). *Proc. Soc. Exp. Biol. Med.* **117,** 252.

Schally, A. V., Kastin, A. J., Locke, W., and Bowers, C. Y. (1967). *In* "Hormones in the Blood" (C. H. Gray and H. L. Bacharach, eds.), 2nd Ed., Ch. XVI, pp. 492–526. Academic Press, New York.

Schneider, H. P. G., and McCann, S. M. (1970a). *Endocrinology* **86,** 1127.

Schneider, H. P. G., and McCann, S. M. (1970b). *Endocrinology* **87,** 249.

Schwartz, N. B., and Waltz, P. (1970). *Fed. Proc. Fed. Amer. Soc. Exp. Biol.* **29,** 1907.

Sgouris, J. T., and Meites, J. (1953). *Amer. J. Physiol.* **175,** 319.

Shelesnyak, M. C. (1954). *Amer. J. Physiol.* **179,** 301.

Shelesnyak, M. C. (1958). *Acta Endocrinol. (Copenhagen)* **27,** 99.

Spies, H. G., and Niswender, G. D. (1971). *Endocrinology* **88,** 937.

Sterental, A., Dominquez, M., Weisman, C., and Pearson, O. H. (1963). *Cancer Res.* **23,** 481.

Talwalker, P. K., and Meites, J. (1961). *Proc. Soc. Exp. Biol. Med.* **107,** 880.

Talwalker, P. K., and Meites, J. (1964). *Proc. Soc. Exp. Biol. Med.* **117,** 121.

Talwalker, P. K., Nicoll, C. S., aid Meites, J. (1961). *Endocrinology* **69,** 802.

Talwalker, P. K., Ratner, A., and Meites, J. (1963). *Amer. J. Physiol.* **205,** 213.

Talwalker, P. K., Meites, J., and Mizuno, H. (1964). *Proc. Soc. Exp. Biol. Med.* **116,** 531.

Tashjian, A. H., Barowsky, N. J., and Jensen, D. K. (1971). *Biochem. Biophys. Res. Commun.* **43,** 516.

Tucker, H. A., and Meites, J. (1965). *J. Dairy Sci.* **48**(3), 403.

Turner, C. W. (1939). *In* "Sex and Internal Secretions" (E. Allen, ed.), 2nd Ed., pp. 740–803. Williams & Wilkins, Baltimore, Maryland.

Valverde, C., and Chieffo, V. (1971). *Proc. 53rd Meet., Endocrine Soc., San Francisco* p. 84.

Vogt, M. (1954). *J. Physiol. (London)* **123,** 451.

Voogt, J. L., and Meites, J. (1971). *Endocrinology* **88,** 286.

Voogt, J. L., Clemens, J. A., and Meites, J. (1969a). *Neuroendocrinology* **4,** 157.

Voogt, J. L., Sar, M., and Meites, J. (1969b). *Amer. J. Physiol.* **216,** 655.

Voogt, J. L., Chen, C. L., and Meites, J. (1970). .*Amer. J. Physiol.* **218,** 396.

Welsch, C. W., Clemens, J. A., and Meites, J. (1968). *J. Nat. Cancer Inst.* **41,** 465.

Welsch, C. W., and Meites, J. (1969). *Cancer* **23,** 601.

Welsch, C. W., and Meites, J. (1970). *Experientia* **26,** 1133.

Welsch, C. W., Clemens, J. A., and Meites, J. (1969). *Cancer Res.* **29,** 1541.

Welsch, C. W., Jenkins, T. W., and Meites, J. (1970). *Cancer Res.* **30,** 1024.

Welsch, C. W., Nagasawa, H., and Meites, J. (1971). *Cancer Res.* **30,** 2310.

Wurtman, R. J. (1970). *In* "Hypophysiotropic Hormones of the Hypothalmus" (J. Meites, ed.), pp. 184–194. Williams & Wilkins, Baltimore, Maryland.

Wuttke, W., and Meites, J. (1970). *Proc. Soc. Exp. Biol. Med.* **135,** 648.

Wuttke, W., and Meites, J. (1971). *Proc. Soc. Exp. Biol. Med.* **137,** 988.

Wuttke, W., Cassell, E., and Meites, J. (1971). *Endocrinology* **88,** 737.

Wuttke, W., Gelato, M., and Meites, J. (1971). *Endocrinology* **89,** 1191.

Wuttke, W., and Meites, J. (1972). *Endocrinology,* **90,** 438.

Yanai, R., and Nagasawa, H. (1970). *Experientia* **26,** 649.

Yoshinaga, K., Hawkins, R. A., and Stocker, J. F. (1969). *Endocrinology* **85,** 103.

DISCUSSION

R. B. Greenblatt: Dr. Meites, you perpetuate the myth that the postpartum lactating woman is protected against conception. In a large measure this may be true, but it is not necessarily so. In French Canada many a peasant woman has 16–18 children without experiencing a menstrual period between pregnancies while constantly nursing an offspring at her breast. In many ancient civilizations intercourse during lactation was taboo; the custom may have arisen from the need to prevent the added burden of another pregnancy while a woman was still lactating. The ancient Hebrews, probably cognizant of this possibility, condoned during lactation the practice of an oft condemned act, i.e., coitus interruptus. Rabbi Eliezer euphemistically stated that at this time "a man may thresh inside and winnow outside."

Now, as to the influence of estrogens on prolactin secretion, you pointed out that estrogens increase serum prolactin levels. This seems to be borne out by the fact that in efforts to arrest abnormal lactation in women harboring prolactin-producing pituitary tumors, physiological doses of estrogens are wholly ineffective; however, massive doses are effective, and once lactation has been arrested, the effect can be maintained by gradually decreasing the dose to physiological levels. What explanation can you offer?

J. Meites: How can large doses of estrogen inhibit lactation? We reported some years ago that large doses of estrogen or estrogen and progesterone administered to lactating rabbits and rats [J. Meites, and J. T. Sgouris, *Endocrinology* **53,** 17 (1953); **55,** 530 (1954)] suppress lactation as they do in humans. But if prolactin is given in sufficient doses it will overcome the inhibitory effects of the estrogens on lactation. We suggested therefore that large doses of estrogen were not interfering with prolactin secretion but with the peripheral action of prolactin on the mammary gland. Large doses of estrogens (and androgens) have a similar inhibitory effect on growth of mammary tumors in rats, and estrogens are used in treating human breast cancer. In a recent experiment in rats, we used large doses of estradiol benzoate (20 μg daily) to suppress growth of mammary cancers, but when we also injected ovine prolactin (1 mg daily with 28 IU/mg) we were able to overcome this suppression by the estrogen [J. Meites, E. Cassell, and J. Clark, *Proc. Soc. Exp. Biol. Med.* (1971) **137,** 1225. Since we have shown that large doses of estrogen do not inhibit pituitary prolactin secretion, I believe these experiments indicate that they inhibit the peripheral action of prolactin on the mammary tissues. It is possible that large doses of estrogen interfere with the binding sites for prolactin in the mammary gland.

There is abundant evidence that the suckling stimulus inhibits gonadotropin secretion. This was first reported by Allen Parks [*Proc. Roy. Soc. Ser.* **B100,** 151 (1926)]. He showed in mice that litter size determined the time at which the mothers resumed cycling in the postpartum lactation phase. Mice with smaller litters started cycling sooner after parturition than mice with larger litters. This has also been demonstrated in rats and cows [see J. Meites, *Neuroendocrinology 1966–1967* **1,** p. 681 (1966)]. The more times per day the cow is milked after parturition, the later is the time at which she resumes cycling. I believe there is some evidence for this in women as well [C. W. Lloyd, *In* "Textbook of Endocrinology" (R. H. Williams, ed.), p. 490. Saunders, Philadelphia, (1968).]

A. R. Fuchs: My question concerns the luteotropic action of prolactin on the corpus luteum in the rat. Dr. Meites, you showed that although small doses of ergocornine prevented the increase in serum prolactin levels that is normally seen following cervical stimulation, the length of pseudopregnancy was only slightly shortened by this treatment. On the other hand, in the earlier experiments of Shelesnyak it was shown that larger amounts of ergocornine did terminate pseudopregnancy in rats and that prolactin given to these ergocornine-treated rats prevented this interruption of pseudopregnancy. Have you studied the effects of larger doses of ergocornine on serum prolactin and LH levels? Do larger doses prevent prolactin secretion completely, as would seem likely on the basis of Shelesnyak's experiments, or do they also interfere with LH secretion?

J. Meites: We have not administered a larger dose of ergocorine than 50 μg per day in the pseudopregnancy experiment. We purposely chose this relatively small dose in order to interfere as little as possible with LH secretion, and you will have noted that serum LH was not lowered by this dose. On the other hand, we have reported that a large dose of ergocornine can reduce serum LH levels, even though it does not interfere with ovulation or cycling [W. Wuttke, E. Cassell, and J. Meites, *Endocrinology* **88**, 737 (1971)]. You are perfectly correct in stating that large doses of ergocornine can terminate pseudopregnancy or early pregnancy as reported by M. C. Shelesnyak [*Acta Endocrinol.* **27**, 99 (1958); *Amer. J. Physiol.* **179**, 301 (1954)]. Presumably this is due to an interference with both prolactin and LH secretion.

A. J. Kastin: My question, prompted by work with MSH, concerns the function of prolactin. Prolactin and MSH share the property of being primarily inhibited by the hypothalamus. Both are released by tranquilizers and possibly stress, and there is some question as to what the target endocrine organ is in each case, particularly in the male. Since there is increasing evidence for an extra pigmentary effect of MSH, I wonder what you consider to be the function of prolactin in males, and whether you have tested the effects of prolactin on tissues other than endocrine organs.

J. Meites: We do not yet know whether prolactin has any functions in males. We have reported that there are no cyclic fluctuations in serum prolactin in male rats [Y. Amenomori, C. L. Chen, and J. Meites, *Endocrinology* **86**, 506 (1970)]. Some investigators have reported that prolactin stimulates growth of the male reproductive tract and acts synergistically with androgens [see J. Meites and C. S. Nicoll, *Annu. Rev. Physiol.* **28**, 57 (1966)]. Insofar as effects of prolactin other than those on the mammary glands and ovaries are concerned, prolactin has been reported to stimulate body growth to some extent and to alter fat and carbohydrate metabolism. In submammalian species, H. A. Bern and C. S. Nicoll [*Recent Progr. Horm. Res.* **24**, 681 (1968)] have brought forth impressive evidence that prolactin influences a great many functions.

R. Guillemin: Relating to the recent paper by A. Tashjian on the effect of TRF on the secretion of prolactin by pituitary tumors maintained in tissue cultures, we have observed that TRF will stimulate the secretion of prolactin from dispersed pituitary cells placed in tissue cultures using cells from a normal pituitary. The effect of TRF on the release of prolactin is even more marked when one used, in the same method, cells from the pituitary glands of thyroidectomized rats. Moreover, and I do not know whether this means anything in terms of PIF activity, we can inhibit all these effects of TRF on the secretion of prolactin by the addition of T4 or T3. Furthermore, addition of T4 or T3 will also inhibit the "paralytic" or spontaneous secretion of prolactin that takes place when the pituitary cells are separated from the hypothalamus.

I would like to hear some clarification on your part about what appears to be a paradox. We all recognize that when pituitary tissues are separated from their hypothalamic connections by transplantation *in vivo* or *in vitro,* they start spontaneously to secrete very large doses of prolactin. In fact, one of the earliest observations to this effect was that of Everett and Nikitovitch-Winer showing persistence of *corpora lutea* in rats bearing pituitary grafts under the kidney capsule. You are now telling us that prolactin has luteolytic activity. How can this be reconciled?

J. Meites: After reading the very interesting report by Dr. Tashjian *et al.* [*Biochem. Biophys. Res. Commun.* **43,** 516 (1971)] on TRH and its stimulatory effect on prolactin release *in vitro,* we immediately tried to repeat this observation. However, instead of using incubations of dispersed cells from pituitary tumors, as Tashjian and you have tried to do, we incubated anterior pituitary halves from normal mature male rats for 4–12 hours with TRH and observed no significant effect on prolactin release. We used the same amounts of TRH reported by Tashjian. We also injected TRH *in vivo* into mature intact male rats and observed no effect on serum prolactin (K. H. Lu and J. Meites, unpublished). I believe you also informed me that you were unable to see any *in vivo* effect of TRH on serum prolactin in rats, and that you obtained some doubtful results when you incubated rat pituitary tissue with TRH.

I am amazed that you observed inhibition of prolactin release *in vitro* when you added thyroxine to a pituitary cell system. Dr. Charles Nicoll and I reported some years ago that addition of thyroxine or triiodothyronine to a culture system containing rat pituitary tissue, stimulated release of prolactin [C. S. Nicoll, and J. Meites, *Endocrinology* **72,** 544 (1963)]. There are many reports in the literature that the thyroid promotes prolactin secretion [see C. L. Chen, and J. Meites, *Proc. Soc. Exp. Biol. Med.* **131,** 576 (1969)]. Thyroidectomy results in a decrease in pituitary prolactin and GH and severely reduces the count of acidophils associated with these two hormones [A. E. Severinghaus, "Sex and Internal Secretions," 2nd Ed., p. 1045. The Williams and Wilkins Co., Baltimore (1939)]. Insofar as the *in vivo* effects of TRH on prolactin release on human subjects are concerned, the possibility has not been eliminated that this may work indirectly through stimulating thyroid function, with the increased thyroid hormones then acting on the pituitary to increase prolactin and possibly GH and ACTH release.

On luteolysis I believe P. V. Malven and C. H. Sawyer [*Endocrinology* **79,** 268 (1966)] established that when prolactin is injected into hypophysectomized rats with fresh corpora lutea, luteal function is maintained. On the other hand, when they injected prolactin several days after the corpora lutea were produced by gonadotropins in the hypophysectomized rats, the corpora lutea disappeared. Apparently the older corpora lutea cannot respond to prolactin by producing progesterone and instead undergo rapid atrophy.

D. B. Bartosik: I would like to respond to Dr. Guillemin's question by emphasizing a conceptual difference between histological luteolysis and physiological luteolysis. A functioning corpus luteum is one which is secreting significant amounts of progesterone. P. V. Malven [*In* "The Gonads" (K. W. McKerns, ed.), p. 367. Appleton-Century Crofts, New York] has shown that rat corpora lutea which persist as nonfunctioning histological units after hypophysectomy undergo histological luteolysis with prolactin administration. Analogously, M. Nikitovitch-Winer and J. W. Everett [*Endocrinology* **62,** 522 (1958)] have demonstrated that the corpora lutea which are histologically present in the rat ovary on the morning of proestrus cannot be induced to function. Such corpora are therefore candidates for the phenomenon of prolactin-induced histological luteolysis.

K. Yoshinaga: I am interested in the functional state of the corpora lutea of your ergocornine-treated pseudopregnant rats. Have you tried to produce deciduoma or measure progesterone levels in these rats?

J. Meites: We have done neither. We do not know whether the corpora lutea of pseudopregnant rats treated with ergocornine secrete progesterone, although it would be difficult to understand how pseudopregnancy can be maintained without progesterone. I am unaware of any definite evidence that the corpora lutea of the rat secrete progesterone during regular estrous cycles.

F. C. Greenwood: Dr. Guillemin said that 80% of an R_f was cleared by the kidney. It is surprising that one can pick up R_f's.

J. Meites: Our laboratory and others have reported the presence of hypophysiotropic factors in the blood of hypophysectomized rats [see J. Meites, *In* "'Hypophysiotropic Hormones of the Hypotholamus" (J. Meites, ed.), pp. 261–281. Williams and Wilkins, Baltimore (1970)], due to removal of pituitary restraint.

A. Bartke: I also wonder whether your work in which several injections of ergocornine were used can be considered as evidence against the role of prolactin in induction of pseudopregnancy. There is very strong evidence that pseudopregnancy or pregnancy does not occur in animals chronically treated wth ergocornine [R. A. Carlsen, G. H. Zeilmaker, and M. C. Shelesnyak, *J. Reprod. Fertil.* **2,** 369 (1961); H. Nagasawa and R. Yanai, *Endocrinol. Jap.* **17,** 233 (1970)] or in animals with hereditary prolactin deficiency [A. Bartke, *J. Reprod. Fertil.* **10,** 93 (1965)]. The role of prolactin in the control of corpus luteum function in rodents has been firmly established, and the mechanism of "luteotropic" action of prolactin in the rat is now well understood [D. T. Armstrong, L. S. Miller, and K. A. Knudsen, *Endocrinology* **85,** 393 (1969); I. Hashimoto and W. G. Wiest, *Endocrinology* **84,** 886 (1969); H. R. Behrman, G. P. Orczyk, G. J. Macdonald, and R. O. Greep, *Endocrinology* **87,** 1251 (1970)].

I believe I can provide a partial answer to Dr. Kastin's question. In hypophysectomized mice and rats treated with LH, prolactin will cause a further significant increase in the production of testosterone *in vivo* and *in vitro* [A. Bartke, *J. Endocrinol.* **49,** 311 (1971); A. A. Hafiez, C. W. Lloyd, and A. Bartke, *J. Endocrinol.* (1972) **52,** 327]. The details of this work were presented last year and this year at the meetings of the Endocrine Society and the Society for the Study of Reproduction.

I. A. Kamberi: I am glad to hear that you have confirmed our results on the effects of serotonin and melatonin on prolactin release. As we reported earlier, LH [I. A. Kamberi, R. S. Mical, and J. C. Porter, *Endocrinology* **87,** 1 (1970)] and FSH release [I. A. Kamberi, R. S. Mical, and J. C. Porter, *Endocrinology* **88,** 1288 (1971)] are suppressed after the intraventricular injection of melatonin and serotonin, whereas prolactin release is increased [I. A. Kamberi, R. S. Mical, and J. C. Porter, *Endocrinology* **88,** 1288 (1971)]. The effect of serotonin or melatonin cannot be seen when these substances are directly infused into the anterior pituitary via a cannulated pituitary stalk portal vessel, which suggests that the effects of these amines on the release of LH, FSH, and prolactin are secondary to the discharge of LH-releasing (LRF), FSH-releasing (FRF), and prolactin-inhibiting (PIF) factor from neurosecretory elements of the hypothalamus. From this result it is clear that the effects of melatonin and serotonin on release of these hormones are opposite to those which we reported for dopamine; namely, small doses of dopamine stimulated the release of LH [I. A. Kamberi, R. S. Mical, and J. C. Porter, *Endocrinology* **87**, 1 (1970)] and FSH [I. A. Kamberi, R. S. Mical, and J. C. Porter, *Endocrinology* **88**, 1003 (1971)], and inhibited prolactin release [I. A. Kamberi, R. S. Mical, and J. C. Porter, *Endocrinology* **88**, 1012 (1971)]. In contrast comparable doses of melatonin and serotonin inhibited the release of LH and FSH, and stimulated the release of prolactin [I. A. Kamberi, R. S. Mical, and J. C. Porter, *Endocrinology* **87**, 1 (1970); I. A. Kamberi, R. S. Mical, and J. C. Porter, *Endocrinology* **88**, 1288 (1970)]. The results of these findings suggest that two mechanisms, viz., a dopaminergic and a serotoninergic mechanism, exist in the brain which have counter effects on the release of LH, FSH, and prolactin. One system stimuates the release of LH and FSH and inhibits the release of prolactin, and the other decreases the release of LH and FSH and increases the release of prolactin.

Dr. Meites, to my knowledge, melatonin has not been found in the hypothalamus. Could you speculate about the mechanism of action of melatonin? How does melatonin exert this effect on prolactin release?

J. Meites: I do not know how melatonin works. It may inhibit PIF release.

D. L. Kleinberg: Have you considered the possibility that the circulating factor that inhibits prolactin release after the administration of *l*-DOPA might be *l*-DOPA itself, without implicating a separate PIF?

J. Meites: Dr. Daughaday and his colleagues [C. A. Birge, L. S. Jacobs, and C. T. Hammer, *Endocrinology* **86**, 120 (1970)] and also Dr. R. M. MacLeod [*Endocrinology* **85**, 916 (1969)] have reported that dopamine, epinephrine, and norepinephrine will inhibit rat pituitary prolactin release *in vitro* when added in small amounts to an incubation system. We repeated their experiments and confirmed their results. However, we also showed that very low doses of norepinephrine and epinephrine could stimulate prolactin release *in vitro* [Y. Koch, K. H. Lu, and J. Meites (1970). *Endocrinology* **87**, 673]. We concluded that catecholamines are not PIF's or PRF's acting directly on the pituitary, but that these are pharmacological effects of doubtful physiological significance. We cannot rule out completely a direct effect of catecholamines on the pituitary, but they have not been detected in the hypophysial portal system and Kamberi *et al.* (1970) observed no effect on prolactin release when dopamine was injected into a single portal vessel in the rat.

G. D. Niswender: You referred to the work of Dr. Spies in which he compared serum levels of gonadotropins in intact and pelvic neurectomized rats following cervical stimulation and found that all three gonadotropins were increased in both groups 20 minutes after mating. It does not seem valid to compare your data regarding FSH levels in mated vs. unmated rats to the data of Spies where he compared FSH levels in intact and pelvic neurectomized rats following cervical stimulation. Obviously, the experimental conditions are not similar. However, Dr. Linkie in our laboratory has shown that prolactin and LH are elevated for at least the first 20 hours after mating of normal rats but that FSH levels are not significantly increased. Following this initial elevation there is no evidence of increased levels of any of the gonadotropins throughout the remainder of gestation.

You commented that you found PIF in the serum of hypophysectomized animals. Dr. Vernon Gay has some unpublished data which clearly suggests that LRF is also present in the serum of hypophysectomized animals. I wonder if you would care to comment about what is unique about the hypophysectomized rat. Do you feel this is evidence for a "short-loop" feedback or is it possible that the pituitary is responsible for altering LRF in some way which prevents its accumulation in serum?

J. Meites: It has been well documented that the anterior pituitary hormones can inhibit their own secretion [see M. Motta *et al. In* "Frontiers in Neuroendocrinology" (W. F. Ganong and L. Martini, eds.), p. 211. Oxford Univ. Press, London and New York, 1969]. This is also true for prolactin. When you remove the pituitary gland, the inhibition on the hypothalamus is removed, permitting the releasing factors to increase to such an extent that they reach the systemic circulation. The disruption of the hypothalamohypophysial portal system also may contribute to this process. This has been shown for practically every hypophysiotropic factor in the hypothalamus including not only PIF, but also GRF, LRF, FRF, and CRF. We have not been able to demonstrate that the rat pituitary inactivates PIF.

C. Ezrin: Suckling produces a rapid rise in prolactin levels that inhibition of PIF seems too sluggish to explain. Is there a stimulating factor involved in the suckling-induced rise in prolactin?

J. Meites: We tested the hypothalami of suckled rats and found a virtual elimination of PIF. We did not find PRF activity. Drs. K. Fuxe and T. Hokfelt [*In* "Frontiers in Neuroendocrinology, 1969" (W. F. Ganong and L. Martini, eds.), pp. 47–96. Oxford Univ. Press, London and New York, 1969.] (1969) reported that suckling causes a marked decrease in catecholamine activity in the hypothalamus, and this is believed to be responsible for the reduction in PIF. Although we cannot definitely exclude the possible involvement of a PRF in suckling-induced release of prolactin, I would remind you of the evidence presented today showing that disruption of the hypothalamopituitary connections by median eminence lesions results in a rapid and sustained rise in serum prolactin levels.

C. Kordon: The antagonistic effects of catecholamines or serotonin precursors on prolactin levels which you have reported appear very interesting in view of the similar situation which exists in the case of LH release regulation. In the latter case, experimentally decreased levels of endogenous dopamine, as well as increased levels of endogenous serotonin, are equally able to block the release of LRF into the hypothalamohypophysial portal system; both regulations seem to occur in the same hypothalamic area (Fig. A) [C. Kordon, *Neuroendocrinology* **7**, 202 (1971)]. In most instances, moreover, an appropriate and simultaneous tuning of both aminergic systems seems to be necessary for proper LH secretion. Do you believe that this also applies to prolactin control?

J. Meites: Drs. Kordon and Glowinski have done some very fine work on the relation of biogenic amines to secretion of gonadotropins. Apparently they have demonstrated that serotonin inhibits LH release. I believe this is yet another illustration of the reciprocal state often found between prolactin and gonadotropin secretion. The catecholamines also have been shown to produce opposite effects on the secretion of these hormones; inhibiting prolactin release but stimulating the release of LH and FSH [I. A. Kamberi *et al. Endocrinology* **88**, 1003 (1971)].

H. G. Kwa: I want to stress the fact that radioimmunoassay of plasma prolactin provides accurate and reliable information as to the secretory activity of the pituitary only when *acute* changes in prolactin release occur; e.g., the proestrous prolactin surge, reaction to suckling, or to tranquilizing agents and 1-DOPA, as has been amply illustrated by Professor Meites in this presentation. Changes in secretory activity of longer duration, however, may not be reflected at all, or

only at the onset of the change. We suspect that this may be the case during
the greater part of pseudopregnancy and also may be the cause that no differences
in prolactin levels in the circulation of DMBA-rats have been found. We have
followed plasma prolactin in various groups of rats with and without isografts
either without estrone or with estrone administered daily for a period of over
400 days [*Eur. J. Cancer* (1971), in press]. Between various groups no consistent
differences in plasma prolactin levels were found by radioimmunoassay, whereas
electron microscopy studies by Dr. C. A. Feltkamp of our institute revealed that
striking differences in prolactin secretory activity existed between such groups of

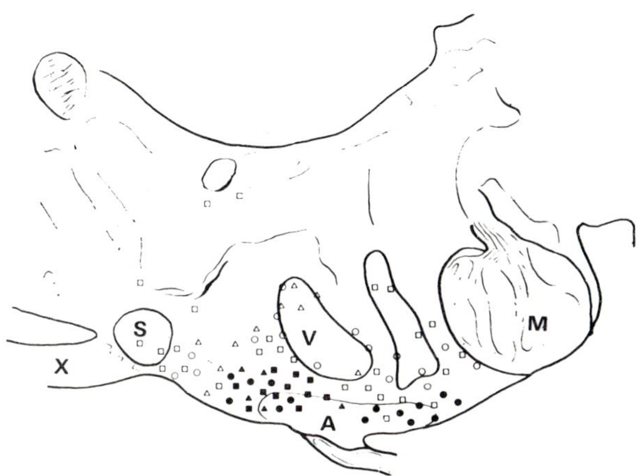

Fig. A.   Sagittal reconstruction of the medial hypothalamus of the rat showing
the effects of microinjected drugs on ovulation. Open symbols, no effect of microin-
jection; filled symbols, ovulation blocked by treatment. Inhibition of catecholamine
release by α-methyl DOPA (α-MD) is indicated by squares (80 ng α-MD) and
triangles (1 μg α-MD); circles concern Nialamid microinjections (75 μg) and the
effect of increased local serotonin levels (since the responses to local Nialamid,
when present, are reversed by previous inhibition of serotonin biosynthesis). Note
the similar distribution of effective sites in both cases. *A*, arcuate nucleus; *M*,
mammillary nucleus; *S*, suprachiasmatic nucleus; *V*, ventromedial nucleus; *X*, optic
chiasm. After C. Kordon, *Neuroendocrinology* **4**, 129 (1969); **7**, 202 (1971).

rats as judged by the presence of Nebenkernen, the development of the endoplasmic
reticulum, etc. This suggested that in rats of such groups the larger amounts
of prolactin secreted by the pituitary (first compartment) into the circulation
(second compartment) must have been compensated by a proportionally large
amount of prolactin leaving this second compartment to enter a third compartment.
At the present stage, however, we cannot as yet commit ourselves as to the nature
of this third compartment; this may either be the capacity of the degradatory
mechanism specific for prolactin, or consumption of the hormone by its target
tissues, or attachment of the hormone to a carrier protein that shields off the
immunoreactive sites.

By taking advantage of the fact that the rat does not discriminate between
endogenous and bovine prolactin, whereas radioimmunoassay does discriminate very

effectively between prolactin of the two species, it should be possible to block the entry of rat prolactin into the third compartment by administering a relatively high dose of bovine prolactin. Any difference in secretory activity should then become reflected in increases of the endogenous prolactin in the circulation.

Figure B shows that this indeed was the case. As can be seen on the left part of the diagram, the relatively high dose of bovine prolactin administered to all 3 groups of rats (at day 270 of the experiment) left the second compartment at the same rate ($\frac{1}{2}t \pm 6$ minutes) over the first 20 minutes of observation. The levels fell from 20 μg/ml at time of injecting the hormone into the tail vein

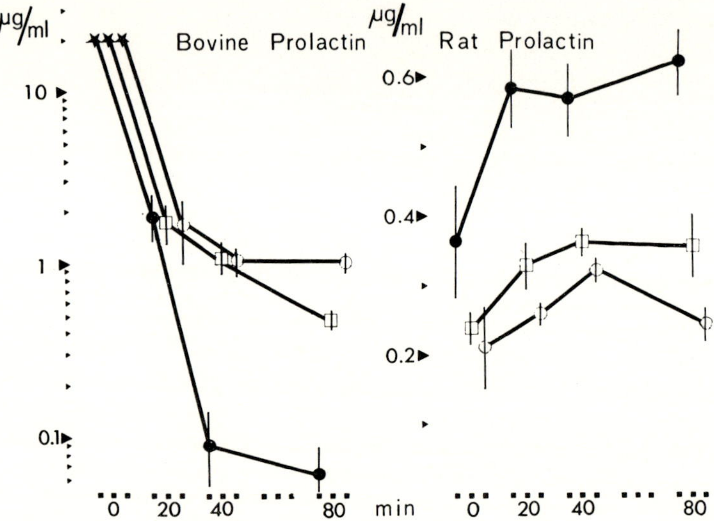

Fig. B. Effect of bovine prolactin on entry of rat prolactin into third compartment in the rat. ○, Castrated rats; □, castrated rats with an isograpft; ●, isografted castrated rats on an estrone regimen.

(time 0, assuming immediate equilibration of the injected dose) to, respectively, 1.82 μg/ml (castrated rats), 1.90 μg/ml (castrated rats with an isograft), and 1.97 μg/ml (isografted, castrated rats on an estrone regimen). In both groups of rats receiving no estrone, the rate of disappearance slowed down considerably in the second 20-minute period ($\frac{1}{2}t \pm 22$ minutes), the levels being 1.06 μg/ml for the castrated rat and 1–195 μg/ml for the isograft-bearing animals. The isograft-bearing rats with estrone showed for the second 20-minute period of observation the same rate of disappearance as for the first 20 minutes ($\frac{1}{2}t \pm 6$ minutes), bringing the level down to 0.089 μg/ml. Over the period of the last 40 minutes of observation, the rate of prolactin entry into the third compartment came to a virtual standstill ($\frac{1}{2}t$ nearing infinity), since the plasma levels were still at 1.045 μg/ml. In the isograft-bearing animal without estrone the rate of disappearance ($\frac{1}{2}t \pm 22$ minutes) was the same as for the second 20-minute period and brought the levels down to 0.495 μg/ml. The isografted rats with estrone showed in the last 40 minutes a further fall in bovine prolactin to 0.064 μg/ml. However, since the slope of these plasma samples differed appreciably in slope (flatter), the assay results in this case should not be considered as valid estimates for intact bovine prolactin, but rather as an indication that mainly immunoreactive fragments of the hormone were present at 80 minutes.

The effects of the effort to overload the third compartment on endogenous prolactin are represented in the right half of the diagram. Even the low secretory activity of the pituitary of the castrated rats resulted in increases of the plasma levels, from 0.211 ± 0.062 μg/ml at time 0 to 0.261 ± 0.018 at 20 minutes, 0.324 ± 0.016 μg/ml at 40 minutes to decline again to 0.246 ± 0.023 at 80 minutes. The isograftbearing rats without estrone showed an endogenous prolactin level of 0.239 ± 0.021 at 0 minute, rising to 0.330 ± 0.034 at 20 minutes, to 0.364 ± 0.021 at 40 minutes, remaining at these levels in the next 40 minutes, 0.358 ± 0.051 μg/ml at 80 minutes), even though the disappearance rate had not been stopped completely. Although the disappearance rate of bovine prolactin indicated that at least over the first 40 minutes it was not appreciably impeded, apparently some slowing down of the prolactin flow from the second into the third compartment was achieved, since the endogenous levels rose from 0.365 ± 0.082 at 0 minute to 0.584 ± 0.057 at 20 minutes, to fluctuate around these levels for the remainder of the observation period (0.568 ± 0.051 at 40 minutes and 0.624 ± 0.047 at 80 minutes).

The results illustrate that high secretory activity of hypophysial prolactin cells may concur with rather low plasma levels due to a rapid flow of prolactin from the (immunoassayable) circulatory compartment into the next compartment.

J. Meites: We have also reported a study on the metabolic clearance of prolactin in the rat and found the half-life for rat prolactin to be about 5 minutes [Y. Koch, Y. F. Chow, and J. Meites (1971). *Endocrinology* **89**, 1303]. The metabolic clearance rate of rat prolactin was not influenced by sex, stage of the estrous cycle, ovariectomy, or hypophysectomy, suggesting that serum concentration of prolactin is determined mainly by the amount of hormone released by the pituitary. Like Dr. Kwa, we found the disappearance curve of injected prolactin was multiexponential.

J. R. Goding: Dr. Meites' paper has marked a turning point for me. No longer can I neglect the rat. We were presented with a mass of data which showed a large measure of agreement between the rat and the sheep. I can think of five situations in which prolactin is released both in the rat and sheep. There is a release at estrus in the cycling ewe, after estradiol in the anestrous ewe, near term in the pregnant ewe, during milking in the lactating ewe, and under conditions of psychological disturbance such as during mating or after the stress of handling [L. R. Fell, C. Beck, J. M. Brown, I. A. Cumming, and J. R. Goding, *J. Reprod. Fertil.* (1971) in press; G. Kann, *C. R. Acad. Sci (Paris)* **272**, 2808 (1971); I. A. Cumming, J. M. Brown, J. R. Goding, G. D. Bryant, and F. C. Greenwood, *J. Endocrinol.* (1971) in press]. We do not see much prolactin secreted during the luteal phase of the cycle, and this may be explained by your experiment in which progesterone administration caused a decrease in prolactin secretion in the rat.

An essential difference between the rat and the sheep is the mechanism causing luteolysis. Prolactin could hardly by involved in the luteolytic mechanism in this species, as luteolysis is already complete several days before the estrous release of prolactin. We still do not know the function of this prolactin surge in the ewe, and we are now reasonably certain that $PGF_2\alpha$ causes luteolysis in the ewe. Among the mammalia, it seems as though the rat, the sheep, and the human all have different mechanisms for bringing about luteal regression.

G. J. Macdonald: It is well recognized that progesterone from the rat corpus luteum can be increased by luteinizing hormone [D. T. Armstrong, *Recent Progr. Horm. Res.* **24**, 255 (1968)]. However, luteinizing hormone does not maintain progesterone secretion in the hypophysectomized nonpregnant rat. This role is dependent on prolactin [G. J. Macdonald, and R. O. Greep, *Perspectives Biol. Med.* **11**, 409 (1968); *Proc. Soc. Exp. Biol. Med.* **134**, 936 (1970)]. Astwood first showed exogenous heterologous prolactin to be luteotropic [*Endocrinology* **28**, 309 (1941)]. Luteotropic ovine prolactin is also antigenic in the rat, limiting the duration of progesterone secretion from the corpus luteum to 17–19 days [G. J. Macdonald, A. H. Tashjian, Jr., and R. O. Greep, *Biol. Reprod.* **2**, 202 (1970)], but if rat prolactin is used the functional life of the corpus luteum is extended to 25 days [G. J. Macdonald, K. Yoshinaga, and R. O. Greep, *Proc. Soc. Exp. Biol. Med.* **136**, 687 (1971)]. Recently, prolactin, not LH, has been shown to maintain the enzymes necessary for cholesterol turnover, a process directly related to progesterone biosynthesis [H. R. Behrman, G. P. Orczyk, G. J. Macdonald, and R. O. Greep, *Endocrinology* **87**, 1251 (1970)]. Thus, without prolactin, progesterone cannot be synthesized by the rat corpus luteum.

I do not agree that the newly formed corpora lutea of the cycle fail to secrete progesterone, for it has been repeatedly shown since the first studies of Eto *et al.* [T. Eto, H. Masuda, Y. Suzuki, and T. Hosi, *Jap. J. Anim. Reprod.* **83**, 34 (1962)] that elevated levels of progesterone are found in ovarian blood during the rat estrous cycle.

Some time ago we found that luteinizing hormone induced estrogen secretion which allowed blastocyst implantation [G. J. Macdonald, D. T. Armstrong, and R. O. Greep, *Endocrinology* **80**, 172 (1967)]. Administration of antiserum developed against luteinizing hormone is able to prevent blastocyst implantation [H. G. Madhwa Raj, M. R. Sairam, and N. R. Moudgal, *J. Reprod. Fertil.* **17**, 335 (1968)], as does ergocornine. Do you think ergocornine acts directly on the ovary to prevent estrogen secretion, which would in turn prevent a prolactin increase?

J. Meites: We have no observations of any direct effects of ergocornine on the ovaries. Although we cannot eliminate the possibility that it may act directly on the ovaries, we have already reported that ergocornine can inhibit prolactin release by a direct action on the pituitary, even under *in vitro* conditions [K. H. Lu, Y. Koch, and J. Meites (1971), *Endocrinology* **89**, 229]. Thus it is not necessary for ergocornine to act via the ovaries to inhibit prolactin release. In fact we observed that ergocornine was effective even in ovariectomized rats.

Certainly progesterone is secreted during the rat estrous cycle, but its source has not been attributed to the corpora lutea but to the ripening follicle [I. Rothchild (1965), *Vit. Horm. New York* **23**, 209.]

Studies on Prolactin in Man[1]

Andrew G. Frantz, David L. Kleinberg, and Gordon L. Noel

Department of Medicine,
Columbia University College of Physicians and Surgeons
and the Presbyterian Hospital, New York, New York

I. Introduction

Until very recently, the steady advances in the field of animal prolactin studies have been in striking contrast to the general uncertainty concerning the status of this hormone in man. After the discovery that human growth hormone preparations possessed strong lactogenic activity (Lyons *et al.*, 1961; Chadwick *et al.*, 1961; Ferguson and Wallace, 1961), the failure to isolate from human pituitaries a prolactin that was definitely distinct from growth hormone (Wilhelmi, 1961; Tashjian *et al.*, 1965; Apostolakis, 1968) led many people to question whether in fact human beings possessed such a hormone (Bewley and Li, 1970). Despite a considerable amount of physiological data favoring its existence, which we have considered elsewhere (Kleinberg and Frantz, 1971), a recent review of the available evidence by Apostolakis (1968) concluded that the existence of a human prolactin separate from growth hormone was still an open question.

Much of the problem lay in the absence of a suitably sensitive and specific bioassay. The various local micromodifications of the original systemic pigeon crop-sac assay (Riddle *et al.*, 1933) have not proved sensitive enough for detection of prolactin in human plasma, and the use of these methods together with extraction and concentration procedures has yielded widely differing results in the hands of different investigators, as well as considerable problems of specificity (Apostolakis, 1968). In the hope of improving both sensitivity and specificity, we began several years ago to experiment with a bioassay employing a mammalian end organ, that of mid-pregnant mouse breast tissue in organ culture. This was a system that had been studied for many years in terms of its hormonal requirements by a number of investigators (Elias, 1957; Lasfargues, 1957; Rivera and Bern, 1961; Juergens *et al.*, 1965; Topper, 1970), but it had not previously been used for assay purposes or been applied to human plasma. We found soon after we began to study this system that by morphological criteria it was capable of detecting prolactin at levels considerably below what had previously been

[1] This work was supported by Grants AM-11294, AM-5397, and CA-11704 from the National Institutes of Health, and by American Heart Association Grant 68-111.

noted (Kleinberg and Frantz, 1969). Later we found that the addition of human plasma, which we had previously thought might be toxic to the cultures, actually stabilized them, increasing the viability of the tissues and enhancing by 2- to 4-fold the sensitivity to prolactin. The findings of high bioassayable prolactin activity in some human plasma samples which had low immunoassayable growth hormone, and the demonstration that anti-human growth hormone antiserum was incapable of neutralizing such activity, whereas it completely neutralized the prolactin activity of growth hormone itself, clearly established the separateness of the two hormones (Frantz and Kleinberg, 1970). In what follows we will describe briefly the assay itself, the evidence it has provided for the existence of human prolactin, its application to physiological investigations in man, and finally a comparison between it and a recently developed radioimmunoassay.

II. Bioassay

The assay, details of which we have given more fully· elsewhere (Kleinberg and Frantz, 1971), employs Swiss albino mice of a local strain, bred in our own laboratory. Animals are used on the eighth or ninth day of pregnancy, the timing having an important effect on the sensitivity of the assay. After sacrifice by ether anesthesia, the breast tissues are cut into small pieces and incubated for 4 days under 95% oxygen, 5% carbon dioxide. The medium 199 is used; to this is added 30% pooled normal male plasma. Such plasma in our experience does not contain detectable prolactin activity. It is present in all dishes containing ovine prolactin standards, which are routinely run at 7 dose levels ranging from 0 to 50 ng/ml. When unknown plasma samples are being assayed, usually at concentrations of 10 and 30%, the normal male plasma is partly or completely replaced by the plasma under test. After fixation and staining with hematoxylin–phloxin–safranine, the microscopic sections are examined by eye. A negative response, which occurs in the absence of prolactin, has tubules whose lumina are empty and devoid of secretory material (Fig. 1). A strongly positive response, occurring with prolactin at a concentration of 50 ng/ml, is shown in Fig. 2. Grading is done visually on a scale of 0 to 4+, based on the number of lumina filled and the degree of filling. The score for each dish is the mean of scores for each fragment in the dish. The results of grading by two independent observers have been in good agreement when, as is always the case, grading is done without the observers' knowledge of what the slide represents. When the scores are plotted together, a standard curve of the form shown in Fig. 3 is obtained, which represents the mean of 60 assays. Also shown in this figure are the points obtained

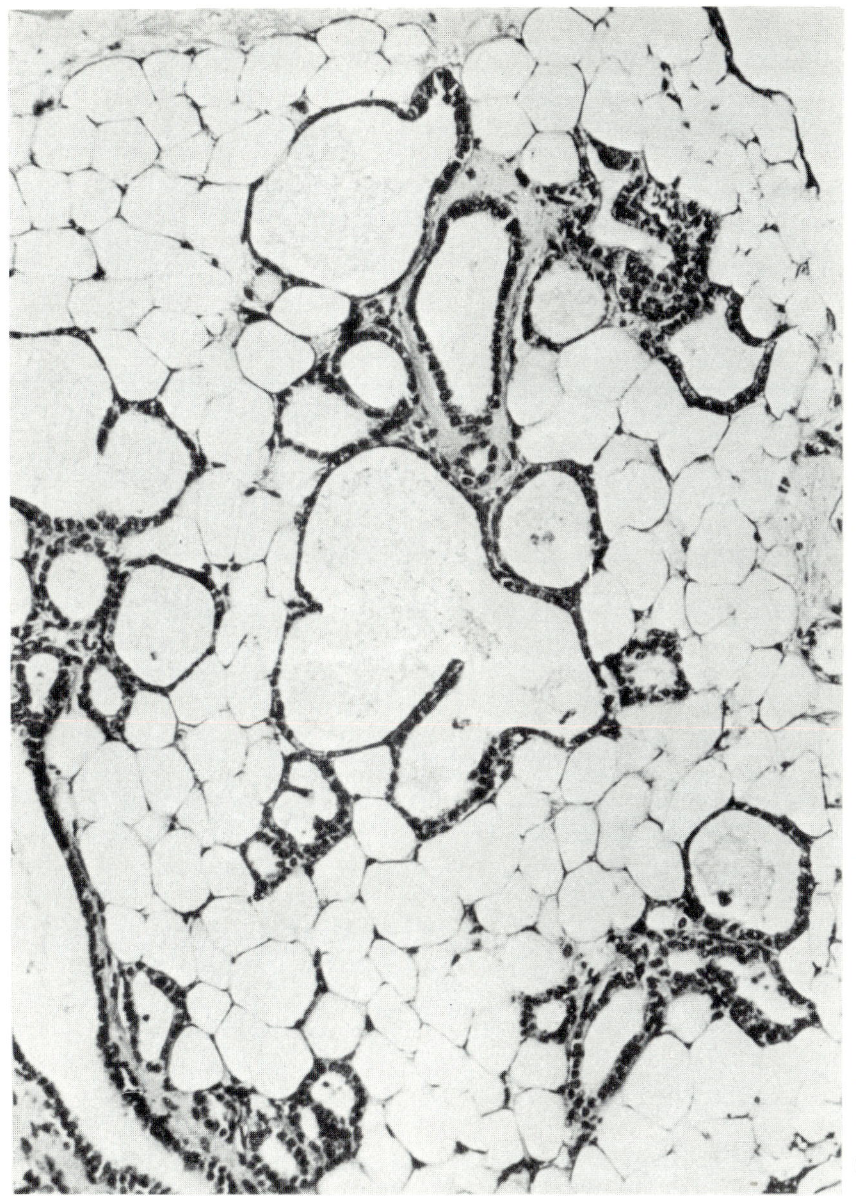

FIG. 1. Mouse breast tissue after incubation in the absence of prolactin. Lumina of tubules are empty. ×150.

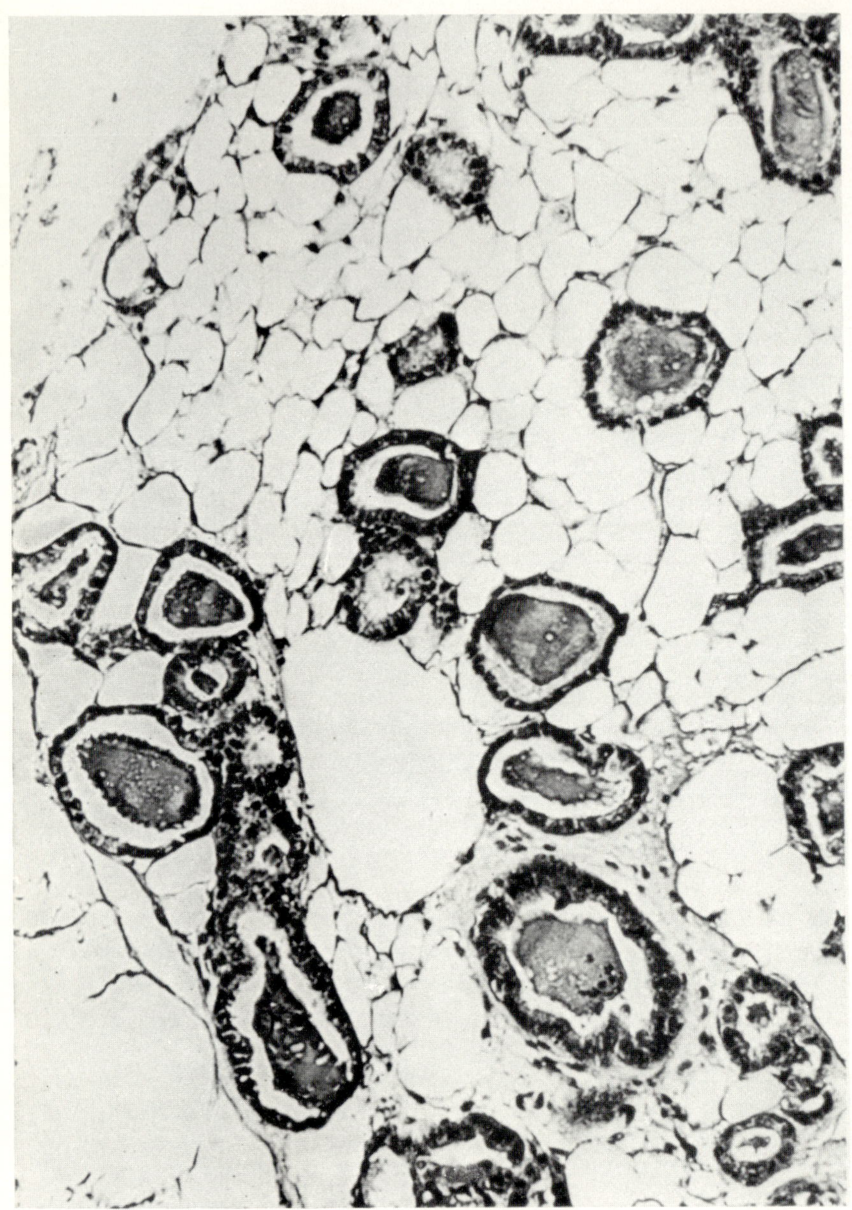

Fig. 2. Mouse breast tissue after incubation in the presence of prolactin, 50 ng/ml. Lumina of tubules are filled with dark red-staining secretory material. ×150.

from a single assay to give some idea of the standard errors involved. We do not regard the precision as adequate unless each sample is run at at least two dose levels in two separate assays, and we consider three assays per sample preferable. Under these conditions, for samples not at the threshold of sensitivity of the assay, measurements obtained will generally have a coefficient of variation of 25–35%. The sensitivity of the assay is high, 5 ng/ml being definitely distinguishable from zero

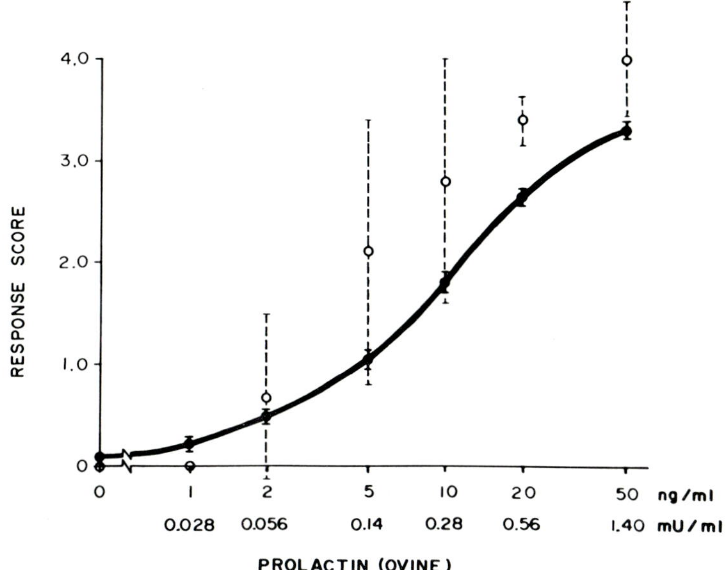

Fig. 3. Standard dose response curve for ovine prolactin. Solid line and bars represent the mean of 60 assays, ± standard error of the mean. Open circles and dashed lines represent the responses derived from a single assay.

in most assays, and sometimes 2 ng/ml being so distinguishable. We usually regard 5 ng/ml as the threshold of the assay; with plasma at a concentration of 30%, there results an effective sensitivity for prolactin in plasma of approximately 15 ng/ml expressed in terms of ovine equivalents.

The specificity of the assay is high. No substance that we have tested that is not known to be a lactogen, including TSH, ACTH, human chorionic gonadotropin, oxytocin, vasopressin, estradiol, progesterone, and testosterone, has given positive responses. Estradiol does not appear to increase or decrease the response to added prolactin standards or unknown samples when added in physiological or supraphysiological concentrations.

Human growth hormone is strongly lactogenic in this assay. Several highly purified preparations have been tested, and all have been found to have activity that ranged from approximately 50% to 80% that of the ovine standard. Precise potency figures are difficult to arrive at in some cases because of the occasional finding that growth hormone appears slightly more potent at lower than at higher concentrations; this tendency to nonparallelism has not been observed in all assays, and is currently being further explored. Human placental lactogen also gives strongly positive responses; the preparations we have tested to date have had approximately one-half to two-thirds the potency of the purest human growth hormone preparations.

III. Evidence for a Separate Prolactin in Human Plasma

As soon as this assay was applied to human plasma it became apparent that a circulating prolactin, distinct from growth hormone, could be detected in a number of conditions. Our initial evidence for the separateness of these two hormones has been presented previously (Frantz and Kleinberg, 1970; Kleinberg and Frantz, 1971) and will only be summarized here:

A. Prolactin Level in Normal vs. Lactating Subjects

Prolactin activity was found to be undetectable, i.e., less than 15 ng/ml, in almost all of a large group of endocrinologically normal males and females, ranging in age from 9 to 67 years. These were healthy, nonhospitalized subjects tested at varying times of day. A very few values were noted at the threshold of detectability of the assay, particularly in a few unusually sensitive assays with thresholds somewhat below 15 ng/ml. In nursing mothers, on the other hand, prolactin was found to be almost always easily detectable with levels occasionally reaching very high values. Figure 4 shows a group of such subjects tested at varying intervals after parturition and at varying times in relation to nursing episodes, a subject to which we shall return shortly. In the nursing mothers shown in this figure, as well as in a number of other patients we shall discuss with elevated serum prolactin activity, growth hormone as measured by radioimmunoassay was at the low end of the normal range and could not possibly have accounted for the prolactin activity observed.

B. Neutralization Experiments with Antihuman Growth Hormone Antiserum

Plasma samples from a number of nursing mothers, as well as from other patients with high prolactin and low growth hormone, were preincubated for 2 hours with a potent rabbit anti-human growth hormone antiserum at a 1:10 concentration and then assayed in parallel with the same samples after preincubation with normal rabbit serum; ovine prolactin standards were also similarly preincubated. In no case was any significant neutralization of prolactin activity observed in the plasma samples, nor was there any neutralization of the ovine standard. By contrast, when human growth hormone was added to normal male plasma in a concentration of 2000 ng/ml and incubated with anti-growth hormone in the same manner, subsequent bioassay of the plasma at the usual 30% concentration indicated complete abolition of all prolactin

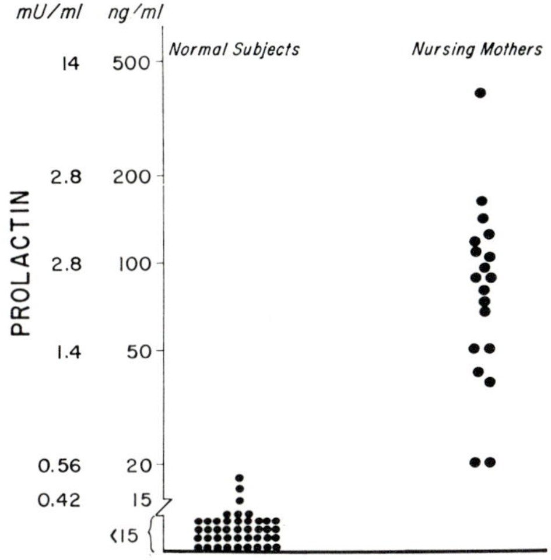

Fig. 4. Bioassay values for plasma prolactin in 42 normal male and female subjects, aged 9–67 years, and 19 nursing mothers 1–150 days post-partum. Samples from mothers were obtained at various times in relation to the last nursing episode.

activity. The prolactin activity associated with high circulating growth hormone levels, both in acromegalic patients and in normal subjects following insulin hypoglycemia, can be largely or entirely neutralized with anti-human growth hormone. These studies indicate that human growth hormone and human prolactin are immunologically distinct molecules, with little or no cross-reactivity demonstrable under the conditions of these experiments by any of the antisera we have used. (Essentially identical results were obtained with each of three antisera raised in rabbits against highly purified Wilhelmi growth hormone preparations: HS 612A, HS 617B, and HS 497C.) The studies also indicate that human growth hormone as it circulates in blood, like the material extracted from pituitaries, possesses intrinsic lactogenic activity (Frantz and Kleinberg, 1970; Kleinberg and Frantz, 1971).

C. Neutralization Experiments with Antiovine Prolactin Antiserum

In contrast to the failure of anti-human growth hormone antiserum to neutralize the prolactin activity in the plasma of nursing mothers and patients with galactorrhea, the data in Table I indicate that preincu-

TABLE I
Neutralization of Bioassayable Prolactin Activity in Human Plasma with Antiovine Prolactin

Patient	Sex	Diagnosis	Growth hormone (ng/ml)	Prolactin ng/ml[a]	mU/ml	Prolactin after incubation with antiovine prolactin[b]
B.J.	F	Galactorrhea: chromophobe adenoma	0.3	140	3.9 ± 1.0^c	Undetectable[d]
E.F.	F	Galactorrhea: hypothalamic glioma	0.8	130	3.6 ± 0.7	Undetectable
M.O.	F	Galactorrhea: chromophobe adenoma	0.3	125	3.5 ± 1.0	Undetectable
L.J.	M	Galactorrhea: possible estrogen withdrawal	1.3	80	2.2 ± 0.9	Undetectable
A.C.	F	Galactorrhea: idiopathic	0.6	75	2.1 ± 0.4	Undetectable
J.W.	F	Galactorrhea: idiopathic	1.2	18	0.5	Undetectable
H.L.	F	Postpartum 3 days	1.0	385	10.8 ± 3.1	Undetectable
E.R.	F	Postpartum 2 days	0.8	165	4.6 ± 1.3	Undetectable
L.F.	F	Postpartum 3 days	0.6	140	4.0 ± 0.5	Undetectable
G.W.	F	Postpartum 2 days	0.3	120	3.4 ± 0.8	Undetectable
C.P.	F	Postpartum 4 days	0.5	100	2.8 ± 1.2	Undetectable
M.S.	F	Postpartum 3 days	1.6	40	1.1 ± 0.3	Undetectable

[a] In terms of NIH-P-S8 ovine standard; 28 units = 1 mg.
[b] Incubated for 2 hours with rabbit antiovine prolactin at 1:10 concentration.
[c] Standard error.
[d] Less than 0.42 mU/ml.

bation of such plasmas with antiserum to highly purified ovine prolactin (NIH-P-S8) under similar conditions completely abolishes detectable prolactin activity. The results of these experiments thus confirm by biological neutralization the immunochemical cross-reaction between ovine prolactin and primate prolactin first noted by Herbert and Hayashida (1970) by means of immunofluorescent techniques.

We have previously reviewed the earlier evidence for a human prolactin (Frantz and Kleinberg, 1970; Kleinberg and Frantz, 1971). Since our own studies were first presented, several papers have appeared offering additional evidence from pituitary incubation studies of the separateness of human growth hormone and human prolactin (Bryant *et al.*, 1971; Hwang *et al.*, 1971a,b; Lewis *et al.*, 1971). Of these, the most important to date appear to be the studies of Dr. Friesen and his colleagues, in which human and monkey prolactin have been isolated in a highly purified form by immunoadsorption techniques and used in the construction of a specific radioimmunoassay, as will be discussed later on.

IV. Physiological Studies of Human Prolactin by Bioassay

A. GALACTORRHEA

Among the first subjects to be studied with our bioassay for plasma prolactin were patients with galactorrhea. In Fig. 5 are shown a group of 47 patients with various forms of galactorrhea whom we have studied in some detail. The great majority of these patients were seen and examined by us at the Presbyterian Hospital. Nine of these patients had pituitary tumors, chiefly chromophobe adenomas, and as a group these tended to have the highest prolactin levels. These patients with tumors, all women, all had amenorrhea and would in general be considered to be examples of the Forbes-Albright syndrome. In contrast to these is the group of patients in the next column of Fig. 5, which we have designated idiopathic. This refers to patients whose galactorrhea is not associated with any tumor or other evident cause, and who have regular menses. Plasma or urinary gonadotropin determinations on these patients, when performed, were within normal range, in contrast to the low gonadotropin levels observed in patients with tumors. The majority of these patients began to lactate after the delivery of a child, and did not cease lactating even though regular menses returned. This group of patients with idiopathic galactorrhea is much larger than is generally recognized clinically, because the women frequently do not complain of the galactorrhea and do not come to see a doctor on account of it. We believe that the largest number of women with galactorrhea in the general population fall into this group. It is striking that the majority of patients within this group have levels of prolactin which were undetectable by bioassay, and therefore not distinguishable from normal. These low values confirm the fact of a relatively mild degree of hypothalamic-pituitary disturbance in these patients, and also indicate that very high levels of prolactin are not necessary for galactorrhea. The group designated drug induced covers a variety of etiological agents and includes patients whose galactorrhea occurred after oral contraceptive withdrawal, on initiation of oral contraceptives, on various tranquilizers including perphenazine, fluphenazine, amitriptyline, isoniazid, α-methyl dopa, and various others. The range of prolactin in these

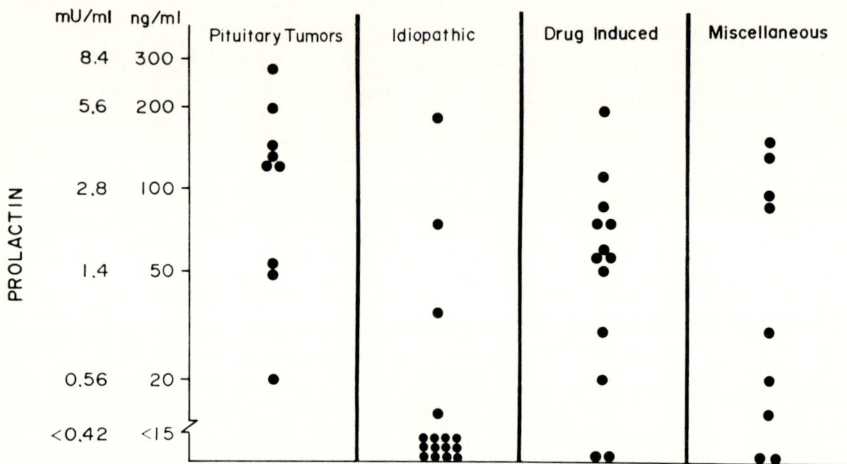

FIG. 5. Bioassay values for plasma prolactin in 47 patients with galactorrhea. Among tumor patients with galactorrhea the highest prolactin level we have yet encountered, not shown in this figure, is 6500 ng/ml (182 mU/ml), occurring in a male. The group designated idiopathic refers to women with galactorrhea without evident cause who are having regular menses. As the figure suggests, this is the largest single group of patients with galactorrhea in our experience. The other columns are explained in the text. All these specimens have also been measured by radioimmunoassay, with results that agree closely with those shown here. In particular, the low levels in the idiopathic group have been confirmed. The 12 patients in this group with prolactin levels by bioassay of less than 15 ng/ml have had levels which by radioimmunoassay ranged from 1.0 to 24 ng/ml, with a mean of 9.1 ng/ml.

subjects is quite wide, extending from normal to considerably elevated levels. The last column in Fig. 5 includes miscellaneous patients not elsewhere classified, including those with hypothalamic lesions, the Chiari-Frommel syndrome, and galactorrhea of uncertain origin associated with amenorrhea. All the patients depicted in Fig. 5 had growth hormone levels within the normal range. A major conclusion from these studies is that prolactin levels in galactorrhea range all the way from normal or close to normal up to greatly elevated values.

B. Pituitary Tumors

As we looked at an increasing number of patients with this assay, it became apparent that elevated prolactin levels were by no means confined to patients with galactorrhea, and that they could be found in a number of conditions where galactorrhea was not present. Figure 6 shows 30 patients with pituitary tumors on whom we have done prolactin measurements. Eighteen of these had chromophobe adenomas either presumed or demonstrated by surgery, and twelve had craniopharyngiomas. Prolactin levels ranged all the way from undetectably low, in

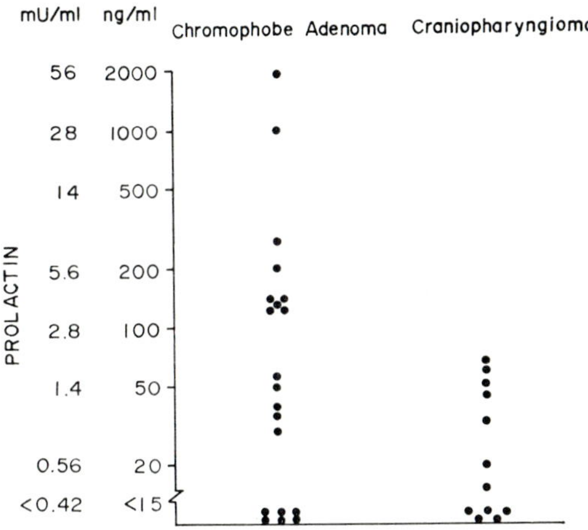

Fig. 6. Bioassay values for plasma prolactin in tumor patients. Many of the patients with elevated prolactin were clinically panhypopituitary. Only 9 of these 30 patients had galactorrhea. An additional three patients, not shown in this figure, have been seen who had prolactin levels in excess of 5000 ng/ml. These values have been confirmed by radioimmunoassay.

a minority, to more than 1000 ng/ml. Less than half of these patients had galactorrhea, and there was no statistically significant difference in prolactin levels between those with galactorrhea and those without. In several patients we have studied pre- and postoperatively, surgery was followed by a marked reduction in circulating prolactin. The patient with the highest levels we have encountered to date, not shown in Fig. 6, was a 32-year-old male with an enlarged sella turcica, presumed due to a chromophobe adenoma, mild gynecomastia, and galactorrhea. There were no signs of acromegaly. A glucose tolerance test was normal. Thyroid and adrenal function were normal, but growth hormone and gonadotropins were low. Prolactin was 6500 ng/ml by bioassay (a later radioimmunoassay was 10,000 ng/ml). Following conventional X-ray therapy to the sella turicica, galactorrhea ceased and prolactin dropped markedly although still remaining elevated. In a number of patients with pituitary tumors, prolactin has been the sole hormone found to be elevated, and the patients have appeared on clinical grounds and by appropriate tests to be hypopituitary with respect to all other anterior pituitary hormones. Our conclusion from these studies, supported by studies on patients with pituitary stalk section whom we shall mention later, is that lesions which tend to interfere with pituitary-hypothalamic connections, either by enlargement of the gland itself or direct involvement of the hypothalamus, may augment prolactin secretion; this presumably occurs through reduction in the influence of the hypothalamic prolactin-inhibiting factor (PIF). Our finding also suggests that measurement of plasma prolactin will be a useful diagnostic test in the evaluation of patients with pituitary disease; in particular, the test may be used to gauge the effects of surgery or radiation therapy in patients with elevated levels of the hormone.

C. Effect of Drugs

Phenothiazine derivatives and some other tranquilizing drugs have long been known to cause breast stimulation and galactorrhea in certain patients, and Sulman (1970) and his colleagues have shown elevation of prolactin in experimental animals receiving these drugs. Pituitary incubation studies indicate that they act via the hypothalamus, not on the gland directly (Danon *et al.*, 1963). Table II shows a group of

TABLE II
Bioassayable Plasma Prolactin During Chronic Tranquilizing Drug Therapy

Patient	Sex	Drug (time in weeks)	Dose (mg/day)	HGH (ng/ml)	Prolactin ng/ml[a]	Prolactin mU/ml
A.E.	F	Perphenazine (78)	12	0.3	111	3.1
M.S.	M	Perphenazine (1)	24	3.3	15	0.42
A.M.	F	Perphenazine (12) + amitriptyline (12)	12 100	0.3	75	2.1
C.G.	F	Fluphenazine (2)	6	<0.3	15	0.42
D.V.	M	Chlorpromazine (4)	2000	0.5	61	1.7
M.M.	M	Chlorpromazine (4)	1500	3.7	20	0.56
T.B.	M	Chlorpromazine (8)	2000	0.4	30	0.84
A.J.	F	Imipramine (4)	150	1.1	20	0.56
M.S.	M	Haloperidol (1)	4.5	0.3	61	1.7

[a] In terms of NIH-P-S8 ovine standard: 28 units = 1 mg.

nine subjects, five males and four females, receiving tranquilizing drugs in high doses for psychiatric purposes. In addition to drugs related to phenothiazine, those administered included imipramine, which is chemically somewhat similar to the phenothiazines, and haloperidol, a butyrophenone derivative. All patients in this table, which includes all whom we have studied on such high doses, had prolactin levels above the threshold of detectability and growth hormone which was at the lower end of the normal range. Only two of these patients, A.E. and A.M., had galactorrhea. It may be noted that chlorpromazine, which raises plasma prolactin, has been reported to inhibit growth hormone secretion

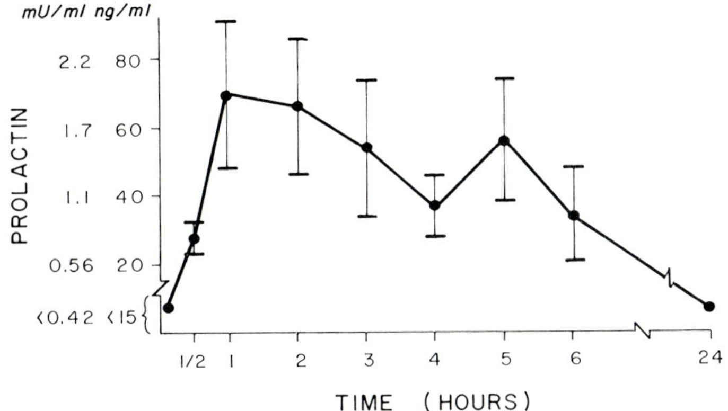

Fig. 7. Bioassayable plasma prolactin in 10 normal subjects who received chlorpromazine, 12.5 or 25 mg intramuscularly, at zero time. Vertical bars represent the standard error of the mean. Prolactin was initially undetectable in all subjects and first became detectable in all at some point between 15 minutes and 2 hours. The apparent dip at 4 hours is not statistically different from the 3- and 5-hour values.

(Sherman *et al.*, 1971). It seems not unlikely that other nonphenothiazine tranquilizers which share with chlorpromazine the property of depleting catecholamines in certain areas of the brain, such as haloperidol, may have suppressive actions on growth hormone as well as a stimulatory action on prolactin.

We next turned our attention to the time course of induction of these high values. Bryant, Connan, and Greenwood (1968) had earlier shown rapid induction of high circulating prolactin values in sheep after acepromazine administration. A group of normal volunteers comprising six men and four women was given chlorpromazine intramuscularly in doses ranging from 12.5 to 25 mg. It can be seen from Fig. 7 that plasma prolactin rose appreciably, becoming detectable for the first time in all subjects at some point between 15 minutes and 2 hours. Individual peak values ranged from 25 to 240 ng/ml and occurred between 30 minutes and 2 hours after injections. The mean peak occurred at 1 hour and there was a slow decline thereafter. Of four subjects tested at 24 hours, prolactin was detectable at low levels in only one; 25 mg of chlorpromazine appeared to produce a somewhat greater response than 12.5 mg, but limited tests with 50 mg revealed no obvious advantage of this higher dose.

Figure 8 shows the results of chloropromazine stimulation in 4 patients with dwarfism secondary to pituitary insufficiency. Two patients classi-

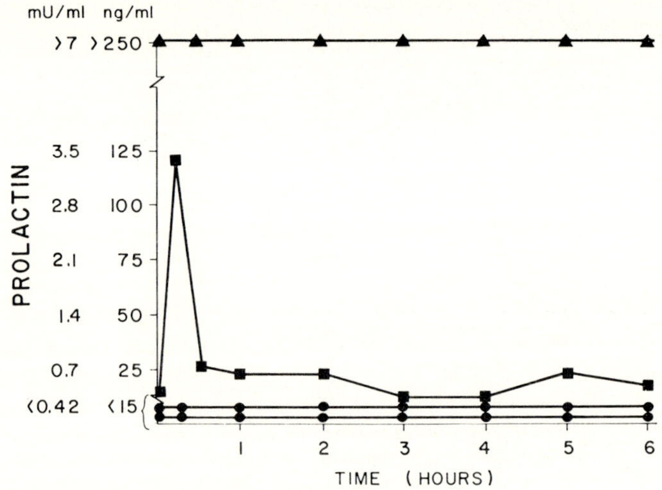

FIG. 8. Bioassayable prolactin in 4 patients with pituitary dwarfism of varied etiology following chlorpromazine, 12.5 or 25 mg intramuscularly, given at zero time. ▲, Chromophobe adenoma; ■, isolated human growth hormone deficiency; ●, idiopathic panhypopituitarism.

fied as having idiopathic panhypopituitarism had undetectable levels throughout the test. One patient with what appeared to be an isolated growth hormone deficiency gave an early but quite definite response. The fourth patient, who had a chromophobe adenoma, had levels which were markedly elevated throughout the test; we have not as yet assayed the samples at sufficiently high dilutions to see whether a change occurred. These studies indicate that prolactin measurement in conjunction with chlorpromazine stimulation affords a valuable means of testing hypothalamic–pituitary function (Kleinberg *et al.*, 1971).

D. SUPPRESSION OF PROLACTIN SECRETION BY L-DOPA

We next wished to see whether L-dopa could antagonize the effect of chlorpromazine. L-dopa, an agent which crosses the blood brain barrier and is decarboxylated to dopamine, has pharmacological actions which are in many ways opposite to those of chlorpromazine. Dopamine has been reported to inhibit prolactin secretion by the pituitary *in vitro* (Birge *et al.*, 1970; MacLeod *et al.*, 1970; Koch *et al.*, 1970) and to do so *in vivo* in the rat by stimulating PIF (Kamberi *et al.*, 1970). L-dopa in humans has been reported to stimulate growth hormone secretion (Boyd *et al.*, 1970). Figure 9 depicts the responses of four

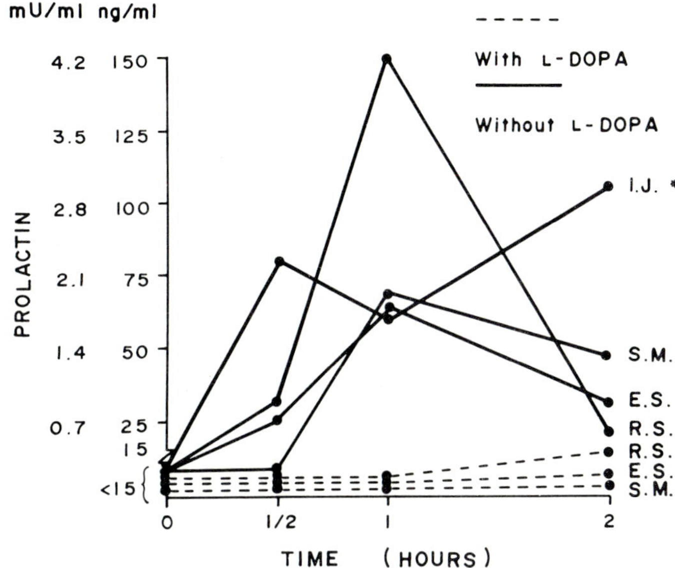

Fig. 9. Bioassayable prolactin in 4 normal subjects who received 25 mg of chlorpromazine (solid lines). The same subjects were given the same dose several days later after pretreatment with L-dopa, 500 mg orally, 0.5 hour before the chlorpromazine (dashed lines). Two of these subjects, R.S. and S.M., also received 250 mg of L-dopa 6 and 12 hours before the test. Prolactin in I.J. was also undetectable at all time intervals after L-dopa, but these values are not shown because of a less sensitive assay in which the threshold of detectability was 30 ng/ml. A marked suppressive effect of L-dopa is evident.

normal subjects to chlorpromazine stimulation tests, together with the responses of the same subjects tested several days later with the same dose of chlorpromazine after pretreatment with 500 mg of L-dopa by mouth 0.5 hour before chlorpromazine injection. A marked inhibition of prolactin response by L-dopa is evident. Growth hormone levels showed no marked changes in any of those tests.

It was also of obvious interest to test the effect of L-dopa on patients with pathologically elevated prolactin levels. Figure 10 depicts the response to 500 mg of L-dopa by mouth in four such patients. Three were women, all of whom had galactorrhea; this was associated with oral contraceptive administration in one, oral contraceptive withdrawal in another, and chlordiazepoxide administration in the third. The fourth patient was a male who had undergone subtotal resection of a prolactin-secreting chromophobe adenoma several weeks previously by transfrontal craniotomy. In all patients prolactin fell rapidly, becoming undetectable

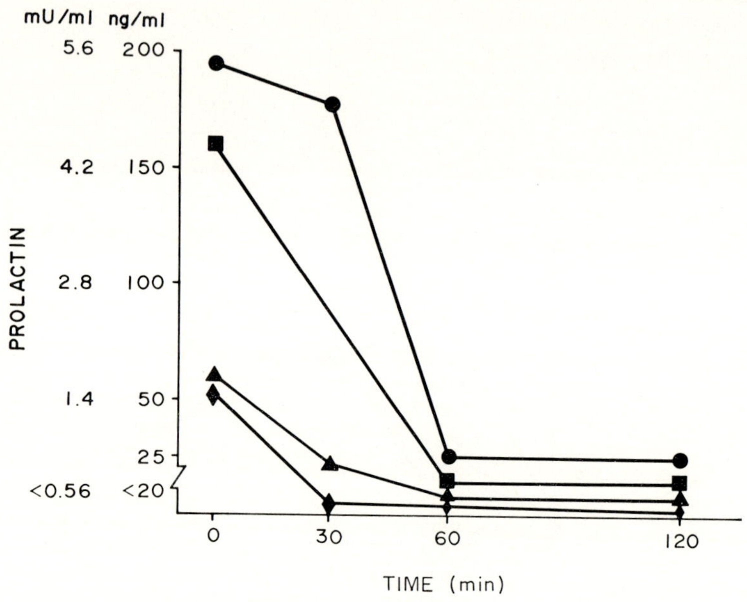

Fig. 10. Bioassayable prolactin following L-dopa, 500 mg orally, at zero time in 4 subjects with initially elevated plasma prolactin: ■, 50-year-old male tested several weeks after incomplete surgical removal of a prolactin-secreting chromophobe adenoma. ●, 25-year-old female with persistent galactorrhea following withdrawal of oral contraceptives; ▲, 29-year-old female with galactorrhea possibly related to chlordiazepoxide administration; ◆, 39-year-old female with galactorrhea beginning after oral contraceptive administration.

or barely detectable by 60 minutes after L-dopa (Kleinberg et al., 1971). The results in the postoperative patient were of particular interest, since the hypothalamic–pituitary connections had been extensively disrupted by surgery. The marked prolactin suppression which occurred in this patient suggests the possibility either of a direct action of L-dopa on the pituitary or else a stimulation of hypothalamic PIF which may have been acting on the pituitary in part at least by the systemic rather than the portal circulation. Ongoing experience in this laboratory with L-dopa in the treatment of galactorrhea associated with high prolactin levels suggests that in some cases L-dopa will abolish both the galactorrhea and the high prolactin levels; in other patients, however, it has produced no change in either of these conditions. While it appears that this agent may prove to be of considerable value in the treatment of certain cases of galactorrhea, its use as a therapeutic agent requires further evaluation.

V. Studies with Radioimmunoassay

A. Development of a Radioimmunoassay for Baboon Prolactin

One of our goals from the beginning of these studies was to use the bioassay in part as a means for the development and validation of a radioimmunoassay for human prolactin. Despite the security it affords

TABLE III
Bioassayable Plasma Prolactin During Perphenazine Treatment of a Lactating Postpartum Baboon

Assay time	Prolactin (ng/ml)[a]	Growth hormone (ng/ml)[b]
Before treatment	25	11.3
1 Day after perphenazine, 20 mg/day	98	12.3
7 Days after perphenazine, 20 mg/day	219	5.3
15 Days after perphenazine, 20 mg/day	750	19.0

[a] In terms of NIH-P-S8 ovine standard.
[b] By radioimmunoassay, in terms of HS 1103C human standard.

that one is measuring a biologically active substance in plasma, the bioassay remains an expensive method of acquiring information, one that is less precise and less well suited than radioimmunoassay to the processing of large numbers of clinical samples. Early in 1970 we began to gather baboon pituitary glands, hoping that baboon prolactin, like baboon growth hormone in the studies of Tashjian, Levine, and Wilhelmi (1965), would show a particularly high degree of immunological cross-reaction with its human counterpart. Over a period of several months we were able to secure four postpartum nursing baboons. These already lactating animals were additionally stimulated with perphenazine, 5 mg intramuscularly, given four times a day for 2 weeks. Plasma from the animals was assayed at intervals during this period; Table III shows that in the postpartum baboon, as in the normal human, perphenazine induced a marked and sustained rise in plasma prolactin; no clear-cut changes in growth hormone were observed. After 2 weeks the animals were sacrificed and the pituitaries were removed under aseptic conditions. The glands were cut up into small pieces and incubated in medium

199 for a period of 7 days. The conditions of the incubation were similar to those used for breast tissue incubation, except that the medium contained no insulin, plasma, or other protein, and was changed daily. Table IV shows that large amounts of bioassayable prolactin were liberated into the medium, together with relatively much lower amounts of growth hormone, over a period of several days. Study of these incubation media by analytical disc gel electrophoresis revealed one to three major bands, all rather closely spaced, and having a mobility very close to that of human serum albumin run in control groups in the same pH 8.3 buffer system.

TABLE IV

Hormone Content of Medium During Pituitary Incubation Postpartum Perphenazine-Treated Baboon

Day	Prolactin (μg/ml)[a]	Growth hormone (μg/ml)[b]
1	500	7.3
2	150	7.2
3	200	4.2
4–5	200	2.5
6	27	1.8
7	11	1.2

[a] By bioassay, ovine standard.
[b] By radioimmunoassay, human standard.

Antibodies were raised in rabbits by several intracutaneous injections of the incubation media with the highest prolactin potency, together with Freund's adjuvant. Because of the shortage of material, the incubation media were used for immunization directly, without attempt at purification at this stage. The antisera so prepared showed precipitin lines against both ovine prolactin and human growth hormone in double diffusion agar gel experiments. The immunizing mixture itself gave precipitin lines both with anti-human growth hormone and with anti-ovine prolactin (Noel and Frantz, 1970).

To prepare a radioactive antigen for radioimmunoassay, the material from one of the incubation media was first dialyzed against phosphate buffer and then labeled with iodine-131 according to the method of Greenwood, Hunter, and Glover (1963). The radioactive proteins obtained from an initial Sephadex column were separated by vertical starch gel electrophoresis at pH 8.6 (Smithies, 1959). Three major bands were identified by radioautography and eluted; the front running band, with a mobility slightly greater than that of human serum albumin, possessed the greatest reactivity with antiovine prolactin as well as with anti-baboon prolactin, and the greatest ability to be displaced from either by rhesus monkey prolactin. A highly purified preparation of the latter hormone, supplied to us by Dr. William Peckham, was used to standardize the immunoassay depicted in Fig. 11; in this assay the labeled baboon prolactin was used together with anti-baboon antiserum. Although offering good sensitivity for rhesus prolactin, this system showed less cross-reactivity than we had expected for human prolactin, with

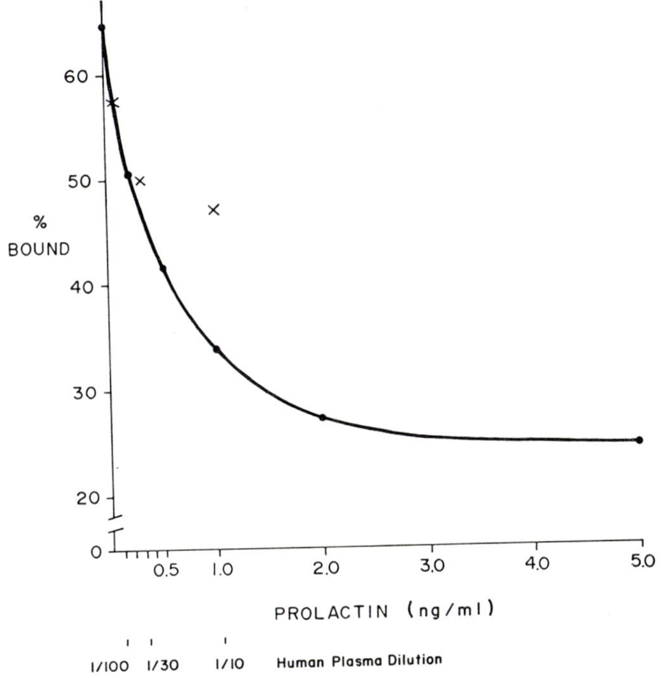

Fig. 11. Standard curve of radioimmunoassay employing baboon prolactin-^{131}I tracer and antibaboon prolactin antibody. See text for details. ●, Rhesus prolactin; ✕, human plasma chromophobe.

marked nonparallelism of the two curves and correspondingly low sensitivity for the latter hormone. Indeed, the sensitivity for human prolactin offered little if any improvement over our bioassay. Ovine prolactin produced no effect in concentrations similar to those of rhesus prolactin; human growth hormone exhibited a slight degree of cross-reactivity which was variable in different assays.

B. COMPARISON OF HUMAN RADIOIMMUNOASSAY AND BIOASSAY RESULTS

In spite of its limitations, we would have continued to explore and use this assay for human purposes had not Drs. Hwang, Guyda, and

Friesen (1971a) recently announced an immunoassay for human prolactin, based on the use of rhesus and, more recently, human prolactin. The isolation of this substance and the use of it in a radioimmunoassay constitute the final piece of evidence confirming the existence in the human pituitary of a separate prolactin. Several weeks before this conference began, Dr. Friesen generously made available to us some of

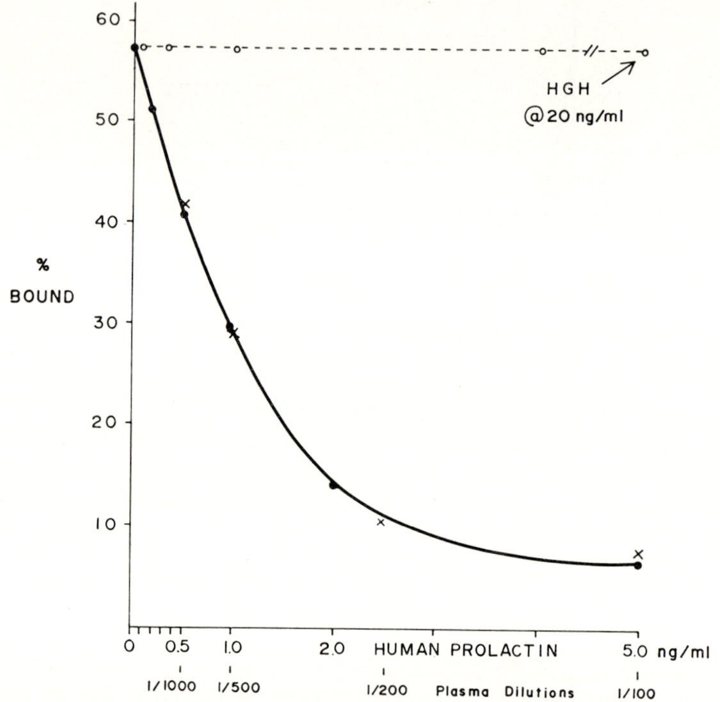

Fig. 12. Standard curve of radioimmunoassay employing human prolactin-^{131}I (Friesen) and antiserum containing both anti-HPr and anti-HGH. See text for details. ●, Human prolactin; ○, human growth hormone; ×, plasma chromophobe.

the human hormone he had prepared together with an antiserum raised against human growth hormone that was discovered to have antiprolactin antibodies as well, presumably because of trace amounts of prolactin in the original immunizing mixture. After radioiodinating Dr. Friesen's prolactin preparation and purifying it slightly by subsequent starch gel electrophoresis, we were able to use it and the antibody to set up the radioimmunoassay shown in Fig. 12. The superiority of this system over our baboon assay for measuring human prolactin is clearly apparent. There is complete coincidence of displacement by the human prolactin standard and the material contained in a human plasma sample with high bioassayable prolactin, as indicated in the figure. Human growth hormone produces no displacement whatever at concentrations anywhere close to what might conceivably be encountered in human plasma, even in the most severe cases of acromegaly.

It has been possible therefore to retest many of our previously bioassayed samples by radioimmunoassay and to correlate the results of

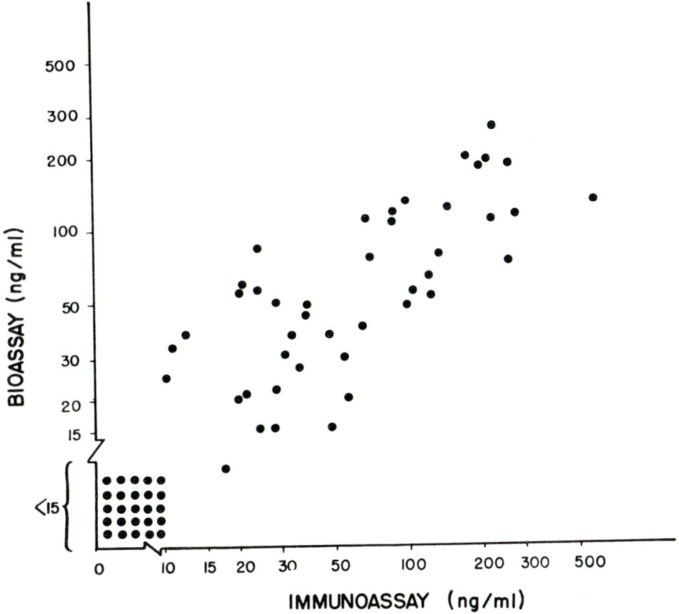

Fig. 13. Scatter diagram showing correlation between results of bioassay and radioimmunoassay for prolactin in human plasma samples. The mean bioassay:immunoassay ratio was 1.09. See text and Table V.

the two measurements. These data are shown graphically in Fig. 13 and numerically in Table V. The results indicate a somewhat higher degree of correlation than we had expected, both at the low as well as the high end of the scale. Samples which were undetectable by bioassay and designated as less than a certain figure were in the great majority of cases found by immunoassay to be in the range designated. Very high values likewise correlated well. Fifty samples with measurable values by both assay methods, representing all such samples tested to date, were compared; in these samples the ratio of bioassay value, expressed in terms of nanograms of the ovine standard, to the radioimmunoassay value, expressed in terms of nanograms of Dr. Friesen's human standard, had a mean of 1.09 (Table V). Two-thirds of all these samples had a bioassay to immunoassay ratio lying between 0.5 and 1.90. A minority of samples, which we plan to examine more closely, had widely divergent potencies by the two assays. Some of these discrepancies may be ascribable to assay variation, and some may possibly be due to the formation of molecules, either by release or peripheral degradation, which possess different ratios of bioassayable to immuno-

TABLE V

Bioassay vs Immunoassay Measurements on Plasma Samples

Number of samples	Bioassay values (ng/ml)[a]	Immunoassay values (ng/ml)[b]	Ratio of values, bioassay:immunoassay
3	29–54	3.5–6.9	4.1 –8.3
8	27–85	9.1–24	2.3 –3.6
13	20–275	20–225	1.00–1.90
20	15–6500	23–10,000	0.50–0.99
6	15–128	49–600	0.21–0.49
40	<15	0.6–15	
5	<15	19–27	

Mean ratio, 50 samples, bioassay:immunoassay: 1.09[c]
95% confidence limits: 0.86–1.39

[a] In terms of NIH-P-S8 ovine standard.

[b] In terms of Dr. Friesen's human prolactin preparation.

[c] Represents geometric mean of individual ratios for all samples tested to date with measurable values by both assays.

assayable potency than the normal prolactin molecule. We are currently in the process of determining the biological potency of Dr. Friesen's human preparation in relation to the NIH-P-S8 ovine standard (see Addendum).

C. Prolactin Levels in Normal Subjects

One of the first and most gratifying confirmations by radioimmunoassay of our earlier bioassay results lay in the category of normal subjects. Figure 14 shows the results of 65 determinations in 54 normal males and 59 determinations in 43 normal females. These were all healthy, nonhospitalized, ambulatory subjects. The mean value for the women, 10.27 ng/ml ± 7.94 standard deviation, was not significantly different from the mean for normal males, 9.21 ng/ml ± 5.65 standard deviation. If one assumes equipotency of human and ovine prolactin, and adds a 10 ng/ml correction factor to our bioassay figures to account for the presumed concentration of prolactin in the normal male plasma which was present with the standard but displaced entirely by the plasma under test at 30% concentration, our previously defined normal range for adults would become less than 25 ng/ml instead of less than 15 ng/ml. In the present series of 97 subjects, only one male and one female

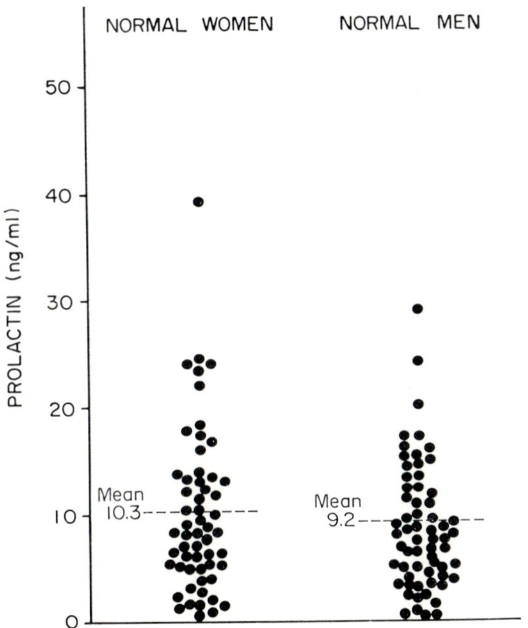

Fig. 14. Radioimmunoassay values for plasma prolactin in normal women and men.

were found to have levels greater than 25 ng/ml. Our present total range for our normal women is 1.0 to 39 ng/ml, and for normal men is 0.6 to 29 ng/ml. These values do not apply to hospitalized subjects or those under stress, in whom, as we shall see later, higher values may be encountered.

D. Nursing Mothers

Table VI shows the effects of suckling in nursing mothers, 2–150 days post-partum. In this table, as in some others that follow, results of bioassay and immunoassay measurements are listed together for comparative purposes. A distinct rise in plasma prolactin following a 30-minute period of nursing can be seen in most of the subjects in this table.

The two subjects who did not exhibit a rise were those with the highest initial prolactin levels, both of whom were tested only 2 days postpartum. No significant rise in growth hormone was noted in any subject. A rise in plasma prolactin due to nursing has also been noted by Hwang, Guyda, and Friesen (1971a). As can be seen from the data in Table VI, the rise in prolactin may be very substantial, as much as 100-fold

TABLE VI

*Plasma Prolactin and Growth Hormone in Postpartum
Women Before and After Nursing*

Name	Days postpartum	Assay	Prolactin		Growth hormone[a]	
			30 Min before nursing	30 Min after nursing	30 Min before nursing	30 Min after nursing
G.W.	2	Bio[b]	124	126	<0 3	0.6
E.R.	2	Bio[b]	213	73		
		Imm[c]	270	290	1.0	0.8
E.R.	8	Bio[b]	75	>200		
		Imm[c]	68	680	0.6	0.6
M.N.	12	Bio[b]	<30	>200		
		Imm[c]	16	1600	14.8	5.9
L.L.	12	Bio[b]	<30	>200		
		Imm[c]	57	475	2.0	1.6
G.W.	17	Bio[b]	50	332		
		Imm[c]	29	406	0.5	0.4
J.P.	90	Bio[b]	20	40		
		Imm[c]	20	64	2.3	4.8
A.B.	150	Bio[b]	60	200		
		Imm[c]	21	180	1.6	2.2

[a] By radioimmunoassay, ng/ml.

[b] By bioassay, ovine standard, ng/ml.

[c] By radioimmunoassay, human standard, ng/ml.

in one of our subjects. The timing of the prolactin rise with suckling was examined in greater detail in several additional subjects by means of multiple samples obtained at closely spaced intervals through indwelling catheters. In two of these subjects the child, obviously eager for feeding, was placed in contact with the mother but denied the breast for a period of 30 or more minutes before being allowed to suckle. In one of these subjects a second study was carried out several weeks later in which the breasts were emptied manually by the mother with the aid of a breast pump, the child not being present. The results in these two subjects are shown numerically in Table VII. Figure 15 also shows

TABLE VII

Plasma Prolactin Before, During, and After Nursing and Breast Pump

Before nursing[a]	−60	−30	−15	−10	−5	0 min
M.B.[b] HPr[c]	92	65	57	50	51	52
M.B. HGH[d]	<0.3	<0.3	<0.3	<0.3	<0.3	<0.3
G.M.[e] HPr[c]	37	24	27	22	18	21
G.M.[f] HPr[c]-breast pump[g]		8.6	10.2			7.5

During nursing	0	+5	+10	+15	+20	+25	+30 min
M.B. HPr[c]	52	50	68	95	270	490	1200
M.B. HGH[d]	<0.3	<0.3	<0.3	<0.3	<0.3	<0.3	<0.3
G.M. HPr[c]	21	22	30	66	91	200	240
G.M. HPr[c]-breast pump[g]	7.5			21			500

After nursing[h]	+30	+40	+45	+50	+60	+75	+90	+105	+120	+150	+180 min
M.B. HPr[c]	1200	850		680	580	450	260	175	125	84	69
M.B. HGH[d]	<0.3	<0.3		<0.3	<0.3	<0.3	<0.3	<0.3	<0.3	<0.3	<0.3
G.M. HPr[c]	240	220		160	93	79	56	25	34	17	21
G.M. HPr[c]-breast pump[g]	500		180		108		44		26	22	17

[a] Children present and in contact with mother during this period; milk let-down occurred around −30 minutes in both mothers.
[b] 28-year-old mother, 23 days post-partum.
[c] Prolactin by radioimmunoassay, human standard, ng/ml.
[d] By radioimmunoassay, human standard, ng/ml.
[e] 26-year-old mother, 25 days post-partum.
[f] Same individual as above, 64 days post-partum.
[g] The child was not present during the breast pump study.
[h] Nursing stopped promptly at +30 minutes.

the prolactin changes during nursing of these plus three additional subjects.

Several significant findings emerge from these studies. First, direct mechanical stimulation of the breast itself appears to be the main stimulus for prolactin secretion in these subjects. No rise in prolactin occurred until after the initiation of suckling or breast manipulation. Thereafter prolactin rose continuously throughout the entire period of nursing, reaching its highest level at the termination of this period. A moderately

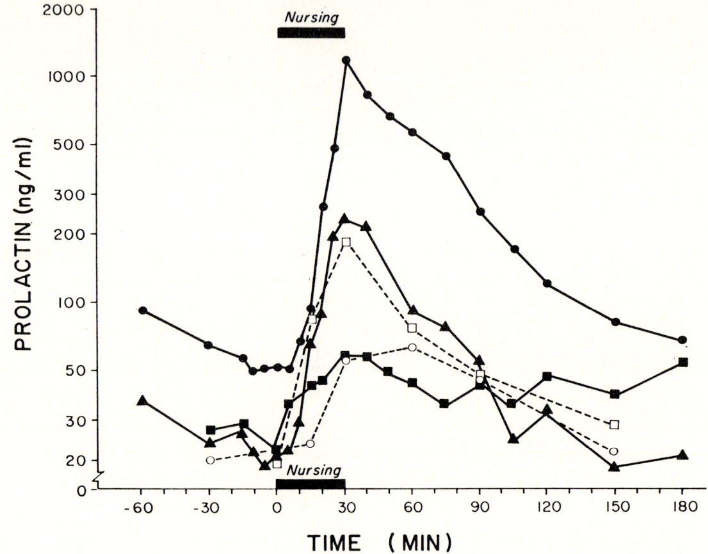

FIG. 15. Plasma prolactin values by radioimmunoassay before, during, and after nursing in 5 mothers, 23–150 days post-partum. In all cases prolactin did not begin to rise until after the onset of suckling, and in all except one the highest value was obtained at the end of the 30-minute nursing period.

rapid fall-off occurred thereafter in most subjects, indicating a maximum half-time of less than 30 minutes for the hormone. Growth hormone remained very low and showed no change whatever throughout the entire period. These studies confirm in humans the importance of the suckling stimulus for prolactin release, first noted in animals a number of years ago by means of pituitary depletion studies (Reece and Turner, 1937; Grosvenor and Turner, 1957).

Second, psychic factors associated with the presence of the infant do not appear to be as important as breast stimulation in prolactin release. Despite the close physical proximity of the child before nursing in the two subjects in Table VII, and despite the fact that milk let-down occurred during this period, no rise in prolactin was observed before actual suckling. In the absence of the infant, the breast pump proved at least equally as effective as suckling in stimulating prolactin release.

Third, the magnitude of the rise suggests that a major portion of the prolactin released by nursing women takes place during actual suckling, and it is probably to a large extent this episodic secretion which maintains the breast in an actively lactating state.

Studies are currently in progress in this laboratory to assess the effect of breast stimulation on prolactin release in nonpostpartum women, both normal subjects and those with galactorrhea or elevated prolactin.

D. Stress

A fall in pituitary prolactin in lactating rats after stress of various kinds was noted by Grosvenor, McCann, and Nallar (1965), and Biyant, Linzell, and Greenwood (1970) have presented evidence that prolactin in goats may be elevated in situations of stress. We have examined the effect of stress in human beings in several different ways (Noel et al., 1971). Table VIII shows prolactin levels in six subjects studied 24 hours before major surgery, during operation, immediately postoperatively, and 24 hours after operation. In all cases a rise in prolactin was noted, the highest values occurring during surgery or immediately thereafter, and in most cases a decline by 24 hours. Growth hormone rose less in some cases than did prolactin, and the timing of rise and fall of the two hormones was frequently not coincident. It is interesting that two of these patients had high prolactin levels 24 hours before the operation. We have tentatively concluded from these and other studies that psychic factors alone, such as the anxieties connected with hospitalization and the anticipation of surgery, can raise prolactin in certain individuals.

The rise which was noted during surgery could conceivably be ascribed to the effects of anesthesia, which was of different kinds in these patients. We therefore turned to another procedure, gastroscopy, where the only medication employed is a standard dose of intravenous diazepam (Valium). Similar amounts of diazepam administered to four normal subjects in control studies produced no change in plasma prolactin over a 4-hour period. As can be seen in Table IX, prolactin levels rose in all of six subjects examined before and after gastroscopy. The magnitude of the rise varied considerably, ranging from less than 2-fold to 10-fold or greater. Growth hormone levels showed no significant change in five patients and a slight rise in one.

In order to avoid the effects of medication completely, we studied the effects in 10 subjects of another procedure, proctoscopy, which is mildly

TABLE VIII

Plasma Prolactin Before, During, and After Surgery

Patient	Sex	Age	Operation	Assay	24 Hours pre-op	During operation	Immediately post-op[a]	24 Hours post-op
R.H.	M	68	Cholecystectomy	Bio[b]	<15	134	74	<15
				Imm[c]	—	100	36	22
				HGH[d]	(1.5)	(10.4)	(3.8)	(1.7)
C.A.	F	36	Cholecystectomy	Bio[b]	80	365	400	132
				Imm[c]	136	>400	>400	400
				HGH[d]	(2.8)	(7.5)	(17)	(2.4)
G.P.	F	29	Thyroidectomy	Bio[b]	<15	23	27	<15
				Imm[c]	13.8	36	40	18
				HGH[d]	(1.3)	(3.5)	(3.2)	(1.9)
A.P.	F	14	Appendectomy	Bio[b]	172	221	168	38
				Imm[c]	104	140	132	43
				HGH[d]	(5.7)	(45)	(2.6)	(3.0)
A.C.	F	38	Tubal ligation	Imm[c]	12.1	96	180	84
				HGH[d]	(2.9)	(2.8)	(18)	(22)
L.S.	F	25	D & C	Imm[c]	17[e]	300	18	76
				HGH[d]	(3.3)	(2.3)	(3.0)	(3.8)

[a] At 15–60 minutes after conclusion of operation.

[b] Prolactin by bioassay, ovine standard, ng/ml.

[c] Prolactin by radioimmunoassay, human standard, ng/ml.

[d] Human growth hormone by radioimmunoassay, ng/ml.

[e] One hour before operation, after premedication.

uncomfortable and involves variable amounts of psychic stress, but for which no permedication is given. As can be seen in Table X, six of the subjects exhibited some degree of rise in prolactin, and four had no change. Growth hormone showed no significant change in any of these subjects.

Since exercise is a known stimulus to growth hormone secretion, we examined the effects of a short period of intense exercise, consisting

TABLE IX
Plasma Prolactin Before and After Gastroscopy

Subject	Sex	Age	Assay	15 minutes before gastroscopy	15 Minutes after gastroscopy
M.R.	F	47	Bio[a]	<30	180
			Imm[b]	60	204
			HGH[c]	(2.0)	(3.0)
R.D.	F	46	Bio[a]	<15	86
			Imm[b]	5.2	>100
			HGH[c]	(3.8)	(8.8)
M.L.	M	74	Bio[a]	<15	<15
			Imm[b]	12.2	20
			HGH[c]	(2.0)	(2.0)
A.P.	F	58	Bio[a]	25	25
			Imm[b]	5.5	46
			HGH[c]	(0.9)	(0.6)
G.R.	M	75	Imm[b]	3.4	6.2
			HGH[c]	(0.3)	(<0.3)
R.G.	F	67	Imm[b]	4.4	42
			HGH[c]	(<0.3)	(<0.3)

[a] Prolactin by bioassay, ovine standard, ng/ml.
[b] Prolactin by radioimmunoassay, human standard, ng/ml.
[c] Human growth hormone by radioimmunoassay, ng/ml.

of running up and down 15 to 25 flights of stairs as rapidly as possible, on nine normal subjects. The results, shown in Table XI, indicate that seven of the nine subjects had some degree of rise in prolactin. The mean for the entire group before exercise was significantly lower than that immediately after exercise ($p < 0.05$) or 15 minutes after exercise ($p < 0.02$). Growth hormone levels in this group as a whole, in contrast to some groups we have previously discussed, rose distinctly more than prolactin levels. Although exercise appears to stimulate growth hormone release by a different and more specific pathway than that of simple stress, the sudden and strenuous exercise in these studies was clearly a stressful experience for the subjects involved. Whether the prolactin rise in these cases is a stress-related response, or whether it is due more specifically to a sudden increased metabolic demand for glucose or other substrates, is not clear from these studies. Such a question might be

TABLE X
Plasma Prolactin Before and After Proctoscopy

Patient	Sex	Age	Assay	Before proctoscopy	Immediately after proctoscopy
L.S.	M	89	Bio[a]	25	36
			Imm[b]	17	24
			HGH[c]	(1.8)	(1.5)
F.V.	F	35	Bio[a]	20	39
			Imm[b]	6.0	10
			HGH[c]	(1.6)	(2.1)
F.U.	F	67	Bio[a]	<15	42
			Imm[b]	9.2	25
			HGH[c]	(1.9)	(1.6)
L.T.	F	67	Bio[a]	<15	<15
			Imm[b]	3.4	<3.0
			HGH[c]	(2.2)	(2.5)
B.M.	F	43	Bio[a]	<15	<15
			Imm[b]	15	14
			HGH[c]	(1.9)	(2.1)
E.A.	F	57	Bio[a]	<15	<15
			Imm[b]	<3.0	<3.0
			HGH[c]	(1.6)	(2.3)
J.A.	M	71	Imm[b]	9.0	46
			HGH[c]	(1.9)	(1.0)
W.J.	M	65	Imm[b]	6.1	7.0
			HGH[c]	(1.0)	(0.3)
A.G.	F	54	Imm[b]	16	27
			HGH[c]	(<0.3)	(0.4)
H.B.	M	—	Imm[b]	8.2	12.8
			HGH[c]	(1.0)	(0.6)

[a] Prolactin by bioassay, ovine standard, ng/ml.
[b] Prolactin by radioimmunoassay, human standard, ng/ml.
[c] Human growth hormone by radioimmunoassay, ng/ml.

answered by glucose administration prior to exercise, since this has been shown to inhibit the growth hormone response to exercise but not to surgery (Glick *et al.*, 1965). Whatever the mechanisms underlying the exercise-induced response may be, we believe it can be concluded from the evidence we have gathered so far that prolactin, like growth hormone, rises in conditions associated with stress. There is often a considerable

TABLE XI
Plasma Prolactin Before and After Exercise

Subject	Sex	Age	Assay	Before exercise	2 Minutes after exercise	15 Minutes after exercise
H.S.	M	30	Bio[a]	<15	48	<15
			Imm[b]	5.0	20	23
			HGH[c]	(1.4)	(1.6)	(4.0)
G.N.	M	30	Bio[a]	<15	—	—
			Imm[b]	13	18	—
			HGH[c]	(2.0)	(7.0)	—
L.F.	M	24	Bio[a]	<20	<20	<20
			Imm[b]	16	18	17
			HGH[c]	(2.0)	(14.2)	(13.3)
A.F.	M	41	Bio[a]	<33	<33	<33
			Imm[b]	8.7	12	13
			HGH[c]	(0.7)	(1.0)	(14.8)
J.C.	M	24	Imm[b]	14.8	24	30
			HGH[c]	(1.8)	(3.8)	(4.9)
L.B.	M	30	Imm[b]	11.6	32	26
			HGH[c]	(2.0)	(9.2)	(14.4)
P.F.	F	22	Imm[b]	24	38	36
			HGH[c]	(36)	(40)	(27)
E.B.	F	23	Imm[b]	10	22	23
			HGH[c]	(<0.3)	(<0.3)	(2.7)
E.S.	M	22	Imm[b]	20	11.2	16.8
			HGH[c]	(0.7)	(3.1)	(16.4)
		Mean, immunoassay:		13.67	21.68	23.10
		SEM:		±2.06	±3.09	±2.86

[a] Prolactin by bioassay, ovine standard, ng/ml.
[b] Prolactin by radioimmunoasay, human standard, ng/ml.
[c] Human growth hormone by radioimmunoassay, ng/ml.

dissociation between the two hormones, however, in the magnitude and timing of their response to a given stimulus.

E. HYPOGLYCEMIA

Last year we reported that insulin tolerance tests in normal subjects induced a rise in bioassayable plasma prolactin activity concomitantly with the rise in growth hormone. This prolactin activity could be entirely neutralized by anti-human growth hormone antiserum in some subjects, and largely, but not entirely, neutralized by such antiserum in others. The failure of complete neutralization indicated that while much of the prolactin effect was due to growth hormone, prolactin in some subjects was probably also stimulated by hypoglycemia (Frantz and

Kleinberg, 1970). Because many of the prolactin levels in our previous group were close to the threshold of sensitivity of the assay, and therefore difficult to measure with precision, we have examined another group of seven normal female subjects by radioimmunoassay before and after insulin-induced hypoglycemia. All had normal growth hormone responses. It can be seen in Fig. 16 that all seven had definite increases in plasma prolactin, the peak being 2 to 19 times the base line value. This rise

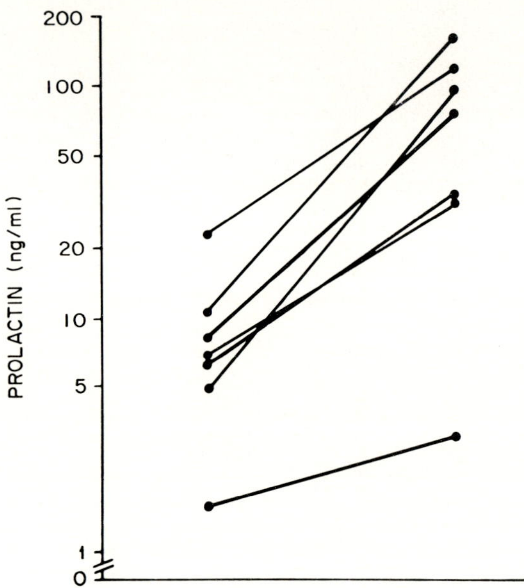

Fig. 16. Plasma prolactin by radioimmunoassay before and 1 hour after intravenous insulin, 0.2 units/kg, in 7 normal women. All had a normal growth hormone response to hypoglycemia.

is distinctly greater than what we had observed in our previous subjects. One possible explanation for the larger rise in these subjects is that they were part of a study in which a higher dose of insulin, i.e., 0.2 unit/kg rather than our customary 0.1 to 0.15 unit/kg, was used. All subjects showed a correspondingly greater than average fall in blood glucose, the mean nadir at 30 minutes being 15.3 mg/100 ml. None experienced any unusual or distressing symptoms during the test. Although the degree of hypoglycemia was greater in these patients than what is frequently experienced in such tests, the conclusion appears justified that hypoglycemia of sufficient magnitude is a definite stimulus to prolactin release, as it is to that of growth hormone.

F. Pituitary Stalk Section

Transection of the pituitary stalk has been associated with lactation in women and in experimental animals (Ehni and Eckles, 1959; Donovan and van der Werff ten Bosch, 1957; Cowie *et al.*, 1964), an effect which has been regarded as being possibly due to increased prolactin secretion from a gland no longer inhibited by hypothalamic PIF. Several years ago we reported the results of testing growth hormone responsiveness to hypoglycemia in a group of diabetic patients undergoing pituitary stalk section for retinopathy (Powell *et al.*, 1966). We have recently reexamined plasma from seven of these patients for prolactin, four of whom were tested preoperatively as well as postoperatively. The plasmas were maintained frozen for periods of five or six years; our present evidence suggests that prolactin, like growth hormone, is stable in frozen plasma for reasonably long periods of time, although the possibility of loss of activity during prolonged storage cannot be ruled out. The results, shown in Table XII, indicate that prolactin rose post-operatively in two patients and in two was unchanged. In no case was there a decrease, even though growth hormone declined considerably and became unresponsive to hypoglycemic testing. Prolactin was unaffected by hypoglycemia before as well as after stalk section; this phenomenon may be related to lesser degrees of hypoglycemia in these patients than in those discussed in the preceding section, or to a generally altered pituitary responsiveness to hypoglycemia in these diabetic patients that was reflected in a diminished preoperative growth hormone response in some cases as well. The relatively short interval, averaging 16 days, between surgery and prolactin testing could well have resulted in a different and perhaps lower prolactin value than would have been obtained after a longer time period. The occurrence of breast enlargement in nine out of eleven women in the original series when they were examined clinically several months after surgery, with galactorrhea in three of these eleven (Powell *et al.*, 1966), suggests the possibility of effects due to increased prolactin. From these limited studies it would appear that pituitary stalk section in human beings, while producing little change in prolactin in some subjects, may in others be followed by significant elevations of this hormone (see Addendum).

G. Craniopharyngioma Patients with Growth after Surgery

A few years ago we reported on the curious phenomenon of the resumption of normal growth following surgery for craniopharyngioma in a selected group of nine children who had exhibited growth arrest prior to operation (Holmes *et al.*, 1968). Growth hormone during the postoperative growth period was undetectable or virtually undetectable in all subjects both before and after insulin hypoglycemia; one patient tested preoperatively also had low growth hormone and failure to respond to insulin. Postoperatively these children were also judged to be hypo-

TABLE XII

Plasma Prolactin and Growth Hormone Before and After Pituitary Stalk Section for Diabetic Retinopathy

Patient	Sex	Age	Operative status	Assay	Insulin tolerance tests			
					Prolactin		Growth hormone	
					0 Min	60 Min	0 Min	60 Min
P.M.	F	28	Pre-op.	Bio[a]	<15	<15		
				Imm[b]	8.6	3.0	3.9	4.0
			15 days	Bio[a]	11	21		
			Post-op.	Imm[b]	10.7	13.9	1.1	2.7
W.T.	M	35	Pre-op.	Bio[a]	<15	6		
				Imm[b]	3.9	4.5	2.6	8.4
			15 days	Bio[a]	10	<15		
			Post-op.	Imm[b]	4.0	5.0	<0.3	1.6
R.C.	F	31	Pre-op.	Bio[a]	<20	23		
				Imm[b]	5.1	3.2	4.2	26.0
			17 days	Bio[a]	44	42		
			Post-op.	Imm[b]	58	47	0.6	2.2
F.S.	F	43	Pre-op.	Imm[b]	5.5	5.6	<0.3	20.0
			16 days Post-op.	Imm[b]	5.4	5.6	0.9	0.9
L.T.	M	45	16 days	Bio[a]	<30	33		
			Post-op.	Imm[b]	25.0	25.0	1.6	0.5
J.H.	M	34	15 days Post-op.	Imm[b]	2.0	6.0	<0.3	<0.3
E.D.	M	40	17 days Post-op.	Imm[b]	2.0	3.0	<0.3	<0.3
Mean pre-op., immunoassay, 4 patients:					5.8	4.1	2.7	14.6
Mean post-op., immunoassay, 4 patients:					19.5	17.9	0.7	1.9
Mean post-op., immunoassay, 7 patients:					15.3	15.1	0.6	1.2

[a] Prolactin by bioassay, ovine standard, ng/ml.
[b] Prolactin by radioimmunoassay, human standard, ng/ml.

pituitary with respect to all other pituitary hormones in addition to growth hormone. The paradox of normal growth in the face of grossly subnormal growth hormone levels was noted, but no obvious explanation was apparent. Kenny and associates (1968) also noted similar findings in three children after operation for craniopharyngioma and speculated on the possibility that the growth after surgery might be due to increased amounts of prolactin functioning as a growth hormone, since in one patient detectable or elevated serum prolactin was suggested by pigeon crop-sac assay. In order to test this possibility we reanalyzed the sera on six of these patients which we had previously studied for growth hormone. The sera had been kept frozen since originally tested 4–6 years previously. The results of these tests are shown in Table XIII. Prolactin

TABLE XIII

Craniopharyngioma Patients with Normal Postoperative Growth

| | | | | | | Insulin tolerance tests | | | |
| | | | | | | Prolactin | | Growth hormone | |
Patient	Sex	Age at surgery	Age when tested	Operative status	Assay	0 Min	60 Min	0 Min	60 Min
E.B.	M	16	16	Pre-op.	Bio[a]	<15	<15		
					Imm[b]	11	14	<1.0	1.6
			16	Post-op.	Bio[a]	<15	<15		
					Imm[b]	8.2	8.6	<0.3	<0.3
R.D.	M	10	15	Post-op.	Bio[a]	37	36		
					Imm[b]	33	47	<0.3	<0.3
A.T.	M	8	10	Post-op.	Bio[a]	45	27		
					Imm[b]	37	35	<0.3	<0.3
R.G.	M	6	12	Post-op.	Bio[a]	31	57		
					Imm[b]	31	20	0.4	0.3
N.C.	F	10	11	Post-op.	Bio[a]	22	<15		
					Imm[b]	28	27	0.3	0.3
V.K.	F	8	10	Post-op.	Bio[a]	<15	<15		
					Imm[b]	3.2	2.2	<0.3	<0.3
				Mean, immunoassay:		25.1	25.6		
4 Craniopharyngioma patients with growth arrest: Mean:						51.0			
				Range:		(5.5–105)			

[a] Prolactin by bioassay, ovine standard, ng/ml.

[b] Prolactin by radioimmunoassay, human standard, ng/ml.

was detectable in all patients, with a mean level of 25.1 ng/ml and no significant change following hypoglycemia. In three individuals the levels were somewhat above our normal range for adults, but two lines of evidence suggest that prolactin in these patients as a group was not the cause of their growth following surgery. First, in the one patient, E.B., who was measured both before and after operation, and who had exhibited complete growth arrest prior to surgery, the preoperative levels of prolactin were similar to, and in fact slightly greater than, the postoperative ones. Second, in four patients of similar ages with craniopharyngiomas who failed to grow before or after surgery, prolactin levels were found to be somewhat higher than in those who grew after surgery, with a mean of 51 ng/ml. Thus, although prolactin appears to be present in normal or elevated amounts in many patients with craniopharyngioma, and may conceivably serve some anabolic function, it appears to us that it is unlikely to be the cause for postoperative growth in the small number of patients in whom such growth occurs.

H. ACROMEGALY

We have previously reported high bioassayable prolactin activity in the serum of 16 acromegalic subjects, the total prolactin activity showing a positive correlation with the growth hormone levels. When some of these sera were tested after incubation with anti-growth hormone antibody, partial or complete neutralization of the prolactin activity was observed (Frantz and Kleinberg, 1970). From these results we concluded that prolactin production did not parallel that of growth hormone in acromegaly, although in some patients the two hormones appeared to be elevated together. We have reexamined this question with radioimmunoassay in another group of acromegalic patients, with results as shown in Table XIV. Of these 15 patients, all with elevated plasma

TABLE XIV
Prolactin and Growth Hormone in Acromegaly

Name	Sex	Prolactin (ng/ml)[a]	HGH (ng/ml)
B.C.	M	95	180
J.N.	M	10	32
G.M.	F	5	26
L.M.	F	50	58
S.G.	F	9	23
D.R.	F	53	28
S.T.	F	8	64
F.M.	F	63	78
A.S.	M	>1000	180
R.H.	M	>1000	105
S.M.	F	28	170
M.P.	F	10	13
J.T.	F	15	80
A.A.	F	25	43
U.A.	F	8	120

[a] By radioimmunoassay.

growth hormone, the majority (9) had normal prolactin levels. Six patients showed elevated prolactin levels of 50 ng/ml or over, and two of these were greater than 1000 ng/ml. There appeared to be no correlation between the levels of the two hormones. Only one of these patients exhibited temporary galactorrhea, which came about immediately after discontinuance of estrogens which had been used as a mode of therapy of his disease. Thus the radioimmunoassay confirms the bioassay conclusion that prolactin is often normal in acromegaly but may in some cases be hypersecreted together with growth hormone.

I. GYNECOMASTIA

Table XV shows prolactin levels in 16 patients with gynecomastia of moderately severe degree. None of these patients was on any medica-

TABLE XV

Plasma Prolactin in Males with Gynecomastia

Patient	Age	Related findings	Prolactin (ng/ml) Bioassay	Immunoassay	HGH (ng/ml)
G.P.	13	None	<30	2.5	3.0
J.G.	16	None	—	12	—
M.F.	19	None	<17	7.2	<0.3
W.V.	19	None	56	24	1.5
U.D.	19	None	27	9.5	0.8
U.T.	56	None	<30	11	<0.3
L.P.	26	Hypogonadotropic hypogonadism	37	12	0.6
M.S.	30	Klinefelter's syndrome	—	6.0	—
G.W.	41	Heroin addiction	<17	19	1.8
J.C.	60	Hepatosplenomegaly, unexplained	29	6.9	<0.3
L.J.	45	Adenocarcinoma of lung	85	24	1.4
B.G.	14	Galactorrhea; normal sella	>200	>500	4.5
M.K.	5	Tumor near third ventricle	<17	22	—
M.D.	61	Enlarged sella	22	22	0.9
U.B.	55	Galactorrhea, enlarged sella	199	220	<0.3
G.S.	32	Galactorrhea, enlarged sella	6500	10,000	—

tion known to be associated with breast stimulation. In six cases there were no associated findings or abnormalities after thorough investigation. All of these had prolactin within the normal range by immunoassay, although in one (W.V.) the hormone was definitely elevated by bioassay and in another (U.D.) it was detectable and therefore considered mildly elevated by bioassay. Three patients (B.G., U.B., and G.S.) had galactorrhea, and all had markedly elevated prolactin by both assays. Two of these patients had pituitary tumors; in the third, B.G., a tumor could not be documented. Two other patients, M.K., and M.D., had evidence of intracranial tumors; in one, M.D., who had an enlarged sella turcica, prolactin was at the upper limit of normal or slightly elevated by bioassay standards. Among the remaining patients, three (C.P., J.C., and L.J.) had elevated levels by bioassay but not by immunoassay. In all, five patients in this group of sixteen exhibited a greater than usual discrepancy between the results of bioassay and immunoassay, with the bioassay giving the higher reading. We have no explanation at present for this finding, which deserves further exploration. Our conclusion from this limited series is that plasma prolactin in gynecomastia seems to be most often within the normal range, with an occasional patient showing somewhat elevated levels. Marked elevation of prolactin tends to be seen if galactorrhea is present, and the presence of galactorrhea or of very high prolactin is suggestive of a pituitary tumor. Elevated prolactin by itself does not appear to be sufficient to produce clinical breast development in the male, since we have encountered a number of patients with high levels of the hormone in association with pituitary tumors in whom there was little or no evidence of gynecomastia.

J. HYPOTHYROIDISM

The very recent discovery by Tashjian, Barowsky, and Jensen (1971) that TRH added to pituitary glands incubated *in vitro* can stimulate release of prolactin, has raised the possibility that prolactin secretion and that of thyrotropin may be linked by a common releasing hormone. Although the status of TRH in primary hypothyroidism is not as yet clear, we felt that it would be of interest to measure prolactin levels in such patients. Table XVI shows the results in a group of 12 patients, all with unequivocal clinical and laboratory evidence of hypothyroidism, demonstrated to be of primary thyroid origin by appropriate tests. Of 12 patients studied, only one, E.Z., had definite elevation of prolactin by both assays. Three others had levels which were detectable by bioassay but within the normal range by immunoassay. One of these, G.W., exhibited a pronounced discrepancy between the two assays with a level of 54 ng/ml by bioassay and 6.5 ng/ml by immunoassay. This was a 12-year-old boy with severe primary myxedema ($T_4 < 1.0$ μg/100 ml), markedly short stature, early pubertal changes, slight gynecomastia, and a normal sella turcica. The patient with the highest level, E.Z., was a 9-year-old girl with primary myxedema (T_4 1.7 μg/100 ml), questionable enlargement of the sella turcica, and no· pubertal changes. This was the only patient in the group with evidence of an enlarged pituitary. Neither this patient nor any of the others had galactorrhea. The association of primary myxedema, enlarged pituitary, galactorrhea, and precocious puberty was described by Van Wyk and Grumbach (1960) in three patients; the galactorrhea, which it was speculated might be due to high circulating prolactin, cleared on treatment with thyroid hormone. A few additional cases of this same syndrome have subsequently been reported. Very recently Forsyth *et al.* (1971) have reported a bioassay similar ours employing breast tissue from pseudo-pregnant rabbits, and have noted one patient with primary hypothyroidism and galactorrhea without evident pituitary tumor who had a prolactin level of 2000

TABLE XVI
Plasma Prolactin in Primary Hypothyroidism

Patient	Sex	Age	PBI or T_4 (μg/100 ml)	Prolactin (ng/ml) Bioassay	Immunoassay	HGH (ng/ml)
E.Z.	F	9	1.7	78	80	<0.3
G.W.	M	12	<1.0	54	6.5	1.4
A.D.	M	65	0.8	33	11	—
J.H.	F	57	1.8	29	3.5	0.7
G.L.	F	80	0.6	<30	9.2	—
F.J.	F	50	—	<30	2.2	0.8
T.T.	M	41	1.4	<15	5.7	—
R.L.	F	63	2.4	<15	8.2	—
H.J.	F	55	1.0	<15	6.9	6.0
E.S.	F	50	1.2	—	2.5	2.0
M.H.	F	37	1.5	—	14.0	<0.5
J.V.	M	55	1.6	—	2.0	<0.3

ng/ml. The prolactin levels reported by these authors in patients with pituitary tumors, as well as the level at which prolactin can be detected in blood, appear to be somewhat higher than our own. The results in the 12 patients we have studied suggest that prolactin in patients with primary hypothyroidism is most often within the normal range, although an occasional high value may be encountered, especially where there is an associated enlargement of the pituitary.

K. Renal Failure

Renal failure appears to be associated with a tendency toward elevation of plasma prolactin. Table XVII lists 13 patients with chronic renal insufficiency due to various causes. More than half of the patients had prolactin levels which by bioassay or immunoassay were considered to be elevated. None of these patients had galactorrhea. Although the administration of. drugs might be invoked to explain the high prolactin values in some of these patients, and although we have previously noted elevated prolactin levels and galactorrhea in a patient receiving α-methyl dopa (Kleinberg and Frantz, 1971), it seems likely to us from these studies that renal failure in some patients can be a cause of elevated

TABLE XVII
Plasma Prolactin in Renal Failure

| | | | | | Prolactin (ng/ml) | | |
| | | | | Associated | | Immuno- | HGH |
Patient	Sex	Age	BUN	drugs	Bioassay	assay	(ng/ml)
R.B.	M	44	80		78	64	6.0
A.N.	M	48	65		<22	17	<0.3
V.V.	F	14	65		20	7.4	11.0
M.E.	F	20	100		65	49	8.7
R.C.	M	19	90		—	102	1.2
C.M.	M	50	140		<33	36	1.3
M.M.	M	28	120	α-Methyl dopa	185	140	1.4
T.G.	M	44	100	α-Methyl dopa	19	36	2.3
A.L.	F	43	35	α-Methyl dopa	33	39	16.8
A.J.	F	28	70	α-Methyl dopa, chlordiazepoxide	53	80	8.6
H.W.	F	36	85	α-Methyl dopa, isoniazid, amitriptyline	204	400	3.0
G.M.	F	32	160	α-Methyl dopa, chlorpromazine	185	140	1.4
J.R.	M	36	115	Chlorpromazine	190	280	5.6

plasma prolactin. Further studies will be necessary to determine the mechanisms by which this effect occurs.

L. Estrogens

Prolactin levels have been measured in 14 male patients receiving chronic estrogen therapy for various conditions, chiefly carcinoma of the prostate, in doses equivalent to 5–50 mg of diethylstilbestrol per day. The subjects were aged 15–73, were active, nonhospitalized, and in fair general health. The mean prolactin level of this group, 20.7 ng/ml, was more than double that of our normal males, and the two means were significantly different ($p < 0.01$). To provide additional evidence of the effect of estrogens, five normal male volunteers, aged 28–35, were given diethylstilbestrol, either 15 or 50 mg/day for 1 week, and prolactin levels were measured at the beginning and end of this period, as shown in the last column of Fig. 17. In all cases administration of estrogen was followed by a rise in prolactin; in three subjects it was relatively slight, whereas in two it was more substantial. The data taken together suggest that estrogen in high doses produces a definite, though modest, elevation in plasma prolactin, being a less potent stimulus in this respect

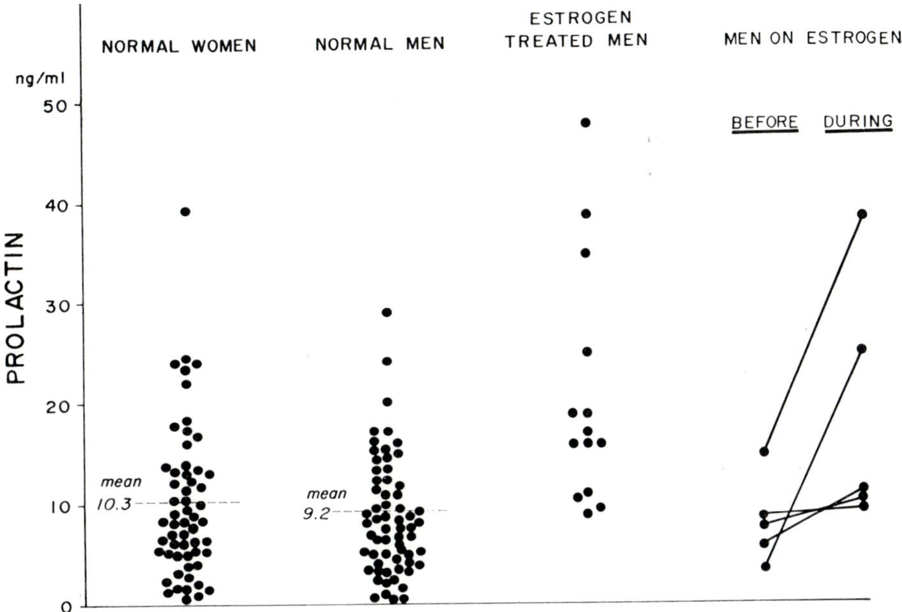

FIG. 17. Plasma prolactin by radioimmunoassay in men on chronic estrogen therapy (5–50 mg diethylstilbestrol per day). The right-hand column represents prolactin in 5 normal males before and after 1 week of diethylstilbestrol, 15–50 mg per day.

than high doses of phenothiazines. We had earlier had definite indications of this effect of estrogens by our bioassay studies, but were able to achieve better quantitation of minimally elevated levels by radioimmunoassay. Estrogens have previously been shown to facilitate growth hormone release (Frantz and Rabkin, 1965). Their effect on plasma prolactin levels in these doses would not appear to be greater in quantitative terms than what we have previously observed for growth hormone. We have not as yet systematically observed the effects of graded doses of estrogen, particularly those closer to physiological levels. The failure to observe major differences in prolactin between men and women suggests that within the normal male and female range of estrogen concentrations, variations in dose levels of this hormone may not produce large effects on plasma prolactin concentrations. Administration of estrogen to animals has previously been considered on the basis of pituitary gland studies to increase prolactin secretion (Meites *et al.*, 1963), and has recently been shown to raise plasma prolactin in the rat by radioimmunoassay (Chen and Meites, 1970).

VI. Discussion and Summary

The identification of a separate human prolactin has opened up a new field of hormone studies in man. It does not appear possible as yet to fit all the data presented here into a unified concept of prolactin regulation. Nevertheless a considerable amount of what we have found confirms in man what had previously been observed in animals (see review by Meites and Nicoll, 1966). In particular this includes the importance of the suckling stimulus in postpartum individuals, a reflex which appears to be mediated by direct sensory stimulation of the breast itself, not by psychic factors. This stimulus seems to be the most specific for prolactin of any yet discovered, and the magnitude of the prolactin response confirms a major role for this hormone in the maintenance of human lactation.

The rise of the hormone seen in conditions of stress is harder to explain from a teleological point of view; it may well be of no functional importance, although it may help to explain the occasional episodes of galactorrhea noted infrequently after various kinds of major surgery. The biological properties of human prolactin preparations on organs other than breast tissue in our own assay system, where it closely resembles ovine prolactin, have not yet been investigated. Ovine prolactin in animals and in man (Beck *et al.*, 1961) has a number of growth hormone-like actions, though to a relatively small degree compared with growth hormone itself. The absence of carbohydrate intolerance or acromegalic changes in one of our patients with a pituitary tumor and prolactin levels of 6500–10,000 ng/ml suggests that human prolactin does not exert growth hormone actions to an important degree. Nevertheless it is entirely possible that the human hormone may possess metabolic or other specific properties which have so far gone unobserved; further study is necessary on this important point.

The control of prolactin secretion in relation to that of growth hormone is a topic of particular interest. Possessing many biological actions in common, and possibly descended from a single ancestral polypeptide (Niall *et al.*, 1971), these hormones show many striking differences in their responses to stimuli as well as some interesting similarities. Suckling produces no stimulation of growth hormone in humans; anatomical lesions which disturb the relation of the pituitary to the hypothalamus, including stalk section and tumors of various kinds, tend to depress growth hormone and elevate prolactin. The same effect is produced by antiadrenergic drugs, such as phenothiazines, which deplete brain catecholamines. L-Dopa acts on both hormones in a manner opposite to phenothiazines. On the other hand, some stimuli, such as stress, hypoglycemia, and estrogens, tend to promote release of both hormones. It is clear that the regulation of prolactin secretion in man, like that of growth hormone, is a complex process and one that remains a challenging field for future research.

The main findings we have presented may be summarized as follows:

1. The mouse breast in organ culture, because of its high sensitivity as well as specificity, is a useful tool for the bioassay of prolactin in plasma.

2. Bioassay as well as immunological studies clearly establish the presence of a human prolactin, separate from growth hormone, which circulates in blood.

3. The act of suckling, in nursing mothers, is a potent stimulus to prolactin release. The reflex is mediated primarily by direct sensory stimulation of the breast, not by psychic factors. A significant part of the total prolactin secreted by nursing mothers is released in association with suckling, and it is probably this episodic secretion which maintains the breast in an actively lactating state.

4. Pituitary tumors of various kinds may be associated with high prolactin levels, often in the absence of galactorrhea; prolactin in these conditions may be high when all other anterior pituitary hormones are low and the patient has clinical panhypopituitarism.

5. Galactorrhea can occur with prolactin levels which range all the way from normal to extremely high. In many cases galactorrhea occurs in association with regular menses and prolactin levels that are within the normal range.

6. Phenothiazines can cause acute as well as sustained secretion of prolactin in normal men and women. Their acute administration can serve as a useful test of hypothalamic–pituitary function.

7. L-Dopa can effectively suppress the prolactin secretion in normal individuals caused by chlorpromazine, and can also suppress prolactin in patients with elevated levels due to a variety of causes.

8. Stress of several different kinds is associated with prolactin release.

9. Hypoglycemia of sufficient magnitude causes prolactin release.

10. Prolactin, whatever the degree of growth hormone-like properties which it may possess, is probably not the agent responsible for the rapid

growth that occasionally occurs following surgery in children with craniopharyngioma.

11. Gynecomastia, unaccompanied by other abnormalities is usually associated with prolactin levels within the normal range. If galactorrhea is also present prolactin will tend to be high, and the presence of a pituitary tumor is suggested.

12. Patients with acromegaly may have elevated prolactin as well as growth hormone, although in the majority of acromegalic patients prolactin is normal.

13. Prolactin is usually normal in primary hypothyroidism, but may be high in an occasional patient, particularly if there is associated enlargement of the pituitary.

14. Renal failure is frequently associated with elevated plasma prolactin.

15. Estrogens in high doses cause a rise in plasma prolactin in adults.

16. Radioimmunoassay of human prolactin, using a human standard recently prepared by Dr. Friesen and his colleagues, has given results that are in good agreement with those of bioassay. Although a minority of plasma samples exhibit possibly significant discrepancies in the two assays, bioassay results indicate that what is measured by radioimmunoassay is biologically active prolactin, and that the human hormone has a potency by weight which is very close to that of the ovine standard.

Addendum

Further work on the bioassay of Dr. Friesen's human prolactin in the mouse breast organ culture indicates that it has a potency by weight of 1.10 (95% confidence limits: 0.89–1.34) times that of the NIH-P-S8 ovine standard, whose potency is 28 U/mg. There is complete parallelism of the response curves for both prolactins at all levels. This figure is in very close agreement with the mean bioassay:immunoassay ratios for 50 plasma samples of 1.09 (95% confidence limits: 0.86–1.39) which we had previously determined (Table V).

Since this work was presented at the 1971 Laurentian Hormone Conference two papers have appeared describing *in vitro* bioassays for prolactin based on the mouse breast organ culture system (Loewenstein *et al.*, 1971; Turkington, 1971). Both use radiochemical rather than morphologic end points, with sensitivity similar to what we have described. The results on plasma are in general agreement with what we had previously reported (Kleinberg and Frantz, 1970; Frantz and Kleinberg, 1970). Turkington's figures for normal human plasma (less than 2 ng/ml) differ significantly from our own, as do the uniformly higher than normal values he has observed in idiopathic galactorrhea. A third paper based on prolactin bioassay which has appeared subsequent to the Conference (Turkington *et al.*, 1971) reports results on 15 patients subjected to pituitary stalk section for metastatic carcinoma of the breast or diabetic retinopathy, 11 of whom had elevated plasma prolactin postoperatively.

ACKNOWLEDGMENTS

The authors are grateful to Dr. Henry Friesen and his collaborators for permitting us to test their human prolactin preparation and for providing the materials that made possible the radioimmunoassay studies. Drs. Anne Forbes, Saul Rosen, and Sander Klein each furnished us with plasma specimens on patients with gynecomastia or galactorrhea. Dr. Wendell Niemann provided invaluable help with the baboon studies. Dr. A. E. Wilhelmi and the NIAMD made available human growth hormone and ovine prolactin. Mr. Robert E. Sundeen, Miss Irene Johnson, and Miss Charity L. Young provided expert technical assistance.

REFERENCES

Apostolakis, M. (1968). *Vitam. Horm. (New York)* **26**, 197.
Beck, J. C., Gonda, A., Hamid, M. A., Morgan, R. O., Rubinstein, D., and McGarry, E. E. (1961). *Metab. Clin. Exp.* **13**, 1108.
Bewley, T. A., and Li, C. H. (1970). *Science* **168**, 1361.
Birge, C. A., Jacobs, L. S., Hammer, C. T., and Daughaday, W. H. (1970). *Endocrinology* **86**, 120.
Boyd, A. E., III, Lebovitz, H. E., and Pfeiffer, J. B. (1970). *New Engl. J. Med.* **283**, 1425.
Bryant, G. D., Connan, R. M., and Greenwood, F. C. (1968). *J. Endocrinol.* **41**, 613.
Bryant, G. D., Linzell, J. L., and Greenwood, F. C. (1970). *Hormones* **1**, 26.
Bryant, G. D., Siler, T. M., Greenwood, F. C., Pasteels, J. L., Robyn, C., and Hubinont, P. O. (1971). *Hormones* **2**, 139.
Chadwick, A., Folley, S. J., and Gemzell, C. A. (1961). *Lancet* **ii**, 241.
Chen, C. L., and Meites, J. (1970). *Endocrinology* **86**, 503.
Cowie, A. T., Daniel, P. M., Knaggs, G. S., Prichard, M. M. L., and Tindal, J. S. (1964). *J. Endocrinol.* **28**, 253.
Danon, A., Dikstein, S., and Sulman, F. G. (1963). *Proc. Soc. Exp. Biol. Med.* **114**, 366.
Donovan, B. T., and van der Werff ten Bosch, J. J. (1957). *J. Physiol. (London)* **137**, 410.
Ehni, G., and Eckles, N. E. (1959). *J. Neurosurg.* **16**, 628.
Elias, J. J. (1957). *Science* **126**, 842.
Ferguson, K. A., and Wallace, A. L. C. (1961). *Nature (London)* **190**, 632.
Forsyth, I. A., Besser, G. M., Edwards, C. R. W., Francis, L., and Myres, R. P. (1971). *Brit. Med. J.* **iii**, 225.
Frantz, A. G., and Kleinberg, D. L. (1970). *Science* **170**, 745.
Frantz, A. G., and Rabkin, M. T. (1965). *J. Clin. Endocrinol. Metab.* **25**, 1470.
Glick, S. M., Roth, J., Yalow, R. S., and Berson, S. A. (1965). *Recent Progr. Horm. Res.* **21**, 241.
Greenwood, F. C., Hunter, W. M., and Glover, J. S. (1963). *Biochem. J.* **89**, 114.

Grosvenor, C. E., and Turner, C. W. (1957). *Proc. Soc. Exp. Biol. Med.* **96,** 723.

Grosvenor, C. E., McCann, S. M., and Nallar, R. (1965). *Endocrinology* **76,** 883.

Herbert, D. C., and Hayashida, T. (1970). *Science* **169,** 378.

Holmes, L. B., Frantz, A. G., Rabkin, M. T., Soeldner, J. S., and Crawford, J. D. (1968). *New Engl. J. Med.* **279,** 559.

Hwang, P., Guyda, H., and Friesen, H. (1971a). *Proc. Nat. Acad. Sci. U.S.* **68,** 1902.

Hwang, P., Friesen, H., Hardy, J., and Wilansky, D. (1971b). *J. Clin. Endocrinol. Metab.* **33,** 1.

Juergens, W. G., Stockdale, F. E., Topper, Y. J., and Elias, J. J. (1965). *Proc. Nat. Acad. Sci. U.S.* **54,** 629.

Kamberi, I. A., Mical, R. S., and Porter, J. C. (1970). *Experientia* **26,** 1150.

Kenny, F. M., Iturzaeta, N. F., Mintz, D., Drash, A., Garces, L. Y., Susen, A. and Askari, H. A. (1968). *J. Pediat.* **72,** 766.

Kleinberg, D. L., and Frantz, A. G. (1969). *Program 51st Meet. Endocrine Soc.* Abstr. No. 32.

Kleinberg, D. L., and Frantz, A. G. (1970). *Clin. Res.* **18,** 363.

Kleinberg, D. L., and Frantz, A. G. (1971). *J. Clin. Invest.* **50,** 1557.

Kleinberg, D. L., Noel, G. L., and Frantz, A. G. (1971). *J. Clin. Endocrinol. Metab.* **33,** 873.

Koch, Y., Lu, K. H., and Meites, J. (1970). *Endocrinology* **87, 673.**

Lasfargues, E. Y. (1957). *Exp. Cell Res.* **13,** 553.

Lewis, U. J., Singh, R. N. P., Sinha, Y. N., and VanderLaan, W. P. (1971). *J. Clin. Endocrinol. Metab.* **33,** 153.

Loewenstein, J. E., Mariz, I. K., Peake, G. T., and Daughaday, W. H. (1971). *J. Clin. Endocrinol. Metab.* **33,** 217.

Lyons, W. R., Li, C. H., and Johnson, R. E. (1961). *Program 43rd Meet. Endocrine Soc.* Abstr. No. 7.

MacLeod, R. M., Fontham, E. H., and Lehmeyer, J. E. (1970). *Neuroendocrinology* **6,** 283.

Meites, J., and Nicoll, C. S. (1966). *Annu. Rev. Physiol.* **28,** 57.

Meites, J., Nicoll, C. S., and Talwaker, P. K. (1963). *Advan. Neuroendocrinol. Proc. Symp., 1961* p. 238.

Niall, H. D., Hogan, M. L., Saure, R.. Rosenblum. I. Y., and Greenwood, F. C. (1971). *Proc. Nat. Acad. Sci. U.S.* **68,** 866.

Noel, G. L., and Frantz, A. G. (1970). Unpublished observations.

Noel, G. L., Suh, H. K., and Frantz, A. G. (1971). *Clin. Res.* **19,** 718.

Powell, E. D. U., Frantz, A. G., Rabkin, M. T., and Field, R. A. (1966). *New Engl. J. Med.* **275,** 922.

Reece, R. P., and Turner, C. W. (1937). *Proc. Soc. Exp. Biol. Med.* **35,** 621.

Riddle, O., Bates, R. W., and Dykshorn, S. (1933). *Amer. J. Physiol.* **105,** 191.

Rivera, E. M., and Bern, H. A. (1961). *Endocrinology* **69,** 340.

Sherman, L., Kim, S., Benjamin, F., and Kolodny, H. D. (1971). *New Engl. J. Med.* **284,** 72.

Smithies, O. (1959). *Advan. Protein Chem.* **14,** 65.

Sulman, F. G. (1970). "Hypothalamic Control of Lactation." Springer-Verlag, Berlin and New York.

Tashjian, A. H., Jr., Levine, L., and Wilhelmi, A. E. (1965). *Endocrinology* **77,** 1023.

Tashjian, A. H., Jr., Barowsky, N. J., and Jensen, D. K. (1971). *Biochem. Biophys. Res. Commun.* **43,** 516.

Topper, Y. J. (1970). *Recent Progr. Horm. Res.* **26,** 287.

Turkington, R. W. (1971). *J. Clin. Endocrinol. Metab.* **33,** 210.

Turkington, R. W., Underwood, L. E., and Van Wyk, J. J. (1971). *New Engl. J. Med.* **285,** 707.

Van Wyk, J. J., and Grumbach, M. M. (1960). *J. Pediat.* **57,** 416.

Wilhelmi, A. E. (1961). *Can. J. Biochem.* **39,** 1659.

DISCUSSION

F. C. Greenwood: I thought it would be helpful to summarize, under the headings present, high levels, rises, and falls, what to my knowledge has been shown by bioassays and radioimmunoassays of human prolactin in plasma, as modified by Dr. Frantz' contribution. There has been no disagreement between the techniques, and Dr. Frantz has shown here a superb correlation.

Under "present," we can put normal children and adults with as yet no sex differences. No discernible pattern is so far seen in the menstrual cycle, but this may require more frequent sampling. Estimates for the half-life of endogenous prolactin are 5 to 15 minutes.

Under "high levels," stalk section, the fetus and newborn can be given unequivocal pluses, but patients with galactorrhea, acromegaly, renal disease, idiopathic hypopituitarism, or myxedema must be rated plus-minus; i.e., high levels are not always found. Under "rises," are suckling, the administration of phenothiazines and butyrophenones, and pregnancy itself. One must single the latter out from this otherwise unattributed catalog, as with Dr. Friesen's finding, particularly as he has shown that this rise does not take place in the monkey during pregnancy. The rises on insulin, previously unequivocal, are now seen as simply dose-dependent, and now conform to the stress response of prolactin previously shown and extended by Dr. Frantz to psychological stress and exercise.

Under the heading of "falls," may be added ergocornine and L-dopa with "no changes" on arginine or in breast cancer.

Is there anyone who would like to argue the point that human prolactin does not exist as a separate entity?

A. Segaloff: If we go back to some of the ancient history and some of the arguments we have had about what you could find in urine by pigeon crop assay, a great deal of what we found and reported coincides with what we have heard today. We have found a mid-cycle rise in active material through quite a few cycles. We also found that stress raises it in some people but not in others, as has been reported today, although many of our patients who were stressed were having radical mastectomies, and that may be a lot more stressful than some of the things reported here. What I would really like to have clarified is whether or not there has been an adequate effort to find a mid-cycle peak of plasma prolactin.

H. G. Friesen: I agree fully with Dr. Segaloff that more work is necessary in this regard. We have simply taken daily samples throughout the menstrual cycle in 20 patients and measured prolactin. There is a considerable degree of fluctuation throughout the cycle in each patient, but certainly no evidence of any systematic change in prolactin levels in relation to the LH peak.

R. B. Jaffe: Together with Drs. L'Hermite and Midgley we have quantitated gonadotropins during and following human pregnancy. In the case illustrated, gonadotropins were quantitated for 295 days after delivery (Fig. A). This subject breast fed. The prolactin concentrations are illustrated above, and the LH and FSH illustrated below. The prolactin was quantitated using homologous ovine-ovine radioimmunoassay and an antibody which Gordon Niswender developed. The episodic variation to which Dr. Frantz referred does occur following delivery in breast-feeding women and, with the advent of weaning, the prolactin levels fall. Following the resumption of normal menses, although there was resumption of a

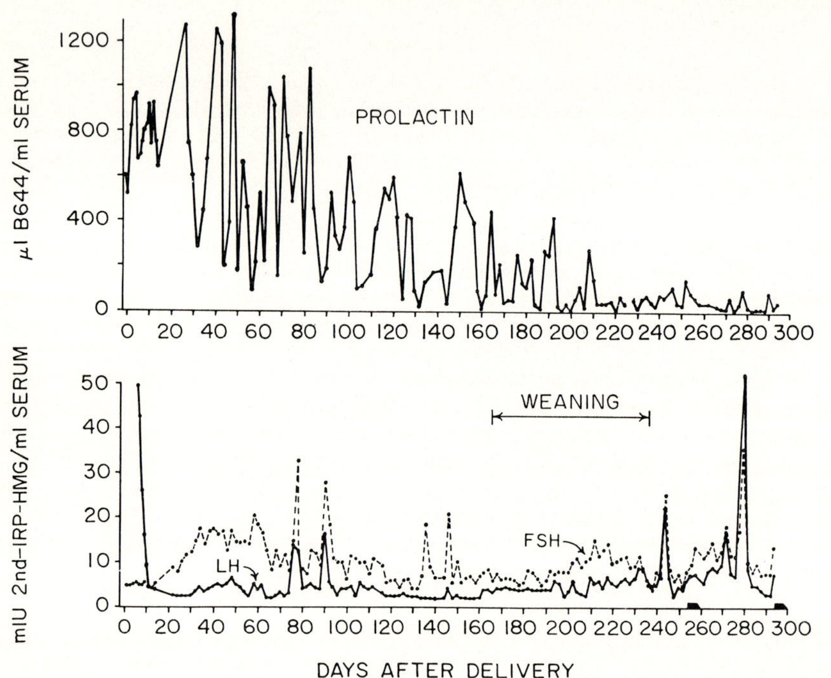

Fig. A.   Hormone concentrations during and after lactation.

cyclic pattern of LH and FSH, we were unable to see any profile of prolactin of a similar cyclic nature. This was also borne out in other studies of cycling nonpregnant women in whom the type of variation to which Dr. Friesen referred was seen, but in whom no cyclic pattern was observed.

G. L. Cohn: I am somewhat confused about what may be discrepancies in the data. After the administration of phenothiazines, which are major tranquilizers, there are striking rises in prolactin. On the other hand, when patients are undergoing operative stress, the rises are also there. Can you explain this discrepancy?

A. G. Frantz: I think we have to assume that the pathways for prolactin stimulation are different in these two cases. Presumably the phenothiazines are acting by inhibition of PIF, since *in vitro* studies in which pituitaries were incubated in coculture with hypothalami showed that hypothalami from perphenazine-treated rats lacked the power of normal hypothalami to inhibit prolactin release. Perphenazine itself did not inhibit the pituitary when added to the medium [A. Danon, S. Dikstein, and F. G. Sulman, *Proc. Soc. Exp. Biol. Med.* **114,** 366 (1963)]. What the stress-mediated pathway is I do not know. It might of course also involve inhibition of PIF in some way, but it is possible to speculate that it might work in quite a different way by stimulating a prolactin-releasing factor such as that recently postulated by Nicoll [C. S. Nicoll, R. P. Fiorindo, C. T. McKennee, and J. A. Parsons, *In* "Hypophysiotropic Hormones of the Hypothalamus: Assay and Chemistry" (J. Meites, ed.), p. 115. Williams and Wilkins, New York, 1970.]. I know of no evidence at present that would support such a speculation, but it seems worth considering.

F. Fuchs: There is no doubt that lactating mothers resume ovulation at a much later time than nonlactating mothers after parturition. Does prolactin in some way interfere with the action of FSH and LH on the ovary and thereby prevent ovulation? I was intrigued by your findings in renal failure because in all of our kidney transplantation patients we have found that they become amenorrheic at a certain stage of their renal failure. Once they have been transplanted, the menstrual cycle returns, and a number of these patients have become pregnant. I wonder if there is a parallel between the inhibition of ovulation postpartum and the amenorrhea found in patients with renal failure. Does venipuncture provide enough stress to cause a rise in prolactin levels in the blood?

A. G. Frantz: We do not have enough data to speculate on your first question about the effect of prolactin on the ovary directly. We do know that there are a number of patients who can certainly menstruate regularly even though they have galactorrhea. Most of these patients do have normal prolactin, but in a few it is elevated. In relation to your last question, in the majority of adult patients venipuncture is probably not a sufficient stress to elevate prolactin because in repeated samplings, where stress is less, levels often remain about the same. There are a few subjects, particularly those who are anticipating some sort of major procedure, who have high values, and I think it is probably the psychic stress that is causing the elevation.

A. C. Carter: Are there differences between the levels of prolactin in the "true" basal state or the "ambulatory" basal state? Are there any changes in prolactin levels during sleep?

A. G. Frantz: We do not as yet have evidence of the kind that we had earlier furnished with respect to growth hormone for differences in prolactin between ambulatory and basal states. We have not yet studied sleep but hope to do so.

N. A. Samaan: Dr. Frantz showed that in idiopathic galactorrhea the prolactin levels in most of these cases were within normal range. What is the mechanism of galactorrhea in these patients? In two of our patients with galactorrhea and amenorrhea, we made similar findings. One of these patients had a prolactin-producing tumor which was removed surgically, the pituitary gland being left intact. The serum prolactin level in this patient was 250 ng per milliliter of serum before surgery and 35 ng postoperatively. Although this patient resumed regular menstruation after surgery, the galactorrhea continued and was of similar degree to that seen before surgery. The other patient was a case of galactorrhea, amenorrhea, diabetes insipidus, and diabetes mellitus. In this patient, we failed to stimulate her growth hormone secretion by insulin hypoglycemia or arginine tests. The prolactin level was only 35 ng, which is again slightly above normal range (normal 5–25 ng/ml serum). The patient's clinical condition deteriorated rapidly, and she developed a florid picture of Letterer-Siwe syndrome in an adult and died.

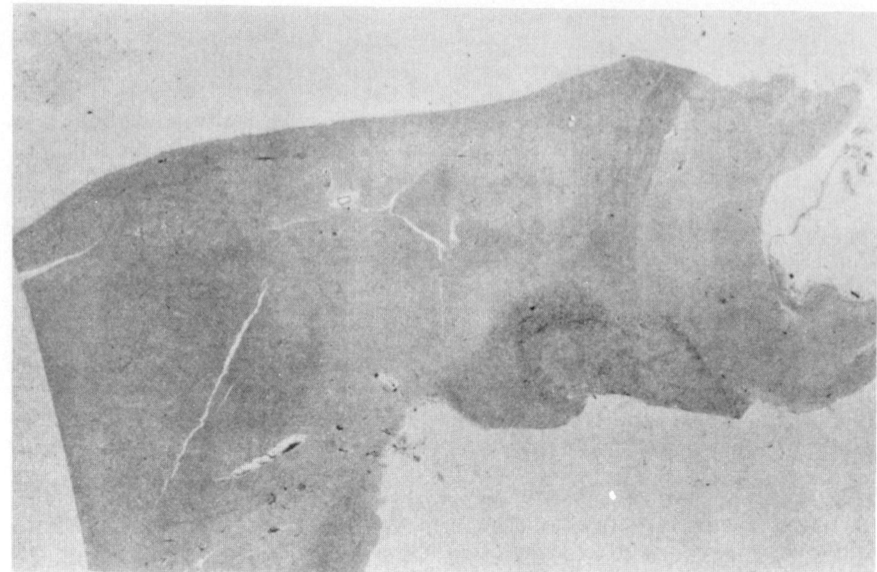

Fig. B. Anteroposterior section of hypothalamus infiltrated with histocytosis X cells.

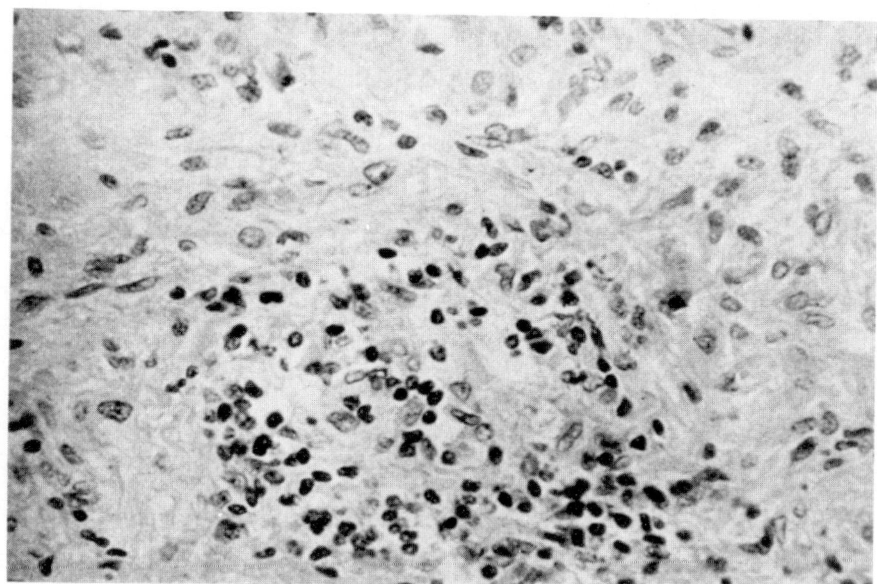

Fig. C. Magnification of Fig. A showing infiltation of the hypothalamus with histocytosis X cells.

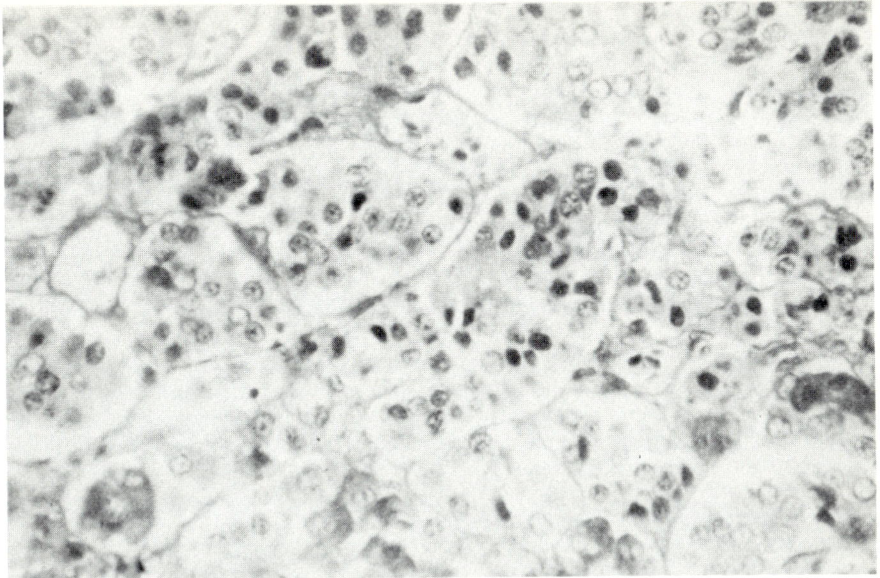

Fig. D.   Pituitary gland tissue stained with Herlant's stain.

Figure B shows an anteroposterior section of hypothalmus infiltrated with histocytosis X cells. Figure C is a magnification of Fig. A, and it shows the infiltration of the hypothalmus with histocytosis X.

Figure D is rather important. It shows the pituitary gland of the patient of Figs. A–C; the Herlant stain was used. With this technique, the prolactin-producing cells appear reddish pink, and the growth hormone-secreting cells stain yellow. I was told by the pathologist that he did not find in this field or other fields growth hormone-producing cells and that all cells seen are prolactin-producing cells. In spite of an abundance of the prolactin-producing cells in the pituitary gland, the prolactin level was slightly elevated. The prolactin level was measured by both Drs. Henry Friesen and W. Daughaday. Here, we have the pituitary gland full of prolactin while the level in the circulation is slightly above the normal range. Why do some of the patients with idiopatic galactorrhea have normal prolactin levels?

A. G. Frantz: In answer to your first question about idiopathic galactorrhea and normal prolactin levels, it certainly does not seem to take very much prolactin to maintain galactorrhea. It may be that a small but significant elevation of prolactin, perhaps occurring only episodically, is sufficient. It may conceivably be that some physical stimulus to the breast itself could function to maintain lactation in certain galactorrhea patients in somewhat the same way that suckling does for nursing mothers. Those are just speculations. The patient shown in Figs. B–D is very interesting. It is perhaps surprising that there should be so many prolactin-secreting cells in the pituitary with such a relatively slightly elevated serum level, but this may be a parallel to the situation in the nursing mother, where there have been shown to be an increased number of prolactin-containing cells in the pituitary, but where in the intervals between nursing plasma prolactin may not be very much elevated.

R. M. Rose: Green and co-workers in Rochester in studying the responses of patients before and during cardiac catheterization found very interesting dissociations in that some patients showed cortisol elevations and some showed cortisol as well as growth hormone responses prior to and during catheterization. We made similar observations of dissociations between growth hormone and cortisol responses during stress in our own laboratory. Do you have any information on the patients you reported where you measured increased prolactin levels in various stressful situations; i.e., whether there were parallel responses in cortisol and/or growth hormone?

A. G. Frantz: Growth hormone did tend to rise along with prolactin in the most stressful situations, that is surgery, but the peaks were not always coincident and the magnitude of the rise, on a percentage basis, sometimes appeared to be less than for prolactin. In the gastroscopy and proctoscopy patients the prolactin response was definitely greater than that of growth hormone, which frequently showed no change. In exercise, on the other hand, the growth hormone rise, in terms of percentage change, was greater. We have not done cortisol determinations as yet on most of these plasmas, but this is definitely something to explore. The cardiac catheterization studies I think are very interesting. We have gathered some plasma samples from patients undergoing cardiac catheterization but have not had a chance to analyze them for prolactin.

F. C. Greenwood: Would you agree with me that prolactins are in fact more volatile in psychic and physical stress than growth hormone, with the exception of exercise?

A. G. Frantz: In a general way, yes, particularly with respect to the milder forms of stress we have looked at. As far as psychic stress goes, we need more data before we could answer that properly.

J. C. Beck: Would you clarify the prolactin content of the growth hormone preparation that you bioassayed? What I am getting at is the undoubted presence of human prolactin in almost all preparations of human growth hormone.

A. G. Frantz: The bioassayable prolactin activity of highly purified human growth hormone preparations, as I mentioned, is of the order of 50–80% that of an equal weight of ovine prolactin in our assay. On the other hand, when human growth hormone was incubated with anti-growth hormone antiserum and then assayed at a final concentration of 600 ng/ml in the incubation medium, all prolactin activity was abolished. The same antiserum under the same conditions produced no neutralization whatever of the prolactin activity in the plasma of galactorrhea and postpartum patients. These experiments indicated that the prolactin activity in the growth hormone preparations we tested was due to the growth hormone itself, not to contaminating prolactin. Human prolactin would have been detected if present in amounts greater than 5 ng/ml, assuming equipotency with the ovine standard. Thus we concluded that human prolactin constituted less than 1% of our human growth hormone preparations. These results correlate fairly well with Dr. Friesen's estimate that the prolactin content of human pituitaries is approximately $\frac{1}{100}$ that of their growth hormone content.

C. Ezrin: I want to illustrate the picture morphologists have had for years of the human pituitary in pregnancy, namely, that it is rich in prolactin cells (see Fig. E). For this reason we have been quietly confident that the game would go our way and that ultimately clinical endocrinologists would accept the fact

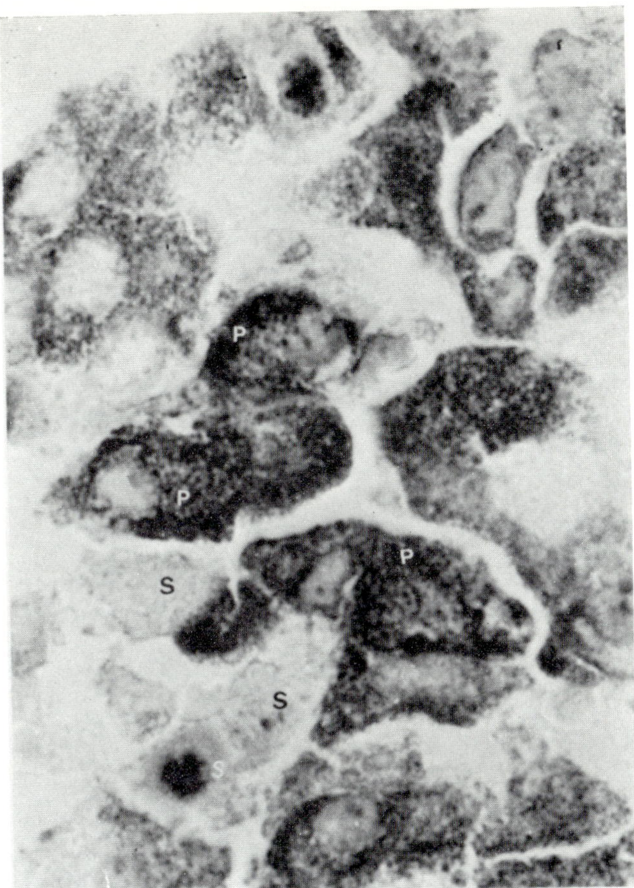

FIG. E.   Anterior pituitary from a pregnant woman at term showing the difference in morphology of the two types of acidophiles. P = prolactin; S = somatotroph or growth hormone cell. Brookes's carmoisin stain. ×1600.

that there is human prolactin. The pituitary of pregnancy shows a large number of prolactin cells only in the second and third trimester; in lactation there is a similar change. In the pituitaries of normal adults or children we do not see heavily granulated prolactin cells. That is not to say that the cell is not there, but rather it is not storing sufficient hormone in granular form to show its characteristic stainability. Drs. Herlant and Pasteels of Brussels were among the earliest to suggest that there are two types of acidophiles in the pituitary of animals and humans. We used a modification of Herlant's technique using carmoisin instead of erythrosin to show the prolactin cells. Dr. Goluboff and I have reported a curious reciprocal relationship between the percentage distribution of growth hormone cells and the prolactin cells in pregnancy [L. G. Goluboff, and C. Ezrin, *J. Clin. Endocrinol. Metab.* **29**, 1533 (1969)]. One wonders whether they both come from a common precursor cell, even though in the main they appear to be distinct adult cell types. Do you have any thoughts about why growth hormone should go down, at least in terms of its storage in the pituitary, while prolactin goes up?

A. G. Frantz: I do not think I can offer any good speculations as to the last point, since it has never been clear to me why there is so much growth hormone stored in the pituitary in normal individuals, in contrast to all other pituitary hormones. The inverse relationship between the number of growth hormone and prolactin cells during pregnancy seems another manifestation of the tendency of many agents to affect these hormones in an opposite way, at least as far as secretion is concerned. Despite the great differences in the amounts present in the gland, it is interesting that the daily production of prolactin, based on plasma levels and half-time estimates, is somewhat greater than that of growth hormone even in nonpregnant and nonpostpartum individuals.

F. C. Greenwood: At a recent Ciba Foundation meeting we asked Professor Herlant about this, and he felt that there is more prolactin in the pituitary of the normal adult than can be accounted for by current histological techniques.

H. G. Friesen: In the last few months we have been turning our attention to the problem of attempting to purify human prolactin from acetone-dried human pituitary powder. Dr. Peter Hwang, working in my laboratory, has had moderate success in his attempts. By a combination of gel filtration and ion exchange chromatography, he has succeeded in obtaining a human prolactin preparation which appears reasonably homogeneous. In Fig. F the prolactin preparation is shown on the right on disc gel electrophoresis at an alkaline pH. Human growth hormone, for comparison, is on the left. The biological potency by the local pigeon crop-sac assay of this preparation is in the neighborhood of 15–20 units/mg. The overall yield of prolactin from the pituitary is very small, perhaps on the order of 50 μg per pituitary as compared to 5 mg per pituitary for growth hormone. The molecular weight of prolactin is very comparable to that of growth hormone, and this makes it difficult to distinguish by gel filtration methods. The electrophoretic mobility of this prolactin preparation has an R_f that is not too dissimilar from that reported by Peake *et al.* [(*J. Clin. Endocrinol.* **29**, 1383 (1969)]. They reported an R_f of 0.44 for prolactin, and we found an R_f of 0.5.

H. D. Niall: I would like to report some very recent studies on the amino acid sequence of human prolactin. We obtained about 1 mg of purified human prolactin from Dr. Friesen. We determined the amino-terminal sequence of the first 20 residues using the automated Edman procedure, in the protein sequenator (Fig. G). The central sequence shown is that of human prolactin. For technical reasons we did not obtain a derivative at step 6, but otherwise we are fairly confident of the identifications. The sequence is shown here in comparison with the revised sequence for human growth hormone (above) and with ovine prolactin (below). Whereas only 4 residues of the first 20 are identical between human prolactin and growth hormone, 13 out of 20 are identical when the two prolactins are compared. This indicates clearly that the human prolactin molecule isolated by Dr. Friesen is chemically quite distinct from human growth hormone in its covalent structure.

Since human growth hormone and prolactin have different amino-terminal residues (phenylalanine and leucine, respectively), we can estimate the degree of contamination of preparations of one by the other, by quantitative Edman amino-terminal end-group analysis. By this criterion we have found that Dr. Friesen's human prolactin preparation has no detectable human growth hormone, and that highly purified human growth hormone preparations (N. I. H. Wilhelmi) have no

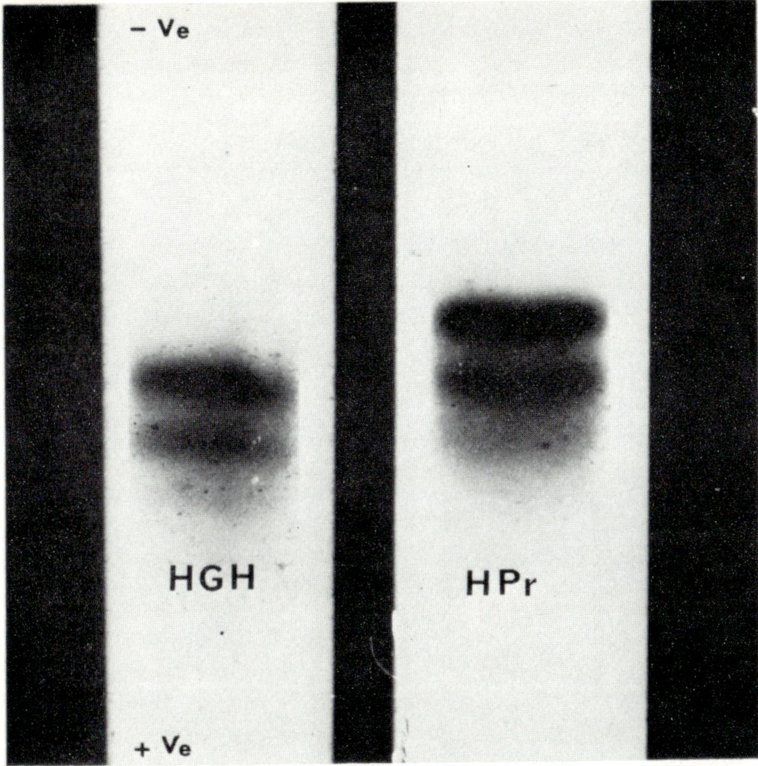

Fig. F. Disc gel electrophoresis upon acrylamide at pH 9.0 of human growth hormone (HGH) and human prolactin (HPr). The anode is at the bottom of the slide. In each case approximately 50 μg of protein was applied. Two protein bands are visible in the HPr preparation. The more anodal band has almost the same electrophoretic mobility as the major band of HGH. The R_f of the two prolactin bands are 0.5 and 0.55, and for HGH 0.55 and 0.62. The HPr preparation contains less than 3% HGH by immunoassay. When the protein in the bands of HPr are eluted from the gel, both cross-react equally in the radioimmunoassay. Presumably the more anodal band is a deamidated form of HPr.

detectable prolactin content. A contamination greater than 2% would be detected by this method.

H. Guyda: I would like to reemphasize what Dr. Frantz has said about the apparent lack of correlation between very elevated levels of prolactin in the blood and apparent physiological effects. We have studied the effect of L-dopa on prolactin secretion in six patients with pituitary disease (Fig. H). Three of these patients had chromophobe tumors. The other three have not been treated as yet, but one does not apparently have a pituitary tumor, whereas the other two patients do. One patient had a prolactin value in excess of 300 ng/ml. She was given 250 mg of L-dopa by mouth, and by 3 hours the value had fallen to 60 ng/ml. A 4-year-old girl with childhood acromegaly and no galactorrhea secreted very high levels of prolactin and also had very high levels of growth hormone. She had serum values of growth hormone in excess of 500 ng/ml. This patient had a paradoxical decrease in growth hormone secretion with L-dopa almost parallel to that seen for the decrease in prolactin. Two patients with intermediate levels of prolactin had a similar fall after L-dopa administration (500 mg p.o.). A male patient (N.C.) had no clinical signs of any pathological effect on his breast tissue.

```
           1                5                10               15               20
HGH    NH₂-Phe-Pro-Thr- Ile -Pro-Leu-Ser-Arg-Leu-Phe-Asp-Asn-Ala-Met-Leu-Arg-Ala -His -Arg-Leu-

HPr    NH₂-Leu-Pro- Ile -Cys-Pro-    -Gly-Ala -Ala -Arg-Cys-Gln -Val-Thr-Leu-Arg-Asp-Leu-Phe-Asp-

OPr    NH₂-Thr-Pro-Val -Cys-Pro-Asn-Gly-Pro-Gly- Asp-Cys-Gln -Val-Ser -Leu-Arg-Asp-Leu-Phe-Asp-
```

FIG. G. Amino-terminal sequences of human growth hormone H. D. Niall [*Nature (London) New Biol.* **230**, 90 (1971); C. H. Li, J. S. Dixon, and W. K. Liu, *Arch. Biochem. Biophys.* **133**, 70 (1969)] human prolactin (present work), and ovine prolactin [Li *et al.*, *Nature* **224**, 695 (1969)].

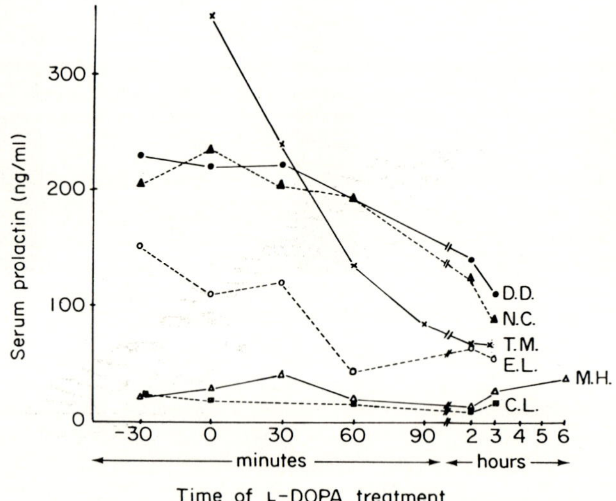

FIG. H. Effect of L-dopa (250–500 mg p.o.) on serum prolactin in patients with pituitary dysfunction (see text).

Another patient (D.D.) had clinical galactorrhea. Two of the patients had normal levels of prolactin (30 ng/ml or less), and both these patients had clinical galactorrhea. Patient M.H. had galactorrhea continuously for two years after removal of a tumor and irradiation. She was given L-dopa (500 mg p.o.). There was very little apparent effect on her prolactin levels; however, she was continued on L-dopa (250 mg t. i. d. by mouth) for 1 week, and within 18 hours there was a dramatic decrease in her galactorrhea and by 48 hours the galactorrhea had disappeared. Patient C.L. also had a relatively normal level of prolactin. She was given L-dopa and also had clinical regression of her galactorrhea within about 1 day of the onset of her daily therapy. These data reveal that L-dopa can decrease serum levels of prolactin and may be associated with regression of galactorrhea.

It is apparent than an elevation in prolactin levels can occur before surgery and just after anesthesia is given (Fig. I). Samples obtained the day before surgery may be low or they may be high.

Eight patients with chronic renal failure undergoing hemodialysis had markedly elevated prolactin levels (up to 1 μg/ml) (Fig. J). Three patients had normal levels. However, in 12 patients who have had renal transplantation, the serum prolactin levels have been normal. It is interesting to speculate on a renal role in the regulation of prolactin secretion.

A. G. Frantz: I think that you have also shown that normal subjects given L-dopa will exhibit a decrease. Were any of your renal failure patients receiving antihypertensive medication?

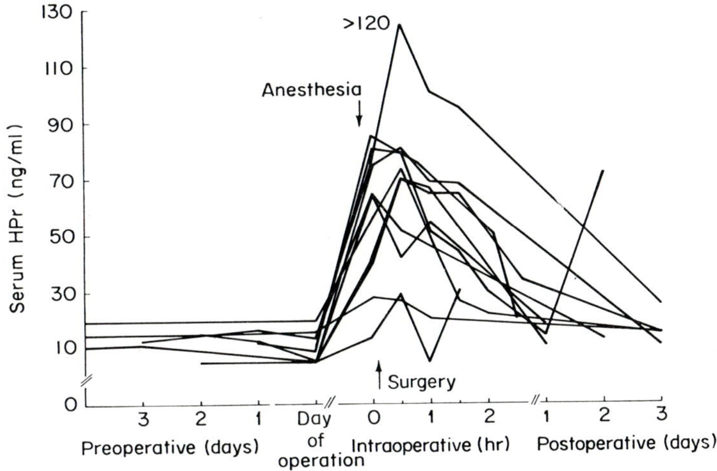

FIG. I. Effect of general anesthesia on serum prolactin (HPr) in 11 patients undergoing abdominal or pelvic surgery.

H. Guyda: Yes, some of them were.

R. B. Jaffe: Dr. Ezrin commented upon the increase in prolactin secreting cells during the latter part of human pregnancy. Prolactin has been measured in women throughout the various stages of pregnancy (Fig. K). Unlike HCG, which diminishes as pregnancy progresses, and unlike FSH, which is suppressed to very low levels throughout pregnancy, prolactin in the human rises as pregnancy progresses, in a manner similar to that Dr. Meites described in the rat and corroborating histologically the findings that Dr. Ezrin mentioned.

R. Jewelewicz: Do you know what happens to prolactin after irradiation of the pituitary? Do you think prolactin is essential for normal lactation? We have had several patients with irradiated pituitary tumors who conceived after ovulation induced by HMG-HCG therapy. Postpartum they had a normal lactation period.

A. G. Frantz: We have had a few patients with chromophobe adenomas and high plasma prolactin whose levels fell, though not to normal, following pituitary irradiation. Among patients previously irradiated for tumors we have encountered both normal and elevated prolactin, but the hormone has always been detectable by radioimmunoassay, i.e., present at levels of 1.0 ng/ml or greater. I do think prolactin is essential for postpartum lactation, since we have never failed to detect it by radioimmunoassay, but the levels may not have to be very high; during non-suckling intervals they may be within the normal range.

J. R. Goding: The graph you presented of human prolactin secretion during pregnancy looked very much like the curves Christine Beck found for human placental lactogen (HPL). Is there any possibility of a cross-reaction between HPL

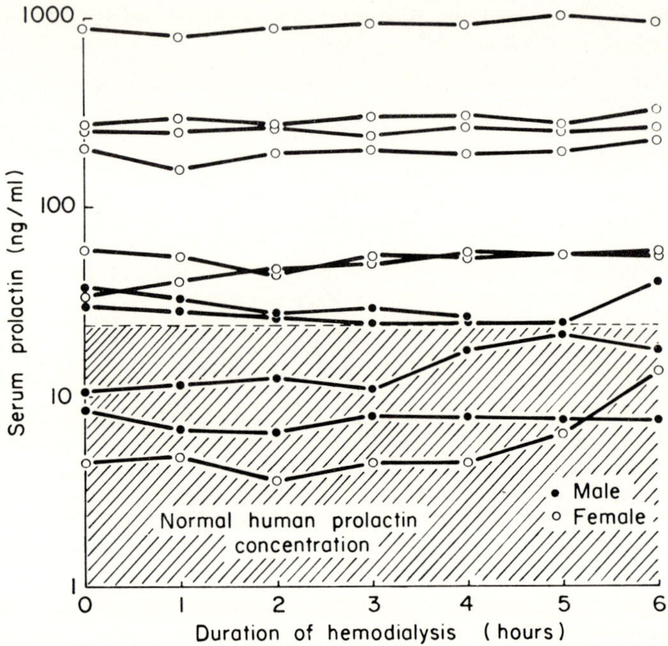

Fig. J. Serum prolactin in 11 patients with chronic renal failure undergoing hemodialysis.

and human prolactin in your assay? On the question of the lability of prolactin, our experience has been that, in sheep, it has been necessary to sample at very frequent intervals. We think that the half-life of endogeneous prolactin in the ewe is somewhat less than 5 minutes [I. A. Cumming, J. M. Brown, J. R. Goding, G. Bryant, and F. C. Greenwood, *J. Endocrinol.* (1971) in press], whereas your data indicated a somewhat longer time. Some of the discrepancies may be resolved by more frequent sampling.

F. C. Greenwood: Dr. Roger Turkington found the half-life of human prolactin by bioassay after hypophysectomy, as I recall, to be 15 minutes. Some of the changes we have seen are more consistent with a half-life of 5 minutes or less. There is the possibility that the half-life might change as a function of the physiological state.

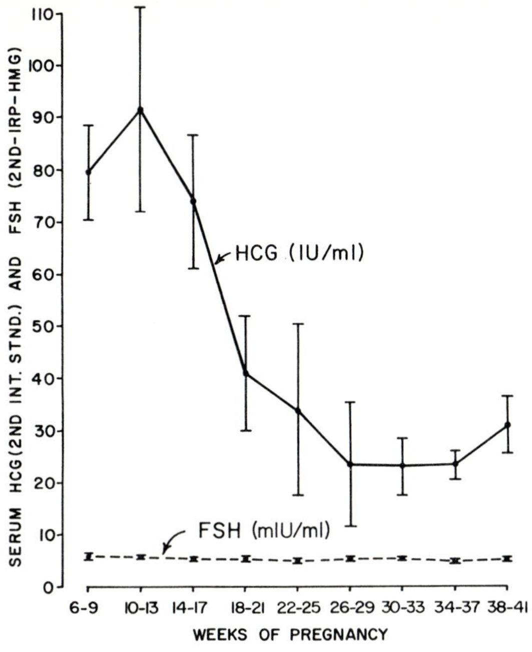

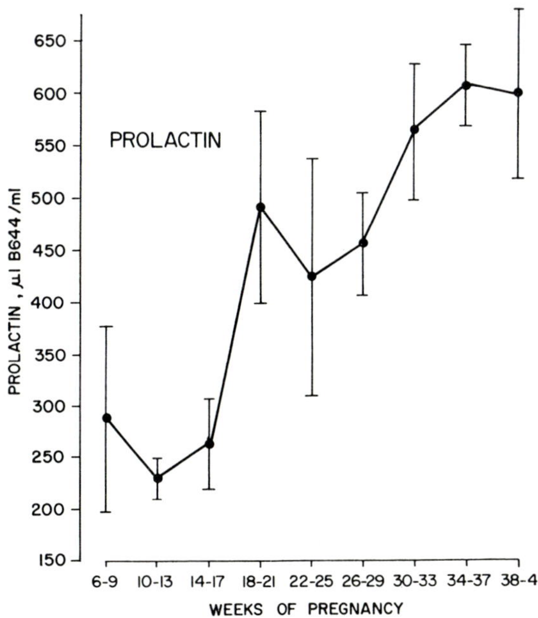

FIG. K. Human growth hormone (HGH), follicle-stimulating hormone (FSH), and prolactin levels during pregnancy.

V. L. Gay: We find that the half-life of prolactin following hypophysectomy of the rat is approximately 13 minutes, but we do observe dynamic changes in the serum prolactin concentrations of intact rats which indicate a half-life of approximately 5 minutes [*Fed. Proc., Fed. Amer. Soc. Exp. Biol.* **29**, 1880 (1970)].

Regarding the failure of women with renal failure to exhibit regular menstrual cycles, we have recently reported that in the rat there is a 4- or 8-fold increase in the half-lives of LH and FSH, respectively, following nephrectomy [*Abstr. The Endocrine Soc.* (1971)]. Serum levels of LH and FSH both increase following nephrectomy, apparently due to this increase in half-life of the hormones. I would suggest that if there are disruptions of the reproductive function during renal failure we should consider the possibility that they may be due to changes in serum LH and FSH as well as changes in prolactin.

W. P. VanderLaan: Dr. U. J. Lewis and I had the chance to examine the pituitary gland of a woman who died during pregnancy U. J. Lewis, R. N. P. Singh, Y. S. Sinha, and W. P. VanderLaan, *J. Clin. Endocrinol.* **33**, 153 (1971). This gland provided us with a new band on polyacrylamide gel electrophoresis; in the crop-sac test it was potent (equivalent to 22 IU/mg). The molecular weight was 21,000.

Using this band to monitor the extraction procedure, Dr. Lewis recovered prolactin from pooled frozen human pituitary glands supplied by the National Pituitary Agency [U. J. Lewis, R. N. P. Singh, and B. K. Seavey, *Biochem. Biophys. Res. Commun.* **44**, 1169 (1971)]. Figure L shows a strong prolactin band from a single fresh pituitary, migrating just behind growth hormone. There is a prolactin band

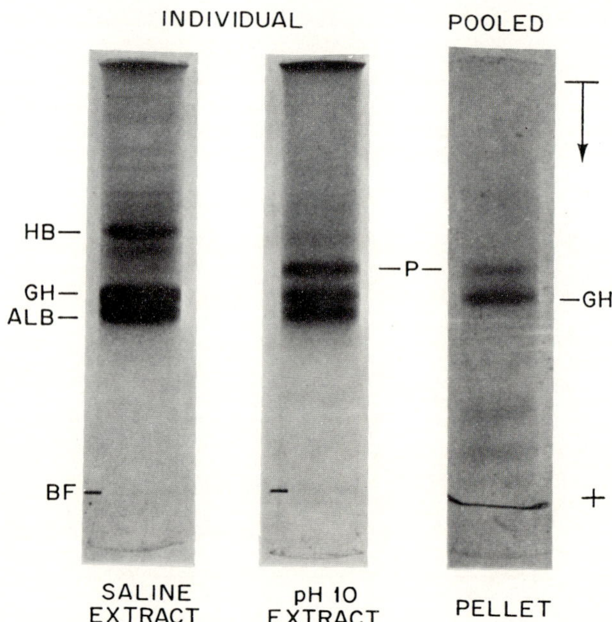

FIG. L. Electrophoretic patterns of pooled frozen human pituitary glands and of individual human fresh pituitary. From Lewis *et al.* (1971).

from the pooled glands; apparently the prolactin deteriorates with storage. A rough estimate would be 500 μg of prolactin in the pituitary of pregnancy, 50 μg per stored gland and somewhat more from the fresh gland.

Figure M shows the prolactin band after purification had been accomplished by standard procedures. A faint band representing deamidated prolactin runs just ahead of the heavier prolactin band. The next gel shows that deamidated prolactin runs to the same point as growth hormone. Could contamination with deamidated prolactin account for the prolactin activity of human growth hormone preparations?

It seems clear that Dr. Friesen and we are talking about the same protein. Figure N shows the electrophoretic pattern of the pituitary of a 14-week-old fetus, and the prolactin band is clearly seen.

R. B. Greenblatt: R. E. Hibbs [*Amer. J. Med. Sci.* **213,** 176 (1947)] and later E. C. Jacobs [*J. Clin. Endocrinol.* **8,** 227 (1948)] reported on a series of severely malnourished prisoners of war who developed temporary gynecomastia on refeeding. Quite a few had colostrum in their breasts; one had frank galactorrhea. The refeeding syndrome following inanition is not new, for in Job (**21, 24**) we find, "his breasts are full of milk and his bones are moistened with marrow." Evidently Job, who had suffered from illness and malnutrition (my bones cleaveth to my skin) was aware of this phenomenon following recovery. Some time ago, Dear Abby, in her newspaper column, replied to an unmarried woman's wish to adopt and nurse a newborn child, that the idea was preposterous. In the correspondence that followed, a young woman wrote in how she had become distressed when her sick sister could not nurse her newborn infant—and soon found that her own breasts

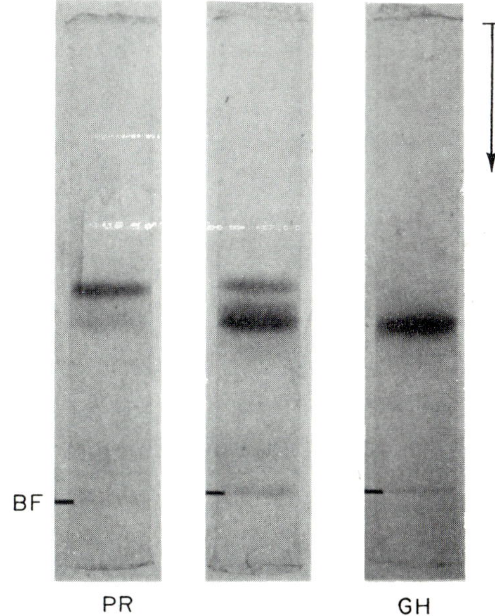

Fig. M. Prolactin band after purification by standard procedures. From U. J. Lewis, R. N. P. Singh, and B. K. Seavey, *Biochem. Biophys. Res. Commun,* **44,** 1169 (1971).

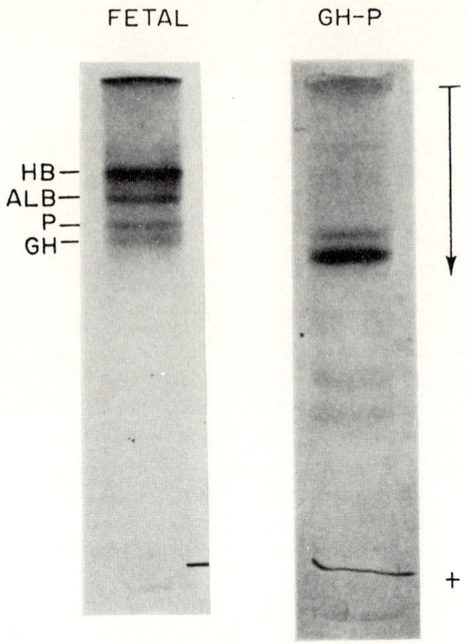

Fig. N. Electrophoretic pattern of pituitary extract from 14-week-old fetus.

filled with milk, enabling her to nurse the child. Evidently psychogenic stress can increase prolactin sufficiently to induce lactation. The Talmud records the story of a man whose wife died at childbirth. He could not afford a wetnurse and so placed the newborn infant to his breasts. Low and behold, a miracle happened and he was able to suckle the child. Sic transit gloria hypothalamicum!

The elegant studies of Dr. Jaffe showed that the ovulatory spurt of FSH and LH was held in abeyance so long as his patient nursed. That may be true in most cases, but it does not necessarily mean that sooner or later an ovulatory response will not occur in many patients. There are too many examples on record to prove that conception does occur in a certain percentage of cases during continuous lactation.

H. G. Friesen: Have you seen galactorrhea in prepubertal or postmenopausal women? Have you looked at prolactin levels in milk itself? We found prolactin concentrations in the milk of patients with galactorrhea to be higher than in the serum. It might be helpful to insist on getting a bit of the milk sample in some of these patients whose serum levels are normal. This may provide a clue to the patients with normal serum prolactin levels, who yet have galactorrhea. It may be that the breast tissue in the patients concentrated prolactin in an unusual manner.

A. G. Frantz: None of our patients with galactorrhea was prepubertal, though some with amenorrhea may have been postmenopausal. We have not looked for prolactin in milk.

C. W. Lloyd: The Incas recognized that pregnancy could interfere with the amount of milk. A woman who was lactating was required to nurse her baby for two years or more. If she became pregnant during the two years after delivery, and therefore had a decrease in milk, she was liable to the death penalty. The Incas recognized that this source of nutrition would be decreased by pregnancy.

Our group found some patients in which it seemed to be quite clear that chromophobe adenomata resulted from a loss of feedback from the gonad. Agnes Russfield proposed this back in the late 1950's. The history suggested intrinsic gonadal failure of long duration. We found five patients who had chromophobes and in whom there was high gonadotropin excretion. Do all of your patients in whom you have seen high prolactin levels have low gonadotropin levels?

A. G. Frantz: We have not encountered any chromophobe adenomas associated with high gonadotropins. The gonadotropins in these patients have been either low, in the majority, or normal, in a few.

A. Wolfson: Have you examined any woman with breast cancer in relation to prolactin levels?

A. G. Frantz: We are in the process of doing so now. We have set up a program at the Presbyterian Hospital to try to screen all women with breast tumors for prolactin. Those few we have studied to date have had prolactin levels within the normal range.

O. H. Pearson: Have you looked for prolactin in the urine?

A. G. Frantz: We are doing so now, but cannot report any definite results. Our preliminary findings are that, if it is present, the amounts seem to be small.

J. R. Marshall: I am bothered by the overall impression left by previous discussants concerning the effectiveness of lactation as a contraceptive. In primitive societies, the median interval between pregnancies is about one year in nonnursing women and two years in nursing women. Thus, from a statistical and ecological point of view, nursing is an effective contraceptive which can significantly diminish the overall birth rate.

C. A. Paulsen: Did your patients with craniopharyngioma have intact ACTH and TSH function? Postoperatively were they treated to induce sexual maturation?

A. G. Frantz: None of the patients was treated to induce puberty, and postoperatively all remained prepubertal. They received thyroid and cortisone postoperatively, since all were panhypopituitary at this time; appropriate pituitary deficiencies had also been corrected preoperatively.

B. F. Rice: Assuming that the breast is an end organ for prolactin, are there any data on prolactin levels in women with bilateral mastectomies? Are there any data on nursing mothers that are blind so that one could try to elucidate some of the mechanisms controlling prolactin secretion in the human?

A. G. Frantz: These data are not available.

F. C. Greenwood: The physiological consequences of these high levels, particularly the transient increases, obviously have to be correlated at some stage with physiological effects. They seem to be related only in lactation as yet, and there is a lot of work to be done in the future.

R. W. Bates: These two papers on prolactin have been of great interest to me. From 1931 to 1941 a group of us in Riddle's laboratory prepared and studied the effects of prolactin mostly in doves, pigeons, and hens, but also in mammals. These studies have been summarized in several places including the first Laurentian Hormone Conference in 1944, the proceedings of which were never published. Few studies were made during the next 20 years. Since then, aided by new techniques, interest and activity in the prolactin field have been accelerating rapidly.

The question was asked: "What does prolactin do in the male?" Prolactin does the same in the male as in the female. Both male and female doves and pigeons develop crop sacs and feed young. This involves a doubling of food intake (an increase in appetite) and an enlarged digestive tract to handle the conversion of the food into raw materials which are carried in the blood to the crop sacs, where crop-sac tissue to the extent of 5% of the parent's body weight is daily fed to the young squab. Male rats with Furth's MtT pituitary tumors have enlarged milk-filled mammary glands, as do females. Relatively too much attention is focused on the effect of prolactin on the mammary glands.

Prolactin is an anabolic hormone which mobilizes raw materials for delivery via the blood to sites where they can be used. These materials are not limited in availability to one target tissue. Also prolactin may be effective at concentrations below those required to stimulate the crop sac or the mammary gland to increase in size. An analogy might be the case of TSH, whose main activity is to mobilize (release) thyroid hormone. TSH also will cause the thyroid gland to increase in weight, but the latter response requires an increase in dosage of nearly 50-fold. A similar situation exists with ACTH and probably with the gonadotropins.

Biological Applications of Electron Ionization and Chemical Ionization Mass Spectrometry

Henry M. Fales, George W. A. Milne, John J. Pisano, and H. Bryan Brewer, Jr.

National Heart and Lung Institute, National Institutes of Health, Bethesda, Maryland

Murray S. Blum, John G. MacConnell, John Brand, and Norman Law

University of Georgia, Athens, Georgia; and Suburban Hospital, Bethesda, Maryland

I. Introduction

This paper deals with a subject that may have rather limited application to hormone research as it is practiced today. Still, when the occasion does arise that mass spectrometry can be used in a problem, there should be no hesitation in doing so since the technique is relatively simple experimentally. Furthermore, interpretation of the data is often straightforward in applications of this sort.

In order to be able to determine when and where the technique may be invoked, it is useful to have some basic understanding of the analytical process. Therefore, it may be helpful to run rapidly through the physics of the method, indicating present limitations and, hopefully, ways that some of these may be overcome in the future.

II. Methods

A basic mass spectrometer system is illustrated by Fig. 1. Essentially, it is composed of a curved, hollow pipe, evacuated to a very high degree, placed in the jaws of an electromagnet. Ions of unit charge formed by various processes at the entrance are accelerated to a constant energy during passage between two plates across which there is a fixed potential, usually of the order of thousands of volts. Although the energy ($mv^2/2$) of all the ions of unit charge is the same, ions of different mass will have different velocities and their momentum (mv) will also vary. A particle of unit charge is deflected in a magnetic field according to its momentum. Even intuitively it will be seen that the more momentum an ion has, the less it will be deflected by a magnetic field of a fixed strength. For this reason, ions of different mass will "fan out" in the magnetic field, and it might be possible to arrange for their detection by allowing them to fall on a photographic plate placed at an appropriate distance from the magnet. In fact, this method of detection is still

591

utilized today in some high resolution instruments where it has the advantage of detecting all ions simultaneously. Thus momentary fluctuations in ion beam intensity due to changes in sample pressure are integrated.

For low resolution work, it is more convenient to vary the strength of the magnetic field by using an electromagnet. The "fan" of ion beams of different mass will increase or decrease their arc as the field is changed, each ion beam falling in succession over a sensitive electrical detector fixed at a location in the ion trajectory. This results in the type of mass spectrum with which most are familiar; a series of peaks

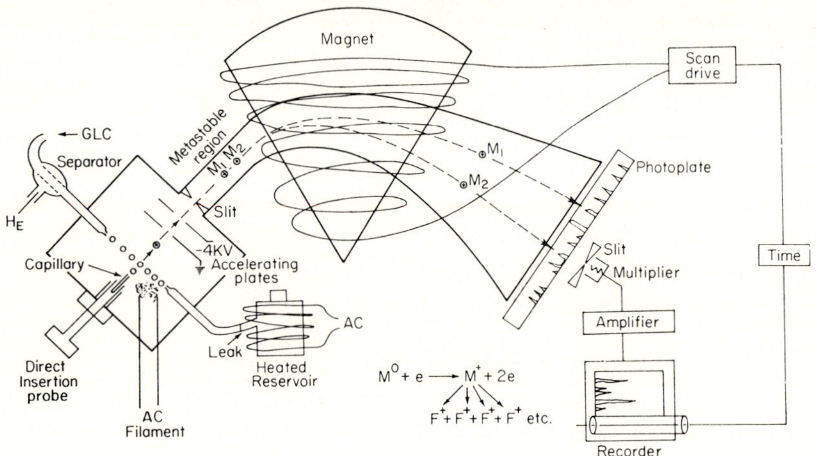

FIG. 1. Schematic diagram of a magnetic deflection mass spectrometer.

arranged according to mass whose height is proportional to intensity of the ion beam. Since the magnet may be scanned very rapidly (1–2 seconds), the entire spectrum is accumulated and a very rapid means of detection and display is required. Although expensive computer options are available today which present data in forms that are easy to comprehend, there are still very real advantages to direct recording, using an oscillograph with 3 beams, attenuated by factors of 1, 10, and 100. One of the strong points about mass spectrometry is the enormous dynamic range of the signal. I know of no other laboratory instrument that routinely uses three pens for recording. In conjunction with this system we use a photomultiplier detector which can detect a single pulse of a few ions. On the other hand, it may take many molecules to produce these ions because of inefficiency in the ionization process. It is in this area that we must look for improvements leading to greater sensitivity. In fact, today all mass spectrometers are very similar in ionization efficiency, and claims of enhanced sensitivity by one manufacturer over another are to be viewed with caution. On the other hand, it is possible to enhance sensitivity by counting ions as we do β-particles. The only disadvantage of this lies in the time required for the process. Time is often very limited in mass spectrometry since it is difficult to predict the optimum temperature for sample volatilization. Choice of too high a temperature results in a strong signal for a very short time, as the sample rapidly evaporates from the probe tip. Gas chromatography aids in visualizing, via the total ion monitor or gas–liquid chromatographic (GLC) detector, the *rate* at which the sample is presented to the ionization chamber, but the typical GLC peak may last only a few seconds.

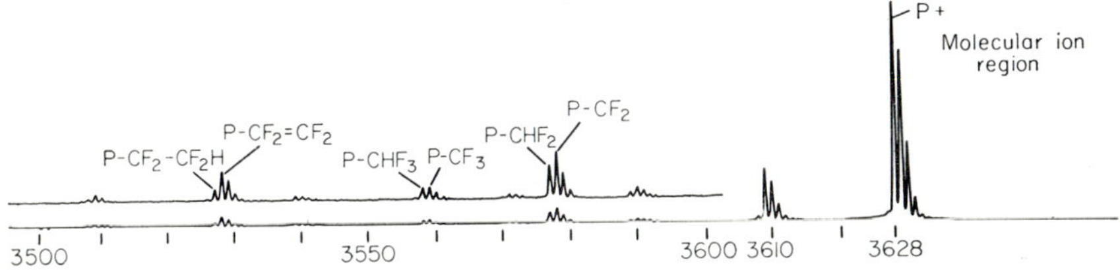

FIG. 2. The mass spectrum of a compound $C_{72}H_{24}O_8F_{128}N_4P_4$ of molecular weight 3628.

There is nothing inherent in the mass spectrometer that limits the molecular weight, or more correctly the ion weight, which may be investigated. The way to analyze higher masses is simply to turn the accelerating voltage down, imparting less momentum to the higher mass ions so that they may be deflected by whatever magnetic field is available. As this is done one tends to lose sensitivity, but it is not here that the true limitation of the mass spectrometer arises, and Fig. 2 shows how we were able to obtain a perfectly good spectrum on a compound $C_{72}H_{24}O_8F_{128}N_4P_4$ of molecular weight 3628 (Fales, 1966). Professor Lederer at Gif (for leading reference, see Vilkas and Lederer, 1968) has successfully analyzed derivatized peptides of well over 1500 molecular weight. Since so many of the conferees are involved in the field of high molecular weight substances such as peptides and proteins, it might be useful to pursue this question by discussing the means by which the molecules are converted to ions. The reason ionization must be brought about is so that the particles may be deflected and sorted according to mass by the various electric and magnetic fields within the spectrometer. Mass spectrometers today are in fact *ion* mass spectrometers. Perhaps someday someone will discover a method for sorting molecules by mass alone, and this would be an authentic *mass* spectrometer.

At the present time, by far the most common method of producing ions from molecules is through electron ionization (Fig. 3). Thus, a high velocity electron approaching a molecule sets up force fields that are strong enough to cause the ejection of an electron from the molecule with the formation of a positive ion. Unfortunately, it is generally very difficult to carry out this process without imparting a good deal of energy (several electron volts) to the resulting ion. Each electron volt is equivalent to 22.4 kcal, and as little as $\sim$ 60 kcal will cause rupture of any C—C bond. Figure 3 also shows some less familiar ways to make ions from molecules. Thus, if a very low velocity electron is used, collision may

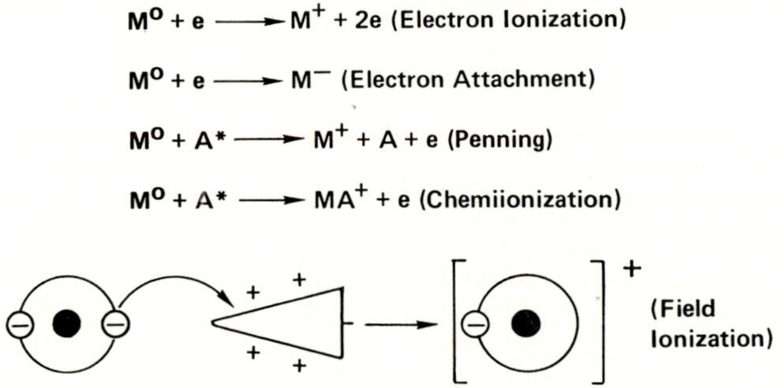

FIG. 3.   Methods of ion formation.

result in attachment of an electron to the molecule resulting in a negative ion. Fragmentation may be minimized, but the technique is experimentally difficult. Or ionization may be brought about by allowing "soft" collisions with atoms in electronically metastable states, an electron being ejected in the process. Since these latter reactions involve only neutral reactants, the process has been called "chemiionization" to draw attention to its relation to condensed-phase chemical reactions. This is not, however, "chemical ionization," which will be discussed later.

Another way to produce ions is to approach the electrons of a molecule with a positively charged tip whose force field is as high as that surrounding an atomic nucleus within the molecule itself. Under these conditions an electron will be withdrawn from the molecule to the tip via an electronic tunneling process, leaving behind a positively charged ion. This is a very gentle process indeed, involving only one or two electron volts of energy, so that one may expect minimum fragmentation; often only the molecular ion is visible. Unfortunately, at present sources built around this principle are difficult to operate and yield rather low intensity ion beams. Again we may look for great improvements in this area in the near future.

A fourth, and theoretically very simple, way to produce ions (Fig. 4), is through an exchange process with other ions formed independently by electron ionization. This is the area of ion molecule reactions that until recently has been the domain of physicists and physical chemists. Today, thanks to the work of Frank Field at Rockefeller University (for leading references, see Field, 1970), Burnaby Munson at the University of Delaware (Munson and Field, 1966), Jean Futrell at the University of Utah (Futrell and Tiernan, 1968), and V. A. Tal'roze in Russia (Tal'rose and Lyubimova, 1952), this method may be utilized by the mass spectrometrist in a very practical way. It is in this area

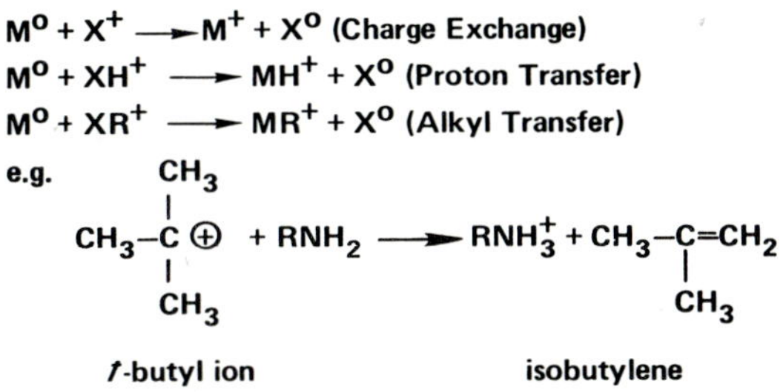

$$M^O + X^+ \longrightarrow M^+ + X^O \text{ (Charge Exchange)}$$
$$M^O + XH^+ \longrightarrow MH^+ + X^O \text{ (Proton Transfer)}$$
$$M^O + XR^+ \longrightarrow MR^+ + X^O \text{ (Alkyl Transfer)}$$

e.g.

$$CH_3-\overset{\overset{\displaystyle CH_3}{|}}{\underset{\underset{\displaystyle CH_3}{|}}{C}} \oplus + RNH_2 \longrightarrow RNH_3^+ + CH_3-\overset{\overset{\displaystyle |}{C}}{\underset{\underset{\displaystyle CH_3}{|}}{=}}CH_2$$

t-butyl ion   isobutylene

Fig. 4. Ion–molecule reactions.

that Dr. G. W. A. Milne and I have been working in recent years, since we feel that the method has particular promise for molecules of biological origin.

In this method an ion may be produced either by electron transfer from another ion or by a proton or alkyl group transfer. The most useful of these processes today is that of proton transfer as exemplified in Fig. 4 by the reaction of the *t*-butyl ion with an amine to form isobutylene and an ammonium ion. The proton derives from one of the methyl groups of the *t*-butyl ion. This reaction will occur with any base whose proton affinity is higher than that of isobutylene and therefore includes compounds containing nearly any heteroatom, aromatic ring, or conjugated olefin. This exemplifies the process of "chemical ionization" so named by Dr. Field to call attention to the essentially "chemical" nature of the bimolecular reactions taking place, and in this respect recalls, but is to be distinguished from, the term "chemiionization" referred to previously.

There are two ways to study ion-molecule reactions. The first (Fig.

5) is by a tandem arrangement of two mass spectrometers, one to prepare
the reagent ion and one in which the actual reaction with another ion
takes place. Such systems are expensive and complex, but the interpretation is straightforward since the reagent ion is precisely known. In the
other system, developed by Field, only one spectrometer is employed.
Here the bimolecular reactions required by the method are brought about
by so constructing the source that pressures of 0.1–2 mm may be toler-
ated. In this case many sorts of ions may be present; interpretation

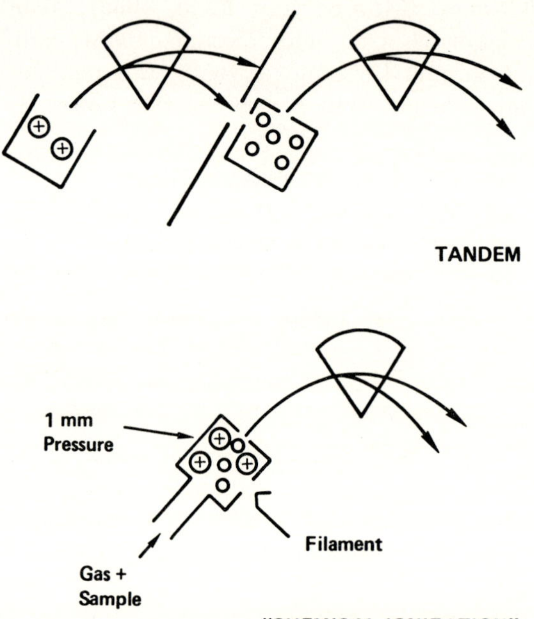

Fig. 5.　Tandem versus chemical ionization spectrometer operation.

is correspondingly complicated, but construction and operation of the
apparatus is much more practical.

From the foregoing discussion it will be noticed that the molecules
have always been assumed to be in the vapor state prior to their conversion to ions. This is a mass spectrometric fact of life that seemingly
bears no relation to the actual production or deflection of the ions. Nevertheless, at the present time it is absolutely essential that the material
be vaporized prior to its conversion to ions. Therefore, in applications
to peptide chemistry, much less protein chemistry, severe limitations
are imposed. For example, the vapor pressure of polystyrene of molecular
weight 14,000 atomic mass units has been calculated (Dole et al., 1968)

to be 10^{-480} atm at room temperature. Nor is it merely a problem of molecular weight; we cannot obtain mass spectra of molecules as simple as the nucleotides because of their high degree of polarity. The situation reaches an extreme for compounds with formal positive and negative charges, such as sodium chloride, which may be held in crystal lattices with electrostatic bond energies at >200 kcal/mole. Furthermore, some polar groups (e.g., hydroxyls) tend to be thermally sensitive, so that decomposition occurs when attempts are made to heat the samples to the temperature required for volatilization. Sometimes we can work around the problem by chemically converting the compound to a less polar and more volatile derivative. The elegant work of Jim McCloskey and his co-workers at Baylor in the field of the nucleic acids (Desiderio *et al.*, 1968) is a prime example. In this case compounds such as adenosine are converted to their polytrimethylsilyl ethers, masking hydroxyl groups and the phosphoric acid in one operation. Satisfactory spectra are obtained (Fig. 6) that should be of great assistance in elucidating the structures of new nucleosides.

Extreme examples of the derivatization technique are probably familiar to you in the work of E. Lederer and his co-workers in France (for leading references, see Vilkas and Lederer, 1968), and A. Kiryushkin and M. M. Shemyakin (Shemyakin, 1968) in Russia, where peptides (even decapeptides) are N-acylated and permethylated using methyl iodide and sodium hydride in dimethylsulfoxide to mask all polar amide NH groups as well as carboxyl groups. Arginine residues are tied up using acetylacetone, etc., and special derivatives must be made of other troublesome amino acids. These methods require a good deal of chemical expertise if the final material is still to reflect the original peptide. When a satisfactory derivative has been prepared, obtaining and interpreting the mass spectrum to yield the amino acid sequence may be relatively straightforward provided sufficient sequence peaks are visible. However, my recommendation to anyone contemplating the use of this technique is to seek firsthand advice from one of the aforementioned groups and to practice derivatization on a reference peptide as similar in amino acid composition as possible to the unknown peptide.

Although derivatization is today the most practical way to overcome lack of volatility, it has always struck me as unreasonable that we must start with a compound that often contains a positive or negative charge, such as a peptide or salt, then chemically modify it so as to *remove* the charge in order to vaporize it, only again to ionize it in the mass spectrometer. For that matter, it is really necessary to vaporize the sample at all? Two interesting experiments have been performed which offer some hope in this area. In the first, Dr. H. G. Beckey,

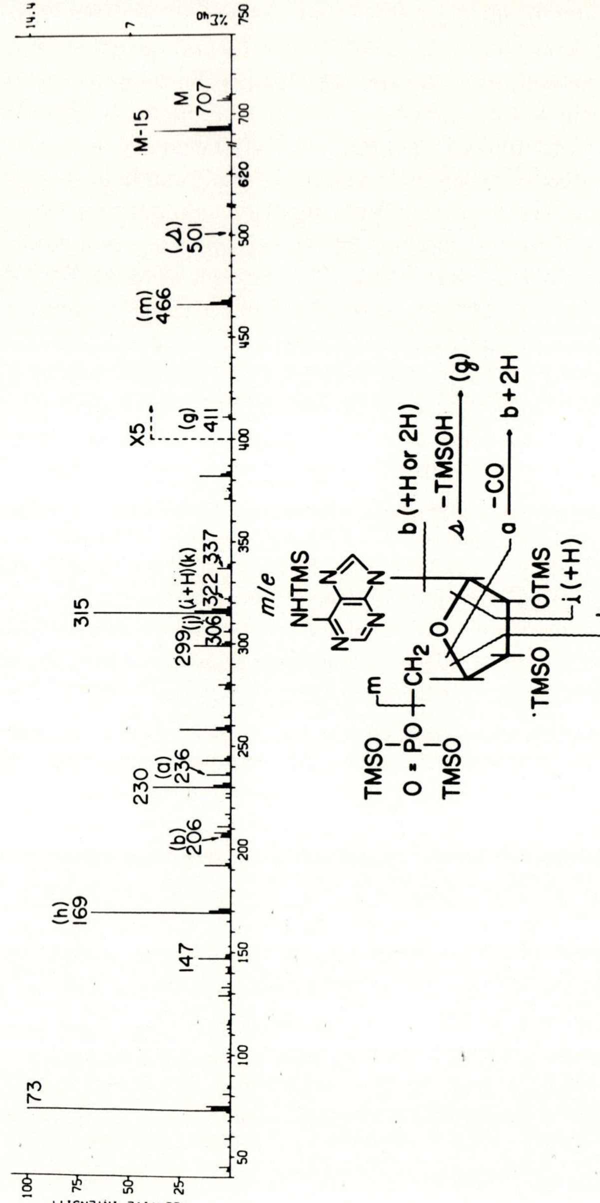

Fig. 6. Mass spectrum of a nucleoside trimethylsilyl ether.

a leader in the area of field ionization, has found that glucose deposited on the tip of a field ionization source may be caused to lose an electron to the tip. When this happens, the resulting positively charged glucose molecular ion will be desorbed from the tip by the mutual repulsion between it and the very highly positively charged tip. In this way he has obtained the very excellent spectrum of glucose (Beckey, 1969), shown in Fig. 7. As will be seen, under ordinary electron ionization conditions, glucose provides no molecular ion. If field desorption methods could be applied to more complex molecules this would be a breakthrough of the first magnitude.

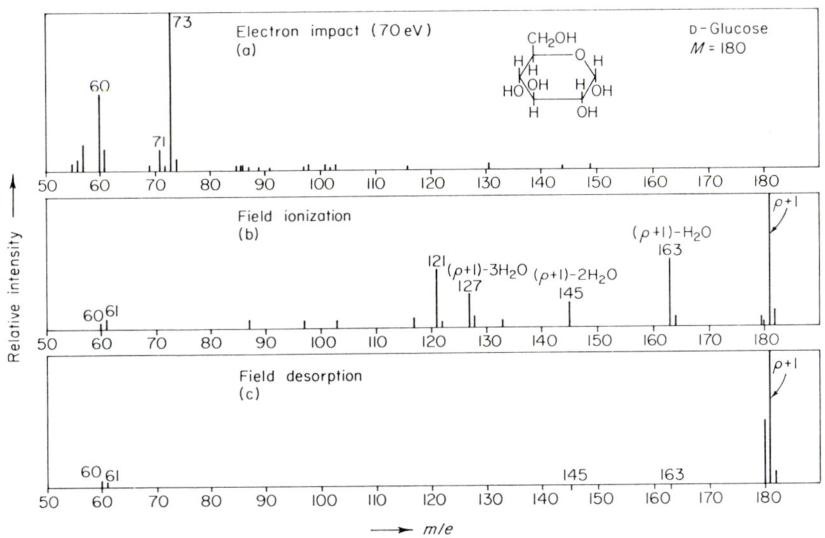

FIG. 7. Glucose mass spectrum via field desorption. From Beckey (1969).

In a second development, Professor M. Dole and his co-workers at Baylor (Dole *et al.*, 1968) have developed a novel mass spectrometer system applicable to compounds of very high molecular weight, such as polymers. The device is in a relatively crude form at present, but the preliminary results are exciting indeed. In this system (Fig. 8) a solution of a polymer is prepared in a solvent at a concentration such that, when it is sprayed in the form of fine droplets from a nebulizer tip, each droplet will contain on the average only one molecule of the polymer. As the droplets are formed at the tip of the nebulizer, they are charged by a potential applied to the tip. The charged, solvated droplet passes into a region where it is mixed with a carrier gas. It proceeds to lose its hydration sphere through collisions with this gas, and the free macromolecular ion is then swept out of an orifice with the gas at ultrasonic velocities. It is then directed toward a collector plate in front of which is placed a charged grid whose potential is adjusted so that ions of increasing energy are selectively repelled. The neat break observed in the resulting plot (Fig. 9) can be calculated

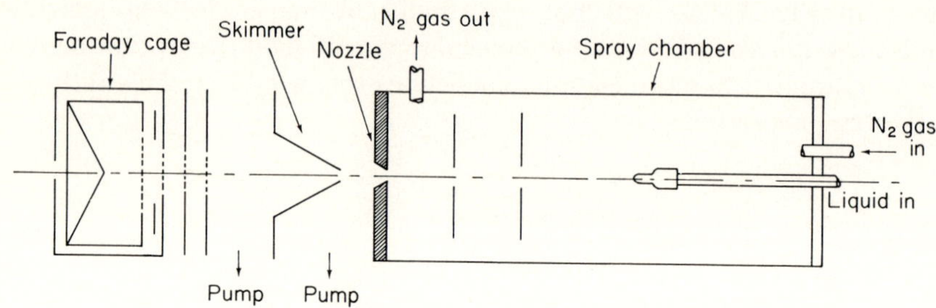

FIG. 8. Electrospray apparatus. From Dole *et al.* (1968).

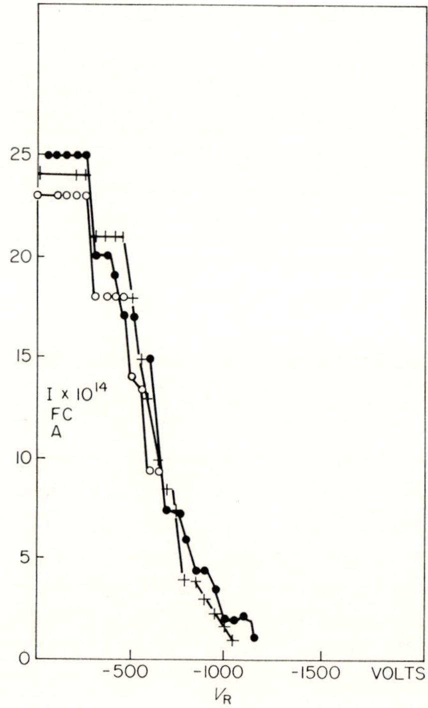

FIG. 9. Current-repeller grid plot from electrospray experiment. Voltage curves for 0.01 wt% solution of 51,000 amu polystyrene; 2, 3, 3 inch spacings. From Dole *et al.* (1968).

to result from a supersonic ion of m/e 50,000. If this eluting stream of macromolecules could be directed toward the entrance port of a specially constructed mass spectrometer, it should be possible to determine the molecular weight with a much higher degree of accuracy than is currently possible. It is my understanding that experiments along this line are already in progress.

III. Results and Discussion

A few examples will follow from current work in this laboratory that may help to indicate the range of problems that may be studied using currently available mass spectrometric systems. In the first place, it is clear that the combination of gas chromatography and mass spectrometry (GC-MS) has had an enormous impact in biochemistry in recent years. Gas chromatography of steroids was discussed some years ago at these sessions by Dr. Evan Horning at Baylor (Horning *et al.*, 1963), and now all of this work has been extended in its utility by the use of GC-MS. The work on steroids by the Finnish and Swedish groups (Luukkainen, Aldercreutz, (1967) and others) demonstrates how valuable it can be in a clinical situation, and recent work of Dr. Engle emphasizes this fact. Recently Professor Röller in collaboration with Dr. C. C. Sweeley showed how the method was applied to the elucidation of the structure of the insect juvenile hormone (Röller and Dahm, 1968).

Entomologists and biochemists studying exocrine secretions and pheromones are indeed in a fascinating field, and I have been privileged to collaborate recently with Drs. Murray Blum, John MacConnell, and John Brand at the Entomology Department of the University of Georgia on related work. For example, we are studying the defensive secretions of the *Chrysomelae interrupta* beetle larvae shown in Fig. 10. This fellow is a local pest, dining on several species of alder trees. When he is attacked by a fire ant or other predator, as shown in Fig. 11, his blood pressure rises forcing inside out small sacs along his back which contain a liquid which is highly offensive to his attacker. When the danger has passed, he simply draws the sacs back along with their droplets, conserving the liquid for future use. (I am indebted to Dr. John Brand for this picture; I understand that such pictures are very difficult to obtain.) The gas chromatogram of extracts of these glands is shown in Fig. 11, and from their mass spectra we have been able to identify the structures of the compounds as β-phenylethyl isobutyrate and β-phenylethyl 3-methylbutyrate. It seems that these materials are unique in the insect world. The mass spectra of the various possible isomers are very similar, and retention time information, in addition to the mass spectra, was required to distinguish among various isomers. These ex-

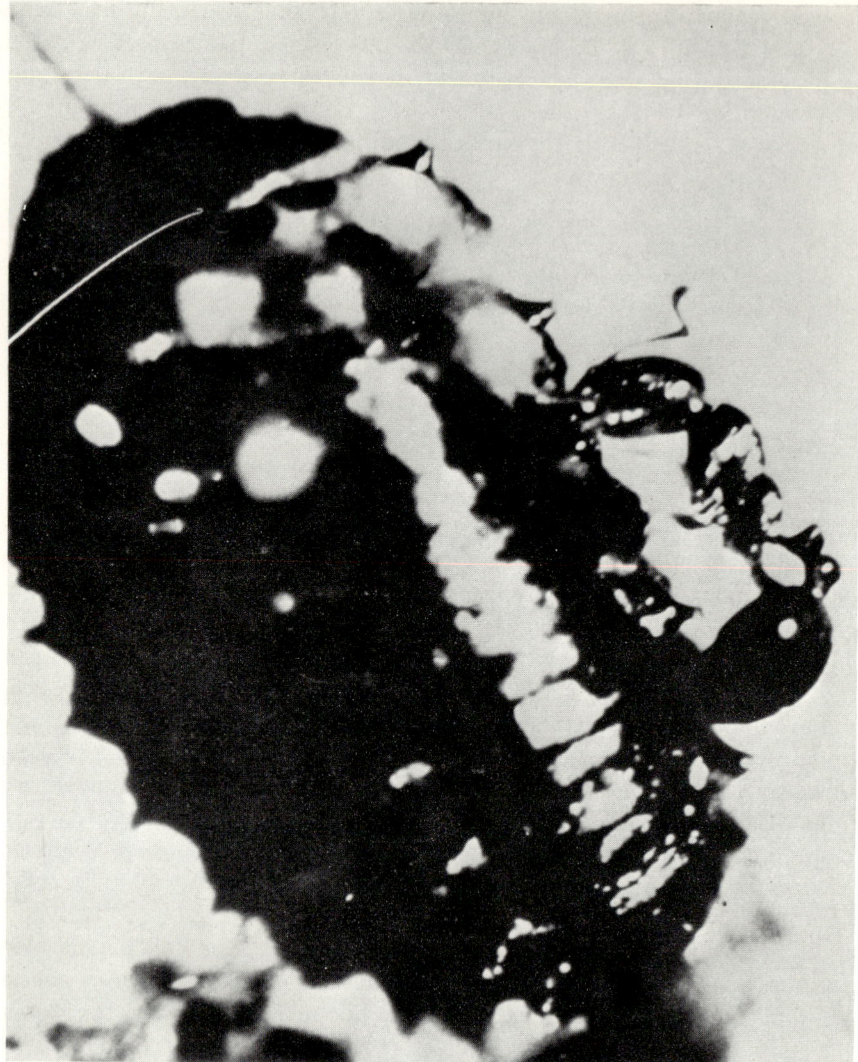

F_{IG}. 10. *Chrysomela* larvae with alarm gland verted being attacked by fire ant.

ocrine secretions are almost uniquely suited for study by GC-MS because such compounds are necessarily volatile so that their signaling action does not persist beyond a reasonable time. Were this not so, considerable confusion would reign in the insect world. The importance of the gas chromatograph cannot be overestimated in this application, since it permits mass spectra to be obtained on biological products without the need for extensive purification. Even more important, it allows diagnostic chemical reactions to be carried out on the micro scale without prior isolation and purification of the product. It is only in the most simple cases such as that described above, that the structure can be ascertained by inspection of a single mass spectrum. For example, we have also been concerned recently with the structures of the compounds found in the venoms of fire ants (e.g., *Solenopsis xylonii*) in the South. The fire ant is a rather serious problem in the southern United States since it is the dominant species in many areas. American taxpayers have spent

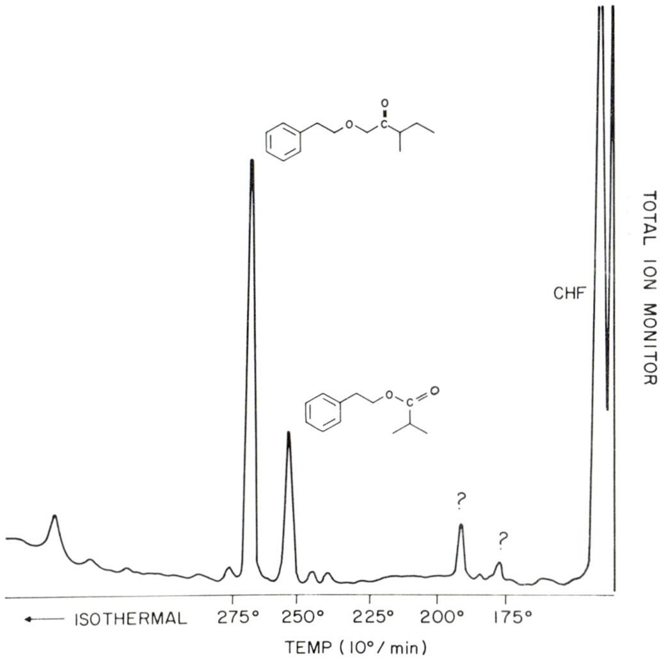

FIG. 11.   Gas chromatogram of *Chrysomela* alarm substances (*Chrysomela* larvae; 10% Carbowax 20M).

around three hundred million dollars on this problem, and suggestions about spraying large areas of the South with insecticides are worrying the environmentalists.

It seems that there are two distinct groups of fire ants: the indigenous species which has been around for many years without causing any particular problem, and the more recently imported species which appears to be the real winner in the battle for survival. Dr. Brand was curious about the possible reasons for its success and felt that the components of its venom might supply a clue. Last year, utilizing GC-MS,

Drs. Blum and MacConnell and I were able to determine their structures (MacConnell *et al.*, 1970, 1971); as Fig. 12 shows, they are simple 2,6-dialkyl substituted piperidines. Now it is interesting to find that the imported fire ant contains almost exclusively the C_{13}- and C_{15}-side chain unsaturated homologs, whose substituents are arranged in a *trans* relation on the ring, whereas the *native* form contains the C_{11}-saturated homolog whose side chains are arranged *cis* as well as *trans* on the

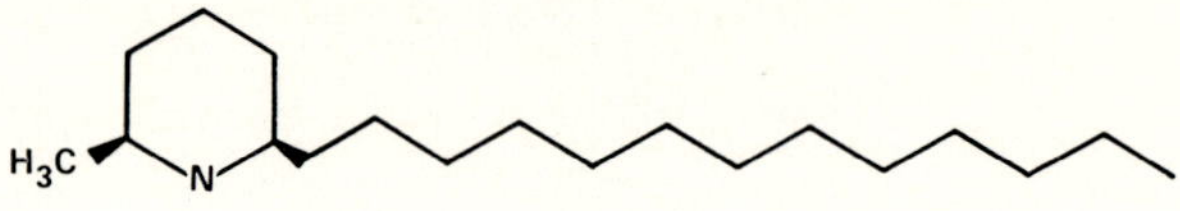

cis–2–methyl–6–n–tridecylpiperidine

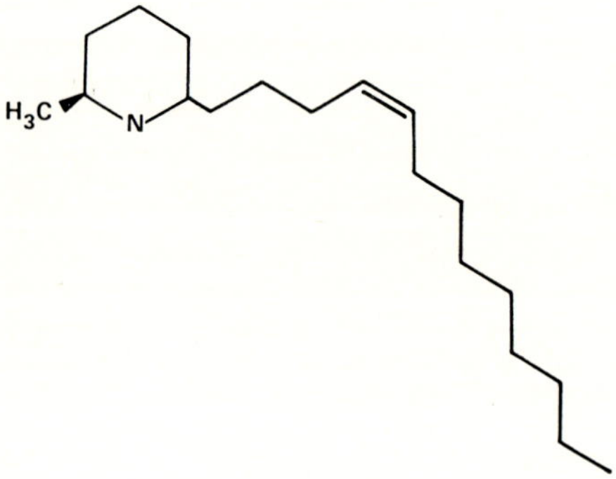

trans–2–methyl–6–n–(cis–4′–tridecylpiperidine)

Fig. 12. Structures of fire ant venom components.

ring. The physical properties of the *cis*- and *trans*-dialkyl piperidines are surprisingly different. In the *cis*-dialkyl piperidine the nitrogen electron pair is not nearly as accessible to various substrates (Fig. 13). For example, on alumina columns the *trans* form interacts and is absorbed much more firmly than the *cis* form.

It is tempting to speculate that the biological activity of the venoms is connected with this *cis-trans* difference, the *trans* being more active. But why is the *cis* form present only in the indigenous species? A recent analysis of this has disclosed the presence of a small peak in the gas chromatogram (Fig. 14) whose mass spectrum (Fig. 15) indicates

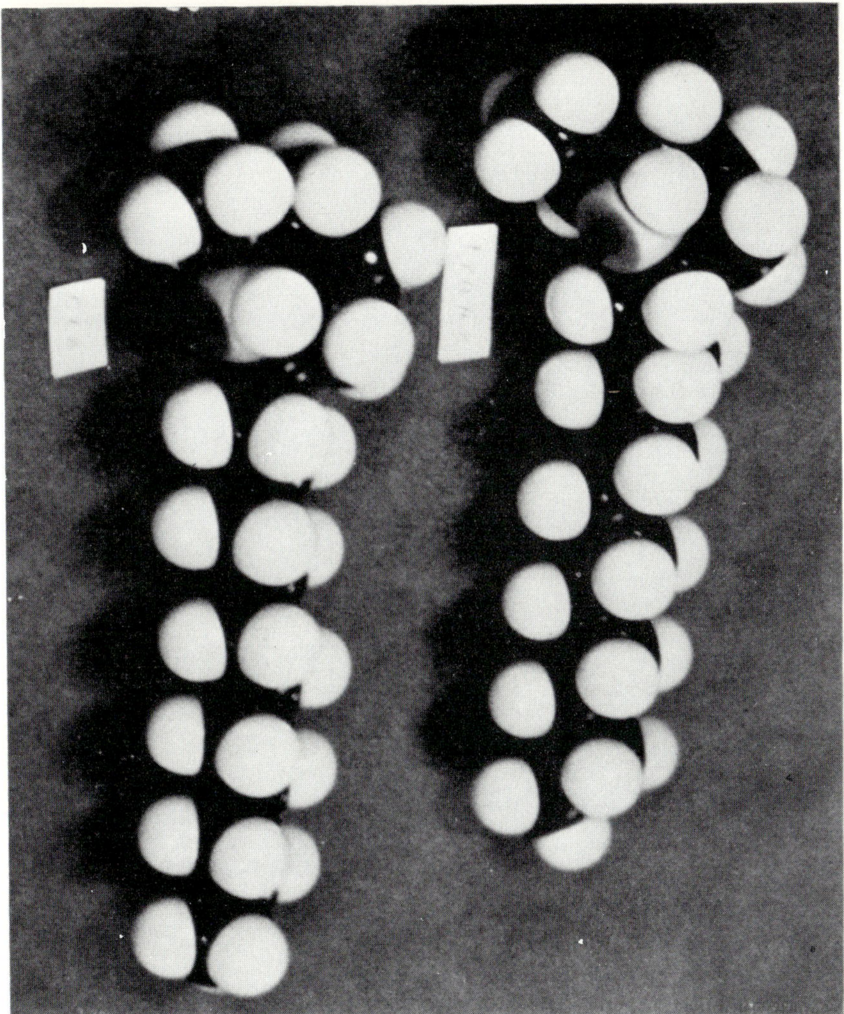

Fig. 13. Models showing accessibility of electron pair on nitrogen.

it to be a $\Delta^{1,2}$-piperideine. Confirming this structure, it undergoes reduction with sodium borohydride, producing the *cis* form almost exclusively. We can then speculate that this $\Delta^{1,2}$-piperideine is a precursor which has been biologically reduced by the imported ant (NADH?) specifically to the *trans* form, while the indigenous ant has an enzyme capable of the borohydride-like *cis* reduction as well. Alternatively, the $\Delta^{1,2}$-piperideine may be an intermediate between the *cis* and *trans* isomers, in which case only the indigenous species has an enzyme capable of oxidiz-

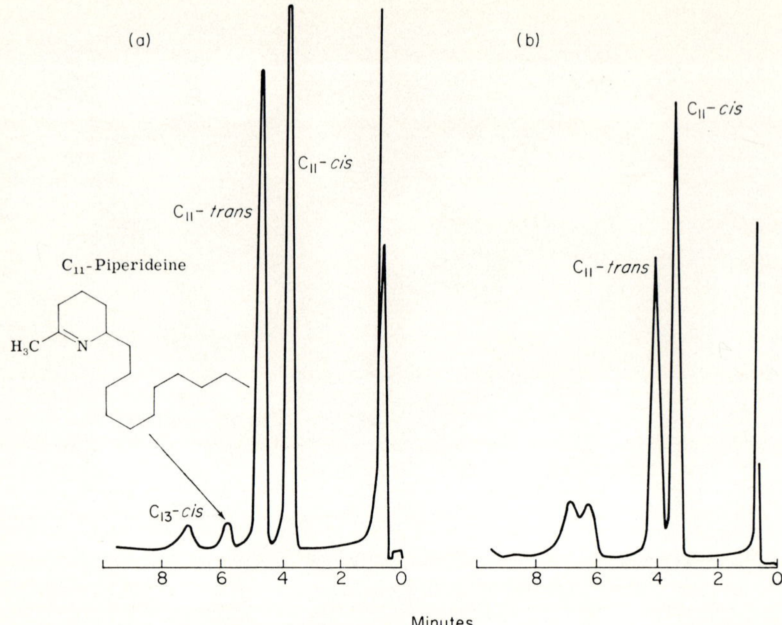

FIG. 14. Gas chromatogram of minor venom component. (*Solenopsis xylonii;* 10% SP-100; 180°C). (a) Before NaBD₄; (b) after NaBD₄.

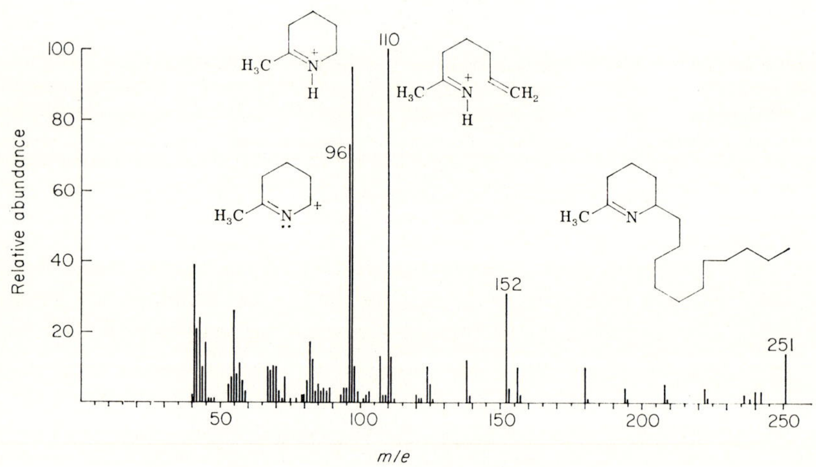

FIG. 15. Mass spectrum of minor venom component.

ing the *trans* form to the $\Delta^{1,2}$-piperideine. I must stress that all this is pure speculation, and biological assay of the various venoms are only now being undertaken. The success of studies such as this on trace amounts of material depends very much on the actual operating condition of the mass spectrometer itself. An instrument in poor condition due to misuse or neglect can be orders of magnitude less sensitive than one properly cared for. We are fortunate at NIH in having the services of Mr. William Comstock, whose expertise was essential to this work.

The fire ant problem was complicated by the fact that the molecular ions of the venom components were extremely weak (Fig. 16) and in fact, an ion that might be mistaken for the molecular ion

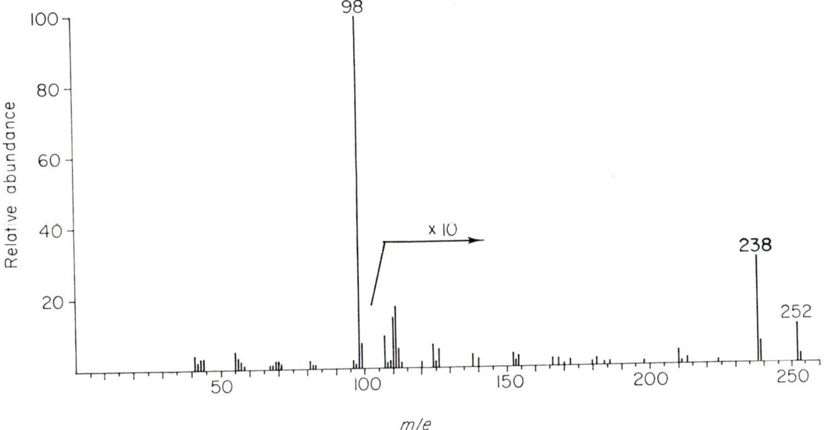

Fig. 16. Electron ionization mass spectrum of *trans* fire ant venom piperidine.

is present at a location 1 mass unit less than expected. This is a common occurrence in the mass spectra of aliphatic amines, and if the compound was truly unknown, one could easily arrive at the erroneous conclusion that the substance was not an amine at all, because its molecular weight apparently would be even. This is a consequence of the so-called "nitrogen rule" in mass spectrometry, which is simply a statement of the fact that compounds with *no nitrogen* or an *even number* of nitrogens are *even* in mass, while those with *one nitrogen* or an *odd* number of nitrogens are *odd* in mass. Because it so often occurs in mass spectrometry that molecular ions (molecular weight ions) are either not present or of such low intensity as to be confused with impurities, we have investigated in some detail the process of "chemical ionization" mentioned previously. For this purpose we have had constructed a special ion source designed by M. Vestal of Scientific Research Instruments (Baltimore,

Maryland) which allows the pressure in the spectrometer to rise to approximately 1 mm of mercury. These are the conditions required for a gas such as isobutane to form almost exclusively the tertiary butyl ion (Fig. 4) under electron bombardment. As mentioned previously, the *t*-butyl ion then transfers a proton to the compound under study, which has been added in trace amounts to the isobutane in the source. Some energy accompanies the proton, and this may result in some fragmentation of the $(M + H)^+$ ion, but in the case of simple bases the only ion observed is at $(M + 1)^+$, i.e., $(M + H)^+$. In the case of the fire ant venom, the intense $(M + H)^+$ ion which results (Fig. 17) has been very helpful in determining its structure since its molecular weight was determined unequivocally. On the other hand, if more fragmentation is desired for

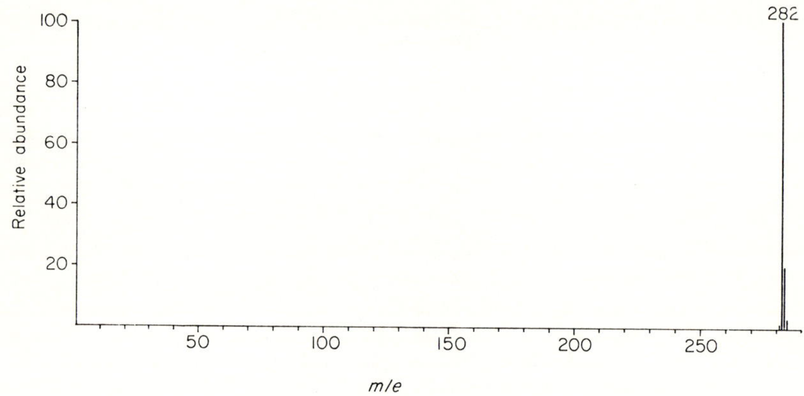

FIG. 17. Chemical ionization mass spectrum of *trans*-C₁₃ fire ant venom piperidine.

structural purposes, one simply employs a more energetic source such as CH_5^+ (protonated methane) formed by electron bombardment of methane at these pressures. An extreme energy source is hydrogen which gives rise to the H_3^+ ion under these conditions. One may actually observe *more* fragmentation in this form of chemical ionization experiment than in electron impact, so the general statement that chemical ionization methods are milder should be slightly modified.

If instead of using a proton-donating reagent gas, we employ nitrogen, that under these conditions gives rise to N_2^+, *charge exchange* will occur and a molecular ion will be observed as in electron ionization rather than a molecular-weight-plus-1 ion. Subsequent fragmentation processes greatly resemble those occurring in electron ionization, since the recombination energy of this ion is similar to the energies used in electron ionization. In the case of simple amines, Field has shown recently that the resulting spectra are virtually indistinguishable (Whitney *et al.,* 1971) from the usual electron ionization spectra. Although this method offers much promise, it is true today that when dealing with an unknown compound it is essential to have its electron ionization spectrum at hand since most of the rules relating fragmentation of the molecular ion to chemical structure have been evolved using this method; furthermore large libraries of such data exist for direct comparison with an unknown. Fortunately one can pass from the chemical ionization to the electron ionization mode at any time by merely turning off the reagent gas and waiting a few seconds for the pressure to drop.

When fragmentation is observed in chemical ionization how are the spectra rationalized as compared with electron ionization? In general,

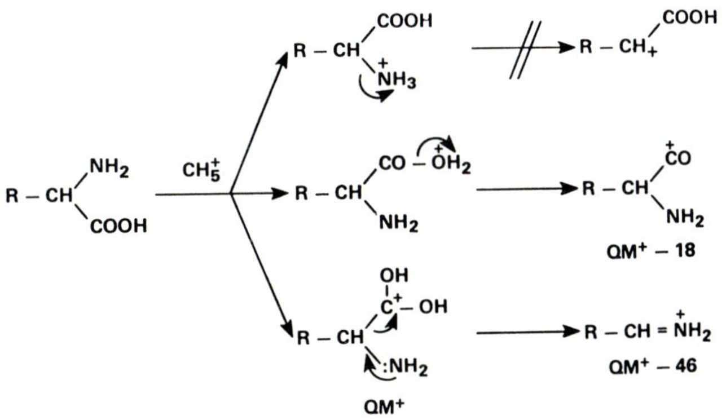

Fig. 18. Rationalization of chemical ionization mass spectra of amino acids.

the fragmentation mechanisms are easier to understand than the relatively high-energy electron ionization spectra. Rearrangements are minimal and fragmentations generally go along pathways very familiar to anyone who has dealt with acid-catalyzed reactions in the condensed state in the laboratory. How simple this can be is illustrated by spectra of the amino acids (Milne *et al.,* 1970) (Fig. 18). In the simple amino acids one can visualize three sites for proton attachment: the amino-group, the hydroxyl, and the carbonyl of the carboxylic acid. Assuming, as seems reasonable, that some protons attach randomly at each possible site, fragmentation will be dictated by the nature of the resulting ion. If the proton attaches to the amino nitrogen it is possible that ammonia will be lost. However, the resulting carbonium ion, being adjacent to a carbonyl group, has no particular stability, especially since its generation is opposed by the dipole of the carbonyl group. On the other hand, if a proton attaches to the hydroxyl of the carboxyl group, loss of water is accompanied by formation of the well-known and very stable acyl carbonium ion. Attachment of a proton to the carbonyl oxygen of the carboxyl group leads to an interesting cleavage with loss of the elements of formic acid ($COOH_2$) and stabilization of the resulting ion by participation of the lone electron pair on nitrogen. The resulting ion is the one which is generally observed at highest mass in electron ionization mass spectrometry of the amino acids and their esters. I think it is interesting to note in passing that all the amino acids except arginine and cystine have finite vapor pressures, so perhaps it is not surprising

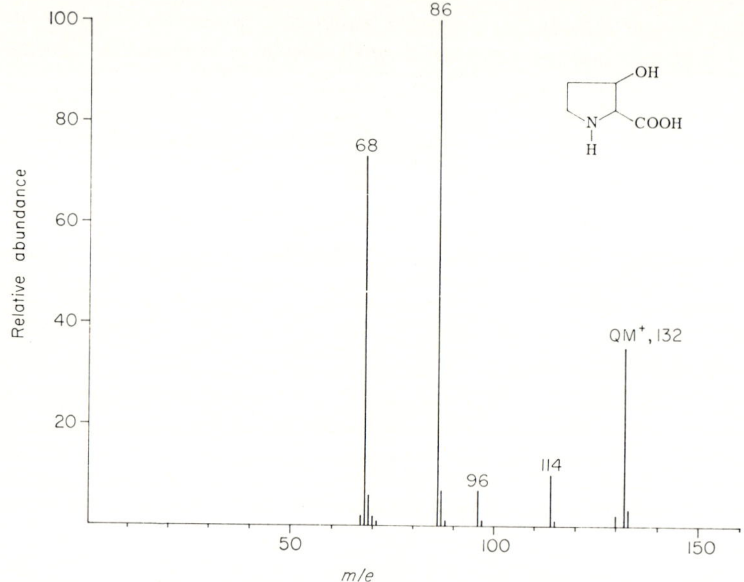

Fig. 19. Chemical ionization mass spectrum of 3-hydroxyproline.

that they all give satisfactory mass spectra under chemical ionization conditions. The advantage of chemical ionization spectra is that one can often discern the functionality of the amino acid in a very simple fashion. For example, 3-hydroxyproline (Fig. 19) displays, besides a peak at molecular weight-plus-1, loss from this ion of water, as well as loss of the elements of formic acid. Water is again lost as a secondary process *after* loss of the elements of formic acid, proving that an alcoholic hydroxyl group is present in the molecule in addition to the hydroxyl group of the carboxylic acid. The technique should be particularly valuable in elucidating the structures of new amino acids found in nature and even perhaps in identifying end-group residues in peptides when they can be removed as ether free or substituted amino acids.

The same technique can be used to elucidate the structures of small peptides (Kiryushkin *et al.*, 1971; Gray *et al.*, 1970) providing they can be volatilized through appropriate derivatization. The processes which occur in this case (Fig. 20) involve simple cleavage of the peptide bond with charge residing on the carbonyl oxygen. In this respect the process strongly resembles the analogous process in electron ionization spectrometry of the peptides. However, using CI it also happens that a proton rearrangement occurs so that the amino side of the peptide is also observed as an ion and one can check the sequence derived by analyzing for both fission products. We have worked out a simple com-

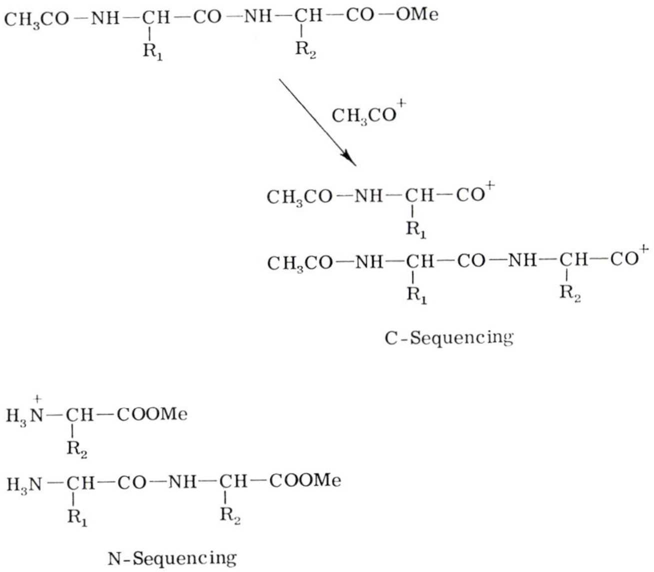

$$CH_3CO-NH-\underset{R_1}{CH}-CO-NH-\underset{R_2}{CH}-CO-OMe$$

$$\searrow CH_3CO^+$$

$$CH_3CO-NH-\underset{R_1}{CH}-CO^+$$

$$CH_3CO-NH-\underset{R_1}{CH}-CO-NH-\underset{R_2}{CH}-CO^+$$

C-Sequencing

$$H_3\overset{+}{N}-\underset{R_2}{CH}-COOMe$$

$$H_3N-\underset{R_1}{CH}-CO-NH-\underset{R_2}{CH}-COOMe$$

N-Sequencing

Fig. 20.   Rationalization of chemical ionization mass spectrum of peptides.

puter program for performing this task (Kiryushkin *et al.*, 1971), but it can be done by hand without much difficulty. Figure 21 gives an example of such an analysis in an admittedly favorable case. As in most published peptide mass spectra, you will notice that the peptide is composed of relatively simple amino acids. Still, the improvement in the spectrum obtained using chemical ionization in this case is obvious. The enhanced intensity in the molecular weight region is important and you may notice that expansion factors of 10 have not been used to expand the higher mass end of the spectra. In all fairness, I must mention that in some other cases the electron ionization spectra have been more impressive than those obtained using chemical ionization.

In the case of the thyroid-releasing factor of Drs. Guillemin and Schally (Fig. 22) brought to our attention by Dr. Dominic Desiderio at Baylor (Desiderio *et al.*, 1971), nature has provided us with both blocking end groups and the spectrum (Fig. 23) obtained on less than 1 μg of material, provides an excellent example of the potential of the method in the area of small peptides that are difficult to analyze by the more usual methods.

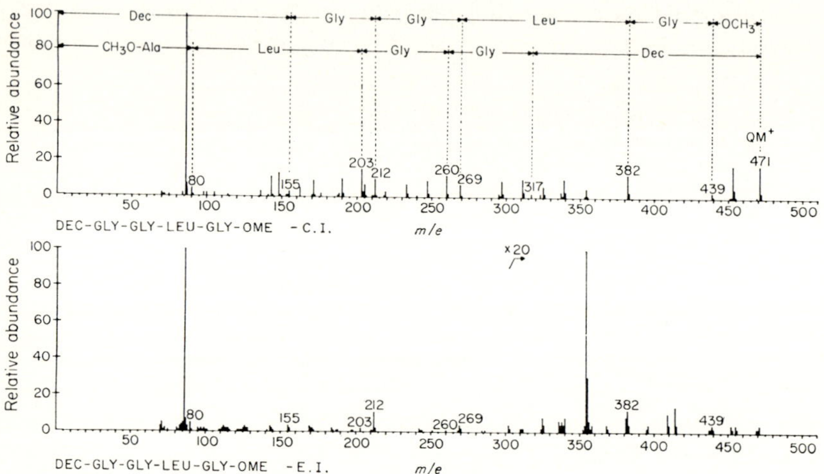

Fig. 21. Chemical ionziation mass spectrum of a typical small peptide.

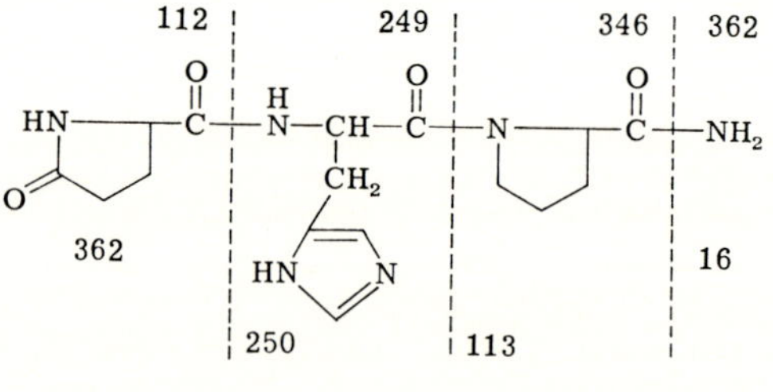

Fig. 22. Structure of the thyroid releasing factor.

From previous remarks, I think you may detect that I am somewhat less than enthusiastic about the general use of mass spectrometry today for the analysis of proteins and larger peptides. This comes about not only from the difficulties associated with the necessary chemical modification, but also from consideration of the very elegant technique of the Edman degradation. Using the automated versions of this method, combined with gas-liquid chromatographic analysis of the phenylthiohydantoins, it is possible to degrade a peptide containing a sequence of 20 amino acid residues within 3–4 days. So rather than compete directly with this powerful method, we decided to see whether mass spectrometry had anything to offer in its behalf. In collaboration with Drs. Bryan Brewer, John Pisano, Yumiko Nagai, and Thomas Bronzert, we have used CI methods to identify and quantitate the phenylthiohydantoins as they are produced by the sequenator (Fales *et al.*, 1971). In chem-

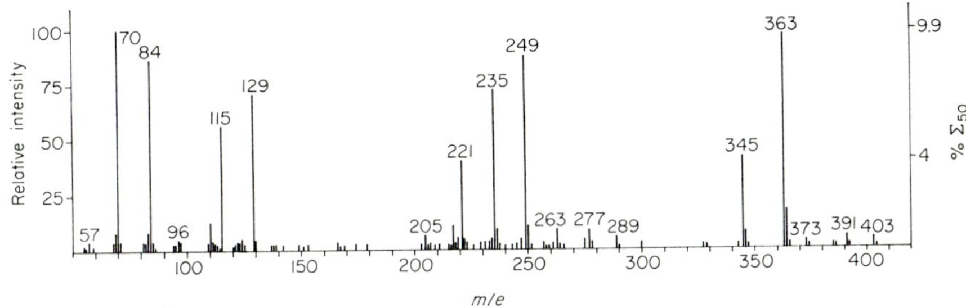

Fig. 23. Chemical ionization mass spectrum of the thyroid releasing factor.

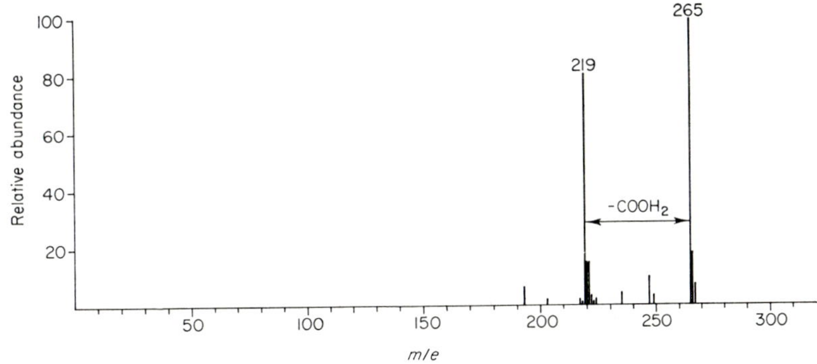

Fig. 24. Chemical ionization mass spectrum of glumatic acid phenylthiohydantoin.

ical ionization using isobutane, nearly all the phenylthiohydantoins provide either a single molecular weight-plus-1 ion or a very easily identified fragmentation peak, as Fig. 24 shows in the case of phenyl-thiohydantoin glutamic acid. A preliminary scan of the sample enables one to determine which phenylthiohydantoin is likely to be present and then addition of a known quantity of pentadeuterio-phenylthiohydantoin internal standard allows more exact quantitation (Fig. 25) by comparison of the ion with its neighbor at 5 mass units higher. Such quantitation is an essential feature late in the sequencing.

Figure 26 shows the results of our first attempt at myoglobin, and the results are seen to be quite satisfactory considering the somewhat impure nature of the myoglobin used in this experiment. We hope to simplify the process by scanning the sample many times, using a quadrupole mass spectrometer as the material evaporates from the direct insertion probe. The scans will be stored in a computer, and subsequently, each scan will be analyzed for the mass (or masses) pertinent to a given phenylthiohydantoin. The sum of these intensities is then an absolute

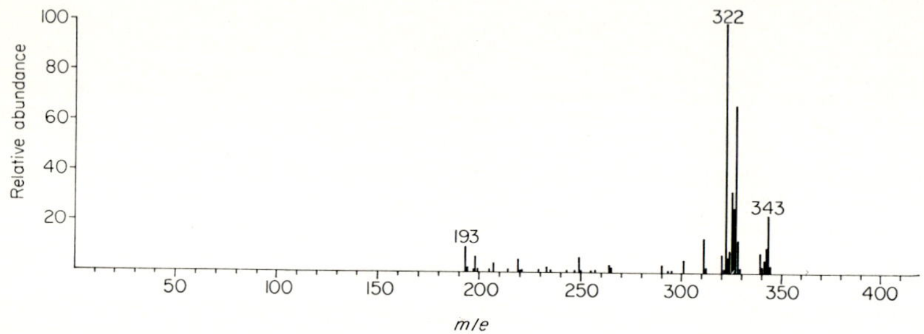

Fig. 25. Chemical ionization mass spectrum of a mixture of phenylthiohydantoins plus internal pentadeuterio standards.

$$
\begin{array}{cccccccccc}
1 & 2 & 3 & 4 & 5 & 6 & 7 & 8 & 9 & 10 \\
H_2N - Val - Leu - Ser - Glu - Gly - Glu - Trp - Gln - Leu - Val - \\
11 & 12 & 13 & 14 & 15 & 16 & 17 & 18 & 19 & 20 & 21 & 22 \\
Leu - His - Val - Trp - Ala - His - Val - Glu - Ala - Asp - Val - Ala - \\
23 & 24 & 25 & 26 \\
Gly - His - Gly - Gln
\end{array}
$$

Fig. 26. Partial structure of sperm whale myoglobin.

measure of the quantity of phenylthiohydantoin present and the only number actually required. A similar system, using electron ionization methods, has been developed by Dr. R. Lovins (Lovins *et al.*, 1971) at the University of Georgia and has proved most satisfactory.

Although alkaloids (Fales *et al.*, 1970) and antibiotics are likely to be of little interest to this group, I cannot resist showing some scans of these basic materials since it is in this area that the chemical ionization method really excels. Figure 27 shows the difference in degree of fragmentation observed when narcotine is investigated by both methods, and it is clear that the molecular weight is much more easily determined from the chemical ionization scan. In the case of deoxyribose (Fales *et al.*, 1969) (Fig. 28) loss of water is observed but the molecular weight + 1 peak is still distinct. Even a molecule such as erythromycin, which contains hemiacetal linkages, survives on protonation (Fig. 29), probably because of the basic nitrogen atom. In general, if a molecule contains an amine function, this is a stabilizing influence in chemical ionization (in contrast to electron ionization) since protonation at this site seldom provides, on cleavage, a carbonium ion with any great stability.

Is chemical ionization useful in the steroid field? Perhaps occasionally, but it has been our experience that those processes which prevent observation of the molecular ion in electron ionization, such as loss of acetic

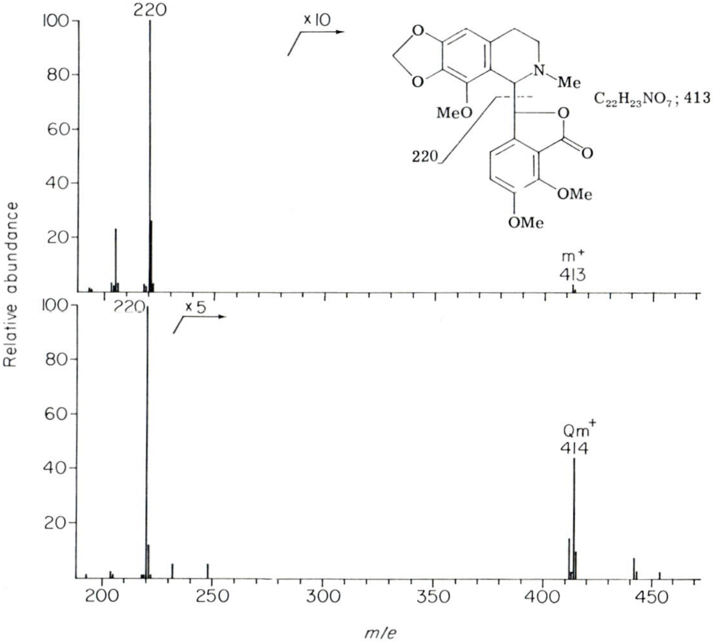

Fig. 27. Chemical ionization mass spectrum (below) of narcotine compared with electron ionization spectrum (above).

acid and water, are even more pronounced in chemical ionization. In fact, it is not so much that chemical ionization is poor for the steroids, but that electron ionization is so very satisfactory, particularly since gas chromatography can also be invoked.

Finally, I would like to discuss a scientifically trivial, but very practical, use of both electron and chemical ionization mass spectrometry in the clinical area. In recent years, Dr. Milne, Virginia Aandahl, and I, in collaboration with Mr. Norman Law at nearby Surburban Hospital (Law et al., 1971, 1972), have found that by using GC-MS we are able

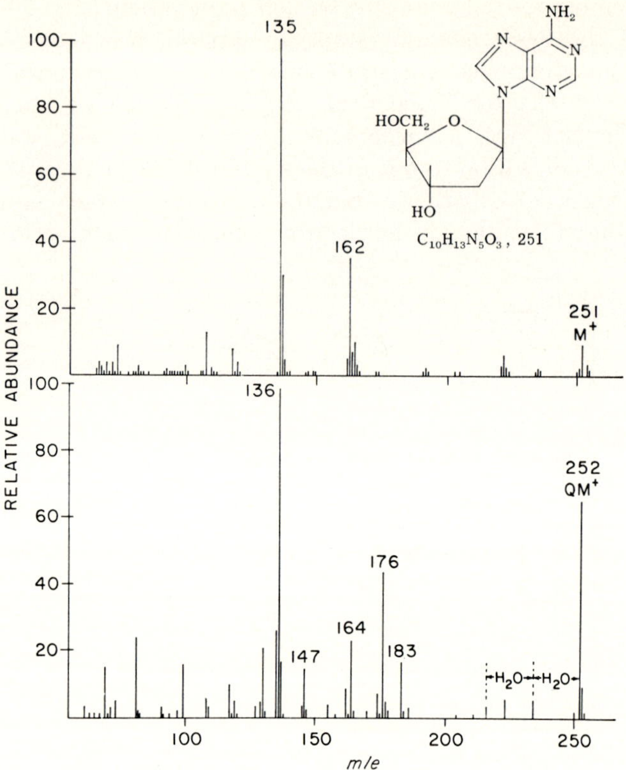

Fig. 28. Chemical ionization mass spectrum (below) of deoxyadenosine compared with electron ionization spectrum (above).

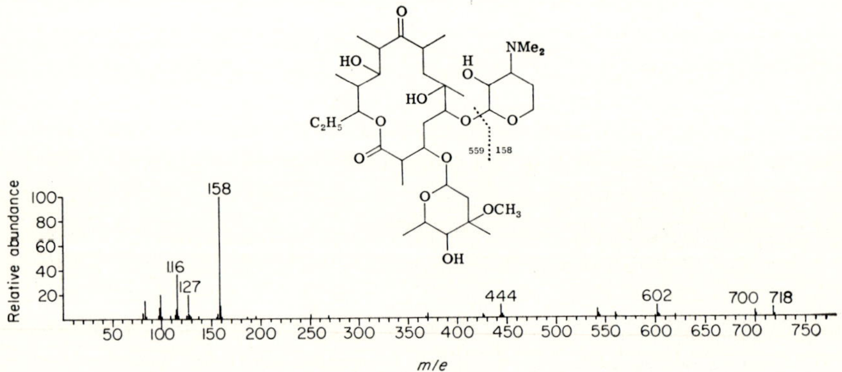

Fig. 29. Chemical ionization mass spectrum of erythromycin.

to identify quite easily the drug (or drugs) responsible for the comatose condition of individuals who had taken overdoses. [A similar technique was demonstrated earlier by Althaus *et al.* (1970) at M.I.T.] This involves a chloroform extraction of serum or gastric lavage, evaporation of the solvent, and GC-MS of the residue. All peaks are scanned as they elute from the gas chromatograph (Fig. 30), and comparison is made either manually or with a computer with a library of about 80 common drug spectra. Often the patient has ingested several compounds and, while this complicates the treatment, the mass spectral identification is straightforward. Even metabolites are picked up, and the potential

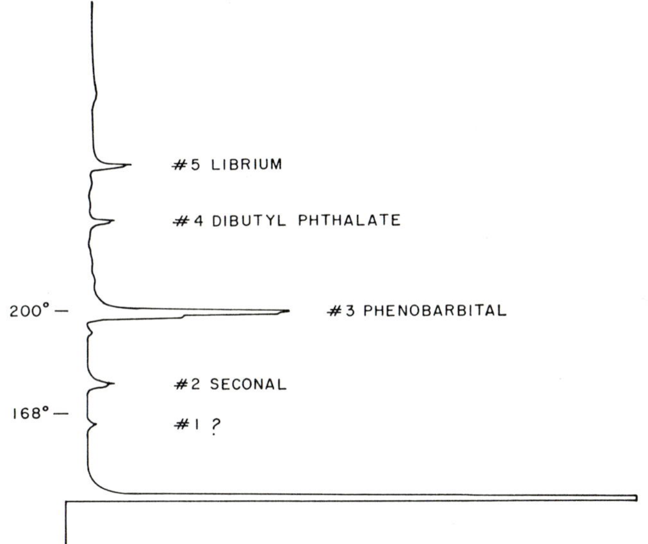

FIG. 30. Gas chromatogram of drugs from gastric lavage in an overdose case.

for drug metabolism studies is obvious. Generally, we can have an unequivocal answer in 0.5–1 hour, and the physicians responsible for decisions about whether or not to dialyze seem to be very happy about the results. We have recently completed 100 cases, and Table I shows the drugs we have identified. Recently, we have wondered whether gas–liquid chromatography might be eliminated by using chemical ionization, as shown in Fig. 31. Most of these drugs (except for the barbiturate) were components of one pill, and even then we missed one trace component (homatropine). Such mixtures are not uncommon, and I think this scan dramatically illustrates the problems facing the clinical toxicologist as long as such combinations are commercially available and prescribed.

TABLE I

*Listing of Drugs Found in Overdose Cases
and Frequency of Occurrence*

Generic name of drug	Number of times identified
Phenobarbital	8
Barbital	1
Amobarbital	8
Hexobarbital	1
Secobarbital	8
Pentobarbital	8
Meprobamate	4
Carbromal	1
Glutethimide	8
Ethchlorvynol	5
Methapyrilene	2
Methaqualone	5
Primidone	1
Trifluoperazine	1
Amitriptylene	2
Chlordiazepoxide	1
Diazepam	6
Meperidine	2
Acetyl salicylate	4
Salicylamide	1
Phenacetin	1
Propoxyphene	2
Methyl salicylate	1
Caffeine	2
Pentazocine	1

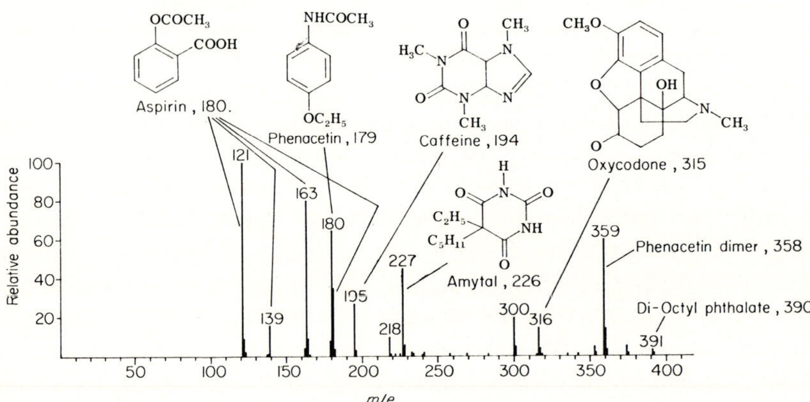

FIG. 31. Chemical ionization mass spectrum of drug sample from overdose case.

ACKNOWLEDGMENTS

I would like again to thank all my collaborators in this work, especially Dr. G. W. A. Milne, whose enthusiasm and skill have been major factors in the success of our applications of chemical ionization, and the National Heart and Lung Institute for allowing us to engage in this only slightly heart-related research.

REFERENCES

Althaus, J. R., Biemann, K., Biller, J., Donoghue, P. R., Evans, D. A., Förster, H.-J., Hertz, H. S., Hignite, G. E., Murphy, R. D., Preti, G., and Reinhold, V. (1970). *Experientia* **26**, 714.

Beckey, H. D. (1969). *Int. J. Mass Spectrom Ion Phys.* **2**, 500.

Desiderio, D. M., Earle, N. R., Krueger, P. M., Lawson, A. M., Smith, L. C., Stillwell, R. N., Tsuboyama, K., Wijtvliet, J., and McCloskey, J. A. (1968). *Sixteenth Annual Conference on Mass Spectrometry and Allied Topics, Pittsburgh, Pa.* p. **228**.

Desiderio, D. M., Burgus, R., Dunn, T. F., Vale, W., Guillemin, R., and Ward, D. N. (1971). *Org. Mass Spectrom.* **5**, 221.

Dole, M., Mack, L. L., Hines, R. L., Mobley, R. C., Ferguson, C. D., and Alice, M. B. (1968). *J. Chem. Phys.* **49**, 2240.

Engel, L. L., and Orr, J. C. (1972). *In* "Biomedical Applications of Mass Spectrometry" (G. Waller, ed.), Chapter 19. Wiley, New York.

Fales, H. M. (1966). *Anal. Chem.* **38**, 1058.

Fales, H. M., Milne, G. W. A., and Vestal, M. L. (1969). *J. Amer. Chem. Soc.* **91**, 3628.

Fales, H. M., Lloyd, H. A., and Milne, G. W. A. (1970). *J. Amer. Chem. Soc.* **92**, 1590.

Fales, H. M., Nagai, Y., Milne, G. W. A., Brewer, H. B., Bronzert, T. J., and Pisano, J. (1971). *Anal. Biochem.* **43**, 28.

Field, F. H. (1970). *J. Amer. Chem. Soc.* **92**, 2672.

Futrell, J. H., and Tiernan, T. O. (1968). *In* "Fundamental Processes in Radiation Chemistry" (P. Ausloos, ed.), Ch. 4. Wiley (Interscience), New York.

Gray, W. R., Wokcik, L. H., and Futrell, J. H. (1970). *Biochem. Biophys. Res. Commun.* **41**, 1111.

Horning, E. C., Luukkainen, T., Haahti, E. O. A., Creech, B. G., and VandenHeuvel, W. J. A. (1963). *Recent Progr. Horm. Res.* **19**, 57.

Kiryushkin, A. A., Fales, H. M., Axenrod, T., Gilbert, E. J., and Milne, G. W. A. (1971). *Org. Mass Spectrom.* **5**, 19.

Law, N. C., Aandahl, V., Fales, H. M., and Milne, G. W. A. (1971). *Clin. Chim. Acta* **32**, 221.

Law, N. C., Fales, H. M., and Milne, G. W. A. (1972). *Clin. Toxicol.* in press.

Lovins, R. E., Craig, J., and Fairwell, T. (1971). *Fed. Proc. Fed. Amer. Soc. Exp. Biol.* **30**, 1241.

Luukkainen, T., and Adlercreutz, H. (1967). *Ann. Med. Exp. Biol. Fenn.* **45**, 264.

MacConnell, J. G., Blum, M. S., and Fales, H. M. (1970). *Science* **168**, 840.

MacConnell, J. G., Blum, M. S., and Fales, H. M. (1971). *Tetrahedron* **26**, 1129.

Milne, G. W. A., Axenrod, T., and Fales, H. M. (1970). *J. Amer. Chem. Soc.* **92**, 5170.

Munson, M. S. B., and Field, F. H. (1966). *J. Amer. Chem. Soc.* **88**, 2621.
Röller, H., and Dahm, K. H. (1968). *Recent Progr. Horm. Res.* **24**, 651.
Shemyakin, M. M. (1968). *Pure Appl. Chem.* **17**, 313.
Tal'rose, V. A., and Lyubinova, A. K. (1952). *Dokl. Akad. Nauk SSSR* **86**, 909.
Vilkas, E., and Lederer, E. (1968). *Tetrahedron Lett.* p. 3089.
Whitney, T. A., Klemann, L. P., and Field, F. H. (1971). *Anal. Chem.* **43**, 1408.

Discussion

J. R. Goding: What happens when you have a mixture—like prostaglandins? Are you able to tell exactly how much of each is present and is the specificity and so on agreeable to people used to working in other fields?

H. M. Fales: The question of quantitation in mass spectrometry is increasingly important, and I would like to answer in some detail.

The mass spectrometer has somehow acquired an undeservedly poor reputation for quantitation. In fact, the spectrometer itself is capable of very high precision and accuracy. There are several ways in which quantitation can be achieved. If the sample is a mixture of gases (or volatile, stable liquids which can be converted to gases below ~250°C), the vapor from a measured quantity may be admitted to the ion chamber through a fixed leak. Under these conditions, assuming trivial loss of sample during the analysis, a certain response will be obtained at each peak in the mass spectrum if all instrumental parameters are fixed. This response should remain constant within a few percent or better over a period of days. The mass spectrometer is very seldom used in this way today since considerable material (~1 mg) is required and few compounds have the required volatility and stability. On the other hand, if a known, small amount of compound is admitted via the direct insertion probe, the spectrum will obviously change intensity as the sample is consumed. Accuracy can still be achieved if the spectra are summed over the total lifetime of the sample, but this is tedious and done conveniently only with a rather expensive computer. This is the approach we are taking to the analysis of phenylthiohydantoin derivatives.

All the instrumental parameters can be eliminated as variables if the method of internal standards is employed. For this purpose, it is usually necessary to synthesize an analog of the material in question labeled with a heavy isotope (e.g., ^{15}N, ^{13}C, ^{18}O, ^{2}H). Because of the natural abundance of such isotopes, the labeling should be at least several percent excess at one location on the molecule. Lower values can be employed, but accurate measurement and subtraction of peak heights become necessary and render the whole procedure less reliable. It is even possible sometimes to substitute a closely related compound (e.g., a dihydro derivative) as an internal standard if it can be established that the volatility of the compound and standard are identical. Admixture of an amount of standard comparable (within a factor of 10–100) with the unknown and evaporation on the direct insertion probe is all that is required for analysis, but handling the volumes involved (1–3 μl) is not as easy as it may sound. In any case, repeated scanning of the spectrometer over the mass range containing the two isotopic species and comparison of the two peak heights gives directly their relative quantities in the sample.

Such analyses can be carried out with high accuracy well below the nanogram range, save for one factor: the instrumental background. A mass spectrometer in general laboratory use will usually show a small peak at every mass unit up to at least m/e 400 under high gain conditions.

Increased confidence in the reliability of a relationship observed between internal standard and unknown peaks can be obtained if their relative intensities remain constant well through the sample lifetime. Clearly, if we are able to compare not just one peak in the spectrum of each compound, but several (or even the whole spectrum), reliability increases still further.

Such comparisons can be done with a computer, but they suffer from the fact that time must be spent (3–10 seconds) in scanning through the whole spectrum. For maximum sensitivity we would like to present the sample to the mass spectrometer in a short burst of approximately the same time period so that only one or two scans could be obtained. This is particularly true when the sample elutes from the gas chromatograph as a narrow peak near the solvent front. One way around this problem involves high-speed switching between several pertinent m/e values by varying the accelerating voltage (the magnet cannot be switched rapidly because of hysteresis) [C. C. Sweeley, W. H. Elliott, I. Fries, and R. Ryhage, *Anal. Chem.* **38**, 1549 (1966)]. Measurements are then made of the relative intensities of unknown and known mass peaks only over that period when their ratio is constant. Alternatively, this system can be used as a GLC detector for one compound without the use of internal standards by focusing on several of its important fragment ions. The total area of any one of the series of peaks is then used as a measure of the sample quantity only while its relationship to the other peaks remains constant [C.-G. Hammer, B. Holmstedt, and R. Ryhage, *Anal. Biochem.* **25**, 532 (1968)]. Measurement in the 50–100 pg range have been claimed for this method [T. E. Gaffney, C.-G. Hammer, B. Holmstedt, and R. E. McMahon, *Anal. Chem.* **43**, 307 (1971)], but it is clear that spectrometer and sample conditions must be ideal to realize this level of detection. When appropriate precautions are observed, there is little reason to doubt the specificity or accuracy of the method, and I am sure that this applies to the prostaglandins as well. It must always be borne in mind, however, that the mass spectrometer does not distinguish stereoisomers or geometric isomers very well, and confusion could arise from this source. Fortunately, GLC often supplies this information. In fact, it appears that most uses of the mass spectrometer in this regard have been to establish that a very weak GLC signal is in fact due to the suspected compound. Quantitation in such experiments is regarded as secondary. The use of mass spectrometers in following changes in biological systems over short time spans where many samples are involved has been minimal. I suppose this is due in some part to the fact that mass spectrometrists do not look kindly upon the required dedication of their instruments to a single project over a considerable period of time. As mass spectrometers become simpler and less expensive, we may expect that its use as a truly quantitative tool will increase.

F. C. Bartter: Have you been able to apply your phenylthiohydantoin method to sequencing of protein hormones?

H. M. Fales: By the time we worked out this method last summer, Dr. Bryan Brewer had already completed his sequencing of the parathyroid hormone [H. B. Brewer and R. Ronan, *Proc. Nat. Acad. Sci. U.S.* **67**, 1862 (1970)], so we had no chance to test it in that case. Since that time we have run some samples for Dr. Brewer when the gas chromatographic analysis is equivocal. One case where it seems to be very useful is in deciding between phenylthiohydantoin-glutamine and -glutamic acid. It is not so useful with phenylthiohydantoin-serine or -threonine because of prior dehydration of the derivatives. We have decided to evaluate the method in depth using a dedicated chemical ionization–quadrupole mass spectrometer–computer system, so we are obviously enthusiastic about its potential.

J. Rudinger: It may be useful in this connection to recall a trick which the late Professor Weygand introduced about three years ago. He proposed the use of variously substituted phenylisothiocyanates in the Edman degradation in such a way that a different isothiocyanate would be used in each of a sequence of degradation steps. The characteristic mass spectrometric peaks of the aryl groups should then make it possible to analyze the sum of the arylthiohydantoins formed in several degradation cycles by a single analysis, or alternatively to distinguish the arylthiohydantoins formed in a given step from those carried over from previous steps and thereby to reduce the "noise" of the determination [F. Weygand, *Fresenius' Z. Anal. Chem.* **243**, 2 (1968); F. Weygand and R. Obermeier, *Eur. J. Biochem.* **20**, 72 (1971)].

H. M. Fales: This is certainly a very valid approach. However, the automated sequenator removes the phenylthiohydantoin derivatives at each step, and there does not seem to be any real advantage in bypassing this feature. Furthermore, it is my understanding that problems do not arise from carrying over the phenylthiohydantoin derivatives from one step to the next, but rather from incomplete or nonselective degradation of protein. If this is true, varying the nature of the phenylisothiocyanate would be of no avail.

B. L. Rubin: I can see that you can tell whether there is a hydroxyl because you have lost that amount of weight, but how do you know where on the molecule it was? What do you do if you have no idea with which to start?

H. M. Fales: Structural information comes out of the mass spectrometer partly as a result of mechanistic considerations gleaned from solution phase organic carbonium ion chemistry [K. Biemann, "Mass Spectrometry, Organic Chemical Applications." McGraw-Hill, New York, 1962] but mostly via liberal use of analogy as in other fields, such as biochemistry and, I suppose, even medicine.

Specifically, it is sometimes quite easy to see where a hydroxyl group is in a steroid because steroids quite often undergo cleavage between the B and C rings localizing the functional groups in either portion of the molecule. At present it is true that some organic chemical training is nearly essential, and experience with many types of compounds greatly enhances one's chances for success. It is possible that in the future we will take much greater advantage of a naive but powerful approach known as the "learning machine." Drs. Isenhour and Jurs [T. L. Isenhour and P. C. Jurs, *Anal. Chem.* **43**, 20a (1971)] are the foremost proponents of the technique in which a computer is fed mass spectral data identified in simple dichotomic terms; i.e., "the molecule contains oxygen—yes, or no?" The computer searches for a "learning" surface in n-dimensions among a set of "training" spectra, which is then used to make decisions in unknown cases.

M. A. Kirschner: At the beginning of your talk you discussed various ways of "brushing the molecules lightly" with electrons or other means in order to delineate the parent ion. Further on you commented that hitting the molecule "hard" causes fragmentation into predictable patterns which can then be used to aid in structural identification. Do you envision a family of mass spectra in the analysis of a compound?

H. M. Fales: Yes; ideally, we would like to have high and low voltage electron ionization spectra, chemical ionization spectra using several reagent gases, and field ionization or field desorption spectra since data from one system often helps to explain unusual features of the other.

G. D. Aurbach: What about using a different type of detection system instead of an ionization chamber? Suppose one were looking for the biosynthesis of a particular compound, for example, from acetate. Would it be possible to hook the spectrometer up to a radiation detector and look only for compounds containing ^{14}C?

H. M. Fales: This has actually been accomplished in a very simple fashion using a graphite-covered photoplate. Thus ions containing ^{14}C will fall at $(M + 2)^+$ rather than M^+, and, since they are radioactive, an autoradiograph will abnormally blacken at this point [H. Knöppel, and W. Beyrich, *Tetrahedron Lett.* p. 291 (1968)]. Alternatively the ^{14}C can be measured directly if it is present at high enough concentration ($\sim 5\%$) [J. L. Occolowitz, *Chem. Commun.* p. 1226 (1968)]. The problem at present is that such detection is very insensitive compared to the usual scintillation counter methods. ^{13}C and other heavy isotopes are attractive alternatives for such studies, but, unfortunately, high-accuracy isotope ratio measurements on large molecules are difficult to obtain. In addition it must be remembered that there is considerable variation in the natural abundance of many of these isotopes. For the biochemist this means that he must achieve very high conversions of substrate. Fortunately, since the spectrometer sensitivity is very high, the experiment need only be performed on very small amounts of material.

J. C. Orr: In conjunction with Drs. Ofner and Engel, [L. L. Engel and J. C. Orr, *In* "Biomedical Applications of Mass Spectroscopy (G. Waller, ed.), Chapter 19, Wiley, New York (1972)] I have been doing some studies of biosynthetic transformation of steroids using gas–liquid chromatography and mass spectrometry. I can see that chemical ionization or other techniques which enhance the molecular ion contribution would be very useful. I took ordinary 17β-hydroxy-5α-androstan-3-one, which has a molecular weight of 290, and mixed it with precisely 43% of 7β-deuterio-17β-hydroxy-5α-androstan-3-one. Natural abundance ^{13}C and deuterium in the unlabeled compound contribute to the peak at m/e 291 $(M + 1)^+$; therefore less than an equimolar amount of the deuterated compound is added to make the ions at m/e 290 and 291 equal in size.

This dihydrotestosterone was incubated with a mince of canine perianal glands. Whatever biosynthetic transformation products you get from this dihydrotestosterone will again have the molecular ion and $(M + 1)^+$ of equal height, or approximately equal height.

In Fig. A on the bottom (merely for comparison) is authentic 5α-androstane-3β,7α,17β-triol. The upper mass spectrum is of the isolated metabolite which contains deuterium in 43% of the molecules. The molecular ions of the triol at m/e 308, 309 are very small; it would have been advantageous to use chemical ionization or some other such technique which causes less fragmentation of the molecule. However, the M-H$_2$O$^+$ ion at m/e 290 and 291 is a double ion and quite prominent enough to reveal this as a metabolite of the dihydrotestosterone. By using GC-MS one can scan a whole lot of mass spectra of the various different GLC peaks, and only those which have something which looks like a "double" molecular ion need you examine further.

In this case, the deuterium being at 7β was extremely fortunate because the 7α-hydroxy was introduced without loss of the label, but on oxidation to 5α-androstane-3,7,17-trione with chromium trioxide under mild conditions the deuterium was lost. This specifically identifies one transformation by the perianal gland as 7α hydroxylation.

H. M. Fales: This is an elegant example of the power of the internal standard method in mass spectrometry. I might add that in my opinion chemical ionization would probably have given even less intensity in the $(M + H)^+$ ion than you observe in the M^+ ion by the electron impact approach since the only basic functions on the molecule are its hydroxyl groups. Probably only $(M + 1 - H_2O)^+$ ions would be visible.

J. Rudinger: Concerning the use of internal standards, ^{15}N is another case in point. By looking for twin peaks in peptides, etc., one can sometimes get the same results as with radioactive labeling since in this case it is much more difficult to localize the site of incorporation.

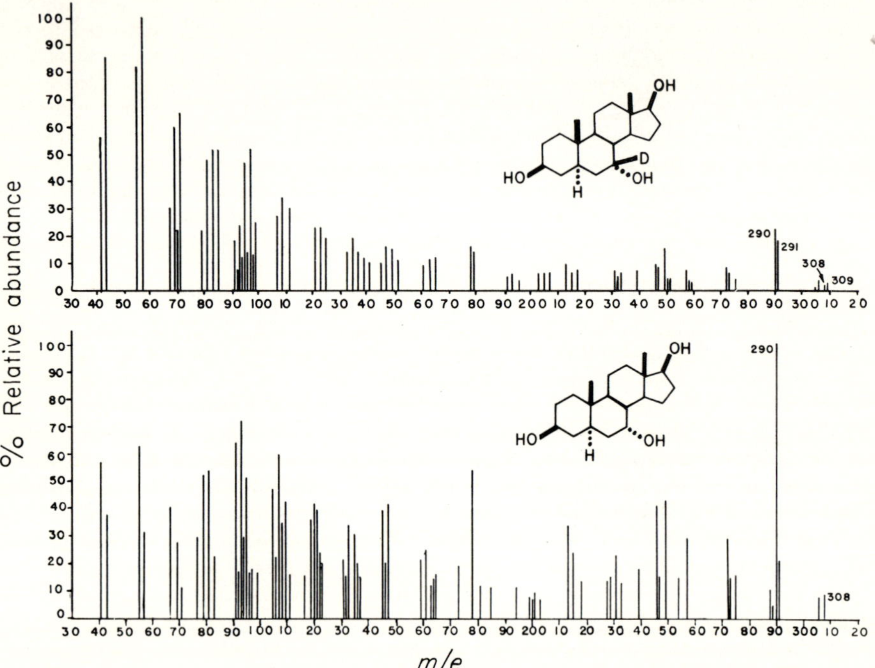

Fig. A. Mass spectra of 5α-androstane-3β,7α,17β-triol (bottom) and of isolated metabolite (top), which contains deuterium in 43% of the molecules.

J. T. Potts, Jr.: Could you amplify further the reasons for the slightly pessimistic scene with regard to peptide sequencing by mass spectrometry? You mentioned that the difficulties were in the derivatization. Is that because material is lost during derivatization or because it is a sort of uneven reaction and therefore the product being analyzed gives confusing results? Is it yield or control of derivatization that is the problem?

H. M. Fales: Most of the methods that have been used have involved prior acylation of the amino end group with something like acetic or propionic anhydride. This reaction is generally quite satisfactory and destroys the very polar zwitterionic character of the peptide providing no arginine or histidine residues are present. All active hydrogens can then be methylated, using methyl iodide and an appropriate base. This is a somewhat unsatisfactory reaction, however, because of the strongly basic conditions that result in some hydrolysis of glutamine residues. Also, incomplete methylation is common. Furthermore, methyl iodide will quaternize any tertiary amines, such as histidine, rendering the protein totally involatile. So, to answer your question, it is a problem of low yield due to inadequate control of derivatization.

J. C. Orr: You mentioned that you got alkyl-substituted methyl piperidines. Were you able to identify which was *cis* and which was *trans* from mass spectrometry?

H. M. Fales: No. Their mass spectra were identical. This differentiation was done strictly on the basis of GLC and synthesis.

R. O. Greep: You have made it appear that these are very useful gadgets to have around and something that no good laboratory should be without. Do they come in different models? Yours is no doubt of the Cadillac variety. Is there something on the order of a Volkswagen available? Is there a model a poor institution could afford and still gain some of the advantages of this useful instrument? Do you have to have a different instrument for electron ionization and chemical ionization? Could you tell us a bit about the practical aspects?

H. M. Fales: The advantage of the Cadillac, or double-focusing, variety is that we can separate ions of mass 100.00 from 100.01 enabling us to calculate directly the formula of the molecule in question. This is due to the very slight difference between the integral mass of the atoms; i.e., CH_4 and O are both nominally 16 but actually differ by an easily measurable 0.0364 mass unit even at m/e 300. Such information is obviously very useful, but such measurements are either very tedious to make or require the use of a very specialized computer system. I cannot believe that this audience would often require the services of such an instrument. At the other extreme, a quadrupole mass spectrometer good to about mass 300–400 can be purchased for about \$28,000. It may serve you very well in much of your work but may be unsatisfactory when you desire to obtain the molecular weight of a trimethyl silyl ether of a pentahydroxylated steroid that may be very easy to see on a GLC run. In between are the so-called medium-resolution, magnetic instruments good to about m/e 1000–1500 mass costing about \$60,000. I feel that if you can afford it, this is by far the best choice for the type of application that most persons here probably have in mind.

Furthermore, if you have a sample you wish to have analyzed using high resolution (or low resolution, for that matter), several commercial organizations still perform the analysis for a fee that may amount to a very small fraction of the cost of upkeep of such an instrument. Currently, the National Institutes of Health are assisting in the analysis of compounds of biological importance; no charge is made if the sample qualifies (contact R. Foltz, Battelle Memorial Institute, Columbus, Ohio). An instrument modified for chemical ionization work can easily be used for electron ionization by just turning off the gas, although it will probably be a shade down in sensitivity. However the reverse is not true due to inadequate pumping in the chemical ionization mode.

F. G. Peron: To what extent can the techniques utilized in your laboratory be used to assess the purity of a particular compound?

H. M. Fales: One of the very strong features of the mass spectrometer is that it is extremely sensitive in disclosing the presence of an impurity, unless it happens to be a very closely related stereoisomer. On the other hand, using fractional evaporation from the probe, impurities may not seriously interfere with one's ability to obtain a useful mass spectrum. High purity is a convenience, but definitely not a general requirement of the method.

L. L. Engel: At a Laurentian Hormone Conference many years ago [R. N. Jones, *Recent Progr. Horm. Res.* **2,** 2 (1948)] the subject of infrared spectroscopy was presented to the endocrinologists present. This physical tool had a profound impact on the development of the field. I think Dr. Fales has described for us a tool which may have an even greater impact on endocrinology.

Techniques for Assessing the Effects of Sex Hormones on Affect, Arousal, and Aggression in Humans[1]

Donald T. Lunde and David A. Hamburg

*Department of Psychiatry,
Stanford University Medical Center,
Stanford, California*

I. Introduction

The combination of techniques and findings from endocrinology and the behavioral sciences has already led to a number of interesting and important observations. For instance, it has been observed that sex hormones, by means of a variety of direct and indirect influences, act on the central nervous system and other organ systems involved in the execution of sexual and aggressive behavior. It is also well known that sexual and aggressive behavior in humans is accompanied by changes in affect; that is, alterations of a subjective nature that are reported as feelings or emotions, such as excitement or anger. Furthermore, recent studies have indicated that variations in female sex hormones during the menstrual cycle and postpartum period may be associated with changes in affect that include anxiety, irritability, and depression (Hamburg *et al.*, 1968). Such phenomena are of great significance to everyday human experience and in extreme cases, such as postpartum psychosis, present challenging problems for the clinician. These and other related observations lead one to question whether or not sex hormones play some role in the arousal to act, the act itself, and the accompanying affect in human sexual and aggressive behavior.

Data from animal studies in this area have definite limitations when extrapolating to the human situation for at least two reasons. First, it is difficult to demonstrate that animals, with the possible exception of some of the primates, experience emotional states that are qualitatively similar to those human experiences that we know as depression, euphoria, anger, and so forth. Second, we are well aware that the effects of sex hormones vary considerably among various animal species. Since the effects of estrogen and progesterone on sexual receptivity, for instance, vary considerably from the rat to the rabbit to the subhuman primates (Kopell, 1969), conclusions drawn from animal studies about the effects of these hormones on human sexual receptivity can only be tentative and are perhaps misleading. In fact, Reichlin (1971) has noted

---

[1] This work was supported by N.I.H. Grant MH-10976 and by the Veterans Administration (PAVAH, Palo Alto, California).

that, on the basis of studies of human subjects, it now seems clear that the primary erotogenic hormones in human females are the androgens of adrenal origin, a conclusion that would not have seemed likely from the many studies of estrus in various animals.

While the value of research with human subjects is perhaps self-evident, the difficulties involved in human research are also apparent. Human subjects for both ethical and practical reasons cannot be subjected to many of the rigorous controls and experimental manipulations that are utilized in animal work. Consequently, there is a need for special techniques that will provide useful data and at the same time prove acceptable and cause no harm to the subjects being studied.

The purpose of this paper is to describe a variety of techniques that we have found useful in exploring the relationship of various hormones to human emotions and behavior. We will discuss briefly our past work, which has involved the assessment of changes in affect associated with variations in estrogen and progesterone, particularly the latter. We will then describe some new EEG techniques that we are currently using to assess arousal to sexual stimuli as well as alterations in arousal in response to hormone administration.

Finally, we will devote the remainder of this paper to a discussion of the rationale for future studies that would explore the relationship of androgens to aggressive behavior in humans. In this context we will emphasize new biochemical techniques that can be used in conjunction with psychological, EEG, and behavioral assessment techniques currently available.

II. Assessment of Changes in Affect

A. During the Menstrual Cycle

Alterations in emotional state during various phases of the menstrual cycle are quite common, particularly during the premenstrual and menstrual phases. Suicides and acts of violence are also more common during these periods when the levels of estrogen and progesterone are rapidly changing (Dalton, 1964; Mandell and Mandell, 1967). In order to document the time course of changes in symptoms and mood over the menstrual cycle, a group of 15 nulliparous women were studied over two consecutive menstrual cycles (Moos et al., 1969). (None of the subjects were using oral contraceptives.) The women were tested on approximately days 2, 7, 14, 19, 24, 25, 26, 27, and 28 of their cycles. On each of these days they completed ratings of their symptoms, moods, and feelings of sexual arousal. Two standardized self-rating scales were used in this study, the Menstrual Distress Questionnaire and the Nowlis Mood

Adjective Check List. Both are tests that are easy to administer and score and are quite reliable. Some of the results are shown in Figs. 1–4.

Figure 1 indicates that pain and water retention symptoms were high in the menstrual phase, decreased rapidly, and then showed a slow but steady increase in the latter part of the cycle. Figure 2 indicates that anxious and aggressive feelings were also high during the menstrual phase and then diminished. Anxiety began to rise again around mid-cycle and was highest on about day 20, after which there was a slight but

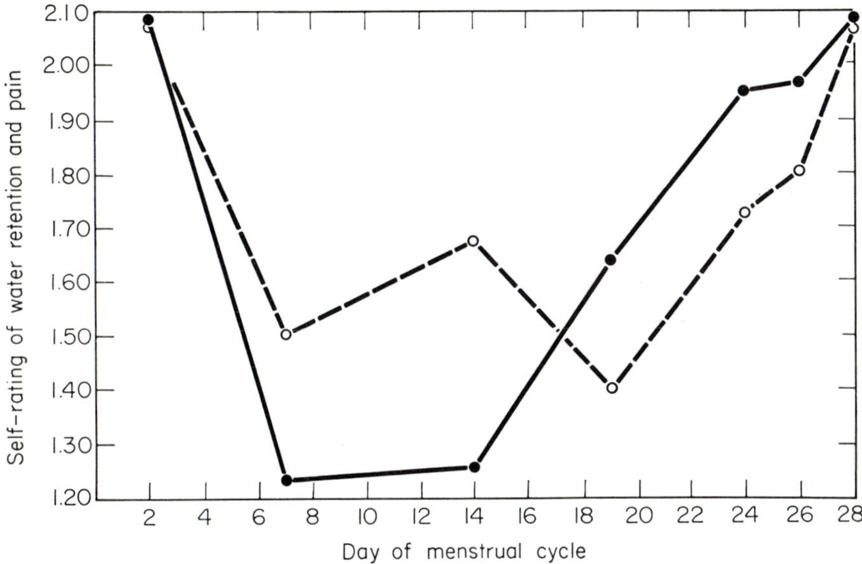

Fig. 1.   Average self-rated pain (○ - - - ○) and water retention (●——●) of 15 women in two consecutive menstrual cycles. Reprinted with permission from Moos *et al. J. Psychosom. Res.* **13,** 37 (1969), Pergamon Press.

not significant decrease on day 28. Aggressive feelings, on the other hand, were highest on day 18' (postovulatory phase) after which there was a slow but steady fall until the end of the cycle. In Fig. 3 one sees that positive affect followed a different pattern from the negative affect variables. Feelings of pleasantness and activation were low in the menstrual phase, high at mid-cycle, and low again in the pre-menstrual phase. Feelings of sexual arousal are shown in Fig. 4. Sexual arousal was low during the menstrual phase but showed a sharp and continuing rise until mid-cycle. After this rise there was a decrease followed by a plateau during the premenstrual phase.

Progesterone studies on this same group of women showed that those who showed rapid clearing of progesterone (i.e., the greatest rate of

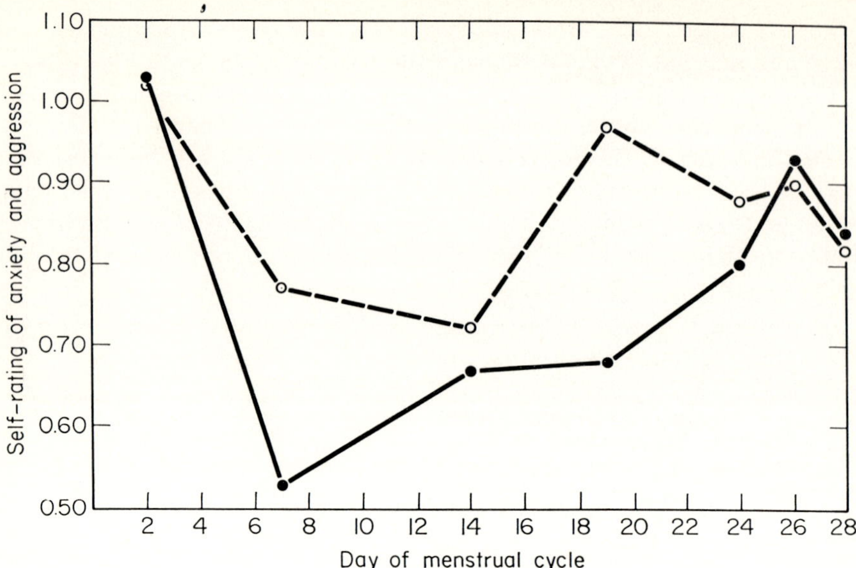

Fig. 2. Average self-rated anxiety (●——●) and aggression (○---○) of 15 women in two consecutive menstrual cycles. Reprinted with permission from Moos *et al. J. Psychosom. Res.* **13,** 37 (1969), Pergamon Press.

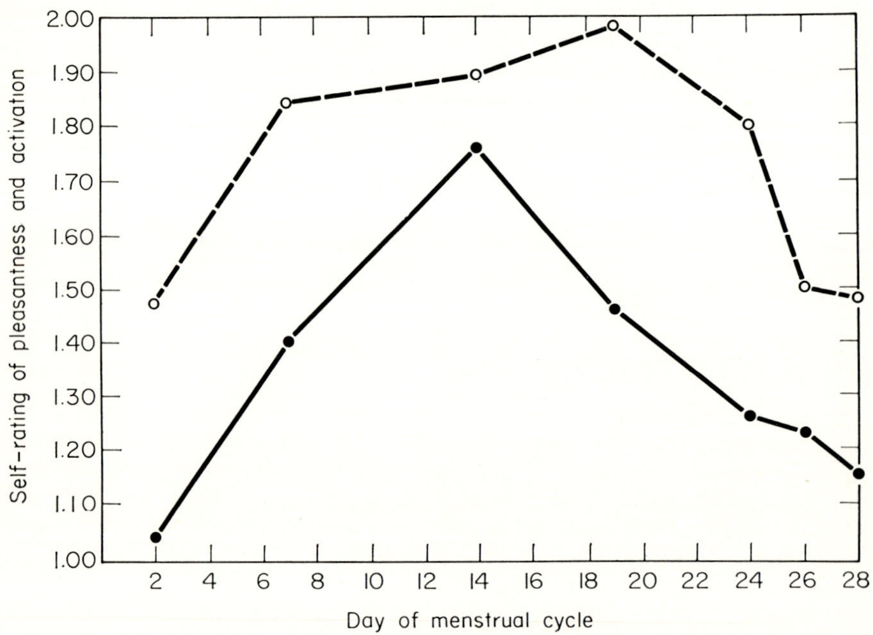

Fig. 3. Average self-rated pleasantness (●——●) and activation (○---○) of 15 women in two consecutive menstrual cycles. Reprinted with permission from Moos *et al. J. Psychosom. Res.* **13,** 37 (1969), Pergamon Press.

change in the premenstrual phase) had fewer complaints of premenstrual and menstrual symptomatology (Moos *et al.*, 1971).

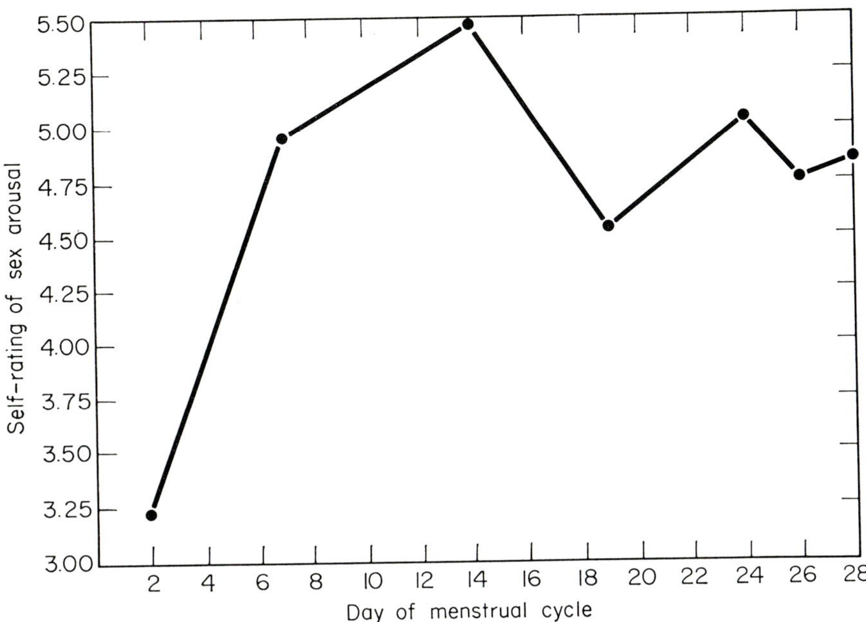

FIG. 4. Average self-rated sexual arousal of 15 women in two consecutive menstrual cycles. Reprinted with permission from Moos *et al. J. Psychosom. Res.* **13**, 37 (1969), Pergamon Press.

B. DURING THE POSTPARTUM PERIOD

A severe affective disorder occurring in about one per thousand women following childbirth has long been recognized. It is characterized by profound depression and known as postpartum psychosis. However, until recently there has been considerable uncertainty as to the incidence of milder depressive episodes occurring immediately postpartum that might not ordinarily come to the attention of the physician. In a study of all the women who were expected to deliver in a given month on a university obstetrics service ($N = 39$), the incidence and degree of postpartum depression were studied utilizing structured interviews, behavioral observations, and psychological tests (Yalom *et al.*, 1968). A summary of the findings from the structured prenatal interview is given in Table I. It should be noted that 73% of our multiparous subjects had experienced postpartum depression with at least one previous delivery. None of these women had sought psychiatric treatment for these reactions, however.

Each of the women was studied for the first 10 days postpartum. She was interviewed at regular intervals and rated by the interviewers on a number of variables related to affective state, including depression, anxiety, irritability, and distractability. The subjects rated themselves on the same variables and also completed the Nowlis Mood Adjective

TABLE I

Summary of Prenatal Interview[a]

	Absent (%)	Mild (%)	Moderate (%)	Severe (%)
Nausea	48.7	38.5	12.8	—
Fatigue	35.0	37.5	20.0	7.5
Overall discomfort	40.5	43.2	13.5	2.7
Fear of labor	43.2	37.8	13.5	5.4
Previous postpartum depression	26.7	30.0	20.0	23.3
Dysmenorrhea	62.9	34.3	—	2.8
Premenstrual tension	54.1	10.8	24.3	10.8
Menstrual irregularity	71.4	8.6	8.6	11.4
Rejection of pregnancy	26.3	29.0	21.1	23.7

Previous mental health

1. Good adjustment	15.4%
2. Mild maladjustment	56.4%
3. Moderate maladjustment	15.4%
4. Moderate emotional disturbance	12.8%
5. Severe emotional disturbance	0

[a] From Yalom *et al.* (1968).

TABLE II

Crying Episodes—Duration and Frequency[a]

	Over 2 hours	1–2 hours	A few minutes to 1 hour	Zero to a few minutes
Number of women having at least 1 episode	5	6	15	13

[a] From Yalom *et al.* (1968).

Check List daily. All episodes of crying were noted and described. The duration and frequency of crying episodes are shown in Table II. The distribution of crying episodes over the 10 day postpartum period is shown in Table III. Two-thirds of the women had episodes of crying lasting 5 minutes or more. The distribution of crying spells was fairly even over the 10 days with a slight tendency toward greater severity in the fifth through tenth days.

The observer ratings, self-ratings, Nowlis MACL scores, and a daily crying score were intercorrelated for each day and a principal-component factor analysis was calculated, yielding a daily depression score for each

TABLE III
Time Distribution of Crying Episodes[a]
(N = 48 Total Episodes)

Duration of episode	Days									
	1	2	3	4	5	6	7	8	9	10
Over 2 hours	1	0	0	1	0	1	2	1	1	1
1 to 2 hours	0	0	2	1	0	1	0	1	2	1
A few minutes to 1 hour	3	6	4	4	3	6	4	4	0	0

[a] From Yalom *et al.* (1968).

TABLE IV
Mean Depression Score[a]

Days	Score
1	187.25
2	235.58
3	100.05
4 in hospital	115.00
4 out of hospital	148.44
5	121.90
6	103.40
7	176.89
8	165.53
9	202.43
10	161.85

[a] From Yalom *et al.* (1968)

woman. The mean depression scores for the group of subjects are shown in Table IV. These data indicate that the terms "third-day blues" or "milk blues," sometimes used to describe mild postpartum depression, are inaccurate. Many of the women were most depressed prior to the third day postpartum and before the onset of lactation.

III. Assessment of Arousal

The concept of CNA arousal is derived from Pavlov's discovery and description of the orientation reaction, which refers to an animal's tendency to reflexly pay attention to novel stimuli (Pavlov, 1927). Arousal refers to a state of alertness that is accompanied by a number of physiological changes including alterations in the electroencephalogram (EEG), respiratory rate, and heart rate. The sense organs become more sensitive to incoming stimuli, and at the same time the capacity to filter out irrelevant stimuli at the precortical level is enchanced. Details of the mechanisms involved in arousal have been reviewed by Lynn (1966).

A. Non-EEG Techniques

There are four relatively simple tests available that are generally considered to reflect state of arousal and that do not involve the more

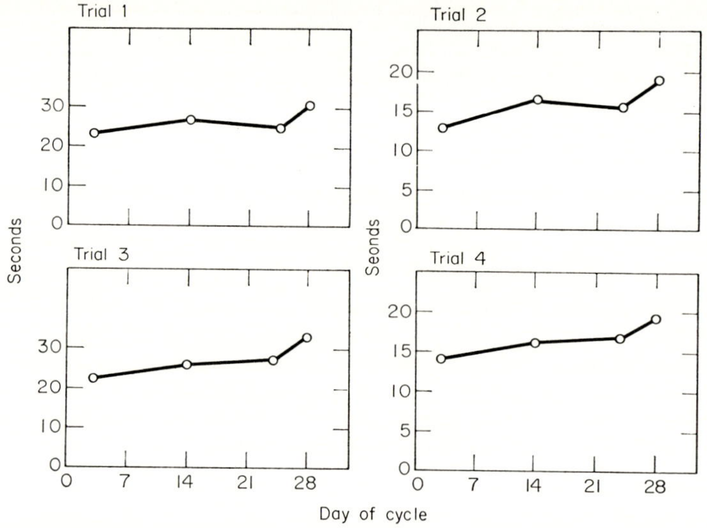

Fig. 5. Time estimation (mean values for 8 women over two consecutive menstrual cycles). From Kopell *et al. J. Nerv. Ment. Dis.* **148** © 1969 The Williams & Wilkins Co., Baltimore.

sophisticated equipment required for direct measurement of cortical activity. These tests are reaction time, galvanic skin potential, two-flash threshold (TFT), and time estimation. These four tests were administered repeatedly over two successive menstrual cycles in a study of eight nulliparous young women (Kopell *et al.*, 1969). Reaction time and skin potential did not show significant variation with the phases of the menstrual cycle in this study, but time estimation showed significant variation over the menstrual cycle with a distinct tendency for the women to estimate the given time intervals as longer during the premenstrual period, as shown in Fig. 5. There are significant correlations between

the results on time estimation and the two-flash threshold. The latter
test involves determining that time interval at which two flashes of
lights are perceived as one flash rather than two. The TFT data are
shown in Fig. 6 and indicate a rise in threshold during the premenstrual
phase.

These findings suggest a diminution of perceptual ability during the
phase of progesterone withdrawal. The time estimation data indicate a
distortion of the basic time sense during the same period and a slowing
of the internal "clock." There is some indication, on the other hand,
of a reversal of these phenomena with the onset of menstruation.

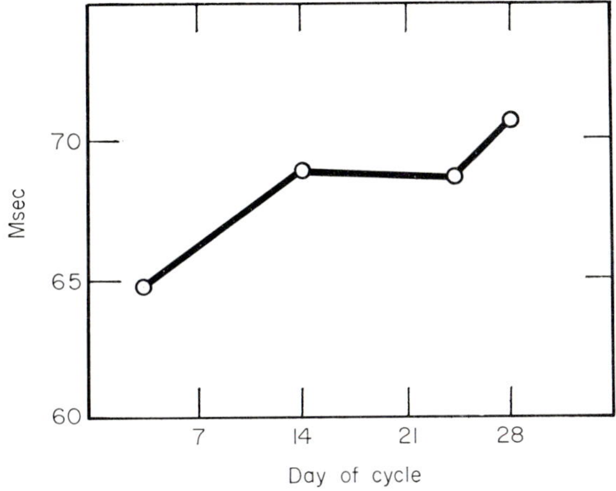

FIG. 6. Two-flash threshold (mean values for 8 women over two consecutive
menstural cycles). From Kopell *et al. J. Nerv. Ment. Dis.* **148** © 1969 The Williams
& Wilkins Co., Baltimore.

B. EEG TECHNIQUES

1. The Averaged Evoked Potential

The technique of computer averaging of the EEG permits the study
of time locked electrical activity of the brain immediately following
discrete sensory stimuli. By averaging a number of EEG segments that
follow the presentation of a series of identical stimuli (visual, auditory,
or other), one can increase the signal-to-noise ratio and obtain a wave
form that is known as the averaged evoked potential (AEP). The characteristic form and standard measurement parameters of the AEP are
shown in Fig. 7.

Utilizing the concept that selective attention is a function of state
of arousal, we have developed a technique that uses the AEP as an
objective measure of the subject's ability to attend selectively to signifi-

cant stimuli while filtering out irrelevant stimuli that are defined arbitrarily during the testing situation. This technique is based on the finding that when a subject attends to a stimulus that is defined as significant or relevant, the amplitude of the accompanying AEP is enhanced (Donchin and Cohen, 1967). Specifically, we have found that when a subject attends to a flash of light rather than to a changing background pattern, the amplitude of the AEP to the flash is increased. (The flash is made "relevant" to the subject by instructing him to push a hand-held button as quickly as he can each time he sees the flash.)

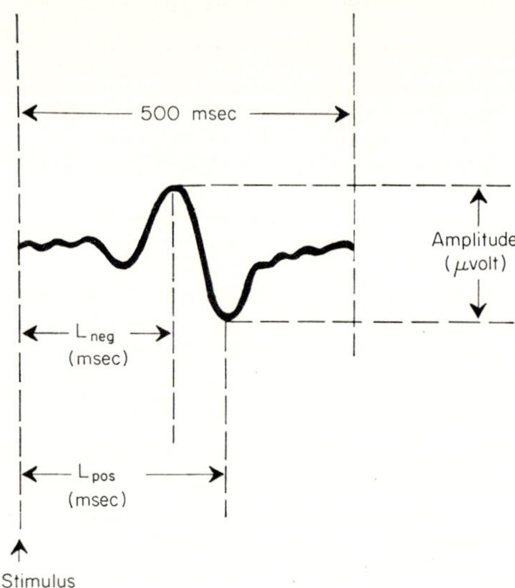

Fig. 7. Average evoked potential to visual stimulus, showing measurement parameters of typical wave form. From Kopell *et al., Psychosom. Med.* **32,** 496 (1970) with permission.

We have conducted two double-blind experiments in which we tested the effects of cortisol and triiodothyronine (T_3) on selective attention as measured by this method (Kopell *et al.,* 1970a,b). These hormones were selected because of their known potential for producing symptoms of mental disorder. Our hypothesis was that some of these symptoms are related to an interference with the ability to filter out irrelevant stimuli, thus allowing the cortex to be bombarded with an overwhelming amount of information that cannot be processed normally.

Some of the data from these studies are summarized in Fig. 8. It should be noted that both hormones tend to reduce the amplitude differences generated by relevant and irrelevant stimuli. Thyroid elevates the amplitude of the AEP to the irrelevant stimuli but does not significantly affect the amplitude of the AEP to the relevant stimuli. Cortisol acts differently. It decreases the amplitude of the AEP produced by the relevant stimuli. In both cases, however, the hormones tend to obliterate the distinction between significant and insignificant stimuli.

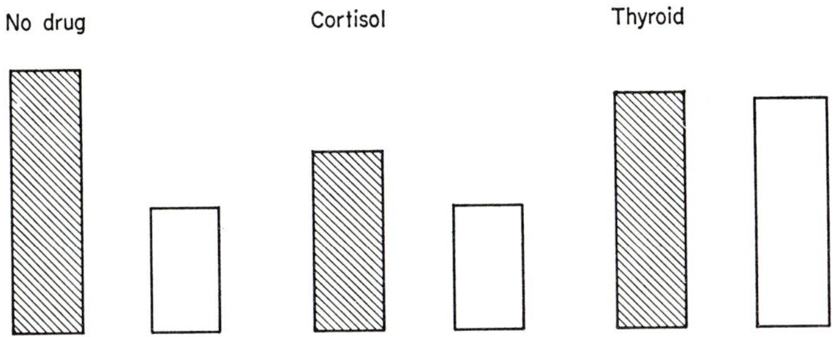

FIG. 8. Relative amplitudes of averaged evoked responses at cortex to irrelevant (□) and relevant (◙) stimuli in three experimental conditions: no drugs, thyroid (T_3), and cortisol.

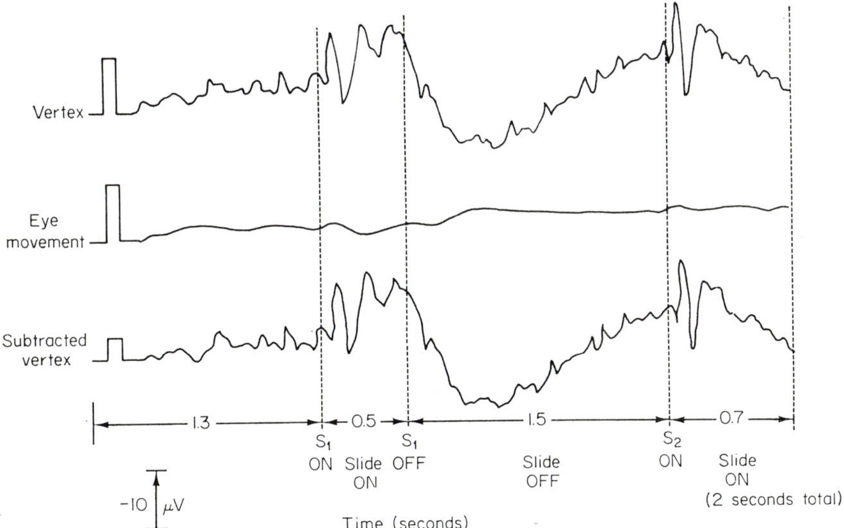

FIG. 9. Typical record of the averaged expectancy wave in the subject anticipating an erotic stimulus (note correction for eye movement artifact).

2. The Expectancy Wave

Walter (1964) has described a negative dc potential shift of the EEG baseline that accompanies a state of psychological anticipation, expectancy, or preparation for a stimulus. This wave form, which like the AEP can be discerned only by means of computer averaging, has been variously named the contingent negative variation (CNV) or the expectancy wave (E-wave). The E-wave typically follows a warning stimulus (S1) that signals a forthcoming major stimulus (S2) (see Fig. 9). It has been shown that the amplitude of the E-wave is enhanced as a function of the subject's interest in or arousal to the major stimulus (Walter, 1965).

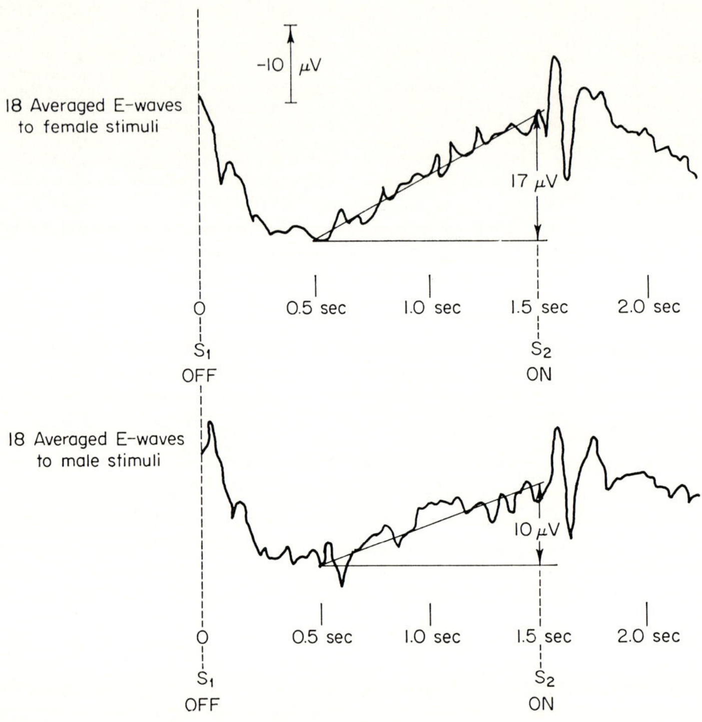

FIG. 10. Differences in amplitude of the averaged expectancy waves of male subject D.D. to female and male stimuli.

In a recent study (Lunde *et al.*, 1971), we utilized the amplitude of the E-wave as a measure of arousal to erotic stimuli. Specifically, we presented a randomized series of slides of male nudes and female nudes to a group of male subjects and a group of female subjects. Our hypothesis was that the amplitudes of the E-waves to the female nudes would be larger for the male subjects and the converse would be true for the female subjects. The amplitudes of the E-waves to the two sets of stimuli were significantly different in both subject groups in the predicted directions. Figure 10 shows a typical record of one of the male subjects. This technique offers an objective means of assessing sexual object preference as well as degree of sexual arousal to a given visual stimulus.

IV. Assessment of Androgens and Aggressive Behavior

The notion that aggressive behavior might be associated in some way with a chemical substance circulating in the blood has intrigued observers of human and animal behavior for many centuries. Hippocrates suggested that an excess of yellow bile was responsible for the aggressiveness and irritability of the choleric individual.

We choose to emphasize in this discussion the possible relationships between androgens and aggressive behavior in humans, recognizing full well that this area is fraught with difficulties for the research scientist.

A. The Concept of Aggression

The problem of defining what one means by aggressive behavior is difficult and may require inferences, particularly in the case of humans. Relationships between sexual behavior and androgens have been a more popular subject for study than aggression, perhaps in part because sexual behavior can be more readily defined. Copulatory behavior, even in humans, can usually be identified and can be safely assumed to be "sexual behavior" when it occurs. On the other hand, aggressive behavior, particularly in humans, is much more elusive. Although behavioral observers generally mean intraspecies fighting or attack behavior when speaking of aggression, it is readily obvious that this definition cannot be easily applied. If one swings his arms behind him intending to stretch and relax and inadvertently strikes another person standing behind him, he has not committed an aggressive act. Yet, to a behavioral rater unaware of the individual's intentions, such an act might be rated as an attack. On the other hand, one might heap verbal abuse on another individual in a most aggressive way or destroy the property of another perhaps as a substitute for a direct attack and not be involved in any fighting behavior per se. Clearly the *intent* of the individual is crucial in defining aggressive behavior. If one includes displaced aggression (attack or inflicting of injury upon someone or something other than the actual "enemy"), the definition becomes more complicated, albeit more encompassing. One final sort of behavior, predation or interspecies fighting and killing, has been considered exempt as a form of aggressive behavior by some investigators. Their argument, simply stated, is that for a carnivore, predatory behavior is simply food-gathering activity. Again, when one looks at the predatory activity of humans (e.g., big-game hunting) or other interspecies forms of fighting (e.g., bullfights), it is clear that there is much more involved than food-gathering needs for the individuals participating in or viewing these activities. In fact, there is no apparent counterpart in the animal world for the phenomenon seen in humans where gratification is experienced by watching animals (e.g., cock fights) or humans (gladiators, prize fighters) fight each other.

Given the difficulties incumbent upon defining aggression in humans, much less doing any reasonably controlled studies involving human subjects, it is not surprising that most research in this area has been confined to lower animals with extrapolations then made to human behavior. In fact, it has been only in the last decade that man's closest relatives, the Old World monkeys and apes, have been carefully studied in this regard (Washburn and Hamburg, 1968).

Lorenz's (1966) widely publicized book, *On Aggression*, advances the thesis that aggressive behavior follows relatively fixed, innate, "instinctual" patterns that have persisted because of their adaptive value over the eons of evolutionary history. Holloway (1968) has criticized the position of Lorenz and other ethologists in their application of an ethological framework to human aggressive behavior. He points out that with man's complex cerebral cortex providing an almost infinite capacity for learning and modifying behavior patterns, one does not see either the motor or sensory (cue) constancy that typifies aggressive behavior in other species. Furthermore, he argues that modern warfare, the most disastrous of man's aggressive activities, cannot be viewed as simply an expanded or cooperative version of the aggressive acts of groups of individuals since war "is an organized activity directed by rational decisions by political bodies whose membership is infinitesimal compared to the aggregates involved in actual combat" (Holloway, 1968, p. 30).

Recognizing these difficulties, it is our contention that aggressive behavior can be identified and studied in humans. Furthermore, among the many variables involved in aggressive behavior, the influence of androgens on the central nervous system and other organ systems may be a significant factor in providing both the capacity and the impulse to act in certain ways, given certain stimuli.

B. Androgens and Behavior-Evidence for a Relationship

There is an obvious way in which male sex hormones influence behavior, and that is in their influence on physical stature, musculature, and strength that in most species gives males a clear-cut advantage over females in fighting. We will not belabor this point but rather will concentrate our attention on the more subtle ways in which androgens may modify the organism so as to facilitate certain patterns of behavior.

1. Early Developmental Influences of Androgens

a. In Small Mammals. The evidence that sex steroids influence the central nervous system in an inductive or organizational way during early development originated and has accrued mainly from the study of laboratory rodents. Levine and Mullins (1966) have reported a series of experiments that showed that administration of a single dose of androgens to a newborn female rat results in a permanent alteration in the sexual behavior of the animal. These females may show acylic sexual receptivity as adults and, depending on the size of the neonatal dose of androgen, varying degrees of aberrant sexual behavior when mature. In addition, administration of androgen neonatally to female rats tends to obliterate the sex differences in amount of open-field activity usually seen in rats (Gray *et al.*, 1965).

With regard to fighting behavior specifically, Edwards (1968) treated female mice with 0.5 mg of testosterone propionate injected subcutaneously within 24 hours after birth. A control group received injections of peanut oil. When mature, these mice were treated with varying doses of testosterone. They were then placed in cages in pairs for 10-minute periods and observed for fighting behavior (biting or persistent attempts to bite). Normal males will almost always fight in this test situation, and 90% of the females who had received testosterone neonatally also fought in the test situation. The amount of fighting among the "virilized" females was proportional to the dose of testosterone received prior to testing. In contrast, only one pair of the females who received oil injections neonatally ever fought. Administering testosterone to adult females who had not received the hormone neonatally failed to result in any fighting behavior in the test situation (see also Edwards, 1969, 1970).

These findings are consistent with the earlier work of Beeman (1947), who found that male mice castrated prior to puberty rarely fought when they reached adulthood. However, if as adults they were given androgen injections, the castrated males would fight when put in a situation where there was another male to attack. The fighting behavior usually ceased when the androgen treatments were stopped. Beeman also noted that if adults were allowed to fight daily and were then castrated, fighting behavior would persist if there were continued opportunity to do so immediately after castration. On the other hand, if the animal were caged separately for several weeks after castration and then placed in a test situation, fighting was much less likely to occur. Thus, without the combination of androgen acting internally and the external stimulus of another male, the fighting habit was seen to extinguish.

One sees in these experiments a model for explaining one possible role of the male hormone in aggressive behavior in such animals; namely, that early exposure to testosterone sensitizes the central nervous system in such a way that subsequent exposure to androgen in adulthood results in fighting behavior *when the animal is confronted with an appropriate external stimulus to fight.* Perhaps there is a lowering of a threshold of a response tendency that is present, but latent, in untreated, castrated adults. Although androgen may affect the "internal urge" to fight, it seems clear that appropriate external stimuli are necessary, if not sufficient, prerequisites for actual fighting to occur. Hinde (1967) has criticized Lorenz' suggesting that aggressive behavior may arise spontaneously from irresistible internal urges. Tinbergen (1968), however, reconciles the "internal vs. external causation" arguments in his discussion of the interplay between the two classes of variables. It seems obvious, and yet one needs to be reminded, that "fighting is started by a number of variables, of which some are internal and some external . . . fighting behavior is not like the simple slot machine that produces one platform ticket every time one three-penny bit is inserted" (Tinbergen, 1968).

An interesting biochemical substance has been used in a number of small animal studies in the past few years and deserves mention here because of its possible application in studies with humans. The compound cyproterone acetate (CA) is known as an "antiandrogen." Neumann and Elger (1966) have reported that this compound blocks the effects of testosterone at the CNS (behavioral) level as well as at the tissue (growth) level. They reported anatomical feminization as well as female patterns of sexual behavior in male rats treated with CA neonatally. Beach and Westbrook (1968) did not find any influence on sexual behavior of CA given to adult male rats, although they did note a decrease in the weight of the seminal vesicles similar to the effect of castration. Neumann *et al.* (1970) reported the findings of their further studies with antiandrogens at the 1969 Laurentian Hormone Conference, and many important studies of these compounds are still in progress.

b. In Primates. A significant series of experiments based on the discoveries in rodents of the effects of early exposure to androgens has been done by Goy, Phoenix, and their colleagues at the Oregon Regional Primate Research Center. Pregnant female rhesus monkeys have been given daily injections of testosterone proportionate during the presumptive period of sexual differentiation (from about day 39 to day 90 of the 108-day gestational period). Pseudohermaphroditic female monkeys have resulted from these pregnancies, and their behavior has been closely studied over a period of years through the various stages of development. Observations of eight such animals through infancy and adolescence have shown that they exhibit behavior in the aggressive sphere (initiating play, threats, rough-and-tumble play, and chasing) which resembles that of the male of the species rather than the female (Goy, 1968). Other forms of behavior that are not sexually dimorphic (e.g., huddling, grooming, withdrawing, and fear-grimacing) are not affected by testosterone treatment.

Sexual behavior is masculinized in these pseudohermaphrodite monkeys with an increase in mounting behavior the most striking difference between normal females. On the other hand, hormonal puberty is female in these animals, and they have menstrual cycles. This is in contrast to the findings in rats and guinea pigs, where early androgen treatment has produced anovulatory, acyclic females. It would appear that the effects of testosterone on the developing central nervous system are not limited to the hypothalamic–pituitary centers as originally suggested but also include other centers or circuits of the brain, which are involved in such activities as play and aggression.

c. In Humans. The preceding experiments in small mammals and subhuman primates raise the question of possible androgen effects in the early development of human females. Studies of two groups of females who were exposed to androgenic compounds *in utero* suggest that sexually dimorphic behavior in humans may be subject to early hormonal influence in a way analogous to the observations in animals. One such group consists of ten young girls ("progestin-induced hermaphrodites") whose mothers received either 17-ethynyltestosterone or 19-nor-17-ethynyltestosterone during pregnancy as a treatment for threatened abortion (Ehrhardt and Money, 1967). Data were gathered from interviews with the girls, their mothers, and psychological tests. Nine of the ten girls were found to be "tomboys," based on the following criteria: (a) preference for boys' toys; (b) preference for and active participation in outdoor, athletic activities; (c) lack of interest (compared with girls in a control group) in feminine clothes, baby care, doll play, and household chores. Again, as with the androgen-treated monkeys, menstruation does not appear to be affected. An unexplained finding was than the average IQ of these girls was 125 (SD 11.8), well above the expected average of 100.

The second group of human females subject to early androgen exposure consists of those women with congenital adrenal hyperplasia (the adrenogenital syndrome). Twenty-three of these women have been studied with regard to their sexual identity and experiences (Ehrhardt *et al.*, 1968a). These women, because of a genetic defect in the biosynthesis of cortisol, were exposed *in utero* and until their condition was recognized and treated, to large amounts of androgens secreted by their adrenal glands.

In this condition, ovaries, oviducts, and uterus develop normally, but there is usually some degree of virilization of the external genitalia, particularly hypertrophy of the clitoris. Data were obtained from medical records kept at Johns Hopkins Hospital as well as from interviews. This was an older group than those previously described, with a mean age of 33. All had received proper treatment and are now anatomically and functionally female in all respects, i.e., cortosone replacement therapy had been instituted resulting in a suppression of adrenal androgen production followed by normal breast development and onset of menstruation. Clitorectomy was performed when indicated.

Eleven of these women reported either homosexual experiences, fantasies, or dreams although none was exclusively lesbian. In addition, most of the women reported sexual stimulation in response to visual and narrative stimuli, tendencies more commonly associated with male than female sexuality according to the Kinsey studies. Intelligence tests were given to twenty of the women and the average IQ was 119, again well above the expected average. This group of women represented cases of late treatment of the adrenogenital syndrome. A group of 15 early-treated females with the adrenogenital syndrome has also been studied (Ehrhardt *et al.*, 1968b) and, particularly with regard to tomboyishness, they resembled the "progestin-induced hermaphrodites."

2. Influences of Androgens on Behavior in Adults

Whereas the organizational or inductive effects of androgens on the developing CNS appear to be relatively permanent and do not require continued presence of the hormone, there are other observable effects on behavior in the adult that seem to vary with the presence or the absence of male sex hormone. Again, these behaviors fall mainly in the sexual and aggressive spheres.

a. In Primates. Dominance orders are established among males of many species by threats and actual fighting. In one series of experiments, Clark and Birch (1945, 1946) studied the effects of sex hormones on dominance order among a pair of cartrated male chimpanzees. The animale was given androgen injections. He subsequently fought and defeated based on fighting over food when only a single item of food was offered. Once the animal was established in the dominant position, the other male was given androgen injections. He subsequently fought and defeated the previously dominant male. Female sex hormones, when administered, seemed to have the opposite effect, i.e., the dominance was reversed. Rose *et al.* (1971) have recently reported a correlation between plasma testosterone levels and dominance rank as well as aggressive behavior in a group of 34 rhesus monkeys.

Aggressive behavior among females is less common and/or severe in most species than in males, and effects of sex hormones on female aggressive behavior have not been well studied. Aggressive behavior can usually be elicited, however, when a mother is caring for and protecting her young. This behavior may be caused by a hormone present in relatively large amounts at this time, e.g., prolactin. Whether or not this applies to primates is presently unknown.

b. In Human Adult Males. Very little is known at the present time about the effects of androgens on the emotions or behavior of human adult males. There is suggestive evidence, however, to support the notion that androgens are involved in sexual and aggressive feelings and behavior.

Persky *et al.* (1970) have recently reported a significant positive correlation between testosterone production rates and feelings of hostility and aggression (measured by a battery of psychological tests) in a group of 18 young men between the ages of 17 and 28.

The effects of surgical castration on sexual and aggressive behavior in males seem to depend upon the age at which castration is performed. Anecdotal data regarding eunuchs suggest that sexual and aggressive behavior persists to varying degrees if the operation is performed after puberty. However, if performed prior to puberty, adult aggressive and sexual behavior patterns usually fail to develop.

"Hormonal castration" is currently being employed on an experimental basis by a number of investigators who are treating hypersexual and sexually deviant men and women with antiandrogen compounds (Laschet *et al.*, 1967; Servais and Hubin, 1968). These drugs seem clearly to diminish sex drive and sexual activity by blocking the activity (rather than the production) of testicular and adrenal androgens. The effect is reversible with cessation of treatment.

Money (1970) has recently reported favorable results in the treatment of eight male sex offenders with an androgen-depleting hormone, medroxyprogesterone acetate. Money's results are particularly intriguing since they involve apparent alterations in sexual object and activity preference as well as sex motivation. In the case of the pedophiliac patient described by Money in some detail, it would be particularly useful to use the E-wave technique for measuring sexual arousal described earlier. If the drug truly altered the patient's arousal to children, one would expect to see a diminution in the amplitude of the E-wave when pictures of children are used as stimuli. If no alteration in the E-wave is seen following treatment, one might attribute the change in the patient's behavior to other factors, such as fear of imprisonment. The E-wave technique could also be employed with homosexual offenders by comparing responses to pictures of men before and after treatment.

c. In Human Adult Females. Androgens seem to have no effect on aggressive behavior in adult female animals. King and Tollman (1956) were unable to elicit fighting behavior in female mice treated with testosterone, and Clark and Birch (1945, 1946) noticed no effects on aggressive behavior when female chimpanzees received androgens. It is unfortunate that so little work has been done with chimpanzees, in view of their especially close relation to man. Human females may react in a distinctive fashion to androgens. Although ovariectomy will result in the elimination of sexual receptivity in almost all subprimate species,[2] ovariectomy in humans has no consistent effects on sex drive or behavior. On the other hand, androgen deprivation by adrenalectomy seems to diminish sex motivation in females (Waxenberg *et al.*, 1959). Furthermore, administration of androgens to adult human females, as in the treatment of carcinoma of the breast, is reported to dramatically increase sexual desire and activity even in seriously ill and debilitated women (Foss, 1951). Whether human females have other distinctive responses to androgens is not yet known.

C. The Measurement of Androgens

Probably the single most promising advance in recent years that will facilitate studies of androgen-behavior relationships is the development of accurate biochemical techniques that allow direct measurement of testosterone levels (or other specific androgens) in plasma or other body fluids. Previously, estimates of androgenic activity were made by bioassay techniques, which were often inaccurate and nonspecific, or by measuring 17-ketosteroids (17-KS) levels in the urine. Dorfman and Shipley (1956) reviewed 276 studies that utilized primarily 17-KS measures and attempted to establish norms for various age groups, sexes, and disease states. In addition, some of the studies attempted to correlate 17-KS levels with behavioral data. Many of these studies were inconclusive.

Aside from the fact that there appear to be certain inherent inaccuracies in 17-KS techniques (Goldheizer and Axelrod, 1962), the measurement of 17-KS levels in various groups of subjects has not yielded any consistent findings, perhaps because the 17-KS group includes metabolites of cortisol and adrenal androgens as well as testosterone.

---

[2] Herbert (1970) has recently reported that the administration of 1 mg/day of testosterone to female rhesus monkeys greatly increases their sexual receptivity (in terms of frequency of presenting sexually to males) but does not increase their attractiveness to the males. Larger doses of 5 and 25 mg/day diminished sexual receptivity but increased aggressiveness.

Testosterone levels can be easily overshadowed by cortisol or less potent andrenal androgen metabolites in the 17-KS fraction. The result is that many studies fail to even find sex differences in 17-KS levels until well after puberty, although it is well known that differences in testosterone secretion rates between boys and girls is a crucial determinant of puberty (cf. Miller and Mason, 1945; Talbot *et al.*, 1943). A recent study of Paulsen *et al.* (1966) employed a modification of the earlier 17-KS techniques in which hydrolysis of the urinary steroids by hot acids resulted in a variable and unpredictable destruction of a certain percentage of the sulfate and glucuronide conjugates. But even using a modified hydrolytic procedure to avoid this problem, no sex differences in 17-KS excretion patterns were found in a sample of 34 normal boys and girls ranging in age from 3 to 16 years.

We shall discuss in this section some of the newer biochemical methods that enable one to make direct, accurate measures of testosterone. These methods, while representing a significant advance over 17-KS measures, are still not perfect. For instance, 90–95% of testosterone is bound to proteins and hence perhaps not physiologically active (Forest *et al.*, 1968). Thus one must assume that the unbound (active) fraction of the total testosterone measured remains constant, an assumption that may or may not be correct.

1. Protein Binding Methods

These methods are based on the discovery that there is in human plasma a protein molecule that binds testosterone quite specifically (Kato and Horton, 1968). This protein is present in small amounts in normal plasma but in elevated amounts in the plasma of men or women undergoing estrogen therapy. It is present in relatively high concentration in the plasma of pregnant women, and consequently the plasma of a pregnant woman is used as an essential ingredient of the assay.

Mayes and Nugent (1968) reported a 64% recovery of the testosterone in their samples using a protein-binding method. Precision and accuracy of the method were quite good over a wide range of concentrations.

The main advantage of the protein-binding method over other methods to be described are that it is less time consuming (3–4 hours), less expensive, and generally less complicated. Samples as small as 2.0 ml plasma for women and 0.2 ml plasma for men can be analyzed by this method and concentrations of testosterone in the range of a few millimicrograms per milliliter can be measured accurately. These practical advantages are significant for behavioral research where it is often necessary to analyze many samples in order to arrive at statistically dependable results.

2. Gas–Liquid Chromatography (GLC)

Electron capture gas–liquid chromatography is a highly sensitive but somewhat complicated method that can be employed for measuring testosterone and other androgens. Most GLC methods involve two or three chromatographic steps (paper and/or thin-layer chromatography) following extraction.

The primary advantage of the GLC method is its sensitivity. Amounts of testosterone in the micromicrogram range can be detected by this method (Exley, 1968). In addition, the method is very precise and accurate. Specificity for testosterone is quite good, although care must be taken in the purification steps to eliminate extraneous peaks from contaminants on the chromatogram that might overlap with the testosterone peak.

The disadvantages of the GLC method are that it is more complex and expensive, more time consuming (24 hours or three working days), and requires larger plasma samples (10–20 ml for women, 2.5 ml for men) than the protein-binding method. For these reasons the GLC method is not widely used in behavioral research at the present time.

3. Other Methods

a. For Plasma Testosterone

(1) *Double isotope dilution derivative methods.* These methods are accurate, sensitive, and reliable but extremely complex and time consuming. In addition to formation of isotope derivatives of testosterone four to six chromatographic steps may be involved. For details regarding the measurement of testosterone and related steroids by this technique, the reader is referred to Gandy and Peterson (1968).

(2) *Fluorescence.* This method involves enzymatic conversion of testosterone to estrogen prior to assay (Forchielli *et al.*, 1963; Lamb *et al.*, 1964). Fluorometric methods have no unique advantage over the methods previously described and are not widely used for androgen assays.

b. For Urinary Testosterone. There are situations, especially in long-term behavioral studies in which urine samples are easier to obtain than blood; or one may have other reasons for measuring testosterone in the urine. Testosterone glucuronide in the urine probably only represents about 1% of the total testosterone secreted, however, the remainder having been metabolized to various 17-ketosteroids (Baulieu and Mauvais-Jarvis, 1964). In addition, urinary testosterone levels in women may appear falsely elevated because of conversion of Δ^4-androstenedione to testosterone prior to excretion (Horton and Tait, 1966).

D. Variations in Androgen Levels with Age, Sex, and Disease States in Humans

1. In Children

Using techniques described in the preceding section, recent studies have demonstrated sex differences in plasma levels of androgens that

TABLE V

Androgen Levels and the Ratio of Androstenedione to Testosterone in Male and Female Babies at the Various Days of Age[a]

Subjects	Sex	Age (days)	Androstenedione (A) (mg/100 ml)	Testosterone (T) (mg/100 ml)	Ratio A/T
1[b]	Male	0 (birth)	113 ± 16 (SD)[b]	43 ± 12 (SD)[c]	2.9 ± 1.2 (SD)[c]
2		1 (5 hr)	438	106	4.1
3		2	236	58	4.1
4		3	32	36	0.9
5		4	306	326	0.9
6		4	54	117	0.5
7		5	20	Undetectable	
8[d]		5	54	58	0.9
9[c]	Female	0 (birth)	81 ± 25 (SD)[c]	57 ± 22 (SD)[c]	1.8 ± 1.2 (SD)[c]
10		2	Undetectable[e]	44	
11		2	Undetectable	22	
12		4	Undetectable	16	
13		5	34	36	0.9

[a] From Mizuno *et al.* (1968) with permission.
[b] Umbilical artery plasma.
[c] Mean ± SD from 5 cases.
[d] A pool of specimens from 3 healthy male babies.
[e] Less than 10 mg/100 ml.

are apparent from birth on. Mizuno *et al.* (1968) found significant differences in androstenedione levels in umbilical vein plasma of 11 newborn male babies compared with 11 females. The mean concentration for males was 118 mg/100 ml and for females, 81 mg/100 ml. Testosterone differences were not significant for the males vs. females at birth, and it is suggested that this might be due to a suppression of testicular production of testosterone in the males resulting from the stress of the delivery. Levin *et al.* (1967) have demonstrated that compounds released by the adrenal medulla and cortex during periods of stress, namely epinephrine and cortisol, can inhibit testosterone production in adult males.

In any event, within the first week of life, sex differences in testosterone levels are reported, and androstenedione in the female babies seems to disappear from the peripheral circulation, suggesting that the

TABLE VI

Plasma Androgen Concentration in Prepubertal Children[a,b]

Sex	Age (years)	Androstenedione (mg/100 ml)	Testosterone (mg/100 ml)
Females	3.7	40	31
	4.0	33	7
	6.0	27	7
	6.3	13	20
	7.0	47	20
	7.0	26	1
	8.3	38	30
	9.3	17	34
		30 ± 4 (SE)	19 ± 8 (SE)
Males	4.0	97	79
	4.7	131	24
	5.0	140	70
	5.0	—	76
	6.0	70	20
	6.7	70	43
	6.7	56	20
	7.0	45	27
	9.3	82	22
		86 ± 12 (SE)	42 ± 9 (SE)

[a] From Frasier and Horton (1966) with permission Holden-Day, Inc.

[b] Values corrected for the mean blank, 15 mg/100 ml androstenedione and 19 mg/100 ml testosterone.

levels of this steroid seen at birth are derived to a great extent from the maternal adrenal glands (see Table V).

Prior to puberty, sex differences in testosterone and androstenedione levels persist, but there appears to be no particular increase in the levels of these androgens during the period from age 3 to 9, according to Frasier and Horton (1966) (see Table VI).

Grandy and Peterson (1968), using a rather elaborate double isotope dilution derivative technique, found sex differences in androgen levels in prepubertal children as well and in addition noted a gradual increase in testosterone levels during the prepubertal years (see Table VII). The

TABLE VII

Peripheral Plasma Concentration of Androgens (µg/100 ml)
in Children[a,b]

Subjects[c]	Age	Testosterone	Androstenedione
Boys	2.5	0.003 (0.001)	0.025 (0.023)
	3.5	0.007 (0.005)	0.046 (0.044)
	4.5	0.003 (0.001)	0.06 (0.058)
	4.5	0.018 (0.016)	0.03 (0.028)
	5	0.016 (0.014)	0.044 (0.043)
	5	0.014 (0.012)	0.04 (0.038)
	5	0.003 (0.001)	0.05 (0.048)
	5.5	0.004 (0.003)	0.065 (0.063)
	7	0.019 (0.017)	0.07 (0.068)
	7	0.03 (0.028)	0.02 (0.018)
	7	0.007 (0.005)	0.05 (0.048)
	8	0.019 (0.017)	0.12 (0.119)
	8	0.014 (0.012)	0.08 (0.078)
	8	0.026 (0.024)	0.13 (0.128)
	8	0.05 (0.048)	0.05 (0.048)
	8	0.06 (0.058)	0.04 (0.038)
	9	0.038 (0.036)	0.039 (0.037)
	9.5	0.03 (0.028)	0.04 (0.038)
	Mean	0.02 (0.018)	0.06 (0.053)
	(SD)	±0.016	0.028 (0.026)
	10	0.17 (0.168)	0.13 (0.128)
	10	0.06 (0.058)	0.05 (0.048)
	10	0.026 (0.024)	0.14 (0.138)
	10	0.028 (0.026)	0.024 (0.023)
	10	0.015 (0.013)	0.11 (0.108)
	11.5	0.18 (0.178)	0.13 (0.128)
	11	0.06 (0.058)	0.04 (0.038)
	11	0.08 (0.078)	
	11.5	0.08 (0.078)	0.03 (0.028)
	11	0.05 (0.048)	0.028 (0.026)
	12	0.09 (0.088)	0.02 (0.018)
	12	0.12 (0.118)	0.04 (0.038)
	13	0.07 (0.068)	
	13	0.38 (0.378)	
	13	0.42 (0.418)	0.11 (0.108)
	13	0.40 (0.398)	0.19 (0.188)
	14	1.42 (1.418)	0.20 (0.198)
	14	0.28 (0.278)	0.09 (0.088)
	15	0.23 (0.228)	0.13 (0.128)
	Mean	0.22 (0.218)	0.09 (0.089)
	(SD)	±0.318	±0.059
Girls	4	0.003 (0.001)	0.04 (0.038)
	5	0.005 (0.003)	0.08 (0.078)
	6	0.014 (0.012)	0.07 (0.068)
	7	0.008 (0.006)	0.08 (0.078)
	7	0.019 (0.017)	0.05 (0.048)
	8	0.009 (0.007)	0.02 (0.018)
	9	0.003 (0.001)	0.11 (0.108)
	Mean	0.009 (0.007)	0.06 (0.052)
	(SD)	±0.005	±0.029

[a] From Gandy and Peterson (1968) with permission.

[b] Values are uncorrected for blank of method; values corrected for blank are cited in parentheses.

[c] Children 2.5–8 years old were sexually immature; those 8–10 years old did not have sexual hair or breast development.

absolute values reported by Grady and Peterson are generally lower than those reported by Frasier and Horton. There also appears to be a great deal of variation in androgen levels for any given age.

Of particular interest are the data in Table VIII for pubertal boys (ages 10–15). Here one sees a dramatic increase in testosterone levels on the order of 10-fold or larger, as well as striking individual differences for a given age level. We shall return to this point later.

TABLE VIII

*Reported Values for Plasma Concentration of
Testosterone (µg/100) in Normal Male Adults[a]*

Number of subjects	Mean	Range
9	0.56	0.01–0.98
11	0.8	0.05–1.1
21	0.74	0.5–1
40	0.71	
11	0.70	0.32–1.7
15	0.74	0.44–1.3
8	0.65	0.44–0.96
10	0.65	0.48–0.79
24	0.73	0.35–1.16
13	0.55	0.37–0.62
8	0.55	
60[b]	0.67	0.28–1.44

[a] Adapted from Gandy and Peterson (1968).
[b] Data of Gandy and Paterson.

2. In Normal Adult Males

A number of recent studies have been directed toward obtaining normative plasma testosterone levels for adult men, usually in the 20–40 age group. Grandy and Peterson (1968) have summarized the data from 12 such studies (including their own), and these data are shown in Table VIII.

Of interest are the observations of Gandy and Peterson and others that testosterone levels in older men in the 60–80 year age group and even older tend to be similar to those found in younger men. This is consistent with the observation that other aspects of testicular function may remain normal well into old age; e.g., normal sperm production has been observed in men as old as 90. Although sexual and aggressive behavior is known to decline in frequency with age, it is not yet clear whether this decline is related to changes in any of the sex hormones or whether it is solely related to such factors as deteriorating health and physical strength.

Blood levels of another androgen, dehydroepiandrosterone, drop off in old age to a level about one-fifth of that seen in younger men (Gandy and Peterson, 1968). It is thought that this androgen is produced primarily by the adrenal glands rather than by the testes.

Although it has long been assumed that the male sex hormone was secreted at a rather constant rate in the adult male,[3] as opposed to the cyclicity of the female sex hormones, there is evidence that testosterone secretion follows a diurnal pattern much like that of cortisol secretion (Resko and Eik-nes, 1966). Testosterone levels tend to be highest in the early morning hours when compared to afternoon or evening levels.[4] Whether this diurnal fluctuation reflects a cyclic secretion of pituitary gonadotropins in the male or simply the diurnal pattern of the adrenal glands (also a source of androgen in both the male and the female) is not yet known.

Plasma testosterone levels are, of course, a function of both the production rate and the metabolic clearance rate of the hormone. It is not surprising that males have been found to have both a higher production rate and a higher clearance rate for testosterone than females (Korenman *et al.*, 1963, Southern *et al.*, 1968). Although the clearance rate appears to be related at least in part to the production rate, other factors are probably involved as well. In addition, the production rate of testosterone in males can be decreased by ACTH stimulation and increased by human chorionic gonadotropin (Rivarola *et al.*, 1966).

3. In Normal Adult Females

Androgens in the female are produced primarily by the adrenal glands. Testosterone levels in the female are, of course, considerably less than those in the male. A summary of testosterone levels reported in 13 different studies of normal women is shown in Table IX.

---

[3] There are a few exceptions to this rule in the animal world; e.g., deer and sheep have a specific "rutting season" during which time androgen secretion is high and the males are sexually aggressively active. In certain wild rodents, the testes are actually withdrawn from the scrotum into the abdominal cavity except during the mating season. During the period the testes are within the abdominal cavity they produce neither androgens nor sperm.

[4] A recent report by Evans *et al.* (1971) indicates that testosterone levels vary during sleep with peaks of plasma testosterone occurring in conjunction with or adjacent to periods of rapid eye movement (REM) sleep. It is of interest that REM sleep is also associated with penile tumescence and erection.

Plasma testosterone levels are elevated in pregnant women, however, as shown in Table X, and these elevations are independent of the sex of the fetus (Rivarola *et al.*, 1968). On the other hand, androstenedione levels tend to be higher when the fetus is a male.

Pregnant women do not usually show signs of virilization despite the elevated testosterone levels, and this is presumably due to the increase in testosterone-binding globulin that was described earlier. This phe-

TABLE IX

Reported Values for Plasma Concentration of
Testosterone (μg/100 ml) in Normal Adult
Females[a]

Number of subjects	Mean	Range
10	0.12	0.02 –0.26
2	0.07	0.059–0.079
12	0.11	0 –0.35
21	0.083	
11	0.18	0.06 –0.31
10	0.07	0.02 –0.12
20	0.11	0.05 –0.29
50	0.054	0.03 –0.1
8	0.12	0.012–0.2
20	0.037	0.014–0.058
9	0.047	0.024–0.07
24	0.017	0.006–0.035
20	0.032	0.002–0.07

[a] From Gandy and Peterson (1968) with permission.

nomenon is probably related to the increase in estrogen during pregnancy, since males undergoing estrogen therapy for carcinoma of the prostrate also show an increase in TBG levels Pearlman *et al.*, 1967).

4. Alterations in Testosterone in Disease States

There is an increasing body of knowledge that indicates that secretion and other aspects of androgen metabolism are disordered in a variety of disease states, some of which have a genetic basis. The following are but a few examples of such disorders.

a. Hypogonadal Men. A variety of conditions in males have been attributed to deficient gonadal production of androgens. The presenting symptoms of the patient with one degree or another of hypogonadism include infertility and deficient or disordered sexual differentiation at puberty with regard to secondary sex characteristics (facial, pubic, and axillary hair, body build, genitalia, breast development, voice change, and libido). There are multiple causes for the symptoms of hypogonadism, including deficient pituitary production of gonadotropins and chromosomal defects such as Klinefelter's syndrome (where the sex chromosomes are XXY rather than the normal male pattern XY) or Turner's syndrome (where the Y chromosome is absent). It has been

TABLE X

Androgens during Pregnancy (ng/100 ml of plasma)[a]

Subjects	Months of pregnancy	Sex of fetus	Testosterone	Androstenedione
E.R.	6	F	101	346
P.G.	5	F	78	265
P.B.	8	F	86	190
A.C.S.	8	F	94	—
M.G.	$6\frac{3}{4}$	F	154	—
N.E.	$8\frac{1}{2}$	F	162	150
S.H.	6	F	145	166
S.K.	7	F	201	256
V.W.	$8\frac{1}{2}$	F	65	99
Pregnant females with female fetus (mean ± SD)			121 ± 46	210 ± 83.5
M.W.	$8\frac{1}{2}$	M	53	—
B.R.	7	M	102	254
S.G.	5	M	110	323
S.S.	6	M	110	312
A.B.	5	M	113	343
J.M.	8	M	121	322
B.O.	6	M	149	308
V.P.	$8\frac{3}{4}$	M	95	156
Pregnant females with male fetus (mean ± SD)			107 ± 27	288 ± 64.4
Nonpregnant females (mean ± SD)			49 ± 13	181 ± 59
Pregnant females, whole group (mean ± SD)			114 ± 38	249 ± 82

[a] From Rivarola *et al.* (1968).

shown that testosterone production is abnormally low in all these conditions (Mauvais-Jarvis *et al.*, 1968; Jeffcoate *et al.*, 1967).

In one condition that involves abnormal sexual differentiation, namely the testicular feminization syndrome, it is significant to note that testosterone secretion is normal, yet virilization does not occur (Mauvais-Jarvis *et al.*, 1968). In these patients the metabolism of testosterone and other androgens is abnormal, with the result that even administration of exogenous testosterone fails to produce virilization. This finding points out another weakness in assuming that plasma testosterone levels necessarily reflect the amount of physiologically active hormone present.

b. Hirsute and Virilized Females. Whereas androgen deficiency produces the most dramatic symptoms in males, androgen excess produces rather distressing problems for females. The source of excess androgens in females may be either the adrenals or the ovaries. The most common presenting symptom in such women is hirsutism, particularly development of excessive coarse facial hair. In addition, there may be symptoms of infertility, menstrual irregularities, and virilization of the genitalia with hypertrophy of the clitoris may occur in some cases.

Urinary testosterone and epitestosterone levels have been found to be elevated in hirsute women, with or without associated known ovarian or adrenal gland pathology (Wieland *et al.*, 1966; de Nicola *et al.*, 1966). It is again worth emphasizing that urinary 17-ketosteroids are not necessarily elevated in these conditions (Lloyd *et al.*, 1966).

Plasma testosterone and androstenedione levels have also been shown to be elevated in women with hirsutism and related symptoms (Horton and Neisler, 1968; Lloyd *et al.*, 1966).

E. Variations in Aggressive Behavior with Age, Sex and Disease States

There appears to be a more than coincidental similarity between those conditions and developmental stages where one sees alterations in androgen levels and those situations when changes in aggressive behavior patterns have been observed to occur, both in animals and in humans. In this section we shall describe some of these observations that have been made relative to variations in aggressive behavior.

1. Animal Studies

We have already mentioned some of the animal studies, particularly in primates, that have shown that fighting behavior as well as related sorts of behavior (rough play, threats, etc.) are sexually dimorphic in many species and are subject to the influence of androgens.

Whereas in many species, androstenedione is the predominant androgen present prior to puberty, testosterone has been shown to emerge as the primary sex steroid in the plasma of adult male guinea pigs, bulls, and rhesus monkeys (Resko, 1967), as it is in humans.

Among most domestic and wild animals, the male of the species is the most aggressive, and castration tends to diminish aggressiveness. One exceptional situation when females may become extremely aggressive, however, is when they are protecting their young. Several authors have suggested that this change in behavior may be related in part to the increase in certain pituitary hormones during nursing in mammals, but no definitive investigations have been done in this area to date.

2. In Children

Sex differences in aggressiveness are readily apparent in humans well before puberty. Kagan and Moss (1962) rated a variety of forms of aggressive behavior in children during the age periods 0–3 years, 3–6 years, 6–10 years, and 10–14 years. The ratings were then compared with ratings of various behavior patterns of adulthood in a longitudinal study that covered a period of 30 years. Whereas aggression toward the mother tended to vary considerably over the years for some children, physical aggression to peers was found to be a very stable behavioral parameter for the first 10 years of life. There was a significant association between physical aggression to peers and mother in childhood (6–10 years) and ratings of competitiveness in adulthood for males, but this association did not hold true for females.

Aggressive behavior in adolescence becomes much more complicated and is subject to more variation and external influences than in the earlier years of life. Both the mode of expression of aggressiveness (verbal, indirect, or direct physical assault) and the recipient of the attack (father, mother, teachers, or peers) are strongly influenced by child-rearing practices and other variables (Bandura and Walters, 1959). It is rare for an adolescent boy to physically attack a parent, but anyone who has worked with juvenile courts or a youth authority is well aware that some adolescent boys are readily capable of destructive behavior, including direct assault, outside of the home. The question whether the striking androgen of male puberty are directly related to the upsurge of aggressiveness in adolescence is one that is ripe for investigation.

3. Adults—Male vs. Female

Aggressiveness as a personality trait as well as specific acts of aggression have throughout history been more characteristic of adult males than of adult females. The expectations of most societies, as reflected by laws and traditions have reinforced such behavioral patterns. Wars, at least the actual fighting phases of wars, are waged by men, and certainly the overwhelming majority of individual acts of physical violence between humans are committed by males. There has been a tendency in recent decades to view these sex differences in behavior as representing strictly learned behaviors, i.e., stemming from differential reinforcement of certain kinds of behavior on the parts of parents and the society in general. But the "heredity vs. environment" controversy has created a false dichotomy in the study of human behavior, since there is considerable evidence that the biological substrate upon which the various environmental influences are brought to bear is not uniform for all mankind and consequently is an important source of variability in the response of the organism to given environmental stimuli. The readiness of human males for learning aggressive patterns early in life deserves systematic investigation.

4. Disease States—The XYY Syndrome

The highly publicized XYY syndrome serves as a useful model for the sort of genetic–hormonal–behavioral interaction that we are suggesting may underlie certain patterns of human behavior and therefore merit further study.

The identification of individuals with two Y (male) sex chromosomes among prison populations led to speculation that the notion of "bad blood" being present in individuals with criminal records might have a genetic basis. Indeed, some courts, notably in France and Australia, have allowed the finding of an XYY karyotype to be admitted as evidence for mitigation in criminal trials.

Most geneticists and behavioral scientists consider this judgment to be premature; nevertheless, the research that has been stimulated in this area may yet prove to be quite significant. There are two excellent review articles that summarize the studies of XYY individuals (Court Brown, 1968; Kessler and Moos, 1970). In brief, there appears to be a higher percentage of XYY individuals in prisons, maximum security hospitals, psychiatric hospitals, and hospitals for the mentally retarded than in the general population. The second general finding is that XYY individuals tend to be taller than comparable XY males. No consistent physical abnormalities or defects in intelligence have been found to date, however, nor is there any clear evidence that the XYY individuals are more aggressive than their counterpart XY inmates when in prison; moreover, it does not appear that they have committed more violent crimes than other prisoners.

Some studies have shown elevated plasma or urinary testosterone levels in XYY individuals, but again the findings are inconsistent. One of the reasons here may be the fact that some of the XYY males are hypogonadal and have undescended, inactive testes, so that mean values for a group of XYY individuals are difficult to interpret.

In any event one can at least speculate about the possibility of a genetic defect (related to the Y chromosome) that leads to excessive production of androgens by the fetal and neonatal testes and in some way alters the early differentiation of the central nervous system. Thus, CNS might become sensitized to certain types of stimuli that might not otherwise provoke aggressive behavior (or at least not to the same degree). Depending on the individual's life experiences, coupled with a greater endowment in terms of size and physical strength (also androgen mediated), such a genetic defect might predispose a person toward violent outbursts or other behaviors not otherwise readily explained.

F. Implications for Further Studies

1. Prenatal Effects of Androgens

The human studies by Money and Ehrhardt already described, as well as the primate work of Goy and his associates, point to the need for larger-scale longitudinal studies to substantiate the long-term effects of prenatal exposure to androgens on aggressive and other behavior patterns. Here, an obvious group that would seem to be ideal for such a study would be XYY individuals identified at birth through a routine screening procedure and then followed into adulthood with both endocrine and behavioral assessments made at regular intervals, using some of the techniques we have been emphasizing.

2. Correlation of Changes in Testosterone Levels and Aggressive Behavior

In parts D and E of this section we have tried to show that in conjunction and certain naturally occurring states, changes in androgen levels and changes in behavior occur simultaneously. The best example of this phenomenon is boys going through puberty. To date, however, endocrine studies and behavioral studies have usually been done quite independent of each other, the former by biochemists, the latter by psychologists. There is a great need here for the combined efforts of an interdisciplinary team that would simultaneously monitor the endocrine, affective, and behavioral changes that occur at puberty and see what correlations might be found between these parameters.

3. Genetic Screening

The identification of genetic defects that might involve concomitant aberrations of androgen metabolism and aggressive behavior would certainly appear to be an area worth pursuing. XXYY individuals have now been identified as well as a variety of mosaics, e.g., 47 XYY/48 XYYY, but further characterization of such individuals as well as longitudinal studies would be of interest.

4. Androgens and Stress

Although we have been discussing primarily the effects of androgens on the CNS and behavior, there is good reason to believe that, as with other steroid hormones, the CNS–androgen relationship is a two-way street; i.e., the CNS can influence androgen secretions as well as be influenced by exposure to circulating androgens. Bliss (1971) has recently completed a series of studies in which he has demonstrated that both acute and chronic stressful experiences will diminish testicular function in rodents, whereas sexual excitation will increase it. Changes in plasma testosterone in the experimental animals were found to occur rapidly in response through the stressful stimuli. Rose (1969), in a study of Vietnam soldiers under conditions of combat stress, has reported significant alterations in urinary testosterone during periods of life-threatening stress. Further studies of alterations in testosterone production under various conditions of stress or other emotional states would be of interest in filling out this other side of the coin.

5. Hormonal Control of Aggressive Behavior

Our intention in this paper has been to emphasize new techniques and new findings that could lead to a better understanding of aggressive and sexual feelings and behavior in humans. The question of a rational way to deal with man's aggressive proclivities certainly hinges on a more rational understanding of the various contributing factors involved. On the other hand, large-scale hormonal manipulation with the goal of altering the aggressive behavior of a society does not seem likely for the reasons already cited by Rothballer (1967): "Because both libido and aggressivity in the human male and female depend upon androgens, their control could never be a popular approach to restraining aggressiveness. Likewise, attempts to manipulate aggressiveness through the perinatal organizing effects of hormones might be fraught with dangerous repercussions, perhaps permanent, in the psychosexual sphere."

From a hormonal as well as a psychological point of view, man's sexuality and his aggressiveness seem to be closely related. Whether they can be separated remains to be seen, but the problems involved in this sphere are increasingly germane to species survival.

REFERENCES

Bandura, A., and Walters, R. H. (1959). "Adolescent Aggression." Ronald Press, New York.
Baulieu, E. E., and Mauvais-Jarvis, P. (1964). *J. Biol. Chem.* **239,** 1579.
Beach, F. A., and Westbrook, W. H. (1968). *J. Endocrinol.* **42,** 379.

Beeman, E. A. (1947). *Physiol. Zool.* **20**, 373.

Bliss, E. L. (1971). Personal communication.

Clark, G., and Birch, H. G. (1945). *Psychosom. Med.* **7**, 321.

Clark, G., and Birch, H. G. (1946). *Psychosom. Med.* **8**, 320.

Court Brown, W. M. (1968). *J. Med. Genet.* **5**, 341.

Dalton, K. (1964). "The Premenstrual Syndrome." Thomas, Springfield, Illinois.

de Nicola, A. F., Dorfman, R. I., and Forchielli, E. (1966). *Steroids* **7**, 351.

Donchin, E., and Cohen, L. (1967). *Electroencephalogr. Clin. Neurophysiol.* **22**, 537.

Dorfman, R. I., and Shipley, R. A. (1956). "Androgens." Wiley, New York.

Edwards, D. (1968). *Science* **161**, 1027.

Edwards, D. (1969). *Physiol. Behav.* **4**, 333.

Edwards, D. (1970). *Physiol. Behav.* **5**, 465.

Ehrhardt, A. A., and Money, J. (1967). *J. Sex Res.* **3**, 83.

Ehrhardt, A. A., Evers, K., and Money, J. (1968a). *Johns Hopkins Med. J.* **123**, 115.

Ehrhardt, A. A., Epstein, R., and Money, J. (1968b). *Johns Hopkins Med. J.* **123**, 160.

Evans, J. I., MacLean, A. W., Ismail, A. A. A., and Love, D. (1971). *Nature (London)* **229**, 261.

Exley, D. (1968). *Biochem. J.* **107**, 285.

Forchielli, E. G., Sorcini, G., Nightingale, M. S., Brust, N., and Dorfman, R. I. (1963). *Anal. Biochem.* **5**, 416.

Forest, M. G., Rivarola, M. A., and Migeon, C. J. (1968). *Steroids* **12**, 323.

Foss, G. L. (1951). *Lancet* **1**, 667.

Frasier, S. D., and Horton, R. (1966). *Steroids* **8**, 777.

Gandy, H. M., and Peterson, R. E. (1968). *J. Clin. Endocrinol. Metab.* **28**, 949.

Goldheizer, J. W., and Axelrod, L. R. (1962). *J. Clin. Endocrinol. Metab.* **22**, 1234.

Goy, R. W. (1968). *In* "Endocrinology and Human Behaviour" (R. P. Michael, ed.), pp. 12–31. Oxford Univ. Press, London and New York.

Gray, J., Levine, S., and Broadhurst, P. W. (1965). *Anim. Behav.* **13**, 33.

Hamburg, D. A., Moos, R. H., and Yalom, I. D. (1968). *In* "Endocrinology and Human Behaviour" (R. P. Michael, ed.), Ch. 6. Oxford Univ. Press, London and New York.

Herbert, J. (1970). *J. Reprod. Fert. Suppl.* **11**, 119.

Hinde, R. A. (1967). *New Society* **9**, 302.

Holloway, R. L., Jr. (1968). *In* "War: The Anthropology of Armed Conflict and Aggression" (M. Fried, M. Harris, and R. Murphy. eds.), pp. 29–48. Nat. Hist. Press, New York.

Horton, R., and Neisler, J. (1968). *J. Clin. Endocrinol. Metab.* **28**, 479.

Horton, R., and Tait, J. F. (1966). *J. Clin. Invest.* **45**, 301.

Jeffcoate, S. L., Brooks, R. V., Lim, N. Y., London, D. R., Prunty, F. T. G., and Spathis, G. S. (1967). *J. Endocrinol.* **37**, 401.

Kagan, J., and Moss, H. A. (1962). "Birth to Maturity." Wiley, New York.

Kato, T., and Horton, R. (1968). *J. Clin. Endocrinol. Metab.* **28**, 1160.

Kessler, S., and Moos, R. H. (1970). *J. Psychiat Res.,* **7**, 153.

King, J. A., and Tollman, J. (1956). *Brit. J. Anim. Behav.* **4**, 147.

Kopell, B. S. (1969). *In* "Metabolic Effects of Gonadal Hormones and Contraceptive Steroids" (A. S. Hilton, ed.), pp. 649–667. Plenum, New York.

Kopell, B. S., Lunde, D. T., Clayton, R. B., and Moos, R. H. (1969). *J. Nerv. Ment. Dis.* **148**, 180.

Kopell, B. S., Wittner, W. K., Lunde, D. T., Warrick, G., and Edwards, D. (1970a). *Psychosom. Med.* **32**, 39.

Kopell, B. S., Wittner, W. K., Lunde, D. T., Warrick, G., and Edwards, D. (1970b). *Psychosom. Med.* **32**, 495.

Korenman, S. G., Wilson, H., and Lipsett, M. B. (1963). *J. Clin. Invest.* **42**, 1753.

Lamb, E. G., Dignam, W. J., Pion, R. J., and Simmer, H. H. (1964). *Acta Endocrinol. (Copenhagen)* **45**, 243.

Laschet, U., Laschet, L., Fetzner, H. R., Glaesal, H. U., Mall, G., and Naab, M. (1967). *Acta Endocrinol. (Copenhagen) Suppl.* **119**, 54.

Levin, J. C., Lloyd, C. W., Lobotsky, J., and Friedrich, E. H. (1967). *Acta Endocrinol. (Copenhagen)* **55**, 184.

Levine, S., and Mullins, R. F. (1966). *Science* **152**, 1585.

Lloyd, C. W., Lobotsky, J., Segre, E. J., Kobayashi, T., Taymor, M. L., and Batt, R. E. (1966). *J. Clin. Endocrinol. Metab.* **26**, 314.

Lorenz, K. (1966). "On Aggression." Harcourt, New York.

Lunde, D. T., Costell, R., Wittner, W. K., and Kopell, B. S. (1971). In preparation.

Lynn, R. (1966). "Attention, Arousal, and the Orientation Reaction." Pergamon, Oxford.

Mandell, A., and Mandell, M. (1967). *J. Amer. Med. Ass.* **200**, 792.

Mauvais-Jarvis, P., Floch, H. H., and Bercovi, J. (1968). *J. Clin. Endocrinol. Metab.* **28**, 460.

Mayes, D., and Nugent, C. A. (1968). *J. Clin. Endocrinol. Metab.* **28**, 1169.

Miller, S., and Mason, H. L. (1945). *J. Clin. Endocrinol.* **5**, 220.

Mizuno, M., Lobotsky, J. Lloyd C. W., Kobayashi, T., and Murasawa, Y. (1968). *J. Clin. Endocrinol. Metab.* **28**, 1133.

Money, J. (1970). *J. Sex Res.* **6**, 165.

Moos, R. H., Kopell, B. S., Melges, F. T., Yalom, I. D., Lunde, D. T., Clayton, R. B., and Hamburg, D. A. (1969). *J. Psychosom. Res.* **13**, 37.

Moos, R. H., Clayton, R. B., Yalom, I. D., and Hamburg, D. A. (1971). Unpublished observations.

Neumann, F., and Elger, W. (1966). *Endokrinologie* **50**, 209.

Neumann, F., von-Berswordt-Wallrabe, R., Elger, W., Steinbeck, H., Hahn, J. D., and Kramer, M. (1970). *Recent Progr. Horm. Res.* **26**, 337.

Paulsen, E. P., Sobel, E. H., and Shafran, M. S. (1966). *J. Clin. Endocrinol. Metab.* **26**, 329.

Pavlov, I. P. (1927). "Conditioned Reflexes." Oxford Univ. Press (Clarendon), London and New York.

Pearlman, W. H., Crepy, O., and Murphy, M. (1967). *J. Clin. Endocrinol. Metab.* **27**, 1012.

Persky, H., Smith, K. D., and Basu, G. K. (1970). *Psychosom. Med.* **32**, 553.

Reichlin, S. (1971). *Medical Aspects of Human Sexuality* **5**(2), 146.

Resko, J. A. (1967). *Endocrinology* **81**, 1203.

Resko, J. A., and Eik-Nes, K. B. (1966). *J. Clin. Endocrinol. Metab.* **26**, 573.

Rivarola, M. A., Saez, J. M., Meyer, W. J., Jenkins, M. E., and Migeon, C. J. (1966). *J. Clin. Endocrinol. Metab.* **26**, 1208.

Rivarola, M. A., Forest, M. G., and Migeon, C. J. (1968). *J. Clin. Endocrinol. Metab.* **28**, 34

Rose, R. (1969). *In* "The Psychology and Physiology of Stress" (P. Bourne, ed.), pp. 117–148. Academic Press, New York.

Rose, R., Holaday, J., and Bernstein, J. (1971). *Nature (London)* **231**, 366.

Rothballer, A. B. (1967). *In* "Brain Function. Vol. V: Aggression and Defense" (C. D. Clemente and D. B. Lindsley, eds.), pp. 135–170. Univ. of California Press, Berkeley, California.

Servais, J., and Hubin, P. (1968). *Encephale* **57**, 333.

Southern, A. L., Gordon, G. G., and Tochimoto, S. (1968). *J. Clin. Endocrinol. Metab.* **28**, 1105.

Talbot, N., Butler, A. M., Berman, R. A., Rodrigues, P. M., and MacLachlan, E. A. (1943). *Amer. J. Dis. Child.* **65**, 364.

Tinbergen, N. (1968). *Science* **160**, 1411.

Walter, G. (1964). *Arch. Psychiat. Nervenkr.* **206**, 309.

Walter, G. (1965). *J. Psychosom. Res.* **9**, 51.

Washburn, S. L., and Hamburg, D. A. (1968). *In* "Primates: Studies in Adaptation and Variability" (P. Jay, ed.), pp. 458–478. Holt, New York.

Waxenberg, S. E., Drellich, M. G., and Sutherland, A. M. (1959). *J. Clin. Endocrinol. Metab.* **19**, 193.

Wieland, R. G., Vorys, N., Folk, R. L., Besch, P. K., Neri, A., and Hamwi, G. T. (1966). *Amer. J. Med.* **41**, 927.

Yalom, I. D., Lunde, D. T., Moos, R. H., and Hamburg, D. A. (1968). *Arch. Gen. Psychiat.* **18**, 16.

[The discussion for this article appears on page 754.]

Neuroendocrine Factors in the Control of Primate Behavior

RICHARD P. MICHAEL, D. ZUMPE, E. B. KEVERNE, AND R. W. BONSALL

Primate Behaviour Research Laboratories, Institute of Psychiatry, Bethlem Royal Hospital, Beckenham, Kent, England

I. Introduction

The view has been quite widely held that, as one ascends the phylogenetic scale, the gonadal hormones become less important in determining the interaction between the sexes: thus, in the higher primates and, particularly in the human female, hormonal factors might be expected to have less prominence. While this generalization certainly has validity, many of the findings obtained in this laboratory during the past few years have tended to indicate that endocrine variables, although sometimes difficult to quantify and assess, do indeed play a crucial role in determining the sexual interactions of male and female rhesus monkeys (Michael, 1964, 1965, 1968a; Michael *et al.*, 1967a; Michael *et al.*, 1968).

The catarrhine monkeys and apes alone among infrahuman mammals have a true menstruation, and in the rhesus monkey the cycle is about 28 days. In recent years, there has been an increasing number of field studies on the social organization and behavior of several anthropoid primates, particularly the baboons and macaques. The females of several primate species appear to have periods of heightened sexual receptivity, usually near mid-cycle, when they are more ready to permit copulation (*Alouatta villosa*, Carpenter, 1934; on vaginal plug evidence in *Ateles geoffroyi Kuhl*, Carpenter, 1935; *Presbytis entellus*, Jay, 1965; *Macaca mulatta*, Ball and Hartman, 1935; Carpenter, 1942a,b; Altmann, 1962; Southwick *et al.*, 1965; *M. fuscata*, Imanishi, 1963; *Papio ursinus* and *P. anubis*, Bolwig, 1959; Washburn and DeVore, 1961; Hall and DeVore, 1965; *P. hamadryas*, Kummer and Kurt, 1963; *Pan troglodytes schweinfurthii*, Young and Orbison, 1944; Goodall, 1965). The pattern of sexual behavior of the rhesus monkey is particularly suited to quantitative study: it consists of a series of separate mounts by the male on the female, each mount generally being associated with an intromission and a variable number of pelvic thrusts. The series of mounts is terminated by a final mount in which ejaculation occurs, and between each mount there is usually an episode of grooming behavior; grooming activity occupies much of the time between successive mounting series (Michael *et al.*, 1966). Our standard observation period is of 1-hour duration, and in this time from zero to as many as 6 to 7 series of mounts can occur, each terminated by an ejaculation: the number of ejaculations depends on the identity of the pair and the stage of the female's cycle.

In these studies, the fallopian tubes of the females were transected and
ligated to prevent unwanted pregnancies, and great care was taken not
to disturb the vascular supply to the ovaries. All the behavior occurring
during an hour's test was recorded both by closed-circuit television and
by observers situated behind one-way vision mirrors; the techniques
for scoring and quantifying the behavioral indices have been fully vali-
dated and described in earlier publications (Michael and Saayman, 1967a
and b, 1968; Michael and Welegalla, 1968).

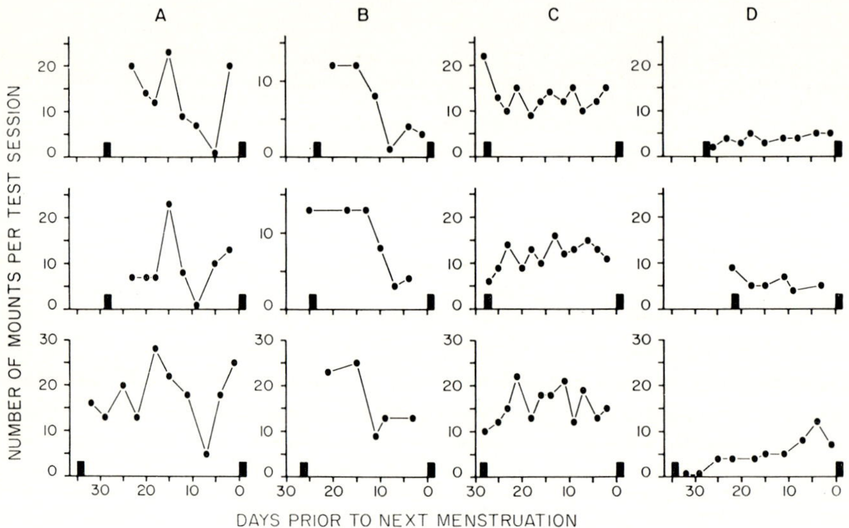

FIG. 1. Changes in the total number of mounts per test in relation to the
menstrual cycle in the rhesus monkey. Twelve pairs showing the four main types
of changes: (A) maxima near mid-cycle and secondary rises before menstruation;
(B) high levels of mounting during follicular phases and low levels during luteal
phases; (C) high levels of mounting throughout the cycles; (D) low levels of
interaction throughout the cycles. Types A and B show rhythmicity. Solid rectangles
indicate menstruation.

II. Influence of Menstrual Cycle on Behavior

When individual pairs of rhesus monkeys were tested together on a
daily basis in the manner described, different pairs showed different
patterns of mounting and ejaculatory behavior in relation to the stage
of the menstrual cycle of the female of the pair. Changes in the number
of mounts per test on successive days of the female's menstrual cycle
are illustrated in Fig. 1 and fall into four main patterns. (A) those
with well-defined maxima near mid-cycle, sharp declines early in the

luteal phase, and a secondary rise immediately before menstruation; (B) those with high levels of activity during the follicular phase, and again sharp declines early in the luteal phase with low levels persisting until the next menstruation; (C) cycles without clear evidence of rhythmic changes but with fairly high levels of mounting throughout the cycle; and (D) cycles again without any rhythmic changes but with low levels of interaction throughout the cycle. Approximately 50% of the pairs and cycles studied showed clear evidence of rhythmicity, and in these cycles neuroendocrine factors appear to be playing a clear-cut role in determining the interaction of the pair (Michael and Zumpe, 1970a). Type C is characteristic of older, more dominant males that show relatively high levels of activity with most partners, and these males seem to be relatively insensitive to the endocrine state of their female consorts. Type D is typical of pairs in which very low levels of sexual interest are shown in the partner. These individual differences in the behavior patterns of different pairs are little understood at the present time, and unraveling the factors that determine them will provide us with a great deal of insight into the mechanisms that underlie primate sexuality. It will be clear from an examination of Fig. 1 that sexual activity in the rhesus monkey can occur throughout the menstrual cycle and that a highly circumscribed period of estrus, such as occurs in many lower mammals, is not found in the rhesus macaque. Furthermore, some rhesus females, but not many, will continue to permit copulation for prolonged periods after bilateral ovariectomy, so that the possible role of non-ovarian factors should also be taken into consideration. The less precise dependence of behavior on hormonal factors, the marked individual differences already described, and the increased importance of neocortical mechanisms (e.g., the effect of past experience on subsequent behavior) all necessitate a more complex treatment of the behavioral aspects than the relatively simple end point of whether or not a positive mating test occurs.

Although different patterns of interaction between the male and female are encountered, it is possible to make a more generalized statement about the changes in the frequency of ejaculation during the menstrual cycle by considering the data from all the 75 menstrual cycles observed in 32 pairs of animals during a 5-year period. These data (mean cycle length 28.3 ± 0.3 days, range 21–34 days) are shown in Fig. 2 (solid line). The mean numbers of ejaculations per test were calculated for all the tests available for each day of the cycle, each point being a mean of 24–30 pairs. There was a maximum incidence of ejaculation 17 days before the first day of the next menstruation, and a progressive decline continued throughout the luteal phase, a low point being reached 2 days

          R. P. MICHAEL ET AL.

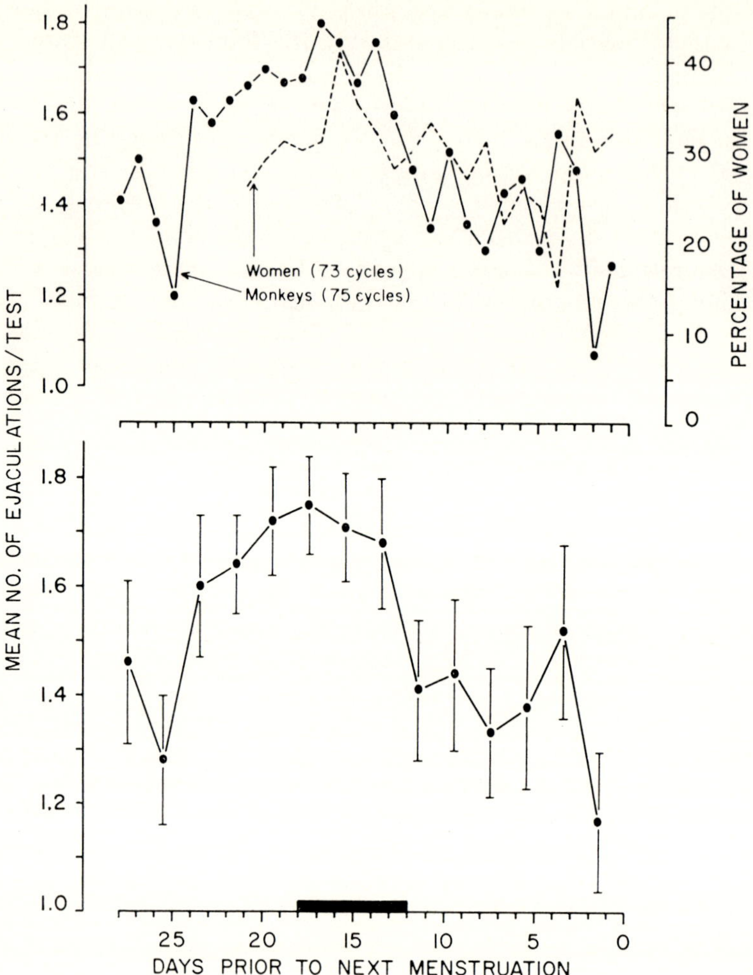

Fig. 2. Top: Comparison of the copulatory activity of rhesus monkey and man in relation to the menstrual cycle, ●———●, mean number of ejaculations per test (32 pairs of rhesus monkeys); - - -, % of women reporting sexual intercourse [40 women (Udry and Morris 1968)]. Bottom: rhesus monkey data smoothed by plotting means of two consecutive days. Vertical bars give standard errors of means. Horizontal bar gives expected time of ovulation (Hartman, 1932).

before menstruation. The lower part of Fig. 2 shows the same data smoothed by plotting the means of 2 consecutive days. The mean number of ejaculations from the 1st to the 12th reverse cycle days was significantly lower than that from the 13th to the 24th reverse cycle days (t test, $P < 0.001$): the high and low plateaus of values in the follicular and luteal phases respectively are separated by the abrupt decline in ejaculation between reverse days 12 and 13: there was a further abrupt fall immediately before menstruation. Since the 75 cycles showed considerable variations in length and included many without any obvious rhythmicity, the highly significant differences between follicular and luteal phases in the over-all data clearly established this pattern of ejaculatory activity as characteristic of the species.

III. Female Sexual Invitations

It is well known that the sexual presentation posture is a frequently observed invitational gesture in many primates including the female rhesus monkey. The frequency with which it is made varies with different females, with the identity of their partners (Michael and Saayman, 1968), with the stage of the menstrual cycle (Michael, 1965; Michael *et al.* 1967b; Michael and Welegalla, 1968) and with their hormonal state (Michael *et al.*, 1968). We have now been able to identify certain other movements and postures which have been shown to express sexual excitement in the female rhesus and which serve as female sexual invitations: these have been named the *hand-reach, head-duck,* and *head-bob* (Michael and Zumpe, 1970b; Zumpe and Michael, 1970a). It is not intended to give a description of these behavior patterns here, but it is important to note them and to take all these specific invitational gestures into consideration when attempting to assess the effects of hormones on the female's motivational state. Female invitational gestures when they are effective tend to elicit male mounting activity, and it might be anticipated that when the female's sexual motivation is high she will make a greater number of invitations. The effects of ovariectomy and of hormone replacement treatments on female invitations and on her refusals of male mounting attempts are shown in Figs. 3 and 4. Ovariectomy clearly decreased female receptivity in that the number of female refusals increased, and this was associated with a decline in female invitational behavior. These effects were reversed by treating females with estradiol, and subsequent administration of progesterone antagonized the effects of estrogen administration. Unfortunately, the prediction so clearly confirmed by these results is not invariably realized because the number and frequency of female invitational gestures depend not only upon her attitude to the male but also upon the male's attitude to her. Thus, in situations where the male is becoming aggressive or is losing sexual interest, female invitations may increase, presumably because of the changes in the behavior of her partner. This illustrates one of the central difficulties in attempting to understand complex primate interactions in terms of one or two behavioral variables. The fact is that it is essential to consider simultaneous changes in several behavioral parameters in order to arrive at adequate interpretations about causation. While a study of the literature cited shows that this

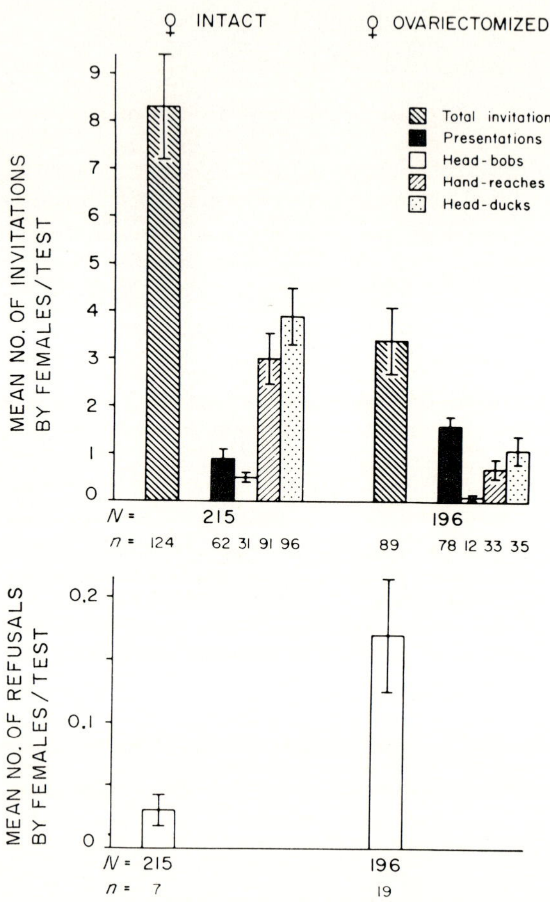

Fig. 3. Effect of ovariectomy on female sexual invitations and refusals. There was a significant decrease in total invitations and increase in refusals ($p < 0.02$). N = number of tests; n = number of tests in which behaviors occurred. Vertical bars give standard errors of means (6 pairs). From Michael and Zumpe (1970b).

has been done, it is not possible to provide all the detailed behavioral data in the context of the present paper, where the emphasis is upon endocrine factors. The fact that behavioral rhythms, shown by several different indices (numbers of mounts per test, mounting rates, numbers

of ejaculations, ejaculation times, etc.), occur in relation to the female's
menstrual cycle and that they are abolished by bilateral ovariectomy
(Michael *et al.*, 1967a), implies that changes in the female's endocrine
status impose changes on the behavioral interactions of the pair. Thus

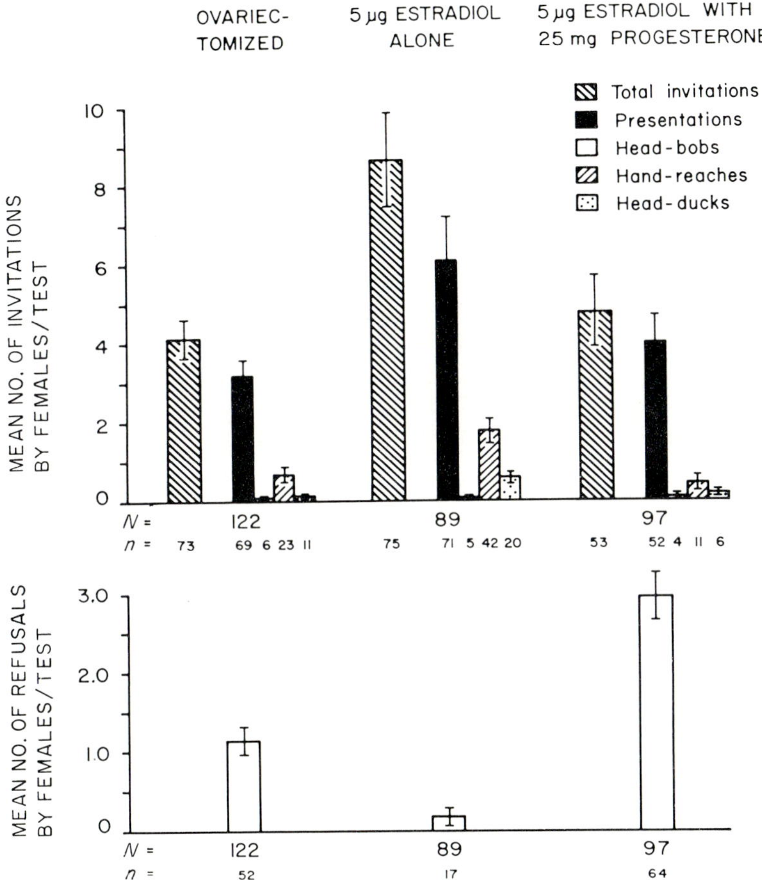

Fig. 4. Effects on sexual invitations and refusals of ovariectomized female rhesus
monkeys of treatments with estradiol and progesterone. Estradiol increased invitations and decreased refusals, and progesterone reversed these effects. N = number
of tests; n = number of tests in which behaviors occurred (8 pairs). From Zumpe
and Michael (1970a).

we are presented with the problem of how information is communicated
to the male partner. We can now consider in more detail the means
by which hormones influence the behavior of the pair, and this may
be done in terms of two major variables, namely, female sexual receptiv-

ity (Michael, 1968a,b) and female sexual attractiveness (Michael *et al.*, 1967a,b).

IV. Sexual Receptivity and Sexual Attractiveness

Sexual receptivity can be distinguished from sexual attractiveness by simple behavioral criteria. These two concepts are, of course, abstractions, but they can be operationally defined and have resulted in useful experimentation. Highly receptive females, for instance, make an increased number of sexual invitations and only rarely refuse the male's mounting attempts. Furthermore, in an operant-conditioning situation (see below) they show high rates of lever-pressing in order to gain access to a male partner. Conversely, unreceptive females make low numbers of sexual invitations and refuse the male partner's mounting attempts more frequently (see other behavior also: clutching reaction, Zumpe and Michael, 1968; withdrawal reaction, Michael and Saayman, 1968; threatening-away behavior, Zumpe and Michael, 1970b). Female attractiveness, on the other hand, is somewhat more difficult to characterize, but when it is high, the male partners make large numbers of mounting attempts and show high rates of lever-pressing in an operant-conditioning situation in order to gain access to the female. It will be obvious that female attractiveness may be high when her receptivity is low and, under these conditions, the male makes many mounting attempts but these will mostly be refused by the unreceptive female. The converse of this situation occurs when a highly receptive female makes numerous sexual invitations but, because her attractiveness is low, the male fails to respond to the majority of these invitations by actually making a mount (Michael *et al.*, 1968). The changing interactions between the pair at different stages of the menstrual cycle can be quite well understood in terms of changes in receptivity and attractiveness, but the situation is simplified and more clear-cut when ovariectomized females are treated with different doses of estrogen and progesterone. Interactions between males and ovariectomized females eventually decline, in most cases to low levels; injecting the females with 5–10 μg of estradiol daily restored male sexual activity in about 3 days, when mounting behavior and ejaculation reappeared (Fig. 5). Estrogen treatment of the female thus restored the attractiveness and receptivity that is lacking in long-ovariectomized females. When females were treated in addition with large doses of progesterone, both the numbers of mounts and of ejaculations made by their male partners were depressed (Fig. 6) (Michael, Saayman and Zumpe, 1967b). This effect of progesterone on behavior is extremely interesting because two mechanisms appear to be responsible for its action in depressing sexual activity. Figure 7 shows that pairs fall into

two groups (A and B) which can be distinguished by the effect of progesterone on the number of mounting attempts made by the males (lower
set of histograms). In group A, mounting attempts remained relatively
constant across treatments whereas in group B, mounting attempts declined conspicuously when 25 mg of progesterone per day was administered to the females (209 tests). In group A pairs, the decline in the
number of mounts made by males (upper set of histograms) with females
on high doses of progesterone was almost entirely due to the marked

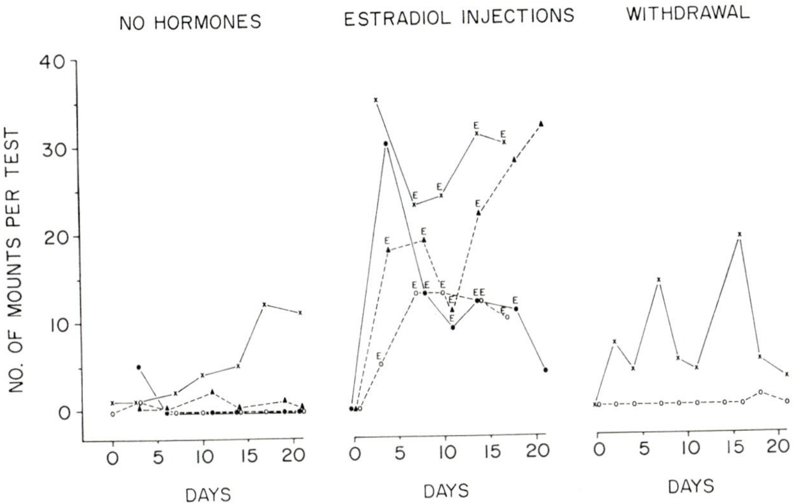

Fig. 5. Effect on the behavior of males when their ovariectomized female partners
were treated daily with estradiol subcutaneously. There was a very rapid
restoration of mounting activity and ejaculation. Two pairs were followed after
withdrawal of hormone, and in one pair behavior remained unstable. E = tests
in which ejaculation occurred (4 pairs). From Michael et al. (1967a).

increase in female refusals (middle set of histograms): in these pairs,
progesterone treatment resulted in a decline in receptivity. However,
the sexual interest of males in these females, as expressed by the number
of male mounting attempts, remained unchanged. In group B pairs, the
situation was quite different. Here, the high dose of progesterone resulted
in a very marked decline in the number of male mounting attempts
but in only a small, nonsignificant increase in the number of refusals
made by the females. The lack of interest of males in progesterone-
treated females was shown by an almost complete cessation of male
mounting activity. Thus we have a situation in which treating females
with progesterone depresses male sexual activity in all cases, but two

quite different mechanisms may underlie this effect, depending upon the identity of the pair. In cases represented in group A, progesterone was presumably acting on a neural mechanism within the brain of the female either to activate a "refusal mechanism" or to inhibit a mechanism

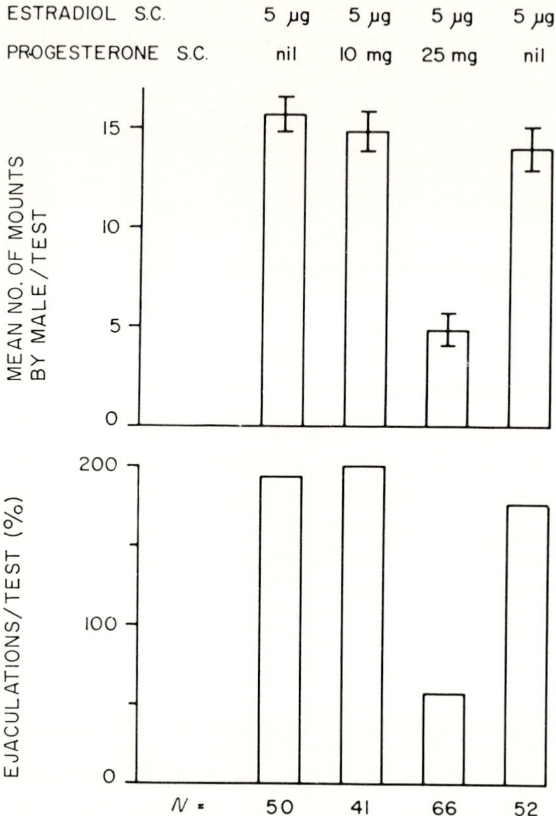

Fig. 6. The suppression of mounting behavior and of ejaculation in male rhesus monkeys when their ovariectomized, estrogen-treated female partners were treated with progesterone. N = number of tests. When values exceed 100%, more than one ejaculation occurred in each test. Vertical bars give standard errors of means (6 pairs). From Michael *et al.* (1968).

responsible for receptivity. However one cares to conceptualize the actions of estrogen and progesterone, the existence and localization of this hypothetical mechanism can conveniently be studied by the application of gonadal steroids and their congeners directly to selected regions of the brain.

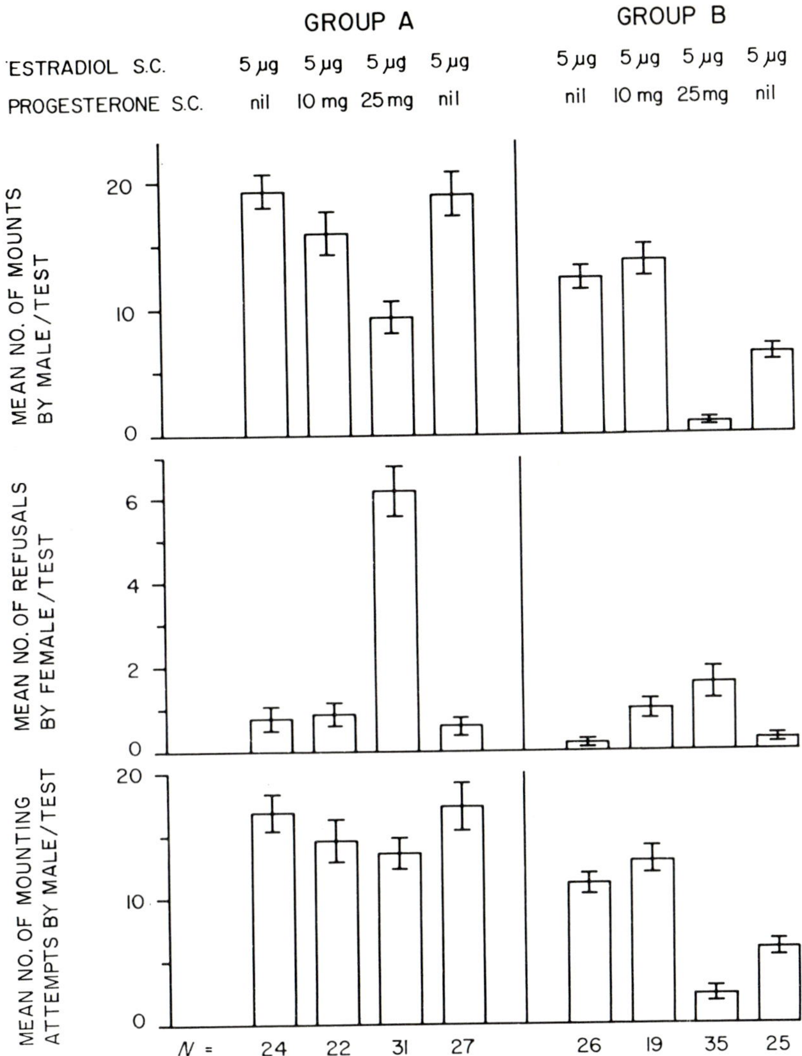

FIG. 7. The two mechanisms responsible for the suppression of mounting activity by male rhesus monkeys when progesterone is administered to their female partners. Group A: three pairs in which there was a marked increase in female refusals. Group B: three pairs in which there was a marked decrease in male mounting attempts. In both groups, a decline in the number of mounts results. N = number of tests. Vertical bars give standard errors of means.

V. Effects of Applying Estrogen to the Brain

Primates have not been systematically investigated with brain implantation techniques until recently (Michael, 1969a, 1971), and, indeed, it was not really feasible to do so until good data became available on behavioral thresholds with the subcutaneous route of administration. The long-term aim is to investigate the diencephalon and related areas of the limbic system of the female rhesus monkey with small, solid implants of estrogen. Initially, attention has been directed to the upper tegmentum and posterior mammillary region using an implantation technique similar to that employed earlier in female cats (Harris *et al.*, 1958; Michael, 1961). Each female rhesus monkey was tested separately with two male partners on a daily basis, the aim being to obtain behavioral changes in the same direction, simultaneously but independently, with each male partner. The effects of the estrogen implant were followed until behavior had again returned to preimplantation baseline and its effects were judged to be exhausted. Four to six months later the same females were reimplanted in the white matter of the frontal subcortex with implants almost identical in shape and weight to those originally introduced into the brain stem. The males with which each female was tested remained unchanged throughout the 12-month period of study. Thus, there was a good opportunity for comparing the behavioral effects of almost identical estrogen implants placed at different neural sites in the same female. It became clear that effects which depended upon the site of the implant, that is differential effects on behavior, were obtained only when small implants were used: larger implants produced similar changes in behavior at whichever site they were implanted. Accordingly, the results from the 12 pairs of animals shown in Fig. 8 are divided into two groups on the basis of implant size. The upper part of the figure shows that, when small implants were used, those in the posterior mammillary–upper tegmental region resulted in three times the number of ejaculations by males paired with these females than when implants of similar shape and weight were placed in the subcortical white matter of the same females. The larger implants also resulted in maximal stimulation of the ejaculatory behavior of the males, but there were no differences in effects at the two neural sites. Figure 8 (lower part) shows that small implants in the brain stem resulted in a marked reduction in the time taken to achieve ejaculation by males paired with these females when compared with the ejaculation times of the same males paired with the same females when they were bearing implants of similar sizes in the subcortical white matter (*t* test, $P < 0.01$). Again, no differential effects were observed when larger im-

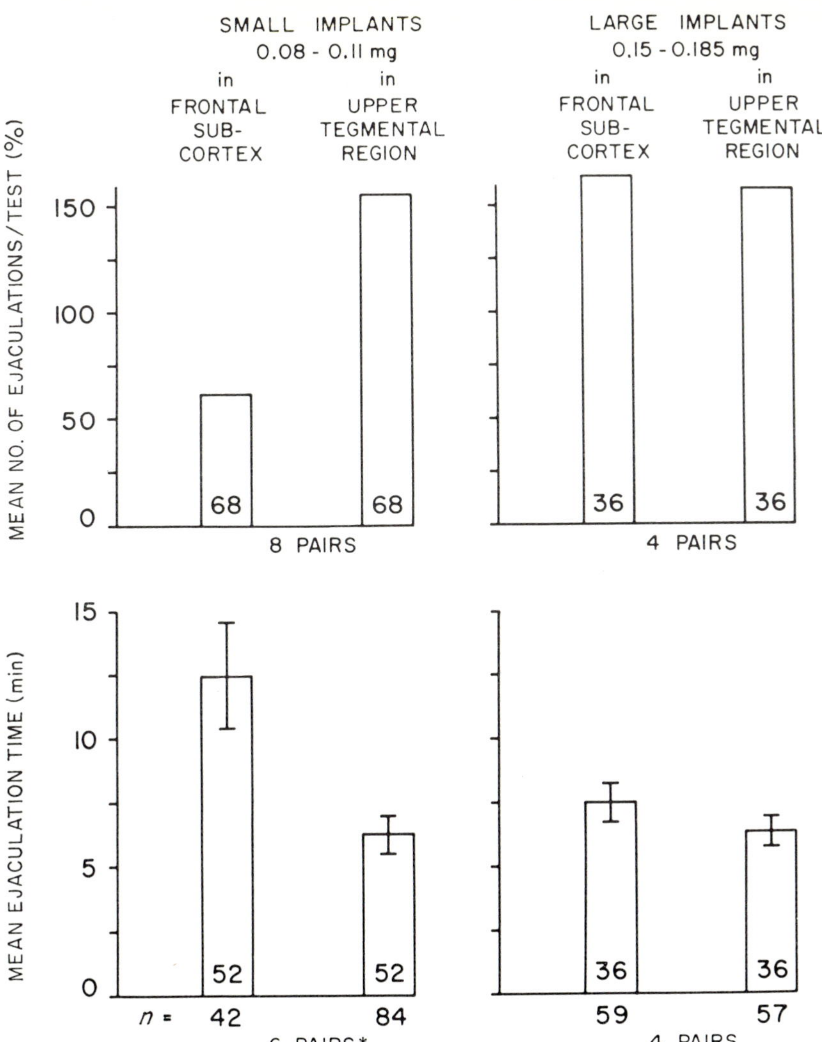

FIG. 8.   Effects on the ejaculation of male rhesus monkeys of implants of estrogen in the brains of female partners. Above: with small implants, behavioral effects depend upon the implant site; three times as many ejaculations resulted with implants in the upper tegmentum-posterior mammillary region. Where values exceed 100%, more than one ejaculation occurred per test. Below: With small implants, those in the brain stem resulted in significantly shorter ejaculation times. * Two pairs excluded because no ejaculations occurred with subcortical implants. Numbers in columns = number of tests. n = number of ejaculations. No differential effect occurred with large implants.

plants of estrogen were used. Other behavioral parameters could also be considered here, particularly those dealing with female receptivity (Michael, 1971), but, simply by using the number of ejaculations per test and ejaculation times, it is clear that the behavior of the male partners is influenced by estrogenizing the deeper, and phylogenetically older, parts of the female's brain. While these preliminary findings are no more than pointers to further study, the fact that differential effects on behavior can be obtained in primates by brain implants of hormone strongly suggests that this line of experimentation will be worth pursuing.

VI. Effects of Applying Androgen to the Brain

All the behavioral studies described above show that estrogen and progesterone have a special importance in determining the sexual activity of female rhesus monkeys. However, there are also data which suggest that androgens, in addition to the main ovarian hormones, may play a role in the manifestation of female libido as, indeed, there is now evidence to suggest that estrogens may contribute to the expression of libido in the male. Women receiving androgen therapy for menstrual disorders and malignant disease have described increased sexual feelings (Loesser, 1940; Greenblatt *et al.*, 1942; Abel, 1945; Carter *et al.*, 1947; Foss, 1951): some of these reports are a little difficult to evaluate but certainly cannot be discounted. Recent studies have shown that androgens may play a part in determining sexual receptivity in the female rhesus monkey (Everitt and Herbert, 1969; Michael, 1971). Using ovariectomized females implanted with 27-gauge stainless steel cannulae in hypothalamus, amygdala, and thalamus, a study was made of the effects of single, small doses of testosterone propionate (0.3–3.0 mg) when administered intracerebrally and intramuscularly in oil. Figure 9 (upper parts) shows a well-marked increase in female invitations with both routes of administration but only a moderate increase in male mounting behavior (Fig. 9, lower parts); indeed, there was only a minimal effect on the mounting behavior of the two males in the pairs on the right-hand side of the figure. Very few mounts were, in fact, accompanied by intromission, and for this reason ejaculation occurred only once in the 50 tests during the first 8 posttreatment days. Thus, the marked effect of testosterone propionate on female invitations was not paralleled by any corresponding increase in the sexual interaction of the pair, although this occurred immediately with the administration of small doses of estrogen.

The effect of androgen on the motivational state of the female was further investigated in an operant-conditioning situation. A twin compartment cage is used, the two sides of which are separated by a lifting partition operated by a servo motor. Females were trained to press a lever for various reinforcements, and doing so 250 times activated the servo motor to give access to the other compartment when a routine mating test followed (Fig. 10). An ovariectomized female would not usually press to criterion in the 30 minutes allowed when the reinforce-

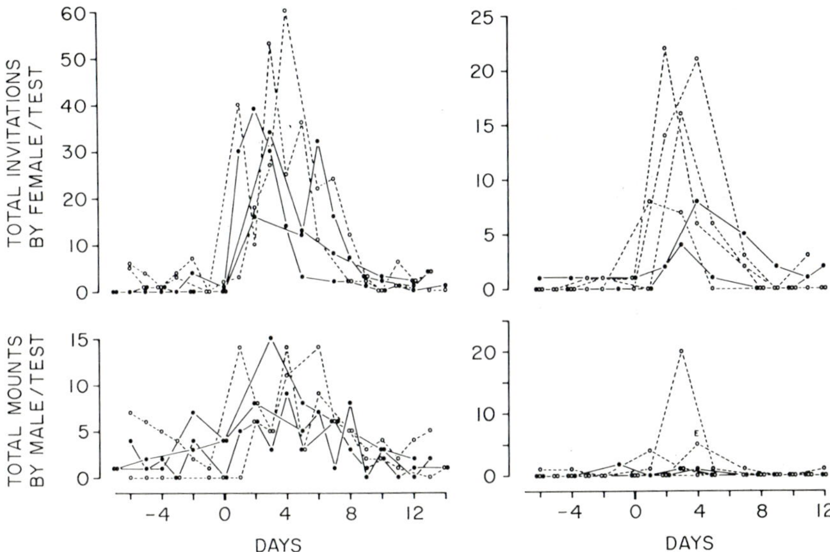

Fig. 9. A well-marked increase in the number of female sexual invitations occurred when they received testosterone proprionate intramuscularly (O---O) and intracerebrally (●——●) (hypothalamus, amygdala, and thalamus). However, there was only a moderate increase in male mounting and a minimal effect on the mounting of the two males in the pairs on the right-hand side. Only one ejaculation occurred. Day 0 = day of injection.

ment was a male partner and, in these circumstances, a routine mating test was carried out at a different time of day in a nonoperant situation to obtain behavioral data. The administration of 0.5 mg of testosterone propionate intramuscularly per day induced rapid lever-pressing, and access to the male was obtained within a few minutes. The increase in operant performance correlated quite well with the increase in the number of female sexual invitations (Fig. 11), but, again, treating the female with androgen failed to have a significant effect on the male's sexual activity, only one ejaculation occurring during the 23 days of treatment. Insofar as the number of invitations and the lever-pressing

Fig. 10. (A) A female rhesus monkey pressing a lever in her cage in order to obtain access to a male partner. (B) After the lever has been pressed 250 times, a servo motor is actuated which lifts a partition between the two sides of the cage, and the female is seen passing to the other side. (C) The male is now seen mounting the female, which has obtained access to him.

performance provide measures of female receptivity, the latter was increased by androgen treatment. However, single doses of 0.5–0.75 mg testosterone propionate administered either intramuscularly or intracerebrally resulted in plasma levels of testosterone during the next 24 hours of between 600 and 2500 ng/100 ml, using a competitive protein binding method (Plant and Michael, 1971). Whether or not they were associated with any changes in female invitational behavior, these

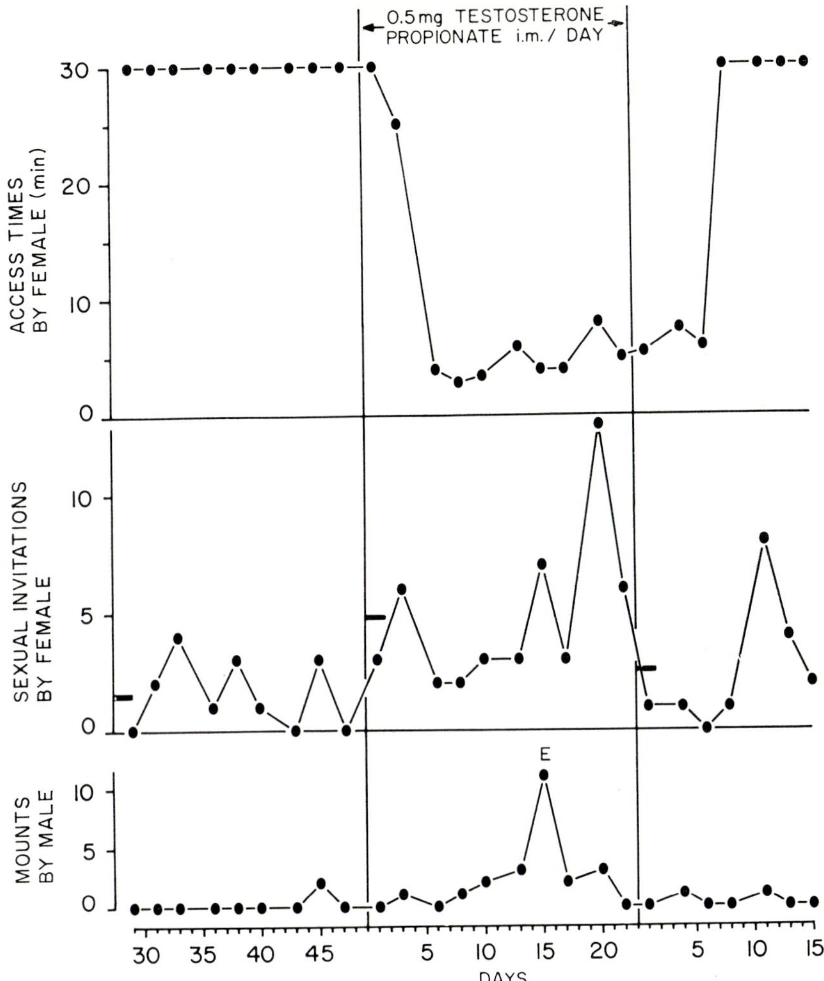

Fig. 11. Effects of androgen on the operant behavior and sexual invitations of an ovariectomized female rhesus monkey, and on the mounting behavior of the male partner. Injections of testosterone propionate induced rapid lever-pressing for access to the male, and female invitations were increased. The male's mounting activity was little affected, and ejaculation occurred only once. Horizontal bars give means.

plasma levels were in the range found in intact, adult male rhesus monkeys. Since the plasma levels of intact female rhesus monkeys are generally below 100 ng/100 ml (Michael and Plant, unpublished results), it seems unlikely that the findings described above have direct, physiological significance.

VII. Effect of Estrogen on Female Sexual Motivation

Assessing the motivational state of the male by his overt behavior with an estrogenized female presents relatively few difficulties: the behavioral initiative is generally with the male and his mounting activity and ejaculatory behavior is readily quantified and scored. The internal motivational state of the female is more difficult to assess: the external signs of female orgasm are more dubious (Zumpe and Michael, 1968), and her participation in courtship and mating activities, as previously mentioned, is greatly influenced by male dominance and the threat of male aggression. However, some interesting findings have emerged from using the free-cage operant conditioning situation in order to provide the female with some measure of environmental control. The twin-compartment cage described above was used, but only the female was trained to lever-press and operate the lifting partition between compartments. The partition lifted to a height that permitted free movement of the female between compartments but obstructed the passage of the larger male. Thus, the female rhesus monkey obtained a greater degree of mastery of the environment than the male by her ability to perform an instrumental response in order to obtain access to, or escape from, her partner. It might be thought that we were excessively naive in our anticipation that the females of this species would use the skill so acquired in a predictably intelligible manner. In fact, there was a marked increase in operant behavior by the female during a rather restricted period in the menstrual cycle near the expected time of ovulation. Figure 12 illustrates the operant behavior of a female for two male partners during the same six successive menstrual cycles. In the case of the more preferred male partner (Fig. 12, lower part) where ejaculations occurred throughout the cycle, the increase in operant behavior was maintained over a longer period than with the less preferred male (Fig. 12, upper part): with the latter, there were fewer ejaculations except near mid-cycle and the increased operant behavior of the female was restricted to this period. In this situation, the female is free to press or ignore the lever as she feels inclined: her increased lever-pressing activity for access to the male near mid-cycle correlates well with the increased excretion of urinary estrone (Hopper and Tullner, 1970) and with the rapid increase in serum levels of estradiol (Hotchkiss *et al.*, 1971) which occur at this time in this species. That these changes in operant behavior were related to the hormonal state of the female was indicated by administering graded doses of estradiol subcutaneously to the ovariectomized female and obtaining a dose-response effect. With the more preferred male (Fig. 13), operant behavior continued for 15 days after

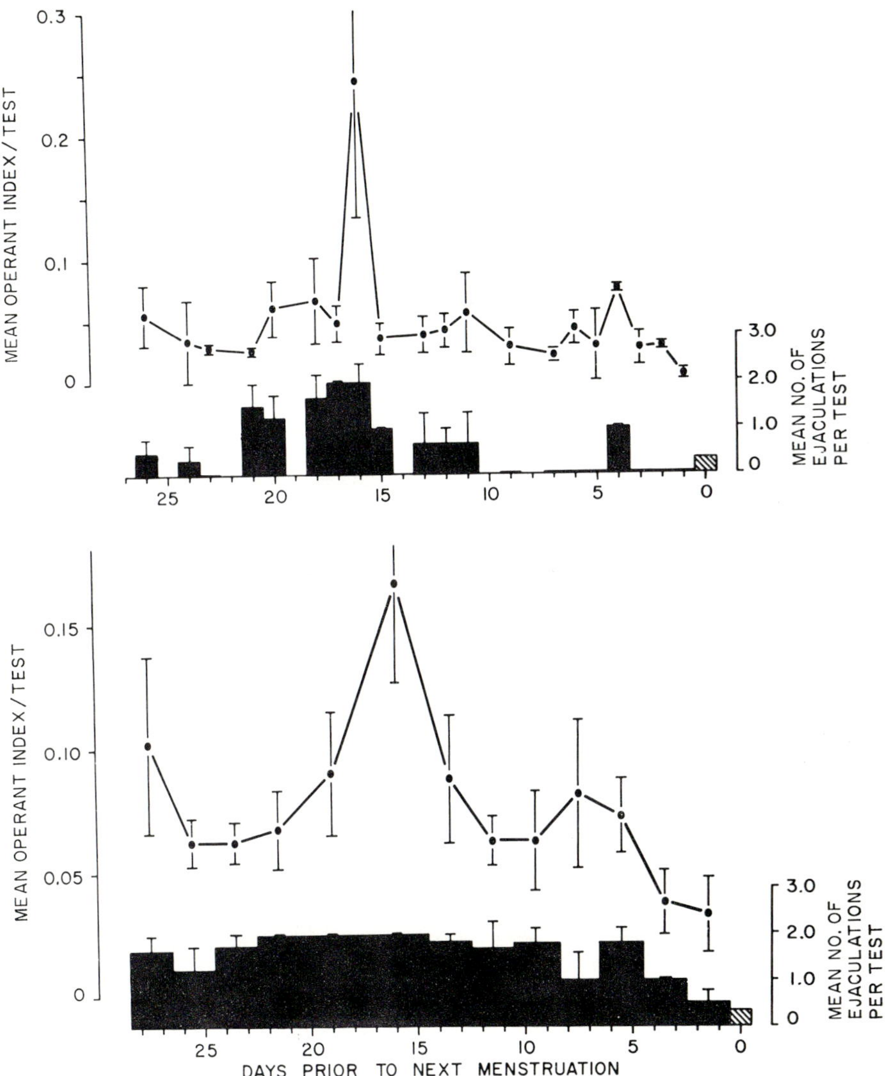

Fig. 12. The operant behavior of a female rhesus monkey throughout the course of six successive menstrual cycles for a preferred male (lower part) and for a less preferred male (upper part). There was an increase in lever-pressing during the middle part of the cycle, but this was for a very restricted period with the less preferred partner. Operant index = reciprocal of product of time to first lever-press and pressing time. Hatched area = menstruation.

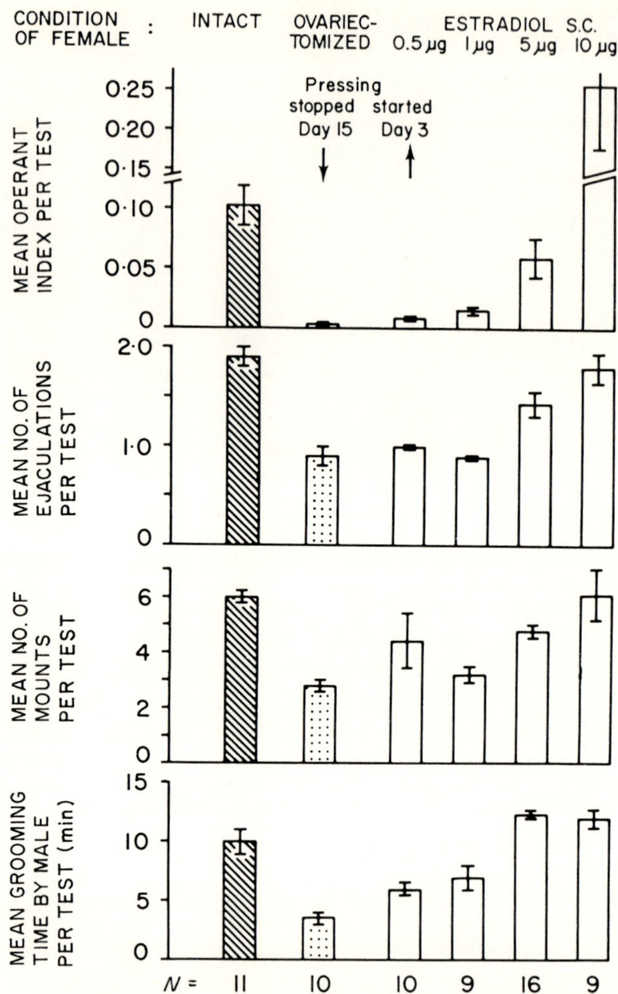

Fig. 13. Administration of graded doses of estradiol subcutaneously to the ovariectomized female produced a graded increase in operant behavior. With this more preferred male partner, lever-pressing continued for 15 days after ovariectomy and was rapidly reinstated with low doses of hormone. N = number of tests.

ovariectomy and was reinstated in only 3 days of treatment with 0.5 μg estradiol given subcutaneously per day. In contrast, with the less preferred male (Fig. 14), operant behavior ceased 4 days after ovariectomy and was restored only 60 days after instituting replacement treatment, for the last 10 days of which the female received 5.0 μg estradiol subcutaneously per day. Although with each male partner the female's

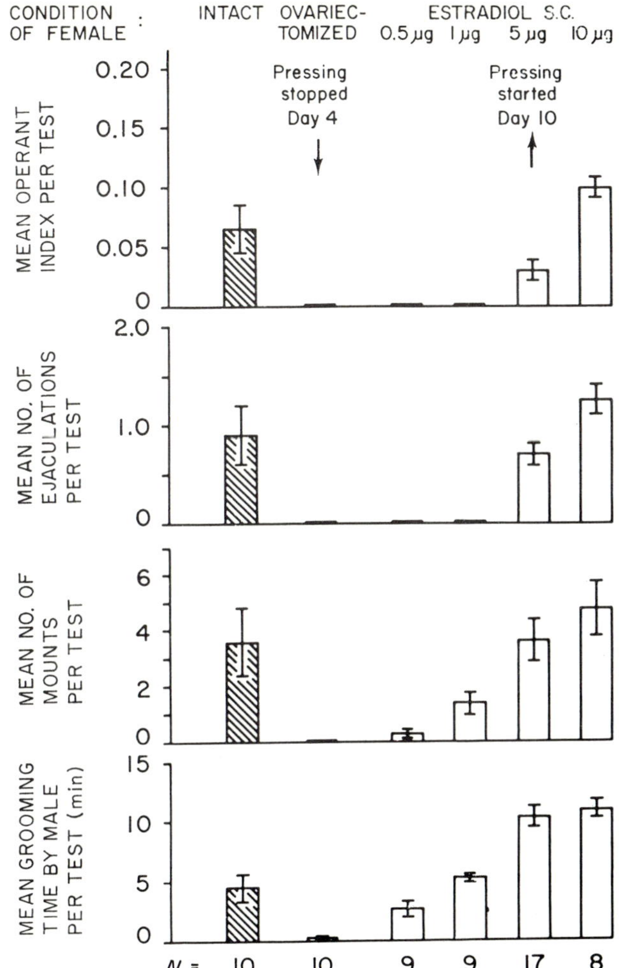

Fig. 14. Administration of graded doses of estradiol subcutaneously to the same ovariectomized female as in Fig. 13 produced a graded response in operant performance. However, with this less preferred male partner, lever-pressing ceased in 4 days after ovariectomy and much larger doses of estradiol were needed to reinstate it. N = number of tests.

operant behavior was related to the dose of estrogen administered, the threshold for the response was determined by the identity of the male. It will be seen that the operant index correlates well with other measures of sexual activity (mean number of ejaculations, mean number of mounts, mean grooming time by male) and would appear to provide a useful, objective measure of female sexual motivation. The use of operant behavior for the assessment of hormonal effects is discussed in Section XII.

VIII. Primate Pheromones and Female Sexual Attractiveness

We have dealt in some detail with the effects of estrogens, progestagens, and androgens on female receptivity, and we have considered some experimental means of assessing the effects of these steroids on the female's emotional orientation toward her male consort. Our earlier work indicated that males lost interest in females receiving high dosages of progesterone and, hence, we used the concept of "loss of attractiveness" when there was no direct evidence available as to the nature of the receptors or the form of distance communication involved. Olfactory communication between individuals by substances with the properties of a pheromone is known to be important in the control of reproductive processes in certain lower mammals, particularly laboratory mice, and this subject has been reviewed recently (Bruce, 1970). The behavioral effects of olfactory stimuli in mammals derive mainly from the breeding of farm animals. Thus, the odor of the boar elicits, in conjunction with other stimuli, the immobilization reflex in estrous sows so that they stand rigidly while being mated (Signoret and du Mesnil du Buisson, 1961; Signoret and Mauleon, 1962). It is now known that the sow detects 5α-androst-16-en-3-one and 3α-hydroxy-5α-androst-16-ene in the salivary secretions of the boar via the sense of smell (Patterson, 1968a). Dogs are attracted over considerable distances by estrous bitches and are specifically attracted by their urine (Beach and Gilmore, 1949) and by their vaginal secretions (Beach and Merari, 1970). Similarly, the urine of estrous mares sexually stimulates the stallion (Berliner, 1959; Wierzbowski and Hafez, 1961). Rams can differentiate between estrous and nonestrous ewes by their scent (Kelley, 1937; Lindsay, 1965). Male laboratory rats differentiate between receptive and nonreceptive females (Le Magnen, 1952; Carr et al., 1966; Stern, 1970). Among primates, the prosimians provide many examples of communication between the sexes by olfaction, and many possess specialized apocrine scent glands used for territorial marking, for self marking, and for marking each other. However, the possible role of olfactory communication between the sexes in higher, anthropoid primates has received little attention until recently. This has perhaps been because the catarrhine monkeys are equipped with an excellent visual apparatus and because, as one ascends the phylogenetic scale from the prosimians, there has been a corresponding reduction in the size of those rhinencephalic brain structures concerned with olfaction. In fact, as Fig. 15 illustrates, the olfactory and limbic brain structures of a typical cercopithecoid monkey quite closely resemble those of man, generally regarded as a microsmatic form. Thus one would not predict with any confidence that olfactory communication would be important in these latter species. However, field observations have been made on the frequent scenting of the female's genital region by males in several species of macaques (*M. sinica, M. radiata, M. fascicularis, M. arctoides, M. nemestrina, M. mulatta*) (see Michael

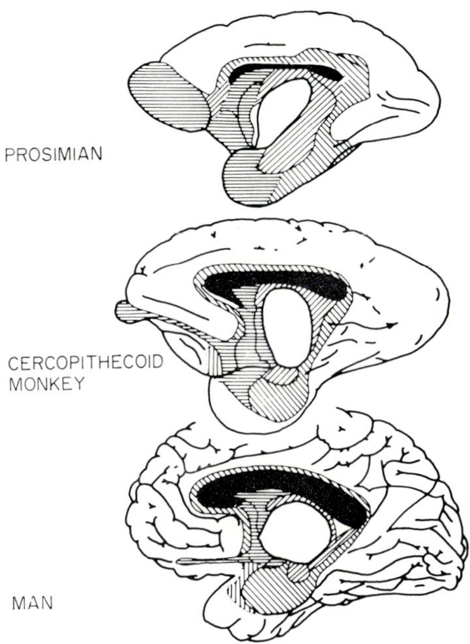

FIG. 15. Illustrates the reduction in the mass of the phylogenetically old brain structures and those associated with olfaction as one ascends the comparative scale from prosimian to man. In this respect, the brain of the rhesus monkey more closely resembles that of man, a microsmatic form, than a prosimian. After Stephan (1963).

and Zumpe, 1971). Similar behavior has been observed under laboratory conditions, and Fig. 16 shows the olfactory examination of a female rhesus monkey's rear by a male during a mating test. The male has seized the female by the tail and his facial expression and protruding lips (cf. "Flehmen" face—van Hooff, 1962) are characteristic of olfactory examination. Use has again been made of an operant-conditioning situation in which male rhesus monkeys were required to press a lever 250 times to obtain access to a female partner. Males were first trained to press a lever for food rewards on a fixed ratio schedule while confined

FIG. 16. Olfactory examination of a female rhesus monkey's rear by the male during a test. The male has seized the female by the tail and pulled her toward him. Note the male's facial expression and protruding lips (cf. "Flehmen" face—van Hooff, 1962). From Keverne and Michael (1971).

in primate chairs. Whe high rates of pressing were consistently achieved, the male was transferred to the free-cage situation and provided with a lever. Reinforcement was initially food and subsequently a female rhesus monkey. When a male pressed to criterion, it obtained access to a female for 1 hour during which their behavior was observed. If a male failed to press to criterion in 30 minutes, the test was terminated and later the same day the pair was brought together in a nonoperant situation for a 1-hour mating test to observe their behavior. In these experiments, unlike the earlier ones described, the females were untrained and not able to operate the servo motor and raise the partition between the cages. When a male rhesus monkey has had a rewarding (or unrewarding) sexual experience with a particular female, it appears to be remembered and may influence behavior in subsequent tests. In order to circumvent this effect, experiments were arranged so that females were initially in an unattractive, unreceptive state. Males did not work consistently in order to obtain access to an ovariectomized, untreated female, but regular, high rates of pressing occurred for access to a female treated with estrogen. After a suitable period of testing (about 30 days) with an ovariectomized, untreated partner, the male was made anosmic by inserting into the nasal olfactory area plugs consisting of gauze impregnated with bismuth iodoform paraffin paste. These plugs were inserted so as to leave a clear nasal airway (Fig. 17) permitting the male to breathe normally but completely blocking his capacity to detect odors. When the male was thus rendered anosmic and the female partner was injected with estrogen, the male's operant behavior remained quite unchanged and he appeared unaware that the hormonal status of his partner had been altered. However, when the nasal plugs were removed and the olfactory acuity of the male returned (which takes a few days), they commenced lever-pressing vigorously and obtained access to the female partner. It appears, therefore, that males are unable to detect the change in the endocrine condition of their female partners when anosmic but do so immediately their sense of smell has been restored. The experiments, briefly outlined here, are difficult to conduct and can be interpreted only when a series of adequate control procedures are undertaken (Michael and Keverne, 1968). It would be inappropriate to go into further details here, but investigations of this type have led us to the view that an olfactory influence emanating from females treated with estrogen can affect the behavior of the male partner. The substances involved would therefore possess the properties of olfactorily acting pheromones (Michael, 1969b).

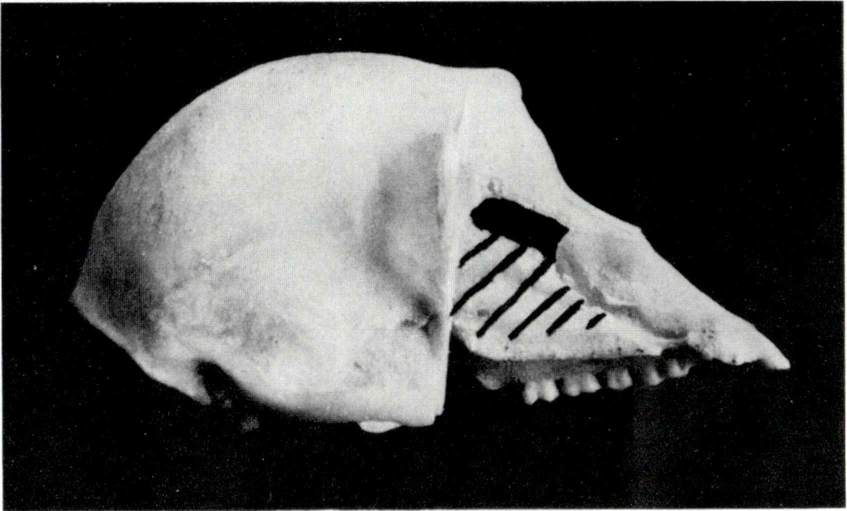

Fig. 17. The skull of a rhesus monkey. The maxilla has been removed revealing the nasal septum, and the area marked in black shows where the nasal plugs are sited. They block the olfactory area of each nostril but leave a clear nasal airway (hatched area) permitting the animal to breathe normally through the nose. When the nasal plugs are removed, the anosmia is reversed.

The possible sites of production of sex-attractant pheromones by females has been studied further by collecting vaginal secretions from ovariectomized, estrogen-treated females (donors) and transferring them to ovariectomized, untreated females (recipients) in order to test the donor secretions for biological activity. Baseline operant and sexual behavior was obtained from six pairs in 71 tests over a 4-week period when normal saline was rubbed onto the areas of the sexual skin of recipient females for control purposes. For the next 4-week period, 2.5 μg of estradiol in oil was rubbed onto the area of the sexual skin, and changes in sexual skin coloration induced by the estrogen treatment were measured by the Munsell system. This latter treatment was withdrawn, and 3 weeks later, donor vaginal secretions collected from estrogen-treated females were then applied to the sexual skin of recipients. During the entire pretreatment period, males pressed for access to recipient females in 11 out of 50 tests, and 4 ejaculations occurred. The application of estradiol to the sexual skin of recipient females produced intense redness but no changes in the male's lever-pressing (9 out of 52 tests) or in sexual behavior (2 ejaculations). During the application of donor vaginal secretions males pressed in 17 out of 37 tests and had 31 ejaculations; these increases were highly significant when compared with the pretreatment condition in each case ($P < 0.001$). The onset of lever-pressing by males within a few minutes of the application of donor secretions to the sexual skins of ovariectomized females for which they had not previously been pressing was a dramatic behavioral change and strongly indicated that estrogen-stimulated secretions possessed sex-attractant properties (Michael and Keverne, 1970). The substance(s) involved, which we have called "copulin," powerfully stimulated both the male's interest in the female, as assessed by his lever-pressing performance, and also overt male sexual behavior, as assessed by his mounting activity and ejaculations. This change occurred in the absence of any appropriate behavioral cues emanating from the female: it should be kept in mind that the recipient females, ovariectomized some 6 months previously, were totally unreceptive and certainly did not encourage mating. These experiments also demonstrated that sexual skin coloration was not an important visual signal in this species since its intense redness failed to stimulate male mounting behavior.

IX. Chemistry of Vaginal Secretions

In order to determine the chemical nature of the substances in estrogen-stimulated vaginal secretions responsible for these powerful behavioral effects, it was necessary to use extraction and fractionation procedures in conjunction with biobehavioral assay methods of the type just described. Vaginal secretions were collected from estrogen-treated, donor females by lavage with a vaginal pipette containing 1 ml of distilled water. After buffering at pH 4.5 with sodium dihydrogen phosphate, samples were extracted with 2 ml of diethyl ether. Two to 3 minutes before a behavioral test, a recipient female was trapped in a net and the 2 ml of ether extract was applied to the area of the sexual skin. Thus, the ether-soluble material present in a single vaginal washing from a donor female was transferred to the sexual skin of a recipient. Diethyl ether alone (2 ml) was applied in the same way for control purposes immediately before tests during the pretreatment period. When these ether extracts were applied to the sexual skin of the recipient, test females, the changes in the males' behavior were quite dramatic (Fig. 18). In all five pairs, a marked and immediate stimulation of the males' sexual behavior resulted. In the 41 pretreatment tests there were 22 mounting attempts compared with 313 in the 33 tests in which ether extracts of secretions were applied [t test, 10.05, degrees of freedom (df) 72, $P < 0.001$]. The onset of ejaculation was more variable: in two pairs, it occurred in the first test with treatment; in two other pairs, during the second test with treatment, and in the fifth pair, although ejaculation was initially impossible (see below), masturbation to ejaculation occurred in the first test with treatment. There were no ejaculations during the 41 pretreatment tests compared with 33 ejaculations during the 33 treatment tests (χ^2 test, 47.75, df 1, $P < 0.001$). The recipient females were unreceptive throughout the entire experimental period (Keverne and Michael, 1971), and the atrophic condition of their genital tracts was demonstrated by the difficulty with which males obtained intromission and the unusually high numbers of nonintromitted mounts. In rhesus monkeys, ejaculation with a female does not occur unless intromission is obtained, and in pair 38–71 (Fig. 18), intromission and ejaculation did not take place until the last treatment test. Despite the severe and prolonged difficulty with intromission (28 nonintromitted mounts), the male's sexual motivation was tremendously enhanced by the application of ether extracts, and ejaculation by masturbation occurred in every test until intromission could be obtained. There can be no doubt from these studies that these extracts contained effective sex-attractant pheromones.

We demonstrated the acidic nature of the behaviorally active constituents in extracts by applying to the sexual skin of recipient females a solution prepared by treating the ether extract with 2 ml of 0.01 N sodium hydroxide, acidifying the alkaline layer, and reextracting into ether: this extract retained behavioral acitivity, a result which suggested

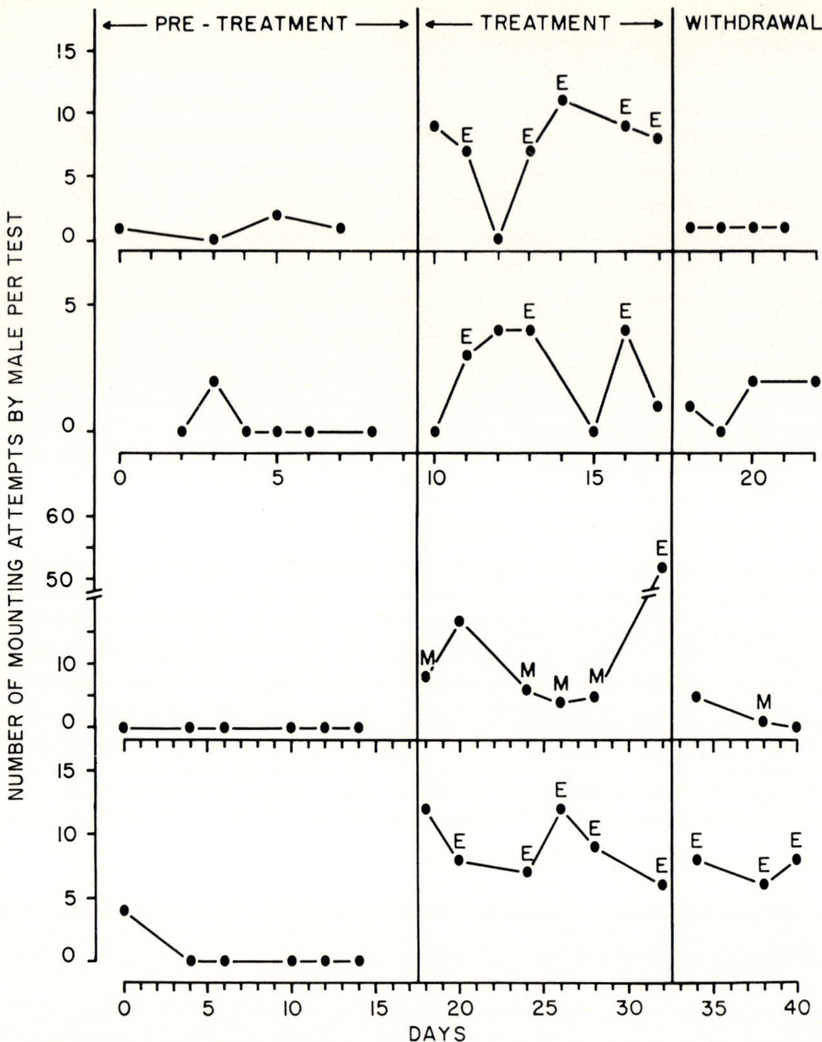

FIG. 18. Changes in the behavior of male rhesus monkeys when ether extracts of vaginal secretions, collected from estrogen-treated donor females, were applied to the sexual skin of ovariectomized, untreated, test females (recipients). There was an increase in male mounting attempts, in ejaculations and in masturbation (5 pairs, 113 tests). ●E = one ejaculation in the test; ●2E = two ejaculations in a test; ●M = masturbation to ejaculation; ○ = tests without ejaculation.

that the active constituents were acidic. Ether-soluble neutral and basic components, notably cholesterol, were removed by this procedure. Fractionation of the acidic components was carried out by ion-exchange chromatography. A pool of 40 vaginal secretion extracts was washed with 5 ml of 0.01 N sodium hydroxide, titrated to pH 7.4 and eluted through a column of DEAE-cellulose (Whatman DE-52) with 50 mM (pH 7.4) sodium phosphate buffer. The eluate was automatically

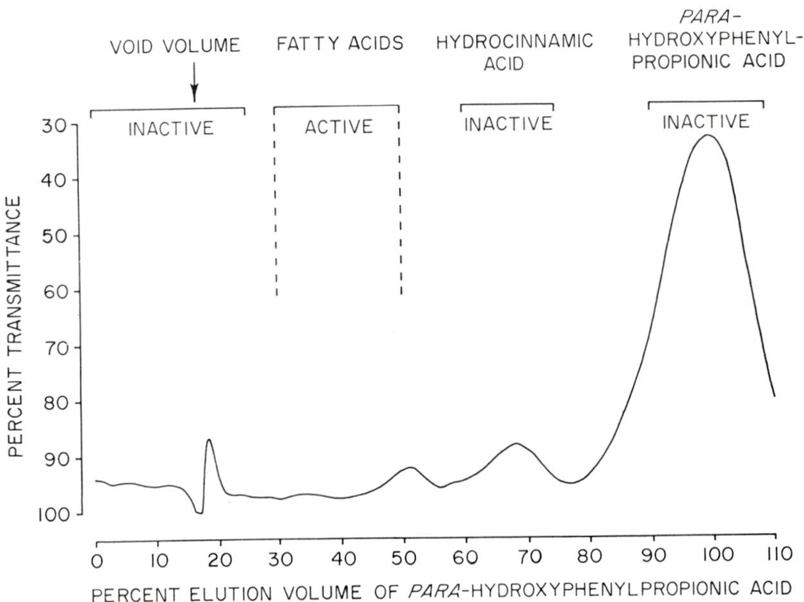

Fɪɢ. 19. Elution profile at 280 nm on DEAE-cellulose of alkaline reextracts of vaginal secretions from ovariectomized, estrogen-treated rhesus monkeys. The fractions indicated above were tested for behavioral activity and only that containing fatty acids was found to be active. This fraction was subjected to further analysis by gas chromatography and mass spectrometry (Curtis *et al.*, 1971).

collected into fractions which were then combined according to their elution volumes relative to that of a phenolic component, detected by continuous monitoring for ultraviolet absorbance at 280 nm, and now identified as p-hydroxyphenylpropionic acid (HPPA). Fractions were acidified, extracted into ether, and tested for behavioral activity: that corresponding to 30–50% of the elution volume of HPPA was found to be highly active (Fig. 19). The elution volume of this fraction corresponded to that of short-chain fatty acids, and ether extracts of this behaviorally active fraction were further examined by analytical gas chromatography. After preliminary screening with different gas–liquid chromatographic media, Carbowax 20M-terephthalic acid columns and temperature gradients between 50 and 200°C were used to study further the acidic, behaviorally active components of secretions. Chromatograms of extracts of individual secretions collected from five ovariectomized, untreated females showed in each case that the amounts of volatile

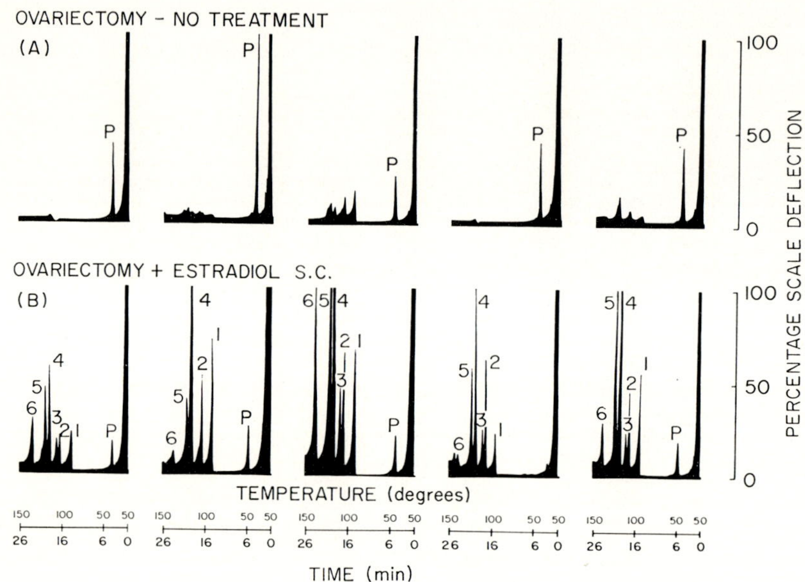

FIG. 20. Gas chromatograms of extracts of rhesus monkey vaginal secretions. (A) Chromatograms of secretions collected from five ovariectomized, untreated females. The volatile acid content of the secretions was low (0–3.3 μg per collection), and the four, from recipient females, that were tested (four at left) were without any behavioral activity (Fig. 21, pretreatment). (B) Chromatograms of secretions collected from five ovariectomized females during treatment with estradiol. The volatile acids in the secretions showed an 8-fold increase (26.0–53.0 μg per collection), and the four from donor females that were tested (four at left) showed marked behavioral activity (Fig. 21, treatment). The right-hand chromatograms (A and B) are from the same animal before and during treatment with estradiol. Peak P, after the solvent front, is authentic n-pentanol added to the extracting ether as a marker. From Michael et al. (1971).

components were low (Fig. 20A). Chromatograms of extracts of individual secretions collected from five ovariectomized females during treatment with estradiol showed that amounts of volatile components (peaks 1–6) were at last eight times greater (Fig. 20B). The righ-hand chromatograms in Fig. 20A and B are from the same animal before and during treatment with estradiol. To determine whether the increased production of these constituents by females during treatment with estradiol was responsible for the changes in the behavioral properties of the secretions, we used an effluent splitting device to trap peaks 1–5 (80–130°C) from the gas chromatograph into ice-cold ether. The material for this trapping procedure was obtained from a pool of 48 vaginal washings collected from three donor females (Michael *et al.*, 1971). The trapped fractions were then applied daily to the sexual skin of four recipient females, each paired with a different male, and an immediate stimulation of the sexual activity of the males resulted in each case (Fig. 21). The four males made a total of ten mounting attempts during 22 tests before treatment compared with 213 mounting attempts during 26 tests with treatment (t test, $P < 0.001$). There were no ejaculations during tests before treatment compared with 14 during tests with treatment (χ^2 test, $P < 0.001$). In three pairs, stopping treatment resulted in an immediate return to the pretreatment baseline, but this did not occur in the remaining pair until the fifth test after stopping treatment; similar carry-over effects have been encountered previously (Michael and Keverne, 1970). By these means it was established that highly active copulins which sexually stimulate these male primates were trapped with peaks 1 through 5. The retention times of the five volatile components, 86, 131, 152, 203, and 251 U were compared with those of the following acids: acetic (87 U), propionic (131 U), isobutyric (152 U), *n*-butyric (203 U), isovaleric (251 U). A synthetic mixture of these five acids coinjected with an equal volume of the natural extract showed no separation between the two mixtures, and confirmation of the identity of the first five components was obtained by preparative gas chromatography and mass spectrometry (Curtis *et al.*, 1971). Five aliquots (2 µl each) of the active ether extract were separately injected onto the same column fitted with a preparative attachment (10:1 effluent split). The effluent corresponding to each of the five peaks was collected individually in melting point U-tubes at −70°C: these were cut and installed in the direct insertion probe of an AEI-MS9 mass spectrometer. The resultant mass spectra were compared with those of authentic samples obtained in identical fashion to avoid differences resulting from the presence of solvent or column impurities. These spectra established the identification of the five volatile constituents of the active fraction. Peak 6 has now been identified as isocaproic acid by analytical gas chromatography.

To ascertain whether the five aliphatic acids in the behaviorally active fraction were indeed responsible for its sex-attractant properties, a mixture of authentic acids was made up to match the concentrations present in a pool of 24 washings collected from ovariectomized, estrogen-treated

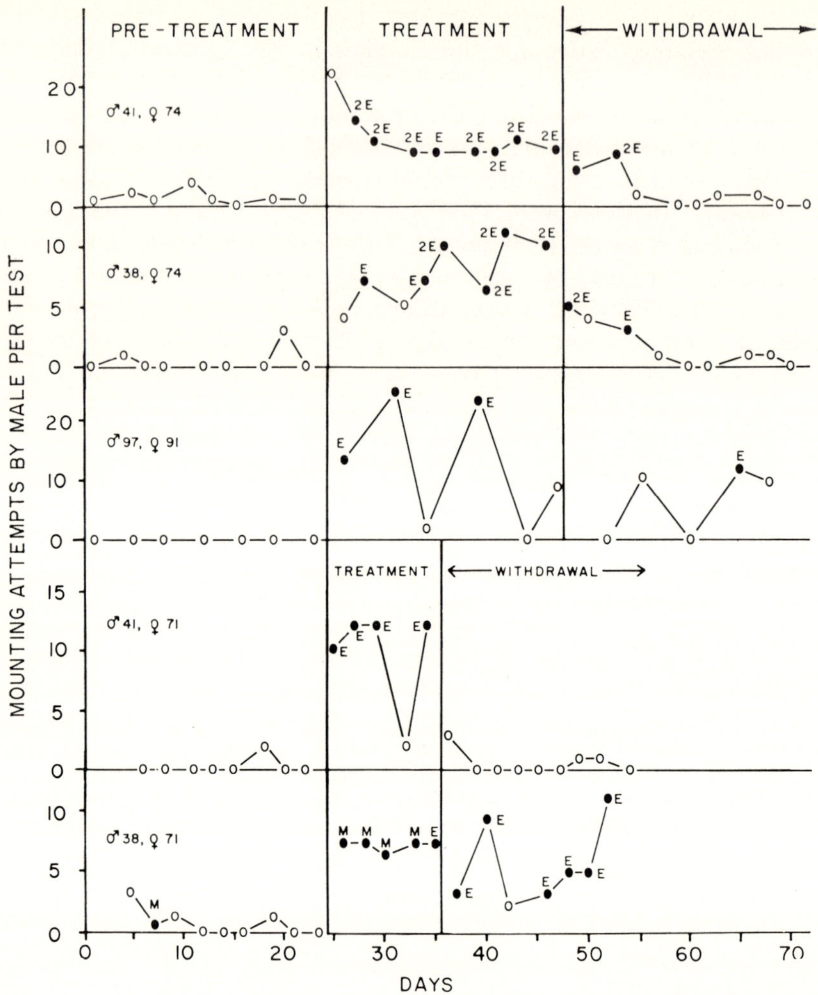

FIG. 21. Sexual stimulation of male rhesus monkeys by components of vaginal secretions fractionated by gas chromatography. In tests with four males each paired with a different ovariectomized (recipient) famale, the application to the latter's sexual skin of material collected by trapping from the gas chromatograph resulted in a marked stimulation of the sexual behavior of their male partners. E = ejaculation; M = masturbation to ejaculation. Time scale of lower two pairs, which were tested on alternate days, is half of upper two pairs, which were tested daily.

female rhesus monkeys. This synthetic mixture contained 9.2 μg of acetic, 8.8 μg of propionic, 4.2 μg of isobutyric, 12.8 μg of *n*-butyric, and 8.3 μg of isovaleric acids per milliliter of ether. A 2-ml sample of the mixture, containing about 1.5 times the aliphatic acid content of a single vaginal washing, was applied to the sexual skin of an ovariectomized female

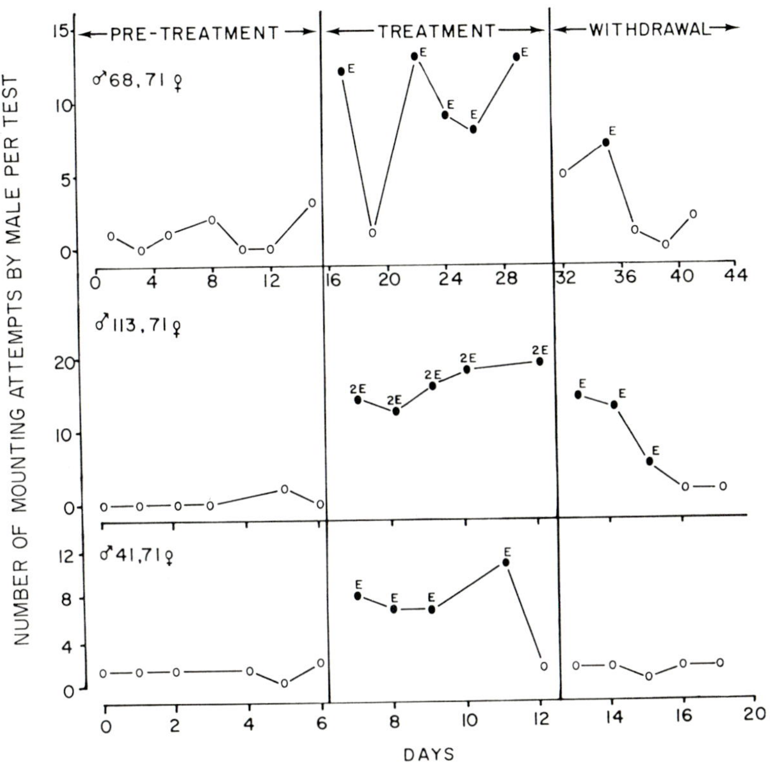

FIG. 22. The sexual stimulation of three male rhesus monkeys by the application of a synthetic mixture of authentic fatty acids to the sexual skin of an overiectomized, recipient female: a significant increase in mounting activity and in ejaculations occurred. ●E = one ejaculation in the test; ●2E = two ejaculations in the test; ○ = test without ejaculation.

and tested for behavioral activity (Fig. 22). The number of mounting attempts made by the three males increased from 15 during 19 pretreatment tests, when ether was applied for control purposes, to 170 during 16 tests when applications of the synthetic mixture of aliphatic acids were made (*t* test, $P < 0.001$): there were no ejaculations during the pretreatment period compared with 19 during treatment (χ^2 test,

$P < 0.001$). The identification of these volatile aliphatic acids as naturally occurring constituents of vaginal secretions in rhesus monkeys, and the demonstration that a synthetic mixture of authentic acids also possesses powerful sex-attractant properties, when taken together, provide rather strong evidence that the attractant properties of estrogen-stimulated vaginal secretions depend, at least in part, on their acid content.

The occurrence of these acids in single vaginal washings, using 1 ml of distilled water, in several primate species including the human is shown in Table I. In the rhesus monkey and anubis baboon, where an adequate number of secretions have been examined, data are based on samples obtained at all stages of the menstrual cycle and include both very low values (<1 μg) and very high values (about 760 μg) of total acids. When the acid content of rhesus monkey samples was examined in relation to the phases of the menstrual cycle, the mean for the middle two quartiles was 166.3 ± 22.8 μg ($n = 45$) compared with the mean for the first and last quartiles of 76.9 ± 19.9 μg ($n = 40$). In anubis baboons, changes in acid content during the menstrual cycle were even more conspicuous (Fig. 23). The mean for secretions collected during the 5 days preceding deflation of the sexual skin was 18.5 ± 5.1 μg, a value that was significantly higher than those obtained just before or just after menstruation.

X. Effects of Female Baboon Secretions on Male Rhesus Monkey Behavior

An examination of Table I indicates that the content of aliphatic acids in vaginal secretions was similar, although by no means identical, in different primate species. It was of interest therefore to ascertain whether interspecies behavioral effects could be obtained by collecting samples from one species, applying them to the sexual skin of another, and thereby influencing their male partners' behavior. Vaginal secretions were collected from two female anubis baboons throughout the course of three menstrual cycles. To test for their sex-attractant effects on rhesus males, ovariectomized rhesus females were trapped in a net and baboon secretions were applied to the sexual skin immediately before a behavioral test with a male rhesus monkey. In the pretreatment period, mounting attempts by males were low (11 in 45 tests) and no ejaculations occurred. During tests when baboon vaginal secretions were applied, mounting attempts increased (363 in 68 tests) and 19 ejaculations occurred. The increase in mounting attempts was highly significant ($\chi^2 = 26.1$, $P < 0.001$), as was the increase in ejaculations ($\chi^2 = 13.9$, $P < 0.001$). These results indicated that baboon secretions were be-

TABLE I

Occurrence of Volatile Fatty Acids in Several Species of Primates[a]

Primate species	No. of samples	Acetic acid	Propionic acid	Isobutyric acid	n-Butyric acid	Isovaleric acid	Isocaproic acid	Total acids
Rhesus monkey	231	36.0 ± 4.1	12.0 ± 0.8	2.7 ± 0.3	35.3 ± 2.0	6.3 ± 0.6	4.6 ± 0.5	96.8 ± 6.8
Anubis baboon	42	6.4 ± 1.5	2.1 ± 0.4	0.6 ± 0.2	4.0 ± 0.9	0.7 ± 0.2	0.4 ± 0.2	14.1 ± 2.8
Patas monkey	5	9.3 ± 1.9	19.5 ± 7.3	26.8 ± 14.7	77.7 ± 47.6	75.5 ± 45.5	9.8 ± 2.4	217.9 ± 116.5
Pigtail macaque	5	29.2 ± 9.5	10.0 ± 4.0	6.3 ± 3.0	34.2 ± 19.0	12.0 ± 6.1	1.6 ± 0.5	93.2 ± 35.0
Macaca fascicularis	6	41.7 ± 13.4	30.0 ± 10.2	10.9 ± 2.4	37.8 ± 8.8	17.4 ± 5.2	6.3 ± 2.2	144.0 ± 34.2
Squirrel monkey	1	36.7	31.0	19.1	2.8	24.8	1.4	115.8
Human	6	9.9 ± 3.1	6.9 ± 2.9	0.6 ± 0.5	6.6 ± 1.5	2.1 ± 1.0	0.6 ± 0.5	26.6 ± 2.1

[a] Values are the means obtained from single vaginal washings using 1 ml of distilled water.

haviorally active in rhesus monkeys and, very interestingly, the rhesus male showed a well-marked mounting cycle with a maximum that corresponded to the time of maximum sexual swelling of the female baboon from which collections had been made. Thus, the behavior of the rhesus males with rhesus females was being driven by the increased aliphatic acid content present in baboon secretions near mid-cycle (Fig. 23)

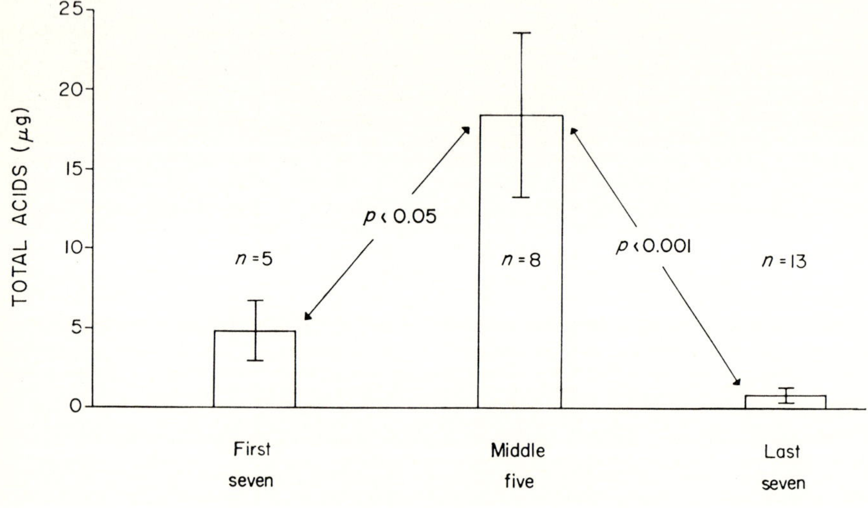

FIG. 23. Changes in the aliphatic acid content of vaginal secretions collected throughout the course of three menstrual cycles of anubis baboons (*Papio anubis*). There was a significant increase in total acids (Mann-Whitney U-test) in the 5 days immediately before deflation of the sexual skin as compared with the periods immediately after and before the first day of menstruation. Total acid = acetic, propionic, isobutyric, *n*-butyric, isovaleric, and isocaproic acids. Vertical bars give standard errors of means.

(Michael, Keverne, and Bonsall, unpublished observations). It would therefore appear to be a practical possibility to use rhesus monkey pairs as a behavioral testing system for sex-attractant pheromones in other primate species including the human.

XI. Source of Vaginal Pheromones

Although these aliphatic acids are readily recovered from the vaginas of primates by lavage, it is not known whether their source is primarily the cervical or vaginal glands, the glands about the introitus or a transudate through the vaginal wall. However, there is evidence that microbial action plays a part in their production. Amounts of behaviorally active, volatile acids in rhesus monkey vaginal secretions increased when they were incubated at 37°C *in vitro*. Concentrations increased as incubation continued in a nonlinear manner typical of bacterial action (Fig. 24). Moreover, autoclaving before incubation and the addition of penicillin, but not of streptomycin, to the media inhibited the production of acids. Production could be restarted in autoclaved

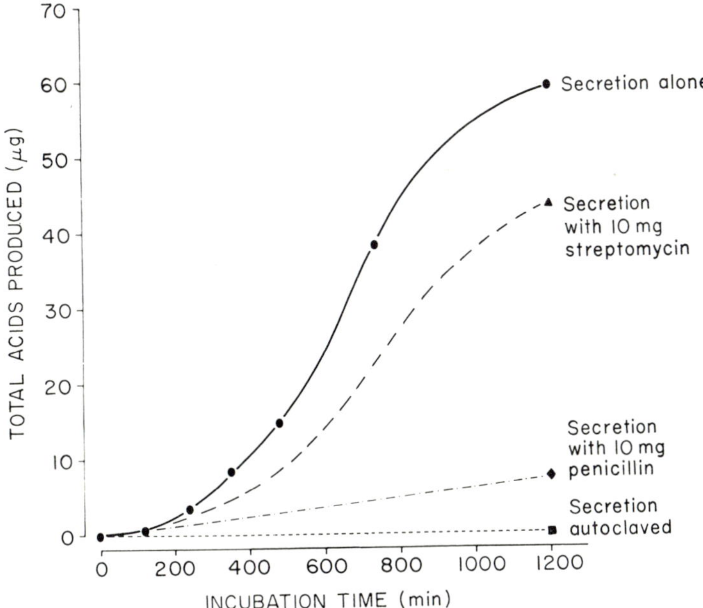

Fig. 24. Production of short-chain aliphatic acids by incubation of vaginal secretions *in vitro* at 37°C. The initial increase in fatty acid content is typical of that of a growing microbial colony. Autoclaving the secretions and also the addition of penicillin, but not of streptomycin, to the media inhibited acid production.

secretions by inoculation with fresh secretions, and acid production could also be obtained from tryptone soya broth by inoculation with fresh vaginal secretions: their continued production following serial inoculations also indicated the growth of microbial organisms (Bonsall and Michael, 1971). The turbid supernatant obtained when secretions were centrifuged at low speeds produced acids from the pellet after this had been autoclaved or extracted with hot ethanol. High speed centrifugation produced a sediment which only synthesized acids when incubated with a heat-stable factor in the supernatant. It seemed probable, therefore, that the production of these acidic pheromones depended upon the bacterial content of the vagina, and that one means by which the ovarian hormones may regulate acid production *in vivo* is by determining the availability of substrate. If this proves to be the case, it indicates the existence of a rather novel symbiotic relationship, but one that might be quite widely represented in nature since similar aliphatic acids are present in the preputial secretions of the boar (Patterson, 1968b), and in the anal gland secretions of cats (Michael—unpublished observations).

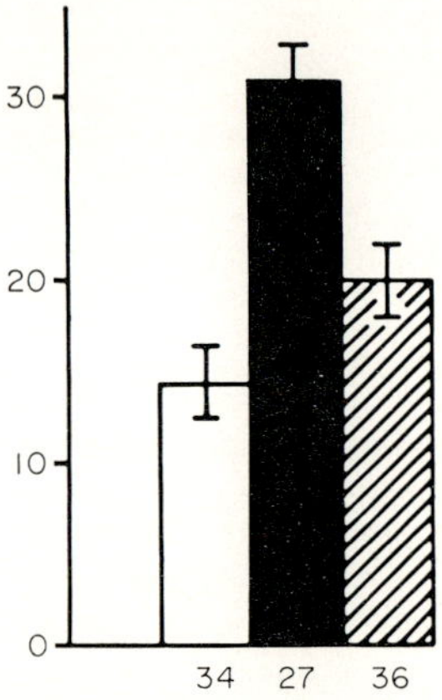

Fɪɢ. 25. Treatment of ovariectomized female rhesus monkeys with ether extracts of vaginal secretions collected from estrogen-treated donors results in a highly significant increase in the time spent by males in grooming them. The increased grooming of intact females by males near mid-cycle may therefore be mediated by pheromones. White column, pretreatment; black column, during treatment; hatched column, after withdrawing treatment. Numbers below columns = number of tests.

XII. Pheromones and Grooming Behavior

Grooming behavior between consorts is intimately related to the sexual interaction of the pair, but it also constitutes an aspect of the social behavior of many primate species. Earlier studies have shown (Michael *et al.*, 1966) that the time spent by the male in grooming the female increases very considerably near mid-cycle compared with the grooming times near menstruation: the changes in the male's grooming activity are a sensitive indicator of his response to the female's hormonal state. An increase in male grooming is one of the earliest behavioral changes observed when ovariectomized females are treated with small doses of estrogen subcutaneously. Figure 25 shows that treatment of ovariectomized females with ether extracts of vaginal secretions also increased male grooming of the females to the same extent (analysis of variance $F = 17.8$, df 2, 94 $P < 0.001$). Thus the increased grooming behavior of intact females by males near mid-cycle may well be mediated by the production of pheromones.

XIII. Conclusions

The complex behavioral interactions of pairs of adult rhesus monkeys of opposite sexes have now been studied under controlled laboratory

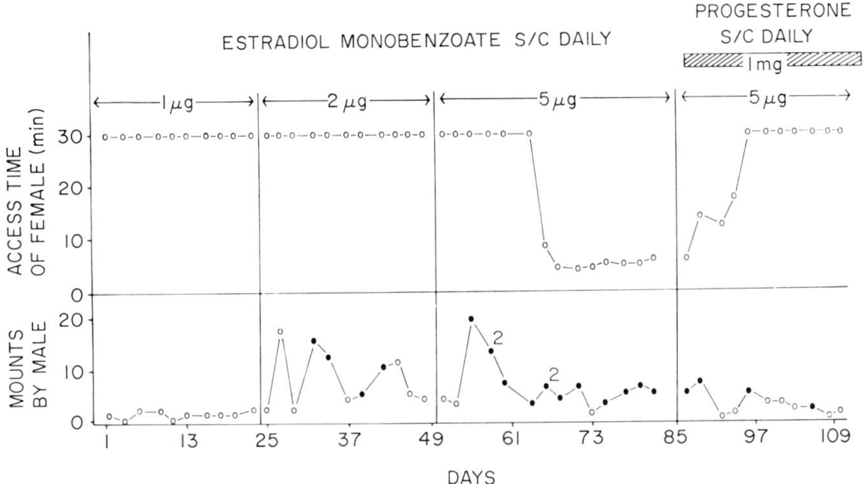

Fig. 26. Effects on female operant behavior and on male mounting of treating an ovariectomized female rhesus monkey with small, increasing doses of estradiol and with a small dose of progesterone: 2 µg of estradiol per day stimulated the male and rendered the female more attractive without increasing her receptivity (lever-pressing for the male). At 5 µg per day estradiol, her operant behavior was stimulated, and the addition of progesterone decreased both her operant behavior and the male's mounting behavior; that is, both receptivity and attractiveness were diminished. ●, test with one ejaculation; ●2, test with two ejaculations; ○, test without ejaculation.

conditions. The sexual activity of the male is clearly influenced by the hormonal status of the female, and the behavioral rhythms that are related to the phases of the menstrual cycle are an example of this. The female's hormonal state influences both her sexual receptivity and her sexual attractiveness and, as a consequence, the level of the male's sexual activity. Figure 26 shows that treating an ovariectomized female with 1 µg of estradiol per day is without effect on the behavior of both male and female; it is below threshold. With 2 µg of estradiol per day, stimulation of mounting behavior by the male occurs without there being any effect on the female's motivation as expressed by her operant behavior. This dose of hormone is sufficient to increase female attractiveness by inducing the production of pheromones that stimulate the male's sexual activity, *without* affecting her receptivity and willingness to work for access to the male. Increasing the dose of estradiol to 5 μg per day stimulated female receptivity, and this change in motivation resulted in the onset of vigorous lever pressing: female attractiveness remained high, and ejaculations occurred more consistently than at the lower dose level. The additional administration of 1 mg of progesterone per day resulted in a rapid decline in female motivation (lever-pressing decreased) and in a marked loss of attractiveness (male mounting activity declined, and ejaculations ceased). Thus, treating ovariectomized females with physiological doses of estrogen and progesterone can reproduce exactly the behavioral events seen during the menstrual cycles of intact females. Although androgens increase sexual receptivity and sexual motivation in female rhesus monkeys, their precise physiological role remains to be established. It has been clearly demonstrated that olfactory signals are important in determining the behavioral interactions of these primates. Their vaginal secretions contain a mixture of short-chain aliphatic acids that stimulate the sexual activty of the males, this effect depends upon olfaction, and these sex-attractants can be regarded as pheromones. Since these acidic pheromones are present in several primate species including the human, it will obviously be of interest to determine their precise role in these species also.

ACKNOWLEDGMENT

Grateful acknowledgment is made to the Medical Research Council, National Institute of Mental Health (MH-10002), Population Council and Foundations' Fund for Research in Psychiatry for grants to support the original work described here. We wish to acknowledge the generous help of Miss Judy Welegalla, Dr. Graham Saayman, Dr. J. Herbert, Dr. Margo Wilson, Dr. T. M. Plant, and Miss Rosemary Evans in the conduct of these experiments. Mr. Michael Sullivan and Mr. Francis Price are thanked for conscientious care of the animals.

REFERENCES

Abel, S. (1945). *Amer. J. Obstet. Gynecol.* **49,** 327.
Altmann, S. A. (1962). *Ann. N.Y. Acad. Sci.* **102,** 338.
Ball, J., and Hartman, C. G. (1935). *Amer. J. Obstet. Gynecol.* **29,** 117.
Beach, F. A., and Gilmore, R. W. (1949). *J. Mammal.* **30,** 391.
Beach, F. A., and Merari, A. (1970). *J. Comp. Physiol. Psychol.* **1,** 1.
Berliner, V. R. (1959). *In* "Reproduction in Domestic Animals" (H. H. Cole and P. T. Cupps, eds.), p. 267. Academic Press, New York.
Bolwig, N. (1959). *Behaviour* **14,** 136.
Bonsall, R. W., and Michael, R. P. (1971). *J. Reprod. Fert.* **27,** 478.

Bruce, H. M. (1970). *Brit. Med. Bull.* **26**, 10.

Carpenter, C. R. (1934). *Comp. Psychol. Monogr.* **10**, 1.

Carpenter, C. R. (1935). *J. Mammal.* **16**, 171.

Carpenter, C. R. (1942a). *J. Comp. Psychol.* **33**, 113.

Carpenter, C. R. (1942b). *J. Comp. Psychol.* **33**, 143.

Carr, W. J., Loeb, L. S., and Wylie, N. R. (1966). *J. Comp. Physiol. Psychol.* **62**, 336.

Carter, A. C., Cohen, E. J., and Shorr, E. (1947). *Vitam. Horm. (New York)* **5**, 317.

Curtis, R. F., Ballantine, J. A., Keverne, E. B., Bonsall, R. W., and Michael, R. P. (1971). *Nature (London)* **232**, 396.

Everitt, B. J., and Herbert, J. (1969). *Nature (London)* **222**, 1065.

Foss, G. L. (1951). *Lancet* **1**, 667.

Goodall, J. (1965). *In* "Primate Behaviour" (I. DeVore, ed.), p. 425. Holt, New York.

Greenblatt, R. B., Mortara, F., and Torpin, R. (1942). *Amer. J. Obstet. Gynecol.* **44**, 658.

Hall, K. R. L., and DeVore, I. (1965). *In* "Primate Behaviour" (I. DeVore, ed.), p. 53. Holt, New York.

Harris, G. W., Michael, R. P., and Scott, P. P. (1958). *In* "The Neurological Basis of Behaviour" (G. E. W. Wolstenholme and C. M. O'Connor, eds.), p. 236. Churchill, London.

Hartman, C. G. (1932). *Contrib. Embryol.* **23**, 1.

Hopper, B., and Tullner, W. W. (1970). *Endocrinology* **86**, 1225.

Hotchkiss, J., Atkinson, L. E., and Knobil, E. (1971). *Endocrinology* **89**, 177.

Imanishi, K. (1963). *In* "Primate Social Behavior" (C. H. Southwick, ed.), p. 68. Van Nostrand, Princeton, New Jersey.

Jay, P. (1965). *In* "Primate Behaviour" (I. DeVore, ed.), p. 197. Holt, New York.

Kelley, R. B. (1937). *Bull. Counc. Sci. Ind. Res. Aust.* **112**.

Keverne, E. B., and Michael, R. P. (1971). *J. Endocrinol.* **51**, 313.

Kummer, H., and Kurt, F. (1963). *Folia Primat.* **1**, 4.

Le Magnen, J. (1952). *Arch. Sci. Physiol.* **6**, 295.

Lindsay, D. R. (1965). *Anim. Behav.* **13**, 75.

Loesser, A. A. (1940). *Brit. med. J.* **1**, 479.

Michael, R. P. (1961). *In* "Regional Neurochemistry" (S. S. Kety and J. Elkes, eds.), p. 465. Pergamon, Oxford.

Michael, R. P. (1964). *In* "Pathology and Treatment of Sexual Deviation" (I. Rosen, ed.), p. 24. Oxford Univ. Press, London and New York.

Michael, R. P. (1965). *Proc. Roy. Soc. Med.* **58**, 595.

Michael, R. P. (1968a). *In* "Endocrinology and Human Behaviour" (R. P. Michael, ed.), p. 69. Oxford Univ. Press, London and New York.

Michael, R. P. (1968b). *In* "Studies in Psychiatry" (D. L. Davies and J. M. Shepherd, eds.), p. 318. Oxford Univ. Press, London and New York.

Michael, R. P. (1969a). *In* "Progress in Endocrinology" (C. Gual, ed.), p. 302. Excerpta Medica Foundation, Amsterdam.

Michael, R. P. (1969b). *In* "Recent Advances in Primatology" (C. R. Carpenter. ed.), Vol. 1, p. 101. Karger, Basel.

Michael, R. P. (1971). *In* "Frontiers in Neuroendocrinology" (L. Martini and W. F. Ganong, eds.), p. 359. Oxford Univ. Press, London and New York.

Michael, R. P., and Keverne, E. B. (1968). *Nature (London)* **218**, 746.
Michael, R. P., and Keverne, E. B. (1970). *Nature (London)* **225**, 84.
Michael, R. P., and Saayman, G. (1967a). *Anim. Behav.* **15**, 460.
Michael, R. P., and Saayman, G. (1967b). *J. Comp. physiol. Psychol.* **64**, 213.
Michael, R. P., and Saayman, G. (1968). *J. Endocrinol.* **41**, 231.
Michael, R. P., and Welegalla, J. (1968). *J. Endocrinol.* **41**, 407.
Michael, R. P., and Zumpe, D. (1970a). *J. Reprod. Fert.* **21**, 199.
Michael, R. P., and Zumpe, D. (1970b). *Behaviour* **36**, 168.
Michael, R. P., and Zumpe, D. (1971). *In* "Comparative Reproduction of Nonhuman Primates" (E. S. E. Hafez, ed.), p. 205. Thomas, Springfield, Illinois.
Michael, R. P., Herbert, J., and Welegalla, J. (1966). *J. Endocrinol.* **36**, 263.
Michael, R. P., Herbert, J., and Welegalla, J. (1967a). *J. Endocrinol.* **39**, 81.
Michael, R. P., Saayman, G., and Zumpe, D. (1967b). *J. Endocrinol.* **39**, 309.
Michael, R. P., Saayman, G., and Zumpe, D. (1967c). *Nature (London)* **215**, 554.
Michael, R. P., Saayman, G., and Zumpe, D. (1968). *J. Endocrinol.* **41**, 421.
Michael, R. P., Keverne, E. B., and Bonsall, R. W. (1971). *Science* **172**, 964.
Patterson, R. L. S. (1968a). *J. Sci. Food Agr.* **19**, 31.
Patterson, R. L. S. (1968b). *J. Sci. Food Agr.* **19**, 38.
Plant, T. M., and Michael, R. P. (1971). *Acta Endocrinol. Suppl.* **155**, 69.
Signoret, J. P., and Mauleon, P. (1962). *Ann. Biol. Anim. Biochem. Biophys.* **2**, 167.
Signoret, J. P., and du Mesnil du Buisson, F. (1961). *Proc. 4th Int. Congr. Anim. Reprod.* p. 171.
Southwick, C. H., Beg, M. A., and Siddiqi, M. R. (1965). *In* "Primate Behavior" (I. DeVore, ed.), p. 111. Holt, New York.
Stephan, H. (1963). *Progr. Brain Res.* **3**, 111.
Stern, J. J. (1970). *Physiol. Behav.* **5**, 519.
Udry, J. R., and Morris, N. M. (1968). *Nature (London)* **220**, 593.
van Hooff, J. A. R. (1962). *Symp. Zool. Soc. London* **8**, 97.
Washburn, S. L., and DeVore, I. (1961). *Sci. Amer.* **204**, 62.
Wierzbowski, S., and Hafez, E. S. E. (1961). *Proc. 4th Int. Congr. Anim. Reprod.* p. 176.
Young, W. C., and Orbison, W. D. (1944). *J. Comp. Physiol. Psychol.* **37**, 107.
Zumpe, D., and Michael, R. P. (1968). *J. Endocrinol.* **40**, 117.
Zumpe, D., and Michael, R. P. (1970a). *Anim. Behav.* **18**, 293.
Zumpe, D., and Michael, R. P. (1970b). *Anim. Behav.* **18**, 11.

[The discussion for this chapter appears on page 754.]

Gonadal Hormones and Behavior of Normal and Pseudohermaphroditic Nonhuman Female Primates

R. W. GOY AND J. A. RESKO

*Wisconsin Regional Primate Research Center and Department of Psychology,
University of Wisconsin, Madison, Wisconsin; and Oregon Regional Primate
Research Center, Beaverton, Oregon and Department of Physiology,
University of Oregon Medical School, Portland, Oregon*

I. Estradiol, Progesterone, and the Behavior of Primates

In Old World monkeys and presumably in apes as well, the relationship between sexual behavior and the ovarian cycle is different from that which exists in lower mammals. In lower mammals sexual receptivity is limited to the brief period of estrus. During estrus the onset of sexual behavior is abrupt and is characterized by the display of highly specialized and stereotyped bodily postures. In rodents the display of these postures is sharply limited to the ovarian condition of preovulatory follicular swelling. The length of time, in hours or days, that sexual behaviors and receptivity are displayed is characteristic for a species, but in all species studied it is generally true with only rare exceptions that sexual behavior is limited to that stage of the cycle characterized by rapid follicular growth and the absence of luteal progesterone.

The endocrine-behavior relationship in lower primates, as represented by the lemurs and galagos, closely resembles the relationship in non-primate mammals. Some years ago a study was made in our laboratory of the sexual behavior and reproductive cycles of *Lemur catta* (Evans and Goy, 1968). Despite the existence of a relatively long ovarian cycle (33–41 days), as judged by recurrent vaginal or behavioral estrus, receptivity to the male is limited to a relatively brief period lasting only about 12 hours. More recently similar investigations have been conducted on *Galago crassicaudatus* (Eaton *et al.*, 1971). In that species vaginal estrus persists for approximately 12 days and can be differentiated easily by morphological criteria from the longer period of diestrus. The duration of vaginal estrus in the galago does not differ greatly from that reported for *Lemur catta*, nor does the overall length of the ovarian cycle. Sexual behavior is displayed throughout much of the period of vaginal estrus in the galago. In *Lemur catta*, however, sexual activity is limited to a single day or less. The relationship of the behavior to the endocrine events throughout the cycle of the galago is shown in Fig. 1. In most of the cycles studied, fertilization was prevented by interrupting copulation as soon as the male had achieved intromission, and there is no

evidence that this limited amount of coital stimulation in any way altered the lengths of the periods of vaginal estrus and diestrus. In both
animals illustrated, as in others studied, the onset of sexual behavior
is abrupt following 3–5 days of endogenous estrogen stimulation. Shortly
after onset, the intensity of sexual behavior reaches its peak, which

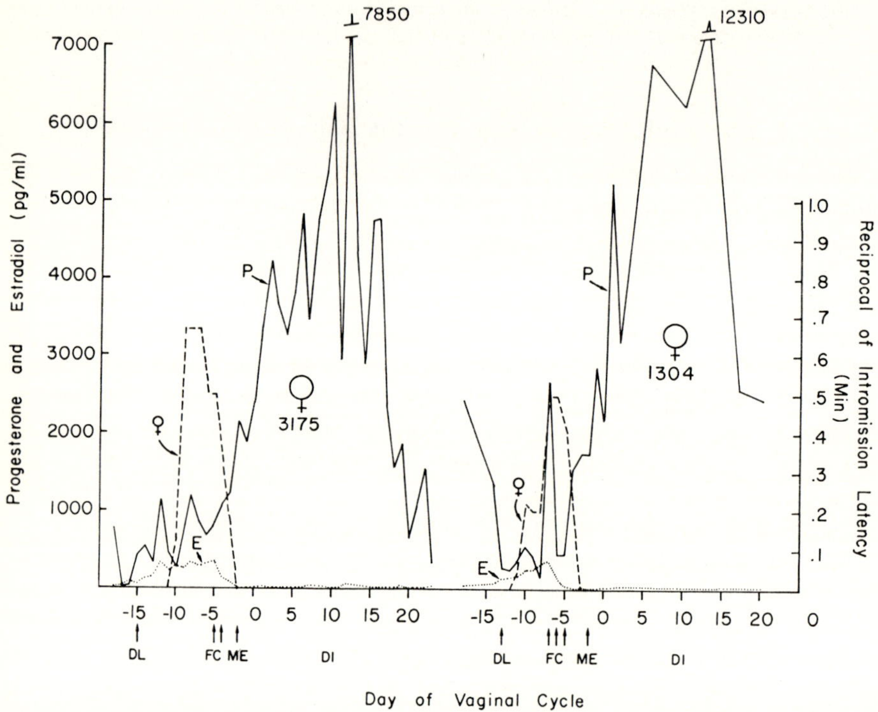

FIG. 1. The mean concentrations of progesterone (P) and estradiol (E) in systemic plasma from the galago and an index of female sexual receptivity (reciprocal
of intromission latency in minutes) across two females' vaginal cycles. Plotted
from the reappearance of leukocytes in the vaginal smears on day 0. DL = disappearance of leukocytes; FC = fully cornified; ME = metestrus; DI = diestrus.
Taken from Eaton *et al.* (1971).

is maintained for 1 or 2 days. The decline in sexual behavior follows,
but is not coincident with the decline in endogenous estrogen concentration. The long diestrum, lasting 23–25 days, is completely concordant
with the life-span of the spontaneously functional corpus luteum and
the associated high levels of progesterone in the systemic plasma.
Throughout the life-span of the corpus luteum, sexual behavior cannot
be elicited from the females by sexually vigorous males.

In contrast to the endocrine-behavior relationship which seems to be characteristic of prosimians as well as nonprimate mammals, the relationship for the Old World monkeys is markedly different. It has been known for a long time (Hartman, 1928; Maslow, 1936; Yerkes and Elder, 1936a,b; Young and Orbison, 1944) that many monkeys and apes copulate throughout the menstrual cycle. This departure from the copulatory patterns typical of lower mammals has been viewed as extreme by some investigators (Rowell, 1963), and the view has been expressed that in the rhesus monkey sexual behavior, far from being limited to the ovulatory period, was displayed at highest frequencies during infertile phases of the cycle. In contrast to this extreme view, the majority of investigators concur that the display of complete copulation, which involves both intromission and ejaculation by the male, increases at mid-cycle and diminishes shortly after ovulation [working with the chimpanzee, Yerkes and Elder (1936b) and Young and Orbison (1944); with the rhesus monkey, Ball and Hartman (1935) and Michael *et al.* (1967); with the pig-tailed macaque, Bullock *et al.* (1968)].

Data from our laboratory have been collected on both the rhesus monkey and the pig-tailed macaque, more completely on the former than on the latter. The general outline of the endocrine-behavior relationship is highly similar, however, in both species. Data on behavioral results in standardized tests with vasectomized males were adapted for present purposes from a previous experiment (Kuehn and Young, 1965). Independently collected data on the endocrine events throughout normal menstrual cycles in combination with these behavioral data are presented in Fig. 2. Results for rhesus monkeys show that there is a marked elevation in the percentage of tests culminating in ejaculation at that time in the cycle when the concentration of endogenous estradiol reaches its maximum. The onset of the luteal phase is characterized behaviorally by a relatively rapid decline in the incidence of full copulatory activity. The inhibitory effect of corpus luteum hormone on sexual behavior persists throughout the duration of the luteal phase. Attention is directed to the fact, however, that this inhibitory action is incomplete, and not as it was found to be in the lower primates and nonprimate mammals. The situation is not different for pig-tailed macaques, as illustrated in Fig. 3.

The most striking feature of these descriptive studies in macaques is that the display of copulatory behavior is not limited to the stage of the cycle characterized by rapid follicular growth and the absence of luteal progesterone. Even at times in the cycle when the concentration of luteal progesterone is high, complete copulation continues to occur, albeit infrequently. Moreover, not all the behavior seen during the follicular phase of the cycle can be correlated with estrogen secretion in either species. Certainly the mid-cycle rise in behavior occurs at a time when estrogen is also rising, but the baseline levels of behavior that remain in the rhesus at other times in the cycle are difficult to explain. The androgens have been suggested as mediators of female sexual behavior in women (Shorr *et al.*, 1938; Greenblatt *et al.*, 1942; Greenblatt, 1943; Salmon and Geist, 1943; Herrmann and Adair, 1946; Waxenburg *et al.*, 1959; Money, 1961), rhesus monkeys (Trimble and Herbert, 1968),

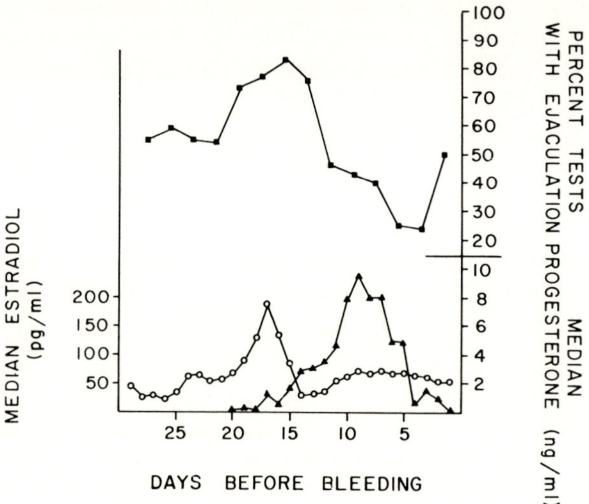

FIG. 2. The relationship of the ovarian steroids, estradiol (O——O, 6 cycles, 4 females) progesterone (▲——▲, 25 cycles, 9 females) to behavior ■——■, 25 cycles, 11 females) throughout the intermenstrual period in the female rhesus monkey. Mating test data are not from the same animals and are plotted according to day before the next menstruation.

and other species (Lindsay and Robinson, 1961, 1964). Preliminary data from our laboratory (Hess and Resko, work in progress), in which testosterone was measured by radioimmunoassay at varying intervals throughout the cycle of four female rhesus monkeys, provide some information about the quantities of this hormone in the circulatory system at various stages of the menstrual cycle. During the follicular phase the concentrations of testosterone ranged from 600 to 900 pg/ml. At mid-cycle the concentrations of this hormone dropped precipitously to about 400 pg/ml. In fact, a reciprocal relationship was found between the concentrations of progesterone and the concentrations of testosterone in systemic plasma. With the decline of the corpus luteum and the secretion of progesterone during the premenstrual period, the concentration of testosterone begins to rise.

The function of androgens during the follicular phase is conjectural but consistent with the hypothesis that they merely provide substrate to enzymes in the nervous system where they are quickly converted to estrogen. Conversions of this type have been reported systemically in women (West *et al.*, 1956; MacDonald *et al.*, 1967; Longscope *et al.*, 1969), but evidence for conversions of this type in nervous tissue

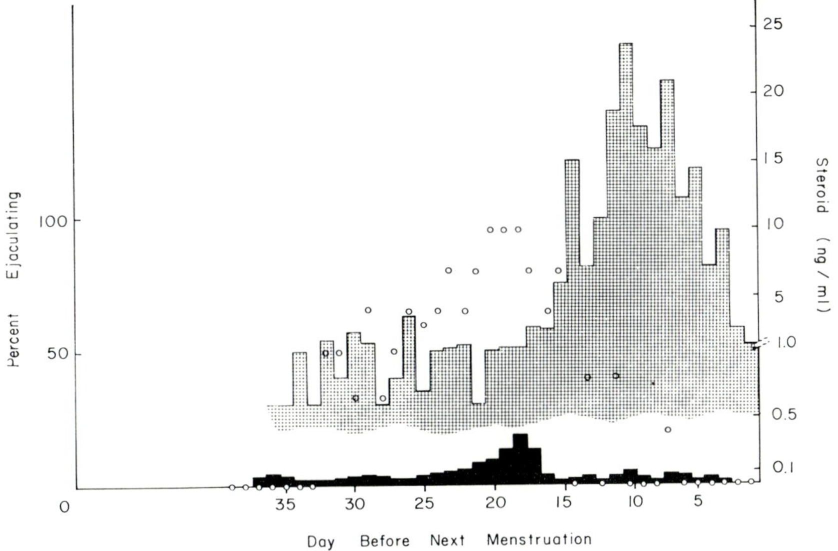

FIG. 3. The relationship of the ovarian steroids, estradiol (solid bars), progesterone (hatched bars), to behavior (open circles) throughout the intermenstrual period in the female pig-tail. Each bar represents a median concentration of steroid in plasma from five animals for estradiol. The progesterone and behavioral data are taken from the same animals and are replotted from data previously published by Bullock *et al.* (1968).

are more difficult to find. Brain tissue of the rhesus monkey is able to convert DHEA to phenolic compounds with the chromatographic mobility of estradiol, estrone, and an unknown compound (Knapstein *et al.*, 1968) but the characterization of these compounds was not made (Knapstein *et al.*, 1968). These data indicate that conversions of androgens to estrogens may take place in nervous tissue. Further support of this point of view can be derived from the recent report by Beyer, Vidal, and Mijares (1970), in which they have shown that only androgens capable of being aromatized induce estrous behavior in the ovariectomized rabbit. A more complete knowledge of the patterns of androgen secretion throughout the cycle of the primate and a better knowledge of the capacity of nervous tissue to metabolize androgen will clarify the position of the male hormone in the mediation of female behavior.

Two other behaviors shown by the female are influenced by the endocrine events of the cycle in a manner similar to that already described for copulatory behavior. These behaviors we have called proximity behavior and fear grimacing. The use of copulatory behavior as a correlate of ovarian activity has the distinct disadvantage that its occurrence

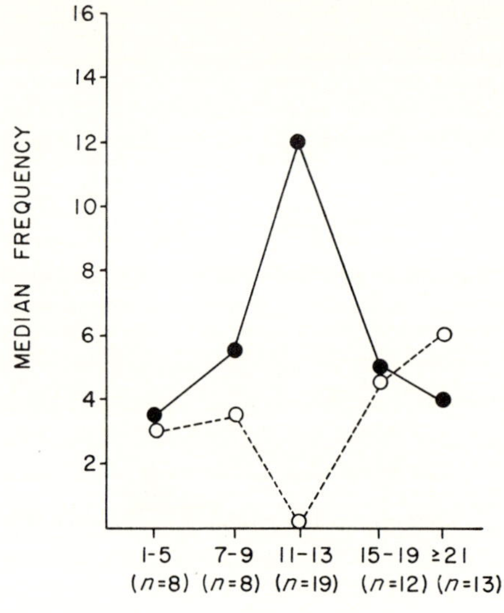

Fig. 4. Changes in selected female behaviors throughout the menstrual cycle of the rhesus monkey. ●——●, Median changes in the frequency of sitting near the male (PROX). ○---○, Median fearfulness of the male (FG) throughout the cycle.

depends as much upon the inclinations of the male as upon the dispositions of the female. This disadvantage is lessened considerably by observing proximity and fear grimacing behaviors, both of which are executed by the female alone. Figure 4 shows the representative changes in these two behaviors when tests were conducted at five different times during the menstrual cycle. It can be seen in the figure that on days 11–13, the time in the cycle when ovulation is most likely to occur (van Wagenen, 1945), the frequency with which the female approaches and sits next to the male, as measured by proximity behaviors, is at a very high level. Conversely, the number of times the female displays fear in response to the approaches or glances of the male is minimal. The increase in the female's motivation to approach and be with or near the male, and her decreased timidity and fearfulness of the male, have been noted previously by a number of field workers for the rhesus (Carpenter, 1942a,b; Altmann, 1962; Conaway and Koford, 1965; Vandenbergh and Vessey, 1968), for the baboon (DeVore, 1965; Hall, 1965; Hall and DeVore, 1965; Kummer, 1968), for the Japanese macaque (Tokuda, 1961), and for the pig-tailed macaque (Bernstein, 1967). What

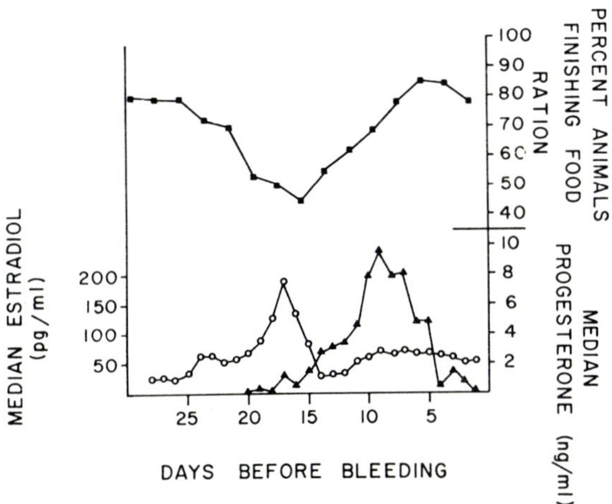

Fig. 5. The relationship of the median concentrations of estradiol (O——O, 6 cycles, 4 females), progesterone (▲——▲, 25 cycles, 9 females) in systemic plasma to eating behavior (■——■, percentage of animals finishing full food ration, 75 cycles, 38 females) in the female rhesus monkey throughout the intermenstrual period. Data are taken from independent groups of animals and plotted according to the day before the next menstruation.

is surprising, in this regard, is that the display of this behavior and its relationship to endocrine events should hold up so well under laboratory conditions. Perhaps the sex-related behaviors which bear no obvious instrumentality to copulation per se are more strictly influenced by the endocrine status of the female than has previously been recognized. Certainly a strict relationship exists between estrogen and eating behavior.

In studies recently conducted by John Czaja, the percentage of females that ate their full daily ration of food declined rapidly at mid-cycle, as shown in Fig. 5. The change in food intake is not gradual, but relatively abrupt, and coincides well with the time of the cycle when endogenous estradiol levels are rising or at a maximum. When a single injection of estradiol benzoate was given to an ovariectomized rhesus female (Fig. 6), a marked and significant decrease in the amount of food consumed within the next 24 hours was observed. The onset of the behavioral change was abrupt, as is ordinarily characteristic of sexual behavior in lower mammals.

Female monkeys which have been rendered pseudohermaphroditic by injecting their mothers with testosterone propionate have been studied in our laboratory for a number of years (Phoenix *et al.*, 1968; Goy, 1968, 1970). Not all the behaviors which are now known to vary cyclically in the normal female have been studied in pseudohermaphroditic

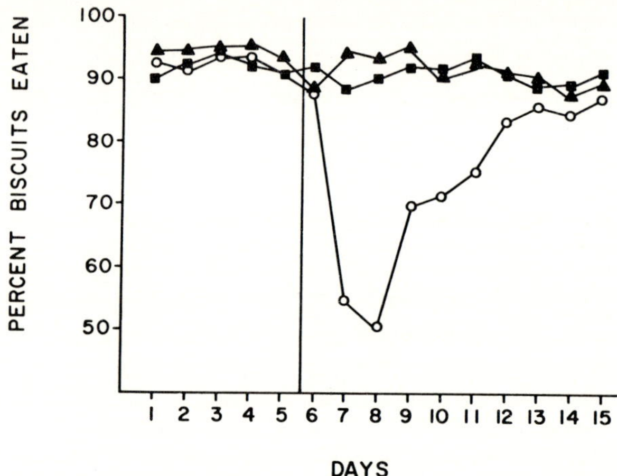

FIG. 6. The effect of estradiol benzoate (○——○, 60 μg, 12 injections, 6 females), progesterone (■——■, 5 mg, 10 injections, 6 females), or oil (▲——▲, 0.5 ml of sesame oil, 12 injections, 6 females), on food intake in the spayed female rhesus monkey. Subjects were given either one or two injections on consecutive days beginning on day 6.

females. In part, the incompleteness is explained by our reluctance to pair pseudohermaphroditic females with adult male studs because of concern for the violent aggression which might occur. Accordingly, what scattered data we have obtained has come from the regularly scheduled tests of pseudohermaphroditic females paired with male peers. These male peers were familiar partners with the pseudohermaphroditic females from the age of 3 months on, but they were not at the time of the tests either as sexually mature or experienced as the stud males used with the normal adult females. Tested under these conditions, the behavior of pseudohermaphroditic females failed to show variations with the stage of the menstrual cycle. Five subjects tested in this manner displayed 3.1 proximity responses per test on the average on tests given 15 to 21 days prior to menstruation and 5.6 responses per test on tests given at other times during the cycle. Six normal females tested similarly, however, showed 2.6 proximity responses per animal per test given 15 to 21

TABLE I

Age at Menarche in Normal and Pseudohermaphroditic Females

Pseudohermaphroditic females			
Animal number	Age at menarche (months)	Days of TP treatment[a]	Total TP (mg)
828	30.4	30	600
829	40.1	30	600
1656	37.8	72	610
1239	34.8	25	625
1664	39.2	82	660
836	33.9	50	750
1616	38.6	50	750
1619	39.0	50	750
1640	37.3	50	750
Mean = 36.8			

Normal females			
Animal number	Age at menarche (months)	Animal number	Age at menarche (months)
822	34.7	1654	30.3
823	24.0	1769	28.5
830	29.1	1838	29.9
831	21.3	2320	31.3
833	31.4	2350	29.7
1252	33.9	2362	29.5
1551	33.3	2551	28.1
1642	29.9	2575	27.4
1649	28.3	2577	24.3
			Mean = 29.2

[a] All treatments began at 38–40 days of gestational age.

days prior to menstruation and no such responses on tests given at other times in the cycle.

Pseudohermaphroditic female monkeys showed a delay in the menarche compared with normal females (Table I). Neither the amount of nor the age at menarche was associated with parameters of the prenatal treatment. Females with short prenatal treatments seemed as affected as those with long treatments. Despite the delay in the menarche, these pseudohermaphroditic females subsequently established menstrual cycles which were as stable and regular as those of normal females at comparable ages. Evidence was obtained, biochemically as well as morphologically, showing that ovulation and normal corpus luteum function occurred in pseudohermaphroditic females (Table II). Direct evidence for ovulation and normal corpus luteum formation was obtained for five of seven pseudohermaphroditic females at the time of ovariectomy performed on day 18 of the cycle. Analysis of ovarian vein blood collected at this time showed that normal quantities of progesterone

TABLE II

Concentrations[a] of Progesterone, Androstenedione, and Testosterone in Plasma from the Ovarian Vein of Untreated and Pseudohermaphroditic Monkeys on Day 18 of the Cycle

Treatment	Number of animals	Source of plasma	Progesterone	Androstene-dione	Testoste-rone
Control ♀	4	Ovary with CL	97.7 ± 28.4	1.0 ± 0.6	ND
		Ovary without CL	40.9 ± 10.9	2.2 ± 0.9	ND
Pseudoher-maphrodite ♀	5	Ovary with CL	95.2 ± 23.2	2.4 ± 1.0	ND
		Ovary without CL	17.8 ± 10.2	2.2 ± 0.6	ND

[a] Values are expressed as nanograms per milliliter of plasma, plus or minus standard error.

were being secreted by the ovary containing the corpus luteum. There was no evidence for secretory abnormality among pseudohermaphrodites in terms of concentrations of either testosterone or androstenedione found in the ovarian veins. Progesterone, however, was deficient in the ovarian vein associated with the ovary lacking the corpus.

We had shown previously that large quantities of progesterone can be found in the ovarian vein of the ovary not having the corpus luteum (Koering *et al.*, 1968). Later this progesterone was shown to have arisen from the ovary with the corpus luteum by connections through the uterus. These connecting vessels, however, were never identified histologically (Riesen *et al.*, 1970). Ovarian venous plasma from control females and the ovary with the corpus luteum contained 97.7 ± 28.4 ng/ml (±SE) of plasma (Table II). The ovary without the corpus luteum contained 40.9 ± 10.9 ng/ml (±SE) of plasma. Pseudohermaphroditic females and the ovary with the corpus luteum contained 95.2 ± 23.2 ng/ml (±SE) but in the ovary without the corpus luteum 17.8 ± 10.2 ng/ml (±SE) were found. The quantities of progesterone in the ovarian venous circulation from the ovary not containing the corpus luteum in the pseudohermaphroditic females were lower than in control females. Very little progesterone was found in ovarian venous plasma from the ovary without the corpus luteum in three of the five pseudohermaphrodites that were analyzed. Two of the remaining animals showed quantities

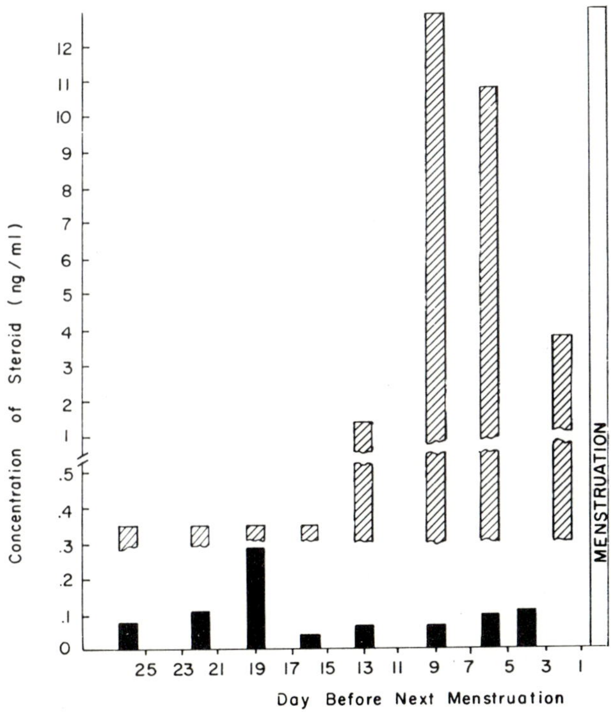

FIG. 7. The concentrations of estradiol (solid bar) and progesterone (hatched bar) in systemic plasma of a pseudohermaphroditic female rhesus monkey (cycle length = 27 days).

of progesterone similar to those found in the normal females. The ratio of the concentration of progesterone (ng/ml) in (ovary with CL) : (ovary without CL) had a range of 1 to 5 in control females and a range of 13 to 25 in the three pseudohermaphrodites mentioned above. The remaining two pseudohermaphrodites had ratios of 1 and 4, respectively. These data seem to indicate that the shunting mechanism whereby progesterone passes from one ovarian vein to another has not developed in three of five rhesus females that were treated with androgen prenatally. The implication is that perhaps one aspect of Müllerian duct development has been altered by prenatal androgen treatment.

Recently one pseudohermaphroditic female has been followed throughout a complete menstrual cycle for the presence of estradiol and progesterone in the plasma (Fig. 7). There is no difference qualitatively in the timing of the estrogen surge or the onset of luteal progesterone between the data obtained for the pseudohermaphroditic female and comparable data obtained for normal individuals (compare Fig. 7 with Fig. 2).

The behavior of the pseudohermaphroditic females was not markedly altered by the realization of cyclical ovarian activity. Those pseudohermaphroditic females which had shown a consistent record of mounting activity prior to the menarche continued to do so afterward. Moreover, there was no evidence that the events of the ovarian cycle influenced the frequency of mounting activity, either negatively or positively. Simi-

Fig. 8. Photograph of pseudohermaphroditic female rhesus monkey (No. 1619) being groomed by a male peer. At this age, 1 year after menarche, the female was capable of displaying erection as illustrated above. The penis is well formed but infantile in size because no testosterone was administered after birth.

larly, phallic erection continued to be displayed by pseudohermaphroditic females following the onset of ovarian activity (see Fig. 8). This means that there is no correspondence between the character of the behavior shown by the pseudohermaphroditic females in adulthood and the nature of the endocrine environment. These observations on the lack of any marked "feminizing" influence of the endogenous estrogens in adult pseudohermaphroditic females complement those observations which demonstrate a lack of any "masculinizing" influence of testosterone on the sexual behavior of adult normal females, as discussed in following sections of this paper.

II. Androgens and the Behavior of Normal and Pseudohermaphroditic Female Monkeys

The hormonal induction of heterotypical behavior in female primates seems to be achieved in very limited ways and in an unpredictable fashion when testosterone is injected at various times after birth, and sexual behavior per se seems highly resistant to any masculinizing influence. For example, in a recent experiment by Dr. W. D. Joslyn, testosterone was injected three times weekly for 8 months into three normal female monkeys, beginning at approximately 5 months of age and ending at 13 months of age. Studies of the behavior of these animals were made beginning with the first injections of testosterone propionate, and continuing throughout the final month of testosterone propionate treatment, and for several months after termination of testosterone treatment. These females became hyperaggressive and almost all their aggressive attacks were directed against the male members of their peer groups. In addition, there were marked and significant increases in yawning behavior, which is ordinarily related to the display of social dominance by males. These behavioral changes induced by testosterone in young females could be considered a kind of masculinization, but it is essential to point out that this masculinizing influence did not extend to other kinds of behaviors which are ordinarily sex-related.

The behavior of young rhesus monkeys in social groups is markedly sexually dimorphic in a number of respects. For example, young male monkeys display sham threat, rough and tumble play, chasing play, and mounting behavior much more frequently than do females in the same groups. Similarly, young males initiate play significantly more often than do females. However, none of these dimorphic behaviors were influenced by testosterone injected for 9 months into the juvenile female monkeys studied by Dr. Joslyn. Their averages for frequency of performance of these activities during the last month of injection are seen in Table III, and their deviations from the standards set by normal control females studied over many years are not large enough to exceed chance expectancy.

The limited masculinization achieved with these postnatal injections of testosterone propionate for a long period of time contrasts markedly with the changes induced in the same patterns of behavior when genetic female monkeys are exposed to androgens prior to birth and rendered pseudohermaphroditic. The patterns of sham threat, rough and tumble play, play initiation, chasing play, and mounting behavior shown by such pseudohermaphroditic females differ significantly from those of normal control females and illustrate the broad scope of the masculinization

TABLE III

Mean Frequencies of Social Behaviors in Normal and Experimentally Treated Rhesus Monkeys Tested for 20 Days at 12 Months of Age

Subjects	N	Play initiation	Rough and tumble play	Threat	Pursuit play	Mounting
Normal males	33	100.4	42.8[b]	55.0	16.2	2.2[b]
Normal females	36	20.8	10.0[c]	13.0	3.6	0.0[c]
Females treated prenatally with TP (pseudohermaphrodites)	8	67.5	30.5	41.4	13.2	2.0
Females treated postnatally with TP[a]	3	2.3	14.3	2.3	0.0	0.0

[a] From experiments by Dr. W. D. Joslyn.
[b] $N = 21$.
[c] $N = 25$.

achieved with prenatal androgenic exposure. Longitudinal and developmental changes in all of these social behaviors are highly similar (Phoenix *et al.*, 1968). For illustrative purposes, changes in the frequency of Play Initiation with age for normal males, normal females, and pseudohermaphrodites are shown in Fig. 9.

The limited effects that androgens have on the behavior of normal females in adulthood can be ascertained from the following set of studies. The first of these studies shows that the behavior of adult rhesus males differs markedly from that of females. When male and female subjects are paired for testing at frequent intervals throughout the menstrual cycle of the female, behaviors bearing no obvious function in reproduction or sexuality, as well as those forming a constant part of the sexual repertoire, differ between the sexes. Data on the frequency of two sexually dimorphic behaviors, "yawning" (see Fig. 10) and mounting,

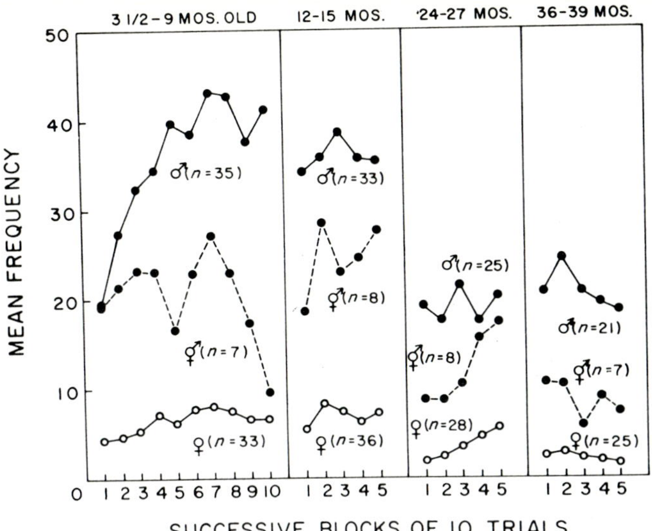

Fig. 9.   The frequency of performance of play initiation shown by normal males (●——●), normal females (○——○), and pseudohermaphroditic female monkeys (●‑‑‑●) during the first 39 months of life.

Fig. 10.   An adult male rhesus monkey yawning. This facial gesture in which the large canines are conspicuously displayed occurs frequently among rhesus in situations requiring mild assertions of social dominance. In tests of oppositely sexed pairs the behavior is most frequently displayed by the male.

were obtained by Stephen Eisele in a recent study of seven pairs of rhesus. In this study the pairs were tested for 12 minutes or until the male ejaculated, if ejaculation occurred sooner than 12 minutes. Under these conditions, as shown in Table IV, the females never yawned or mounted, regardless of the stage of the cycle when the tests were given. Males, in contrast, displayed both behaviors with reasonable consistency, and there was a tendency for the frequency of performance to be higher when tests were given on the female's follicular days than when the female partner was in the luteal phase of her cycle.

TABLE IV

Sex Differences in Yawning and Mounting Behavior for Normal Adult Rhesus Tested in Heterosexual Pairs

| | N (Animals) | Mean Frequency of Behavior per Test | | | | Per Cent of Tests on Which Behavior was Displayed | | | |
| | | Follicular Phase[a] | | Luteal Phase[b] | | Follicular Phase[a] | | Luteal Phase[b] | |
		Yawn	Mount	Yawn	Mount	Yawn	Mount	Yawn	Mount
Males	7	1.9	5.9	1.2	3.2	61.9	90.4	41.3	45.6
Females	7	0	0	0	0	0	0	0	0

[a] Tests given 17 to 20 days prior to menstruation were classed as Follicular Phase. The number of tests given each pair varied from 2 to 4 during this period, and the total number of tests was 21.

[b] Tests given 3 to 10 days prior to menstruation were classed as Luteal Phase. The number of tests given each pair varied from 4 to 8 during this period, and the total number of tests was 46.

In part, the dimorphism observed in Mr. Eisele's study of heterosexual pairs can be attributed to the insufficiency of endogenous androgens in the female compared with the male, and in part it is due to the psychological influence of the partner. When females are tested with another female rather than with a male, a higher frequency of yawning and mounting typically occurs. In short, in the pair-testing situation, the probability of performance of yawning and mounting is increased by the use of a female as partner regardless of whether the performer is male or female.

A study carried out in our laboratory in collaboration with C. H. Phoenix showed that even when this psychological factor was taken into account, the behavior of adult females injected with testosterone propionate was only partially masculinized. Of some 28 behavioral characteristics recorded during standardized 10-minute tests of females tested with female partners, only one behavior, which we call "yawning" and which is normally characteristic of the adult male tested under similar circumstances, was influenced. The effects of testosterone propionate administered for 4 weeks and in two different dosages on both yawning and mounting behavior in adult females are summarized in Table V. The reader can observe in Table V that in female–female pair tests both yawning and mounting are displayed prior to the injection of testosterone

TABLE V

Differential Effect of Testosterone Propionate on Mounting and "Yawning" Behavior of Adult Rhesus Females

Behavior	Number of animals	Prior to TP treatment	During TP treatment		After TP treatment
			(5.0 mg/day)	(10.0 mg/day)	
Average frequency of mounts per animal per test	5	0.6	1.1[a]	0.2[a]	0.1[a]
Average frequency of "yawns" per animal per test	5	0.1	3.9[a]	5.6[b]	2.0[c]

[a] Not significantly different from pretreatment score ($P > 0.05$, dependent t).

[b] Significantly different from pretreatment score ($P < 0.05$, dependent t).

[c] Significantly different from 10 mg/day TP treatment score ($P < 0.05$, dependent t).

propionate. This differs from the results obtained for females when adult males were used as partners (cf. Table IV). When testosterone propionate was injected, the frequency of mounting did not increase significantly and remained substantially below that for normal males. In contrast, the frequency of yawning behavior not only increased significantly during injections of testosterone propionate, but it was substantially higher than that for normal males tested with females. Again, as in the previous example with juvenile females discussed in an earlier section of this article, testosterone propionate in the adult female, although it induced a 20-fold increase in yawning behavior, failed to induce any significant increases in homosexual mounting activity. We interpret these results to mean that testosterone has little or no ability to "masculinize" the sexual behavior of normal female rhesus even though some influence of this hormone on social behaviors can be measured. Although the hamster is more sluggish in its response than other rodents (Swanson and Crossley,

1970), the results of similar experiments in females from lower mammalian species contrast markedly with our results on primates. In the former species mounting behavior is regularly induced or augmented with injections of testosterone (Ball, 1939; Beach, 1942; Gassner, 1952; Phoenix *et al.*, 1959; Goy *et al.*, 1967; Pfaff, 1970; McDonald *et al.*, 1970).

Clitoral hypertrophy was marked in these adult females injected with testosterone propionate, and an increase in the erotic value of clitoral stimulation was obvious from the behavior of one female excluded from the previous table. This particular female had an idiosyncratic method of achieving clitoral stimulation with her submissive partner. The method consisted of climbing upon the partner's back, sitting down upon the partner's rump, and rubbing the clitoris vigorously back and forth while maintaining the seated posture (Fig. 11). When testosterone propionate was injected into this female monkey, the frequency with which she performed this behavior increased 5-fold and was maintained at this level for 1 week following the cessation of testosterone propionate injections (Fig. 12). It is important to point out a limitation of hormone action illustrated by the response of this idiosyncratic behavior pattern to exogenous testosterone propionate. The injections of testosterone into

Fig. 11. Photograph of idiosyncratic behavior of female rhesus monkey for clitoral stimulation. Androgen treatment intensifies this behavior but does not change it to a male pattern.

this female did not modify this preexisting behavior into a male-type pattern of mounting. Apparently for sexual activity, at least, the androgens serve in adulthood only to bring out or intensify the pattern of behavior that is already there. In this regard, it is well to recall the discussion in an earlier section of this paper dealing with endogenous testosterone and the behavior of female rhesus during early follicular stages of the cycle. We have hypothesized on the basis of our experi-

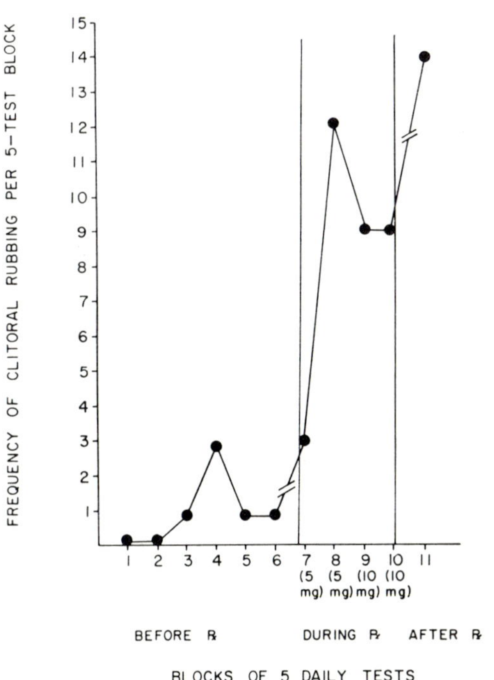

Fig. 12. The effect of testosterone on the frequency of clitoral rubbing behavior of female No. 821 during tests with a female partner.

ments, as others have done using alternative approaches (Trimble and Herbert, 1968), that endogenous testosterone present during the follicular phase enhances feminine copulatory behavior rather than induces masculine characteristics.

When testosterone propionate is injected into pseudohermaphroditic females in adulthood the effects are not limited to changes in the frequency of yawning behavior, as is the case in normal females. Recently, with Dr. Gray Eaton, we have examined the effects of androgens on the behavior induced in adult pseudohermaphroditic monkeys. The results of these experiments varied with the specific prenatal treatment as well as with the individual. Pseudohermaphroditic monkeys produced by prenatal testosterone treatments lasting only 30 days (from day 39 to day 69 of gestation) did not show further augmentation of masculine sexual behavior when given testosterone in adulthood. None of these females, in tests during daily administration of testosterone propionate, displayed either intromission or ejaculation. Three pseudohermaphrodites produced by prenatal treatments lasting for 50 days (from day 39 or 40 to day 89 or 90 of gestation) showed on the average more marked but variable changes in behavior. One of these three pseudohermaphrodites (No. 1640) developed the complete pattern of male copulatory behavior and a cinematographic record of her performance has been made. The yawning response of pseudohermaphrodite No. 1640 is illustrated in Fig. 13. The reader may compare this figure with Fig. 10 and note the conspicuous difference in the size of the canines of the pseudohermaph-

Fig. 13. Yawning behavior of pseudohermaphroditic female No. 1640 induced by administration of testosterone propionate in adulthood. Note the development of the canine teeth, which appear to be intermediate in size between those characteristic of genetic males and females.

rodite and the normal male. None of the prenatal androgen treatments which we have given have produced canines in females comparable to those of normal male monkeys. When genetic males are castrated at birth, however, the development of the large canines is not prevented, so the possibility exists that longer prenatal treatments of genetic females would produce masculinization of the structure of the canines. In Figs. 14 and 15 the mounting postures of a normal adult male and pseudo-hermaprodite No. 1640 are illustrated. The postural differences which are apparent involve the degree of flexion of the knee. Flexion is greater for the normal male and expresses the accommodation required to achieve intromission when the male's feet clasp the partner's legs just below the knee. The shorter legs of the pseudohermaphrodite do not require the same degree of flexion of the knee for intromission to occur.

The variation encountered in response of pseudohermaphrodites to testosterone in adulthood is in part related to the amount and duration of the exposure to androgens during fetal development. The length of

Fig. 14. Mounting behavior of a normal adult male rhesus monkey showing the characteristic clasping of the partner's legs by the male's feet.

Fig. 15. Mounting behavior of pseudohermaphroditic female No. 1640. The pattern does not differ in any way important to theoretical issues from that displayed by normal males.

time we treated our experimental females was based only upon the guidelines established by experimental embryology for the induction of complete masculinization of genital morphology (van Wagenen and Hamilton, 1943; Wells and van Wagenen, 1954). The possibility exits, however, that the differentiation of complete male psychosexual orientation requires different endocrine parameters than those of the distal genital tract. Our own data on the character of the androgens during fetal and embryonic development of rhesus monkeys show that testosterone is present throughout most of prenatal development. Using radioimmunoassay procedures for androgens, we have shown that testosterone and dihydrotestosterone are present in the umbilical circulation in both males and females. We have analyzed the plasma from 12 individual fetuses (six males and six females ranging in age from 60 to 141 days of gestation) for testosterone and dihydrotestosterone. The data are shown in Table VI. A significant difference was found in the concentration of testosterone in mixed umbilical vessel plasma from the male compared to the female fetus. No overlap was found between the two groups. Concentrations among individual females were relatively uniform. In contrast, wide variation was found in the concentration of testosterone in plasma from individual fetal males. We are not certain whether this is due to the age of the fetus, or whether some males are exposed to much

TABLE VI

Concentrations of Testosterone and Dihydrotestosterone in Mixed Umbilical Vessel Plasma from the Fetal Rhesus Monkey

Day of gestation	Testosterone (pg/ml plasma)	Dihydrotestosterone (pg/ml plasma)
Male fetus		
60	>1500[a]	840
80	1017	478
80	1191	647
81	879	—
110	875	—
141	1736	715
Mean ± SE	1200 ± 144[b]	670 ± 75
Female fetus		
79	185	—
106	395	824
111	404	561
115	338	586
115	145	—
118	483	355
Mean ± SE	325 ± 54	582 ± 96

[a] A value of 1500 was used for statistical analysis.

[b] Quantities of testosterone, male *vs* female, were significantly different by a *t* test. $P < 0.01$; dihydrotestosterone male *vs* female were not significant.

more prenatal androgen than others. If the latter should prove to be the case in future research, the possibilities for endocrine explanations of male psychosexual and morphological variation would be greatly enhanced over the present limited value of endocrine data on the adult. The concentrations of dihydrotestosterone, unlike those for testosterone, do not differ between the sexes (Table VI). Accordingly, dihydrotestosterone may not be crucial to sexual differentiation in the rhesus, or, alternatively, its concentrations in target tissues rather than in plasma are more related to its importance in development. We interpret these findings, along with those obtained previously showing no difference between the sexes in the concentration of umbilical androstenedione (Resko, 1970), as supportive of two hypotheses. First and probably of greater importance, testosterone is the major hormonal principle accounting for sexual differentiation. Second, because the period of endogenous testosterone stimulation is protracted in long-gestation mammals, the opportunity exists for processes of neural differentiation to be continuously supported by the presence of testosterone throughout most of gestation.

We have not aimed our experiments with rhesus monkeys toward the identification and demarcation of a critical period for induction of the whole set of male behaviors. Perhaps testosterone support throughout the entire period of gestation would be required to achieve this result with regularity and relative uniformity from individual to individual. However, the performance of our three "high-dosage" pseudohermaphrodites compares favorably with what can be expected from genetic males which presumably had adequate, if not identical, prenatal endocrine environments. In studies still in progress we have found that of six males castrated within 3 or 4 months of birth and injected with testosterone in adulthood, only two showed the development of ejaculatory behavior, one showed occasional intromissions without the development of ejaculation, one showed augmentation of mounting only, and the remaining two failed to show any significant changes even in the frequency of mounting behavior during the period of injections.

The results of our experiments with primates show that the effects of testosterone given before birth are not greatly different from those of similar treatments in nonprimate mammals. However, such a conclusion seems premature, and the manner in which the hormones act to accomplish masculinization may be, in fact, quite different in these two mammalian groups. The influence of experience on the development of sexual and social behaviors is small in lower mammals compared with higher primates. For the former, it may be possible to postulate, as Dorner (1970) has done, that the androgens present early in development act upon separate neural circuits for the expression of masculine and feminine components of sexual behavior. This seems all the more plausible since very small brain lesions and very discrete placements of steroid hormones in the brains of castrate animals have been shown to abolish or restore, respectively, full sexual behaviors in a variety of nonprimate animals. When one considers, however, the important contribution which experience may make to the expression of sexual and social behaviors in primates (Rosenblum, 1961; Mason, 1963; Mason and Sponholz, 1963; Harlow *et al.*, 1963; Harlow, 1965) it does not seem reasonable to postulate a simple manner or a discrete locus of androgen action. If in higher primates a large part of sexual and sex-related behavior is experientially determined, then it seems reasonable to infer that androgens present before birth act upon the neural circuits mediating the effects of experience as well as upon the neural circuits for masculine and feminine behavior. Our experiences with the rhesus monkey suggest that one might turn one's attentions instead to the effects of androgen on the development of neural systems underlying drive and reward, i.e., those systems which permit experience to exert its ultimate influences on behavior.

ACKNOWLEDGMENTS

Special thanks are due C. A. Paris, J. Czaja, and K. Wallen for helpful discussions and diligent assistance with various parts of the manuscript and Nancy Arnold for expert secretarial assistance.

We also wish to acknowledge the financial support of several grants which have helped not only in the conduct of the work but also in the preparation of the manuscript itself: Grant MH-21312, Grant FR-00163 to the Oregon Regional Primate Research Center, and Grant FR-00167 to the Wisconsin Regional Primate Research Center. Publication number 11-027 of the Wisconsin Regional Primate Research Center.

REFERENCES

Altmann, S. (1962). *Ann. N.Y. Acad. Sci.* **102**, 338.

Ball, J. (1939). *J. Comp. Psychol.* **28**, 273.

Ball, J., and Hartman, C. G. (1935). *Amer. J. Obstet. Gynecol.* **29**, 117.

Beach, F. A. (1942). *Endocrinology* **31**, 673.

Bernstein, I. S. (1967). *Primates* **8**, 217.

Beyer, C., Vidal, N., and Mijares, A. (1970). *Endocrinology* **87**, 1386.

Bullock, D. W., Paris, C. A., Resko, J. A., and Goy, R. W. (1968). *Proc. 6th Int. Congr. Anim. Reprod. Artif. Insem., Paris* **2**, 1657.

Carpenter, C. R. (1942a). *J. Comp. Psychol.* **33**, 113.

Carpenter, C. R. (1942b). *J. Comp. Psychol.* **33**, 143.

Conaway, C. H., and Koford, C. B. (1965). *J. Mammal.* **45**, 577.

DeVore, I. (1965). *In* "Sex and Behavior" (F. A. Beach, ed.), pp. 266–289. Wiley, New York.

Dorner, G. (1970). *Deut. Med. Wochenschr.* **15**, 108.

Eaton, G. G., Slob, A., and Resko, J. A. (1971). *Anim. Behav.* (in press).

Evans, C. S., and Goy, R. W. (1968). *J. Zool.* **156**, 181.

Gassner, F. X. (1952). *Recent Progr. Horm. Res.* **7**, 165.

Goy, R. W. (1968). *In* "Endocrinology and Human Behaviour" (R. P. Michael, ed.), pp. 12–31. Oxford Univ. Press, London and New York.

Goy, R. W. (1970). *Phil. Trans. Roy. Soc. London, Ser. B* **259**, 149.

Goy, R. W., Phoenix, C. H., and Meidinger, R. (1967). *Anat. Rec.* **157**, 87.

Greenblatt, R. B. (1943). *J. Amer. Med. Ass.* **121**, 17.

Greenblatt, R. B., Mortara, F., and Torpin, R. (1942). *Amer. J. Obstet. Gynecol.* **44**, 658.

Hall, K. R. L. (1965). *In* "The Baboon in Medical Research" (H. Vagtborg, ed.), Vol. 1, pp. 29–42. Univ. of Texas Press, Austin, Texas.

Hall, K. R. L., and DeVore, I. (1965). *In* "Primate Behavior: Field Studies of Monkeys and Apes" (I. DeVore, ed.), pp. 53–110. Holt, New York.

Harlow, H. F. (1965). *In* "Sex and Behavior" (F. A. Beach, ed.), pp. 234–265. Wiley, New York.

Harlow, H. F., Harlow, M. K., and Hansen, E. W. (1963). *In* "Maternal Behavior in Mammals" (H. L. Rheingold, ed.), pp. 254–281. Wiley, New York.

Hartman, C. G. (1928). *J. Mammal.* **9**, 181.

Herrmann, J. B., and Adair, F. E. (1946). *J. Clin. Endocrinol. Metab.* **6**, 769.

Knapstein, P., David, A., Chung-Hsiu Wu, D. F., Archer, D. F., Flickinger, G. L., and Touchstone, J. C. (1968). *Steroids* **11**, 885.

Koering, M., Resko, J. A., Phoenix, C. H., and Goy, R. W. (1968). *Anat. Rec.* **160**, 378.

Kuehn, R. E., and Young, W. C. (1965). *Amer. Zool.* **5**, 687.

Kummer, H. (1968). "Social Organization of Hamadryas Baboons: A Field Study." Univ. of Chicago Press, Chicago, Illinois.

Lindsay, D. R., and Robinson, T. J. (1961). *Nature (London)* **192**, 761.

Lindsay, D. R., and Robinson, T. J. (1964). *J. Reprod. Fert.* **7**, 267.

Longscope, C., Kato, T., and Horton, R. (1969). *J. Clin. Invest.* **48**, 2191.

MacDonald, P. C., Rombant, R. P., and Siiteri, P. K. (1967). *J. Clin. Endocrinol. Metab.* **27**, 1103.

McDonald, P. G., Vidal, N., and Beyer, C. (1970). *Horm. Behav.* **1**, 161.

Maslow, A. H. (1936). *J. Genet. Psychol.* **48**, 310.

Mason, W. A. (1963). *Percept. Motor Skills* **16**, 263.

Mason, W. A., and Sponholz, R. R. (1963). *Psychiat. Res.* **1**, 1.

Michael, R. P., Herbert, J., and Welegalla, J. (1967). *J. Endocrinol.* **39**, 81.

Money, J. (1961). *In* "Sex and Internal Secretions" (W. C. Young, ed.), Vol. 2, pp. 1383–1400. Williams & Wilkins, Baltimore, Maryland.

Pfaff, D. (1970). *J. Comp. Physiol. Psychol.* **73**, 349.

Phoenix, C. H., Goy, R. W., Gerall, A. A., and Young, W. C. (1959). *Anat. Rec.* **133**, 323.

Phoenix, C. H., Goy, R. W., and Resko, J. A. (1968). *In* "Reproduction and Sexual Behavior" (M. Diamond, ed.), pp. 33–49. Indiana Univ. Press, Bloomington, Indiana.

Resko, J. A. (1970). *Endocrinology* **87**, 680.

Riesen, J. W., Koering, M. J., Meyer, R. K., and Wolf, R. C. (1970). *Endocrinology* **86**, 1212.

Rosenblum, L. A. (1961). Ph.D. Thesis, Univ. of Wisconsin, Madison, Wisconsin.

Rowell, T. E. (1963). *J. Reprod. Fert.* **6**, 193.

Salmon, U. J., and Geist, S. H. (1943). *J. Clin. Endocrinol.* **3**, 235.

Shorr, E., Papanicolaou, G. N., and Stimmel, B. R. (1938). *Proc. Soc. Exp. Biol. Med.* **38**, 759.

Swanson, H. H., and Crossley, D. A. (1970). *In* "Proceedings of the International Conference on Hormones and Development" (M. Hamburgh and E. J. W. Barrington, eds.). Sponsored by Nat. Found. Appleton-Century-Crofts, New York.

Tokuda, K. (1961). *Primates* **3**, 1.

Trimble, M. R., and Herbert, J. (1968). *J. Endocrinol.* **42**, 171.

Vandenbergh, J. G., and Vessey, S. (1968). *J. Reprod. Fert.* **15**, 71.

van Wagenen, G. (1945). *Endocrinology* **37**, 307.

van Wagenen, G., and Hamilton, J. B. (1943). *In* "Essays in Biology" (T. Cowles, ed.), pp. 581–607. Univ. of California Press, Berkeley, California.

Waxenburg, S. E., Drellich, M. D., and Sutherland, A. M. (1959). *J. Clin. Endocrinol. Metab.* **19**, 193.

Wells, L. J., and van Wagenen, G. (1954). *Contrib. Embryol.* **35**, 93.

West, C. D., Damast, B. L., Sarro, S. D., and Pearson, O. H. (1956). *J. Biol. Chem.* **218**, 409.

Yerkes, R. M., and Elder, J. H. (1936a). *Comp. Psychol. Monogr.* **13**, 1.

Yerkes, R. M., and Elder, J. H. (1936b). *Proc. Nat. Acad. Sci. U.S.* **22**, 276.

Young, W. C., and Orbison, W. D. (1944). *J. Comp. Psychol.* **37**, 107.

[Discussion for this chapter appears on page **754**.]

Gender Dimorphic Behavior
and Fetal Sex Hormones[1]

JOHN MONEY[2] AND ANKE A. EHRHARDT

Department of Psychiatry and Behavioral Sciences and Department of Pediatrics The Johns Hopkins University School of Medicine, Baltimore, Maryland; and Department of Pediatrics and Department of Psychiatry, State University of New York Children's Hospital, Buffalo, New York.

I. Introduction

Animal experimentalists have demonstrated delayed effects of fetal hormones on behavior dimorphism. The only way of testing these observations in human beings is not by planned experiments but by experiments of nature, spontaneously occurring as human clinical syndromes with a known history of fetal hormonal anomalies.

II. Fetally Androgenized Genetic Females

A. THE SYNDROME OF PROGESTIN-INDUCED HERMAPHRODITISM

The closest approximation in human beings to the planned experimental masculinization of genetic female animals before birth is the syndrome of progestin-induced hermaphroditism. This condition of prenatal masculinization was, two to three decades ago, inadvertently induced in a few genetic female fetuses by pregnancy-saving hormones given to the mother to prevent a miscarriage. The hormones belong to a recently synthesized group of steroids which, although related in chemical structure to androgens, are, in biological action, substitutes for pregnancy hormone (progesterone) and hence are named progestins. When they were first synthesized, it was not known that certain of them would exert a masculinizing influence, in rare instances, on a daughter fetus. Thus, before this effect was discovered, and before the use of the hormones was discontinued, there were in the 1950's a few rare mothers who gave birth to a daughter with masculinization of the clitoris. Usually the masculinization was restricted to enlargement of the clitoris plus or minus a certain amount of labial fusion, but in rare instances it became a complete masculinization, producing a penis and empty scrotum.

[1] See also, "Man and Woman, Boy and Girl: The Differentiation and Dimorphism of Gender Identity from Conception to Maturity," by Money and Ehrhardt, The Johns Hopkins University Press, Baltimore, Maryland, in press.

[2] Research support by Grant 5K03-HD18635 and Grant 2R01-HD00325, U.S. Public Health Service.

The more complete the degree of masculinization, the deeper within, and closer to the neck of the bladder, was the opening of the vagina. But the internal organs of the female were always present. The internal organs of the male were not differentiated, apparently because of a timing effect. The masculinizing effect of the progestinic medication has proved to be confined to the fetal period of external sexual differentiation.

Because the influence of synthetic progestin pills taken by the mother ceases at birth, and because there are no other telltale signs or side effects to arouse diagnostic suspicion, the baby born with a penis would, naturally enough, be regarded as a boy with undescended testes, and brought up as a boy. The baby with only incomplete masculinization of the clitoris, however, would more likely be subject to diagnostic evaluation leading to the decision that, because the genetic, gonadal, and internal sex were female, the sex of assignment should be as a girl. Upon completion of surgical feminization of the external genitalia, no further surgical or hormonal treatment would be required. At the age of puberty, the girl's own ovaries would function normally, totally feminizing the body and inducing menstruation, though possibly a little late.

B. The Female Adrenogenital Syndrome

The hermaphroditism and development of genetic females with the progestin-induced condition has a parallel in genetic females with the adrenogenital syndrome. The parallel exists only if the adrenogenital child is hormonally regulated from birth onward by treatment with cortisone to prevent continuance of developmental masculinization, postnatally.

The adrenogenital syndrome is so named because the adrenal glands have a defect of function, beginning in fetal life, which in turn causes a defect of the genital anatomy, if the fetus is female. The primary defect is a genetic one, transmitted as a genetic recessive, which prevents the adrenal cortices from synthesizing their proper hormone, cortisol. They release instead a precursor product which is, in biological action, a male sex hormone—an androgen. This androgen enters into the blood stream of the fetus too late to induce extensive masculinization of the internal reproductive ducts (the Wolffian ducts), but in time to masculinize the external genital anlagen. As in the progestin-induced syndrome, the result may be complete masculinization to form a penis and empty scrotum, or less complete, to form a grossly enlarged clitoris that resembles a hypospadiac penis, with partial fusion of the labia majora. In either case, the internal female reproductive organs will have differentiated. The vaginal orifice will be near its expected location if masculinization has been minimal. Otherwise, its opening will be into the wall of the urethra, internally, and will need to be brought to its normal external location, surgically.

Such is the nature of adrenal malfunction in the adrenogenital syndrome that some babies born with this condition are also lacking in correct salt and fluid balance in the body, and some lack in blood-pressure regulation. In consequence, they may become desperately sick and rapidly die unless the condition is recognized and cortisone regulation promptly established. Especially because of the telltale symptoms associated with salt loss, even the baby masculinized to the extent of having a full penis is likely to be diagnostically recognized soon after birth as a genetic female with two ovaries and a uterus internally. There are some who escape recognition and grow up successfully as boys; and in times past there have even been a few born with an incomplete penis who, because of an incomplete diagnostic workup, were surgically corrected as boys, and also grew up to have a masculine gender identity.

The adrenogenital female hermaphrodites of special interest in this paper, however, are those whose diagnosis is promptly established neonatally so that they are assigned to grow up as girls. They are given at least the first stage of surgical feminizing repair of the genitalia within the first weeks or months of life. Sometimes additional, minor vaginal surgery may be needed in teenage. Their early surgery taken care of, the girls grow up to see themselves as girls, as well as to be seen as girls by the people who take care of them or otherwise see their genitalia.

Because the abnormality of adrenocortical function does not correct itself postnatally, it is necessary for children with this condition to be regulated on cortisone therapy throughout the growing period and, indeed, in adulthood also. This regulation is imperative to prevent growth and maturation in childhood which is too rapid, too early, and too masculine. Otherwise, the physique would simulate that of normal male puberty. Of course, masculine puberty in a child living as a girl is unsightly and wrong. In later years, a girl also does not desire to be hairy, deep-voiced, and masculine in appearance, hence hormonal regulation on cortisone must be maintained. Properly treated, she will develop a feminine physique at the expected time of puberty. Her menses may be about a year late in appearing. In a few instances they may be excessively delayed (Jones and Verkauf, 1971). Otherwise she can expect to be able to conceive, though perhaps not as expeditiously as she might desire. She can expect to give birth to a normal child, but proper cortisone regulation is imperative to prevent miscarriage. Lactation is also possible.

Cortisone therapy for the adrenogenital syndrome was discovered in

1950. It was just prior to this time that the first babies with progestin-induced hermaphroditism were born. Thus, the oldest adrenogenital girls treated with cortisone since babyhood, and the oldest of the progestin-induced group are now young adults. In the years 1965–1967 when our special follow-up study of their behavioral development and psychosexual identity was undertaken, the oldest was sixteen.

Not Additionally Androgenized Postnatally: Behavioral Sequelae

The purpose of this follow-up study was to compare some aspects of sexually dimorphic behavior in fetally androgenized, genetic females of the human species in childhood and adolescence with that of matched controls, in order to see whether prenatal androgens may have left a presumptive effect on the brain, and hence on subsequent behavior. The sample comprised a census of all ten available genetic females with progestin-induced hermaphroditism, all given early corrective surgery, if needed, and all reared as girls; and an unbiased selection of fifteen early-treated genetic female adrenogenital hermaphrodites of suitable age, hormonal history, surgical history, and geographic proximity. They ranged in age from 4 to 16 years, the majority being in middle childhood.

For each of these 25 fetally androgenized girls, a normal girl as a matched control was found through the courtesy of local school authorities. Matching was on the basis of age, IQ, socioeconomic background, and race. All fifty girls and their mothers were interviewed with a standard schedule of topics, and were tested with sex-role preference procedures. Further details of methodology are in Ehrhardt and Money (1967), Ehrhardt *et al.* (1968a), and Ehrhardt (1969).

Since the findings on the two diagnostic groups closely paralleled one another, they are presented together in the sections and tables that follow.

1. Tomboyism

Table I shows that fetally androgenized girls differed from their matched controls in regarding themselves as tomboys, a status of which they were proud. There were actually 9 of the 10 girls with the progestin-induced syndrome and 11 of the 15 with the adrenogenital syndrome who claimed they were tomboys. This status was confirmed by the mother, and was recognized and accepted by playmates and others. A statistically unimportant number of girls in the matched control sample identified themselves also as tomboyish, but in most cases for only a limited episode, whereas for the patient groups tomboyism was a long-term way of life.

Tomboyism did not necessarily include explicit dissatisfaction with being a girl, although the difference between patients and controls on this criterion reached the 5% level of significance in the case of the adrenogenital syndrome. Some girls in the patient groups said they would rather have been born a boy, had there been a choice. Others were ambivalent in the sense that they would not make up their minds whether it would be better to be a boy than a girl. None of the girls actually wanted to change her sex, as some hermaphroditic children do, and none had entertained a conception of sex reassignment.

TABLE I

Tomboyism, Energy Expenditure, and Clothing-Adornment Preference in Fetally Androgenized Girls versus Their Matched Controls, Plus Girls with Turner's Syndrome versus Their Matched Controls[a]

Criteria	PI vs. C	AGS vs. C	C vs. TS
Evidence of tomboyism			
1. Known to self and mother as tomboy	$p \leq 0.05$	$p \leq 0.01$	0
2. Lack of satisfaction with female sex role	0	$p \leq 0.05$	0
Expenditure of energy in recreation and aggression			
3. Athletic interests and skills	$p \leq 0.05$	$p \leq 0.10$	$p \leq 0.05$
4. Preference of male versus female playmates	$p \leq 0.05$	$p \leq 0.01$	0
5. Behavior in childhood fights	0	0	$p \leq 0.05$
Preferred clothing and adornment			
6. Clothing preference, slacks versus dresses	$p \leq 0.05$	$p \leq 0.05$	0
7. Lacking interest in jewelry, perfume, and hair styling	0	0	$p \leq 0.05$

[a] Symbols and abbreviations: PI = progestin-induced hermaphroditism $(N = 10)$; AGS = adrenogenital syndrome $(N = 15)$; TS = Turner's syndrome $(N = 15)$; C = matched controls; 0 = no significant difference.

2. Energy Expenditure in Recreation and Aggression

The common denominator of many tomboyish activities in girls is a high level of physical energy expenditure, especially in the vigorous, outdoor play, games, and sports commonly considered the prerogative of boys. Such activities correspond, it would appear, to the rough-and-tumble play of prenatally masculinized female rhesus monkeys. Table I shows that the girls of both diagnostic groups tended to outstrip their matched controls in preference for athletic energy expenditure, and definitely to have more interest in joining with boys in their energetic play. Team games with a ball, especially neighborhood football and baseball games, received frequent mention. Some of the girls in the diagnostic groups liked to play with girls as well as boys, although some preferred boys as playmates. Most of the control girls preferred to play with other girls rather than boys, if they had a choice.

In the Oregon studies, experimentally masculinized female monkeys showed not only an elevated incidence of rough-and-tumble play, but also of threat in play (Goy, 1970). One might speculate, therefore, that fetally androgenized girls might have evidenced more aggression against their playmates than did their matched controls. In point of fact, this did not prove to be true. The fetally androgenized girls were well able to take up for themselves if attacked, but they were not rated by themselves or their mothers as aggressive children who liked to pick fights. In this respect they did not differ from their matched controls (Table I).

This lack of predisposition to aggressive attack suggests that aggressiveness per se is the wrong variable on which to expect gender-dimorphic behavior, despite popular stereotypes to the contrary. It is more likely that the correct variable is dominance assertion, and striving for position in the dominance hierarchy of childhood. Although this variable was not identified in advance, so that specific data pertaining to it were not collected, in retrospect it appears that the girls of the two diagnostic groups did not strive for dominance in competition with boys, and they were not interested in the rivalries of other girls. Quite possibly, they may have exempted themselves from competitive rivalry with boys, sensing that, because they were socially identified as girls regardless of their skills and accomplishments, they were obliged not to trespass on the culturally defined right of male superiority. Such trespassing might have resulted in their being evicted by the boys from their recreational groups. Under the circumstances, they were better off to avoid dominance rivalry than to be rejected.

3. Clothing and Adornment

In keeping with their energetic outdoor recreations shared with boys, girls of the two diagnostic groups preferred the utilitarian and functional in clothing, rather than the chic, pretty, or fashionably feminine. Usually they chose to wear slacks or shorts and skirts rather than dresses. They were not, however, compulsively averse to dressing up on special occasions as a genuinely transvestic or transsexual child might be. They simply preferred more practical clothes, and in this respect were significantly different from their matched controls. The same trend carried over to accessories, namely jewelry, perfume and hairstyling, although there was a sufficient frequency of moderate interest in these personal adornments to make the contrast with frequency of interest in the matched control group not statistically significant.

4. Childhood Sexuality

Sexual play among the primates in childhood serves a rehearsal function in preparation for the reproductive behavior of adulthood. It is normally observed in human juveniles, provided it is not socially tabooed and inhibited. In rhesus monkeys, there is no absolute difference between the sexual play patterns of males and females, but there is a relative difference: males mount more than females do, the females being more often mounted. Moreover, males progress from keeping both feet on the ground while they mount to using their feet to grasp the shanks of the female. Normal female monkeys do not practice this skill in their play, but fetally androgenized females do.

TABLE II

Sexual Play of Childhood in Fetally Androgenized Girls versus Their Matched Controls, Plus Girls with Turner's Syndrome versus Their Matched Controls[a]

Criteria	PI vs. C	AGS vs. C	C vs. TS
Evidence of childhood sexuality			
1. Attention to genital morphology	0	0	0
2. Masturbation	0	0	0
3. Shared genital inspection and play	0	0	0
4. Shared copulation play	0	0	0

[a] Symbols and abbreviations: PI = progestin-induced hermaphroditism ($N = 10$); AGS = adrenogenital syndrome ($N = 15$); TS = Turner's syndrome ($N = 15$); C = matched controls; 0 = no significant difference.

Among the fetally androgenized human subjects and their matched controls, reports of manifest childhood sexual activity did not differ statistically (Table II). There were only a few girls for whom manifest sexual behavior was reported. Presumably all of them, subjects and controls, had established obedience to the cultural norms and either avoided sexual play or kept it private. The important point is that a history of fetal masculinization did not per se bring about an observable change in the sexual play of childhood. Nor did it bring about a greater degree of verbal curiosity or interest. The amount of sex education received was more complete for the diagnostic than for the control groups, because what they learned at home was augmented in discussions at the hospital.

5. Maternalism

With respect to gender dimorphism of behavior, there is a certain unity to energetic recreation, dominance assertion, relative indifference to personal adornment, and, possibly, masculine positioning in sexual play, all of which belong to the stereotype of boyishness. Conversely, there is a unity that belongs to the stereotype of girlishness. It pertains to rehearsals in childhood play and fantasy of maternalism, marriage, and boyfriend romance. Developmentally, maternalism appears first, in connection with its rehearsal in doll play.

Table III shows that fetally androgenized girls differed from their matched controls in the preferred toys of childhood. They were indiffer-

TABLE III

Anticipation and Imagery of Maternalism, Marriage, and Romance in Fetally Androgenized Girls and Their Matched Controls, Plus Girls with Turner's Syndrome versus Their Matched Controls[a]

Criteria	PI vs. C	AGS vs. C	C vs. TS
Maternalism			
1. Toy cars, guns, etc. preferred to dolls	$p \leq 0.05$	$p \leq 0.05$	0
2. Juvenile interest in infant care lacking	0	$p \leq 0.001$	0
3. No daydreams or fantasies of pregnancy and motherhood	0	$p \leq 0.05$	0
Marriage			
4. Wedding and marriage not anticipated in play and daydreams	0	$p \leq 0.05$	0
5. Priority of career versus marriage	$p \leq 0.05$	$p \leq 0.05$	0
Romance			
6. Lack of heterosexual romanticism in juvenile play and daydreams	0	0	0
7. Lack of adolescent (age 13–16) daydreams of boyfriend and lack of dating relationships	0	0	0
8. Lack of homosexual fantasies reported	0	0	0

[a] Symbols and abbreviations: PI = progestin-induced hermaphroditism ($N = 10$); AGS = adrenogenital syndrome ($N = 15$); TS = Turner's syndrome ($N = 15$); C = matched controls; 0 = no significant difference.

ent to dolls, or openly neglectful of them. They turned instead to cars, trucks, guns, and other toys that traditionally belong to boys.

Lack of interest in dolls later became a lack of interest in infants, that is, in doing things for their care, or in expecting to do such things, even as a paid baby sitter, in the future. The discrepancy between diagnostic groups and their controls reached statistical significance on the criterion of baby care in the androgenital syndrome only. Some girls in this group distinctly disliked handling little babies and believed they would be awkward and clumsy. By contrast, many of the control girls rated high in enthusiasm for little children; they adored them and took every opportunity to get in close contact with them.

All control girls were sure that they wanted to have pregnancies and be the mothers of little babies when they grew up, whereas one-third of the fetally androgenized girls with the adrenogenital syndrome said they would prefer not to have children. The remainder, as well as the ten girls with a history of fetal progestin, did not reject the idea of having children, but they were rather perfunctory and matter-of-fact in their anticipation of motherhood, and lacking the enthusiasm of the control girls.

6. Career versus Marriage

When queried about the priority of career versus marriage in the future (Table III) the majority of fetally androgenized girls subordinated marriage to career, or else wanted an occupational career other than housewife concurrently with being married, occupational and marital status both being regarded as equally important. Among the control girls, the emphasis was in favor of marriage over nonmarital career. For the majority of these girls, marriage was the most important goal of their future.

At first glance, it might appear that preferences for a career over housewife in the fetally androgenized girls was related to their generally high IQ level (Money and Lewis, 1966; Ehrhardt and Money, 1967; Ehrhardt et al., 1968a; Money, 1971). This finding of a trend toward IQ elevation in children exposed to an excess of fetal androgen (it occurs also in genetic males similarly exposed) was a serendipitous one, and one concordant with a high level of attainment academically. It raises questions, as yet unanswered, regarding the relationship of gonadal steroids to development of the cerebral cortex in the fetus, as well as questions regarding the relationship of high IQ to tomboyism or vice versa in genitally normal females (Maccoby, 1963). Obviously high IQ per se is not the determinant of tomboyism, for the fetally androgenized girls of the present sample were matched with the controls on the criterion of IQ; and the controls were not persistent tomboys.

7. Romanticism, Boyfriends, Homosexuality

The fetally androgenized girls did not carry forward the same interest in romance and boyfriends from their play and daydreams into adolescent dating as did the control girls. Those few who were already adolescent lagged behind their agemates in beginning their dating life and venturing into the beginnings of love play. Although the sample of adolescents was too small for any definitive conclusion, subsequent observations on additional teenaged patients bears out the impression that they are late in reaching the boyfriend stage of development, and in getting married.

In view of all the foregoing signs of tomboyism, it is of considerable importance that there were no indications of lesbianism of erotic interest in the fetally androgenized girls, or in their controls (Table III). Thus, it appears that in the fetally androgenized girls it may have been that the biological clock for falling in love was in arrears, but not that it was set to respond to a member of the same sex. It may, indeed, be easier for a fetally androgenized girl than for her control to grow up with a lesbian's biography, but so far this has not been observed, either in the present sample or in other adolescent patients who have subsequently augmented it.

8. Comment

The most likely hypothesis to explain the various features of tomboyism in fetally masculinized genetic females is that their tomboyism is a sequel to a masculinizing effect on the fetal brain. This masculinization may apply specifically to pathways, most probably in the limbic system or paleocortex, that mediate dominance assertion (possibly in association with assertion of exploratory and territorial rights) and, therefore, manifests itself in competitive energy expenditure. Fighting and aggression are not primarily implicated.

Masculinization of the fetal brain may apply also to the inhibition of pathways that will eventually subserve maternal behavior. More correctly, one might say partial inhibition of these pathways, for normal males are capable of paternalism, much of which is identical with maternalism, both being manifestations of parentalism or caretaking.

The noteworthy lack of masculinization in fetal androgenization pertains to those pathways of the brain that subsequently will mediate love and eroticism in response to a mating partner. Evidently, the deciding factor as to the characteristics of the sexual mate as male or female operates postnatally, not prenatally. It remains to be explained why the timing of dating, romance, and the first love affair appeared to be delayed in fetally androgenized girls, until later than in their agemates of either sex.

In the lower species, fetal androgenization may automatically reverse gender dimorphic behavior by prenatal hormonal decree, so to speak. In human beings there is no such automatic decree. So much of gender-identity differentiation remains to take place postnatally, that prenatally determined traits or dispositions can be incorporated into the postnatally differentiated scheme, whether it be masculine or feminine.

Additionally Androgenized Postnatally:
Behavioral Sequelae

Prior to 1950, when cortisone therapy was discovered, adrenogenital females had no choice except to grow up severely virilized by chronically high levels of adrenocortical androgen. They were not only prenatally masculinized, but postnatally also. In teenage and adulthood, they could report on erotic and sexual arousal before their excessively high androgen levels were brought down to normal under the influence of cortisone therapy, after 1950.

The study of 23 of these older adolescents and adults living as virilized women (Ehrhardt *et al.*, 1968b) corroborated, in general, the findings on the early-treated group reported in the foregoing. The majority of the 23 reported having been tomboys throughout childhood, and in adulthood their first preference was for a career other than full-time housewife.

Among the 23 there was a relatively high incidence of homosexual as well as heterosexual imagery in dreams and fantasies, namely in ten individuals. Three of these ten, plus one other, reported bisexual experience also. No woman, however, considered herself to have been erroneously assigned as a female, and none had entertained the idea of a sex reassignment.

A complicating factor in the bisexuality of imagery and experience in these women is that several of them still had a hypertrophied clitoris or clitoral stump capable of erecting under the influence of excess androgen, but not after the excess had been cortisone controlled. Their sexual behavior may have been indirectly influenced also by their image of their bodies under the influence of postnatal virilization, even if the virilization had been partially corrected.

Among the 23 women, 13 are now known to have married, all but two of them after being feminized on cortisone. Five are known to have had at least one pregnancy and delivery. One had three. The babies are nonhermaphroditic and usually healthy. In some instances, the mothers worried about the wisdom of their becoming pregnant, for they felt insecure and feared they would be clumsy with small infants and would fail in mothering them. Actual practice with the baby enabled them to overcome their misgivings. At least one mother was able to breast feed her infants, of whom she had three (Money and Raiti, 1967).

Among the 23 women there were 11 (52% of the 21 for whom sufficient data were available) who stated that having a child usually would not enter their thoughts, fantasies, or dreams. In the files of 12 patients there was also information about how they expected they might feel, prior to actual experience, if they had to hug and cuddle a tiny infant. Only two of them (17%) expressed a positive and genuine desire to be affectionate with small children. The other ten were noncommittal and preferred children who were at least at the toddler age.

There were some late-treated women with the adrenogenital syndrome who, even prior to cortisone treatment, manifested relatively little behavior that might be classified as more typically masculine than feminine. These women show that postnatally elevated androgen levels, persisting into adulthood, do not dictate a masculine gender role or gender identity. They also show that, if there is a permanent prenatal hormonal effect on a part of the central nervous system that mediates sexually dimorphic behavior, then it obviously is in some manner selective. The selectivity may pertain to the timing and/or amount of fetal androgen exposure. An alternative explanation may be that postnatal gender-identity differentiation may be capable of overriding prenatal precursors, or at least of modifying them to an extensive degree.

III. Fetal Lack of Gonadal Hormones

TURNER'S SYNDROME

The antithesis of a genetic female masculinized *in utero* is not a genetic male feminized but, more accurately, a genetic male nonmasculinized. In the absence of androgen, the genetic male differentiates as a female. In the absence of estrogen, the genetic female differentiates as a female. In the absence of gonads and their hormones in entirety, the fetus develops as a female, regardless of genetic sex.

The fetus with Turner's syndrome is one that has no gonadal hormones whatsoever, for its gonads are primitive streaks of tissue instead of being fully formed and differentiated. The etiology of the gonadal defect is cytogenetic, for individuals with the condition most commonly have only one sex chromosome, an X, the other having been lost either prior to, or immediately after, fertilization.

The baby born with Turner's syndrome will look like a girl genitally and so will be assigned and reared as one. Lack of gonadal hormones will be without known effect during the years of childhood, until the time arrives for the beginning of puberty. Then it will be necessary to give estrogen by pills, or possibly by injection. Without this substitution therapy in puberty, it will be extremely difficult for the girl to establish the same degree of psychosocial maturity as her agemates, or to be treated by them as their psychosocial equals, all the more so because of her typically short stature.

The biography of a girl with Turner's syndrome is of special interest with respect to gender-identity differentiation, insofar as it can be expected to show whatever effect, if any, may appear on account of prenatal absence of gonadal hormones. For this reason, and in connection with the study on fetally androgenized girls already presented in the foregoing sections, we undertook a comparison of 15 girls with Turner's syndrome and their matched controls, on the basis of the same criteria as those used to compare the fetally androgenized girls and their matched controls (Ehrhardt *et al.*, 1970). The girls ranged in age from 8 to 16.5 years, with a mean and a median of 12.5 years. Only the oldest Turner girl was menstruating on cyclic estrogen substitution therapy when last seen at the time of the study. The patient sample was selected from a larger group seen in the pediatric endocrine clinic and the psychohormonal research unit without known sampling bias.

In Table I, it can be seen that girls with Turner's syndrome did not differ from their matched controls on four counts, being equally feminine as the normal girls in these respects. On the three other counts they were even more extremely feminine than their controls: as a group they manifested a lesser incidence of athletic interest and skill, a lesser incidence of childhood fighting, and a greater interest in personal adornment.

Table II shows that girls with Turner's syndrome were similar to their matched controls in their limited manifestation of childhood sexuality. Table III shows that they also resembled their matched controls in childhood manifestations and anticipations of maternalism, marriage, and romance. Despite the handicap of their stature and infertility, which all the older Turner girls knew about, all but one explicitly hoped to get married one day. They all reported daydreams and fantasies of being pregnant and wanting to have a baby to care for one day. All but one had played with dolls exclusively, and the one preferred dolls even though she played with boys' toys occasionally. Twelve of them had a strong interest in taking care of babies, tending to their younger siblings, or babysitting for another person; two had a moderate interest in such maternalistic activities, and for the one remaining girl information was missing.

Taken as a whole, these findings indicated that girls with Turner's syndrome differentiate an unequivocally feminine gender identity. Thus one may infer that, in order to differentiate postnatally as feminine, gender identity is not dependent on prenatal gonadal hormones (estrogen and/or androgen) acting presumptively on the brain. Nor is a feminine gender identity dependent on the presence of a second X chromosome. A feminine gender identity can differentiate very effectively without any help from prenatal gonadal hormones that might influence the brain and perhaps, in fact, all the more effectively in their absence.

IV. Fetally Nonandrogenized Genetic Males

A. ANDROGEN-INSENSITIVITY SYNDROME

The individual with Turner's syndrome is not a genetic male who feminized *in utero*. She is not, therefore, the genetic male counterpart of the genetic female who masculinized *in utero*. The human clinical syndrome that most closely approximates the antithesis of fetal androgenization of the genetic female is the syndrome of androgen insensitivity (testicular feminization) in the genetic male. That the antithesis is not total is due to the fact that the syndrome represents not fetal estrogenization of a genetic male, but failure of androgenization of the genetic male. Without androgen, the genetic male fetus, like the 45,X fetus of Turner's syndrome, differentiates morphologically as a female. The same feminization occurs if the fetus is supplied with androgen but is unable to use it, which is precisely what happens in the androgen-insensitivity syndrome.

The defect in androgen utilization in the testicular feminizing syndrome is known to be a genetically transmitted trait that travels down the generations among some of the fertile females of an affected pedigree. Two modes of genetic transmission are possible, an X-linked recessive or a male-limited X or autosomal dominant, and it cannot yet be decided which applies. In either case, whether a genetic male offspring will be affected or not will depend on which one of the pair of maternal chromosomes, the carrier or the noncarrier, happens to get into the egg when the pairs of chromosomes separate so that the egg receives only one-half of the total number of 46 chromosomes. If the egg does carry the affected chromosome, then the resultant baby will manifest the clinical defect only if it is fertilized by a Y-bearing sperm from the father. If the sperm bears a second X chromosome, then the baby will be a fertile female, and a covert carrier of the genetic defect, capable of transmitting it in the succeeding generation.

The nature of the biochemical defect responsible for inability to utilize androgen in the testicular feminizing syndrome has not yet been identified. It is probably enzymatic. The site of action is presumed to be within the cell and to affect virtually every cell in the body that should be responsive to androgen. The exception may be the cells of pubic and axillary hair follicles which, in the normal female, are believed to be stimulated by adrenocortical androgens. Even these cells may be unresponsive in the individual with the androgen-insensitivity syndrome, for there are some patients who have no pubic or axillary hair in adulthood. Head hair is not affected, but hair on the face and body does not appear.

The testes in the androgen-insensitivity syndrome secrete androgen into the blood stream in amounts normal for a male. They also are the source of the much lesser amount of estrogen normal for a male. This estrogen proves sufficient at the age of puberty to bring about complete feminization of the bony structure and outer contours of the body, including normal feminine growth of breasts.

The effect of androgen insensitivity in fetal life is suppression of masculine differentiation of the Wolffian ducts and of the anlagen of the external genital organs. The suppression is so complete that many affected babies are born indistinguishable in genital appearance from normal females. Since it is not usual to do a detailed pelvic examination on a newborn girl, it is not discovered that the the vagina may be represented only by a dimple, or by a shallow cavity that ends blindly and which will need surgical lengthening in (or later than) mid-teenage. Such a vagina has no connection with a cervix or uterus. The uterus itself is not properly formed, but is a cordlike structure without an interior cavity. The embryological explanation for this defect of the uterus is that in early stages of sexual differentiation the testes released not only their androgenic substance, which the body failed to use, but also their Müllerian-inhibiting substance, which the body did not fail to use. Thus the Müllerian ducts began the process of becoming vestigial, as expected in a male, instead of enlarging and growing into the uterus and Fallopian tubes of the female. This circumstance of prenatal development accounts for the fact that, at adolescence, even though feminization is otherwise complete, there is no menstruation.

Although there are some babies with the androgen insensitivity syndrome whose anomaly easily passes completely undetected in infancy and childhood, there are some who present the telltale sign of lumps in the groin or in the labia majora, which are actually the feminizing testes trying to descend from their original abdominal position. If they are then surgically removed, the child will need replacement therapy with estrogen at the age of puberty.

Another telltale sign, present at birth, may be a slight enlargement of the clitoris, insignificant in itself, but enough to lead the alert physician to make further diagnostic investigations. In a very small number of recorded cases, the clitoral organ may be large enough that, in combination with the fact that the testes can be palpated in the groins, the physician may decide the baby should be declared a boy, and surgically corrected as much as possible as a boy. From puberty on, these boys are imprisoned in a body that develops like that of a female and absolutely refuses to masculinize despite all therapeutic efforts, however heroic. The breasts can be removed surgically, of course, but it is impossible to induce deepening of the voice and masculine hair distribution. A particularly disheartening feature of failure to masculinize is that it also creates the impression of failure to age in appearance and look mature enough for one's age. As a husband, if a man with this condition surmounts the obstacles of his physique and sexual anatomy and gets married, he may scarcely be able to cope with the indignity of being mistaken for his wife's son.

It is fortunate for most babies with the androgen insensitivity syndrome that they are so completely feminized in genital appearance that, even if their testes are palpated neonatally, they are assigned to be reared as girls. As these patients grow up, their biographies are of special value as test cases of whether sexually dimorphic behavior will have been influenced by either the genetic or the gonadal status of the fetus as a male, in the absence of fetal hormonal masculinization.

B. Behavioral Sequelae of Androgen-Insensitivity Syndrome

Although it is exceptionally rare, and also exceptionally difficult to explain embryologically, it may happen that a genetic male is born with the external genitals differentiated as a female, and with two undescended testes that will not feminize but masculinize the body at the time of puberty. For this reason, it is preferable that psychological studies of androgen-insensitivity pertain to individuals who, even though they have been followed through childhood, are old enough to have shown the signs of testicular feminization at puberty. There were ten patients in this older age range, at the time we undertook a survey of our clinical data (Money *et al.*, 1968; Masica *et al.*, 1969, 1971), who had been born with normal-appearing female sex organs. Since then, there have been four more patients, on whom the findings confirm those of the other ten. The psychohormonal files on these ten patients were abstracted and tabulated according to the same categories as those used for the fetally androgenized and the Turner-syndrome girls.

With respect to marriage and maternalism, the girls and women with the androgen-insensitivity syndrome showed a high incidence of preference for being a wife with no outside job (80%); of enjoying homecraft (70%); of being resigned to their permanent incapacity for pregnancy (70%); of having dreams and fantasies of raising a family (100%); of having played primarily with dolls and other girls' toys (80%); of having a positive and genuine interest in infant care, even though they had to forfeit the care of the newborn (60%); and of high or average affectionateness, self-rated (80%). Two of the married women each had adopted two children, and they proved to be good mothers with a good sense of motherhood.

On the criteria of sex and eroticism, when the findings on the 10 androgen-insensitive patients were compared with those of the 23 late-treated adrenogenital women (see above) by way of contrast, the androgen-insensitive stood out as strongly feminine. The findings can be summed up by saying that nine of the androgen-insensitive women pretty much conformed to the idealized stereotype of what constitutes femininity in our culture. The tenth, a teenaged girl with an exceptionally troubled home background and adverse surgical and medical history, was going through a troubled period of adolescent development in which she rejected herself, her sex, her family, and her religion.

The group incidence of exclusive heterosexual relations among the 10 androgen-insensitive patients was 80%, and the incidence of adult homosexual experience was nil, with only one case of homosexual activity and bisexual adolescent fantasy (not admitted until several years later), the remaining girl being sexually inexperienced. Six of the women rated themselves as having an average level of libido, having orgasm most of the time, and being reserved and passive in coitus; two rated themselves as above average in libido, always having orgasm, and predominantly initiating sex. The remaining two had not begun their sex lives. Seven considered themselves conservative with respect to coital positions, using one or two only. The breasts and clitoris were regarded as erotic zones by 80%, the vagina by 80%, with no information from the one inexperienced girl. Sensory erotic arousal was predominantly through the sense of touch (80%).

The majority (90%) of androgen-insensitive women rated themselves as fully content with the female role, only one being ambivalent. By contrast, only 47% of adrenogenital girls (the 15 early-treated ones are used here for comparison) so rated themselves, 33% being ambivalent and 20% preferring to have been born male. Only one (10%) androgen-insensitive patient, the disturbed adolescent girl, disliked feminine clothing style, and 90% favored it. The corresponding percentages in the adrenogenital girls were 60% and 40%, respectively. A strong interest in personal adornment was declared by 80% of the androgen-insensitive group as compared with 13% of the adrenogenital group.

These various percentages clearly tell a story of women whose genetic status as males was utterly irrelevant to their psychosexual status as women, as also was the histology of their gonads. Their behavior and outlook were feminine, concordantly with the feminine hormonalization of their bodies at puberty and thereafter. Insofar as there had been any prenatal hormonal influence on pathways in the brain that would subsequently mediate gender dimorphic behavior, including sexual behavior, one must infer that it had been a nonmasculinizing influence. In these cases the presumptive prenatal influence was thus nicely congruent with the postnatal influences of feminine rearing and feminine gender-identity differentiation. This congruence stands in contrast with what is the presumed incongruous prenatal androgenization effect when genetic females with either of the prenatal androgenization syndromes are reared as girls and differentiate a tomboyish version of a feminine gender identity.

Incongruence of a different type is manifested when a baby with the androgen-insensitivity syndrome is declared a boy because of palpable testes and a slight enlargement of what could otherwise be considered to be a clitoris. In this case, the boy differentiates a male gender identity in spite of the limits set by the obstacle of an absence of presumed fetal masculinizing effects on the brain, and absence of postpubertal androgen effects. The impairment of his masculinity is particularly noticeable when he talks about his erotic arousal, sensations, and feelings, as well as of the sexual functioning of his genitalia (Money, unpublished data). Just as a color-blind person cannot talk from the vantage point of color-seeing, so also such a man cannot talk from the sexual vantage point of the ordinary male.

For the systematic accumulation of knowledge, it would be ideal if there existed a clinical syndrome in which lack of, or insensitivity to, androgen in fetal life could be fully corrected at birth. Such a syndrome would correspond to the two syndromes of fetal androgenization both of which can be corrected at birth by surgical feminization of the external genitalia, with hormonal correction by means of cortisone added in cases of the adrenogenital syndrome. There is no such nonandrogenization syndrome which is fully correctable, for the problems of plastic surgery in attempting to construct a simulated penis from the patient's own skin grafts are virtually insurmountable, and promise to remain so at least until the technical problems of a genital-organ transplant have been solved. In addition to the surgical challenge, the hormonal challenge still remains unmet in the androgen-insensitivity syndrome. There is no known way of correcting the basic cellular resistance to androgen, either at birth or subsequently.

Lacking all the evidence one might desire, one must be satisfied with such evidence as is available. The human clinical syndromes reviewed in this report suggest that there is in human beings a counterpart of experimental animal data on the influence of prenatal hormones on gender behavior. Nonetheless, as compared with the lower species, much that pertains to human gender-identity differentiation remains to be accomplished after birth, not in a developmental vacuum, so to speak, but, like language, in interaction with the social environment. The prenatal determinants of gender identity can be perhaps not entirely overridden, but they can be and are incorporated into the postnatal program of differentiatibn.

The importance of postnatal, social-developmental determinants on gender-identity differentiation can be conclusively demonstrated in pairs of matched hermaphrodites (Money, 1970). In such pairs, both patients are concordant for chromosome count, sex of gonads, fetal hormonal history, and external genitalia at birth, but discordant for sex of assignment, one being female and the other being male. Provided certain conditions are met, such as no doubt and no ambiquity in the parents' minds as to the appropriateness of the sex of rearing, early gender-appropriate surgical repair of the external genitalia, and administration of gender-appropriate sex hormones at the usual age of puberty, then each member of a matched pair of hermaphrodites typically differentiates a gender identity in concordance with the sex of assignment and rearing.

Summary

Genetic females prenatally androgenized with progestin-induced hermaphroditism or adrenogenital hermaphroditism, but not subject to further androgenization postnatally, differentiate a gender identity as tomboys when they are assigned and reared as females. Genetic females with the adrenogenital syndrome who grew up prior to 1950 without the benefit of cortisone therapy and were, therefore, also postnatally androgenized, manifested a frequency incidence of bisexual imagery and/or experience not encountered in a comparison group of women with a diagnosis of testicular-feminizing, androgen-insensitivity syndrome. Girls with the latter syndrome are reared as females and differentiate a female gender identity, as do also girls with Turner's syndrome. Prenatal hormonal exposure may influence gender-dimorphic behavior traits in later development, presumably by way of an influence on pathways in the limbic system of the brain; but prenatal hormones do not predestine gender identity differentiation in its entirety. Postnatal influences are equally or more important.

REFERENCES

Ehrhardt, A. A. (1969). Dissertation zur Erlangung des Doktorgrades der Philosophischen Fakultät, Universität, Düsseldorf.

Ehrhardt, A. A., and Money, J. (1967). *J. Sex Res.* 3, 83.

Ehrhardt, A. A., Epstein, R., and Money, J. (1968a). *The Johns Hopkins Medical Journal* 122, 160.

Ehrhardt, A. A., Evers, K., and Money, J. (1968b). *Johns Hopkins Med. J.* **123**, 115.

Ehrhardt, A. A., Greenberg, N., and Money, J. (1970). *Johns Hopkins Med. J.* **126**, 237.

Goy, R. W. (1970). *Phil. Trans. Roy. Soc. London, Ser. B* **259**, 149.

Jones, H. W., Jr., and Verkauf, B. S. (1971). *Amer. J. Obstet. Gynecol.* **109**, 292.

Maccoby, E. E. (1963). *In* "The Potential of Woman" (S. M. Farber and R. H. L. Wilson, eds.). p. 33, McGraw-Hill, New York.

Masica, D. N., Money, J., Ehrhardt, A. A., and Lewis, V. G. (1969). *Johns Hopkins Med. J.* **124**, 34.

Masica, D. N., Money, J., and Ehrhardt, A. A. (1971). *Arch. Sexual Behav.* **1**, 131.

Money, J. (1970). *Eng. Sci.* **33**, 34.

Money, J. (1971). *Progr. Brain Res.,* **32**, 295–304.

Money, J., and Lewis, V. (1966). *Bull. Johns Hopkins Hosp.* **118**, 365.

Money, J., and Raiti, S. (1967). *J. Amer. Med. Women's Ass.* **22**, 865.

Money, J., Ehrhardt, A. A., and Masica, D. N. (1968). *Johns Hopkins Med. J.* **123**, 105.

DISCUSSION FOR ARTICLES BY LUNDE AND HAMBURG, R. P. MICHAEL *et al.* GOY AND RESKO, AND MONEY AND EHRHARDT

C. W. Lloyd: Could you give us a brief description of the meaning of the way you score the postpartum depression? What is the significance of these different parameters?

D. Lunde: We used a number of ways of rating and scoring depression in these women and then intercorrelated the various measures because at the time we were not sure which would be the most useful. Consequently, we used a number of tests. One was the Nowlis Mood Adjective Check List, which was also used in the study of women during the menstrual cycle. This is a well-standardized test used in psychological research. We also used a self-rating sheet that we devised ourselves; this included parameters that we thought were important for such things as depression, anxiety, irritability, and distractibility. These were 9-point rating scales that the women themselves filled out twice a day. Then we did behavioral observations. We interviewed each woman twice a day and rated the women on a series of variables, including some of the same ones—how anxious they appeared, how irritable, etc. We did some standard tests of concentration and tests that are used to pick up symptoms of delirium because some people have reported a sort of organic brain syndrome occurring in some women very temporarily, postpartum. The tests we used there were some standard memory kinds of tests that are very simple to do; such as, asking the subject what she had for breakfast that day. We also did digit span—how many numbers they can remember forward and backward. In addition to that we also drew bloods for measurement of progesterone and cortisol.

We found crying correlating highly with depression but not synonymous with depression in these women. We eliminated some crying episodes from the depression data—those episodes where a woman said she really did not feel sad or depressed. A few women cried when they said they felt very happy. Most of them did not feel exactly depressed in the usual psychiatric sense either. Most of the women who cried (and many cried for long periods of time—1 or 2 hours a day) at some point in the first 10 days would say afterward when we interviewed them that it was a very puzzling, strange experience. They knew they should be happy, the baby was healthy, and so forth. None of the women had a very difficult delivery, all the babies were normal, and yet they were crying. We asked, "Why are you crying?" and some said, "I don't know—you tell me!"

A. J. Kastin: I was very interested to hear your report that cortisol diminished the difference in evoked potential between relevant and irrelevant stimuli. Since melanocyte-stimulating hormone seems to have the opposite type of effect on the somatosensory-evoked response in man, it is possible that our findings were due to the release of cortisol. Your results, as well as our measurements, seem to rule this out. Can you tell us more about the adrenal status of the patients to whom you administered the cortisol?

D. Lunde: They were not patients. They were college students—18 normal, healthy male volunteers who were screened on the basis of having no discernible adrenal pathology. They had no histories of ever having been hospitalized, no serious illness of any sort, and also no family history of adrenal problems per se or hypertension or anything else that might have been related. The group of patients I referred to in which we do see psychiatric difficulties related to steroids is the heart transplant patients who are sometimes getting 100–150 mg/day of prednisone for a long period of time. Some of them became quite psychotic [D. T. Lunde, *Amer. J. Psychiat.* **126,** 369 (1969)].

E. D. Bransome: You referred to abrupt changes in progesterone levels, one in the premenstrual period and one in the postpartum period. I am curious as to why you mentioned a possible causal relationship for this one variable when there seemed to be so many. Does this mean that you have some prospective data on the point?

D. Lunde: What I meant to imply is that progesterone changes may be an important variable that has been previously overlooked. Actually the hormonal influence probably has something to do with the ratio of progesterone to estrogen, particularly in the menstrual cycle phenomenon. Postpartum, I am not so sure that the ratio is as important, since the progesterone drop at that time is very dramatic in the first day or two and may completely overshadow the estrogen. I am not saying there is a causal relationship between the depression one sees and the progesterone phenomenon, but I would say that the progesterone phenomenon is more striking in the postpartum period than in any other period in a woman's lifetime. I did not mean to imply, certainly, that progesterone is the only variable involved in the mood and behavioral changes that we have observed in the premenstrual and postpartum period.

G. L. Cohn: Did you observe any hormonal changes during postpartum psychosis?

D. Lunde: We drew bloods from the women each time they were interviewed or tested so we have a considerable amount of blood waiting to be analyzed. Most of it has been run, but we do not have the data yet. Let me clarify that what we were observing was not postpartum psychosis. These were normal women who became depressed, but not psychotic. The basic finding was that 2 of 3 normal women, with a normal pregnancy, labor, and delivery experience a mild form of postpartum depression. It is transient and does not usually require treatment.

H. Persky: We have done studies of a more systematic nature on the relationship of testosterone production to aggression in young men. We have found a relationship between four objective measures of hostility and aggression with testosterone production rates yielding a multiple correlation coefficient of 0.90. Because of this high degree of relationship, we felt that this would enable us to predict androgenic activity from psychological assessment using the instruments that we have employed. We have recently undertaken a systematic study of a similar group of subjects and attempted to manipulate the psychological variables using suggestions of hostility and aggressiveness administered under hypnotic trances to these subjects. We were able to raise their hostility scores by more than 100% with some degree of increase in the production rate of testosterone. Conversely, we attempted in another similar group of young men to determine the opposite side of the coin; namely, what effect does raising the circulating levels of testosterone have on their psychological performances, at least with respect to hostile moods. To date, in one study where we only modestly raised the plasma level of testosterone 10-fold, we observed no changes in their measures of hostility and/or aggressivity over a brief period of time of less than 1 hour.

R. B. Greenblatt: May I comment on the slogan "Make Love Not War"? Hitler made little love, much war. He had only one testicle and I suspect that it may not have been functional. On the other hand, Napoleon's prowess on the battlefield could be equated with his priapic activity in the bedroom. When at age 42 he became impotent, his capacity for quick decision and bold maneuver faltered and so too his success in battle.

D. Lunde: Then Napoleon does follow along the line of what I was saying. The success of his love life paralleled the success of his battle career.

It appears that one influence of androgens on subsequent behavior is related to the presence of a crucial amount of androgen (testosterone or some other androgen) present in the early developmental stages *in utero* and while the brain is still differentiating. Early exposure to androgens certainly seems to have some effect on adult sexual behavior in primates. I think for males and females in adulthood there is probably only a minimal maintenance amount of testosterone (or other androgen) that is essential to maintain the physiological aspect of sex drive. Sex hormones in no way seem to influence the preference of sexual partner or sexual activity. The evidence indicates just that these hormones do have some effect on sex drive or motivation per se. The use of antiandrogens seems to indicate that if you block androgens you get diminution of sex drive. When you administer androgens to human females, where normally there is little circulating androgen, and you are creating a significant increase from almost negligible blood levels, there do seem to be dramatic effects.

Dr. S. Reichlin [*Medical Aspects of Human Sexuality* 5(2), 151 (1971)] stated that the erotogenic hormones in human females are the adrenal androgens. Clinical studies do seem to indicate that androgens may be more active in this regard than the female sex hormones in humans.

H. A. Robertson: For the study of reproduction in the sheep, there is a basic requirement to monitor the love life of the ram and the ewe. We do this by setting up a harem of 10 ewes and 1 ram. When the ram mounts the ewe he leaves a mark on the rump of the ewe. Unfortunately for the ram there is a period of 6 months when the ewes do not breed; nevertheless, we find that a fair proportion of these ewes get marked by the ram several times during anestrum. It is conceivable that the environmental conditions produce a state of frustration or boredom in the ram, and as a consequence all the sexual advances come from the ram.

In an attempt to clarify the problem, we have tested ovariectomized ewes (ovariectomized two years earlier) and find that they likewise get mounted. The frequency of mating is greater during the normal breeding season than during anestrum. This may be related to the seasonal changes in the libido of the ram or perhaps to some cyclicity of the adrenals of the ewes. The question to Dr. Michael is: If mating does occur, would you classify this as rape by the ram or as a laissez-faire situation, or even prostitution by the ewe?

I should like to ask Dr. Michael a further question. What happens if a mature female monkey, ovariectomized before puberty, is presented to a mature but sexually inexperienced male? Does mounting occur?

R. P. Michael: I am sure your ewes are marked by rams during anestrum because the males retain a higher level of sexual activity during this period than the females, and are relatively less sensitive to exteroceptive factors. However, even under relatively well-controlled laboratory conditions, male rhesus monkeys show seasonal changes in sexual activity [R. P. Michael and E. B. Keverne, *J. Reprod. Fert.* **25**, 95 (1971)]. As to the question of rape by the ram or seduction by the ewe, a competent reply can come only from the ewe. With regard to Dr. Robertson's final question, we have not done this experiment, but my guess would be that there would be very little sexual interaction between the pair.

A. L. Southren: I wonder whether you measured the diurnal pattern of plasma testosterone in the male monkey and whether this has an effect on his sexual behavior. We know a diurnal pattern of plasma testosterone in the human male.

R. P. Michael: Yes, we have measured changes in plasma testosterone levels four times during the 24 hours. Intact adult males showed a progressive decline in values from 08:00 to 16:00 hours—a variation that was similar to that reported by you and by others for the human male. However, in about half our animals this variation was not observed. The most consistent feature of our study was the finding that testosterone levels in plasma collected at 22:00 hours were nearly double those observed at other times (1665 ng/100 ml). The exact mechanism responsible for the high level at night remains to be elucidated, but it would appear to depend on changes in testicular secretion or metabolic clearance, since it was not observed in castrates. I hesitate to call this a diurnal variation until we have sampled more frequently during the 24 hours and determined the relationship of the high values at 22:00 hours to paradoxical sleep.

J. Kowal: How often do these monkeys become pregnant, and what is the effect of pregnancy on their sexual behavior?

R. P. Michael: The females described in these studies do not become pregnant since we ligate their fallopian tubes and remove a portion of each tube. The effect of pregnancy on sexual behavior is extremely interesting because males continue to show a lot of sexual interest in pregnant females under our test conditions. The females, however, become quite unreceptive and reject the males with a display of considerable aggression. In fact, this seems to be one of the few circumstances in which female rhesus monkeys can be aggressive toward males and get away with it [R. P. Michael and D. Zumpe, *Anim. Behav.* **18**, 1 (1970)].

K. Sterling: One of your slides showed the number of mounts, the number of refusals, and the number of male attempts. The progesterone dose given seemed identical in the two groups. I could not see the significance of the division into two experimental groups using the one dose of progesterone.

R. P. Michael: The significance of the division of pairs into two groups was this. The females of one group increased their refusals under the influence of progesterone, and this accounted for the decline in the pairs' copulatory activity.

The females of the other group did not refuse the sexual advances of the male, but the males' interest in these females showed a decline in any case. We feel the mechanism responsible for this loss of male interest depended upon the inhibition of attractant pheromone production in this group of females by progesterone. In this group too, the copulatory activity of the pairs declined—but for a different reason.

C. H. Rodgers: Dr. Michael, you presented data showing that males bar-press more for an estrous female than for a castrated female. Have you observed whether males will bar-press for other males?

R. P. Michael: If two dominant, sexually active males are placed in either side of the twin-compartment operant cage, they spend most of their time threatening each other through the wire partition. I would prefer not to allow them access to each other as, under our experimental conditions, serious aggression would occur. In fact, this has happened by accident.

F. C. Greenwood: Can your synthetic mixture turn sexually ill-assorted pairs into the rather more active-pair behavior?

R. P. Michael: Indeed, yes. Dr. Margo Wilson working in my laboratory has shown that the least preferred of three females by three males (9 pairs) will rise substantially in the popularity stakes when vaginal secretions collected from an estrogen-treated donor was applied before tests with these males. Pheromones may very well be important in determining partner preferences and individual differences in sexual activity.

R. Horton: Plasma testosterone levels in the mature male rabbit are quite low, and, although there is a seasonal variation, the exhibition of a mature female results in a striking and immediate rise in testosterone levels [*Endocrinology* **82,** 627 (1968)]. An obvious extension of this information is to ask the question, does the vaginal substance act like a releasing factor, causing changes in testosterone levels after exposure to the pheromone?

R. P. Michael: We have not investigated the effect on plasma testosterone levels of exposing the male to pheromones: it would be an extremely interesting study to do. Our data on the effects of castration and subsequent testosterone replacement treatment [M. Wilson, T. M. Plant, and R. P. Michael *J. Endocrinol.* **52,** ii (1972).] indicate, however, that changing the levels of plasma testosterone may not have an immediate behavioral effect. After castration, changes in plasma testosterone values occurred very abruptly but the behavioral decrement took several weeks to appear. Conversely, injections of testosterone propionate restored the plasma values in a matter of hours, but the restoration of behavior occurred gradually, again over several weeks.

R. M. Rose: In the discussion of brain-hormonal interrelations, as suggested by Dr. Horton's studies in the rabbit, it is important to recall that we are dealing with two sides of a feedback system. One is the possible action of sex steroids on behavior, and the second relates to the influence of psychological and social variables on the regulation of testosterone secretion. That is, the brain acts as a modulator on the secretion of these behaviorally active hormones.

In collaboration with Dr. Irwin Bernstein at the Yerkes Regional Primate Center, we studied a group of adult male rhesus monkeys living in a large outdoor compound, one-third of an acre in area. We found interesting correlations between the frequency of various aggressive behaviors and plasma testosterone in this group of 34 males. In an attempt to isolate the effects of sexual behavior, females were excluded. We also observed a positive correlation between plasma testosterone and the dominance rank order of the animals. Animals who were more aggressive or more dominant tended to have higher plasma testosterone levels. Following these studies, we took animals who were isolated from females and then exposed them to receptive females. There was a very rapid rise in testosterone one day after exposure to the females. The testosterone levels began to fall after this during the 2-week period the males were in consort with the females, although they were still actively copulating. We then took these animals and placed them in a group with strange animals, which was a very stressful experience, as the recently introduced males were actively rejected by the group. During this period there was a dramatic fall in plasma testosterone levels secondary to the stressful confrontation. In two animals plasma testosterone was observed to rise once again when the males were removed from the group and placed with receptive females.

These findings parallel a study we did in Officers Candidate School at Fort Benning, Georgia. Fifty-eight men were studied on two occasions, early in training, which was potentially very stressful, and 1 week prior to graduation when the men were relaxed and confident they had passed the course. Early in training there was a 25–40% decline in plasma testosterone levels, supporting the interpretation of stressful stimuli inhibiting testosterone secretion. It is therefore possible that testosterone secretion may be activated or inhibited by social and environmental factors, and that these alterations in testosterone levels may in turn be effective in influencing behavior.

C. W. Lloyd: Drs. Klaiber and Braverman at the Worcester Foundation have performed a study showing that with a prolonged infusion of testosterone into normal men the normal degradation of performance of certain psychological tasks (serial subtraction) does not occur. In other words, the administration of testosterone over a period of 8 hours tends to keep this performance from getting worse as normally occurs with only an infusion of saline. You might find things if you gave it for a considerably longer period of time.

R. Horton: Does vaginal instillation of a locally active estrogen also result in formation of the active pheronome?

R. P. Michael: Estradiol is locally active when applied intravaginally, and this results in the formation of pheromones in considerable quantities.

J. M. McKenzie: I would like to hear more about your studies with the pheromone, Dr. Michael, particularly since you indicated that synthetic material is being studied. Also, your brief reference to a certain lack of zoological specificity was interesting. Can you say anything about crossing with other nonsimian species?

R. P. Michael: There is no doubt that a synthetic mixture of authentic fatty acids made up to match those present in a pool of rhesus monkey vaginal secretions exerts a strong sex-attractant effect on the rhesus male. I am not certain, however, whether we have found the optimal concentrations and proportions of acids as yet. We have a lot more experiments to do, and it may be that some of the phenolic compounds in vaginal secretions which are not active in themselves can have the effect of enhancers.

J. R. Marshall: You administered testosterone on days 38 or 40 of gestation. Have you studied the effects of androgen given earlier in pregnancy, and have you been able to duplicate any of the hypothalamic effects of administered androgens as described by Gorski in the rat?

R. W. Goy: We have not begun injections earlier. I am not sure that it would be of any advantage to do so if we can extrapolate from results with the guinea pig. We have done extensive experiments on the so-called critical period in the guinea pig. In that species the effects of injected testosterone in inducing masculinization correspond very well to Dorothy Price's data on the same species for the periods of development when the fetal testis is normally active. In addition, psychological masculinization of the female guinea pig appears to be accomplished at an earlier development age than that for modification of the hypothalamic–pituitary–gonadal axis. In the rhesus monkey the best data we have on the age at which the fetal testis becomes active estimate that age to be 45 days, and for the monkeys reported on in that paper injections were begun slightly ahead of that time. We have not been able to induce anovulatory sterility in the rhesus monkey, but the range of treatments that we have tried is very small indeed. I would not be surprised if in the rhesus monkey, as in other species, variations in timing and dosage are critical determinants of the specificity of end points influenced. Changes in behavior and psychosexual status can be influenced separately from changes in ovarian function. I assume, for the time being, that we simply have missed the critical period in development for blockade of ovulation. The possibility should not be discounted, however, that the sex difference in hypothalamic–pituitary–gonadal axes is not conditioned by early developmental endocrine factors in the rhesus as it is in the rat, guinea pig, and hamster. At present, there is no theoretical basis for predicting which sexually dimorphic characters depend upon conditioning actions of the embryonic, fetal, or larval hormones and which are conditioned solely by the type of hormone present in adulthood. In the present paper, for example, we have shown that "yawning" behavior is sexually dimorphic in the adult, but the dimorphism depends only on the presence of testosterone during adulthood. In contrast, the display of mounting, intromission, and ejaculation require the presence of testosterone both before birth and in adulthood.

A. Bartke: If a treated pregnant female were pregnant with a male fetus, would the resulting male be normal?

R. W. Goy: As far as we can tell, the resulting males are normal.

J. M. Davidson: I think it is rather striking that you find a delay in menarche in the monkeys treated with prenatal androgen. Rodents show a different effect; i.e., earlier vaginal opening in rats treated neonatally with androgen. Do you have any explanation for this?

R. W. Goy: We do not have an explanation. I do not agree entirely with your statement about rodents. What you say is true for the rat, but as Professor Keith Brown-Grant has shown, it is not true for the guinea pig. The guinea pig, like the monkey, when it is given testosterone prior to birth shows a delay in the onset of the first estrus compared to normal animals. Perhaps some of the difference is attributable to differences in the end point used to indicate puberty.

R. P. Michael: You test your males for 10 minutes; we test routinely for an hour. You get a distribution of ejaculation in relationship to the menstrual cycle. Is it possible that you have selected fast ejaculators?

R. W. Goy: The stud males from the rhesus that we have vasectomized and used in the experiments described were selected on the basis of their tolerance for females, not on the basis of speed of ejaculating performance. A few of the eleven stud males had been purchased from other laboratories because they had histories which recommended them as "good breeders." The rest were purchased as feral males, and only two of these were excluded on the basis of a high incidence of aggressive attacks upon the females.

R. B. Greenblatt: Dr. Goy suggested that androgens do not change the behavior pattern of normal adult women. I can support this statement because we have administered androgens to hundreds of women in an effort to improve their sexual desire or overcome relative frigidity. We noted no change in the orientation of their sexual drives.

Dr. Michael pointed out that progesterone was indeed anaphrodisiacal. We have learned to treat nymphomanical women with continuous doses of progestogens. With a dampening of their frenetic search for sexual gratification, many were able to make an adjustment in their way of life.

C. W. Lloyd: There are many data on progesterone as an anaphrodisiac in the male. It works both by suppressing testosterone production and probably also by a central nervous system effect which might be important in the inhibition of aggression.

A. S. Goldman: A group of subjects in Norway that might complement your feminized males has been studied as a result of a suggestion we made several years ago [A. S. Goldman and A. M. Bongiovanni, *Ann. N.Y. Acad. Sci.* **142,** 755 (1967)]. We noted a couple of cases of hypospadias in boys whose mothers had been treated with large doses of progestin. We wondered whether, in addition to the well known virilizing effects of maternal administration of progestins on the female fetus, such treatment might prevent masculinization of males in a fashion similar to our experimental model of congenital adrenal hyperplasia due to inhibition of 3β-ol-dehydrogenase. Thus we warned that maternal progestin treatment may also feminize males. Aarskog has reported several cases of hypospadias in Norway and found five cases of boys with hypospadias whose mothers had been treated with progestin during the critical period of penis formation [D. Aarskog, *Acta Paed. Scan. Suppl.* **1,** 203 (1970)]. As we had predicted, the closer the progestin treatment was started to the beginning of differentiation of the penis, the more marked was the hypospadias. Thus, I think a hypospadic male with a history of maternal progestin treatment during such a critical period may be an additional group for your studies.

G. T. Bryan: Were normal female siblings of the girls with adrenogenital syndrome or of the girls from progestin-treated mothers used as control subjects? Would not these siblings make better control subjects than matched controls from outside of the family?

A. A. Ehrhardt: At the time of Dr. Money's and my original study, we saw only a few of the siblings. At present, I am involved in Buffalo in a sibling and parent study of patients with the adrenogenital syndrome. The study was mainly designed to shed more light on the still unexplained finding on elevated IQ in patients with the adrenogenital syndrome [J. Money and V. Lewis, *Bull. Johns Hopkins Hosp.* **118,** 365 (1966)]. At the same time, we are in the process of collecting data on sexually dimorphic behavior in these siblings to be compared with the patients.

J. C. Orr: You reported that the girls had a higher level of intelligence than was average for the population. Did you compare the average intelligence of their parents with that of the general population, and, if so, was this also higher? You mentioned that none of them showed lesbian tendencies. Has there been any converse study on the hormonal status of lesbians?

A. A. Ehrhardt: I am testing the parents of patients with the adrenogenital syndrome in Buffalo right now. To my knowledge, there has been very little work done on the endocrine status of lesbians. I am not aware of any report on established differences between lesbians and normal controls.

S. L. Cohen: A paper presented at the Hamburg meeting last year reported that testosterone, dihydrotestosterone, estrogen, and pregnenediol excretion were all followed in lesbians and normals. There were no differences between the lesbians and the normals.

F. C. Bartter: Have you studied any of those patients with congenital adrenal hyperplasia who, either by a decision of the doctor earlier, or because of an unusually complete fusion of labia, have been brought up as males? I have in mind a patient we are studying who, with complete fusion and a penile urethra, reports that he is happily married, and troubled only by hematuria once a month. The karyotype is, of course, XX.

A. A. Ehrhardt: Yes, we have seen several cases of genetic female patients with the adrenogenital syndrome raised as males. We also recently saw a patient (genetic female) whose mother had been treated with progestins and who had been raised consistently as a male. He has been followed by Dr. Kupperman in New York. The patient is now a teenager and has clearly identified as a male. The same holds true for the patients with the adrenogenital syndrome who were raised as males. They follow the rule that gender-identity differentiation usually is consistent with the sex of rearing, provided there is no confusion and ambiguity in the patient's mind and the environment's mind as to the sex of assignment.

C. Ezrin: This question concerns the statement that the assigned sex or "sex rearing" is more important than the hormonal influence in determining the ultimate psychogender. While I would agree that this is probably a justifiable conclusion to draw from studies of patients who have the adrenogenital syndrome, is it also true of male pseudohermaphrodites who have been raised as females? I know of one such patient who had considerable emotional disturbance when virilism appeared at puberty; this led to a request to be transferred back to the male sex.

A. A. Ehrhardt: The conclusion that postnatal environment influences have an extraordinary effect on gender-identity differentiation is also true for male pseudo-hermaphrodites. In fact, a genetic male baby with a microphallus with no good chance for adequate surgical repair has a much better prognosis if the sex of rearing is female. It is essential though that certain conditions are met in such cases—namely, that the decision of the sex of assignment is made as soon as possible after birth and that the parents are interviewed and receive counseling so that no ambivalence and confusion is left in their minds that they are raising a girl. Surgical feminization of the genitalia should be performed as early as possible and female hormonal replacement therapy should be administered at the time of puberty. If all these conditions are met, we have found that these patients typically identify as females, consistent with the sex of rearing and in contrast to their genetic, gonadal, and fetal hormonal sex.

J. Weisz: Could you tell us what the individual profiles were like? For instance, did the girls who scored high on those behavioral parameters that are considered to be expressions of high energy expenditure score high also on the aspects of maternal behavior? Did the different behavioral expressions parallel each other, or could they be dissociated?

A. A. Ehrhardt: We did not do a correlation analysis of the different variables. From inspection of our data, it appears that the girls who had a high energy expenditure level and liked to play predominantly with boys, had typically a low interest in maternal behavior. However, there were also some girls who did not follow this pattern and were tomboys in some respect, but still were rehearsing the maternal role in fantasy and play.

R. B. Greenblatt: Dr. Ehrhardt, have you ever treated any of your patients with testicular feminization with large doses of androgens and noted their response as to their sexual drive? I ask this question because many years ago, before we knew of the enzyme defect common to these individuals, we attempted, following the removal of the abdominal testes, to induce sexual hair growth with large doses of testosterone. There were no alterations in any of the parameters of androgen activity, such as growth of sexual hair, voice changes, acne, oiliness of scalp and skin, or clitoral enlargement. One patient volunteered that her libido was greater following androgen than estrogen therapy. Was this effect cerebral?

A. A. Ehrhardt: I do not know any patient with testicular feminization who complained about her low sex desire. The patients who were old enough to have an active sex life did not differ from other females in their level of libido.

Author Index

Numbers in italics refer to the pages on which the complete references are listed.

Meyer, R. K., 105, 106, 121, *128*, 716, *732*

Meyer, W. J., 185, *199*, 653, *662*

Mical, R. S., 265, *276, 281, 282, 283*, 501, 504, 506, 510, *513, 520, 521*, 541, *572*

Michael, R. P., 627, 643, *661*, 665, 666, 667, 669, 670, 671, 672, 673, 674, 676, 678, 680, 681, 682, 687, 688, 689, 690, 691, 693, 694, 695, 701, 702, *704, 705, 706*, 709, 714, *731, 732, 757, 758*

Michel, I., 1, *42*

Michelakis, A. M., 316, *339*

Midgley, A. R., Jr., 174, 175, 178, 181, 182, 185, 186, 187, 191, 192, 193, 194, *198, 199*, 203, 204, 205, 206, 207, 208, 209, 210, 211, 213, 214, *216, 217*, 473, 481, 484, 504, *513, 515*

Mielke, J. E., 399, *458*

Migeon, C. J., 647, 653, 654, 655, *661, 663*

Mijares, A., 711, *731*

Milhaud, G., 184, 187, *198*, 402, 450, *458*

Miller, H. H., 399, 405, 414, 415, 443, 448, *458*

Miller, J. M., 399, *457*

Miller, M., 146, *164*

Miller, M. C., 204, 205, 208, 211, 213, *217*

Miller, L. S., *520*

Miller, N. E., 273, *276*

Miller, S., 647, *662*

Milne, G. W. A., 604, 609, 611, 613, 614, 615, *619*

Minaguchi, H., 472, 477, 501, *512, 515*

Minkin, C., 414, 418, 423, 424, 425, 439, 451, *459*

Mintz, D., 560, *572*

Mishkinsky, J., 502, *515*

Misuraca, G., 98, 104, *128*

Mitnick, M., 249, *276*

Mittler, J. C., 173, 174, *199*, 201, 209, *217*

Miyazaki, A., 299, *337*

Mizuno, H., 490, *516*

Mobbs, B. G., 2, *43*

Mobley, R. C., 596, 599, 600, *619*

Moertel, C. G., 399, 405, 436, 450, *459*

Mohla, S., 23, *43*

Mole, B. J., 66, *71*

Moll, J., 233, *277*

Möllendorf, W., *279*

Molnar, G. D., 399, *458*

Monahan, *226*

Monchkton, G., 145, *163*

Money, J., 643, 644, 645, *661*, 710, *732*, 738, 743, 745, 747, 750, *753, 754, 761*

Monroe, S. E., 484, *515*

Montagna, W., 98, 101, 109, *128*

Montegut, M., 141, *163*

Montemurro, D. G., 488, *512*

Montgomery, R. G., 54, *71*

Moor, R. M., 57, 58, 59, 67, *71, 73, 78, 79*

Moos, R. H., 627, 628, 629, 630, 631, 632, 633, 634, 635, 636, 658, *661, 662, 663*

Mooz, E. D., 258, *276*

Morgan, M. S., 3, 12, 13, 17, *42*

Morgan, R. O., 568, *571*

Morain, W. D., 406, *459*

Morii, H., 381, *391*

Morrell, R. M., 249, *276*

Morris, B., 66, *72*

Morris, N. M., 668, *706*

Mortara, F., 678, *705*, 710, *731*

Moses, A. M., 146, *164*

Moss, H. A., 657, *661*

Motta, M., 242, 273, *275, 276*, 288, 292, 306, 330, *337*, 507, 508, *513, 522*

Moudgal, N. R., *526*

Moukhtar, M. S., 184, 187, *198*, 450, *458*

Mousseron-Canet, M., 3, 22, 23, *41*

Mowles, T. F., 288, *339*

Moyer, F. H., 116, 119, 125, *128*

Mueller, G. C., 1, 3, 11, 12, 13, 14, 15, 16, 17, 18, 19, 20, 22, 23, 24, 26, 28, 29, 30, 31, 32, 33, 34, 35, 36, 37, 39, *41, 42, 43, 44, 45*

Muhlbock, O., 488, *515*

Muller, E. E., 173, *199*

Mullins, R. J., 641, *662*

Munson, M. S. B., 595, *620*

Munson, P. L., 358, 368, 370, 374, 379, 383, *390, 391*, 405, 438, *457, 458*

Murad, F., 412, *459*

Muramatsu, M., 2, *42*

Murasawa, Y., 649, *662*

Murphy, B. E. P., 24, *44*, 291, *339*

Murphy, M., 654, *662*

Murphy, R., 640, *661*

Murphy, R. D., 611, *619*